FLORA OF GLAMORGAN

D1419751

To: Eleanor Vachell, Arthur E. Wade, Eileen and Maria

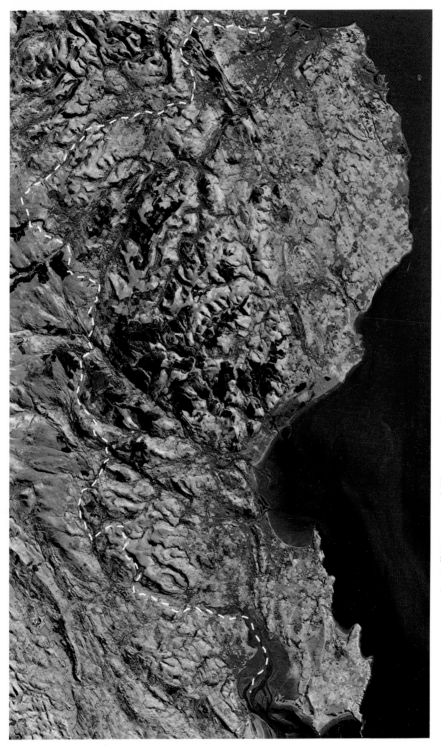

A satellite image of Glamorgan. Large areas of moorland and lowland heath show a mosaic pattern of plant communities with contrasting dominant species, overlaid by recent conifer plantations in the hills. Near the coast, duneland and salt-marshes are prominent, with a narrow strip of limestone cliffland in south Gower. Urban and industrial areas are intermingled with farmland and moorland.

THE NATURAL HISTORY MUSEUM

FLORA OF GLAMORGAN

A. E. Wade[†], Q. O. N. Kay and R. G. Ellis
and the NATIONAL MUSEUM OF WALES

with contributions by
M. G. Bassett, A. J. Morton, A. Orange, R. M. Owens,
A. R. Perry and P. S. Wright

LONDON : HMSO

A catalogue record for this book is available from the British Library

ISBN 0 11 310046 9

Cover designed by Mike Morey

Produced by Hugh Tempest-Radford

Typeset in Palatino by Galleon Photosetting, Ipswich

Printed in Great Britain by St Edmundsbury Press

HMSO publications are available from:

HMSO Publications Centre
(Mail fax and telephone orders only)
PO Box 276, London, SW8 5DT
Telephone orders 071-873 9090
General enquiries 071-873 0011
(queuing system in operation for both numbers)
Fax orders 071-873 8200

HMSO Bookshops
49 High Holborn, London, WC1V 6HB
(counter service only)
071-873 0011   Fax 071-873 8200
258 Broad Street, Birmingham, B1 2HE
021-643 3740   Fax 021-643 6510
Southey House, 33 Wine Street, Bristol, BS1 2BQ
0272 264306   Fax 0272 294515
9–21 Princess Street, Manchester, M60 8AS
061-834 7201   Fax 061-833 0634
16 Arthur Street, Belfast, BT1 4GD
0232 238451   Fax 0232 235401
71 Lothian Road, Edinburgh, EH3 9AZ
031-228 4181   Fax 031-229 2734

HMSO's Accredited Agents
(see Yellow Pages)

and through good booksellers

Contents

Dedications

This work is dedicated to the memory of two botanists, Eleanor Vachell and Arthur Wade, who between them contributed over 100 years of work in the flora of the county, and also to our wives Eileen Kay and Maria Ellis for putting up with 'the book' for so many years. Eleanor Vachell (1879–1948) was Secretary of the previous Flora Committee set up in 1900 to oversee the preparation of Trow's *The Flora of Glamorgan*, and was an active worker on the Glamorgan flora all her life. As mentioned below (p. viii) she made a small bequest towards the publication of a new *Flora*, and also left an annotated copy of Trow's *Flora* and many other notes and records to be used in the new compilation. Arthur Edward Wade M.Sc. (1895–1989) the senior author of this book, joined the staff of the National Museum of Wales in 1920 and, as has been detailed elsewhere, spent the rest of his long life working on the Welsh flora. He was the guiding light behind this title, carried out a vast amount of fieldwork, and wrote preliminary accounts of about a third of the book prior to his emigration to New Zealand in 1983. It is a source of continual regret and sorrow that he did not live to see the publication. This also seems to be a fitting place to pay tribute to Mary Percival who was responsible for the innovation of the inclusion of details of floral biology including insect visitors to many of the more widespread species found within the county. Sadly, she too did not live to see this publication. Biographical notes on these and other renowned Glamorgan botanists will be found in Chapter 1.

Acknowledgements

A work such as this is of necessity a co-operative effort and many people have helped in its compilation. Detailed acknowledgements to these are found below and a list of those who contributed records will be found in Appendix II.

Most of the systematic survey of plant distribution was carried out between 1970 and 1985 by a relatively small number of fieldworkers (in the east and north of the county, J. P. Curtis, R. G. Ellis, M. E. Gillham, G. Hutchinson, A. M. Pell, F. H. Perring, A. R. Perry, A. B. Pinkard, B. Scotter and A. E. Wade; and in the west, C. R. Hipkin, H. Hipkin and Q. O. N. Kay), but many others have contributed valuable records or helped in other ways. These include the farmers, site operators and landowners whose friendliness, tolerance and helpfulness made fieldwork more enjoyable. Our thanks are also due to several organisations which have supported the work, especially Cardiff Naturalists' Society, the Glamorgan Wildlife Trust and the Nature Conservancy Council, and to the Botany Departments of the University College of Swansea and the National Museum of Wales for providing essential facilities and financial support.

Further thanks are due to the specialists who have looked at and identified many specimens belonging to critical groups, especially C. R. Fraser-Jenkins (*Dryopteris*); A. Newton (*Rubus*); Rev. A. Primavesi and Mrs I. M. Vaughan (*Rosa*); P. F. Yeo and A. J. Silverside (*Euphrasia*); C. Howarth and A.J. Richards (*Taraxacum*), J. Bevan and P. D. Sell (*Hieracium*) and A. O. Chater (*Carex*). Thanks are also due to Dr D. Edwards for information on fossil plants for the Geology chapter, and to the Director General of the Ordnance Survey for permission to base many of the maps in this book on Ordnance Survey maps. Many of the portraits of botanists, scenes of Glamorgan landscapes and colour portraits of Glamorgan plants are reproduced from negatives or transparencies in the possession of the authors or the National Museum of Wales. Credit for each plate is acknowledged in its caption, but mention should be made of Dr Mary E. Gillham and Mr Jeff P. Curtis for allowing their slides to be used.

Figs. 2.1 to 2.4 and maps 1 and 2 were drawn by Mr H. M. Reynolds, and figs 3.1 to 3.4 by Mrs D. G. Evans and Mrs L. C. Norton, all of the National Museum of Wales.

Finally, thanks are due to Mrs P. Knapman and Mrs S. Mynott of the Department of Geology who prepared the chapter on Geology; to Mrs O. M. Evans, Mrs M. J. McKenzie and especially Miss H. E. Fraser of the Department of Botany who between them prepared most of the typescript; and to Dr G. Hutchinson who bullied, cajoled and encouraged the various authors to finish their work; prepared the first drafts of the Bibliography and gazetteer appendices and who co-ordinated the final push to get the *Flora* ready for the printers.

Thanks are due to the National Museum of Wales for sponsorship of the colour plates, and to the Countryside Council for Wales for sponsorship of the production of this book.

Introduction

The Flora of Glamorgan Committee was formed in 1969 with the aim of preparing a new *Flora* of the county. The existing Glamorgan Floras – A. H. Trow's *The Flora of Glamorgan*, which appeared in instalments from 1906 to 1911, and H. J. Riddelsdell's *A Flora of Glamorganshire*, printed as a supplement to the *Journal of Botany* in 1907 – were out of date and long out of print. Trow's *Flora* was never completed by the addition of an ecological section, as had originally been intended, although an ecological account of the flora of the county by R. C. McLean eventually appeared in Volume 1 of the *Victoria County History of Glamorgan*, published in 1936.

Glamorgan has been spared the wholesale destruction of rural habitats that has taken place in some English counties, and many of the localities that were described by Trow and Riddelsdell still exist. Others have, inevitably, been destroyed or changed radically by urban and industrial development and by new techniques in agriculture, forestry, and mining. Many species have changed in abundance, especially in the urban areas and lowlands of the county where a surprising number of species that were rare or absent from the county eighty years ago are now conspicuous and ecologically important, for example, *Buddleia davidii*, *Epilobium ciliatum*, *Reynoutria japonica* and *Senecio squalidus*. In the uplands afforestation and open-cast mining have led to great changes, with fewer gains.

By the late 1960s it was clear that a new *Flora* of the county was badly needed. The stimulus for the founding of the Flora of Glamorgan Committee came from two sources. One was the preparation of the first edition of the *Atlas of the British Flora*, published in 1962, which had involved an extensive collaborative survey co-ordinated by the Botanical Society of the British Isles. This both increased interest in plant distribution among botanists and led to increased awareness of the gaps in our knowledge of the flora of the county. The other source was the continuing tradition of botanical work in Glamorgan, stemming directly from Trow's group and maintained chiefly by the Cardiff Naturalists' Society and the National Museum of Wales. Eleanor Vachell, who had herself contributed many records to Trow's *Flora* and had continued active work and publication until after the Second World War, made a small bequest to the Society to support work toward the publication of a new *Flora of Glamorgan*. The Flora of Glamorgan Committee that was founded with the aid of this bequest, linked the Society with groups from the Botany Departments of the National Museum of Wales and the University College of Swansea who were already carrying out preparatory recording work for a new *Flora*. The records in this book are restricted to vice-county 41, corresponding to the old county of Glamorgan, excluding the parts of the modern counties of Mid-Glamorgan and South Glamorgan that are north or east of the old county boundary.

The idea of producing a new *Flora of Glamorgan* was first mooted in 1968 by R. G. Ellis and a Flora of Glamorgan Committee was set up to oversee the project. The Committee first met in May 1969 and fieldwork by members commenced in that year although the special recording cards, instructions to fieldworkers and keys to critical groups were not available until 1970.

1

A Short History of Botany and Botanists in Glamorgan

R. G. Ellis and A. E. Wade

The earliest botanical visitor to Glamorgan of whom we have any record is that famous naturalist John Ray, who passed through the county on his second tour through Wales in 1662. We have no record of any botanist living in the county either then or before that date.

Ray in the company of his friend Francis Willughby set out from Cambridge on 8 May 1662 on what was to prove his most rewarding tour. They reached North Wales about 10 days later; after travelling south by way of Aberdovey, Cardigan, Pembrokeshire, and Kidwelly they came to Glamorgan.

Unfortunately, Ray and his companion seem to have been in a great hurry, for although numerous records for other counties are given in his published *Itinerary* (Lankester, 1846), the only plant they recorded for Glamorgan was *Senecio congestus* (Marsh Fleawort), which Ray says they saw growing – 'Near Aberavon, on the sandy meadows by the sea-side.' Even this solitary record is subject to considerable doubt, since no plant resembling the Marsh Fleawort has ever been seen there by anyone else, and the only reliable records for it are from the fens of East Anglia. Riddelsdell (1907) states – 'Ray afterwards says that the locality is in Merionethshire; but I cannot trace an Aberavon there.' If the locality is in Merionethshire (the Fleawort has never been recorded there either), then Ray entered it in a strange place in his journal, sandwiched between Kidwelly, Carmarthenshire on the one side and Caerwent, Monmouthshire on the other.

In 1668 a certain Richard Kayse of Bristol visited Dinas Powis. Little is known of him and only one other record, from Clifton near Bristol stands to his credit. He found in Ray's words 'on a rock in a wood near Dinas Powis Castle, Glamorganshire in Wales' *Polypodium cambricum* 'Cambricum' (Laciniated Polypody). Since Ray in his *Synopsis* (Ray, 1690) gives the record without authority it has been mistakenly credited to him. There is little doubt that the locality is the steep rocky slope in Cwm George. John Storrie in his *Flora of Cardiff* (Storrie, 1886) records that the fern was completely exterminated by a dealer in 1876, who uprooted and sold hundreds of plants at one shilling each. Storrie mentions that he bought one plant himself. It would be interesting to know if any of these still survive in cultivation. However the fern was not quite exterminated for the Rev. H. J. Riddelsdell re-discovered it in 1908, but it was not until some 70 years later in 1977 that it was found again, and then only by an intrepid local naturalist Mr Malcolm Wood abseiling down the cliff face. The Laciniated Polypody is a rare sterile variety of the Southern Polypody and has the frond much divided into segments.

At about the same time or slightly later Dr James Newton made an interesting discovery on an island in the Severn Estuary. Ray in his *Historia Plantarum* (Ray, 1688 p. 1126) records *Allium ampeloprasum* (Wild Leek) – 'In parva insula *Holms* dicta supra

Bristolium in Sabrinae aestuario copiose provenit'. J. E. Dandy refers to the same record in *The Sloane Herbarium* (Dandy, 1958 p. 172) – 'In Petiver's "Hortus siccus Anglicanus" . . . is a specimen of *Allium ampeloprasum* L., a species added to the British flora by Newton . . ., labelled, "Dr Newton found this in Holms Isles". . .'. It has always been assumed that 'Holms Isles' refers to Steep Holm in Somerset, but as the island was not named, it is interesting to conjecture that the specimen might have come from its close neighbour, Flat Holm in Glamorgan, where the Wild Leek is also very abundant.

Nearly thirty years pass before we come across any more Glamorgan plants, but in 1698 another fern for which Glamorgan is famed – *Adiantum capillus-veneris* (Maidenhair Fern) – was discovered by the famous Welsh antiquarian and botanist Edward Llwyd.

In 1695 Edward Llwyd was invited by a group of people in Glamorgan to write a 'Natural History of Wales', for which purpose they offered him £10 a year for seven years. Llwyd did not think very highly of the offer and complained of the indifferent financial support for natural history as compared to that given for his *Dictionary* (unpublished) and *Archaeologia* (Llwyd, 1707). Nothing came of the project, but between 1697 and 1701, Llwyd made his 'Great Tour' of Wales, Ireland, Scotland, Cornwall and Brittany, to collect materials mainly of an archaeological nature, for Gibson's proposed revised edition of Camden's *Britannia* (Camden, 1695). The first two years of the 'Tour' were spent in Wales and it was then that Llwyd discovered the Maidenhair Fern and wrote in a letter to Dr Richard Richardson dated 19 September 1698 – 'I was surprised at none so much as to see the [Maidenhair Fern] growing very plentifully out of a marly incrustation, both at Barry Island and Parth Kirig [Porthkerry] in Glamorganshire, and out of no other matter; . . .' (Gunther, 1945). It was thought to be extinct on Barry Island, but was rediscovered there in 1983 by Peter Jones who, in the same year, also recorded it at 43 other sites between Barry Island westwards to Dunraven. It always occurs in wet sheltered places where calcareous tufa is formed and the colonies are usually well out of reach, high up on cliff ledges. Llwyd, in the same letter to Richardson, expresses surprise that *Anaphalis margaritacea* (Pearly Everlasting) '. . . should grow on the banks of Rymny [Rhymney] river for the space of at least 12 miles, begining near the fountain-head in a mountain in this county; and yet not a plant of it to be seen elsewhere throughout Wales.' (Gunther, 1945). Pearly Everlasting is a North American plant which was originally introduced as a garden flower. How it came to be so abundant as a naturalized plant in the Rhymney valley, before it was found elsewhere in Britain is difficult to explain. It has since escaped from cultivation in various parts of Britain and is now fairly widespread on railway banks, walls, sand dunes and waste places. Edwin Lees gives a clue to its possible origin (Lees, 1851); after mentioning having seen the plant in seven or eight places in the Rhymney valley, on the margins of woods and in waste places, he writes – 'In South Wales the Pearly Everlasting is much used as a fit emblem to adorn the graves of departed friends, as a simple but intelligible "in memoriam".' This custom may have had some symbolic significance, apart from the adorning of graves, since it is one of the so-called everlasting flowers. If this practice was more or less confined to the Rhymney valley, at the time of and before Llwyd's visit to the district, it could account for the abundance of Pearly Everlasting as a naturalized plant in that area and not elsewhere.

Among the plants listed by Llwyd in 'Camden's Britannia' was another fern *Asplenium ceterach* (Rustyback), which he found on the walls of Bonvilston Church. Rustyback is a conspicuous feature on limestone walls in Glamorgan, a fact which strikes the visiting botanist, since away from the west of Britain it is much less common. He also saw *Centranthus ruber* (Red Valerian) on the walls of Margam Abbey. This species was introduced into cultivation from southern Europe and soon became naturalized on old walls. It is now a common sight (in white, red, and pink colour forms) as it grows in great abundance on sea-cliffs, walls, railway banks and waste ground etc., throughout the county. Also mentioned were *Aquilegia vulgaris* (Columbine) near Margam; *Antirrhinum majus* (Snapdragon) on walls of Margam Abbey; *Salix repens* (Creeping Willow) on sands, sea coast; *Erodium maritimum* (Sea Stork's-bill) seashore; and *Centaurium pulchellum* (Lesser Centaury) seashore.

Llwyd, in his search for information for his revision of the Welsh sections of Gibson's edition of Camden's Britannia, had sent out 'parish-by-parish Queries' to local correspondents. One of these was Isaac Hamon of Bishopston, Gower, and he sent to Llwyd an extensive description of Gower in July 1697. The manuscript is preserved in MS. Carte 108 at the Bodleian Library, Oxford.

F. V. Emery (1965), mentions the contribution of Isaac Hamon to our knowledge of Gower plants in the 1690s with the following words – 'The botanical value of his tally of forty-two plants, the first Gower Flora, is hard to over-estimate, several of them in fact having become rarities since he wrote.' In view of the importance of this list it is given here in full, with modern Latin and English names added in square brackets, where a name is in doubt it is prefixed with a ?.

The Sea Cost [sic]

The South pt of Gowersland . . . being in length from Swanzey to Worms head about 12 miles is for the most corn ground with store of limestones, & limestone cleeves, wherein are many great holes or Caves, . . . here are these sorts of sea hearbes, as scurvie grasse [*Cochlearia officinalis* (Common Scurvygrass)], Sampire [*Crithmum maritimum* (Rock Samphire)] & lavar [*Porphyra umbilicalis*] of Rock herbs, Cetrack [*Asplenium ceterach* (Rustyback)], maiden hair [*Asplenium trichomanes* (Maidenhair Spleenwort)], walrue [*Asplenium ruta-muraria* (Wall-rue)], & in the p[ar]ishes of Bishopstown, Pennard, & Oystermouth there is plenty of Juniper [*Juniperus communis*] & some buckthorn [*Rhamnus catharticus*]

Of field hearbs (especially in the said 3 pishes) Agrimony [*Agrimonia eupatoria*], wild carret [*Daucus carota* (Wild Carrot)], mulleyn [*Verbascum thapsus* (Great Mullein)], Dandelyon [*Taraxacum officinale* (Common Dandelion)], Pelamountain [*Thymus praecox* subsp. *arcticus* (Wild Thyme), mallows [*Malva* sp.], Burdock [*Arctium minus* (Lesser Burdock)], Tutsan [*Hypericum androsaemum*], Eybright [*Euphrasia officinalis* (Eyebright)], Bettony [*Stachys officinalis* (Betony)], Elecampane [*Inula helenium*], Foxfingers [*Digitalis purpurea* (Foxgloves)], yellow & blue Kay-roses [?*Primula veris* (Cowslip) and cultivated 'Polyanthus'], Rames or Ramsey [*Allium ursinum* (Ramsons)], Centry [?*Centaurium erythraea* (Common Centaury)], Yarrow [*Achillea millefolium*], Adders tongue [*Ophioglossum vulgatum* (Adder's-tongue)], vervain [*Verbena officinalis*], St Johns wort [*Hypericum* sp. (St John's-wort)], Cancker wort [?*Papaver rhoeas* (Common Poppy)], Devilles bit [*Succisa pratensis* (Devil's-bit Scabious)], Ragwort [*Senecio jacobaea* (Common Ragwort)], mugwort [*Artemisia vulgaris*], Breakestone-psley [*Aphanes arvensis* (Parsley-piert)], Larks bill [?*Consolida ambigua* (Larkspur) or *Aconitum napellus* (Monk's-hood)], plantane [*Plantago* sp. (Plantain)], Pimpnell [?*Anagallis arvensis* (Scarlet Pimpernel)], Fumitory [*Fumaria* sp.], Burnet [?*Sanguisorba minor* (Salad Burnet) or *Rosa pimpinellifolia* (Burnet Rose)], Botchwort [?]

of hearbs in some waterie places, as water cresses [*Nasturtium officinale*], Rosa solis [*Drosera rotundifolia* (Round-leaved Sundew)], Lungwort [*Pulmonaria officinalis*] Liver wort [a Liverwort] . . .

The Corn & grain that growes in all pts of Gowersland & thereabts are wheat [*Triticum* sp.], barley [*Hordeum* sp.], Ry [*Secale cereale* (Rye)], oates white & gray [*Avena* sp.], pease [*Pisum* sp.], & fitches [vetches, *Vicia* sp. and *Phaseolus* sp.], & in divers pts of Swanzey hund[re]d there are much Clover [*Trifolium*] grasse & seede especially at & near Bpstown, . . .

After Llwyd's visit to Glamorgan, a long night seems to have settled on botanizing in the county, for it is not until 75 years later, in 1773, that any additions were made to the handful of species recorded. In that year the Rev. John Lightfoot, in the company of Sir

Joseph Banks, made a botanical tour of Wales. The Rev. John Lightfoot was born at
Newent in Gloucestershire, he became chaplain to the Duchess of Portland and later
lived at Uxbridge near London. He was one of the founders of the Linnean Society and
a Fellow of the Royal Society.

Sir Joseph Banks is the more famous of the two companions, and is especially
remembered for his collections having formed the foundation of the British Museum
(Natural History). Charles Lyte in *Sir Joseph Banks* (1980), tells us something of Bank's
botanizing methods while travelling (slowly) by coach in 1774; he quotes from the
reminiscences of George Colman, then a 12 year old lad:

> Unwearied in botanical research, he travelled with trunks containing voluminous
> specimens of his hortus siccus in whitey-brown paper; and large recepticles for further
> vegetable materials, which he might accumulate in his locomotions . . . we never saw a tree
> with an unusual branch, or a strange weed, or anything singular in the vegetable world,
> but a halt was immediately ordered. . . . Many articles "all agrowing, and agrowing",
> which seemed to me no better than thistles, and which would not have sold for a farthing
> in Covent Garden Market, were pulled up by the roots, and stowed carefully in the coach,
> as rarities.

Banks and Lightfoot set out with the intention of following Ray's itinerary, but in the
reverse direction. Starting from Bristol they reached Glamorgan on 2 July 1773. Much
of the ground they covered in West and North Wales coincided with Ray's itinerary
and many of their records repeated those of Ray and Llwyd, but in journeying through
Monmouthshire and Glamorgan they opened new ground; consequently this part of
their journey was the most original and the most rewarding.

They visited Cwm George, but failed to find the laciniated polypody although they
could not have spent much time looking for it, since on the same day they travelled to
Porthkerry and saw *Adiantum capillus-veneris* (Maidenhair Fern) growing on the cliff
face. Their most interesting discovery here was *Buglossoides purpurocaeruleum* (Purple
Gromwell) growing among the bushes on the cliff top. This is one of the rarest British
wild flowers, confined to Glamorgan, Monmouthshire and Denbigh, and to a few
counties in southern England. Other plants of special interest that Banks and Lightfoot
recorded from Porthkerry were *Linum bienne* (Pale Flax) and *Ophrys apifera* (Bee
Orchid).

The following day they went to Flat Holm where they found, among other plants,
Allium ampeloprasum (Wild Leek) growing by the landing stage, and *Ophrys apifera* (Bee
Orchid) and *Erodium maritimum* (Sea Stork's-bill) all over the island. They also visited
Steep Holm and again saw the wild leek and several other plants but made no mention
of *Paeonia mascula* (Peony) for which the island is famed, a fact which suggests that it
might be a comparatively recent introduction there. The wild leek still thrives on Flat
Holm and John Storrie in the last century compared it with the cultivated leek from the
culinary angle and found that they differed much in flavour, the taste of the wild leek
being particularly abominable, lingering in the mouth for days!

The travellers then went to the saltings at Rumney, where they saw the somewhat
rare grass *Alopecurus bulbosus* (Bulbous Foxtail), which is still a feature of the locality.
Lightfoot in his diary goes on to record that they saw at Rumney near Cardiff large
crops of *Plantago maritima* (Sea Plantain), called by the inhabitants 'Gibbals', and refers
to the then custom of turning pigs out on to the saltings so that they could feed on the
plantain, which – 'They rout up the Roots as we saw, and grow fat upon them, as were
assured.'

After passing through Cowbridge, visiting St Donats Castle and Nash Point, they
came to Briton Ferry and there found *Matthiola sinuata* (Sea Stock), on a sandy bank.
The Sea Stock was apparently quite common early last century on the sand hills

between Swansea and the Mumbles and also grew more sparingly at Crymlyn Burrows; then for some unexplained reason it seemed to disappear from the county for over a hundred years until it was rediscovered in dunes on both sides of the Neath estuary in the 1960s. It is now thriving again on Jersey Marine. The related species *Matthiola incana* (Hoary Stock), which curiously enough Lightfoot and Banks failed to notice, is a conspicuous feature on inaccessible cliff ledges at Nash Point, but perhaps like the Paeony on Steep Holm it is a relatively recent arrival.

In the caves and crevices of the rocks at Nash Point they found *Asplenium marinum* (Sea Spleenwort): this and *Adiantum capillus-veneris*, are the only British ferns more or less confined to sea-cliffs.

At Briton Ferry they visited the sand dunes and recorded several of the typical sand dune plants, including 'A yellow flowered violet which seemed to be only a variety of *Viola tricolor*'. It was not until 1886 that the violet mentioned was recorded for the county under its correct name of *Viola curtisii* (now *V. tricolor* subsp. *curtisii*).

On 8 July 1773, the two companions were:

> On our right Hand 2 miles from Breton ferry and about ¼ of a mile from the Sea, [where] we observ'd a marshy Lake extending a mile in length, call'd Coars Crym Lyn [Crymlyn Fen] in which we found plenty of the *Typha angustifolia* [Lesser Bulrush], *Schoenus mariscus* [*Cladium mariscus* (Great Fen-sedge)], *Nymphaea alba* [White Water-lily], *Hydrocharis Morsus Ranae* [*H. morsus-ranae* (Frogbit)].
>
> Upon the marshy Tufts in the Lake, *Osmunda regalis* [Royal Fern], *Myrica gale* [Bog-myrtle], *Hypericum elodes* [St John's-wort], *Carex pseudo-Cyperus* [*C. pseudocyperus* (Cyperus Sedge)] *Carex paniculata* [Greater Tussock-sedge], *Geranium cicutarium* [*Erodium cicutarium* (Common Stork's-bill)].

Crymlyn Fen has suffered much from industrialization, and the pool and fen are now much reduced in size. Some of the plants which grew there when Lightfoot and Banks visited the area such as *Osmunda regalis* maintain a precarious hold, but others have long since disappeared. Chief among these are *Drosera anglica* (Great Sundew) and *Andromeda polifolia* (Bog-rosemary). At long last Crymlyn Fen has been declared a National Nature Reserve and it is to be hoped that this will help prevent the loss of any more species. H. J. Riddelsdell published a transcript of Lightfoot's Diary (Riddelsdell, 1905d), and it is from this that the above quotations are taken.

The Rev. Sir John Cullum (1733–1785), who is, perhaps, best known botanically for his discovery of *Veronica verna* (Spring Speedwell) in East Anglia, visited the Pontypridd area of the county sometime in the latter half of the eighteenth century. Two of his records appear in Turner & Dillwyn (1805). These were for *Wahlenbergia hederacea* (Ivy-leaved Bellflower) which Sir John found 'plentiful in moist places about Pont y Pridd', and *Sibthorpia europaea* (Cornish Moneywort) 'under a damp shady wall on the left about 200 yards before you come to Pont y Pridd from Cardiff.'

Up to the end of the eighteenth century very few botanists, with the notable exceptions of those already mentioned, had visited Glamorgan, and resident botanists seem to have been non-existent. This lean period was in some degree a reflection of the state of botanical affairs generally. There were few botanists in Britain at the time and most of these were professional botanists or medical men. This is, however, hardly surprising when one recalls that practically all botanical works were written in Latin!

Ray's *Synopsis Methodica Stirpium Britannicarum* (*Synopsis of British Plants*, Ray, 1690), was the standard flora for over 70 years, until Hudson's *Flora Anglica* (Hudson, 1762), also in Latin, appeared. Furthermore, during the first half of the eighteenth century, natural history did not figure very much in the general culture of the period – a period when the proper study of mankind was considered to be man himself, – when a devotion to what Ray described as 'the rich array of springtime meadows' was quite unforgivable.

The later half of the eighteenth century, however, witnessed a revival of interest in natural history, and a love of the out-of-doors. It was the period with which one associates the name of Gilbert White of Selbourne; at this time, too, arose the English School of water colour painting and of landscape painting generally. This Romantic Revival was accompanied by a love of wild flowers, for their form and beauty, and a desire to know them for their own sakes.

A natural consequence of this revival was a desire for botanical books, especially those written in English rather than Latin. William Withering's 'flora', (*A botanical arrangement of all the vegetables naturally growing in Great Britain . . .*) which first appeared in 1776 and ran to several editions under slightly different titles (see Withering, 1776), met a widespread want. Other works and small handbooks began to appear in increasing numbers so that – to use Withering's own words – botany became '. . . familiar to those who are unacquainted with the Learned Languages.'

A further factor was the introduction of the Linnean system of classification which, although highly artificial, greatly simplified the identification of plants, based as it was on the number of stigmas and stamens in the flower; but what was equally important was Linnaeus's binomial system of nomenclature. As was once remarked, the opportunity of calling the Lesser Periwinkle, *Vinca minor*, instead of 'Clematis daphnoides minor, seu Vinca pervinca minor' must have proved most attractive.

Thus began a new era; that of the accumulation of data culminating in the production of local lists and floras.

The first such list for Glamorgan was produced by Dr William Turton for the *Swansea Guide* (Oldisworth, 1802). Dr Turton, whose interests were mainly the study of shells, was born in Gloucestershire in 1762 and eventually went to Swansea as a medical practitioner. His collection of shells is now in the United States National Museum at Washington; *Turtonia*, a genus of bivalve molluscs, was named in his honour in 1849. It has been remarked that Turton was not always to be relied on in his published statements; and this is especially true of his botanical records, for the list of plants he supplied for the *Swansea Guide* contains an undue number of very dubious records. Incidentally this list seems to have given rise to a rumour that Turton published a pocket flora, if he did no record of it has been traced! Amongst the plants correctly recorded by Turton were *Monotropa hypopitys* (Yellow Bird's-nest), from Margam woods.

To John Lucas of Stout Hall near Swansea, who, with William Turton and Charles Collins, assisted the author of the *Swansea Guide*, the Rev. John Oldisworth, belongs the credit of having discovered *Draba aizoides* (Yellow Whitlowgrass) near Worm's Head in 1796. This plant, a member of the cabbage family, with rosettes of rather stiff leaves, fringed with bristles, and bright yellow flowers which are rather large for so small a plant, grows on Pennard Castle, and on the cliffs between there and Worm's Head, and nowhere else in the British Isles. Since its nearest other locality in Europe is in the Alps of central and south-eastern Europe, its occurrence in Glamorgan provides one of the puzzles of plant distribution; it has been considered by some botanists to be a naturalized garden escape, but its distribution in Gower scarcely supports this view.

Outstanding amongst these early nineteenth century botanists in the county was Lewis Weston Dillwyn, born in 1778 in Ipswich, son of William Dillwyn, who was descended from an old Breconshire family. Lewis Weston was sent by his father in 1803 to take charge of the Cambrian Pottery at Swansea which he had bought. But Dillwyn is remembered not so much as a potter, but as a scientist, author of an important work on British algae (*British Confervae*, 1802–1809), and joint author with Dawson Turner of the *Botanists' Guide*. Dillwyn was elected a Fellow of the Royal Society at the unusually early age of 25 and was but 27 when the *Botanists' Guide* was published in 1805. This work consisted of lists of the rarer plants of Britain arranged under

Fig. 1.1 Lewis Weston Dillwyn (1778–1855)

county headings. The Glamorgan list was partly drawn from published sources (26 records), some records came from correspondents, especially his friends Edward Forster and Joseph Woods (41 records), but the majority were made by Dillwyn himself (79 records). Amongst the plants he recorded were: *Cladium mariscus* (Great Fen-sedge) from Crymlyn Bog near Swansea, where it still survives in one of only two sites in the county; *Lobelia dortmanna* (Water Lobelia) from the lakes of Craig-y-llyn, the only Glamorgan site and the most southerly in the British Isles; *Geranium pratense* (Meadow Crane's-bill) from Pont-nedd-fechan, a local plant in Glamorgan, although quite common in the neighbouring counties of Breconshire and parts of Monmouthshire; and *Juncus acutus* (Sharp Rush), 'common on the sand hills about Swansea'.

In 1840 Dillwyn had privately printed his *Contributions towards a History of Swansea*, which included an eight page alphabetical list of 117 species or varieties of the rarer flowering plants that had been found within 20 miles of Swansea. Apparently the list was compiled in great haste and 300 copies printed for sale at a Bazaar in aid of Swansea Infirmary. As Carter (1955) wrote – 'Perhaps this is the only occasion upon which a Flora has been compiled with such a charitable end in view.'

In 1848 the British Association held its Annual meeting in Swansea, and Dillwyn, who occupied the Chair of the Natural History Section, published privately, and dedicated to the President and Council of the British Association, an important contribution to the botany of Glamorgan with the modest title of *Materials for a fauna and flora of Swansea and neighbourhood*. The 'floral' part was an enlargement of his 1840 list and included localities for 267 species of flowering plants, ferns and stoneworts. Among the contributors to the list were: James Bicheno, Edward Forster, J. W. G. Gutch, Edward Hawkins, A. Henfrey, Dr Joseph Hooker, Mr & Mrs Dillwyn Llewelyn, John Lucas, Thomas Milne, Matthew Moggridge, John Motley, Rev. John Montgomery Traherne, William Turton, and Joseph Woods. It is in this work that we find the first mention of *Aconitum napellus* (Monk's-hood) growing in Glamorgan. Dillwyn, who thought it to be a naturalized garden escape, found it in Nicholaston Wood and in woods at Penllergaer. Monk's-hood is now better known in its apparently native habitats along the River Ely in the Vale of Glamorgan.

Other interesting plants which Dillwyn records in his book are: *Helleborus foetidus* (Stinking Hellebore), from Park Mill, Gower, it still grows in the area; *Paris quadrifolia* (Herb-Paris), in the woods at Nicholaston and in other Gower woods; *Ranunculus lingua* (Greater Spearwort), from Crymlyn Bog and Kenfig Pool, still found at the former site but probably extinct at Kenfig and all but one of its former sites in the eastern part of the county.

About *Cardaria draba* (Hoary Cress), Dillwyn wrote:

> In 1802, I found this plant on the banks of the Tawey [R. Tawe], above Swansea, where ballast had been formerly deposited; and in 1840, as the old habitat had been covered by the Hafod wharfs, I requested Mr. Moggridge [who was Hon. Curator of the Museum of the Royal Institute of South Wales] to search for it a little higher up the river, and he found it growing there plentifully. I do not suppose that any ballast can have been deposited so high up the river as where the plant is found for much more than half a century, and as it has so long continued to ripen the seed, it must be at least perfectly naturalized, and become as much of a native as many of the plants which appear in the British Flora.

Dillwyn's 1802 record was the first for *Cardaria draba* in Britain, it is a native of S.E. Europe and Asia Minor, and his words proved prophetic; there have been subsequent introductions, and, since it spreads rapidly by its underground runners and sets seed freely, it is now a frequent weed of arable and waste land throughout Great Britain.

Dillwyn also records another introduction, *Narcissus* × *medioluteus* (Primrose-peer-less), on the authority of James Bicheno, 'it grows in abundance in Kenfig village, not far from the Church.' It thrived in the same locality until 1976 when the field in which it grew was ploughed up and planted with barley!

James Ebenezer Bicheno (1785–1851) lived for a time at Pyle. He was an accomplished botanist and collected many plants in the area, his extensive herbarium was acquired by the Royal Institution of South Wales at Swansea (Riddelsdell, 1902a), as was that of another very good botanist, who lived for a time at Aberavon, James Motley (1821?-1859) (Riddelsdell, 1902b).

Matthew Moggridge (1803–1882) was the Honorary Curator of the Museum of The Royal Institution and contributed to the Report of the Institution a list of plants seen in flower in the neighbourhood of Swansea for each month of the year (Moggridge, 1844a). He was a son-in-law of L. W. Dillwyn and the father of John Treharne Moggridge who wrote the *Flora of Mentone* published in 1867.

Many of Dillwyn's other relations were also active in natural history in the county. His elder son, John Dillwyn-Llewelyn (1810–1882) besides collaborating with Wheatstone in his work on the electric telegraph, and Fox Talbot (a relative by marriage) in photography, was also 'a sedulous botanist' (Lloyd & Jenkins, 1959). His son (L. W. Dillwyn's grandson) (Sir) John Talbot Dillwyn-Llewelyn (1836–1927) was an important public figure and a keen gardener, who became vice-president of the Royal Horticultural Society. L. W. Dillwyn's younger son, John Llewelyn Dillwyn (1814–92) married a daughter of that noted geologist and friend of the family Henry de la Beche, and was himself a competent geologist. In Carter (1955) we find that:

> Connected by marriage was the Traherne family, notable among whom was Rev. John Montgomery Traherne, F.R.S. (1788–1860) of Coedrhiglan, Chancellor of the diocese of Llandaff, who formed a collection of plants including some of the earliest Glamorganshire specimens [now in the herbarium of the National Museum of Wales, including one of *Pilularia globulifera* which Traherne found at Pysgodlyn Mawr near Welsh St Donats] . . .
>   Other related members of this family were: Sir Thomas Mansel Franklen, an authority on Glamorgan ferns; Mrs. George Montgomery Traherne, who contributed a list of Glamorgan plants to the Cardiff Naturalist Society Transactions, and Capt. G. Traherne who contributed about 113 records . . . to Trow's *Flora of Glamorgan*.

Since Lewis Weston Dillwyn's father was a member of the Society of Friends (Quakers), he received his early education at a Friends' school at Tottenham, where he first met Joseph Woods. Dillwyn and Woods later went to Dover where they met up with Edward Forster and all three remained firm friends for life; while in Dover, Dillwyn wrote a flora of the area (Dillwyn, 1802).

We know that both Forster and Woods visited Glamorgan in or before 1805, probably at the invitation of Dillwyn, since some of their records for the county appear in the *Botanists' Guide*.

Edward Forster (1765–1849) was born at Walthamstow, Essex, and became one of the foremost amateur botanists of the late eighteenth and early nineteenth centuries. He was so keen that he is said to have risen early in order to work on his herbarium, before going to the bank where he was employed; his evenings he devoted to reading and further botanical work. He left notebooks recording his fieldwork, and from these we know that he visited Glamorgan several times between 1805 and 1810. The areas he visited most were in Gower, and around Swansea and Pont-nedd-fechan in the west, and around Quaker's Yard and Caerphilly in the east. He was one of the first to explore thoroughly the wild and treacherous wastes of Crymlyn Bog near Swansea. Among the many plants he recorded were *Rhynchospora fusca*, *Drosera anglica* and *Andromeda polifolia*, all unrecorded elsewhere, and now believed extinct in the county. He was also the first to find *Atropa belladonna* in the cathedral graveyard at Llandaff and *Crepis paludosa* at Pont-nedd-fechan, although most of his recording in that latter area was in Breconshire. He found *Anaphalis margaritacea* 'undoubtedly wild in a wood between Caerphylli and Quakers Yard', but we now know that this species is native to North America not Britain, and is only naturalized here.

Joseph Woods (1776–1864), was born at Stoke Newington, Middlesex, and became an architect, but not a particularly good one, and eventually retired from that profession to devote himself to botany. He is perhaps best known for his book, *The Tourist's Flora*, a *Descriptive Catalogue of the Flowering Plants and Ferns of the British Islands, France, Germany, Switzerland, Italy, and the Italian Islands*, published in 1850.

As mentioned above, Woods was a personal friend of Dillwyn and visited the county several times, but, unlike Forster, he left no field notebooks giving details of the plants he saw. From the records in Turner & Dillwyn (1805) it would appear that he and L. W. Dillwyn were the first botanists to explore the Pont-nedd-fechan and Craig-y-llyn areas of the county. Among the plants he recorded from Pont-nedd-fechan were: *Geum rivale* (Water Avens), *Saxifraga hypnoides* (Mossy Saxifrage), and *Trollius europaeus* (Globeflower), this latter plant of moist pastures once extended down the River Rhymney as far south as Michaelston y Fedw but is now confined to the north of the county. He also recorded one of our less common orchids from this area, *Neottia nidus-avis* (Bird's-nest Orchid), now sadly extinct in that part of the county, but still occasionally met with in woods in Gower and around Cardiff, especially those at Castell Coch.

From Llyn Fach, Craig-y-llyn, Woods recorded the small aquatic fern *Pilularia globulifera* (Pillwort); it survived there until the end of the nineteenth century but is now thought to be extinct in this and its only other Glamorgan site. He collected plants during his visits to various parts of Britain and these were given to his friend James Bicheno, who in turn presented them to the Royal Institution of South Wales, Swansea, when he left Britain for the Far East in 1839. (Riddelsdell, 1902a).

Woods published an account of 'a visit to Glamorgan and Monmouthshire, in the latter part of July and beginning of August, 1850' in that semi-popular botanical journal, *The Phytologist* (Woods, 1850). The major part of this account is taken up with his searches for brambles (*Rubus* spp.) but he does list a few other plants that he came across. It is interesting to note that he found *Brassica oleracea* (Wild Cabbage) plentiful on Barry Island; in the late nineteenth century this plant extended from Barry to Southerndown, but has now become very rare, if not extinct, east of Llantwit Major. Another interesting comment is about *Matthiola sinuata* (Sea Stock), Woods writes:

> When Dillwyn's *Botanist's Guide* was first published, [it] used to be found in several places among the sand-hills of the shore. Afterwards it disappeared. Two or three years ago it again showed itself, but has again disappeared. The corporation of Swansea, it seems,

sometimes take turf from the sand-hills and replace the soil with some they want to get rid of from the neighbourhood of the town, and in these spots I noticed Calendula officinalis [Pot Marigold], Koniga maritima [*Lobularia maritima* (Sweet Alison)], Delphinium con-solida [*Consolida regalis* (Forking Larkspur)], a cultivated Pimpinella and other garden plants. This could hardly be the origin of the Matthiola; but I suspect that it has produced the D. consolida, noticed by Mr. Lees, at least I neither saw nor heard of the plant except in such situations.

Matthiola sinuata 'disappeared' again from Crymlyn Burrows between Woods' *c.*1848 record and the 1960s, but is now plentiful there, and apparently spreading to other dune systems. In the same paper, Wood's refers to:

A more curious instance of incipient naturalization was exhibited a few years ago in the *Yucca gloriosa* [Spanish Dagger], a root of which was cast on the shore of Cromlyn burrows. It established itself there, sending out suckers, and for some years seemed quite disposed to be permanent, till a storm buried it under a heap of shingle.

One of the rarest plants in the British flora was discovered by Daniel Sharpe on rocks at Port Eynon in 1828. This was *Ononis reclinata* (Small Restharrow), which elsewhere in Britain is known as a native only in Pembrokeshire, South Devon, and the Channel Isles. It still occurs at Port Eynon. Sharpe's specimen came into the possession of Joseph Woods, and it was he who first identified it as *Ononis reclinata* (H. & J. Groves, 1907).

The study of the distribution of British plants was to become a major study of British botanists and the pioneer of this study was an amateur botanist, Hewett Cottrell Watson (1804–1881). In 1835 he published *The New Botanist's Guide to the Localities of the Rare Plants of Britain*. This followed the same general plan as Turner & Dillwyn's *Botanist's Guide*, but excluded the lower plants. For Glamorgan, only 9 plants were listed that were additional to those in the *Botanist's Guide*; among these was *Consolida regalis* (*Delphinium consolida*), Forking Larkspur, 'Truly wild on the sandy shores of Swansea Bay. E. Lees mss.', which, as mentioned above, was still present in 1850.

Watson considerably elaborated his views on geographical distribution in his *Cybele Britannica, or British Plants and their Geographical relations*, published in four volumes between 1847 and 1859, and they reached their final form in his classic work *Topographi-cal Botany* published in 1873–74, with a second edition in 1883. In this work, Watson divided the British Isles into 112 vice-counties using arguments in their favour which are still as valid today as they were then. Ellis (1974) gave these arguments in full and continued:

Indeed in the last 100 or more years, all the county Floras and lists published have been based on Watson's vice-county system. It is important to remember that although county boundaries may change with the passage of time, the boundaries of vice-counties are fixed for all time to the position they held in the 1840s when Watson was formulating his system.

In Glamorgan we are fortunate that the boundary of the vice-county almost exactly coincides with that of the old administrative county, but it is however quite different from those of the new administrative counties of Mid, South, and West Glamorgan (see map). In the second edition of *Topographical Botany*, some 737 species were recorded from Glamorgan; thirty-six persons were named as having contributed records for 126 of these, with the remaining 611 species being listed without authority. Of the 36 contributors, only 4 were credited with more than 10 species: Miss A. M. Barnard (27), Matthew Moggridge (21), T. J. Dyke (11), and L. W. Dillwyn (11).

By the middle of the nineteenth century botany had become not only a serious popular pursuit, but also one of the polite accomplishments, fostered by innumerable popular works, often very sentimental and freely sprinkled with verses of poetry. Anne Pratt's *Wild Flowers* is a familiar example of the better kind of popular work. The lowest depths seem to have been reached by Edwin Lees, although he was a very accomplished botanist, having given up business to devote all his time to the subject. He wrote *The Botanical Looker Out among the wild flowers of England and Wales*. This work, which first appeared in 1842, is a semi-autobiographical account of botanical excursions arranged according to the months of the year. Lees paid several visits to Glamorgan and some of the plants he met with are referred to in this work. A much enlarged second edition contained many additional records for Glamorgan. It is odd that these records were overlooked by A. H. Trow, editor of *The Flora of Glamorgan* (1911); he may have been put off, as were many botanists, by Lees' amiable twitterings. Thus, referring to *Geum rivale* (Water Avens) in the waterfall district in the north of the county he wrote – 'Shall we dash among the wild cataracts of Glamorgan, beauteous with the crimson drooping blossoms of Water-avens . . .'. Lees also visited Crymlyn Burrows, which he says:

> . . . extend for some miles along the coast, and bounding them inland is a great morass [Crymlyn Bog] intersected with drains and pools covered with the white-water lily [*Nymphaea alba*] and the great Spearwort [*Ranunculus lingua*]. A canal now cuts through the morass amidst reeds and tall aquatics, . . .

In addition to the two plants mentioned above, he also recorded from the Bog both *Drosera intermedia* (Oblong-leaved Sundew) and *D. rotundifolia* (Round-leaved Sundew), *Baldellia ranunculoides* (Lesser Water-plantain), *Cladium mariscus* (Great Fensedge), both species of *Typha*, *T. angustifolia* (Lesser Bulrush) and *T. latifolia* (Bulrush), and *Eriophorum angustifolium* (Common Cottongrass), and from the Burrows *Glaucium flavum* (Yellow Horned-poppy), *Cakile maritima* (Sea Rocket), *Diplotaxis tenuifolia* (Perennial Wall-rocket), *D. muralis* (Annual Wall-rocket), *Oenanthe lachenalii* (Parsley Water-dropwort), *Erigeron acer* (Blue Fleabane), *Erodium cicutarium* (Common Stork's-bill), *Samolus valerandi* (Brookweed), and *Rosa pimpinellifolia* (Burnet Rose), one of our most delightful wild flowers. Among the other places visited by Lees in Glamorgan was Oxwich Bay, which he said 'presents a silent scene of beauty to the explorer'. Today the beauty remains but alas not the silence.

A name which occurs frequently in Trow's *Flora of Glamorgan* is that of John Wheeley Gough Gutch. Gutch, a surgeon and Queen's Messenger, was born in Bristol in 1809. He lived in Swansea for some years, and whilst there published two lists of Swansea plants. The first, *A list of some of the rarer plants met with in the neighbourhood of Swansea*, appeared in 1839 in a medical journal (Gutch, 1839). In it Gutch remarks: 'I am well aware that the foregoing list is comparatively of no value without the insertion of the various habitats.' This defect was remedied in the second of his lists, *A List of Plants met with in the neighbourhood of Swansea, Glamorganshire*, which was published, in parts, in *The Phytologist* between 1841 and 1842 (Gutch, 1844a).

In the short introduction to this latter list, Gutch refers to the extensive herbarium which had been presented to the Royal Institution of South Wales, in Swansea by J. E. Bicheno in 1839, and writes that he looks 'forward with confidence to the time when a complete list of all the plants found in the neighbourhood, with a specimen of each species, will be deposited in the Institution'. The latter wish has not been accomplished so far as the Royal Institution of South Wales is concerned; it remained for another Institution to put it into effect, for the policy adopted by the Department of Botany of the National Museum of Wales in building up its Herbarium has carried out Gutch's idea and, moreover, extended it to the whole of Wales.

Gutch's list of some 550 flowering plants naturally contains many of outstanding interest. Amongst these are *Iris foetidissima* (Stinking Iris), a familiar feature of bushy places on limestone, especially on the coast; *Helianthemum canum* (Hoary Rock-rose) at Cockett, near Swansea; *Utricularia minor* (Lesser Bladderwort) from Crymlyn Bog –

where it still occurs; and *Silene vulgaris* subsp. *maritima* (Sea Campion), a fairly common plant on the Gower coast, which was formerly quite common about Barry, but has now become very rare in the east of the county. *Crambe maritima* (Sea Kale), a very rare plant, now apparently extinct in Glamorgan, he found on the sandy shore between Neath and the Afan Rivers. The last confirmed record of this species appears to be that from Sully made in 1900, when a seedling was found, but in the 1930s a leaf was seen by a student from University College, Cardiff on Worm's Head, and it may have been present in an unspecified Gower locality in the 1950s.

Gutch's list was swiftly supplemented by T. B. Flower and Edwin Lees (Flower & Lees, 1842), and by T. Westcombe (1844), who published additions in later pages of the same volume of *The Phytologist*. T. B. Flower (1817–1899), a surgeon and competent amateur botanist from Wiltshire, had previously published 'A Catalogue of Swansea plants' in *Loudon's Magazine of Natural History* (Flower, 1839).

During the latter half of the nineteenth century the number of botanists who resided in or visited Glamorgan and recorded their finds rapidly increased; many are mentioned in Carter (1955), and most of them confined their attention to the western half of the county, principally Gower and the sand dune areas; consequently, prior to the arrival of John Storrie, we find very few plants recorded for the eastern half.

John Storrie, born in Lanarkshire in 1843, started his working life as a printer, and it was in that capacity that he came to Cardiff to join the staff of the *Western Mail*. In 1877 he became part-time curator of the Cardiff Museum at 9d per hour, but at the end of a few months he resigned. His two successors proved unsatisfactory and Storrie was re-appointed in 1881 and remained curator for the next 11 years. His collection of plants, together with that of Charles Conway, formed the foundation of the Cardiff Museum herbarium, out of which the Herbarium of the National Museum of Wales has grown.

John Storrie's botanical monument is his *Flora of Cardiff* published by the Cardiff Naturalists' Society in 1886. The area covered went beyond the confines of Cardiff and actually comprises some 530 square miles. The majority of the records were made by Storrie himself, and a considerable number of species were recorded for the county for the first time including the following:

> *Gymnadenia conopsea* (Fragrant Orchid), formerly common in a marsh near Llanishen, now rather scarce in Glamorgan; *Platanthera chlorantha* (Greater Butterfly-orchid) from about Llandough and Penarth; *Anacamptis pyramidalis* (Pyramidal Orchid) from Porthkerry and Rhiwbina; *Lathraea squamaria* (Toothwort) from Cooper's Fields, Cardiff; *Polygonatum multiflorum* (Solomon's-seal) from Llwyn-y-grant, Coedriglan, and Llanedeyrn, it still grew in the latter locality until about 1921 and can still be found about Castell Coch and Draethen; *Impatiens capensis* (Orange Balsam) was first recorded in error as *Impatiens noli-tangere* (Touch-me-not Balsam) from

Fig. 1.2   John Storrie (1843–1901)

Llandough, near Cowbridge, where it still grows in abundance, it is closely related to *Impatiens glandulifera* (Indian Balsam), so common now along river banks throughout the county, although surprisingly absent from Gower.

The spread of Cardiff has naturally resulted in the disappearance of many interesting plants Storrie knew, for example *Butomus umbellatus* (Flowering-rush) on Leckwith and Canton Commons, where it grew in abundance up to the 1920s. Other examples are *Hydrocharis morsus-ranae* (Frogbit), also on Leckwith Common; *Trollius europaeus* (Globeflower) from St Fagans; *Crambe maritima* (Sea-kale), from the Leys, Aberthaw; *Geranium pratense* (Meadow Crane's-bill) from Sully and Blackweir; *Parnassia palustris* (Grass-of-Parnassus), from Heath near Llanishen. The locality where this grew was also the home of many other interesting plants, including *Dactylorhiza incarnata* (Early Marsh-orchid), *Genista anglica* (Petty Whin), *Potentilla palustris* (Marsh Cinquefoil), *Epipactis palustris* (Marsh Helleborine), *Pedicularis palustris* (Marsh Lousewort) and *Drosera rotundifolia* (Round-leaved Sundew).

During the nineteenth and the early years of the present century, when field botanists were almost all amateurs, a remarkable number of clergymen found the pursuit of botany a relaxation, although many put so much into it that relaxation is hardly the word. Outstanding amongst these was the Reverend Augustine Ley, born at Hereford in 1842. He became vicar of Sellack in 1887. He was a very close friend of Dr W. A. Shoolbred of Chepstow, whose magnificent herbarium was bequeathed to the National Museum of Wales in 1928. Ley explored considerable areas of Wales, especially Breconshire and the Wye Valley, often on horseback, for other forms of transport were then almost non-existent. Unfortunately he did not visit Glamorganshire very often, preferring the uplands of Breconshire, where the hawkweeds, in which he specialised, are so well represented. It was during one of his rare visits to the county that he discovered, in the upper part of the Rhondda Valley 'near a waterfall', the rare *Circaea alpina* (Alpine Enchanter's-nightshade). This interesting record was strangely enough overlooked until A. E. Wade came across the specimens in Ley's herbarium, now at Birmingham University. *Circaea alpina* is a more delicate, smaller plant than the familiar *Circaea lutetiana* (Enchanter's-nightshade), and is usually found in shaded rocky places and woods in mountainous districts. Its main area of distribution in Britain is Scotland; it becomes very rare as one travels southwards, and like a few other species occurring in the north of Glamorgan, it would appear to be at its most southerly limit here. Ley's record was the first and last for the county as it has not been seen since, despite repeated searches. However, the description of the locality on the specimen's label is very vague and covers a considerable area, so it is just possible that the species still survives, unnoticed, in its original locality.

Two clergymen brothers, the Reverends E. F. and W. R. Linton, who, like Ley, specialized in the genus Hieracium, spent some time botanizing in the county, especially in the Swansea area, in the latter part of the nineteenth century. Details of their records were published in the *Journal of Botany*, and E. F. Linton also contributed records to Watson's *Topographical Botany* and Trow's *Flora of Glamorgan*.

Another botanical clergyman was the Reverend H. J. Riddelsdell. He was resident first at Aberdare and later at Llandaff and explored considerable areas of the county; he published a *Flora of Glamorgan* as a supplement to the *Journal of Botany* for 1907. It was he who discovered *Liparis loeselii* (Fen Orchid) at an undisclosed Glamorgan locality in 1905, and *Poa bulbosa* var. *vivipara* (Bulbous Meadow-grass) at Cold Knap near Barry in 1907. The Fen Orchid has since been found at Kenfig, Margam, Baglan, Crymlyn, Oxwich and Whiteford Burrows, occurring in the damp hollows of the dunes, usually where water has stood during the winter.

So many botanists have contributed to our knowledge of the flora of Glamorgan in the present century that only those who have made a major contribution can be mentioned here, details of some others may be found in Carter (1955).

Henry Harris of Ystrad Rhondda, compiled for the Rhondda Naturalists' Society, *The Flora of Rhondda*, published in 1905. From the preface we learn that – 'The list of plants ... is by no means intended to be complete, but is an attempt to collect and classify the plants found in the Rhondda Valleys – Porth to Mardy, Pontypridd to Blaenrhondda.' Harris also acknowledges 'his indebtedness to the work of the late J. Storrie, Esq.'. Nineteen ferns and 368 flowering plants are mentioned but no grasses, sedges or rushes.

Some tribute must be paid to the botanical investigations of A. H. Trow who added much to our knowledge of the Glamorgan flora and who edited *The Flora of Glamorgan* published by the Cardiff Naturalists' Society in parts between 1906–1909, and as a separate book in 1911.

Fig. 1.3   Rev. Harry Joseph Riddelsdell (1866–1941)

Born at Newtown, Montgomeryshire in 1863, Trow trained first at the University College of North Wales, Bangor, and later under Professor Strasburger at Bonn in Germany. In 1883 he took up a lectureship in the Department of Biology at University College, Cardiff, became Professor of Botany when that department was created in about 1888, and later was appointed Principal of the College.

Trow's most important botanical find was probably that of *Orthilia secunda* (Serrated Wintergreen) on Craig-y-llyn, which he first discovered there in 1892. This is a plant of woods and damp rocky ledges of Northern Europe and the Alps. Apart from the record from the Wyndcliff near Chepstow, the Glamorgan site was the most southerly in Britain, but sadly *Orthilia* has not been seen there for many years; perhaps, with *Circaea alpina*, it still survives unnoticed.

Amongst his many other discoveries, mention should be made of *Hippocrepis comosa* (Horseshoe Vetch) from Pennard Castle in 1892 and subsequently from Ogmore and Ewenny Down, and *Hypericum montanum* (Pale St John's-wort) from three separate localities in the Barry district.

A Mudwort collected by Professor Trow

Fig. 1.4   Prof. Albert Howard Trow (1863–1939)

in 1897 at Kenfig Pool was thought at first to be a narrow-leaved variety of *Limosella aquatica*, but in about 1930 it was identified by Dr Gluck of Heidelburg as *Limosella australis* (Welsh Mudwort), which previously had been known only from Eastern North America. It was later found near the River Glaslyn and Dysynni Broadwater in North Wales. It is very erratic in appearance at Kenfig, coming up in abundance one year and then disappearing for several years until conditions are again suitable; the seeds appear to be capable of remaining dormant in the soil for many years.

One whose knowledge of the county flora surpassed that of any previous worker was Miss Eleanor Vachell. Born in Cardiff in 1879, she was introduced to field botany at the age of 10, by her father Charles Tanfield Vachell, a prominent Cardiff physician. At the age of 12 Miss Vachell was given a botanical diary by a family friend, Mrs George Montgomery Traherne, and on 10 August 1891 she wrote in it details of the discovery of Enchanter's Nightshade (*Circaea lutetiana*), Tutsan (*Hypericum androsaemum*), Wood Sage (*Teucrium scorodonia*) and Creeping Jenny (*Lysimachia nummularia*) in 'Coed Rhyd-y-glyn woods', her first records of Glamorgan plants. At more or less the same time she and her father started their herbarium, which was at first housed 'in one cardboard portfolio' in the linen press of her father's dressing room. A few years later she set herself the task of 'painting her Bentham'; this involved seeing, in situ, every species recognised as British, collecting it if allowable, and recording it by colouring the illustration in Fitch's *Illustrations of the British Flora* and later in Butcher and Strudwick's *Further Illustrations of British Plants*. This led her to almost every corner of the British Isles, and on her death (in 1948), only 13 species remained uncoloured out of about 1800.

She embarked on her life-long association with the flora of Glamorgan by becoming, with her father, joint recording secretary for the *Flora of Glamorgan*, published in the *Transactions of the Cardiff Naturalists' Society* from 1906 to 1909. This work, as mentioned previously, was edited by A. H. Trow and published in book form in 1911.

Miss Vachell also kept a botanical journal that contains accounts of her botanical trips and all sorts of interesting anecdotes. One interesting and amusing episode she records took place in May 1926. Mr R. L. Smith, a local botanist with a great interest in the alien flora of Cardiff and joint author with A. E. Wade, of the 'The Adventive Flora of the Port of Cardiff'; 'Additions to the Adventive Flora of the Port of Cardiff'; and 'Notes on the Adventive Flora of the Cardiff District', published in the *Botanical Society and Exchange Club, Reports* for 1925, 1926 and 1938 respectively, discovered *Roemeria hybrida* (Violet Horned-poppy) growing in an allotment in Splott, Cardiff. This very rare casual he showed to Miss Vachell who, after confirming its identity, immediately set about informing her numerous botanical friends throughout the country of this exciting find. Botanical 'twitchers' started pouring into Cardiff and Mr Smith and Miss Vachell were kept extremely busy conducting them from the station to the allotment. Mr O'Sullivan, who worked the allotment where the poppy appeared, made a small fortune in tips, and his 'little blue flower' as he called it, also got him a job; his long period of unemployment ceased after Miss Vachell had commended him to a prospective employer.

During her lifetime Miss Vachell made many important botanical discoveries, none more so than the four mentioned below.

In about 1930 a Scandinavian botanist working on the microspecies of *Taraxacum* (dandelions), H. G. A. Dahlstedt, received some specimens from her. He decided that one represented a new, undescribed, microspecies which he named *Taraxacum vachellii*, an honour which Miss Vachell looked upon as a humorous episode rather than one of any great scientific importance; a justifiable attitude for *T. vachellii* was, for a time, regarded as conspecific with *T. hispanicum*. But *T. vachellii* has recently been recognised as a 'good' microspecies once more (Haworth, 1988); it grows in several counties in South Wales but has yet to be found in Glamorgan.

It was also in 1930 that she made another important discovery, that of the rare Arum Lily, *Arum italicum* subsp. *neglectum*, in Cwm George, Dinas Powis. This plant is similar to our common Lords-and-Ladies (*Arum maculatum*), but differs chiefly in having slightly different shaped leaves that appear in November, and a larger inflorescence with a greenish lemon-yellow spadix rather than purple. The Dinas Powis and an adjacent locality are still its only native Welsh sites.

In 1937 Miss Vachell attended a meeting of the British Association for the Advancement of Science at Nottingham. She took with her an exhibit of various forms of *Limosella* (Mudwort) that she had found growing at Morfa Pools near Port Talbot. Dr K. Blackburn took a keen interest in the exhibit and was sent further specimens for cytological examination. Chromosome counts proved the existence of two species, *Limosella aquatica* and *L. australis*, together with the previously unknown hybrid between them. The locality of Morfa Pools is now submerged under British Steel's Eglwys Nunydd Reservoir, and it is feared that this hybrid, which has been recorded nowhere else in the world, is now extinct.

Miss Vachell summarised her intimate and unique knowledge of Glamorgan plants in two major publications: 'A List of Glamorgan Plants' in the *Botanical Society and Exchange Club Report* for 1933; and 'Glamorgan Flowering Plants and Ferns' in the *Glamorgan County History*, Vol. I, *Natural History*, 1936.

On her death in 1948, her botanical library (including her books, records, diaries and journals) and herbarium was bequeathed to the National Museum of Wales. The C. T. & E. Vachell Herbarium can still be seen as a separate entity in the Department of Botany, the only collection to be afforded this special treatment. Miss Vachell also made a bequest of £500 to the Cardiff Naturalists' Society for the furtherance of amateur botanical research, and for the publication of a new *Flora of Glamorgan*, providing that her name and records were associated with it. This present *Flora* fulfils these provisions and the authors are proud to dedicate it to the memory of Miss Vachell.

Mr R. Melville, who later found fame as an expert on the genera *Rosa* and *Ulmus* while working at the Royal Botanic Gardens, Kew, received his botanical education while exploring the docks, railways and industrial waste places of the Cardiff area, usually in the company of Arthur Wade and Ronald Smith; many of Melville's records appear in Wade & Smith (1926 and 1927).

1934 was a momentous one for Welsh botany, it saw the publication of a book devoted solely to listing the species of flowering plants found in each of the Welsh vice-counties. The book was *Welsh Flowering Plants*, and the authors, H. A. Hyde and A. E. Wade, Keeper of Botany and assistant respectively at the National Museum of Wales. As Ellis (1974) wrote: 'Welsh botany owes much to the devoted services of these two men; neither was of Welsh origin, but both, in their different ways, advanced the progress of botany in Wales.'

Hyde had previously written *Welsh Timber Trees* published in 1931, and in 1940, the first edition of *Welsh Ferns* by Hyde and Wade appeared; both these books provided keys to, and descriptions of, the taxa included, as well as distributional data. A second edition of *Welsh Flowering Plants* was published in 1957, a 4th edition of *Welsh Timber Trees* (revised by S. G. Harrison) in 1977, and a 6th edition of *Welsh Ferns* (revised and expanded by S. G. Harrison) in 1978. In the first edition of *Welsh Flowering Plants*, 886 native and established introductions were recorded for the county, by the time the second edition appeared some 23 years later, this total had risen to 1144.

Harold Augustus Hyde (1892–1973), besides being author or co-author of the books mentioned above, had wide botanical interests and distinguished himself particularly in the field of aeropalynology. In 1941 he began the first day-to-day survey of atmospheric pollen grains and spores to be undertaken in Europe (Harrison, 1974). This work developed into research on asthma and the forecasting of pollen counts.

Fig. 1.5 Dr Harold Augustus Hyde (1892–1973) with Hirst pollen trap on roof of the National Museum of Wales, Cardiff. Photo: *Western Mail & Echo* © 1970

John Arthur Webb (1886–1961) of Mayals, Swansea, was a schoolmaster who, during school holidays and later when he retired, travelled the length and breadth of Wales recording and collecting specimens which he sent to the National Museum of Wales. The first specimens arrived in 1921 and the last in 1956 when he was over 70 years of age. In all he contributed over 4,000 specimens, many of which were from Glamorgan.

In collaboration with A. E. Wade, Webb wrote manuscript floras of four Welsh counties. Complete copies of those for Radnorshire and Montgomeryshire are held in the Library of the National Museum of Wales, whilst partial copies of those for Breconshire and Gower have recently been discovered in the Library of University College Swansea. Many of Webb's Glamorgan records of flowering plants were published in the *Proceedings of the Royal Institution of Swansea* from 1914–1926, and from 1927 onwards in 'Reports of Phanerogamic Botany with the Pteridophyta and Charads' published in *Proceedings of the Swansea Scientific and Field Naturalists' Society*. In the same Journal in 1929 Webb wrote on 'The presumably extinct plants in West Glamorgan, with notes on dubious and erroneous records' (Webb, 1929c), in 1932 'On *Polygonum cuspidatum* (Japanese Knotweed) in West Glamorgan' (Webb, 1932c), and, in 1956, the first (and only published) part of *The Flora of Gower* (Webb, 1956b). In 1948 a new publication – *Journal of the Gower Society* – was launched; Webb contributed many articles, notably *Naturalists in Gower* (Webb, 1948, 1949), *Gower's own plant – Draba aizoides montana* (Webb, 1950), *The Trees and Shrubs of Gower* (Webb, 1954, 1955b), and *The Ferns and Fern Allies of Gower* (Webb, 1957, 1958).

In the 1950s a new impetus was given to botanical recording throughout Britain by the Botanical Society of the British Isles when it set out to record the presence, or absence, of each species of vascular plant in every one of the 10 kilometre squares of the British Isles. 'With the help of about 1,500 botanists, both amateur and professional, sufficient records, a total of some 1.5 million, were received from most of the 3,500 10 km squares, within the five years allowed for the task. Publication of *The Atlas of the British Flora* edited by F. H. Perring and S. M. Walters took place in 1962 and, for the first time, botanists were able to see, at a glance, the detailed distribution of any species within the British Isles.' (Ellis, 1974).

Some 50 botanists contributed records from Glamorgan; most worked in only one or two squares and made fewer than 500 records, but there were notable exceptions. Beverley A. Miles, a very able botanist who specialised in the critical groups of *Hieracium* and *Rubus*, contributed over 1,300 records from four squares in the Bridgend area. Miles met a tragically early death in 1970 at the age of 33; Gordon T. Goodman contributed some 1,750 records from 6 squares in Gower and the Swansea area; but pride of place goes to Arthur Wade who visited 22 squares throughout the county and contributed over 3,200 records, besides vetting every Welsh record for the *Atlas* scheme and recording in many of the other Welsh counties.

Ellis (1974), has remarked that 'the initiation of the mapping scheme seemed at last to provide the stimulus that had been sought for so long in Wales and work was started on the preparation of Floras in many Welsh counties'. In Glamorgan it was not until 1968 that the idea of starting work on a new Flora was first mooted, but this did not mean that local botanists had been idle. In 1960, Dr M. S. Percival contributed a chapter on botany in the Cardiff region for a book that was published for a meeting of the British Association for the Advancement of Science, held that year in Cardiff (Percival, 1960). Dr Percival was very interested in the floral biology of plants and made a detailed study of the insect visitors to flowers; the results of this research is summarized, for Glamorgan, after the accounts for many of the species in this Flora.

Dr M. E. Gillham, author of several papers on the ecology of Pembrokeshire islands, joined the staff of the Extra-Mural Department of University College Cardiff in the early 1960s and immediately started investigating the flora of the county. The numerous papers and several books that appear under her name in the Bibliography (Appendix IV), testify to her extraordinary energy and enthusiasm.

Mr A. E. Wade retired from the National Museum of Wales in 1962 and his successor as Assistant Keeper, Dr Brian A. Seddon, initiated a 'Lake Flora Survey' which aimed to investigate the vegetation of all the Welsh lakes. This was more or less completed by the time Dr Seddon moved to Reading University in 1966 and some of the results were published in Seddon (1972).

As mentioned above, the idea of starting work on a new 'Flora of Glamorgan' was first mooted in 1968 and a committee, set up to oversee the project, held its first meeting in May 1969. It was chaired by the then Keeper of Botany at the National Museum of Wales, Mr S. G. Harrison and was set up under the auspices of the Cardiff Naturalists' Society; Mrs A. B. Pinkard (N.M.W.) was appointed Secretary, A. E. Wade (ex N.M.W.) Treasurer, and Dr Quentin Kay (U. C. Swansea) and R. Gwynn Ellis (N.M.W.) recorders.

Over 50 volunteer field workers attended an evening meeting held in the Department of Botany, National Museum of Wales on 15 April 1970, eventually well over 100 persons expressed a willingness to help, and over 90 provided some records.

Of the 52 field-workers who contributed more than 100 records, only 10 recorded in more than 6 squares and 5 of these were members of the Flora Committee. Most of the remaining field-workers 'adopted' one or a few 5 km squares, usually around their homes, and many covered their squares very thoroughly indeed. Thirty-eight persons contributed fewer records (1–89), and a further 35 sent in records of cowslips following a newspaper article about the Flora project.

From these figures it would seem that the main burden of recording fell on members of the Flora Committee, but the very valuable contributions made by many of the

Fig. 1.6   Sidney Gerald Harrison (1924–1987)

other field-workers must not be overlooked and those who contributed significantly to the project are listed in Appendix II.

Since botanists have been exploring Glamorgan for at least 325 years, and as some 1,190 native or introduced flowering plants, 550 casuals, 140 microspecies and 140 hybrids have been recorded (approximate numbers), one might be forgiven for thinking that the pursuit of floristic botany in the county was no longer worthwhile. But apart from the fact that there are still relatively neglected parts of the county, the flora is not static, whilst some species become rarer and a few become extinct, others extend their range. From time to time newcomers also appear and a number settle down as permanent members of the flora. In the 80 years since Trow's *Flora of Glamorgan* was published, over 100 species have been added as more or less permanent members of the flora. The latest as recently as 1987 when *Trifolium occidentale* (Western Clover) was discovered by Miss J. Dunn growing on cliff-top turf in Gower.

The importance of recording newcomers as soon as noted is thrown into relief by the fact that the Pineappleweed, *Matricaria matricarioides*, is dismissed in Trow's *Flora of Glamorgan* in a few lines as something that, although widespread, might have been introduced in chicken food. It is now one of the commonest weeds of cultivation and waste ground, not only in Glamorgan but throughout the British Isles.

It may be appropriate to close this account of the history of botanical recording in Glamorgan with reference to a few of the more interesting plants that have been added to our flora during the present century and have since become permanent parts of it.

A notable addition was the occurrence early this century of *Senecio squalidus* (Oxford Ragwort), first seen on railway embankments, it having spread along the Great Western Railway from Swindon. This plant, a native of Sicily, seems to have been grown in the Botanic Garden at Oxford in the eighteenth century; later it was seen growing on old walls and waste places in the town. There it appears to have stayed until the construction of the railway, when it spread along the embankment and cuttings, reaching first Reading and then Swindon whence it continued its migration along the G. W. R. Prior to the First World War it was, away from the G. W. R., a rare plant, but is now abundant in many parts of Britain.

Impatiens glandulifera (Indian Balsam), although now so abundant on river banks throughout the county (Gower excepted), has appeared, in abundance, only since the First World War. It is not mentioned at all in Trow (1911), possibly because it was common in cottage gardens and escapes, if they occurred, were regarded as ephemeral occurrences. Riddelsdell (1907), gives only two, casual, occurrences; at Merthyr Mawr and Peterstone Moor. R. L. Smith first noticed it, in 1910, on the banks of the River Taff just below Radyr, by 1923 it had become very common all along the lower reaches of the Taff and remains so today. Indian Balsam is of course a foreigner – a native of the Himalayas, which was introduced as a garden plant. It escaped, and is now established as a permanent element of our flora. It finds the silt left by river flood water an ideal habitat which in some respects is similar to that favoured by the species in its native Himalaya.

Several other foreign plants have become established within recent years: notably *Epilobium ciliatum* (American Willowherb), *Epilobium brunnescens* (New Zealand Willowherb), *Veronica filiformis* (Slender Speedwell, from Asia Minor), *Reynoutria japonica* (Japanese Knotweed), and *Rumex frutescens* (Argentine Dock).

Epilobium ciliatum was introduced accidentally from North America and, since the 1950s, has become a common weed of waysides and waste places throughout the county, and has been recorded from almost every 5km square during the present survey. It was first recorded in 1943 by Eleanor Vachell in the grounds of Rockwood Hospital, Cardiff.

Epilobium brunnescens, a native of the gravelly margins of streams in New Zealand, was first recorded in the county in 1931 as a garden weed at Rhiwbina, Cardiff (Harrison, 1968). It is now well established over a wide area of upland Glamorgan especially by streams and wet rocky places and looks for all the world like a true native.

Veronica filiformis, introduced from Asia Minor as a rockery plant, is so invasive that it soon becomes a pest. As seed is very rarely produced in Britain, garden throwouts have undoubtedly helped this plant to colonise grassy waysides and river banks in lowland parts of the county.

Reynoutria japonica, a native of Japan, Taiwan and Northern China, was first introduced to Britain as a garden plant in 1825. The first record of it establishing itself in the wild in Britain was provided by John Storrie who in his Cardiff Flora of 1886 referred to it as being – 'Very abundant on the cinder tips, near Maesteg'. Ann Conolly in her masterly paper 'The distribution and history in the British Isles of some alien species of *Polygonum and Reynoutria*' (Conolly, 1977), describes the spread of this species throughout Britain; by 1907 it was reported as 'already common almost throughout Glamorgan . . .', and has been recorded from almost every 5 km square in the county during the present survey.

Rumex frutescens was perfectly at home both on Kenfig Burrows, where it was found in 1932 by Miss Mary Thomas and the foreshore at Swansea, where it was found by Eleanor Vachell in 1940. When first discovered it was abundant at both sites but is now quite rare.

Among native plants, probably the most interesting addition to the flora was the discovery by Mr John Lord in 1938 of *Pyrola rotundifolia* subsp. *maritima* (Round-leaved Wintergreen) on Kenfig Burrows. It has since been found on Crymlyn Burrows, Crymlyn Fen, Oxwich Burrows and Whitford Burrows. Prior to 1939 it was known in Wales from only one locality in Flintshire where it was first recorded in 1926, subsequent discoveries have been made at Newborough Warren, Anglesey (1954); Tywyn Burrows, Carmarthenshire (1964); and Morfa Dyffryn, Merioneth (1981) (Kay, Roberts & Vaughan, 1974). The Round-leaved Wintergreen thus provides an interesting example of a nationally rare species which actually appears to be increasing in numbers.

Other natives or presumed natives that have been discovered in the county in the last 30 years or so include the following species.

Mibora minima (Early Sand-grass), a diminutive, early flowering, annual grass, which was found on Whiteford Burrows, Gower, by E. Duffy and G. T. Goodman in 1964. It is now known to be fairly abundant in an area covering a couple of hectares.

Frankenia laevis (Sea-heath), was found by Steve Waldren growing on a salt-marsh at Merthyr Mawr in 1981. From the size of the population it must have been present, but overlooked, for many years. Coincidentally, both *Mibora* and *Frankenia* find their only other Welsh sites in Anglesey.

As mentioned previously, the most recent addition to the native flora of Glamorgan is *Trifolium occidentale* (Western Clover). It was discovered by Miss J. Dunn in 1987, growing in cliff-top turf in Gower and is now known to be locally frequent along some 1.5 km of cliff.

Who knows what the future will bring?

2
Soils of Glamorgan

P. S. Wright

Introduction

Soil investigations began in Glamorgan before the 1939–45 war under G. W. Robinson (Robinson *et al.*, 1930; Robinson & Hughes, 1936). Between 1958 and 1966 C. B. Crampton of the Soil Survey of England and Wales mapped the soils of the Vale of Glamorgan at 1:63,360 (Crampton, 1972) and made an unpublished reconnaissance survey of Glamorgan, Brecon and Monmouth. Relevant parts have been summarized by Bridges (1967) and Bridges & Clayden (1971), and further fieldwork on Gower has been done by Bridges (1973 & 1976). In the uplands the Forestry Commission has made unpublished detailed soil maps of their forests. This information, with further fieldwork in south Wales by the present author has been incorporated into the 1:250,000 Soil Map of England and Wales (Soil Survey staff, 1983).

Soil Forming Processes

Soil is the upper layer of the earth's crust and the product of a number of complex interacting processes.

Ice, wind and water have been responsible for the accumulation of an unconsolidated debris, or *regolith* mantling the rocks. The regolith in which soils develop is the *parent material* and this has undergone differentiation through biological, physical and chemical processes which add or remove soil constituents. The chemical and physical processes causing disintegration of rocks and minerals are known as *weathering* and in south Wales the main processes are solution, hydration, oxidation and hydrolysis. Physical weathering, principally involving freeze–thaw cycles has become less important since the end of periglacial conditions.

Soil water is the main agent of weathering; rain falls on the surface as a weak acid and as it moves through the soil it dissolves and decomposes minerals carrying away with it bases released, especially calcium, magnesium and sodium, a process called *leaching*. In this way upper soil horizons are continually depleted of bases and without additions of lime and fertilizers can become acid and impoverished.

Eluviation is the process by which clay is mechanically moved in suspension by water percolating to lower levels (argillic horizons) where it is plastered as coatings on structure faces, in pores and round stones. Wetting of dry clay soils also disperses clay particles, and soils displaying this phenomenon are most common in the Vale of Glamorgan where clays are widespread and the climate is driest.

Under heath, permeable coarse textured acid soils commonly have bleached eluvial and dark coloured or ochreous illuvial horizons or an ironpan. These horizons are thought to result from reactions involving the release of aluminium and iron from soil

21

minerals and organic matter, and their translocation and solution within profiles. These and allied processes are known as *podzolization*.

Soils which are periodically, or more or less permanently, waterlogged develop characteristics which result from the reduction and reoxidation of iron oxides. Brown and reddish ferric oxides and hydroxides are unstable in reducing conditions and change to grey or bluish grey ferrous compounds; in contrast ferrous compounds are unstable in oxidizing conditions and tend to change to brown or reddish brown ferric compounds. Thus, periodically waterlogged soils (alternating reducing and oxidizing conditions) are mottled in grey and ochreous colours, whereas persistently water-logged soils (reducing conditions) are grey or bluish grey. The process of reduction of ferric to ferrous conpounds is known as *gleying*. Ferrous iron compounds are relatively soluble and can be removed from soil horizons either partly or completely. Manganese compounds are similarly mobilized in intermittently waterlogged horizons and can be redeposited, as manganese dioxide, either in black concretions or coatings to ped faces.

Plants grow in soil, extracting nutrients needed for growth, and returning their roots and debris to it. Soil organisms living in this organic matter either transform it *in situ* or mix the ingested material with mineral matter over a limited depth. The rate at which plant remains decompose depends on environmental conditions: in base-rich aerated soils it is usually swift and mineral and organic components are thoroughly mixed to form *mull*; in strongly acid soils the rate of decomposition is slow and the organic horizons, more or less clearly separable from the mineral soil, are termed either *moder* or *mor*; *peat* is characteristic of more or less permanently waterlogged sites where decomposition is also slow. Roots and larger fauna like earthworms and ants increase porosity and permeability, and gums and mucilage produced by soil microflora and fauna aggregate soil particles.

The pressures exerted by roots forcing their way into the soil, the swelling and shrinking of clays and the activities of soil fauna produce structural units or *peds* – a process which destroys or modifies lithological structure.

The overall effect of the processes described above is to produce in the soil parent material layers or *horizons* approximately parallel with the ground surface, and distin-guished by such properties as texture, structure, colour, consistence, stoniness, organic matter content and degree of root development. The vertical sequence of soil horizons, approximately to the lower limit of plant activity, constitutes the *soil profile*.

Soil Forming Environment

The character and thickness of the regolith, together with relief, climate, vegetation, man and time determine the characteristics of soil from place to place, since they control the kind of processes which operate and the rate of differentiation of horizons.

Relief
Inland from the coastal marshes and dunes, Gower and the Vale of Glamorgan are low plateaux rising gently inland to about 120m, the former having some higher prominences around 180m on Cefn Bryn, Rhossili Down and Llanmadoc Hill. A dramatic change to upland relief occurs north of a line which runs from Pontardulais to Aberkenfig and Caerphilly (Fig. 2.1), following the outcrop of Carboniferous Pennant sandstone. This forms a large upland area consisting of a series of broad plateaux rising from 300m in the south to 600m in the north, and dissected by deep, steep-sided valleys. North of the steep scarp of the Pennant outcrop, which is most dramatic at Craig-y-llyn (G.R. SN9103), shales of the Middle and Lower Coal Measures are marked

by a broad east–west trough at 200–300m through which runs the former county boundary between Rhymney and Brynamman.

Climate
The climate is controlled by the pattern of relief and exhibits corresponding contrasts. Precipitation ranges from 1000mm in the Vale and Gower to nearly 3000mm on the highest summits (Fig. 2.1). This range is reflected in the moisture balance or potential water deficit, which is the average excess of rainfall over evapo-transpiration in any year. Estimates of water deficit range from about 80mm in the Vale and Gower to zero on the highest summits (Bendelow & Hartnup, 1980). There is a similar topographic effect on temperature. The warmth of the growing season can be expressed by the accumulated temperature above a base of 5.6°C, which is a threshold for grass growth, and values range from 1925 day/°C in the mild areas of the Vale and Gower, to around 1000 day/°C on the highest ground (Bendelow & Hartnup, 1980). These climatic contrasts not only control land use, but also affect the development and distribution of the soils, as is shown below, with regard to the formation of acid peat and podzols on the high ground, resulting from strong leaching and slow rates of decay of organic matter; in contrast to the predominance of less strongly leached, near neutral, brown earths on Gower and the Vale.

The Regolith (Superficial Geology)
A description of solid geology is given in 'The Geological History of Glamorgan' (pages 40–54). In Glamorgan, as in most of Britain, the relationship of soil to solid geology has been considerably modified by the presence of Quaternary superficial deposits (drift), most of which accumulated under glacial or periglacial conditions. A number of glacial periods occurred in the last 2 million years but only the last (Devensian) and penultimate (Wolstonian) glaciations are significant in providing soil parent materials. The Wolstonian glaciation was the most extensive, covering the whole of Wales, ice from the Irish Sea over-riding Gower and the Vale of Glamorgan as far as Cardiff. Another ice mass originating in the Brecon Beacons and the Pennant sandstone uplands entered the Vale from the valleys. The subsequent Ipswichian interglacial period was characterized by a warm temperate climate, and certain reddish soil horizons on limestone may have originated at that time (p. 33). The Devensian glaciation was less extensive than the Wolstonian and its approximate limit is shown in Fig. 2.2. Ice probably did not reach Glamorgan from the Irish Sea, but was entirely of Welsh provenance; tongues extended southward down the coalfield valleys, and combined to form a small piedmont glacier where they debouched onto the Vale and Gower. Severe periglacial conditions existed beyond the ice limit, and over the whole of south Wales, once the ice had retreated about 12000–10000 B.P. The nature and distribution of the various drift deposits of Glamorgan are shown in Fig. 2.3, in relation to the derived soils.

Till
Till or boulder clay occurs mainly within the Devensian ice limits and is mostly of that age and of Welsh provenance. It is accordingly derived from Carboniferous sandstone and shale with some material from the Old Red Sandstone, and is typically an unsorted deposit of rounded and angular stones and boulders in a clay loam matrix of high bulk density. Frequently, there is an extremely compact subsurface 'indurated' layer thought to have formed under permafrost conditions (Fitzpatrick, 1956). Periglacial processes, principally solifluxion, have reworked much of the till and moved it downslope to lower ground so that the thickest drift deposits are in the valleys and depressions over Coal Measures shales.

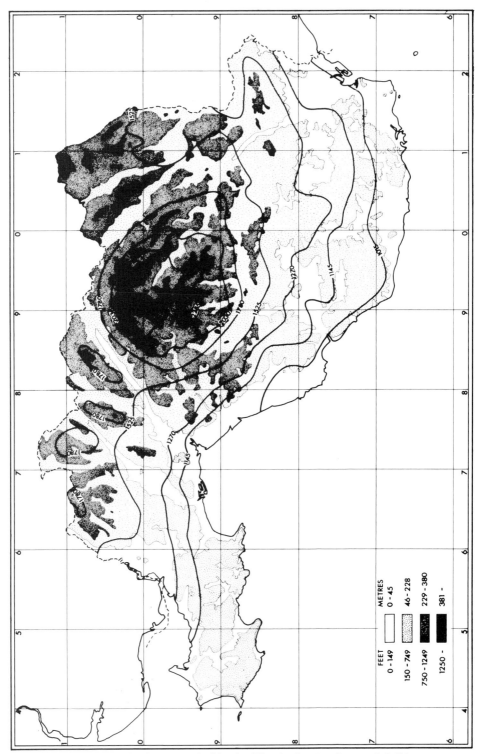

Fig. 2.1   Map showing isohyets and altitude. The isohyets show rainfall in mm.

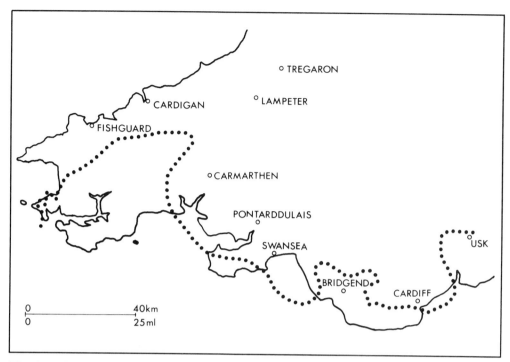

Fig. 2.2   Late Devensian ice limit (after Bowen, 1981).
. . . Ice limit

Soils in this material are usually slowly permeable and strongly gleyed as a result of prolonged waterlogging. The material is also of low base-status and the soils are acid and poor in nutrients.

Till and Glaciofluvial Deposits
In south and west Gower beyond the assumed limit of Devensian ice there are mixed deposits of till and glaciofluvial material. Bowen (1971) believes the till is Wolstonian, and if so it has been extensively reworked during several periglacial episodes, and mixed with glaciofluvial material from the wasting Devensian ice mass on north Gower. Similar deposits, around Cardiff, are termed 'morainic drift' by the Geological Survey (Squirrell & Downing, 1969), and probably compromise the moraines and outwash deposits left by a piedmont glacier. They are derived partly from the adjacent Devonian sediments, which give them a reddish colour, and are sometimes ranging over short distances from compact till to stratified sand and gravel. Soils in these materials are varied in their drainage characteristics from permeable and well drained to slowly permeable and strongly gleyed. The low base-status of the parent materials results in acid soils, poor in nutrients.

Glaciofluvial Deposits
These are the products of meltwater flowing beneath and issuing from ice fronts, and are found in the north of the Vale around the limits of Devensian ice. They consist of rounded stones in a sandy loam or sandy matrix, and often show stratification. In places, most noticeably in Hensol Forest, there are kames and intervening enclosed hollows giving 'kame and kettle' topography. Elsewhere the terrain is hummocky, as

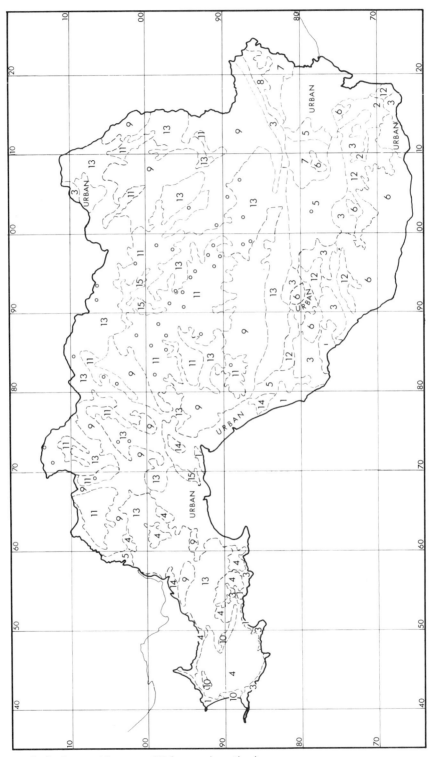

Fig. 2.3   Soils (see table page 27 for explanation)

Fig. 2.3 Soils

Map Unit	Geology	Dominant soil group	Associated soil groups	Soil Characteristics
1	Dune sand	Raw sands	Sand-pararendzinas Humic-sandy gley soils	Calcareous windblown sand; thin topsoils present only on stabilized dunes. Waterlogged soils with humose surface in hollows.
2	Triassic mudstone	Argillic pelosols	Calcareous pelosols Stagnogley soils	Slowly permeable, often reddish clayey soils, usually calcareous at depth, sometimes throughout.
3	Carboniferous limestone and Triassic limestone breccia	Brown earths	Paleo-argillic brown earths Rankers	Well drained loamy or loamy over clayey soils over limestone, shallow in places especially on steep slopes and crests.
4	Till and glaciofluvial deposits	Brown earths	Stagnogley soils	Deep well drained loamy soils and some slowly permeable, seasonably waterlogged soils.
5	Glaciofluvial deposits	Brown earths	Cambic gley soils Humic gley soils	Deep well drained loamy soils, often over sand and gravel. Waterlogged soils with peaty surface in hollows.
6	Lias limestone	Argillic brown earths	Argillic pelosols Rendzinas	Well drained loamy over clayey soils over limestone with some slowly permeable clayey soils. Non-calcareous except some shallow soils.
7	Till and glaciofluvial deposits	Argillic brown earths	Stagnogley soils Brown earths	Deep reddish loamy soils with slowly permeable subsoils and slight seasonal waterlogging. Some slowly permeable, seasonably waterlogged soils and some deep well drained loamy soils.
8	Devonian mudstone and sandstone	Argillic brown earths	Brown earths	Well drained reddish loamy soils over mud-stone. Some similar soils with slowly permeable subsoils and slight seasonal waterlogging and some well drained loamy soils over sandstone.
9	Carboniferous sandstone	Brown podzolic soils	Stagnohumic gley soils Brown earths	Well drained loamy soils over sandstone, usually on steep slopes. Some slowly permeable, seasonably waterlogged soils with a peaty surface usually on valley floors.
10	Devonian conglomerate and sandstone	Podzols	Brown earths Brown podzolic soils	Well drained very acid sandy soils with a bleached subsurface horizon over conglomerate. Associated with less acid well drained reddish loamy soils over sandstone and siltstone.
11	Carboniferous sandstone	Stagnopodzols	Stagnohumic gley soils	Very acid loamy permeable upland soils over sandstone with a wet peaty surface horizon and bleached subsurface horizon. Some soils with thin ironpan.
12	Lias shale and limestone	Stagnogley soils	Argillic pelosols	Slowly permeable seasonally waterlogged clayey soils.
13	Till	Stagnohumic gley soils	Stagnogley soils Disturbed soils	Very acid slowly permeable seasonally waterlogged loamy soils with a peaty surface horizon. Some similar non-peaty soils and some disturbed by opencast coal mining.
14	Estuarine alluvium	Alluvial gley soils	Raw sandy gley soils Unripened gley soils	Loamy alluvial soils with high groundwater. Calcareous and saline in unreclaimed marshes. Sand or soft mud on intertidal flats.
15	Blanket and basin peat	Raw peat soils		Thick, very acid perenially wet peat in the uplands. Eutrophic peat in Crymlyn Bog.

○ Peat bogs; ⌒⌒ Soil boundaries; (URBAN) Principal urban areas. Scale 1:4,000,000

around Margam. Gravel deposits too small to show on the map extend up most of the valleys but have mainly been built over.

Soils in these deposits are permeable, but low-lying areas are affected by a fluctuating groundwater table which may form ponds in winter. The gravels are deficient in bases and stongly leached owing to the porosity, and the soils are naturally acid and poor in nutrients.

Head Deposits
In the cold periglacial climate, frost shattering of bare rock outcrops produced the angular regolith which is characteristic of the hill slopes. This has mostly been moved downslope by solifluxion, resulting in unsorted and crudely sorted 'Head' deposits which are thickest on lower slopes and thinner on upper slopes and summits. Locally derived Head is the principal parent material of soils shown on Fig. 2.3 in which a rock formation is named in the table of geology. Soils in such deposits are usually permeable and well drained. Base status of soil parent material is extremely variable from Carboniferous sandstones in the uplands which produce acid, nutrient deficient soils, to limestones and calcareous mudstones in the Vale and Gower producing base-rich soils with near neutral pH values.

Dune Sand
The extensive dune systems of Glamorgan are of relatively recent (Holocene) origin but are derived from the vast amount of sand washed into what is now the Bristol Channel by glacial meltwater. Archaeological evidence (Higgins, 1933) suggest that dune development began in the Neolithic period with subsequent phases of stabilisation, alternating with periods of great dune migration, as in the mid-Iron Age and the fourteenth century. The continual accretion and movement of sand means that dune soils have little or no profile development.

Alluvium
The most extensive area of alluvium is Llanrhidian Marsh in the Burry Inlet. The alluvium has been transported in post-glacial times by the river Loughor from the hinterland of mainly Carboniferous rocks and deposited under tidal conditions. Alluvial deposits, too small in extent to show in Fig. 2.3, occur in all the river valleys and are generally loamy, frequently overlying gravel and becoming finer textured downstream. In the south of the Vale the alluvium is clayey.

Vegetation and Man

The plant communities of Glamorgan are a product of human activities and so vegetation and man must be considered together in assessing their impact on soil development. Useful summaries of post-glacial vegetation and related soil changes of relevance to Glamorgan are given by Kay (1979) and Taylor (1974).

Glamorgan became free of ice by between 12 000 and 10 000 B.P., and the climate rapidly improved, bringing recolonisation by trees, so that by the mid-Boreal period around 8500 B.P. there was a nearly continuous forest cover of oak, birch, pine and elm with an understory of hazel; the hardier birch and pine predominating at higher altitudes. The spread of trees was aided by a climate that was warmer than at present with smaller temperature lapse rates, and in this environment the soils in the uplands were mostly brown earths. However, Mesolithic man was affecting even this forest, principally by burning, as shown by the pollen record from Cwmllynfell bog (Trotman, 1963) which had an increase in grass pollen by the late-Boreal (7500 B.P.) period.

In Neolithic times pressure on the forest increased with widescale clearance, creating large, fluctuating areas of heath in the uplands. This coincided with the wetter conditions of the Atlantic period (7500–5000 B.P.) which resulted in increased leaching of soils no longer protected by a forest cover. Growth of blanket peat was initiated in Wales at this period. The basal layer of blanket peat at Glaslyn Bog near Plynlimon, in a similar topographic position to the Glamorgan blanket peats gave a radiocarbon date of 4220 B.P. and the underlying soil contains pollen of open-habitat species (Taylor, 1974).

The Bronze Age saw a return to a warmer drier sub-Boreal climate. The concentration of barrows of this age on the Pennant sandstone uplands and on Gower prominences such as Rhossili Down suggests an open habitat, and this is supported by the pollen record at Cwmllynfell (Trotman, 1963) which shows high grass and bracken levels, with some cereals. The increased population of man and domesticated grazing animals would have further hindered forest regeneration.

By Iron Age and Roman times the proportion of tree pollen in the total pollen rain in Wales was as low as at the present day (Moore, 1977), and around 2500 B.P. began the sub-Atlantic period of climatic deterioration to cooler, more humid conditions. Without the protective deciduous forest cover, leaching of soils, initiated in the earlier Atlantic period, was greatly increased. The wet heathy vegetation which replaced forest of the uplands gives rise to an organic surface layer which is resistant to decay and tending to hold water, so initiating peat development. The acid products of decomposition intensified the leaching process, leading to podzolization, giving the peaty podzols which cover the upland surfaces today. Where factors of relief and climate produce waterlogged conditions needed for peat growth, blanket peats have formed, the principal sites being the high plateaux above Craig-y-llyn.

On hillsides, forest clearance and ploughing was accompanied by soil erosion, evidenced by the thick, almost stoneless topsoils which are a common feature of lower slope soils in Wales (Wright, 1980). Much of the deep, stoneless alluvium in the valleys of south Wales, as in other parts of Britain, may result from this accelerated soil erosion (Hazelden & Jarvis, 1979).

In the twentieth century, technology has brought about further wide scale changes to the soil. Land drainage was begun in the last century but few old schemes have remained effective. The Second World War saw the introduction of grant-aided schemes and since then there has been a great increase in the amount of land drained, and a revolution in equipment and techniques. Thus, much old wet pasture and numerous springs and flushes have been drained and reseeded. Many of the upland plateaux have been traversed by closely spaced, deep, open drains for forestry and these, together with the increased evapo-transpiration rates from a forest cover, have reduced the wetness of the peaty soils. In addition, grant-aided reclamation of rough grazing land has resulted in ploughing and reseeding of large areas of upland soils. Finally, the availability of cheap fertilizers, particularly basic slag from local steelworks has raised the pH and base status of agricultural soils.

Soil Classification

The soil classification is that of the Soil Survey of England and Wales (Avery, 1980). Brief definitions of the classes of soils found in Glamorgan with their main distinguishing features are given below. However these definitions are incomplete and reference to the original is advised for the full classification of soils.

The soil groups are differentiated primarily by observable or measurable characteristics of the soil profile, including distinctive surface and subsurface horizons resulting from alteration of the original material by pedogenic processes, and proportions of organic matter, calcium carbonate and differently sized mineral particles within specified depths.

Soil materials containing more than 20–30 per cent organic matter, depending on clay content, are classed as *organic*. Mineral or organo-mineral soil materials, containing less organic matter, and less than 70 per cent by volume of stones (2–200mm), are classed as sandy, clayey or loamy according to mass-percentages of sand (2,000–60μm), silt (60–2μm) and clay sized particles (<2μm) in the inorganic fraction <2mm, as follows:

> *Sandy*: percentage silt + twice percentage clay less than 30.
> *Clayey*: more than 35 per cent clay.
> *Loamy*: other materials of intermediate composition.

Raw Soils. These have no distinct pedogenic horizons other than a superficial organo-mineral or organic layer less than 7.5cm thick. They are usually sparsely vegetated and of negligible value for agriculture or forestry.

> **Raw Sands**. Raw sandy soils, chiefly dune sands.
> **Raw Sandy Gley Soils**. Raw sandy soils with a gleyed subsurface horizon, chiefly on inter-tidal flats or saltings.
> **Unripened Gley Soils**. Raw gley soils having a soft, wet, unripened mineral horizon starting within 20cm depth, chiefly on intertidal flats and saltings.

Lithomorphic Soils. These have a distinct organo-mineral or occasionally organic surface horizon and bedrock or little altered unconsolidated material at 30cm or less.

> **Rankers**. Non-calcareous soils, normally well drained and usually shallow, over non-calcareous rock or massive limestone.
> **Rendzinas**. Calcareous soils, normally well drained and usually shallow, over shattered limestone, chalk, or soft, extremely calcareous (>40 per cent $CaCO_3$) material.
> **Sand Rankers**. Non-calcareous soils in sandy deposits other than recent alluvium.
> **Sand Pararendzinas**. Calcareous soils in sandy deposits other than recent alluvium.

Brown Soils. Well drained or moderately well drained soils with an altered subsurface horizon, usually brownish, that has soil structure rather than rock structure and extends below 30cm depth.

> **Brown Earths**. Loamy soils with a brown or reddish, friable, non-calcareous subsurface horizon, normally similar in texture to the topsoil, in materials other than recent alluvium.
> **Argillic Brown Earths**. Loamy or loamy-over-clayey soils with a subsurface horizon of clay accumulation, normally brown or reddish.
> **Paleo-Argillic Brown Earths**. Loamy, loamy-over-clayey, or clayey soils with a strong brown to red subsurface horizon of clay accumulation, normally attributable to pedogenic alteration of the original material before the last glacial period.

Podzolic Soils. Well drained to poorly drained soils with black, dark brown or ochreous subsurface horizons, often partly cemented, in which aluminium and/or iron have accumulated in association with organic matter. An overlying bleached horizon depleted of iron, peaty topsoil, or both, may or may not be present.

Brown Podzolic Soils. Loamy or sandy soil, normally well drained, with a dark-brown or ochreous, friable subsurface horizon and no overlying bleached horizon or peaty topsoil.

Podzols. Sandy or coarse loamy soils, normally well drained, with a bleached horizon and/or a dark brown or black compact subsurface horizon enriched in humus, and no immediately underlying grey or mottled (gleyed) horizon, or peaty topsoil.

Stagnopodzols. Podzolic soils, usually loamy, with a peaty topsoil, a periodically wet (gleyed) bleached horizon, or both, over a thin ironpan and/or a brown or ochreous, relatively permeable subsurface horizon.

Pelosols. Clayey soils, usually calcareous or over a calcareous substratum, in materials other than recent alluvium. They crack deeply in dry seasons and have a brown, greyish or reddish, blocky or prismatic subsurface horizon, often slightly mottled. These soils are slowly permeable when wet, but a strongly mottled (gleyed) horizon is absent or occurs only below 40cm depth.

Calcareous Pelosols. These have a calcareous subsurface horizon normally similar in texture to the topsoil.

Argillic Pelosols. These have a subsurface horizon of clay accumulation, usually decalcified.

Gley Soils. Soils with grey-and-brown mottled or uniformly grey (gleyed) subsurface horizons in which the original material has evidently been altered by reduction, or reduction and segregation, of iron caused by periodic or permanent saturation with water in the presence of organic matter. Horizons characteristic of podzolic soils are absent or incompletely developed.

1. Gley Soils Without A Humose Or Peaty Topsoil. These are seasonally wet in the absence of effective artificial drainage.

Alluvial Gley Soils. Loamy or clayey soils in recent alluvium, affected by fluctuating groundwater.

Cambic Gley Soils. Loamy or clayey soils in materials other than recent alluvium, with a relatively permeable substratum affected by fluctuating groundwater.

Argillic Gley Soils. Loamy or loamy-over-clayey soils with a subsurface horizon of clay accumulation and a relatively permeable substratum affected by fluctuating groundwater.

Stagnogley Soils. Non-calcareous loamy, clayey or loamy-over-clayey soils in which drainage is impeded at moderate depths by a relatively impermeable subsurface horizon or substratum.

2. Gley Soils With A Humose Or Peaty Topsoil. These are normally wet for most of the year in the absence of effective artificial drainage.

Humic Gley Soils. Loamy or clayey soils in materials other than recent alluvium, affected by a high groundwater-table.

Stagnohumic Gley Soils. Non-calcareous loamy, clayey or loamy-over-clayey soils in which drainage is impeded at moderate depths by a relatively impermeable subsurface horizon or substratum.

Peat Soils. These have an organic layer at least 40cm thick, starting at the surface or at less than 30cm depth.

Raw Peat Soils. Peat soils that remain permanently waterlogged (unripened) and/or contain more than 15 per cent of recognizable plant remains within the upper 20cm.

The soil map (Fig. 2.3) is based on the recently compiled 1:250,000 Soil Map of England and Wales and the 1:63,360 soil map of the Vale of Glamorgan (Crampton, 1972); it is highly generalized and not intended as a substitute for the originals. The soils are described below in the order of the map units on the legend of Fig. 2.3.

Alluvial soils are only shown in the Burry Inlet and Margam Moors, but they also occur in narrow strips along all river valleys, and show complex patterns of particle-size classes and drainage states. The broad particle-size range is summarized on p. 28.

Soils in industrial waste, notably those in the lower Swansea valley are not discussed as the problems of soil chemistry, in particular heavy metal toxicity are extremely complex and already well documented (Bridges *et al.*, 1980).

Map Unit 1

Calcareous sand dunes composed largely of quartz grains and shell fragments occur at intervals around the coast. Except on the inland portions where they have been fixed by vegetation they are unstable and show no profile development below the top few centimetres; they are classed as raw sands. On fixed dunes there may be more than 5cm of topsoil passing to calcareous sand within 30cm and these are sand-pararendzinas. Humic-sandy gley soils form in hollows where the water-table is near the surface, often forming ponds in winter. They have a surface horizon rich in organic matter over gleyed sand. Stabilized dunes on cliff tops as at Pennard, where there is little accretion of fresh sand, are sometimes decalcified to below 30cm and the resulting soils are sand-rankers.

Dune surfaces are very unstable and fresh migration of sand has frequently buried old topsoils causing layering. Increasingly, trampling by visitors is having a serious effect, destroying the vegetation cover and producing extensive areas of loose sand, causing 'blow-outs' which hinder recolonisation by vegetation.

Map Unit 2

These are clayey soils with near neutral to slightly alkaline pH values of 6.5–7.5 derived from grey, green and red Triassic mudstone that are subject to periodic waterlogging. Most common are argillic pelosols which are slowly permeable soils, clayey throughout, with an increase in clay in the subsoil resulting from illuviation, giving a clay enriched argillic horizon. They are seasonally waterlogged and this has resulted in some gley mottling in the upper 40cm, which may be masked in soils derived from red mudstone (Keuper Marl). Many profiles are calcareous, throughout or at depth, especially those derived from grey and green mudstones, and calcareous nodules are sometimes present. More poorly drained soils, found on lower ground are similarly clayey, with strong evidence of gleying in the upper 40cm and these are stagnogley soils.

Map Unit 3

Well drained fine loamy and silty brown earths predominate on Carboniferous limestone and Triassic limestone breccia. They commonly overlie rock at 40–60cm and those over the Trias are reddish. Limestone fragments are common but fine material has been decalcified. The calcareous parent material and relatively low rainfall of the Vale and Gower, together with additions of lime and basic slag on agricultural land result in these soils commonly having a near neutral pH value of 6.5–7, although it may be lower in the deeper profiles. Many profiles are silty and this together with mineralogical data (Crampton, 1972) indicates that wind-blown material of loessial origin is an important constituent of these soils.

On plateaux on Gower and the Vale there are similar but deeper soils, which beneath a silty loessial cover pass to reddish clay directly above and infilling fissures in the limestone bedrock. The bisequal nature of these profiles and similarity of the subsoils to reddish limestone soils of Mediterranean lands suggests that lower horizons are relicts of interglacial soil formation (Ball, 1960; Clayden, 1977) and such profiles are termed paleo-argillic brown earths.

Soils shallower than 30cm are common, especially on cliffs and even these have been mostly decalcified in their fine material and are classed as rankers, although calcareous equivalents, called rendzinas, also occur.

Map Unit 4

Brown earths are the main soils in the complex till and glaciofluvial deposits occurring in south and west Gower and in small areas north of Swansea. These are agriculturally important, deep, loamy soils with rounded stones throughout. Many profiles have gley mottles in the subsoil, especially where compact till occurs at depth or in hollows in the undulating landscape. Also common, especially in north-west Gower, are stagnogley soils which are poorly drained with slowly permeable subsoils and exhibiting strong gleying within 40cm of the surface.

Although these soils are derived from Carboniferous rocks of low base-status their pH has been raised by the application of lime and basic slag and commonly ranges from moderately acid to neutral, around 5.5–7.0.

Map Unit 5

Soils in glaciofluvial deposits are coarse loamy, containing well rounded stones, frequently becoming gravelly and sandier with depth. Well drained brown earths predominate in the characteristically hummocky topography. In Hensol Forest, enclosed hollows or 'kettle holes' commonly have humic gley soils with a peaty topsoil over a gleyed subsoil resulting from the high ground-water-table. Elsewhere, areas of low relief are occupied by cambic gley soils, strongly gleyed within 40cm of the surface, but lacking an organic topsoil.

These soils have a parent material of a low base-status and they are also strongly leached, owing to their permeability and coarse texture. Their pH values are commonly strongly acid, around 4.5 under forest and old pasture increasing to slightly acid around 6.0–6.5 on improved land.

Map Unit 6

Well drained argillic or paleo-argillic brown earths form most of the important agricultural land over the Lias limestone of the Vale. They are silty over a clayey argillic horizon which may be locally reddish, as in the paleo-argillic brown earths over

Carboniferous limestone. They are seldom more than 60cm deep and the fine material has been largely decalcified, although hard limestone fragments occur. Moderately well drained soils which are clayey to the surface but still exhibit a clay increase with depth are also common and these are argillic pelosols. Rendzinas which are calcareous and less than 30cm deep also occur.

The calcareous parent material, low rainfall and frequent additions of lime and basic slag give these soils near neutral to slightly alkaline pH values of around 6.5–7.5.

Map Unit 7

Argillic brown earths predominate in the reddish 'morainic drift' north east of Cardiff and in a similar area of reddish drift north of Peterston-super-Ely. These soils are loamy with an increase in clay content down the profile and often exhibit gley mottling, especially where a slowly permeable subsoil is present. On lower ground there are stagnogley soils with slowly permeable subsoils which are poorly drained and strongly gleyed. Also in this map unit are well drained coarse loamy brown earths in glaciofluvial deposits.

The parent material of these soils is of low base-status but their pH values have been raised by applications of lime and basic slag, and range from moderately acid to neutral, around 5.5–7.0.

Map Unit 8

The main Devonian outcrop is north east of Cardiff with sandstone forming a ridge and mudstone the footslopes. Well drained, shallow, stony brown earths occupy the sandstone ridge. Over the more extensive mudstones there are fine loamy or silty argillic brown earths with an increase in clay content in the subsoil. Many of these soils experience seasonal waterlogging owing to slow subsoil permeability but gleying is masked by the reddish colour.

The soils on sandstone are usually strongly acid with pH values of 4.5–5.0, whereas those over mudstone are commonly moderately acid to neutral, around 5.5–7.0.

Map Unit 9

Brown podzolic soils are widespread in Head deposits on valley-side slopes in the Pennant sandstone uplands. They are well drained coarse loamy soils, distinguished from the duller coloured brown earths which also occur by their bright orange brown subsoil horizons which contain larger proportions of hydroxides of iron and aluminum. They are common on steep slopes in the humid climate of western Britain where intense leaching can result in losses of silicon and bases, causing an *in situ* accumulation of iron and aluminium (Crompton, 1960). Under old rough pasture, or woodland, usually on the steepest slopes, the dark coloured topsoils are commonly no more than 5cm thick. Soil depth varies considerably but most profiles pass to extremely stony Head or bedrock between 60–80cm.

Valley floors too narrow to show in Fig. 2.3 contain stagnohumic gley soils in till, described on p. 36. Their relationship to brown podzolic soils within the landscape is depicted in Fig. 2.4.

Except on recently improved pasture, these soils have extremely, to strongly acid pH values of 4.0–5.0.

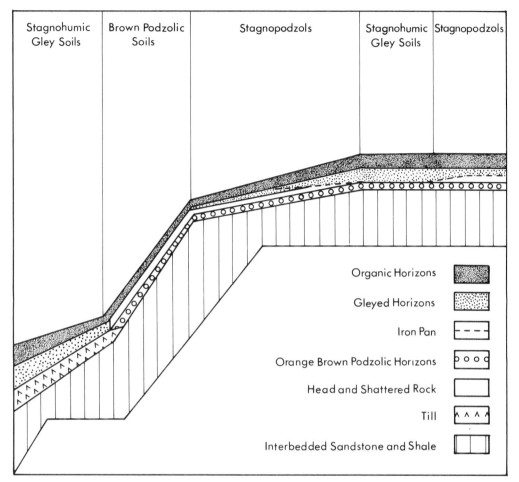

Fig. 2.4   Topographic sequence of soils of the Pennant Sandstone Uplands.

Map Unit 10

The crests of Cefn Bryn, Rhossili Down and Llanmadoc Hill on Gower are composed of Devonian conglomerates and coarse sandstones, and soils are mainly rocky, coarse loamy podzols. They are found under heath vegetation, and below a surface accumulation of acid plant debris there is a thick bleached sandy horizon which overlies a dark zone of humus, iron and aluminium accumulation. This passes to a yellowish red horizon enriched with hydroxides of iron and aluminium.

Associated fine loamy brown earths and brown podzolic soils are mainly under *Festuca-Agrostis* grassland with bracken and their distribution is related to the lithology of the underlying rocks. On Rhossili Down for example, they extend to the summit on the western side of the ridge where the substrata are fine sandstones and siltstones of the Brownstone group. Similar soils over Brownstones occupy the col on the summit of Llanmadoc Hill, flanked by podzols on conglomerate.

The podzols are extremely acid with pH values of less than 4.0, and the associated brown earths and brown podzolic soils are only slightly less acid, with pH values around 4.5.

Map Unit 11

Coarse loamy stagnopodzols predominate on the summit plateaux of the Pennant sandstone uplands. Strong leaching and the low base status of the parent material makes these soils extremely acid, with pH values around 4.0. They have a peaty topsoil, which is commonly around 20cm thick, but becomes thinner on convex sites as at plateaux edges. Below this is a grey strongly leached and often very stony horizon with evidence of gleying succeeded by a thin dark reddish brown zone of iron, aluminium and humus accumulation. In places this takes the form of thin ironpan no more than a few millimetres thick. The ironpan is rarely continuous, however, and in sections along forest roads it can be seen to disappear and reappear over distances of a few metres. These upper horizons commonly overlie a freely draining orange brown podzolic horizon similar to that of brown podzolic soils on adjacent slopes, indicating that the stagnopodzols may have resulted from a superficial modification of brown podzolic soils following the spread of heath vegetation. Most pass to little altered, very stony Head or sandstone rock between 40–80cm depth, but some are shallower, perhaps as a result of soil erosion after forest clearance, in which case an ironpan is locally developed in rock debris.

Stagnohumic gley soils also occur and may predominate in some areas, as illustrated in Fig. 2.4. They resemble the associated stagnopodzols in having a peaty topsoil, around 20cm thick overlying a grey gleyed layer depleted of iron, but the underlying horizons are also gleyed, with ochreous mottles, and commonly merge into rock debris at around 50–90cm depth. In both types of soil, the peaty topsoils and associated surface wetness are induced by the combination of high rainfall, relatively low evaporation, and gentle relief which restricts lateral water movement. They therefore differ in their water relations from stagnohumic gley soils in till, described below, that have impermeable subsoils.

Map Unit 12

Poorly drained, clayey stagnogley soils occupy depressions in the Vale over Lias and Rhaetic shales and limestones. They are strongly gleyed owing to prolonged waterlogging which results from their slow permeability and low topographic position. The soils are non-calcareous but limestone fragments are common throughout the profile.

Argillic pelosols, likewise clayey, occur as intergrades to the argillic brown earths on the Lias limestone plateau. They are better drained than stagnogley soils owing to the stronger relief, with few gley mottles in the upper 40cm.

The pH value of these soils is very variable depending on the level of management and amount of limestone present, commonly ranging from moderately acid to neutral, 5.5–7.0.

Map Unit 13

Stagnohumic gley soils predominate on the till which mantles the low ground of the Coal Measures shale outcrop and the upland valley floors (Fig. 2.4). The till is dense and impermeable often containing an extremely compact or indurated horizon within 100cm of the surface. Subsoil impermeability, together with gentle slopes and the humid climate, cause these soils to be wet for much of the year. They have a humose or peaty topsoil around 15–30cm thick over a grey gleyed, loamy horizon depleted of iron, which passes into a dense gleyed loamy subsoil horizon with grey and ochreous mottles. The profile usually becomes more dense and less gleyed with depth and merges into little altered till between 100cm and 150cm. Stagnogley soils that lack a humose or topsoil but are otherwise similar are common in agricultural land on Gower and the southern coalfield.

Large areas where these soils occur have been disturbed by opencast coal working, which involves complete removal of the over-burden and its restoration for agriculture. The quality of restoration depends on whether care is taken to preserve and replace topsoil. In many schemes, the replaced surface layer consists largely of compacted till and shale, giving a poor rooting medium.

Unimproved soils are extremely acid with pH values around 4.0. On improved land pH values are higher, commonly moderately to slightly acid, around 5.5–6.5.

Map Unit 14

Salt-marshes occur along the Burry inlet, the most extensive being Llanrhidian Marsh. The alluvium varies in particle size and the depositional pattern is complex, but according to Bridges (1976) silty and loamy soils predominate, with coarser deposits on the levees of creeks that traverse the marsh and on flats towards the centre of the estuary. In its fresh state the alluvium is saline and calcareous with a pH value around 8.

All the soils are affected by a high water-table and gleyed to the surface and are classed as alluvial gley soils, with the exception of fresh sand and mud deposits on intertidal flats showing no profile development. The degree of profile development is influenced by slight differences in elevation and four zones are recognised, each with a distinctive vegetation community. Between 3.3m and 4.1m O.D. there is an extensive area of saltmarsh-grass (*Puccinellia*) where soils show structural development in the upper 20cm and some decalcification. At 4.1m to 4.3m O.D. under red fescue (*Festuca rubra*) the soils have well developed subsoil structures and are almost completely decalcified in the upper 20cm. The highest zone of the marsh is the landward edge containing sea rush (*Juncus maritimus*), where some fresh-water flushing occurs and the soils are completely decalcified with a pH value around 5.5 in the upper 50cm.

On Margam Moors, south of the steelworks and behind the coastal dunes, there is an area of old estuarine alluvium which has been drained and reclaimed for agriculture, although the drainage system is no longer fully effective and parts are reverting to reed swamp. Non-calcareous silty and clayey alluvial gley soils predominate with cal- careous sandy soils adjacent to the dunes.

Map Unit 15

Peat soils are those in peat deposits more than 40cm deep. The most extensive area of upland peat is the blanket bog on the highest parts of the Pennant sandstone uplands above Craig-y-llyn. In the lowlands the largest is Crymlyn Bog, occupying a deep glacial trough behind the dunes of Swansea Bay. The approximate location of other peat bogs larger than 10ha is shown by a symbol on Fig. 2.3 and their exact location may be obtained from Soil Survey and Geological Survey maps.

Few of the Glamorgan bogs have been intensively drained or cultivated and conse- quently bear raw peat soils on which soft wet peat extends to the surface, and the crumbly 'earthy' topsoil characteristic of cultivated peat soils is lacking. There are few data concerning Glamorgan upland peats, but observations in the course of soil survey suggest that blanket and basin bogs of the uplands are similar in composition. All are extremely acid with pH values less than 4.0 and most are strongly decomposed and amorphous with few horizons of more fibrous material. The commonest visible plant remains are dead roots of *Molinia*, which is also predominant in today's vegetation.

In contrast, Crymlyn Fen is in a lowland situation behind the dunes of Swansea Bay and is eutrophic over much of its area, with pH values of 6.0–7.0 (Meade, 1982). Its position makes it heavily dependent on the quality and trophic status of the water flowing in from the surrounding urban and industrial areas.

References

Avery, B.W. (1980). *Soil Classification for England and Wales.* Soil Survey Technical Monograph No. 14. Harpenden.

Ball, D. F. (1960). Relic-soil on limestone in South Wales. *Nature,* **187**, 497–498.

Bendelow, V. C. & Hartnup, R. (1980). *Climatic Classification of England and Wales.* Soil Survey Technical Monograph No. 15. Harpenden.

Bowen, D. Q. (1971). The Quaternary succession in south Gower. In: *Geological excursions in south Wales and the Forest of Dean* (Ed. by D. A. Bassett & M. G. Bassett). Cardiff.

Bowen, D. Q. (1981). The 'South Wales End-Moraine' : Fifty years after. In: *The Quaternary in Britain* (Ed. by J. Neale & J. Flenley). Oxford.

Bridges, E. M. (1967). The soils of Gower. *Gower,* **18**, 66–72.

Bridges, E. M. (1973). *An investigation into the origin and development of selected soils of the Gower Peninsula.* Ph.D. thesis. University College of Swansea.

Bridges, E. M. (1976). Soils of the alluvial lowlands of the Burry Inlet. In: *Problems of a small estuary* (Ed. by A. Nelson-Smith & E. M. Bridges). Burry Inlet Symposium, **2**: 1/1–1/15. Swansea.

Bridges, E. M. & Clayden, B. (1971). Pedology. In: *Swansea and its Region* (Ed. by W. G. Balchin). Brit. Assn. Adv. Sci. 77–84, Swansea.

Bridges, E. M., Chase, D. S. & Wainwright, S. J. (1980). Soil and plant investigations in the lower Swansea valley since the 1967 Report. In: *Dealing with dereliction* (Ed. by R. D. F. Bromley & G. Humphrys). Swansea.

Clayden, B. (1977). Paleosols. In: Studies in the Welsh Quaternary (Ed. by D.Q. Bowen). *Cambria,* **4**, 84–95.

Crampton, C. B. (1972). *Soils of the Vale of Glamorgan* : Sheets 262 and 263 Memoir of the Soil Survey of England and Wales. Harpenden.

Crompton, E. (1960). The significance of the weathering/leaching ratio in the differentiation of major soil groups, with particular reference to some very strongly leached brown earths on the hills of Britain. *Trans. 7th Int. Congr. Soil Sci.,* **4**, 406–412.

Fitzpatrick, E. A. (1956). An indurated horizon formed by permafrost. *Jour. Soil Sci.,* **7**, 248–254.

Hazelden, J. & Jarvis, M. G. (1979). Age and significance of alluvium in the Windrush valley, Oxfordshire. *Nature,* **282**, 291–292.

Higgins, L. S. (1933). An investigation into the problem of sand dune areas on the south Wales coast. *Archaeol. Cambrensis,* **88**, 197–213.

Kay, Q. O. N. (1979). The post-glacial history of commonlands in West Glamorgan. In: *Problems of commonland. The example of West Glamorgan* (Ed. by E. M. Bridges). Commons symposium, **3**: 1/1–1/19. Swansea.

Meade, R. (1982). Private communication, Nature Conservancy Council, Oxwich, Gower.

Moore, A. D. (1977). Vegetational History. In: *Studies in Welsh Quaternary* (Ed. by D.Q. Bowen). *Cambria,* **4**, 73–83.

Robinson, G. W., Hughes, D. O. & Jones, B. (1930). Soil Survey of Wales progress report 1927–1929. *Welsh Jour. Agric.,* **6**, 249–265.

Robinson, G. W. & Hughes, D. O. (1936). The soils of Glamorgan. In: *Glamorgan County History. Vol. 1 Natural History.* Cardiff.

Soil Survey Staff (1983). *1:250,000 Map, Soils of England and Wales: Sheet 2, Wales.* Harpenden.

Squirrell, H. C. & Downing, R. A. (1969). *Geology of the South Wales Coalfield, Part 1. The country around Newport (Mon.).* Mem. geol. Surv. U.K.

Taylor, J. A. (1974). Organic soils in Wales. In: *Soils in Wales* (Ed. by W. A. Adams). Welsh Soils Discussion Group Report No. 15, 30–43.

Trotman D.M. (1963). *Data for Late-Glacial and Post-Glacial history in South Wales.* Ph.D. thesis, University College, Swansea.

Water Resources Board (1972). South Wales rainfall map : Sheet 12. Reading.

Wright, P. S. (1979). Soils of the West Glamorgan commons. In: *Problems of commonland. The example of West Glamorgan* (Ed. by E. M. Bridges). Commons symposium, **2**: 3/1–3/17. Swansea.

Wright, P. S. (1980). *Soils in Dyfed IV : Sheet SN62 (Llandeilo).* Soil Survey Record No. 61. Harpenden.

3

The Geological History of Glamorgan

M.G. Bassett and R.M. Owens

Introduction

The present-day distribution of rocks, the relief, and the patterns of drainage of Glamorgan, which are together crucial in determining the nature of the soils and flora, are the products of thousands of millions of years of continuous change. From a study of the rocks and their contained fossils it is possible to interpret the nature and changes of ancient environments, climates and geographies, and from such interpretations it is known that the area now occupied by Glamorgan has experienced such varied conditions as subtropical coral-rich seas, coastal swamps, arid deserts and ice-covered landscapes. One major reason for these changes is that this region of the Earth's crust has moved its latitudinal position through time, since the crustal plate to which it belongs has drifted progressively northwards from south temperate latitudes over the 425 million years represented by the geological record in Glamorgan. It is worth emphasising here that this span of time represents less than one-tenth of the total age of the Earth, which is estimated to be approximately 4600 million years (the oldest known rocks on Earth are some 3800 million years old; the oldest in Wales are about 700 million).

All the rocks preserved in Glamorgan originated as sediments; that is, as muds, silts, sands, gravels and limy oozes – the products of the erosion of pre-existing rocks – or as accumulations of organic remains. Ultimately, they were compacted to form the mudstones, siltstones, sandstones, conglomerates, limestones and coals that we see today exposed in cliffs, valley sides, cuttings, quarries and mines. Many contain abundant fossils, among which there are large numbers of fossil plants at some localities. These plants afford evidence not only of past floras, climates and habitats in Glamorgan, but also important clues in plant evolution; and they set in context the present day flora, which must be seen as only a temporary phase in the long vegetational history of the region.

Within its boundaries, the county embraces the central portion of the South Wales Coalfield, which is a broadly east-west trending basinal structure. Its core is occupied by a wide outcrop of Upper Carboniferous Pennant sandstones (Fig. 3.1) that belong to the upper part of the Coal Measures, bounded in descending geological succession to the south, east and north (the so-called South, East and North crops) in turn by the coal-rich shales, mudstones and sandstones of the Lower and Middle Coal Measures, the Millstone Grit, and finally the Carboniferous Limestone, which forms the prominent escarpments as a rim around the Coalfield. The East Crop and parts of the North Crop lie outside Glamorgan. The South Crop, defining the southern margin of the Coalfield, extends westwards to Gower. The Vale of Glamorgan is underlain for the most part by a flat-lying blanket of younger Triassic and Jurassic strata, breached in

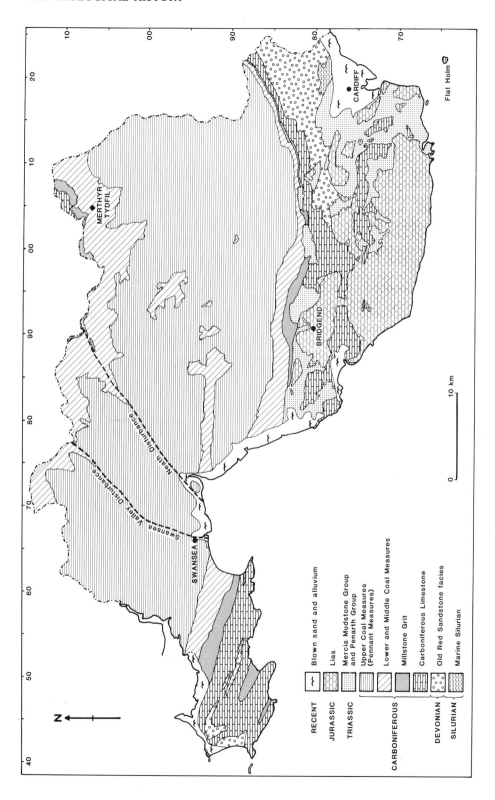

Fig. 3.1   Simplified geological map of Glamorgan (many minor structures are omitted).

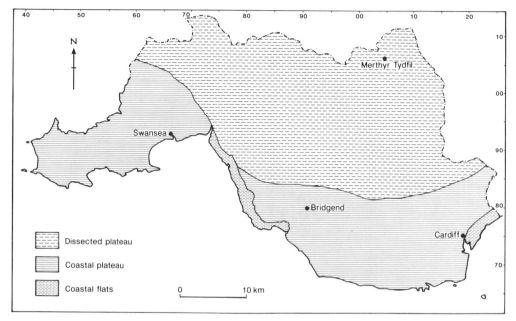

Fig. 3.2   Major relief regions of Glamorgan (after Brown, 1960, fig. 1).

places by the upfolded Carboniferous Limestone and Devonian (Old Red Sandstone) sediments. Finally, in the south east corner of the county, Old Red Sandstone and Silurian sediments crop out in the Cardiff district. This broad distribution of strata in turn largely controls the major patterns of relief within the region (Fig. 3.2). In summary, these geomorphological divisions comprise the dissected plateau of the

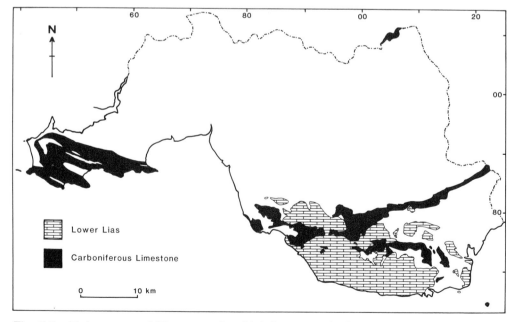

Fig. 3.3   Major areas of Glamorgan in which limestone is the dominant bedrock.

ERA	PERIOD	AGE (millions of years before present)	DOMINANT ROCK TYPES IN GLAMORGAN	SUMMARY OF ENVIRONMENTS AND EVENTS
CENOZOIC	HOLOCENE	0.01	Peat, alluvium, estuarine mud, blown sand, 'head' and tufa	Oscillating sea levels and drowning of coastal areas
	PLEISTOCENE (QUATERNARY)	c. 0.275	Glacial till, sand and gravel	Glacial and milder interglacial periods
		195	mid and late Jurassic, Cretaceous and older Cenozoic absent (including early Pleistocene)	Alpine Earth Movements: uplift and erosion
MESOZOIC	JURASSIC	205	Blue Lias: mudstones and limestones	Offshore and nearshore marine
	TRIASSIC	210	Penarth Group: mudstones and sandstones	Brackish and restricted marine
		220	Mercia Mudstone Group: red and green mudstones, siltstones, sandstones, breccias, limestones and evaporites	Warm, semi-arid, with lakes and coastal flats
			latest Carboniferous, Permian and early Triassic absent	Variscan Earth Movements: uplift and erosion
PALAEOZOIC	CARBONIFEROUS	300	Coal Measures { Upper or Pennant Measures: massive sandstones with coals / Lower and Middle: mudstones, many coals, marine bands	Fluviatile and deltaic, coastal plains and swamps; short-lived marine incursions
		310	Millstone Grit: sandstones and shales, marine bands	
		325	Carboniferous Limestone { Limestones, dolomitized in places, oolites and thin mudstones, with shales at bottom and top of succession	Marine: warm, shallow carbonate-rich sea; some local uplift and erosion
	DEVONIAN	355	Upper Old Red Sandstone: sandstones, siltstones, mudstones, conglomerates, calcretes	Fluviatile, warm climate
		370	Middle Devonian absent	Caledonian Earth Movements: uplift and erosion
		390	Lower Old Red Sandstone: mudstones, siltstones, sandstones, conglomerates, calcretes	Fluviatile, warm climate
		410		
	SILURIAN	420	Ludlow Series: mudstones, thin siltstones and sandstones	
			Wenlock Series: mudstones, sandstones and siltstones, prominent grit horizon	Offshore and nearshore marine

Fig. 3.4   Geological succession and dominant rock types in Glamorgan.

central coalfield: the coastal plateaux of Gower, the Swansea hinterland and the Vale of Glamorgan; and lastly the coastal flats between Baglan and Kenfig Burrows, Merthyr Mawr and the Wentloog levels to the east of Cardiff.

Solid Geology

Marine Silurian
The oldest sediments that now crop out in Glamorgan were deposited in the intertidal to shallow subtidal areas of a marine platform during mid and late Silurian times (about 420 to 410 million years ago). The shore-line lay very close to the present-day Cardiff area, aligned roughly east–west, with a land mass to the south known as Pretannia occupying most of the Bristol Channel region and beyond over south-west England. From the shallow seas across southern Wales, the platform deepened northwards towards a relatively deep marine basin that occupied most of central Wales. The Silurian sediments of Glamorgan comprise a succession of silty mudstones, sandstones and occasional thin limestones with abundant marine fossils. By about 410 million years ago, before the end of the Silurian Period, the sea retreated from the Glamorgan area. This change was brought about by the final closure of the Iapetus (or 'proto-Atlantic') Ocean as two crustal plates moved together and collided in northern Britain. One result of this collision was the development of the Caledonian Mountain range, which extended from present day Scandinavia through Scotland, the Lake District, north and central Wales, northern Ireland and into the Appalachian Mountains of the eastern U.S.A. It should be noted that this was not an 'instantaneous' event, and different parts of Iapetus closed at different times throughout the Silurian and early Devonian periods. This meant that the uplift, mountain building, initiation of non-marine conditions and development of erosional processes took place at different times in different areas – for example they began earlier in Pembrokeshire than in south-east Wales.

Marine Silurian rocks are limited to a small area of south-east Glamorgan, where they crop out in isolated exposures in the Penylan-Rumney district of Cardiff. These deposits yield occasional plant remains, including spores and the enigmatic *Prototaxites storriei*, which comprises bundles of cylindrical tubes about 0.02mm across. This latter plant may be an alga, perhaps related to brown algae such as the giant sea kelps, although it has also been suggested that *Prototaxites* may be of terrestrial origin, representing an early experiment in colonisation of the land. It is from rocks of mid Silurian age elsewhere in South Wales that some of the earliest vascular land plants have been found.

Old Red Sandstone facies (late Silurian-Devonian)
It is convenient to describe the latest Silurian rocks, deposited between about 415 and 410 million years ago, with those of the succeeding Devonian Period, because together they comprise a group of sediments known as the Old Red Sandstone. This succession of generally red-coloured mudstones, siltstones and sandstones was deposited in dominantly southward-flowing rivers and across a broad floodplain that lay between the Caledonian mountain chain to the north and the sea to the south. However, pebbles in one lower Devonian formation, the Llanishen Conglomerate, also suggest an intermittent southerly source, perhaps from Pretannia in the Bristol Channel area. The environment of deposition was generally hot and humid, as witnessed, for example, by the numerous thin calcareous soil horizons occurring within the sequence. These palaeosols, or *calcretes*, form today under such conditions and in the Old Red Sandstone they commonly bear evidence of rootlets. Important early land floras evolved during

the late Silurian and early Devonian periods, including precursors of the lush floras of the later, Carboniferous coal swamps. Localities in the Brecon Beacons, just beyond the northern borders of Glamorgan, have yielded *Dawsonites*, a possible forerunner of a Middle Devonian complex of plants from which the ferns and gymnosperms probably evolved, and *Drepanophycus*, which may be associated with the ancestry of the lycopods. These plants from the Beacons are also accompanied by other rich floras.

In the uppermost Old Red Sandstone at Tongwynlais to the north of Cardiff, where the beds are probably of earliest Carboniferous age, floras include spores and stalked seeds which are the oldest known seeds in Britain. In beds of comparable age in the Swansea valley, deposited probably in fresh or brackish water, there are poorly preserved specimens of charophytes.

The Old Red Sandstone comprises two major divisions, the older Lower Old Red Sandstone of late Silurian to early Devonian age (*c.*415 to 390 million years old), and the younger Upper Old Red Sandstone of late Devonian to earliest Carboniferous age (*c.*370 to 350 million years old). The intervening period is unrepresented in South Wales, where there was a period of uplift and erosion. The Old Red Sandstone crops out in the northern suburbs of Cardiff, and extends in a narrow belt westward towards Cowbridge; small, isolated outcrops occur in west Cardiff and in the Cowbridge area. There are also further outcrops on the Gower, which form the high ground of Cefn Bryn, Llanmadoc Hill and Rhossili Down. Although outcrops are relatively few, the presence of Old Red Sandstone is commonly evident immediately from the distinctive red colour that it gives to the overlying soil.

Dinantian (lower Carboniferous)
Evidence for a gradual return to marine conditions occurs in the highest beds of the Old Red Sandstone, where marine fossils, principally brachiopods, are present. During the Dinantian Epoch, in the first half of the Carboniferous Period (350 to 325 million years ago) Glamorgan lay under a shallow, subtropical sea, with land (St George's Land) to the north and open sea to the south. After an initial period when predominantly muddy sediments were deposited, the greater bulk of this time interval was characterised by the deposition of limy oozes, which together with debris from calcareous shells of marine organisms ultimately formed the Carboniferous Limestone. Recent studies have shown that the deposition of the various kinds of limestone embraced within this thick group of rocks was controlled by eustatic changes in sea level. There were several advances (with intervening retreats) of the sea from the south, and gradually a carbonate pile built up under a very shallow sea that was rich in corals and brachiopods. From time to time there was subaerial erosion, as witnessed by breaks in the succession, by the presence of rootlet horizons in widespread calcretes, and by palaeokarst. Collapsed breccias suggest the presence of former evaporites – salts that formed by evaporation of sea water under subaerial conditions in a hot climate. Other deposits suggest lagoons and oolite shoals, comparable with those forming today in areas such as the Bahamas Banks. At some levels there are algal tufts and mats, which formed as calcareous layers built up by lime-secreting algae that favoured shallow, warm calcareous seas.

The Carboniferous Limestone crops out around the rim of the South Wales Coalfield, but apart from a small area around Morlais near Merthyr Tydfil, only the southern rim falls within the confines of Glamorgan. The outcrop is narrow in the vicinity of the Taff Gorge, due to the high inclination of the strata, but it is broader between Llantrisant and Bridgend, and between Bridgend and Porthcawl. The Carboniferous Limestone is particularly well displayed in the Gower, with fine outcrops in the cliffs along its southern coast. Smaller isolated outcrops occur to the west of Cardiff, and in the Barry area, generally as upfolded or faulted remnants. In some areas, notably to the north and north-west of Cardiff, the original calcium carbonate of the limestone has been replaced by magnesium carbonate to produce dolomite. It is believed that this chemical process took place soon after deposition of the original sediments.

Namurian (middle Carboniferous)
Environments changed markedly with the commencement of deposition (about 325 million years ago) of sediments now forming the Millstone Grit. Carbonate deposition ceased, and was replaced by alternating non-marine and short-lived marine episodes, with sands and muds as the dominant sediments. Swampy coastal plains and river systems were developed across Glamorgan. When the sea retreated far to the south, rivers ran southwards from St George's Land and from the east, from an uplifted area in the Usk district. Sedimentation was continuous from late Dinantian times in the Gower and in the Bridgend area, but in the north-east part of the coalfield there was upwarping and erosion which removed the higher parts of the Carboniferous Limestone prior to deposition of the Millstone Grit sediments. On the South Crop the Millstone Grit sequence is predominantly shaly, with subordinate sandstones, but on the North Crop sandstones dominate, with a thick Basal Grit (which crops out widely in the Merthyr district) at the bottom of the sequence, passing upwards into a shale member. Floras that developed across the coastal plain and deltaic environments of the Namurian Epoch represent an initial phase in the build-up of vegetation that was to become dominant in the immediately succeeding period.

Westphalian (upper Carboniferous)
The environmental conditions of Namurian times continued into the latest Carboniferous, with extension of the coastal plain swamps and river systems. Sediments were again derived mainly from the north and east. The Lower and Middle Coal Measures are predominantly mudstones and shales, and contain a large number of coals with several marine intercalations. They also contain ironstones, especially in the lower part, formed probably by penecontemporaneous segregation of iron-rich clays into nodules or bands ('pins') parallel to the bedding. These ironstones were important in the early industrial development of the area, especially in the Merthyr district. The coal seams originated as peat in low-lying swamps on the coastal plains, only slightly above sea level. Coal formation may be linked to sea level changes. As sea level rose and caused a rise in the water table, this in turn would have impeded drainage and prevented oxidation of accumulating organic matter, thus encouraging the formation of peat. It has been estimated that a layer of plant debris between 4 and 6 metres thick would have produced a coal seam 0.5m thick. Repeated rapid rises in sea level produced a large number of coal seams in the Lower and Middle Coal Measures, and also led to repetition of similar sequences of associated sediments, including freshwater shales with mussels, seat-earths with rootlets, and, when the sea invaded temporarily, marine bands. The Coal Measures swamps are generally considered to have formed in subtropical or tropical zones, analogous to modern mangrove swamps, but there has been a recent suggestion (see Ramsbottom, 1984) that they might have originated in cooler climates; the greatest thickness and extent of modern peats is in temperate or cool climates, and the suggestion that the late Carboniferous swamps formed in such climatic zones is supported to some extent by oxygen isotope studies.

Lower and Middle Coal Measures generally form the lower ground adjacent to the Pennant escarpments on the North and South crops; they are also brought to the surface by a faulted upfold in the Maesteg area, and are exposed in the floor of the Rhondda valleys. The upper part of the Coal Measures is dominated by massive beds, the Pennant sandstones, which crop out over the central part of the coalfield, and are up to 1800 metres thick. They comprise thick feldspathic sandstones and grits, and detailed studies have shown that they originated in some areas in river channels, such as those near Briton Ferry. It is also apparent that in contrast to the underlying parts of the Coal Measures, the sediments were derived mostly from a land mass lying to the south of Wales, perhaps roughly in the position of the present Bristol Channel (Kelling, 1974). Although coals are present within the Pennant Measures, there are no marine bands, and sea level seems generally to have remained low.

The freshwater and brackish swamps, marshes and lagoons of the Coal Measures supported luxuriant forests and undergrowth, and the presence of large numbers of fossils of fern-like plants (the seed ferns) together with true ferns gives rise to the term *Age of ferns*, which is sometimes applied to this period of geological time. One of the most striking features of Coal Measures plants was the occurrence of large, tree-sized forms in groups that are now represented only by small herbaceous types. The most spectacular examples are in the lycopods and horsetails. Such large plants invariably broke up on transportation or burial, resulting generally in fragmentary fossils consisting of discrete pieces of stem, roots, leaves, seeds, or sporangia. As with all fossil plants of different geological periods, these individual organs that once made up a parent plant are given separate names.

There is a wealth of genera and species of plants known from the Coal Measures of Glamorgan, which has thus been a region of considerable study for various aspects of ecology and taxonomy (see Bassett & Edwards, 1982; Thomas, 1986).

Variscan Earth Movements (late Carboniferous-Triassic)
Rocks of latest Carboniferous age are not represented in Glamorgan. At about this time a prolonged period of uplift and erosion affected much of Wales and adjacent regions, lasting through the Permian Period and well into Triassic times – an interval of some 80 million years. During this period the Coal Measures and older strata were folded into the present basinal structure that dominates the coalfield. At the same time any Coal Measures that were deposited to the south of the coalfield were removed by erosion, so that in the Vale of Glamorgan post-Carboniferous sediments rest directly upon the Carboniferous Limestone. This unconformity is particularly well seen, for example, at Ogmore-by-Sea, Barry Island (Friars Point) and on Sully Island. It is worth emphasizing here that immensely long periods of time can be represented by gaps in the sequence, and this interval with no rock record is considerably longer than that (*c.* 55 million years) represented by all the Carboniferous strata.

Triassic
Late in the Triassic Period southern Wales lay in northern tropical latitudes and the climate was hot and semi-arid. From about 220 to 210 million years ago a variegated complex of red muds and silts with scattered nodules and bands of gypsiferous evaporite was deposited, succeeded by green and grey mud with abundant nodular evaporites; together these sediments now comprise the Mercia Mudstone Group. The mudstones are largely of lacustrine and alluvial plain origin, and the evaporites indicate rapid evaporation and prolonged periods of subaerial exposure. Local sheets of breccia and conglomerate represent coarse gravels and boulders brought down in rivers and wadis by occasional flash floods. Close to the eroded remnants of former Carboniferous Limestone 'islands', the mudstones and siltstones pass laterally into a mixed association of conglomerates, breccias, sandstones, limestones and the dolomites – the so-called marginal facies of the Triassic. At the Bendricks, near Barry, the red sandstones and siltstones have yielded numerous dinosaur footprints, and algal laminae are not infrequent. Throughout this interval, the fluvial systems running generally southwards across Glamorgan were draining into a lake or lakes whose northern shore lay close to the present Bristol Channel. In some areas the Triassic

sediments penetrate and fill karstic fissures in the Carboniferous Limestone, and a few localities have yielded bones of small dinosaurs and some of the earliest known mammals.

Sediments of the Mercia Mudstone Group underly much of the low ground (much of it drift and alluvium covered) in the Cardiff district, and forms the low red cliffs in the Penarth and Barry areas. There are also numerous scattered outcrops around Cowbridge and Bridgend. The marginal facies tends to be harder and more resistant to erosion, and is particularly well seen at Ogmore-by-Sea.

During the final 5 million years of the Triassic Period (210–205 million years ago) marine conditions gradually returned to South Wales. Sediments of the Penarth Group are mostly mudstones and sandstones with thin limestones, representing a succession of brackish, lagoonal and restricted marine environments. Close to the base is the so-called 'Rhaetic bone bed', which is probably a winnowed strand-line deposit, packed with disarticulated vertebrate bones and teeth and with phosphatic nodules.

As with the older Triassic deposits, some of these younger sediments also fill fissures in the Carboniferous Limestone, and some have yielded interesting plant remains. Near Bridgend, cones, bark, wood and shoots of the conifer *Hirmerella* have been found as uncompressed charcoal, which preserves fine details of the cellular structure, and there are also remains of the leaf *Pterophyllum*. The charcoal provides evidence of a late Triassic forest fire in which the more substantial timber was probably only scorched whilst the burnt twigs, leaves and other debris were washed into the fissures and preserved (see Bassett & Edwards, 1982; Ager & Edwards, 1986).

At Cornelly, a clay pocket within a sandstone known as the Quarella Stone yielded leaves of the fern *Clathropteris* in the workings for the M4 Motorway. From the fragments collected, it has been estimated that the leaf might have been up to 2 metres wide. Present day members of the family to which *Clathropteris* belongs grow in thickets in the tropics; in Glamorgan this fern probably grew in damp, humid areas close to rivers, where it was associated also with an *Equisetum*-like plant (Ager & Edwards, 1986).

Jurassic

Above the Penarth Group, the distinctly bedded, alternating mudstones and limestones of the Blue Lias represent the advent of fully marine conditions by early Jurassic times (205 million years ago). Marine transgression across Wales probably continued throughout the Jurassic Period, but only the lowermost beds are now preserved. The Blue Lias was deposited as muds and lime muds in a relatively quiet sea, and the beds now cover much of the Vale of Glamorgan, including the horizontally bedded cliffs occupying much of the coast in the Lavernock area and from Barry westwards to Llantwit Major and Southerndown. Close to the islands of Carboniferous Limestone around which the early Jurassic sea lapped, the typical Blue Lias gives way to purer, lighter coloured limestones of shallow water origin.

The Blue Lias contains some drifted plant remains, including pieces of wood from coniferous trees. Occasional fronds of the palm-like gymnosperm *Otozamites* also occur. These plants presumably grew on land areas bordering the Blue Lias sea, lying perhaps not far to the north, as well as on the Carboniferous Limestone 'islands'.

The Blue Lias rocks represent the youngest bedrock sediments in Glamorgan; however, younger Jurassic sediments crop out on the floor of the adjacent Bristol Channel to the south. Many authorities have long claimed that the whole of Wales was covered originally by thick layers of such younger Jurassic rocks, plus succeeding deposits of Cretaceous age, including the Chalk. No evidence of these younger rocks remains today, but by consideration of distributions and thicknesses elsewhere in southern Britain it seems likely that progressive Mesozoic marine transgression did affect most of Wales. The Chalk, for example, may well have covered much of the country, with the exception of perhaps small areas of mid and north Wales; if so, it is likely that this late Cretaceous deposit was removed by erosion within only some 30 million years of its deposition. Given such a history, it is improbable that a Chalk cover played much part in the shaping of the present landscape of Wales (see Cope, 1984).

Alpine Earth movements (late Mesozoic-Cenozoic)
During the Cenozoic Era, before the onset of the Pleistocene glaciations, Wales was subjected to a prolonged period of pulsatory uplift and erosion, which resulted in the development of a series of planation surfaces. These surfaces impart a remarkably flat profile to the hill tops of upland Wales, but whether they are of subaerial origin, as proposed by Brown (1960), or are the product of marine erosion, as proposed by George (1970) remains a matter of conjecture. Evidence for their dating is circumstantial, but since all the surfaces cut across pre-existing structure in the rocks, it is safe to assume that the erosional events postdate the structures. By analogy with other parts of southern Britain, the folding of the youngest strata in Wales, the Lias, took place during Miocene times, between about 24 million and 5 million years ago; the likelihood is that planation took place during the succeeding Pliocene Period, between 5 and 1.6 million years ago. It was during this interval that the present day drainage pattern began to emerge.

Three major planation surfaces have been distinguished across Wales (Brown 1960), each of which is represented by relief within Glamorgan:

High Plateau (510m–570m)
Middle Peneplain (360m–480m) (most commonly 360m–420m and/or 450m)
Low Peneplain (210m–330m) (most commonly 210m–270m and/or 300m–330m)

The Middle Peneplain slopes in accordance with the regional trend of the drainage, and the Low Peneplain always slopes in the direction of the major rivers crossing it. In all cases these features are much dissected. The High Plateau in Glamorgan is restricted to a 12km long segment at a height of 510 to 540m running southwards from Craig y Llyn to Mynydd Llangeinor, between the headwaters of the rivers Afan and Rhondda; on average it is 3 to 5km wide and descends steeply on all sides. The Middle Peneplain is well developed on the summit surfaces on the crests of the divides between the Rhonddas, the Cynon, the Taff and the Rhymney, and two outliers at Mynydd y Glyn and Twyn Howel overlook the Low Peneplain near Pontypridd. Farther west, the Middle Peneplain can be identified at Mynydd Maesteg, Mynydd Maendy, Mynydd Marchywel, Hirfynydd and Resolven Mountain. West of the Tawe there is only one remnant probably referable to the Middle Peneplain, Penlle'rcastell, a flat-topped monadnock rising above the Low Peneplain.

In the area of the Loughor, Amman and Tawe rivers, summits between 300 and 322m belong to the Low Peneplain, remnants of which occur also on Mynydd Drumau between the Tawe and Dulais, and Carn Caca and Cefn Morfudd in the Neath valley. This surface cuts across the south rim of the coalfield between the Ogmore and Ebbw valleys, and can be identified in numerous areas including the Caerphilly district, Gelli-gaer Common, Mynydd Troed y Rhiw between the Rhonddas, Llwynypia Mountain, and Mynydd Eglwysilan.

On Gower, in the Swansea hinterland, and in the Vale of Glamorgan there is a series of stepped, but highly dissected coastal platforms at levels of 60m, 90m, 120m and 180m, and these are considered to be wave-cut. Like the upland peneplains, they truncate pre-existing geological structures, and presumably represent successive changes in sea level. Remnants of the 180m platform are represented on Gower by the Old Red Sandstone ridges of Cefn Bryn, Llanmadoc Hill and Rhossili Down. Similar remnants occur near Swansea. Much of the Gower is at a lower level, and like the Vale of Glamorgan extends up to and a little over 120m. The 120m platform is not well seen, because of encroachment of the 60m level, but it can be identified near Penclawdd, Three Crosses and Clyne Common. The south edge of the coalfield flanking the Vale of Glamorgan represents a platform cut into the south-facing scarp.

The present-day drainage pattern is much modified from that initiated in Pliocene times, with river capture a common phenomenon, caused by a combination of sea level changes and the effects of glacial erosion. Wind gaps, such as that between Rhigos and Hirwaun bear testimony to old river courses, and in the Neath and Tawe valleys many right bank tributaries are matched by left bank wind gaps, indicating routes prior to capture. Fine examples of river capture are to be seen in the Neath and Tawe headwaters, and in the Cynon, which via the Rhigos–Hirwaun gap formerly included what are now the Neath and Tawe headwaters. The courses of the Tawe and Neath have been adapted secondarily to the weak structural zones of the Neath and Swansea valley disturbances (Fig. 3.1) that they follow so conspicuously today.

Superficial Geology

Pleistocene

The whole of Glamorgan was affected by the Pleistocene glaciations, although as over much of Britain the associated landforms can be related generally only to the last two episodes. Thus only the latter part of the Pleistocene Period is represented, dated as extending from 275 000 years ago to approximately 10 000 years ago. It could be that many of the erosional surfaces described immediately above date from the early Pleistocene instead of from the Pliocene, and in particular some of the coastal platforms. The present landscape of Glamorgan is not considerably unlike the preglacial morphology, although it is modified in detail. Glacial erosion has overdeepened the valleys, and changed their profile from V-shaped to U-shaped; some have rock floors over 45m below the present ground level (e.g. the Neath, subsequently filled with alluvium), and many valleys have oversteepened hill slopes. Glacial deepening has led to rejuvenation, with the development of waterfalls in the upper reaches of rivers, whilst the oversteepening is a causal factor of landslips in the central coalfield valleys. Other notable erosional features include corries on the north-facing Pennant scarp at Craig y Llyn, and at Cwmparc in the Rhondda, which developed on slopes facing from the north-northwest to the east, i.e. directions of low insolation. Evidence for ice movement is present as striations, which are well developed on hard rocks such as the Basal Grits (Namurian) of the North Crop, and also high on valley sides in the Neath and Tawe valleys. Ice-polished and roughened surfaces and roches moutonnées are also best seen on hard rocks. Meltwater channels formed in certain areas during deglaciation, such as at Briton Ferry (now followed by the present River Neath); and parts of the Afan, Llynfi, Garw and Ogmore Valleys were initiated or at least strongly modified by meltwater. Depositional features include boulder clays or tills (rounded or angular boulders in a sandy or clayey matrix), outwash sands and gravels produced by meltwaters from the ice sheets, plus moraines and drumlins. Well-developed drumlins near Hirwaun and Coelbren have their long axes oriented in the direction of ice movement.

Deposits from the penultimate glaciation (formerly known as the 'Older Drift') are widespread in the Gower and in the Vale of Glamorgan; these areas were covered by ice from the Irish Sea, whereas ice sheets in districts to the north were fed by glaciers from Breconshire and by local valley glaciers. Till from the Irish Sea includes erratics from west and north Wales and from Scotland, including the celebrated Ailsa Craig microgranite. Some is calcareous and clay-rich, and contains shells picked up from the sea bed. The till at Pencoed and Ewenny belongs to this type, and at the latter locality has been used for the manufacture of pottery since Roman times.

As ice from the penultimate glaciation melted, sea levels rose to some 6 to 9m above that at present, as witnessed by the Raised Beach deposits of the Gower, such as the *Patella* Beach. These deposits formed during the Ipswichian interglacial, between 130 000 and 120 000 years ago. Associated with them are the well-known cave deposits of the Gower, such as Minchin Hole, which have yielded bones of vertebrates such as elephant, hippopotamus and rhinoceros and which provide evidence of a climate warmer than today.

Many of the features of the penultimate glaciation were destroyed by the last, Devensian glaciation, which began about 120 000 years ago and was responsible for most of the glacial landforms preserved today, and for the deposition of the 'Newer Drift'. However, this episode was not so extensive geographically, so that south-west Gower and the Vale of Glamorgan for example remained ice-free, and experienced periglacial conditions. All the ice was derived from Wales; in the central coalfield, corrie glaciers developed on the Pennant escarpment, as at Craig y Llyn. A local ice cap to the south of Craig y Llyn fed glaciers in the Neath, Afan, Ogmore, Rhondda and Cynon valleys; other areas were supplied from Breconshire and mid Wales. It is likely that much of the Pennant plateau in the central coalfield was not covered by ice, although most was probably snow-covered. Limits of the ice are indicated by morainic belts, such as those at Radyr Golf Course, Talbot Green and between Margam and Pyle. Retreat of the ice was pulsatory, and valley moraines indicate temporary readvances. Good examples of these are seen at Glais in the Tawe valley, and at Talbot Green in the Ely valley, and in both cases the rivers have been deflected to flow along the valley sides.

During the last glaciation, the climate ameliorated at times, and periglacial conditions prevailed, producing deposits of 'head', which comprises crudely sorted, frost-shattered angular rock fragments in a scree-like matrix that have moved downslope by solifluxion. The 'Red Lady of Paviland' (actually the bones of a male) dates from about 18 000 years ago, immediately before the last major ice advance. Cold climate faunas, including woolly mammoth, woolly rhinoceros and ermine have been found in some of the Gower cave deposits. Periglacial conditions also prevailed immediately following the last glaciation, and subsequently the climate improved gradually.

Recent or Holocene
At the end of the Pleistocene Period (10 000 years ago) sea levels stood well below the present, at least 22.5 metres from evidence from peat deposits around the Glamorgan coast. Such peats have been found in many areas, including Crymlyn Bog, Swansea Bay, between Port Talbot and Kenfig, and in Cardiff and Barry docks. The Crymlyn Bog section records some 8 000 years of postglacial history, and pollen analysis shows a succession of climatic changes. Peat horizons are intercalated with marine, freshwater and estuarine clays, and in addition to pollen they have yielded stumps, branches and leaves of such trees as oak, hazel and birch, plus remains of beetles and flint implements, the latter giving evidence of early human activity in the district. Early in the postglacial period much of the present day Bristol Channel formed wooded coastal flats. As the ice sheets retreated and the ice melted, sea levels gradually rose during the Flandrian transgression, and at its greatest extent the sea covered all the coastal flats of Glamorgan. Blown sand piled up by prevailing winds now covers most of the flats at Merthyr Mawr, the stretch between Porthcawl and Baglan Moors, Crymlyn Burrows, and Oxwich Burrows. The sand dunes have impeded drainage, giving rise to the areas of marshy ground behind them. Other Recent processes include river and coastal erosion and the deposition of alluvial and estuarine muds, all of which continue to modify the superficial morphology of Glamorgan.

Fig. 3.5 Silurian-Carboniferous fossil plants from Glamorgan and adjacent areas. **a**, longitudinal and **b**, transverse sections through a petrified cylindrical axis of *Prototaxites storriei*, Silurian (Wenlock Series), Rumney, Cardiff (×100 approx.); **c, d** spores: *Artemopyra granulata* and *Emphanisporites protophanus*, Silurian (Wenlock Series), Rumney borehole, Cardiff (×1000 approx.); **e** *Cooksonia* sp., Silurian (Ludlow Series), Cwm Graig ddu, Powys (×4); **f** stalked seed (×4); **g** group of sporangia containing microspores, cf. *Xenotheca* sp. (×2); **h** spore, *Retusotriletes incohatus* (×400); **i** spore, *Hystricosporites* cf. *delectabilis* (×100), all basal Carboniferous, Upper Old Red Sandstone, Tongwynlais, near Cardiff; **j** alga, *Garwoodia* sp. (×6); **k** alga (×1); **l** stromatolite (×0.75), all from Carboniferous Limestone, Llanelly Formation, Blaen Onneu, near Crickhowell, Powys; **m** seed fern, *Neuropteris loshi*?, Coal Measures, Danygraig, Swansea (×0.4); **n** true fern, *Pecopteris plumosadentata*, Coal Measures, Rhigos, near Hirwaun (×0.6) (by permission of the Trustees, Natural History Museum).

Fig. 3.6 Carboniferous (Coal Measures) fossil lycopods from Glamorgan and Gwent. **a** terminal branches of *Lepidodendron acutum* with pendulous cone (arrowed), Gilfach Goch (×0.3); **b** spore, *Densosporites*, of probable lycopod origin (×500 approx.); **c** lycopod spore, *Lycospora* sp. (×500 approx.); **d** *Lepidodendron* cf. *lanceolatum*, part of flattened trunk, Porth (×0.5); **e** cone, *Lepidostrobus ornatus*, Abercarn, Gwent (×0.75); **f** root, *Stigmaria ficoides*, Gyfeillion (×0.6); **g** root and rootlets, *Stigmaria ficoides*, Pontycymmer (×0.5); **h** part of flattened trunk, *Sigillaria scutellata* (×1).

Fig. 3.7 Carboniferous (Coal Measures) horsetails (a–e) and cordaitales (f, g) from Glamorgan. **a** pith cast of *Calamites approximatus* (×0.3); **b** leaves of *Calamites* : *Asterophyllites equistiformis* (×0.7); **c** leaves of *Calamites* : *Annularia radiata* (×1.5); **d** herbaceous horsetail, *Sphenophyllum cuneifolium* (×0.8); **e** attached cone (*Bowmanites*) of same species (×0.75); all from Rhigos, near Hirwaun; **f** strap-shaped leaf of *Cordaites principalis*, Wernddu, Caerphilly (×0.75); **g** pollen grain, *Florinites* sp., probably from *Cordaites* (×500). **b–e** by permission of the Trustees, Natural History Museum.

Fig. 3.8 Mesozoic plants from Glamorgan. **a** leafy shoot of conifer *Hirmerella muensteri*? (×20); **b** fragment of conifer wood (×350 approx.), both Cnap Twt, near Bridgend; **c** part of frond (×1.25) and, **d** detail of veins in leaf (×0.8) of fern *Clathropteris meniscoides*, Triassic (Rhaetian), Pyle; **e** conifer shoot (×1), **f** part of palm-like bennetitales frond *Otozamites* cf. *obtusus* (×1), both Jurassic (Lower Lias), Penarth.

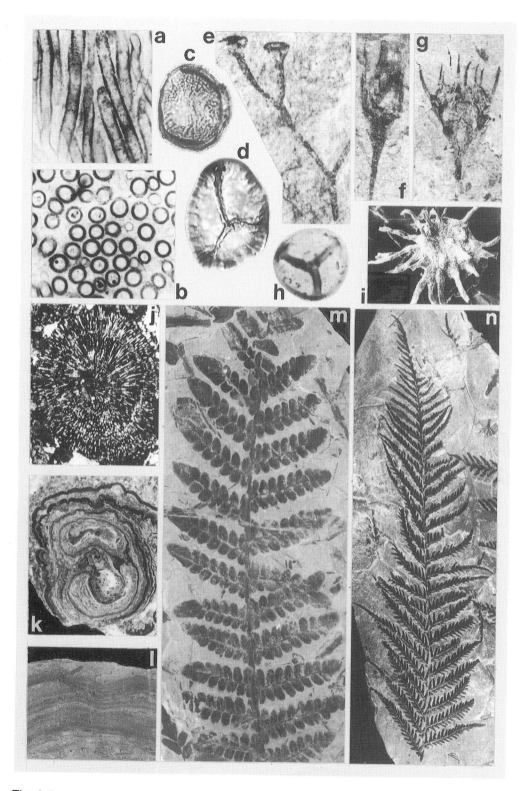

Fig. 3.5

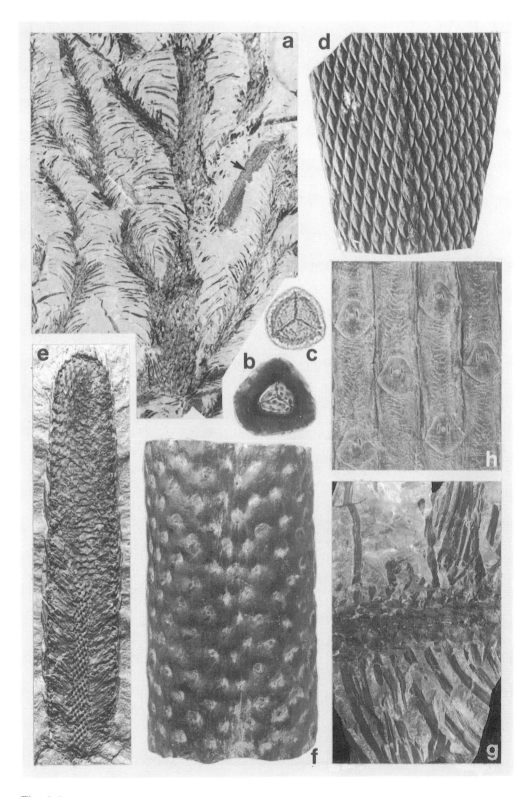

Fig. 3.6

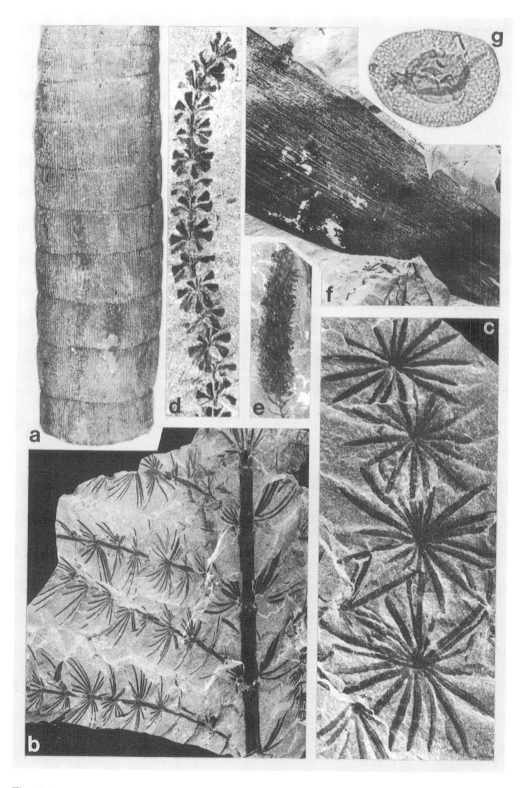

Fig. 3.7

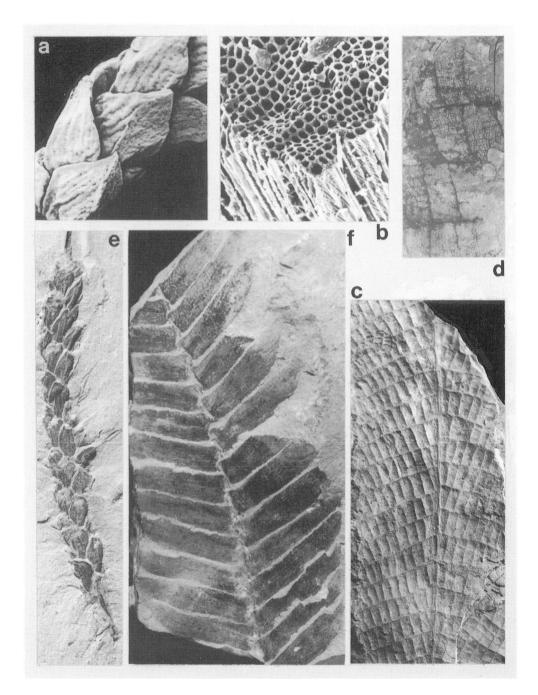

Fig. 3.8

References

Ager, D.V. & Edwards, D. (1986). The fauna and flora of the Rhaetian of South Wales and adjacent areas. *Nature in Wales* (*n.s.*), **4** (1 & 2), 71–9.

Allen, J.R.L. (1974). The Devonian rocks of Wales and the Welsh Borderland. 47–84 *in* Owen, T.R. (Ed.) *The Upper Palaeozoic and post-Palaeozoic rocks of Wales.* University of Wales Press, Cardiff.

Barclay, W.J. 1989. Geology of the South Wales Coalfield. Part II, the country around Abergavenny (3rd edition). x + 147pp., 16 pls. *Mem. Br. Geol. Surv.,* Sheet 232 (England and Wales).

Barclay, W.J., Taylor, K., and Thomas, L.P. 1988. Geology of the South Wales Coalfield. Part V, the country around Merthyr Tydfil (3rd edition). x+ 52pp., 10 pls. *Mem. Br. Geol. Surv.,* Sheet 231 (England and Wales).

Bassett, M.G. & Edwards, D. (1982). Fossil plants from Wales. Geological Series No. 2, National Museum of Wales, Cardiff. 42pp.

Bowen, D.Q. (1970). South-east and central South Wales. 197–227 *in* Lewis, C.A. (Ed.) *The glaciations of Wales and adjoining regions.* xv + 378pp., 33 pls. Longman, London.

Brown, E.H. (1960). *The relief and drainage of Wales. A study in geomorphological development.* xii + 186pp., 24 pls. University of Wales Press, Cardiff.

Cope, J.C.W. (1984). The Mesozoic history of Wales. *Proc. Geol. Ass.,* **95,** 373–85.

George, T.N. (1970). *British Regional Geology. South Wales.* (3rd ed.). xii + 152pp., 13 pls. H.M.S.O. London.

Kelling, G. (1974). Upper Carboniferous sedimentation in South Wales. 185–224 *in* Owen, T.R. (Ed.) *The Upper Palaeozoic and post-Palaeozoic rocks of Wales.* University of Wales Press, Cardiff.

Owen, T.R. (1973). *Geology explained in South Wales.* 211pp., David & Charles, Newton Abbot.

Owen, T.R. (1984). Upper Palaeozoic Wales – a review of studies in the past 25 years. *Proc. Geol. Ass.,* **95,** 349–64.

Ramsbottom, W.H.C. (1984). Developments from faunal studies in the Carboniferous of Wales. *Ibid.,* **95,** 365–71.

Squirrell, H.C. & Downing, R.A. (1969). Geology of the South Wales Coalfield. Part I, the country around Newport (Mon.) (3rd edition). xiii + 333 pp., 12 pls. *Mem. Geol. Surv. U.K.,* Sheet 249 (England and Wales).

Thomas, B.A. (1986). In search of fossil plants: the life and work of David Davies (Gilfach Goch). Geological Series No. 8, National Museum of Wales, Cardiff. 54pp.

Waters, R.A. and Lawrence, D.J.D. 1987. Geology of the South Wales Coalfield. Part III, the country around Cardiff (3rd edition). x + 114pp., 10 pls. *Mem. Br. Geol. Surv.,* Sheet 263 (England and Wales).

Wilson, D., Davies, J.R., Fletcher, C.J.N. and Smith, M. 1990. Geology of the South Wales Coalfield. Part VI, the country around Bridgend (2nd edition). viii + 62pp., 20 pls. *Mem. Br. Geol. Surv.,* Sheets 261 and 262 (England and Wales).

Woodland, A.W. & Evans, W.B. (1964). Geology of the South Wales Coalfield. Part IV, the country around Pontypridd and Maesteg (3rd edition). xiv + 391pp., 8 pls. *Mem. Geol. Surv. U.K.,* Sheet 248 (England and Wales).

1:50 000 GEOLOGICAL MAPS (S: solid, D: drift).

230   Ammanford (S) (D) 1977
231   Merthyr Tydfil (S) (D) 1979
232   Abergavenny (S & D) 1990
246   Worms Head (D*) 1960
247   Swansea (S) 1977 (D*) 1972
248   Pontypridd and Maesteg (S*) 1963 (D) 1975
249   Newport (S) 1975 (D*) 1969
262   Bridgend (S & D) 1990
263   Cardiff (S) 1986 (D) 1989

* denotes available only in 1:63,360 edition

4

Land Cover of Glamorgan: A Satellite Image

Alan J. Morton

In a detailed study of the flora and vegetation of a large area such as a county or vice-county, it is often difficult to form an overall or 'synoptic' view of that area. Vegetation maps may be available, but often they are not, or they are rather out of date. This latter point is particularly important in modern times with the large changes in our natural environment which are brought about by coniferous afforestation, agricultural improvement and urban development. Recent technological developments have, however, provided us with an opportunity of gaining this synoptic view of our area of interest, in the form of satellite imagery.

The main source of this imagery is the North American Landsat series of satellites, and the most recent of these which has provided the imagery presented in this chapter is Landsat 5 which was launched in 1984. These satellites do not record pictures in the way that a camera does, they systematically record digital data which is transmitted back to Earth for analysis and re-assembling as an image. Landsat 5, for example, orbits the earth from Pole to Pole at an altitude of 706km (440 miles), scanning the Earth's surface with its sensors. The most advanced sensor system on Landsat 5 is the Thematic Mapper which records radiation reflected and emitted from the Earth's surface in each of seven separate wavebands. Three of these wavebands are in the visible part of the spectrum (i.e. 'light') whilst the other four are in the infra-red or thermal part of the radiation spectrum. As the Thematic Mapper scans the earth's surface, it systematically records the intensity of this radiation from areas on the ground which are approximately 30m × 30m (33yds × 33yds) in size, and this is sometimes referred to as the 'spatial resolution' of the data or image. For some purposes, this resolution is rather coarse, but for obtaining a synoptic view of a large area it is adequate and is much better than earlier scanner systems because, at the improved resolution of 30m, features such as field boundaries, roads and tracks are clearly visible.

The Frontispiece is a satellite image which includes the vice-county of Glamorgan. The colouring of the image is not natural and is termed 'false colour', and is in fact a 'false colour composite' of three wavebands, one visible and two infra-red. The use of the infra-red bands and false colouring gives rise to an image which shows sharp contrasts between land cover types. The image is an autumn scene, recorded on 16 October 1986, and is particularly good for discriminating between vegetation types. The main land types which are readily discernible are as follows:

1. Vegetation

1.1 Improved pastures, leys and meadows (orange)
   At the time of year of the imagery, both pastures and meadows give the spectral

response of healthy, active vegetation. It is not possible to separate hay meadows from pasture in late autumn, but in the mid-summer hay-cutting period, satellite images show a sharp contrast between these grasslands.

1.2 Moorland and rough pasture (light greenish blue to green)
In contrast to the improved pastures and meadows, the moorland grasslands are largely senescent, giving rise to a very different spectral response. Heather moorland also has a distinctive spectral response, but there is too little of it to show on the imagery; most of the moorland being acid grasslands with such species as *Nardus stricta* and *Molinia caerulea*. The lightest tones on the image are likely to be the *Molinia* grasslands.

1.3 Arable land (bright blue)
Although this has a similar colour to some of the moorland, it can be readily distinguished by its contextual appearance and location. Instead of the extensive, irregular shapes of the moorland vegetation, it is seen as a patchwork of arable fields and adjacent pastures (see, for example, the Gower Peninsula). Much of the arable land is bare soil at the time of year of the image, giving rise to this contrast with the adjacent pastures and leys.

1.4 Conifer forest (dark reddish brown)
This is possibly the most obvious vegetation type seen on the image. Conifers have a distinctive spectral response due to the characteristics of conifer leaves and canopy structure, and this combined with the sharply defined edges of conifer plantations sets them apart from other vegetation types.

1.5 Broadleaved woodland (light reddish brown)
Very little of this is evident on the image, but small remnants can be detected in Gower and in the Vale of Glamorgan.

2. Non-Vegetation

2.1 Urban areas (grey-blue)

2.2 Colliery wastes and opencast mines (dark blue)

2.3 Reservoirs and docks (black)
Clear, deep water strongly absorbs both visible and infra-red radiation, and hence appears very dark on the image. Note the contrast between the relatively clear water of the docks and reservoirs and the sediment-laiden coastal waters.

If we apply the preceding interpretation to the whole image of Glamorgan, we clearly see the contrasting regions of the vice-county; the heavily industrialized and urbanized areas of Swansea, Neath, Port Talbot, Bridgend, Cardiff and the Rhondda and Cynon Valleys; the extensively conifer-afforested uplands of West and Mid Glamorgan; and the rural areas of Gower and the Vale of Glamorgan. A more detailed look at the vice-county is described in four regions.

Gower

The Gower Peninsula is seen as a very diverse area in terms of land cover types. It is predominantly a patchwork of pasture and arable land on the well drained loamy soils, but many other land cover features are also seen and these include:

The rough grassland and moorland plant communities which occur on the conglomerate and sandstone outcrops of Rhossili Down, Llanmadoc Hill and Cefn Bryn, and also at Pengwern, Fairwood and Clyne Commons. Some of these areas have an abundance of *Agrostis curtisii* (Bristle Bent) and this is particularly obvious in the light tones of the southern slope of Cefn Bryn.

Broadleaved and mixed woodlands are seen near Penrice and to the east of Llanrhidian, and also in the Ilston and Parkmill valleys. The colour of these woodlands on the image is intermediate between the light orange tones of the pastures and the dark tones of conifer forest.

The sand dunes at Broughton and Whiteford Burrows can be detected by their light green colour and the characteristic texture of the dune systems. The conifer plantations on Whiteford Burrows are clearly visible, contrasting strongly with the adjacent dunes. Dune systems are also visible at Oxwich and Pennard Burrows.

Extensive estuarine silt deposits are seen in the Burry Inlet and these form the salt marshes and creeks which lie between Whiteford Point and Loughor. The open water of Broad Pool and the pools in Oxwich Marsh appear as small dark spots on the image.

The rugged limestone cliffs between Worm's Head and Mumbles Head are an obvious feature, as also are the north and west facing cliffs and steep slopes which are characterised by the linear shadows which they cast.

The bog and fen vegetation of Crymlyn Bog can clearly be seen as a green coloured area adjacent to the sharply conrasting blue tones of the oil refinery complex between Swansea and the River Neath.

The Western Uplands

This area, to the north of Swansea, shows a gentle transition from pasture on glacial deposits to extensive unafforested moorland on sandstone above the Lliw Valley. This area of moorland contrasts strongly with the extensively afforested moorland of the upland further east. The reservoirs in the Lliw Valley are clearly visible. On the north side of the western uplands, waste tips and open-cast mines are seen as dark blue areas south of Brynamman, between Gwaun-cae-gurwen and Ystalyfera. The dark linear features in the uplands are steep west-facing slopes which are in shadow at the time of overpass of the satellite, for example on the east side of the Afon Tawe between Pontardawe and Ystalyfera.

The Uplands East of the Afon Tawe

This area is characterised by extensively conifer-afforested moorland on peaty and podsolic soils overlying sandstone. Rheola Forest is among the larger of these forests. Only a small proportion of unafforested moorland remains, mainly to the west of Rhondda Fawr and in the Blaen Rhondda – Hirwaun Common area and even this shows some evidence of young conifer plantations. The Lluest-wen reservoir shows clearly on the image, because of its contrast with the adjacent moorland vegetation, but other small lakes in the area, such as Llyn Fawr and Llyn Fach, are more difficult to discern because of shadowing by adjacent hills.

The extensive waste tips and opencast mines between Onllwyn and Glyn Neath, and also around Hirwaun, dominate the northern edge of the uplands. Further to the east, the opencast mine at Dowlais Top between Merthyr Tydfil and Rhymney also makes its mark on the landscape.

Leading away from the uplands, the valleys of the Rhondda, Cynon and Rhymney are seen as urbanized strips separated by the largely pastoral hills.

The Vale of Glamorgan

Apart from the large conurbations of Cardiff, Bridgend, Barry and Port Talbot, this area is seen as largely agricultural, with pasture and meadow predominant in the northern part (Ogwr and Taff-Ely) and a mixture of pasture and arable on the better drained loamy soils of the southern Vale of Glamorgan. Evidence of mining activity is confined largely to the northern part of the Vale, notably near Llanharry and east of Pyle, and also the limestone quarry to the north-west of Cardiff.

Some areas of broadleaved woodland can be seen in the Vale, for example the strips of woodland in the valleys around Llancarfan, and the limestone beechwoods between Radyr and Taff's Well.

The sand dune systems at Kenfig Burrows and Merthyr Mawr Warren stand out clearly and are seen as green areas on the image, similar to some of the moorland areas. This is due to a covering of largely senescent grass at the time of year of the imagery. Kenfig Pool is prominent within Kenfig Burrows, and to the north of this is the largest freshwater lake seen on the image, Eglwys Nunydd Reservoir.

5

The History, Ecology and Distribution of the Flora of Glamorgan

Q. O. N. Kay

Glamorgan is a relatively small county in total area, but its long and varied coastline, its geographic position, and its geological and topographical diversity have given it an exceptional variety of plant habitats and plant communities. These range from the sunny, south-facing limestone slopes and sand-dunes of the Gower Peninsula, where several Mediterranean and continental-southern plants approach the northern and western limits of their European ranges, to the high Pennant Grit escarpment in the north of the county, where some arctic-alpine and montane species find their southernmost refuges in Britain on the bleak north-facing cliffs of Craig-y-llyn and in the cold waters of Llyn Fach.

Ecological contrasts and diversity are greatest in southern and western Gower. Here, at Oxwich or Whiteford, one can walk in an afternoon from sand-dunes with rich orchid communities in slacks fringed by the Mediterranean rush *Juncus acutus*, past salt-marshes and *Phragmites australis* fen, across limestone cliffland with a rich mosaic of cliff-ledge, grassland and scrub habitats, through limestone woodland on steep slopes, past hedged fields where increasingly scarce annual weeds such as *Kickxia elatine*, *Fumaria capreolata*, and *Lamium hybridum* still grow, to emerge on acid moorland where *Drosera rotundifolia*, *Osmunda regalis* and *Thelypteris limbosperma* grow on bryophyte-rich streamsides and the moorland community of *Calluna vulgaris*, *Molinia caerulea*, *Erica tetralix*, *Juncus squarrosus* and *Scirpus caespitosus* resembles that of the high uplands.

The postglacial history of the flora of Glamorgan

At the height of the Devensian glaciation, about 20 000 to 18 000 years ago, ice-sheets covered most of Glamorgan apart from the southern Vale and south-western Gower. The ill-drained boulder clay that was deposited along the edge of the ice cover underlies the chain of commons that lie between northern and southern Gower – Welsh Moor, Pengwern, Fairwood and Clyne Commons – and spreads across the northern Vale. Severe periglacial conditions must have existed beyond the limits of the ice-sheet, and on Rhossili Down and in the limestone clifflands of southern Gower there is much evidence of ice-shattering of the rocks. The sea-level was much lower than at present, perhaps more than 100 metres below its present level. The vegetation of the lowlands to the south of the ice-sheet was probably similar to that of modern Arctic tundra, with arctic-alpine species such as *Dryas octopetala*, *Salix herbacea*, *Rhodiola rosea* and *Saxifraga oppositifolia*.

As the ice retreated, at first slowly and intermittently, then more rapidly with one sharp but final oscillation of temperature about 10 000 years ago (the Allerød warm period followed by the 500-year Loch Lomond cold period) grass and sedge-dominated plant communities of a temperate rather than arctic type spread northwards in western Europe. Many of the plants that are now found in heathland, moorland or limestone grassland in Glamorgan grew in these open communities of the Late-glacial and early Post-glacial periods, but they were rapidly replaced by woodland plants in most of our area as birch and pine forests also spread northwards. After the Loch Lomond cold period, juniper scrub may have been dominant for a time in the hills before birch forest became established there too (Walker, 1982).

In previous interglacial warm periods, large grazing and browsing mammals prob-ably maintained park-like rangeland conditions in much of Europe by the intensity of their grazing pressure, which prevented the development of closed forest conditions. But these large herbivores – for example extinct species of elephant and rhinoceros, as well as wild horses and cattle – appear to have been exterminated or drastically reduced in numbers in their refuges in southern Europe during the later stages of the Devensian glaciation (Martin, 1967). As a result, the forests that spread northwards into the British Isles at the end of the Devensian were closed forests without clearings, and most plants of open habitats were unable to survive in these forests.

By the middle of the Boreal period, about 8500 years ago, closed forests of oak, pine, birch and elm with an understorey of hazel probably covered most of modern Glamorgan, with pine and birch forests extending over the summits of the coalfield hills. Plants of open habitats found inland refuges only in a few bogs (where several of the important dominant species of heathlands and moorlands are able to grow), and on cliffs and riversides. Riverside gravels are likely to have been a particularly important open habitat that is also thought to have been occupied by the bands of Mesolithic hunters and food-gatherers who lived in the forests. These Mesolithic people probably made clearings and tracks, where some weeds (plants of man-made habitats), for example *Chenopodium album*, may have been able to grow.

More refuges for plants of open habitats were found on the coast, where windswept sea-cliffs and sand-dunes remained free of trees; these seaside habitats were probably especially favoured by the Mesolithic tribal groups that are known to have fished and gathered shellfish. These people, too, may have created habitats suitable for some weeds around their occupation sites. The sea-level was rapidly rising during the Boreal period (Culver, 1976) and was now 15–20 metres below its present level, with the shoreline only a few kilometres beyond its present position.

We are able to follow the changes in vegetation during the post-glacial period mainly by using the technique of pollen analysis, which depends upon identification and counting of the pollen-grains that are preserved in peats, lake muds or other suitable deposits. The early pollen analysts thought that Man had virtually no influence upon vegetation until the Neolithic period, which began about 5700 years ago in our area, and that even then climatic factors were more important than human activities until the Iron Age, which began only 2500 years ago here. We now believe that Man's influence upon vegetation began very much earlier than this, and that it may even, as Evans (1975) has suggested, have been locally important during the Hoxnian interglacial, more than 150 000 years ago. Recent work has shown that Mesolithic people were probably responsible for deforestation in many upland areas of Britain, from the Pennines and the Yorkshire Moors to central Wales and Dartmoor, relatively early in the postglacial and well before the arrival of Neolithic farming communities. When farming communities did arrive, it is now thought that the clearances they made were much more extensive and permanent than was once believed. Habitats for moorland and grassland communities may thus have existed quite widely within our area from a relatively early stage, after a fairly short period of continuous forest cover.

Mesolithic people were present in quite small numbers and probably influenced vegetation mainly through the use of fire, both to make clearings around their camp-sites and, more extensively, to clear underbrush and perhaps to drive game. Repeated burning of the forest would lead to an increase in fire-resistant pioneer species like pine and hazel, and it has been suggested that the known abundance of hazel during the Boreal period may represent a fire-climax, which would possibly benefit Mesolithic food-gatheres by increasing the crop of hazel nuts. The more open conditions produced by burning would also tend to increase the carrying capacity of the forest for grazing animals, and it is possible that during Mesolithic times some of these (even including red deer) may have been semi-domesticated. Burning would however speed up the rate of leaching and loss of soil nutrients (a process which is still continuing on repeatedly burned heathland and moorland in Glamorgan) and would produce increasing acidity of the soil. In marginal upland habitats this could eventually lead to the replacement of the forest by moorland blanket-peat vegetation with *Sphagnum*, as seems to have happened on the Yorkshire Moors about 7000 years ago.

Although radiocarbon dating of blanket peat from drier parts of the Craig-y-Llyn plateau has given a date of only 3625 ±80 years (late Neolithic) to the mineral soil/peat transition there, analyses of pollen diagrams from upland South Wales (Trotman 1963, Kay 1978, Chambers 1983) and the abundance of Mesolithic microliths in the Glamorgan uplands (Green & Stanton, 1984) do suggest that fire-using Mesolithic hunters may have influenced the vegetation and soils of our area in these ways during the Boreal and succeeding Atlantic periods, especially from the late Boreal (about 7500 years ago) onwards, when there was a large temporary increase in hazel and grasses at the Cwmllynfell site studied by Dorothy Trotman, bracken appeared in some quantity, and pine (which probably grew mainly at the upper margins of the forests) declined and was perhaps replaced by *Sphagnum* moorland in wetter sites. The high grass pollen levels in the late Boreal at Cwmllynfell are particularly striking; they are higher than at any subsequent stage, apart from a brief peak in the recent past which may date from the boom in upland arable farming during the Napoleonic wars. These high Boreal grass levels are followed by very high levels of hazel during the early Atlantic period at Cwmllynfell, suggesting partial regrowth of the forest through a scrub phase. Thus the agricultural clearances made by the Neolithic farmers who arrived at the end of the Atlantic period may, as Smith (1970) has suggested, have been clearances of secondary forests, and may have been aided by the presence of open heathland created by earlier Mesolithic clearances.

Neolithic farming communities are now known to have arrived in our area considerably earlier than was once supposed. The age of the Sweyne's Houses megalithic graves on Rhossili Down is uncertain; similar tombs in Pembrokeshire probably date from the Early Neolithic, and may have been constructed by 3700 B.C., but Sweyne's Houses might be more recent than this. The age of the Parc Cwm tomb in Gower, and of similar Severn-Cotswold tombs in other parts of southern Glamorgan, is more firmly established. These probably date from the Middle Neolithic period of about 3200 B.C. to 2800 B.C.

Farming originated in the Fertile Crescent of the Middle East before 8000 B.C., and then spread gradually westwards and northwards across Europe. Our Neolithic cultures were essentially farming societies whose crops, farm animals and techniques nearly all stemmed from the Middle East and south-eastern Europe. Although some transitional Mesolithic/Neolithic and early Neolithic groups in our area may have been entirely pastoral, most Neolithic farmers appear to have been mixed farmers, who grew cereals (mainly emmer wheat, with some einkorn wheat, bread wheat and barley) and also kept cattle, pigs and sometimes (as for example in Breconshire) sheep. Early Neolithic farmers may have collected leafy branches of trees, especially the elm, for animal fodder, and this has been suggested to be the cause of the decline in elm pollen that can often be seen in pollen diagrams from Britain at about 3400 to 3200 B.C. Alternative explanations of the elm decline are selective clearance of trees from the better, base-rich soils preferred by elm, or spread of Dutch elm disease when farmers disturbed the forest ecosystem.

One of the most characteristic artefacts of the Neolithic period, in Glamorgan as in other parts of the British Isles, was the polished stone axe. In Glamorgan, axes were made either of flint (29% of finds) or of igneous rock from the Welsh axe factories at Graig Lwyd and in north-western Pembrokeshire (Grimes, 1984). Axe finds are widely distributed but concentrated in the lowlands; in the west of the county, for example, Rutter (1948) reported that single polished stone axes had been found at Llandewi, Paviland, Ammanford, Ystradgynlais and Briton Ferry, and three axes at different sites in Oystermouth. An axe found recently at Aberafan retained part of its birchwood handle (Grimes, 1984). Danish experiments have shown that these axes are surprisingly effective in tree-felling; a ten-inch spruce tree was felled in twenty minutes, and 600 square yards of birch forest were felled by three men in four hours, with one axe-head lasting for a hundred trees. Thus Neolithic farmers were well able to clear the forest (perhaps ring-barking larger trees). On fertile soils in the lowlands, permanent clearances for arable cultivation seem to have been made, and although firm evidence from pollen diagrams is lacking it is likely that this happened in parts of Gower and the Vale. On less fertile soils elsewhere, temporary clearances may have been made, followed by regrowth of the forest or the development of heathland or moorland conditions when the fertility of the soil had declined to a point at which it was no longer worthwhile to grow crops. These temporary clearances (landnam phases) are distinctive in pollen diagrams, and repeated landnam phases can be seen in pollen diagrams from other parts of western Britain, including west-central Wales (Turner, 1964; Moore & Chater, 1969).

It seems likely that the relatively infertile uplands of northern Glamorgan would have been unattractive to Neolithic farmers. The pollen diagram from Cwmllynfell (Trotman, 1963) suggests that the uplands retained their forest cover in the valleys and on hill-slopes during Neolithic times, but that there were fairly large and fluctuating areas of heathland that had first appeared during the Mesolithic period. The pollen evidence indicates that there were only three temporary clearances during the Neolithic and early Bronze Age at this site (Kay, 1978).

We do not have continuous pollen records for Gower and the Vale, so we must rely on other sources of evidence and on comparisons with other parts of the British Isles to reconstruct the likely course of events there during the Mesolithic and Neolithic periods. The existence of the Mesolithic flint chipping site on Burry Holms, close to Rhossili Down and Llanmadoc Hill, and the similarity in soils, topography and vegetation between these western hills of Gower and the hills of the Uplands, suggests the possibility that there may have been Mesolithic fire-clearances in Gower which could have enabled heathland vegetation to spread to the western hills from its refuges on exposed coastal sites. Podsolization might have followed, and parts of Rhossili Down could have remained as heathland subsequently.

There is less uncertainty about the possible course of events after the arrival of the Severn-Cotswold people who built the megalithic tombs that survive at Parc Cwm, Nicholaston, Penmaen Burrows, St Lythan's and Tinkinswood. Their settlements were probably nearby and other occupation sites may have existed in other parts of Gower and the Vale. These tombs adjoin some of the best arable land in Gower and the Vale, on well-drained gentle slopes over limestone, and it is likely that the land was permanently cleared, as was the case on the fertile coastal strip of south-western Cumberland (Pennington, 1969) where both the soil and the climate are less favourable than in Glamorgan. Limestone woodland would probably persist on the steep and uncultivable slopes in Gower and the Vale where it still grows today, but in the small cornfields of the cleared areas weeds like *Lamium* spp. (dead-nettles), *Sinapis arvensis* (charlock) and *Papaver* spp. (poppies) could become established after being introduced with crop-seed, with *Rumex* and *Plantago* spp. on fallows and in pastures on cleared land.

The sea-level, which had been rising throughout the Mesolithic period, was close to its present level in Neolithic times, and sand-dunes and saltmarsh probably fringed Oxwich Bay, within a few minutes' walk of Nicholaston and Penmaen, and salt-marsh may also have extended up the Pennard valley towards Parc Cwm. These are naturally open habitats, and salt-marshes provide nutritious and productive grazing. Evans (1975) has suggested that coastal pastures may have been important even for Mesolithic people because of their attraction for wild cattle and deer in the winter when there was little to browse in the forest, and it is reasonable to suppose that Neolithic farmers exploited them just as farmers do today. The grazing potential of fixed-dune grassland is smaller, but the Neolithic farmers may have extended the area of permanent pasture by clearing dune woodland, which would be unable to regenerate. This seems to have happened at similar settlement sites in Cornwall, the Outer Hebrides and the Orkneys (Evans, 1975).

Four of the large inland commons of Gower (Cefn Bryn, Welsh Moor, Pengwern Common and Fairwood Common) lie within two or three miles of Severn-Cotswold tombs. With the exception of the well-drained but extremely acid central ridge of Cefn Bryn, these commons occupy soils that are mainly poorly drained, infertile and very heavy, and it seems unlikely that Neolithic farmers would have undertaken the labour of felling the oak-birch-alder forest that probably grew on these soils, when fertile brown-earth woodland was available in other parts of Gower. Nevertheless it is possible that there were temporary clearances on some parts of these commons, and grazing pressure and perhaps also fires may have prevented regeneration of the woodland there. The concentrations of Bronze Age barrows and cairns on the ridge of Cefn Bryn and, to the west, on Rhossili Down and Llanmadoc Hill, suggest that these areas had been cleared by the end of the Neolithic period, or at least by the early Bronze Age. Fox (1923) pointed out that the siting of barrows in situations where they are clearly visible from the valleys below indicates that the environment was open when the barrows were built. Many Bronze Age barrows also survive on the hill-tops in the uplands of western Glamorgan, where the pollen record suggests that open heathland may have existed since the Mesolithic period.

The types and numbers of domestic animals that were kept by prehistoric farmers are valuable indicators of the character of the landscape. Forests can support relatively large numbers of pigs, but few cattle, because cattle require considerable areas of woodland for browsing – perhaps 10 hectares per head – if no grassland is present. A high ratio of pig to cattle bones at a prehistoric site therefore indicates forest conditions. Much larger numbers of cattle can be supported by grassland, and sheep are unable to live in woodland, so large numbers of cattle bones, particularly when they are accompanied by sheep bones, indicate the presence of a significant amount of open grazing land. If sheep bones predominate open heathland or upland grazing is likely to have been the chief pastoral resource. At the eastern Neolithic Severn-Cotswold sites in the Cotswold Hills, a mixture of sheep and cattle bones has been found, but in the outlying group of Severn-Cotswold sites in southern Breconshire, for example Ty-isaf, sheep bones predominate. These sites are probably contemporary with the Severn-

Cotswold sites in Gower and the Vale; at Tinkinswood, sheep or goat bones also outnumber cattle bones among the small number that have been found (Grimes, 1984). Weobley (1976) has suggested that there may have been extensive pastoral activities with transhumance from lowland bases to upland summer grazings during this period. Some support for this theory is provided by the recent discovery of a Late Neolithic house on Cefn Glas ridge above the Rhondda Valley at a height of 500 m (Grimes, 1984) which could be interpreted as a hafod site occupied for summer grazing, although several finds of Neolithic arrowheads from the Uplands suggest that hunting was also important. If pastoral transhumance was practised by the Severn-Cotswold people of the Neolithic sites in Gower and the Vale, grazing pressure from wintering herds and flocks in the lowlands could have led to deforestation, perhaps even without any deliberate clearance of the woodland on the less fertile soils. Similar patterns of land-use, with settled land cultivation on better soils in the lowlands but also pastoral transhumance from the lowlands to the open uplands, are likely to have existed throughout the Bronze Age and the Iron Age, and continued through medieval times into the relatively recent past (hendre and hafod). Several groups of upland Iron Age huts in east Glamorgan have been suggested to be summer-occupied hafodtai by Savory (1984). Knight (1984) has suggested that one of the tasks of the Roman troops stationed at the upland forts at Caerphilly, Gelligaer and Merthyr may have been to police the tribespeople by control of their seasonal cattle movements. Pastoral trans-humance still continues in a modified form on a reduced scale today, but with the flocks and herds being transported by road and the shepherds and herdsmen no longer living with their flocks.

Several studies have shown that the Neolithic period apparently passed fairly smoothly into the Bronze Age in Wales (Savory, 1984). 'Beaker' pottery and burials, which are usually taken to indicate the presence of early bronze-using people, appeared in our area in strength about 4000 years ago, but scattered examples date from about 4500 years ago. It has often been suggested that these people were primarily nomadic shepherds, with herds of cattle and sheep, who also hunted for food (Houlder & Manning, 1966). The pollen record and the distribution of their burial mounds shows that there was much open grazing land inland and that their occupation sites were apparently more evenly dispersed through Wales, especially in the uplands, than those of the preceding Neolithic farmers, whose settlements were apparently concentrated into a few lowland areas. Several recent studies have shown that there were repeated temporary clearances of woodland in the uplands during the Bronze Age (Turner, 1970) and that some of these clearances are associated with field systems and evidence of cereal cultivation (Feachem, 1973). The Cwmllynfell pollen diagram shows *Plantago* and bracken pollen through much of the probable Bronze Age period at levels that are higher than any occurring in the Neolithic, with scattered cereal and weed pollen grains, indicating clearance and cultivation of parts of the uplands in our area.

Rainfall was lower and summer warmth greater in the Bronze Age than they are now, and arable cultivation in the uplands may have been abandoned as the climate worsened to its recent state with the onset of the Sub-Atlantic period around 500 B.C. However, the Bronze Age farmers are also likely to have grown cereal crops in lowland areas and may have had field systems there. If this was so, these field systems are unrecognised or have been obliterated by later cultivation in Glamorgan, although they survive in a few areas such as Salisbury Plain (Evans, 1975) and the Isles of Scilly (Ashbee, 1974). Some boundaries of Bronze Age field systems may however survive to the present day in our area; the original limits of some of the Gower commons are likely to be among the oldest surviving boundaries, and it would be extremely interesting to date them.

During the Bronze Age, agricultural implements seem to have been made mainly of wood or stone, as they were in the Neolithic period, and agricultural techniques probably showed little change. Bronze was scarce and expensive, and was at first used mainly for axes, spear-heads, swords and other weapons. As bronze became more abundant it was used for woodworking tools, and finally, during the late Bronze Age, bronze sickles were made; there were two bronze sickles in a late Bronze Age hoard (dating from after 750 B.C.) found near Cardiff, and there were several bronze sickles in the Llyn Fawr hoard, which also includes a few iron objects and presumably dates from the end of the Bronze Age. Iron-working was introduced into Wales in the sixth century B.C. Iron was far superior to bronze as a metal for weapons, tools and agricultural implements, and was also far cheaper; iron ore is widely distributed in South Wales, and wood and charcoal for smelting could be readily obtained from the forests and woodlands that were still widespread.

The availability of iron implements gradually transformed agriculture. It was now far easier to clear woodland, and it had become possible to cultivate the heavy but fertile soils that had defeated wooden ploughs. The pollen record shows extensive clearances, on a much larger scale than the relatively small and usually temporary clearances of the Neolithic and the Bronze Age, in west-central Wales around 400 B.C. (Turner, 1965), in the Somerset Levels around 300 B.C., and in many other areas of Britain. The partial occupation of Wales by the Romans is also associated with forest clearance in the uplands of west-central Wales (Moore & Chater, 1969) which in some cases has lasted until the present. At Cwmllynfell, a marked peak in bracken pollen with a simultaneous increase in the pollen of Ericaceae and decrease in tree pollen, but no corresponding increase in grass pollen, may date from this period and suggests fire-clearance for non-agricultural purposes. The Roman road leading from Neath to Brecon along the crest of Hirfynydd is likely to have followed open country that already existed, but the Romans may have carried out fire-clearances near other roads and around their forts at Coelbren Gaer (just within the boundary of Glamorgan) and Llwchwr.

In Gower and the Vale, several large hill-forts and many smaller fortlets survive from the Iron Age and indicate that the population was denser than in the Bronze Age and Neolithic periods. Many of these defensive structures are close to modern farms and villages, which may thus have existed at or near their present sites since the Iron Age and Romano-Celtic periods. The pattern of small enclosed fields near the village and shared grazing on common land that still survives in parts of Gower may also date from this period, modified by the partial imposition of the manorial system and later enclosures (Rackham, 1986).

In the uplands, the Cwmllynfell pollen record shows some increase in trees during the Dark Ages, followed by a gradual and rather uneven increase in grass, bracken and *Plantago* pollen through the Middle Ages and recent times, reaching a peak that may represent a combination of the climax of cereal-growing in the uplands during the Napoleonic wars, when high grain prices coincided with a run of good summers (Thomas, 1963) and the cumulative effects of wood-cutting for tanbark and charcoal during the seventeenth and eighteenth centuries, when intensive coppicing may have reduced tree pollen production. After this peak, grass and *Plantago* decrease rapidly and tree pollen shows some increase, reflecting the cessation of woodland management for charcoal and tanbark production, and some spread of woodland as a result of the progressive abandonment of upland holdings that has continued until very recently. It is interesting to see how the pollen spectrum of about 1900 was similar to that of the late Bronze Age at Cwmllynfell. Since the early years of this century, extensive afforestation with alien conifers, urbanization, the spread of coal-tips and in the last thirty years very large open-cast coal workings, and the recent grant-aided revival in upland agriculture have brought about substantial changes in the vegetation of much of the Uplands. But the upland commons of north-western Glamorgan have escaped most of these changes so far, preserving a pattern of vegetation, soils and land-use that may be more than five thousand years old and may have had its origin in clearances made during the Mesolithic period by men who lived more than seven thousand years ago.

Plant communities and plant distribution in modern Glamorgan

Gower and the lowlands from Swansea to Gorseinon
The Gower peninsula and the lowlands to its north form part of the southern rim of the South Wales Coalfield syncline. To the south of a line from Llanrhidian to Oyster-mouth, there is an undulating plateau of Carboniferous Limestone, about 60–69 m above sea-level, with stepped and sloping limestone cliffs where it meets the sea. North of this line the limestone dips below Millstone Grit and Coal Measures sandstones and shales which rise to 120–130 m in the peninsula and to 180 m at Kilvey Hill east of Swansea. Inliers of Devonian sandstone and conglomerate rise above the limestone plateau of Gower to form the conspicuous rounded hills of Cefn Bryn, Rhossili Down and Llanmadoc Hill, about 180 m above sea-level, and Millstone Grit occurs in synclines at Oxwich and Port Eynon.

In the limestone area, much of the plateau and some of the gentler slopes on the sea-cliffs are covered by acid glacial drift of varying thicknesses, and, on the plateau, calcareous soils, often in permanent pasture around limestone outcrops, are limited in extent. The most interesting plant communities of calcareous soils are found on the uncultivated slopes of steep-sided valleys, and on the sloping, terraced sea-cliffs. Well-developed limestone woodland occurs in the valleys and on the sheltered north-facing cliffs above the salt-marshes of the northern coast. These woodlands are dominated by ash (*Fraxinus excelsior*) and oak (*Quercus robur*) in varying proportions; until recently, wych-elm (*Ulmus glabra*) was a third co-dominant, but it has been reduced drastically by Dutch elm disease. Small-leaved lime (*Tilia cordata*) and field maple (*Acer campestre*) are frequent in some woods. Introduced sycamore (*Acer pseudo-platanus*) is locally dominant and is, as in other parts of the British Isles, spreading rapidly and often replacing the native tree species. Hazel (*Corylus avellana*) is usually common, spindle (*Euonymus europaeus*), privet (*Ligustrum vulgare*), blackthorn (*Prunus spinosa*) and dogwood (*Cornus sanguinea*) are frequent. Buckthorn (*Rhamnus catharticus*) is scarce in the shrub layer and at the woodland margin, with the hedgerow species hawthorn (*Crataegus monogyna*) and elder (*Sambucus nigra*) most frequent in open and disturbed areas of the woods. The field layer has a rich flora, usually dominated by *Allium ursinum* and *Mercurialis perennis*, with *Asplenium scolopendrium* and *Polystichum setiferum* very abundant on shaded slopes. *Brachypodium sylvaticum*, *Bromus ramosus*, *Galium odoratum* and *Sanicula europaea* are frequent. The more notable species include *Daphne laureola*, *Euphorbia amygdaloides*, *Lathraea squamaria*, *Paris quadrifolia* and *Ruscus aculeatus* (Kay & Page, 1985), all of which are most abundant in southern Britain and are near the western or north-western limits of their British ranges in Gower. Patches of scrub-woodland, dominated by *Fraxinus excelsior*, *Corylus avellana* and *Prunus spinosa* but with *Rhamnus catharticus* and *Cornus sanguinea* in places, survive here and there on the exposed limestone cliffs of southern Gower; they were probably more extensive in the past, and have been reduced and replaced by gorse and cliff-heath as a result of repeated fires and grazing by sheep. Good examples can be seen on the southern side of Oxwich Point, where *Buglossoides purpurocaerulea* and *Hypericum montanum* grow at the scrub margins, and in Ramsgrove Valley, where *Sorbus porrigentiformis* grows in the scrub. *Sorbus porrigentiformis* also grows at Crawley Cliff overlooking Oxwich Marsh, with the taller rock whitebeam (*S. rupicola*) and wild service-tree (*S. torminalis*). Here there is a patchwork of species-rich cliff-scrub and limestone woodland with *Tilia cordata* and *Quercus robur*. Juniper (*Juniperus communis*) is still well-established in cliff scrub and around limestone outcrops in a few places on the southern cliffs.

The limestone cliffland of the southern coast of Gower has a flora of outstanding interest. There is an intricate mosaic of habitats here. The gentler slopes are mantled by glacial drift on which surprisingly acidic soils (pHs of 5 or less are frequent) bear acidic heath vegetation, often within a few metres of limestone grassland, scrub and scree communities around the limestone outcrops and bluffs. The limestone crevices and cliff ledges are floristically especially rich. On the lower part of the cliffs, halophytic spray-zone plants are found, often showing particularly clear zonation, but except on the more exposed headlands the community of the upper slopes is not halophytic.

In the spray-zone, *Armeria maritima* typically grows nearest to the sea, followed by *Crithmum maritimum* and often *Inula crithmoides* and *Limonium procerum*. Above these more species appear successively; first, *Plantago maritima* and the glaucous, compact spray-zone forms of *Agrostis stolonifera* and *Festuca rubra*, often with *Leontodon taraxacoides*, *Anthyllis vulneraria*, *Desmazeria marina* and *Centaurium erythraea* and sometimes with *Carex distans* and, rather surprisingly, *Danthonia decumbens*. The transition between the spray-zone and the limestone grassland or cliff-heath community above is sometimes gradual but is often marked by an eroded step, above which *Daucus carota* subsp. *gummifer* frequently grows and *Festuca rubra* or, on acid drift, *Erica cinerea* typically form a belt. On the more exposed cliffs of western Gower halophytic *Festuca rubra* or *Plantago maritima* sward sometimes extends some distance up the cliff-slope and also grows on the cliff crests, with *Armeria maritima*, *Silene vulgaris* subsp. *maritima* and other halophytes. *Trifolium occidentale* grows in cliff-top sites around Fall Bay and Mewslade Bay in the far west. *Spergularia rupicola* is common on Worm's Head, where halophytic communities are particularly well-developed, but very scarce elsewhere. A robust perennial form of *Cochlearia officinalis* characteristically grows in crevices on higher vertical cliffs exposed to salt spray, and the diploid cytotype of *Leucanthemum vulgare* (Atherton, 1975) and a fleshy-leaved form of *Centaurea scabiosa* (Valentine, 1980) grow in similar but less exposed sites.

The cliffs above the spray-zone have a rich flora. Yellow whitlowgrass (*Draba aizoides*), which is locally common in crevices on the upper cliffs between Pwll-du Head and Rhossili, is an Alpine species which grows nowhere else in Britain and has its nearest Continental localities in the Belgian Ardennes and on the Côte d'Or near Dijon; it is of great phytogeographical interest (Kay & Harrison, 1970). *D. aizoides* is an extremely attractive plant when in flower but can easily be overlooked at other times of year, and was not discovered in Gower until 1795. Its conspicuous spikes of yellow flowers, rising from compact, saxifrage-like rosettes of leaves, are at their best in March and sometimes appear in late February. Several other uncommon limestone species are relatively abundant in crevices and on shallow soils near limestone outcrops in Gower, for example *Helianthemum canum*, *Potentilla tabernaemontani* and *Geranium sanguineum*. A number of calcicole grassland species, including *Acinos arvensis* and *Hippocrepis comosa*, approach the western limits of their British ranges here. The absence of a number of species that are common in chalk and limestone grassland in England is striking; *Brachypodium pinnatum*, *Cirsium acaule*, *Filipendula vulgaris*, *Avenula pratensis* and *Plantago media* are all absent. In contrast to this, some of the rarest species found in Gower have strikingly disjunct distributions, occurring at a number of widely separated sites on the limestones of western Britain. One example is *Aster linosyris*, which is found at a few sites in shallow gorse west of Port Eynon. Elsewhere in Britain it grows in only four other localities, all on limestone in the west, from Torbay to Lancashire. *Veronica spicata*, also found west of Port Eynon in a few sites on cliff ledges, has a slightly wider distribution on the limestones of western Britain, and also occurs in the breckland of East Anglia; and *Helianthemum canum* grows on Carboniferous Limestone in Gower, Pembrokeshire, North Wales, western Ireland, and at two sites in northern England. Plants with this type of distribution have probably grown at their present refuge sites since the early post-glacial period, often in very small populations that have been isolated from each other for several thousand years.

Another species that is extremely rare in Britain, the small annual *Ononis reclinata*, was first found in 1828 in the west of the peninsula. It is still well-established in the same small area, and also grows at one locality on limestone in Pembrokeshire, and on the Torbay limestone in Devon. It may have spread to Gower more recently than the 'refuge' species, as may a group of scarce annual species of the Gower cliffs that, unlike *O. reclinata*, were once more frequent as weeds of cultivated land. These include *Euphorbia exigua* and *Kickxia elatine* (which usually appear only after gorse fires, with *Anagallis arvensis*) *Gastridium ventricosum* (locally quite abundant in good years in open grassland at several sites from Pwll-du Bay to Mewslade; Gower is one of its remaining strongholds in Britain) and *Anthemis arvensis*, which grows on the crest of the Inner Head of Worm's Head (Kay, 1971b). Some of these may be growing in their original natural habitats, but others are likely to have established themselves from the richer weed populations of cultivated land in the past.

The western extremity of the Gower limestone forms the long promontory of Worm's Head, a tidal island which can be reached on foot over a rocky causeway at low tide. The Inner Head, which until recently was grazed by sheep, is floristically very similar to the mainland cliffs, with *Asperula cynanchica*, *Geranium sanguineum*, *Helianthemum canum*, *H. nummularium*, *Koeleria macrantha*, *Scabiosa columbaria* etc., growing on the shallow soil of the upper south-facing slopes above mixed *Festuca rubra* grassland and *Ulex europaeus* thickets. The more sheltered parts of the Middle and Outer Heads, which are inaccessible to sheep, are covered with a deep, springy turf of *Festuca rubra*, although this is no longer 'largely pure' as McLean described it in 1935 (McLean, 1935), but mixed with *Cirsium arvense*, *Poa pratensis*, *Rumex acetosa* and other weedy species, perhaps as a consequence of reseeding after a fire here in the 1950s. Although McLean attributed the development of the deep *F. rubra* turf of the Middle Head to the absence of grazing, the occurrence of similar *F. rubra* turf on parts of the grazed Inner Head with a similar aspect casts some doubt on this interpretation. *Scilla verna*, which is abundant on exposed headlands on the south coast of Gower, grows to an exceptionally large size on the Middle Head, and the halophytes of the spray-zone are also exceptionally fine; *Inula crithmoides*, *Silene vulgaris* subsp. *maritima* and *Spergularia rupicola* are very abundant. The herring-gull colony on the crest of the Outer Head had an interesting flora including many weeds (Gillham, 1964); after the nesting gulls had left, in July, the summit plateau was sometimes purple with the flowers of *Malva sylvestris*. *Lavatera arborea* is, rather surprisingly, not found here, but does grow in the kittiwake and herring-gull colony on the magnificent cliffs between Mewslade and Fall Bay.

Acid soils are widespread in Gower, on glacial drift, on the Devonian sandstones of Cefn Bryn and the other Gower hills, and on the Millstone Grit and Coal Measures of the north-east. The large areas of common land which are such a feature of the peninsula have mixed open heath vegetation, in which *Calluna vulgaris* and *Ulex gallii* form a patchwork of purple and gold flowers in August, interspersed with patches of marsh and bog and with a few surviving woods of oak, birch, sallow and alder around their fringes. The commons usually occupy the most infertile, stony or ill-drained areas of acid soil. Their vegetation is interesting, especially in marshy and boggy areas, where *Carum verticillatum, Dactylorhiza maculata, Drosera rotundifolia, Menyanthes trifoliata, Narthecium ossifragum, Myrica gale* and *Osmunda regalis* are locally abundant, and *Baldellia ranunculoides, Drosera intermedia, Eriophorum vaginatum, Potentilla palustris* and *Vaccinium oxycoccos* are well-established in a few sites. *Agrostis curtisii*, which is locally dominant on the driest and most acid soils on the crest and southern slopes of Cefn Bryn and Fairwood Common, and also grows in cliff-heath on Oxwich Point and at Longhole Cliff west of Overton, is at the north-western limit of its European range in Gower. The abundance of *Eriophorum angustifolium, Juncus squarrosus, Molinia caerulea, Nardus stricta, Scirpus caespitosus* and locally *Vaccinium myrtillus* on several commons gives them a surprisingly 'upland' character, despite their moderate elevation.

The natural vegetation of the commons was probably a mixed forest of sessile oak (*Quercus petraea*) and birch (*Betula pubescens* and some *B. pendula*) with hazel (*Corylus avellana*) and holly (*Ilex aquifolium*) as understorey trees and bluebell (*Hyacinthoides non-scripta*), *Holcus mollis* and *Lysimachia nemorum* in the field-layer in drier sites. Alder (*Alnus glutinosa*) and sallows (*Salix cinerea, S. caprea* and *S. aurita*) would grow in wetter sites, with guelder-rose (*Viburnum opulus*) and some alder buckthorn (*Frangula alnus*). A few *Sphagnum* bogs probably existed in the wettest sites from an early stage, but open heath and moorland vegetation may at first have been restricted to the exposed slopes of Rhossili Down and perhaps Cefn Bryn. The commons are maintained in their present open state by constant grazing and burning. The abundance of oak seedlings (perhaps growing from acorns dropped by jays) on many of the commons indicates that they would rapidly revert to woodland if grazing and burning ceased. The surviving fragments of oak-birch woodland have, where they are least disturbed, a varied flora again including a number of 'upland' species (Kay, 1971a). For example, the upper part of Penrice Wood by Mead Moor (SS 483834) had until recently a luxuriant bryophyte flora including *Dicranum majus, Hylocomium splendens, Plagiothecium undulatum* and *Rhytidiadelphus loreus* under an open canopy of oak and birch; and in an open oak-birch wood with much hazel and some alder buckthorn in the upper part of Clyne Woods (SS 598910) *Equisetum sylvaticum, Melampyrum sylvaticum* and *Oreopteris limbosperma* are abundant.

There are several fine sand-dune systems in Gower, and some smaller areas of dunes. At Pennard, Penmaen, Nicholaston and Llangennith Burrows the sand has been blown inland over Carboniferous Limestone. Whiteford Burrows and Oxwich Burrows (both of which are National Nature Reserves) are noted for their extensive systems of slacks, usually dominated by *Salix repens*, which have rich floras with many scarce species, for example fen orchid (*Liparis loeselii*), the sedge *Carex serotina*, the horsetail *Equisetum variegatum, Gentianella uliginosa, Monotropa hypopitys* and *Pyrola rotundifolia* subsp. *maritima. Vulpia fasciculata* is abundant on more mobile dunes. *Botrychium lunaria* is still found in fixed dune grassland at Llangennith and on Great Tor to the south of Oxwich Bay; *Ophioglossum vulgatum* is locally abundant in dune-slacks at Whiteford, Oxwich and Crymlyn Burrows.

In north-eastern Gower, sand-dunes form the coast or cover the coastal slopes above low cliffs for a distance of about eight kilometres, from Whiteford Point to Hillend Burrows. Although afforestation has destroyed part of the large area of dune grassland and slacks at Whiteford Burrows, the majority survive or are being reinstated and are now in the care of the National Trust and the Nature Conservancy Council. The slacks are varied and species-rich; orchids, including *Dactylorhiza incarnata, D. praetermissa, Epipactis palustris* and *Listera cordata* are abundant and *Liparis loeselii* is locally frequent. *Gentianella uliginosa* grows in low open vegetation in a few slacks, often, as at Oxwich, near paths or in lightly trampled areas. Its associates in a typical low slack near the transition zone at Whiteford (G.R. SS 448558, 24 May 1972, Domin cover-abundances) were; *Agrostis stolonifera* 3, *Anagallis tenella* 3, *Bellis perennis* 1, *Carex arenaria* 2, *C. flacca* 3, *Dactylorhiza fuchsii* 1, *Eleocharis quinigueflora* 3, *Epipactis palustris* 2, *Equisetum variegatum* 4, *Festuca rubra* 3, *Gentianella uliginosa* 3, *Hydrocotyle vulgaris* 2, *Leontodon taraxacoides* 3, *Lotus corniculatus* 3, *Ononis repens* 1, *Pilosella officinarum* 1, *Polygala vulgaris* 1, *Prunella vulgaris* 2, *Salix repens* 3, *Sonchus arvensis* 1, *Taraxacum laevigatum* agg. 1, *Trifolium fragiferum* 1 and *Trifolium repens* 3.

Another noteworthy plant of the Whiteford dunes is the diminutive grass *Mibora minima*, which is known only from an area of about 1.5 ha of low fixed dunes where it grows in association with other annual grasses and *Hornungia petraea*, often near rabbit-scrapes in bryophyte-rich *Festuca rubra* / *Thymus praecox* turf. *M. minima* flowers from early February to the end of March. Although it has to be searched for carefully its purple-tinged spikelets are very distinctive. *Thalictrum minus* grows with *Geranium sanguineum* and *Rosa pimpinellifolia* on fixed dunes near limestone outcrops at Cwm Ivy Tor and Hills Tor. *Anacamptis pyramidalis* and *Primula veris* are especially abundant below Cwm Ivy Tor, colouring the slopes with their flowers in favourable years, and scattered plants of *Saxifraga granulata* grow in the same area.

At Broughton and Llangennith Burrows there are a few slacks on the landward side with *Epipactis palustris*, but these systems consist mainly of high, mobile or partly fixed dunes over a raised limestone base. As at Whiteford, *Cerastium diffusum*, *C. semidecandrum*, *Hornungia petraea*, *Vicia lathryroides*, *Saxifraga tridactylites* and other small annuals are locally abundant in open turf and on mossy slopes.

The dunes between Port Eynon Point and Horton have no slacks but show a range of dune habitats within a relatively small area, including some fine fixed dunes with *Geranium sanguineum* between Port Eynon and Horton, and an interesting area of low dune exposed to salt spray, with *Halimione portulacoides*, *Inula crithmoides* and other halophytes, on Salthouse Point. Several introduced species have become locally naturalised on the landward side of the dunes, including *Reseda alba* and *Foeniculum vulgare*. Successful dune conservation work – fencing and the provision of boardwalks – has saved these dunes, which are precariously situated between a series of holiday caravan sites and a popular beach, from the trampling and erosion that threatened to destroy the system in the early 1970s. Similar successful conservation work has been carried out by Swansea City Council at Broughton and Pennard Burrows, and by the Nature Conservancy Council at Oxwich Burrows.

Oxwich Burrows resemble Whiteford Burrows in having a fine series of slacks rich in orchids, with *Gentianella uliginosa* locally frequent in favourable years. At Oxwich there are interesting transitions from dune to dry salt-marsh and from low dunes or dune-slacks to woodland with birch (*Betula pubescens*), sallow and oak (Goodman, 1963). Oxwich Burrows were used for invasion exercises during the 1939–45 war and became temporarily seriously eroded as a result. The abundance at Oxwich of *Oenothera* species, both *O. erythrosepala* and the recently described *O. cambrica* (*O. novae-scotiae*), may be a consequence of this disturbance, as may the genetic structure of the common milkwort (*Polygala vulgaris*) populations at Oxwich. Each slack or group of slacks tends to be occupied by a distinctive, genetically homogeneous milkwort population, probably as a consequence of recolonisation from a few scattered surviving plants after the war (Lack & Kay, 1987). Nicholaston Burrows, on the northern side of Oxwich Pill, have some relict areas of fixed dune where *Geranium sanguineum* and *Rosa pimpinellifolia* are conspicuous with, rather surprisingly, an understorey of ivy (*Hedera helix*) and its parasite ivy broom-rape (*Orobanche hederae*); these also grow on Crawley Cliff behind the dunes. Open *Euonymus europaeus* scrub is well-developed on Nicholaston Burrows.

Penmaen Burrows on the western side of Nicholaston Pill and Pennard Burrows to the east resemble one another in being high dune systems extending over the plateau above the cliffs, and are largely fixed. Like the similar dune-system at Merthyr Mawr, they are archaeologically interesting, with buried or partly exposed ancient sites, including a Neolithic megalithic tomb on Penmaen Burrows and a Norman castle on Pennard Burrows. The older parts of the dunes are leached and acidic on the surface, with much bracken (*Pteridium aquilinum*); *Vaccinium myrtillus* also grows at Penmaen, and there is still some fine but increasingly eroded *Calluna* dune-heath rich in lichens and bryophytes (including *Ptilidium ciliare* and *Racomitrium lanuginosum*) on Pennard Burrows. Younger parts of the dunes, and areas where base-rich sand has come to the surface again, are covered by moss-rich grassland communities dominated by *Festuca rubra*, often with much *Rhinanthus minor*. Dry open parts of the dunes have a range of annuals including *Trifolium arvense*, *T. striatum* and *Vicia lathyroides*. *Rhynchosinapis monensis* is locally abundant on young dunes and sandy slopes above Pobbles Bay, growing with *Ammophila arenaria* and *Ononis repens*.

Crymlyn Burrows, on the eastern side of Swansea, are surrounded by industrial sites and crossed by roads and pipelines, but preserve much of what must have been a outstandingly interesting series of habitats between Crymlyn Fen and the sea (Kay, 1974a). Several orchid slacks with *Epipactis palustris* and *Dactylorhiza* spp. survive near the road, some with a few plants of *Liparis loeselii* and one with abundant *Pinguicula vulgaris*. *Juncus acutus* is a conspicuous feature of these slacks, as in other western Glamorgan dune-systems. *Equisetum hyemale* is locally abundant both in some slacks and on low sandy ground near the road. *Matthiola sinuata*, which is probably native here (but has recently appeared on several other dune-systems in Glamorgan as a suspected introduction, for example at Port Eynon) grows in some abundance on drier dunes around the salt-marsh and towards the sea. *Parentucellia viscosa* still grows in a few places in dune grassland near the golf-course at Jersey Marine. There are some particularly interesting established introductions here which may originate from dock-side ballast at the nearby Briton Ferry or Swansea Docks, for example *Artemisia campestris* and *Lathyrus tuberosus*. *Hirschfeldia incana*, *Rhynchosinapis cheiranthos* and *Senecio squalidus*, which grow mainly as weeds of urban waste ground in Glamorgan, are abundant throughout the dune-system.

Freshwater communities – marsh and fen, with some open water – are well-developed behind sand-dunes at several sites, especially at Oxwich Marsh and Crymlyn Fen. Freshwater marshes and other habitats for aquatic plants also occur in some Gower valleys, especially along the Burry Pill and in the valley leading from Blackpill to Gowerton (now, inevitably, much reduced by drainage and tipping), on some commons, and above the landward fringes of the salt-marshes in the Burry Inlet and Llwchwr valleys. Freshwater habitats of a different type are found in the many ponds that survive in Gower.

Crymlyn Fen, behind Crymlyn Burrows, is also rather misleadingly named Crymlyn Bog; bog plant communities in the strict sense are restricted to a small relict area at the northern end, where *Drosera rotundifolia*, *Eriophorum vaginatum*, *Narthecium ossifragum* and *Utricularia minor* are abundant, and a fringe along the eastern margin. The greater part of the wetland area is in fact occupied by fen communities in which many rare or local plants are found. Some of the wettest parts are dominated by an association of two handsome and scarce species – lesser bulrush (*Typha angustifolia*) and marsh cinquefoil (*Potentilla palustris*). Most of the remaining area is covered by common reed (*Phragmites australis*) fen, which is species-rich in places, especially north-west of Gelli'r-allor where *Phragmites* is co-dominant with *Cladium mariscus*, and *Carex elata*, *Osmunda regalis* and *Ranunculus lingua* are abundant, with *Sparganium minimum* in pools. Fen carr scrub is best developed on the eastern side, and is dominated by *Betula pubescens*, *Alnus glutinosa* and *Salix cinerea*, with locally abundant *Frangula alnus*, *Osmunda regalis*, *Viburnum opulus*, and some *Myrica gale*. *Dryopteris carthusiana* is fairly common on *Carex paniculata* tussocks here. *Schoenus nigricans* is found only on the western side, where it grows in a tongue of *Molinia caerulea* extending into the reed fen below Cwm-bach. Three even rarer species, *Carex limosa, Eriophorum gracile* and *Carex dioica,* grow in similar sites along the western margin where acid water drains into the reed fen. This is the largest and botanically richest area of acid marsh and fen in south Wales. Species like *C. elata, C. limosa* and *Sparganium minimum* – all three are at the extreme south-western limit of their British ranges here – and *Carex dioica, C. curta* (also found on the western margin), *Cladium mariscus, Ranunculus lingua, Schoenus nigricans* and *Typha angustifolia* are becoming rarer year by year as the fens and marshes in which they grow are drained, polluted or filled in, and they now survive in very few localities indeed in south Wales.

Oxwich Marsh, described by Goodman (1963), consists mainly of *Phragmites australis* fen, with belts of fen carr scrub on the landward margins, and some areas of open water. The fen itself is relatively base-rich but is floristically much less diverse than Crymlyn Bog. *Carex pseudocyperus* and *Typha angustifolia,* which also grow at Crymlyn, are among the relatively small number of noteworthy species recorded from Oxwich; others include *Butomus umbellatus,* which is scarce and easily overlooked, and the submerged aquatics *Utricularia australis, Hippuris vulgaris* and *Ceratophyllum submersum.*

Salt-marsh communities are well represented in Gower. The greatest extent of salt-marsh is in northern Gower, where there are 16–20 square kilometres of salt-marshes in the Burry Inlet and the upper estuary of the Afon Llwchwr (Kay & Rojanavipart, 1976). These range from fully maritime salt-marshes near Whiteford Point, where there are large beds of *Zostera noltii* on the lower shore beyond the marshes, to brackish transitional high-level marsh near Pontarddulais. Much of the marsh in the Burry Inlet is grazed intensively by sheep and ponies. Grazing has reduced the impact of *Spartina anglica,* which was introduced near Whiteford Point in 1931 and had spread to the entire estuary by the 1950s. *S. anglica* forms pure stands only in some little-grazed areas, for example east of Salthouse Point at Penclawdd. Elsewhere it is a relatively minor component of the vegetation, predominating only along muddy creeks and in some of the pioneer low-level marsh.

Accretion is taking place rapidly at present on the southern shore of the Burry Inlet, where open stands of *Salicornia* spp. and *Spartina anglica* are succeeded by *Puccinellia maritima* hummocks which coalesce to form a grazed general salt-marsh community with *P. maritima, S. anglica, Aster tripolium, Cochlearia anglica, Halimione portulacoides, Spergularia media* and *Triglochin maritima. Limonium vulgare* is absent from most of the grazed areas here but is locally common near Berges Island on Whiteford Point, and in a fenced area of salt-marsh south of Llwchwr. At higher levels *Armeria maritima, Festuca rubra, Glaux maritima, Juncus gerardi, Plantago coronopus, P. maritima* and *Triglochin maritima* form a closely grazed turf, with *Agrostis stolonifera, Alopecurus bulbosus, Parapholis strigosa* and *Trifolium repens* locally frequent immediately below the *Juncus maritimus* belt in which *Carex extensa, Leontodon autumnalis* and *Oenanthe lachenalii* are common. Annual *Salicornia* spp. and *Suaeda maritima* are found in open muddy areas throughout the marsh. In high grazed marshes near the transition zone, for example between Llanrhidian and Wern-halog, *Althaea officinalis* and *Artemisia maritima* are locally abundant. Shallow brackish pools here contain *Ranunculus baudotii, Ruppia maritima,* and *Zannichellia palustris,* with *Scirpus maritimus, Schoenoplectus tabernaemontani* and *Phragmites australis* in brackish ditches. The transition zone between salt-marsh and sand-dune, well-developed at Whiteford Burrows, has several characteristic species including *Centaurium pulchellum* and *Sagina maritima* (both of which grow on the anthills which are a surprising feature of the upper salt-marsh), *Apium graveolens, Juncus acutus* and *Limonium binervosum* (Kay & Woodell, 1976).

The other main salt-marshes in the western part of Glamorgan are at Oxwich, along Pennard Pill, and behind Crymlyn Burrows and in the estuary of the River Neath. These show general similarities to the North Gower marshes, although the generally lower levels of grazing in several of these marshes (Pennard Pill and the Neath marshes excepted) appear to have reduced the complexity of the community, with single species tending to be dominant at each level and stage of development of the salt-marsh. This is particularly well shown at Crymlyn Burrows where single species, e.g. *Festuca rubra* and *Parapholis strigosa*, tend to form pure stands.

The drift-lines, on the magnificent but much-visited sandy beaches of western Glamorgan, are often so disturbed that few plants are able to persist there. On some less-trampled sandy beaches, for example at Whiteford and Crymlyn Burrows, *Crambe maritima* and *Salsola kali* can form a broad belt on the drift-line in favourable years and *Honkenya peploides* is locally frequent, with scattered plants of *Atriplex* species; *A. laciniata* grows at both Whiteford and Oxwich. Most other drift-line plants, for example *Beta vulgaris* subsp. *maritima* and *Rumex crispus* var. *trigranulatus*, require a more stable substratum and are found only where sand is stabilised by shingle, or where conditions are locally suitable above salt-marshes or below cliffs. Isolated plants of *Raphanus maritimus* are not infrequent on drift-lines in the county, but it is well-established only on shingle-beaches at Llwchwr and Sully. *Atriplex littoralis* is also locally common at Llwchwr, but scarce or absent elsewhere. *Polygonum oxyspermum* subsp. *raii* has probably always been scarce in the county but has been found recently on drift-lines at Three Cliffs Bay, where there is an interesting shingle beach on which yellow horned-poppy *Glaucium flavum* still grows. *G. flavum* has also been seen recently at Caswell Bay, where it grows on the undercliff. *Galeopsis angustifolia* grows on stabilised shingle (mainly originating from nineteenth century limestone quarry waste and barge ballast) at Pwll-du Bay.

The Uplands and the Border Ridges
Coal Measures sandstones and shales, with some Millstone Grit and Carboniferous Limestone in the extreme north-east and south, underlie most of this area.

The resistant Pennant Grit sandstone forms a high plateau in the north, rising from about 260 m near Pontardulais to 600 m at Craig-y-Llyn above Hirwaun. The plateau is dissected by deep and usually steep-sided valleys. The hilltops and higher slopes have a rather limited moorland flora, with *Calluna vulgaris*, *Erica cinerea* and *E. tetralix*, *Vaccinium myrtillus*, *Deschampsia flexuosa*, *Festuca ovina*, *Nardus stricta*, *Juncus squarrosus*, *Molinia caerulea*, *Scirpus caespitosus* and *Eriophorum angustifolium* in varying proportions. Flushes and streamsides on the slopes of the hills provide more variety and interest, often with creeping stems of the delicate ivy-leaved bellflower (*Wahlenbergia hederacea*), bog pimpernel (*Anagallis tenella*) and the introduced New Zealand willowherb (*Epilobium brunnescens*) growing among mosses and liverworts. *Succisa pratensis* and *Cirsium dissectum* are often abundant in marshy grassland at lower altitudes, with *Carum verticillatum* in the west. Oak (*Quercus petraea*) or oak-birch woods, in which ash (*Fraxinus excelsior*) and rowan (*Sorbus aucuparia*) are often common survive here and there on the sides of the valleys and in ravines. Alder (*Alnus glutinosa*) is common on stream-banks and in wet woods. Some of the best oakwoods are found in the deep valley of the Afon Pyrddin west of Glyn Neath, where the bryophyte flora is particularly rich and *Crepis paludosa*, *Trollius europaeus* and other rare species grow on rocky streamsides in the wood. In the upper oakwoods of Cwm Dimbath, near Blackmill, the filmy-ferns *Hymenophyllum tunbrigense* and *H. wilsonii* still grow in ravines. However, although old records suggest that *Hymenophyllum wilsonii*, *Trollius europaeus* and other upland species which are now very rare in Glamorgan were fairly widespread in the past, most of the remaining valleyside woods now have a limited flora of bryophytes and vascular plants. Great expanses of conifers have been planted on the valley slopes and lower hills in the last sixty years, and more recently have been extended over some of the higher tops, even including Craig-y-Llyn. Although young plantations often pass through a phase in which they are attractive to wildlife, successful conifer plantations can eventually destroy or drastically modify the native flora. The dense, closed stands of evergreen conifers that develop as the plantations mature exclude and destroy native vegetation, and, as increasingly strong evidence shows, acidify the soil and the water of the streams that pass through the plantations, bringing about changes that may sometimes be irreversible. Realisation of these problems is gradually bringing about changes in woodland management, which may return to modern versions of the coppicing techniques by which high and sustainable yields were obtained from native deciduous woodland in the past (Rackham, 1986).

Most valleys in the uplands were industrialised during the nineteenth and early twentieth centuries. There is now little heavy industry left in the valleys, but the valley floors are still heavily urbanized, with tips of slag and colliery waste occupying remaining open spaces on the valley floors and lower slopes. Redistribution and landscaping of these tips in the last few years, although undertaken with the best of intentions, has sometimes reduced the variety and interest of the plants and wildlife in the valleys, both by destroying the vegetation of the tips themselves (sometimes unusual and fascinating in its own right, especially on basic slag) and by obliterating surviving fragments of natural vegetation and farmland around the tips. Fortunately, the widespread acceptance of the value of wildlife and habitat diversity to urban communities has been reflected in recent schemes.

Although the upland valleys are at last recovering from the scars left by steelmaking, coal-mining and their associated heavy industries, the last phase of the exploitation of the uplands by industry is still developing and has been perhaps the most drastic of all in its destructive effects on the environment and vegetation. Open-cast coal-mining on a huge scale, often not visible from the valleys below, has devastated the more accessible parts of the northern plateaux and high slopes in recent years, permanently and irreversibly destroying the natural landscape and habitat. Open-cast mining especially threatens the few surviving enclaves of species-rich hay-meadow, pasture and hedgebanks in the patches of old farmland above the main valleys, for example at Blaen-gwrach above Glyn Neath, where *Epipactis helleborine* grows on the laneside banks, and at Tairgwaith above Gwaun-cae-gurwen.

The high north-facing Pennant Sandstone escarpment of Craig-y-Llyn above Hirwaun, with its two small cwm lakes Llyn Fach and Llyn Fawr, is one of the most interesting localities in the Uplands. Several northern and arctic-alpine species have been found on the cliffs above the lakes, and there is a rich bryophyte flora in ravines and flushes. *Cryptogramma crispa, Dryopteris oreades, Hymenophyllum wilsonii, Lycopodium selago, Vaccinium vitis-idaea, Sedum rosea, Thalictrum minus,* the mosses *Andreaea rupestris* and *Bartramia ithyphylla* and many other montane species are still well-established here, but *Orthilia secunda* has not been seen recently. Some of these species also grow on similar though less extensive mountain cliffs around the head of the Rhondda Valley. The mountain lake species *Lobelia dortmanna* and *Isoetes echinospora* (Kay, 1974b) are abundant in Llyn-fach, where *Sparganium angustifolium* grows in its southernmost site in the British Isles.

Ravines provide some of the best woodland, streamside and cliff habitats for both vascular plants and bryophytes in the Uplands, especially below waterfalls. In the Pyrddin valley west of Pont-nedd-fechan, *Festuca altissima, Meconopsis cambrica,* and *Rubus saxatilis* grow in the ravines below the waterfalls of Sgwd Gwladys and Sgwd Einon Gam. At Melin-y-cwrt near Resolven *Osmunda regalis* and *Phegopteris connectilis* grow on the rocks beside the waterfall and *Hymenophyllum tunbrigense* and *Tilia cordata* in the ravine below the falls.

A variety of native and introduced species grow on river-banks and riverside gravels in the Uplands, even in the most industrialized valleys. Monkeyflower, *Mimulus guttatus* or *M. guttatus × luteus* is often abundant, for example in the Rhondda valleys. *Ranunculus penicillatus*, the perennial water-crowfoot of the fast-flowing upland rivers, is another attractive species which flowers in the shallows and in pools on gravel-banks exposed in the summer. Indian balsam (*Impatiens glandulifera*) is locally abundant by rivers both here and in the Vale. *Rorippa sylvestris* and *Elymus caninus* are characteristic riverside species in the Uplands; *R. sylvestris* often forms a belt on sloping gravelly banks. Many annual weeds grow on riverside gravels, for example *Galeopsis tetrahit*, *Polygonum persicaria* and *Senecio viscosus*.

Coal-tips often have an interesting flora including some unexpected species. Pearly everlasting (*Anaphalis margaritacea*), an early introduction from America first recorded in Glamorgan before 1729, is a conspicuous plant of coal-tips on which it is locally very common in some valleys, for example near Ystalyfera. Small cudweed (*Logfia minima*) is much less conspicuous but is a very characteristic and often abundant plant of coal-tips and restored opencast sites. The normally calcicole biennial *Carlina vulgaris* is another characteristic though less common coal-tip plant, as are annual *Trifolium* species, most often *T. dubium* but occasionally other species including *T. ornithopodioides* and *T. striatum*. For example, the following species were recorded on a coal-tip north-east of Gwaun-cae-Gurwen (SN 713123) in August 1974 (Domin cover-abundances); *Aira caryophyllea* 1.3, *Arenaria serpyllifolia* s.l. 2, *Cerastium fontanum* 2, *Cirsium arvense* 3–4, *Cirsium vulgare* 1, *Epilobium ciliatum* 1, *Festuca rubra* 3, *Filago minima* 3, *Hieracium* sp. 2, *Holcus lanatus* 2, *Hypochoeris radicata* 3, *Lathyrus pratensis* 4, *Leontodon taraxacoides* 3, *Poa pratensis* 2, *Senecio viscosus* 1, *Tragopogon pratense* 1, *Trifolium dubium* 4, *Trifolium striatum* 3, *Vulpia bromoides* 2.

Another industrial habitat, now less widespread than it once was but still important in some valleys, is railway ballast. *Vulpia myuros* and *Chaenorrhinum minus* are characteristic species which are rarely found in any other habitat in this area. Calcicoles like *Anthyllis vulneraria* sometimes grow where basic steel slag or limestone chippings have been used as ballast. A few moderate calcicoles also grow on some old basic slag-heaps, for example *Linum catharticum* and *Thymus praecox* on the slag-heaps by the Tawe below Ystalyfera. Most of the Carboniferous Limestone on which the ironworks of the heads of the Glamorgan coalfield valleys depended outcrops to the north of the county boundary, in Breconshire, but there is a minute area of limestone on the Glamorgan side at Bwa Maen north of Hirwaun – *Botrychium lunaria* and *Cornus sanguinea* grow in a patch of limestone grassland and scrub there – and a much larger area at Morlais Castle Hill and in the Taf Fechan gorge north of Merthyr Tydfil. This has been much disturbed by quarrying and tramway construction but is still full of interest, and is especially rich in limestone bryophytes. Mountain everlasting (*Antennaria dioica*), *Asplenium viride*, *Cardamine impatiens*, *Galium sterneri*, *Gymnocarpium robertianum*, *Melica nutans* and *Saxifraga hypnoides* are among the scarce or local montane limestone species that have been recorded here. Most were found in small numbers and some have not been seen recently; fortunately, nearly all are still locally frequent not far away, on the Breconshire limestone.

Along the southern margin of the coalfield, Millstone Grit, Carboniferous Limestone and to the east of Cardiff Old Red Sandstone outcrop to form the Border Ridges. Beech (*Fagus sylvatica*), which appears to reach the western limit of its native distribution in Britain in eastern Glamorgan (Hyde, 1960) still forms some fine woods on Carboniferous Limestone and Old Red Sandstone north and east of Cardiff, although much woodland has been lost or impoverished by felling, disturbance or quarrying, as for example at Garth Woods. As in other woods on basic soil in Glamorgan, the commonest species in the field layer are *Mercurialis perennis* and *Allium ursinum*. Several scarce limestone woodland species also occur here and in similar woods in the eastern Vale, for example *Convallaria majalis*, *Monotropa hypopitys* and *Neottia nidus-avis*. Acid soils on Millstone Grit, glacial drift and lower Coal Measures elsewhere on the Border Ridges provide a complete contrast; here marshy heathland resembling the Gower commons is found, for example at Kenfig Hill where *Agrostis curtisii* and *Genista anglica* are locally common, and at Llantrisant Common where *Vaccinium oxycoccos* and other bog species grow.

The Vale of Glamorgan
The southern part of the Vale of Glamorgan is an undulating plateau of Lias limestone rising from about 50 m near the sea to 100 m in the north, meeting the sea in vertical cliffs with few ledges. Most of this area is cultivated farmland, but, as in Gower, woods and some limestone grassland survive in narrow, deeply incised valleys, especially near the sea. To the north and east Triassic sandstones and marls as well as Lias limestone overlie Carboniferous Limestone and the countryside is more varied and broken, with wide valleys and rounded hills rising towards the Border Ridges, and considerable thicknesses of periglacial deposits on which acidic, poorly drained soils have developed. Where the Carboniferous Limestone is exposed there are some steep-sided rocky valleys, for example Pant St Brides south of Bridgend and Cwm George near Dinas Powis, and there are also some steep-faced minor escarpments of Lias limestone on which limestone woodland has survived, for example Pencoed Wood (SS 952810). Along the coast, salt-marshes have formed in the estuaries of the Taff, Thaw, Ogmore and some smaller streams. The western coast is bordered by a broad belt of sand-dunes from the Ogmore estuary northwards. The belt of dunes is about 1.5 km wide at Merthyr Mawr and reaches a maximum of 3 km inland at Kenfig, but only 0.5 km or less on the seaward side of the alluvial flat extending to the Neath Estuary at the foot of the steep escarpment behind Port Talbot.

Several conspicuous limestone species reach the western limit of their British ranges in the Vale of Glamorgan, and are absent from Gower, or grow there only as scattered individuals. Examples include wayfaring-tree (*Viburnum lantana*), which is a common and conspicuous hedgerow shrub in the Vale, woolly thistle (*Cirsium eriophorum*), and some of the most characteristic species of limestone grassland in the Vale; *Bromus erectus*, *Cirsium acaule*, *Hypericum hirsutum*, *Plantago media*, *Silaum silaus*, and the scarcer *Campanula glomerata*. The differences between the limestone grassland communities of the two areas are increased by the absence or scarcity of some typical Gower limestone plants in the Vale. For example, *Acinos arvensis* and *Helianthemum canum* are absent from the Vale, *Geranium sanguineum* was always scarce and may be extinct there, *Scabiosa columbaria* has been found at only two sites and *Asperula cynanchica* is rare. There are fewer differences between limestone woodland communities in Gower and the Vale, partly because beech (*Fagus sylvatica*), which is native only in the eastern Vale, has been introduced widely in the west and in Gower, and perhaps also because most of the remaining woodland on limestone in Gower and the Vale is so fragmented and disturbed. Coed Garnlwyd (ST 058712), between Llancarfan and Bonvilston in the east-central Vale, is a good example of a mixed limestone woodland dominated by ash and oak, with hazel (*Corylus avellana*), field maple (*Acer campestre*), spindle (*Euonymus europaeus*), hawthorn (*Crataegus monogyna*) and wayfaring-tree (*Viburnum lantana*) making a diverse shrub layer and Herb-Paris (*Paris quadrifolia*) and early-purple orchid (*Orchis mascula*) growing in a rich field layer with *Anemone nemorosa* and other widespread species, most of them also present in Gower limestone woods.

Coastal scrub and the narrow strip of limestone grassland that survives above the unstable sea-cliffs of the Vale provide important refuges for limestone plants, some of which were once more widespread inland. *Buglossoides purpurocaerulea, Daphne laureola, Helleborus foetidus, Iris foetidissima* and *Ruscus aculeatus* grow at the edge of coastal scrub on the southern cliffs. Good examples of coastal limestone grassland survive at Nash Point (SS 914685) and Cwm Nash (SS 904702). Here *Campanula glomerata, Cirsium acaule, Avenula pubescens, Hippocrepis comosa, Koeleria macrantha, Sanguisorba minor* and *Ranunculus parviflorus* grow on south-facing slopes and *Cirsium tuberosum* grows both in cliff-top grassland with *Silaum silaus* near the lighthouse and on the cliffs below. Hoary stock (*Matthiola incana*) also grows on the cliffs here; other noteworthy species of the Vale cliffs are wild cabbage (*Brassica oleracea*), which grows in the spray-zone with *Festuca rubra* and other halophytic species and is particularly fine at Southerndown, and maidenhair fern (*Adiantum capillus-veneris*) which grows on inaccessible tufaceous seepages at several places.

At Ogmore Down and Old Castle Down on opposite sides of Pant St Brides (SS 89.76.) some of the finest remaining stretches of limestone grassland in the Vale and an extremely interesting area of intermingled limestone and acid grassland survive. Here Carboniferous Limestone is partly mantled by acid loessic soils. On deep acid soil, communities of calcifuge species (*Agrostis curtisii, Calluna vulgaris, Erica cinerea, Ulex gallii* and others) grow. These also grow on shallower, apparently neutral soil over limestone, but intermingled with calcicoles (*Cirsium acaule, Helianthemum nummularium, Hippocrepis comosa, Poterium sanguisorba,* and even the extreme calcicole *Asperula cynanchica*). Earlier workers interpreted this as a consequence of soil layering, with deep-rooting calcicoles penetrating to calcareous soil below an acid surface layer, but recent work suggests that the mixed community may be maintained by burning and grazing on an unlayered neutral soil (Etherington, 1981).

Kenfig Burrows on the western coast of the Vale is in many respects the finest dune-system in Glamorgan, rivalled only by Whiteford Burrows in the extent and floral diversity of its slacks. *Liparis loeselii* and other orchids are particularly abundant at Kenfig, which has some species that are rare or absent from other Glamorgan dunes, for instance *Herminium monorchis, Orchis morio* and *Gymnadenia conopsea*. As at Whiteford Burrows, *L. loeselii* grows in low, relatively open vegetation in the wetter slacks, often in association with *Equisetum variegatum* and *Salix repens*, and varies greatly in apparent abundance from year to year. In unfavourable years many plants apparently fail to produce aerial rosettes, but persist underground as mycorrhizal pseudobulbs. *Pyrola rotundifolia* is also abundant at Kenfig, where it was first recorded in the 1930s. The dune flora is richest in and around the series of large slacks between Kenfig Pool and the sea, but some scarce species grow on the dry, leached dune grassland towards Sker, for example *Ornithopus perpusillus* and *Omalotheca sylvatica*. Kenfig Pool and some of the wettest slacks have a rich aquatic flora with several scarce *Potamogeton* species; waterside plants here include *Limosella aquatica* (mudwort) and the extremely rare *L. subulata* (not seen recently), *Littorella uniflora* and *Baldellia ranunculoides*. The reens draining the marshy meadows between Kenfig Burrows and Margam also have an interesting aquatic flora, with *Butomus umbellatus, Carex disticha, Hydrocharis morsus-ranae* and *Sagittaria sagittifolia*. Both *Limosella* species used to grow on the muddy, cattle-trampled shores of Morfa Pools at Margam, but have not been seen since the construction of the Margam steelworks in the late 1940s (Vachell, 1950).

References

Ashbee, P. (1974). *Ancient Scilly, from the first farmers to the early Christians.* David & Charles, London.

Chambers, F.M. (1983). Three radiocarbon-dated pollen diagrams from upland peats north-west of Merthyr Tydfil, South Wales. *Journal of Ecology,* **71**, 475–487.

Culver, S.J. (1976). The development of the Swansea Bay area during the past 20,000 years. *Gower,* **27**, 58–63.

Etherington, J.R. (1981). Limestone heaths in southern Britain; their soils and the maintenance of their calcicole-calcifuge mixtures. *Journal of Ecology,* **69**, 277–294.

Evans, J.G. (1975). *The environment of early man in the British Isles.* Paul Elek, London.

Feachem, R.W. (1973). Ancient agriculture in the highlands of Britain. *Proceedings of the Prehistoric Society,* **39**, 332–353.

Fearn, G.M. (1972). The distribution of intraspecific chromosome races of *Hippocrepis comosa* and their phytogeographical significance. *New Phytologist,* **71**, 1221–1225.

Fox, C. (1923). *Archaeology of the Cambridge Region.* Cambridge University Press.

Gillham, M.E. (1964). The vegetation of local coastal gull colonies. *Transactions of the Cardiff Naturalists' Society,* **91**, 23–33.

Goodman, G.T. (1963). *Plant Life in Gower.* Gower Society, Swansea.

Green, H.S. & Stanton, Y.C. (1984). The Old and Middle Stone Ages. In: *Glamorgan County History. Volume II. Early Glamorgan* (Ed. by H.N. Savory), pp. 11–121. Glamorgan County History Trust, Cardiff.

Grimes, W.F. (1984). The Neolithic Period. In: *Glamorgan County History. Volume II. Early Glamorgan* (Ed. by H.N. Savory), pp. 123–153. Glamorgan County History Trust, Cardiff.

Houlder, C. & Manning, W.H. (1966). *Regional Archaeologies – South Wales.* Cory, Adams and Mackay, London.

Kay, Q.O.N. (1971). Botany. In: *Swansea and its Region* (Ed. by W.G.V. Balchin), pp. 85–100. University College, Swansea.

Kay, Q.O.N. (1974a). Botany. In: *Neath and District* (Ed. by Elis Jenkins), pp. 305–318. Elis Jenkins, Neath.

Kay, Q.O.N. (1974b). Diploid *Isoetes echinospora* in Britain. *Fern Gazette,* **11**, 56–57.

Kay, Q.O.N. (1975). J.A. Webb's account of 'The presumably extinct plants of West Glamorgan' reconsidered after forty-five years. *Transactions of the Cardiff Naturalists' Society,* **96**, 23–28.

Kay, Q.O.N. (1978). The post-glacial history of commonlands in West Glamorgan. In: *Problems of Commonland* (Ed. by E.M. Bridges), pp. 3:1/1 – 3:1/19. University College, Swansea.

Kay, Q.O.N. & Harrison, J. (1970). Biological Flora of the British Isles. *Draba aizoides* L. *Journal of Ecology,* **58**, 877–888.

Kay, Q.O.N., Roberts, R.H. & Vaughan, I.M. (1974). The spread of *Pyrola rotundifolia* L. subsp. *maritima* (Kenyon) E.F. Warb. in Wales. *Watsonia,* **10**, 61–67.

Kay, Q.O.N. & Rojanavipart, P. (1976). Saltmarsh ecology and trace-metal studies. In: *Problems of a small estuary* (Ed. by A. Nelson-Smith), pp. 2.1–2.16. University College, Swansea.

Kay, Q.O.N. & Woodell, S.R.J. (1976). The vegetation of anthills in West Glamorgan saltmarshes. *Nature in Wales,* **15**, 81–87.

King, T.J. (1981). Anthill vegetation in acidic grasslands in the Gower peninsula, South Wales. *New Phytologist,* **88**, 559–571.

Knight, J.K. (1984). Sources for the early history of Morgannwg. In: *Glamorgan County History. Volume II. Early Glamorgan* (Ed. by H.N. Savory), pp. 365–409. Glamorgan County History Trust, Cardiff.

Lack, A.J. & Kay, Q.O.N. (1987). Genetic structure, gene-flow and reproductive ecology in sand-dune populations of *Polygala vulgaris. Journal of Ecology,* **75**, 259–276.

Martin, P.S. (1967). Prehistoric Overkill. In: *Pleistocene Extinctions - the search for a cause* (Ed. by P.S. Martin & H.E. Wright), pp. 75–120. Yale University Press, New Haven.

McClean, R.C. (1935). An ungrazed grassland on limestone in Wales. *Journal of Ecology,* **23**, 436–442.

Moore, P.D. & Chater, E.H. (1969). The changing vegetation of west-central Wales in the light of human history. *Journal of Ecology,* **57**, 361–379.

Pennington, W. (1969). *The history of British vegetation*. English Universities Press, London.

Percival, M.S. (1960). Botany. In: *The Cardiff Region* (Ed. by J.F. Rees), pp. 45–57. University of Wales Press, Cardiff.

Rackham, O. (1986). *The History of the Countryside*. J.M. Dent, London.

Rutter, J.G. (1948). *Prehistoric Gower*. Welsh Guides, Swansea.

Savory, H.N. (1984). The Early Bronze Age in Glamorgan (*c.* 2500–1450 B.C.); The Later Bronze Age in Glamorgan (*c.* 1450–500 B.C.); Early Iron Age Glamorgan (*c.* 500 B.C.–A.D.. 100). In: *Glamorgan County History. Volume II. Early Glamorgan* (Ed. by H.N. Savory), pp. 155–275. Glamorgan County History Trust, Cardiff.

Simmons, I.G. (1969). Evidence for vegetation changes associated with mesolithic man in Britain. In: *The domestication and exploitation of plants and animals* (Ed. by P.J. Ucko & G.W. Dimbleby), pp. 111–119. Duckworth, London.

Smith, A.G. (1970). The influence of Mesolithic and Neolithic man on British vegetation; a discussion. In: *Studies in the vegetational history of the British Isles* (Ed. by D. Walker & R.G. West), pp. 81–96. Cambridge University Press, London.

Swan, M.C. (1981). The orchids of Gower. *Bios (Swansea)*, **7**, 41–45.

Thomas, D.A. (1963). *Agriculture in Wales during the Napoleonic Wars*. University of Wales Press, Cardiff.

Trotman, D.M. (1963). *Data for late glacial and post glacial history in South Wales*. Ph.D. Thesis, University College of Swansea.

Turner, J. (1964). The anthropogenic factor in vegetational history. 1. Tregaron and Whixall Mosses. *New Phytologist*, **63**, 73–90

Turner, J. (1965). A contribution to the history of forest clearance. *Proceedings of the Royal Society, B*, **161**, 343–353.

Turner, J. (1970). Post-Neolithic disturbance of British vegetation. In: *Studies in the vegetational history of the British Isles* (Ed. by D. Walker & R.G. West), pp. 97–116. Cambridge University Press, London.

Vachell, E. (1950). The disappearance of Morfa Pools. *Transactions of the Cardiff Naturalists' Society*, **79**, 40–42.

Walker, M.J.C. (1982). Early and Mid-Flandrian Environmental History of the Brecon Beacons, South Wales. *New Phytologist*, 91, 147–165.

Webley, D.P. (1976). How the west was won: prehistoric land-use in the Southern Marches. In: *Welsh Antiquity* (Ed. by G.C. Boon & J.M. Lewis), pp. 19–35. National Museum of Wales, Cardiff.

6

Flowering Plants and Ferns of Glamorgan

Arrangement of the Catalogue

The sequence of families, genera and species is mainly that of *Flora Europaea*, edited by T. G. Tutin *et al.*, and published in five volumes between 1964 and 1980. Some of the less common introduced taxa are grouped together at the end of a genus, section, or family.

The entry for each species follows the same general pattern. **Latin name** (in bold type, followed by the authority and synonym if given, in italics), English name and Welsh name are shown on the first line, followed by a paragraph describing the status, frequency, habitats, general distribution etc. of the species in Glamorgan, and in a few cases abbreviated information on floral biology (see page 82). A map shows the detailed distribution of the species in the 5km squares of the Ordnance Survey Grid within the county. Rarer species are not mapped; instead, individual records are listed by 5km squares. All these are explained in detail below.

Latin name: The name adopted here is that given in *Flora Europaea* unless more recent research has shown this to be wrong, or inappropriate. Synonyms are only given where the currently accepted names differ from those in widely used recent Floras. In the case of some rare aliens, the name under which the original record was published is given either as the currently accepted name or as a synonym.

English name: In general only one English name is given, and for native and naturalized plants this has been taken from *English Names of Wild Flowers*, 2nd ed. by J. G. Dony, S. L. Jury and F. H. Perring, published by the B.S.B.I., 1986. Names of other taxa have been taken from a variety of sources. Occasionally a second English name is given if the authors felt that that was equally well known in Glamorgan.

Welsh name: In general only one Welsh name is given, and this is the recommended Welsh name taken from a manuscript list of Welsh names of wild plants, edited by Dafydd Davies, and hopefully to be published by the National Museum of Wales. This lists all the Welsh names known to have been given to plants and indicates which name is recommended for general use. Occasionally a second Welsh name is given if the authors felt that that was equally well known in Glamorgan, and this too has been taken from the same source.

Status: Only three categories of status are employed.

> Natives: Species believed to have been in Wales before man, or to have immigrated without his aid by natural means of dispersal, or to have arisen naturally here.

> Introduced: Species believed or known to have been introduced by the intentional

or unintentional agency of man and to have more or less established themselves in the county.

Casuals or Aliens: Species known to have been introduced by the intentional or unintentional agency of man, which do not persist.

Frequency, habitats and general distribution: A longer or shorter account, depending on the rarity or interest of the taxon in question, is given of its frequency, habitats and general distribution within the county; details of its history may also be mentioned.

Floral biology: For many native and some of the more widespread introduced species, months of flowering and fruiting, as observed within the county, are given, often followed by a list, in abbreviated form, of the insects which have been noted visiting the flower and an indication where possible if the insect was gathering pollen or nectar. A more detailed account is given below.

Distribution: The distribution of plants within Glamorgan (v.c. 41) has been mapped using the 5km square as the base unit, and the county contains all or part of 132 of these squares. The 5km square distribution is given in the following ways.

(a)   For taxa recorded from fewer than 10 squares full details, as set out below, are provided for each locality (there may be several localities in one 5km square, in which case all are treated separately). The entries are listed in numerical order of 10km squares.

5km square: Although four 100km square units are found within the vice-county, their code letters or numerals have been omitted as they do not give rise to any ambiguous 10km square numbers. The 5km squares are denoted by their 10km coordinates subdivided into NW (A), NE (B), SW (C) and SE (D) quarters. Occasionally the letter may be omitted, especially for old records, and for records in Perring and Walters (1962) or Perring (1968), when the exact locality may be uncertain.

Locality: The place name given is usually that of the nearest feature named on the appropriate 1:50 000 or 1:25 000 O.S. map, but in the case of very rare plants it may be made deliberately vague. The correct Welsh spelling of place names is only used when it is unlikely to cause confusion. For some rare aliens or hybrids, the locality name is omitted for reasons of space.

Four figure map references to most place names are given in the gazetteer in Appendix III.

Recorder's name or its abbreviation: Recorders, whose names occur on only a few occasions, are given in full but usually without title (Mr, Mrs, Miss, Dr, etc.). Recorders whose names occur more frequently are given in abbreviated form, and a key to these may be found in Appendix I.

The full or abbreviated names of the three authors, R.G. Ellis, Q.O.N. Kay and A.E. Wade, are rarely given; they are indicated instead by the symbol ! usually after the date (e.g. 1979!). Records made before 1965 which have this symbol are almost invariably those of A.E. Wade. When the record was made in the company of another recorder, the name of the latter is given (full or abbreviated) with ! following the date (e.g. 'ABP 1976!', indicates a record made by Mrs A.B. Pinkard and one of the authors in 1976).

A complete entry enclosed in parentheses indicates that the taxon is known to have been introduced in that locality. This convention is only used when the taxon is native elsewhere in the county.

Date: When only an approximate date is known this is indicated by *c.* (circa). In some cases all that is known of the date of a record is that it was made before or after a certain year; this is indicated by using the prefix pre- or suffix + respectively.

Symbols following date: The use of the exclamation mark ! is explained above. Single or double asterisks (*/**) following the date indicate that a specimen is deposited in **NMW** or **UCSA** respectively.

The entries for casual species may be more concise, often with related species amalgamated into collective accounts to save space. Individual records are nevertheless listed where possible because of the considerable interest of the adventive and casual flora of Glamorgan.

(b) Taxa recorded in more than 10 5km squares are mapped. On all the maps, an open circle (○) indicates a pre-1960 record; a solid dot (●) indicates a 1960 or later record; a cross (×) indicates an introduction, pre-1960 and a plus sign (+) indicates an introduction, post-1960 (these latter symbols are only used when the species is native elsewhere in the county). An open circle or cross in the centre of a 10km square indicates records which it has not been possible to assign to a 5km square and which are the only known records for that 10km square (mainly ex *Atlas of the British Flora*).

Floral Biology

Phenology and insect visitors

Surprisingly little is known of the floral biology of many species, and the information that can be found in botanical literature is often derived from scattered observations made during the late nineteenth century. We have therefore incorporated our own records of the periods of flowering of some of the species in Glamorgan and, for insect-pollinated species the nectar- or pollen-seeking visitors that we have observed in the county, as brief summaries at the end of the accounts of some species. The main periods of flowering (fl.) and fruiting (fr.) are described first, with the months numbered from January (1) to December (12). Lists of the insects that have been observed visiting the flowers of each species in Glamorgan follow; (n) indicated that they were apparently seeking nectar, (p) that they were seeking pollen. A key to the abbreviated names that we have used follows. Most observations were made by Dr Mary Percival; those made by the authors (mainly QONK) are preceded by (!). Unidentified species in a particular group are denoted by a query preceding the name of the group, for example ?Bb means an unidentified Bumblebee.

Nomenclature follows Kloet, G.S. & Hincks, W.D. *A Checklist of British Insects* (2nd ed.) The abbreviations used for the names of common flower-visitors are as follows:

Honey-bee	
Apis mellifera	Hb
Bumblebees	Bb
Bombus	B
B. hortorum	B.hort
B. lapidarius	B.lap
B. lucorum	B.luc
B. monticola	B.mont
B. pascuorum	B.pasc
B. pratorum	B.prat
B. terrestris	B.terr

(*B. magnus* was not distinguished from *B. lucorum* in the field).

Solitary bees	Sb
Hoverflies (Syrphidae)	Syr
Eristalis s.l.	E
E. arbustorum	E.arb
E. intricarius	E.int
E. tenax	E.ten
Helophilus pendulus	H.pen
Leucozona lucorum	Leu.luc
Rhingia campestris	Rh.cam
Scaeva pyrastri	Sc.pyr
Sphaerophoria scripta	Sp.scr
Syrphus s.l.	S
S. balteatus	S.balt
S. cf. *luniger*	S.lun
S. cf. *ribesii*	S.rib
Syritta pipiens	Sy.pip

Other Diptera	Dip
Calliphora	Call
C. erythrocephala	Call.ery
Empis tesselata	Em.tess
Lucilia	Luc
Polietes lardaria	Pol.lar
Pollenia rudis	Pol.rud
Sarcophaga carnaria	Sa.car
Scatophaga stercoraria	Sc.ste
Beetles (Coleoptera)	Col
Lepidoptera (Butterflies)	Lep
Aglais urticae	Ag.urt
Anthocharis cardamines	An.car
Callophrys rubi	Ca.rub
Celastrina argiolus	Ce.arg
Cynthias cardui	Cy.car
Gonepteryx rhamni	Go.rha
Inachis io	In.io
Lycaena phlaeas	Ly.phl
Maniola jurtina	Ma.jur
Ochlodes venata	Oc.ven
Pararge aegeria	Pa.aeg
Pieris brassicae	Pi.bra
Pieris napi	Pi.nap
Pieris rapae	Pi.rap
Polygonia c-album	Po.c-al
Polyommatus icarus	Po.ica
Pyrgus malvae	Py.mal
Pyronia tithonus	Py.tit
Vanessa atalanta	Va.ata
Moths	
Plusia gamma	Pl.gam

Summary Explanation of Abbreviations and Symbols used in the Flowering Plants and Ferns Catalogue

1. An entry enclosed in square brackets [ ] indicates an error.
2. Name of species (see also p. 80)
   Bold type: accepted Latin name
   Italics: synonyms.
   A cross (×) indicates a hybrid.
3. Floral biology (see also p. 82)
   A key to the abbreviations of insect names is given above.
   fl. = flowering.
   fr. = fruiting.
   1–12 indicates months of the year (January–December).
   (n) = seeking nectar.
   (p) = seeking pollen.
   (!) preceding an abbreviated name indicates a record made by QONK, all other records by Dr M. Percival.

4. Distribution
    (a) For species with fewer than 10 records (see also p. 81).
        5km square A,B,C,D, indicates NW, NE, SW and SE quarters of 10km square.
        Recorders' initials are given in Appendix I.
        ! following date: record made by one of the authors (see also p. 81).
        * following date: specimen in **NMW** (National Museum of Wales).
        ** following date: specimen in **UCSA** (University College, Swansea).
        Entry enclosed in parentheses ( ) taxon introduced in that locality although
        native elsewhere in the county.
    (b) Species with more than 10 records are mapped (see also p. 82).
        ○ – pre-1960 record (to 1959)
        ● – post-1960 record (1960 onwards)
        × – known introduction pre-1960 (to 1959)
        + – known introduction post-1960 (1960 onwards)

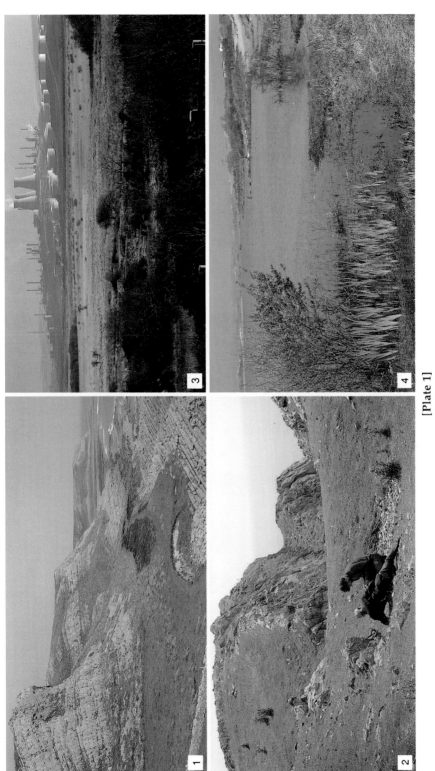

[Plate 1]

Habitats. 1–2, Limestone cliffland in south-western Gower, the habitat of *Aster linosyris*, *Ononis reclinata* and *Veronica spicata*. **1,** Overton Cliff and Port Eynon Point. **2,** Clifftops near Ramsgrove, with A. E. Wade and Q. O. N. Kay.

3–4, Major wetland sites. **3,** Crymlyn Bog near Swansea, now a National Nature Reserve; the BP Llandarcy oil refinery overlooks the site. **4,** Kenfig Pool. (Photographs: **1,** Q. O. N. Kay; **2,** H. R. W. Herbert; **3,** J. P. Curtis; **4,** Mrs A. C. Evans).

[Plate 2]

Upland cliffs, refuges for montane species at the southern limit of their ranges. **1,** Graig-fawr above Cwm Parc, Rhondda. **2,** Craig-y-llyn above Hirwaun (*Hieracium* Study Group meeting, 1985). **3,** Sgwd Einion Gam waterfall on the Afon Pyrddin near Pontneddfechan. (Photographs: 1, Q. O. N. Kay; 2–3, R. G. Ellis).

LYCOPSIDA

LYCOPODIACEAE

HUPERZIA Bernh.

Huperzia selago (L.) Bernh. ex Schrank & Mart.   Fir Clubmoss. Cnwpfwsogl Mawr.
Native. Rare; found only on damp ledges on high north-facing cliffs in the northern Uplands.
90C, Craig-y-llyn, THT 1891, above Llyn Fach, AJES 1962**, 1983!, above Llyn Fawr, 1970!*; 99A, Cwm Parc, AHT *c*.1911, Crug-yr-afan, EV 1927*, Graig-fawr, PJo 1983, PJo and GH 1985.

LYCOPODIUM L.

Lycopodium clavatum L.   Stag's-horn Clubmoss. Cnwpfwsogl Corn Carw.
Native. Very rare; found on mountain heaths and old colliery tips in the Uplands, with only a few plants in each locality.
70A, above Rhyd-y-fro, H. Jones 1970; 89C, SW bank of reservoir in Cwm Wernderi, EV & EMT 1941*; 99D, shale tip in Clydach Vale, JTa 1972.

Selaginella kraussiana (G. Kunze) A. Braun (Selaginellaceae) was found growing on a hedgebank in Mill Lane, Mayals (69C) by J.A. Webb in 1944!*; it was probably a greenhouse escape and apparently did not persist.

ISOETACEAE

ISOETES L.

Isoetes lacustris L.   Quillwort. Gwair Merllyn.
Native. Found only in Llyn Fach and Llyn Fawr below Craig-y-llyn in the northern Uplands. **I. lacustris** was recorded from Llyn Fach by Matthew Moggridge in the late 1840s (Dillwyn, 1848) but he probably did not distinguish it from **I. echinospora**. Both **I. lacustris** and **I. echinospora** grew in Llyn Fawr in abundance during the 1890s, but have not been seen there since its conversion to a reservoir.
90C, Llyn Fach, EV 1927, 1969!*, Llyn Fawr, AHT 1892*, 1899.

Isoetes echinospora Durieu   Spring Quillwort. Gwair Merllyn Bychan.
Native. Found only in Llyn Fach and Llyn Fawr below Craig-y-llyn, with recent records only from Llyn Fach, where it is abundant. Plants from the Llyn Fach population have a chromosome number of 2n=22, the usual number in **I. echinospora** (Kay, 1974b).
90C, Llyn Fach, HJR 1901*, EV 1938*, 1947*, AJES 1962**, 1983!, Llyn Fawr, AHT 1892, 1899.

SPHENOPSIDA

EQUISETACEAE

EQUISETUM L.

Equisetum hyemale L.   Rough Horsetail. Marchrawn y Gaeaf.

Native. Very scarce, with recent records only from Crymlyn Burrows near Swansea, where it is locally abundant in dune slacks, on low lying disturbed dunes, and in ditches near the main road.
18C, damp woods near Tongwynlais, JS *c*.1886; 18D, railway embankment at Llanishen, EV 1927*, HAH 1942*; 69A, Penllergaer, M. Moggridge *c*.1842; 79C, Crymlyn Bog, M. Moggridge *c*.1842, rediscovered on Crymlyn Burrows 1974!**.

Equisetum variegatum Schleicher ex Weber & Mohr   Variegated Horsetail. Marchrawn Amrywiol.
Native. Locally abundant in dune-slacks from Whiteford to Porthcawl. H.J. Riddelsdell (1907) and A.H.Trow (1911), both record E. **variegatum** only from Crymlyn Burrows (79C) where it was first found by W.R. Linton in April 1886, growing 'in enormous quantity, filling the ditches on both sides of the railway' (W.R. Linton, 1886). It is unlikely that it could have been overlooked in the other Glamorgan dune systems by Riddelsdell and Trow, and it thus appears that E. **variegatum**, which is at the south-western limit of its European range in Glamorgan, has spread to several new localities in the county since the early part of this century. It is now an abundant and characteristic member of the dune-slack community at Whiteford, Oxwich and Kenfig. In the following list of records the earliest ascertainable record at each site is given.
49A & C, Whiteford Burrows, HAH 1924*, 1983!; 58B, Oxwich Burrows, BSBI maps scheme 1958, 1983!; 78D, Kenfig Burrows, EV *c*.1925, 1967!*, 1982!; 79C, Crymlyn Burrows, W.R. Linton 1886, HJR 1905*, 1983!; 87A, Porthcawl, EV 1925*; 87B, Merthyr Mawr Warren, 1972!, 88C, Kenfig, 1972!.

Equisetum fluviatile L.   Water Horsetail. Marchrawn yr Afon.
Native. Common in marshes, fens and shallow water in the northern and western parts of the county, but more local in the Vale and central Uplands. Map 1.

Equisetum palustre L.   Marsh Horsetail. Marchrawn y Gors.
Native. Fairly common in marshes and ditches in Gower, but more local in the Vale and the Uplands. Map 2.

Equisetum sylvaticum L.   Wood Horsetail. Marchrawn y Coed.
Native. Rather local, growing on acid soils in marshy places on heaths and in woods; most frequent in the Uplands and in northern Gower. Map 3.

Equisetum arvense L.   Field Horsetail. Marchrawn yr Ardir.
Native. Very common and widespread, growing in damp grassland, hedgebanks and waste ground, and in cultivated land where it is sometimes a troublesome weed. Map 4.

Equisetum arvense × E. **fluviatile** (E. × **litorale** Kuhlew. ex Rupr.) has been recorded from Oxwich

(58A, 1974!) and Crymlyn Burrows (79C, 1974!), and is abundant on Kenfig Burrows (C.R. & C.D. Fraser-Jenkins 1969, **BM**, P. Taylor 1976, P. Jepson 1980).

Equisetum telmateia L. Great Horsetail. Marchrawn Mawr.
Native. Locally common in hedgebanks and damp woodland in western and southern parts of the county, scarce elsewhere. Map 5.

FILICOPSIDA
OPHIOGLOSSACEAE
OPHIOGLOSSUM L.

Ophioglossum vulgatum L. Adder's-tongue. Tafod y Neidr.
Native. Now very scarce in damp grassland inland, but locally abundant in dune-slacks at Whiteford and Kenfig and also growing in dune-slacks at Oxwich and Crymlyn Burrows. It has been lost from many localities in the Vale where it grew in old pastures which have been destroyed by ploughing. A.H. Trow (1911), described it as a widespread and abundant plant in the Vale, growing in pastures with cowslips and **Orchis morio**. Map 6.

Ophioglossum azoricum C. Presl Small Adder's-tongue. Tafod y Neidr Bach.
Plants resembling this species have been recorded from Whiteford Burrows (49A) but more field studies and cytological work are necessary to help establish the status of this difficult taxon in Britain (A.M. Paul, 1987 and in litt.)

BOTRYCHIUM Swartz

Botrychium lunaria (L.) Swartz Moonwort. Crib y Ceiliog.
Native. Dry grassland, mainly on limestone or on fixed dunes. **B. lunaria** has been lost from several sites in the Vale where it grew on field banks or in old pastures which have been destroyed by ploughing, and has also been lost from a number of dune localities where, as at Oxwich, it grew on fixed dunes which have been destroyed by erosion. It is now almost confined to coastal dunes and to base-rich grassland in the northern Uplands. Map 7.

OSMUNDACEAE
OSMUNDA L.

Osmunda regalis L. Royal Fern. Rhedyne Gyfrdwy.
Native. Damp heathland, ditches, fens and bogs. Locally frequent in the western part of the county, for example in Crymlyn Bog, on Clyne Common, and near Rhyd-y-fro, but now rare or absent in the central and eastern Uplands and the Vale. Map 8.

ADIANTACEAE
ADIANTUM L.

Adiantum capillus-veneris L. Maidenhair Fern. Briger Gwener.
Native. Locally fairly abundant in sheltered crevices on the Lias Limestone sea-cliffs of the southern Vale from Nash Point (96A) to Porthkerry (06B), where it usually grows in places where there is some seepage of ground water from below. It was first recorded in the county by Edward Llwyd in 1698 'growing very plentifully out of a marly incrustation both at Barry Island and Parth Kirig, ... and out of no other matter, ...' (Irvine, 1855–1856; Gunther, 1945). A recent survey by P.S. Jones (1983a) has shown populations in 06A, 06B, 16A, 16B, 87D, 96A and 96B. Recorded in the past as a garden escape in Penarth (17D).

CRYPTOGRAMMACEAE
CRYPTOGRAMMA R.Br.

Cryptogramma crispa (L.) Hooker Parsley Fern. Rhedynen Bersli.
Native. High screes of Pennant Grit in the northern Uplands; rare, but with recent records from several sites in or near the northern Rhondda valleys. Young (1856) reports it as only known at Aberdare in South Wales.
00C?, mountain above Aberdare, JS c.1890; 09A, Llanwonno Forest, MG 1970; 18A, Mynydd Meio, THT c.1890, EV 1940; 90C, above Llyn Fawr, AHT 1905*, Craig-y-bwlch, BAW c.1907, 1969!*; 90D, Cwm Dâr Rocks, BAW 1932; 99A, Graig Fach, 1970!, Graig-fawr, Cwm Parc, AHT 1898*, Craig Blaen Rhondda, D.P.M. Guile 1950*; 99B, Taren Maerdy, D.P.M. Guile 1950*, JTa 1970, Ferndale, JTa 1970; 99C, Craig Ogwr, JTa 1971, GH 1987.

HYPOLEPIDACEAE
PTERIDIUM Scop.

Pteridium aquilinum (L.) Kuhn Bracken. Rhedynen Gyffredin.
Native. Very common and widespread on heathland and hillsides, in open woodland and hedgebanks, in abandoned or neglected pastures, and on old fixed dunes. Map 9.

HYMENOPHYLLACEAE
HYMENOPHYLLUM Sm.

Hymenophyllum tunbrigense (L.) Sm. Tonbridge Filmy-fern. Rhedynach Teneuwe.
Native. Shaded banks in moist ravines. Very rare; now known from only three localities, Melincourt in the Vale of Neath, Cwm Dimbath (Darren Dimbath) near Blackwood, and above Llyn Fach at the head of the Rhondda Valley.
79B, Bryn-cous above Cadoxton, LWD c.1840; 80B, near Pont-nedd-fechan, LWD c.1805 (possibly in Breconshire); 80C, in ravine below fall at Melincourt, T. Westcombe & E. Young c.1850, 1975!**; 90A,

Pen-cae-drain, AHT 1892*–1905* (site destroyed by quarrying *c*.1910); 90C, above Llyn Fach, 1968!; 98B, Cwm Dimbath, HOB Booker *c*.1910, EMT 1932*, 1974!.

Hymenophyllum wilsonii Hooker   Wilson's Filmy-fern. Rhedynach Teneuwe Wilson.
Native. Shaded rocks, especially on high north-facing cliffs. Very scarce, but with recent records from several localities in the northern Uplands.
80B, Pyrddin Valley, E. Young *c*.1850, Scŵd Einion Gam, MG 1975; 80C, Melincourt Glen, E. Young *c*.1850; 90A, Pen-cae-drain, AHT 1892*–1905*; 90C, Craig-y-llyn, AHT 1892*, 1908*, above Llyn Fawr, 1920!*, above Llyn Fach, 1965–1983!, Craig-y-bwlch, BAW 1937*; 98B, Cwm Dimbath, HOB *c*.1910, 1974!; 99A, Cwm Selsig, 1970!, Tarren Rhiw-maen, ARP 1974, Graig-fawr, PJo and GH, 1985; 99C, Craig Ogwr, ARP 1975*.

THELYPTERIDACEAE
OREOPTERIS Holub

Oreopteris limbosperma (All.) Holub Lemon-scented Fern. Marchredynen Arogl-lemwn.
Native. Steep hillsides and streambanks, cliffs and screes, and rocky places in open woods on acid soils. Common in the Uplands but rather scarce in Gower and the northern Vale, and absent from the southern Vale. Map 10.

THELYPTERIS Schmidel

Thelypteris palustris Schott (*Thelypteris thelypteroides* Michx subsp. *glabra* Holub).   Marsh Fern. Marchredynen y Gors.
Formerly native, but now apparently extinct; once recorded from two sites near Swansea, Cwmbwrla Bog north of Townhill, which was drained in the mid-nineteenth century, and Sketty or Singleton Bog in the Clyne Valley west of Swansea, now almost completely destroyed by a mid-twentieth century rubbish tip. **T. palustris** was also recorded by H.J. Riddelsdell from Llyn Fach (90C), an unlikely locality. The Clyne Valley site was probably ecologically reasonably suitable for **T. palustris** although it was an isolated and outlying locality near the south-western limit of the species in Britain; the Cwmbwrla site, apparently close to Townhill Bog in which **Drosera anglica** grew, would appear to have been less suitable.
69A, Cwmbwrla, JWGG *c*.1842; 69C, Sketty or Singleton Bog, EF *c*.1800, E. Young (1856), EV 1919*, T.J. Foggitt 1940? (Ms note E. Vachell, n.d.4 'a very damp corner of ''Killay Common'' ' – either the upper part of Clyne Valley in 69C or part of Fairwood Common in 59D?).

PHEGOPTERIS (Presl) Feé emend. Ching

Phegopteris connectilis (Michx) Watt (*Thelypteris phegopteris* (L.) Slosson).   Beech Fern. Rhedynen y Graig.

Native. Locally frequent on damp cliffs and shaded streamsides or woodland banks around the heads of the Neath, Afan, Rhondda and Taff valleys in the Uplands, and in a few sites further west. Map 11.

ASPLENIACEAE
ASPLENIUM L.

Asplenium marinum L.   Sea Spleenwort. Dueg-redynen Arfor.
Native. Fairly frequent in crevices, overhangs and caves on limestone sea-cliffs on the coast of Gower and the southern Vale, thinly scattered along the coast; most frequent in Gower. Map 12.

Asplenium trichomanes L.   Maidenhair Spleen-wort. Duegredynen Gwallt y Forwyn.
Native. Common and widespread on walls, especially on old mortar, and in rock crevices on basic substrata; particularly abundant on limestone in Gower. These comments refer to subsp. **quadrivalens** D.E. Meyer emend Lovis. There is one record of subsp. **trichomanes** from acidic rocks in the north of the county (99A, Graig-fawr, PJo & GH 1985*). Map 13.

Asplenium viride Hudson   Green Spleenwort. Duegredynen Werdd.
Native. Very scarce, growing in limestone crevices or base-rich seepages on high crags or near waterfalls at a few localities in the Uplands.
00A, Morlais Castle Hill, AHT 1890*, 1923!*; 00B & 01D, Morlais quarries, GH 1986*; 28A, near Draethen, TGE 1973*; 80B near Pontneddfechan, JWGG *c*.1840; 80C, Melincourt Falls, 1972!; 99A, Crug yr Afan above Nant-y-moel on coal seam, T.M. Franklin & EV 1927*.

Asplenium billotii F.W. Schultz   Lanceolate Spleen-wort. Duegredynen Reiniolaidd.
Native. Very rare, with recent records only from one site in Gower, where it grows on acid soil in the crevices of a field retaining wall of Devonian conglomerate above the sea.
09D, Pontypridd, AHT 1891*, THT 1894* (probably a garden escape); 58A, Penmaen, EFL *c*.1900, 1974!**; 69C, Townhill, JWGG *c*.1840.

Asplenium adiantum-nigrum L.   Black Spleen-wort. Duegredynen Ddu.
Native. Widespread on moist walls, in rocks crevices and in rocky hedgebanks, but not common; most frequent in Gower. Map 14.

Asplenium ruta-muraria L.   Wall-rue. Dueg-redynen y Muriau.
Native. Common and widespread on walls and in rock crevices. Map 15.

Asplenium ceterach L. (*Ceterach officinarum* DC.). Rustyback. Rhedynen Gefngoch.
Native. Limestone crevices, walls and rocky hedge-

banks. Locally common in Gower and the Vale but less frequent in the Uplands, where it usually grows on mortared walls. Map 16.

Asplenium scolopendrium L. (*Phyllitis scolopendrium* (L.) Newman).   Hart's-tongue. Tafod yr Hydd.
Native. Common, and often abundant, in woods and hedgebanks on limestone or base-rich soils in Gower and the Vale; widespread but relatively scarce and usually confined to damp mortared walls in the Uplands. Map 17.

ATHYRIACEAE
ATHYRIUM Roth

Athyrium filix-femina (L.) Roth   Lady-fern. Marchredynen Fenyw.
Native. Common and widespread in woods, on damp moorland and in bogs and marshes, and on streambanks and hill-slopes, usually on acid or base-poor soils; absent from the southern Vale. Map 18.

CYSTOPTERIS Bernh.

Cystopteris fragilis (L.) Bernh.   Brittle Bladder-fern. Rhedynen Frau.
Native. Moist shady rocks and walls; fairly frequent in the northern Uplands but scarce or absent elsewhere. Map 19.

ASPIDIACEAE
POLYSTICHUM Roth

Polystichum aculeatum (L.) Roth   Hard Shield-fern. Gwrychredynen Galed.
Native. In woods and on hedgebanks and rocky slopes on base-rich soils, and on old walls. Very local in the Uplands and the Vale, and rare in Gower. **P. aculeatum** is sometimes confused with **P. setiferum**, which is common in lowland Glamorgan, and **P. aculeatum** may consequently have been over-recorded in the past. In Upland populations, for example in the Pyrddin and Taf-fechan valleys, some young plants have entire pinnae resembling **P. lonchitis** and only develop pinnatifid pinnae with just one or two basal pinnules – previously called var. *cambricum* (S.F. Gray) Hyde & Wade – but now considered only a growth form (O.C. Lloyd, 1979), recorded from 00A, 80B, 90D and 99A. Map 20.

Polystichum setiferum (Forskål) Woynar   Soft Shield-fern. Gwrychredynen Feddal.
Native. In woods and on hedgebanks on base-rich soils. Very common in Gower and parts of the Vale where it is often, after **Asplenium scolopendrium**, the most abundant fern in limestone woodland, but rare in the Uplands where it has been found only in sheltered woodland in the upper Neath valley. Map 21.

DRYOPTERIS Adanson

Dryopteris filix-mas (L.) Schott   Male-fern. Marchredynen Wryw.
Native. Common and widespread in woods, hedgebanks and rocky places and on damp walls. Map 22.

Dryopteris affinis (Lowe) Fraser-Jenkins (*D. pseudomas* (Wollaston) Holub & Pouzar, *D. borreri* Newman).   Scaly Male-fern. Marchredynen Euraid.
Native. Woods, hedgebanks, streamside banks and screes. Locally common in Gower and the Uplands where it is often more abundant than **D. filix-mas**; less frequent in the northern Vale, and uncommon in the southern Vale. Map 23.
Three subspecies are now recognised in the British Isles (Derrick, Jermy & Paul, 1987) all of which have been recorded from Glamorgan (specimens determined mostly by C.R. Fraser-Jenkins).

Dryopteris affinis (Lowe) Fraser-Jenkins subsp. **affinis**, and **D. affinis** (Lowe) Fraser-Jenkins subsp. **borreri** (Newman) Fraser-Jenkins are by far the commonest in the county and can sometimes be found growing together in woods in the presence of **D. filix-mas**. Further field-work is required to ascertain their full distribution in the county, but provisional maps are given below.
The third subspecies, **Dryopteris affinis** (Lowe) Fraser-Jenkins subsp. **cambrensis** Fraser-Jenkins (*D. affinis* subsp. *stilluppensis* sensu Fraser-Jenkins), has been recorded from upland Wales and occurs in the north of the county.
00A, ditch, Cyfarthfa Castle, Merthyr Tydfil, GH 1990*; 99C, stream gully, Bwlch y Clawdd, C.R. Fraser-Jenkins 1986, **Hb. CRFJ**.

Dryopteris affinis subsp. **affinis**. Map 24.

Dryopteris affinis subsp. **borreri**. Map 25.

Dryopteris affinis × **D. filix-mas** (**D.** × **complexa** Fraser-Jenkins (*D.* × *tavelii* sensu auct.)).
Native. Under-recorded and has been confused with robust forms of **D. affinis**.
07A, Hensol Forest, C.R. Fraser-Jenkins 1970, **BM**; 16B, Cogan Plantation, GH 1986*, det. C.R. Fraser-Jenkins; 18C, Tongwynlais, GH 1988*; 58B, Ilston Cwm, GH 1974*.

Dryopteris oreades Fomin (*D. abbreviata* (DC.) Newman).   Mountain Male-fern. Marchredynen y Mynydd.
Native. On north-facing cliffs of Pennant Grit; rare, recorded from the cliffs about Craig-y-llyn where it was first found in 1974 above Llyn Fach (90C!**, det. A.C. Jermy), and then in 1981 above Llyn Fawr (90C*, GH, det. A.C. Jermy & J.M. Camus); also in 1989 at Graig-fawr where it is at its southernmost extant site in the British Isles (99A*, J. Bevan & GH, det. C.R. Fraser-Jenkins).

Dryopteris carthusiana (Vill.) H.P. Fuchs   Narrow Buckler-fern. Marchredynen Gul.
Native. Widespread but very scarce, growing in wet woods, marshes and fens; perhaps overlooked in

some sites and thus under-recorded to some extent. Map 26.

Dryopteris carthusiana × D. dilatata (D. × deweveri (Jansen) Jansen & Wachter) has been recorded from Mynydd-y-glew (07A) by C.R. Fraser-Jenkins in 1967*, and from Mynydd Radyr (18C) by H.J. Riddelsdell in 1907 (det. A.C. Jermy) **BM**.

Dryopteris dilatata (Hoffm.) A. Gray (*D. austriaca* (Jacq.) Woynar). Broad Buckler-fern. Marchredynen Lydan.
Native. Very common in most of the county, growing on acid soils in woods, rocky places, marshy heathland and bogs; absent from some limestone-soil areas in the southern Vale. Map 27

Dryopteris aemula (Aiton) O. Kuntze   Hay-scented Buckler-fern. Marchredynen Aroglys.
Native, perhaps extinct. Reported from Melincourt Waterfall (80C) (Young, 1856). Young apparently collected a large number of specimens there, and it has not been subsequently refound; it was thought to be probably extinct in the county by A.H. Trow in 1910. **D. aemula** is often confused with **D. dilatata** in the field, and Trow excluded a number of dubious field records by Storrie for some coastal sites and for Craig-y-llyn, where a form of **D. dilatata** grows which closely resembles **D. aemula** in some characters. Eleanor Vachell (1934) and J.A. Webb (1947d) reported finding **D. aemula** at Cwm Dimbath (98B), a site where, as at Melincourt, **Hymenophyllum** spp. grow, in sheltered humid conditions that appear to be suitable for **D. aemula**. Recent searches of both localities (1974!, 1975!) have been unsuccessful, but it is possible that the species may still grow in Glamorgan.

GYMNOCARPIUM Newman

Gymnocarpium dryopteris (L.) Newman Oak Fern. Llawredynen y Derw.
Native. Shaded rocks in the northern Uplands, usually in boulder-scree or near waterfalls. Always scarce and now very rare in the county, perhaps as a result of over-collecting in the past.
07C, St. Hilary (J. Storrie, 1886); 08, Pontypridd, A. Jones 1952*; 28A, near Draethen, Mr Birkenhead (Trow, 1911); 50D, near Pontarddulais, HJR *c*.1907; 80B, near Pont-nedd-fechan, LWD *c*.1805, Pyrddin Valley, C.M. Thompson, 1879*, 1974!; 90C, Craig-y-bwlch, 1969!*; 99A Graig Fach, HJR *c*.1907.

Gymnocarpium robertianum (Hoffm.) Newman   Limestone Fern. Llawredynen y Calchfaen.
Native. Limestone rocks, screes and old walls. Very rare in Glamorgan (v.c. 41), although it is locally abundant on limestone screes below Darren-fawr 4km north of Merthyr Tydfil, now in the administrative county of Mid-Glamorgan but in v.c. 42 (Breconshire) for plant and animal records.
00A, Merthyr, AHT 1890*, Morlais Castle Hill, G.R. Willan 1906*, limestone rocks in Morlais Glen, R.B.

Ivimey-Cook 1954*, wall at Cyfarthfa Castle, 1969!*; 00B & 01D, Morlais Quarries, GH 1986*; 18B, tunnel spoil heap on Cefn Onn, 1924!*; 28A, near Draethen, AHT *c*.1910; 87B, Old Castle Down, AHT *c*.1910; 90A, stonework of a bridge at Hirwaun, 1938!*.

BLECHNACEAE
BLECHNUM L.

Blechnum spicant (L.) Roth Hard Fern. Gwib-redynen.
Native. Common in open woods, in rocky places, and on hedgebanks and streamside banks on acid soils in Gower, the Uplands and the northern part of the Vale, but absent from the limestone areas of the southern Vale. Map 28.

POLYPODIACEAE
POLYPODIUM L.

Polypodium cambricum L. (*P. australe* Feé). Southern Polypody. Llawredynen Gymreig.
Native. Limestone rocks and walls in Gower and the Vale, very local. Small forms of **P. interjectum** are common in similar habitats. Map 29.

Polypodium cambricum L. var. **cambricum** Welsh Polypody, Laciniated Polypody.
Native. Only known from one site in Glamorgan, at Cwm George, Dinas Powys (17C) growing on a limestone cliff where it was first discovered by Richard Kayse of Bristol in 1668 (Ray, 1690 and specimen of Ray's at **BM**). John Storrie (1886) states that the fern was completely rooted out by a dealer in 1876, but the Rev. H.J. Riddelsdell re-discovered it in 1908 and it was refound by Malcolm Wood in 1977!* (Wood, 1978) and seen by GH in 1985.

[**Polypodium cambricum** × **P. interjectum** (**P. × shivasiae** Rothm.) has been recorded in error.]

Polypodium vulgare L.   Polypody. Llawredynen y Fagwyr.
Native. Acid rocks and walls, less often on trees. Scarce in Gower, rather local in the Vale (where it may have been confused with **P. interjectum**, which is the commonest polypody in lowland Glamorgan), but commoner than **P. interjectum** in the Uplands. Map 30.

Polypodium interjectum Shivas   Intermediate Polypody. Llawredynen Rymus.
Native. On trees, walls, rocks and fixed dunes. Very common in Gower where it is by far the most abundant member of the **P. vulgare** aggregate; common in the Vale, but rather scarce and usually confined to sheltered valleys in the Uplands. Map 31.

Polypodium interjectum × **P. vulgare** (**P. × mantoniae** Rothm.).

Under-recorded due to confusion with the parent species.
17C, Cwm George, GH 1983*, det. R.H. Roberts; 49D, Llanrhidian, C.R. Fraser-Jenkins 1971, **BM**; 78D, Kenfig Dunes, GH 1985*, det. R.H. Roberts.

MARSILEACEAE
PILULARIA L.
Pilularia globulifera L.   Pillwort. Pelenllys.
Native. On shelving banks or in shallow water at the edges of ponds and lakes with acid water; always rare, and now perhaps extinct. It has been found at two sites in the county, Pysgodlyn Mawr near Welsh St. Donats (07A), where it was last seen in 1971, and has not been found in recent careful searches of the sites where it once grew, being probably exterminated by dredging, and Llyn Fach (90C) where it was last seen in 1894. In both sites the first record was made at the beginning of the nineteenth century.
07A, Pysgodlyn Mawr, J.M. Traherne 1807*, HJR 1904*, EV 1911*, 1920!*, 1971!*, not seen since 1972; 90C, Llyn Fach, J. Woods *c*.1805, JS 1894*.

AZOLLACEAE
AZOLLA Lam.
Azolla filiculoides Lam., Water Fern, Rhedynen y Dwr, was found in the moat of Cardiff Castle (17B) by J.W. Davies in 1971. It formed a complete carpet over the surface of the moat; the cover was so good that a child nearly drowned trying to walk on it as a result of which the moat was drained in *c*.1972 and the **Azolla** exterminated.

CONIFEROPSIDA
PINACEAE
ABIES Miller
Abies grandis (D.Don) Lindley   Giant Fir. Ffynidwydden Wych.
Introduced, mainly as a specimen tree or in parks.

Abies alba Miller   European Silver-fir. Ffynidwydden Arian.
Introduced, mainly as an ornamental tree.

PSEUDOTSUGA Carrière
Pseudotsuga menziesii (Mirbel) Franco   Douglas Fir. Ffynidwydden Douglas.
Introduced. Forestry plantations on relatively fertile soils, sometimes regenerating.

TSUGA (Antoine) Carrière
Tsuga heterophylla (Rafin.) Sarg.   Western Hemlock-spruce. Hemlog y Gorllewin.
Introduced. Occasionally planted for timber; sometimes regenerating.

Tsuga canadensis (L.) Carrière   Eastern Hemlock-spruce. Hemlog y Dwyrain.
Introduced, mainly in parks.

PICEA A. Dietr.
Picea abies (L.) Karsten   Norway Spruce. Spriwsen Norwy.
Introduced. Commonly used for forestry plantations, sometimes regenerating.

Picea sitchensis (Bong.) Carrière. Sitka Spruce. Spriwsen Sitka.
Introduced. The chief species used for forestry plantations on poorly drained peaty soils, especially on the hills of the Uplands.

LARIX Miller
Larix kaempferi (Lamb.) Carrière (*L. leptolepis* (Siebold & Zucc.) Endl.).   Japanese Larch. Llarwydden Japan.
Introduced. One of the chief species in forestry plantations on better-drained soils; sometimes regenerates well.

Larix decidua Miller   European Larch. Llarwydden Ewrop.
Introduced, mainly in older plantations on moderately fertile soils; sometimes regenerating and more or less naturalized. Now seldom planted.

PINUS L.
Pinus contorta Douglas ex Loudon   Lodge-pole Pine. Pinwydden Camfrig.
Introduced. Sometimes used in forestry plantations, especially in exposed sites on the higher hills in the Uplands.

Pinus radiata D.Don   Monterey Pine. Pinwydden Monterey.
Introduced. Sometimes planted near the sea, as for example at Whiteford Burrows (49C).

Pinus nigra Arnold subsp. **laricio** (Poiret) Maire   Corsican Pine. Pinwydden Corsica.
Introduced. Forestry plantations on a range of soils; often planted on sand-dunes. Sometimes regenerates.

Pinus sylvestris L.   Scots Pine. Pinwydden yr Alban.
Introduced? Commonly planted in woods and plantations; often regenerating and sometimes naturalized. Map 32.

CUPRESSACEAE
CUPRESSUS L.
Cupressus macrocarpa Hartweg   Monterey Cypress. Cypreswydden Monterey.
Introduced. Occasionally used in plantations; sometimes regenerating.

CHAMAECYPARIS Spach

Chamaecyparis lawsoniana (A. Murray) Parl. Lawson Cypress. Cypreswydden Lawson.
Introduced. Occasionally used in plantations; sometimes regenerating.

THUJA L.

Thuja plicata D. Don ex Lamb. Western Red-cedar. Cedrwydden Goch.
Introduced. Occasionally planted for timber on good soils; sometimes regenerating.

JUNIPERUS L.

Juniperus communis L. subsp. **communis** Juniper. Merywen.
Native. Limestone cliffs near the sea in southern Gower from Langland to Rhosili. A relict population which has probably survived there since the Late-glacial period. Rapidly decreasing as a result of cliff-fires in eastern Gower, where it is now absent or reduced to a few isolated and ageing bushes in sites where it was fairly frequent in the early 1960s. Still locally frequent between Paviland and Mewslade.
48A, between Paviland and Rhosili, 1970!, 1988!; 48B, Horton, HAH 1929*, 1986!; 48D, Overton Cliff, JAW 1946; 58A, Crawley Cliff, 1972!; 58B, Knap Cliff, Pwlldu, JAW 1921*, 1948*, 1970!; 58C, Oxwich Point, 1975!; 68A, W. Langland Cliff, JAW 1945*, 1970!.

TAXACEAE
TAXUS L.

Taxus baccata L. Yew. Ywen.
Probably native in the Vale and perhaps on limestone in Gower and formerly on Morlais Castle Hill, but introduced elsewhere. Its native range is hard to determine, because it is often bird-sown and naturalized from trees anciently planted in churchyards, and it is thought to have been planted in sacred sites even in pre-Christian times. Map 33.

ANGIOSPERMOPSIDA
DICOTYLEDONES
SALICACEAE
SALIX L.

Salix pentandra L. Bay Willow. Helygen Beraro-glaidd.
Probably introduced; rare, with no recent records. Recorded in the past from 00C, Aberdare, HJR 1907; 07B, St George's and Peterston-super-Ely, JS c.1886; 17A, St. Fagans, JS c.1886; 88A, Twmpath Valley, WWB c.1909.

Salix fragilis L. Crack Willow. Helygen Frau.
Probably native in part of the Vale but introduced elsewhere; often planted by streams and rivers, and fairly common in suitable localities. Map 34.

Salix alba × fragilis (**S. × rubens** Schrank) has been recorded from four localities in the past and one recently (00C, Aberdare, HJR 1907; 17A, Ely, AHT c.1907; 17B Pontcanna, AHT c.1907, EV 1933*, Leckwith Bridge, GH 1989*, det. R.D. Meikle; 69D, Swansea, JWGG 1842) and may be more frequent than these records indicate.

Salix alba L. White Willow. Helygen Wen.
Perhaps native in parts of the Vale but introduced elsewhere; fairly often planted by streams and rivers, but generally less common than **S. fragilis**. Map 35.

Salix triandra L. Almond Willow. Helygen Drigwryw.
Probably introduced as an osier and persisting by streams and ditches in a few localities; apparently rare.
00C, Aberdare, HJR c.1906; 17D, Leckwith, Cardiff, JS c.1876; 28C, Coed-y-gores, HJR 1909; 49D, Burry Pill, JAW 1947*, Cheriton 1972!; 69A, Waunarlwydd, H & CH 1974; 69D, Swansea, JWGG c.1840; 78D, Kenfig Pool, JM 1846; 87, Porthcawl to Bridgend area, BAM 1951⁺; 88C, between Pyle and Port Talbot, E.S. Marshall & WAS 1901; 97D, Cowbridge, HJR c.1906.

Salix triandra × S. viminalis (**S. × mollissima** Hoffm. ex Elwert) has been reported from two localities (17A, Ely Bridge, EV 1904* & 1934*(Meikle, 1984); 88C?, between Pyle and Port Talbot, E.S. Marshall & WAS 1901) and may, like some other taxonomically difficult **Salix** hybrids, be more frequent than the records indicate.

[Old records of **Salix nigricans** Sm. from 00C and 18D are now regarded as doubtful.]

Salix cinerea L. subsp. **oleifolia** Macreight (*S. atrocinerea* Brot.). Rusty Willow, Common Sallow. Helygen Olewydd-ddail.
Native. Very common and widespread in damp woodland and hedges, on heathland, in marshes and fens, by streams and pools, in dune-slacks, and on roadsides, waste ground, tips and railway embankments. Fl. 3–4, fr. 5; (!)Hb (p,n), B.terr (p,n), B.luc, ?Sb (p), E.ten, S.ste, ?Dips (and also Blue Tit, *Parus caeruleus*, (n), frequently; see Kay, 1985b). Map 36.

Salix cinerea × S. purpurea (**S. × sordida** A. Kerner) has been recorded once, from Taff's Well near Cardiff (18C, HJR c.1907).

Salix cinerea × S. viminalis (**S. × smithiana** Willd.) was recorded from several localities by Trow (1911) and Riddelsdell (1907) but has been seen recently, in only three. Map 37.

Salix aurita L. Eared Willow. Helygen Grynglystiog.
Native. Common in damp woodland and marshes, by ditches and on marshy moorland. Map 38.

Salix aurita × S. caprea × S. viminalis (**S. × stipularis**

Sm.) has been recorded from Swansea (69, W.R. Linton 1891), **Salix aurita × S. repens (S. × ambigua** Ehrh.) from Aberdare (00C, HJR 1911*), and **Salix aurita × S. viminalis (S. × fruticosa** Döll) also from Aberdare (00C, HJR 1911*).

Salix aurita × S. cinerea (S. × multinervis Döll) was recorded from several localities by AHT & HJR but has been seen recently in only two. Map 39.

Salix caprea L. Goat Willow, Great Sallow. Helygen Grynddail Fwyaf.
Native. Common in woods and hedges and on roadsides, waste ground and tips; usually less frequent than **S. cinerea** in the countryside but more frequent in towns and industrial habitats. Map 40.

Salix caprea × S. cinerea (S. × reichardtii A. Kerner) has been recorded from several localities, mainly in the west and may be more frequent than records indicate. Map 41.

Salix caprea × S. aurita (S. × capreola A. Kerner ex N.J. Andersson) has been recorded from 00C (HJR c.1906, KA 1973* det. c.f. R.D. Meickle) and 18C (HJR c.1908).

Salix repens L. Creeping Willow. Corhelygen, Helygen y Cwn
Native. Very abundant in dune-slacks and on low fixed dunes, where **S. repens** often forms 'hedgehog' dunes; locally common on dry heathland, especially in the western part of the county. (The dune form has in the past been treated as a species (*S. arenaria* L.), but is now considered to be, at most, a variety of **S. repens** (var. **argentea** (Sm.) Wimmer & Grab.)). Fl. 4–5, fr. 5; (!) Hb (p,n), Bb(p,n), Sb incl. *Colletes* sp. (p). Map 42.

Salix viminalis L. Osier. Helygen Wiail.
Perhaps native, but usually planted. Fairly common by rivers, streams and pools and in small marshes near farmland. Map 43.

Salix purpurea L. Purple Willow. Helygen Gochlas.
Probably introduced; occasionally planted as an osier in small marshes or by streams, lakes and ponds, but much less frequently than **S. viminalis**. Map 44.

Salix purpurea × S. viminalis (S. × rubra Hudson) was found in two localities by H.J. Riddelsdell c.1907 (00C, Aberdare; 17B, Leckwith Wood) and may still occur in these localities and elsewhere.

[The following hybrids have been recorded in the past but all are probably errors: **Salix alba × S. triandra**, (87B, WAS 1901), **Salix caprea × S. repens (S. × laschiana** Zahn), (49A, AHT 1908), **Salix fragilis × S. triandra (S. × speciosa** Host.), (00C, HJR 1905), **Salix cinerea × S. nigricans** (00C, HJR c.1907).]

POPULUS L.
Populus tremula L. Aspen. Aethnen.
Native, but often planted. Fairly frequent in woods and hedges, but not common. Map 45.

Several species of **Populus** are planted for timber or as ornamental trees.

Populus alba L. White Poplar. Poplysen Wen.
Introduced; more or less naturalized on damp soils in a few sites in the Vale or near the coast; spreading vegetatively after planting.
06A, Aberthaw, MG 1974; 17B, Cardiff, 1978!*; 18C, Whitchurch, GH 1980; 48B, Stouthall Woods, JBe 1982; 78B, Port Talbot, 1974!; 79C, Jersey Marine, 1974!; 87B, Merthyr Mawr, JAW 1956*, 1974!.

Populus alba × P. tremula (P. × canescens Aiton. Sm.). Grey Poplar. Poplysen Lwyd.
Introduced; like **P. alba**, more or less naturalized on damp soils in a few sites in the Vale or near the coast; spreading vegetatively after planting. Map 46.

Populus candicans Aiton (*P. × gileadensis* Rouleau). Balm-of-Gilead.
Introduced; occasionally planted and sometimes persists in a similar fashion to the two previous taxa, as, for example, at Merthyr Mawr Warren (87B, 1972!). Map 47.

Populus deltoides × P. nigra (P. × canadensis Moench). Italian Poplar. Poplysen Ddu Ffrengig.
Introduced; frequently naturalized, though often planted originally. Map 48.

Populus nigra L. *s.l.* Black Poplar. Poplysen Ddu.
Introduced; frequently planted, especially along roadsides and sometimes naturalized. Two distinct forms occur; **P. nigra** cv. 'Italica', Lombardy Poplar, Poplysen Lombardy, and **P. nigra** *s.s.*, less often planted and apparently confused with **P. × canadensis** by Trow (1911). Map 49.

MYRICACEAE
MYRICA L.
Myrica gale L. Bog-myrtle. Helygen Fair.
Native. Locally common in bogs and on wet heathland in western Gower, especially on Cefn Bryn to the south of Broad Pool (49D & 59C, 1964!, 1983!); now apparently very scarce elsewhere, with few recent records. Map 50.

JUGLANDACEAE
JUGLANS L.
Juglans regia L. Walnut. Coeden Gnau Ffrengig.
Introduced. Limestone cliff and woods. Often planted for its edible fruit.
49D, Stembridge, T. Davies 1985; 58B, Widegate, JAW c.1925; 06A, St. Athan, HJR c.1906.

BETULACEAE
BETULA L.

Betula pendula Roth Silver Birch. Bedwen Arian.
Native. Fairly common in woods on well-drained soils, on old colliery tips, and on hillsides in the Uplands, but scarce or absent in much of Gower, where it is replaced by **B. pubescens**, and in the southern Vale. Map 51.

Betula pubescens Ehrh. Downy Birch. Bedwen Lwyd.
Native. Common in woods on acid or neutral soils, on heathland and hillsides, and sometimes on old colliery tips and waste ground. Usually the commonest birch, especially in the west, but absent from parts of the southern Vale. Map 52.

Betula pubescens subsp. **pubescens** has been recorded from several sites in the eastern Uplands and eastern Vale, but is often not clearly separable from subsp. **carpatica** (Willd.) Ascherson & Graebner in Glamorgan populations; subsp. **carpatica** has been recorded from two localities in the northern Uplands (10C, Troed y Fwch, 1974!*; 90A, Cwm Hwnt, 1969!*), and one in the Vale (78D, Kenfig 1956!*)

Betula pendula × B. pubescens (**B. × aurata** Borkh.) has been recorded twice (00A, Taf Fechan, 1969!*; 98A, Cwm Dimbath, ARP 1974!*)

ALNUS Miller

Alnus glutinosa (L.) Gaertner Alder. Gwernen.
Native. Common by streams and rivers and in marshes and wet woodland; sometimes locally abundant in dune-slacks. Map 53.

Alnus incana (L.) Moench Grey Alder. Gwernen Lwyd.
Occasionally planted and sometimes regenerates. 06B, Mill Wood, Porthkerry, 1975!*; 18D, Ty Celyn, Cardiff, GH 1985; 49C, Whiteford plantation, 1977!.

CORYLACEAE
CARPINUS L.

Carpinus betulus L. Hornbeam. Oestrwydden.
Occasionally planted as an ornamental tree, and recorded in woods and hedges in several localities. It does not appear to spread or regenerate. Map 54.

CORYLUS L.

Corylus avellana L. Hazel. Collen.
Native. Very common in woods, cliff-scrub and hedges on most soils, often forming a scrub layer in deciduous woodland. Map 55.

FAGACEAE
FAGUS L.

Fagus sylvatica L. Beech. Fawydden.
Probably native in the east of the county but introduced in the west. It is locally dominant and appears to be fully native in woods on the Old Red Sandstone and Carboniferous Limestone ridges north of Cardiff, for example on Caerphilly Mountain (18B), around Tongwynlais (18C), on Craig Llanishen (18D), near Draethen (28A) and in Coed-y-goras (28C). Elsewhere it is fairly common in woods and hedges, and sometimes regenerates, but is certainly or probably planted. Map 56.

CASTANEA Miller

Castanea sativa Miller Sweet Chestnut. Castanwydden.
Fairly widely planted in woods, plantations and parkland; it occasionally produces good crops of seed (for example after the hot summer of 1976) but usually does not regenerate and spread. Map 57.

QUERCUS L.

Quercus petraea (Mattuschka) Liebl. Sessile Oak. Derwen Ddigoesog, Derwen Ddaildigoes.
Native. Very common in woods and hedgerows in the Uplands and on acid soils in Gower, but rather local in the Vale. Map 58.

Quercus robur L. Pedunculate Oak. Derwen Goesog.
Native, but often planted in woods and in parkland. Common in the Vale and fairly common in Gower in woods and hedgerows on basic and neutral soils; occasional in the Uplands. Map 59.

Quercus robur × Q. petraea (**Q. × rosacea** Bechst.) is widespread and probably occurs whenever the parent species come together; under-recorded. Map 60.

Several species of **Quercus** have been planted as specimen trees in woods or parkland.
Quercus cerris L., Turkey Oak, Derwen Dwrci, is the commonest of these and is more or less naturalized in a few places, for example near the Taf-fechan Gorge north of Cyfarthfa Castle (00A). Map 61.

Quercus ilex L., Evergreen Oak, Derwen Fythwyrdd, is also commonly planted and has become more or less naturalized in one or two sites, for example in the lower Caswell valley on the Gower coast (58B). Map 62.

Quercus rubra L. Red Oak, Derwen Goch, has been recorded from 17A, 79C and 88A as planted trees.

ULMACEAE
ULMUS L.

Ulmus glabra Hudson Wych Elm. Llwyfen Lydanddail.
Native. Once common in woods and hedges on

base-rich soils, but much reduced by Dutch elm disease after 1976. Map 63.

Ulmus procera Salisb.   English Elm. Llwyfen Gyffredin.
Probably introduced. Once fairly common in hedges except in the central Uplands, where it was rare; now, like other **Ulmus** species, much reduced by Dutch elm disease and persisting only as suckers except in a few isolated outlying localities (for example the Pitton area in south-western Gower (48A) where some trees still survived in 1983). Map 64.

Ulmus minor Miller   Small-leaved Elm. Pren Llwyf.
Introduced. Once fairly frequent in hedges and as planted trees; now persisting, if at all, mainly as suckers, but with a few trees still surviving in isolated sites. **U. minor** is a taxonomically difficult and confusing group of clones, some of which are distinctive and were often planted. The herbarium specimens of **U. minor** and related taxa and presumed hybrids in the National Museum of Wales, many of which were collected by J.A. Webb in the 1930s and 1940s, have mostly been identified by R. Melville. Typical **U. minor** (*U. carpinifolia* auct.) has the recorded distribution shown on the map, but was more frequent than the small number of records indicates. Map 65.

Ulmus angustifolia (Weston) Weston has been recorded from only three squares (17B, Llandaff, HJR 1912*, Cardiff, EV 1931*; 17C, near Cwrt-yr-ala, 1921!*; 18D, Rhiwbina, HAH 1925*). **U. plotii** Druce, was recorded by J.A. Webb from Sketty (69C) in 1926*, and **U. minor × U. plotii** (**U. × viminalis** Loddiges) from the same area between 1932* and 1952*. **U. glabra × U. plotii** was recorded from Gower (without locality) by H.A. Hyde in 1930.

Ulmus minor × U. glabra × U. plotii (**U. × hollandica** Miller) (*U. major* Sm.).   Dutch Elm.
Introduced. Once fairly frequent in hedges in the lowlands and probably more frequent than the records indicate; recorded only twice recently. Map 66.

Ulmus minor × U. glabra (**U. × vegeta** (Loudon) A. Ley) (*U. carpinifolia × glabra*) has been recorded from six squares (00C Abernant park, HJR 1904*; 17B, Cardiff, 1930!*; 68A, West Cross, JAW 1942*; 69C, Sketty, JAW 1932*; 79B, Cadoxton, JAW 1943*; 80C, Resolven, 1972!*), and yet another hybrid, **U. angustifolia × U. glabra** from two (17C, Dinas Powis Common, 1929!*; 69C, Mayals, JAW 1932* & 1940*).

MORACEAE
FICUS L.

Ficus carica L.   Fig. Ffigysbren.
Introduced; occasionally naturalized on river banks, cliffs, railway embankments and retaining walls, and waste ground. Established trees live for many years and may reach a very large size, even in the Uplands. Map 67.

CANNABACEAE
HUMULUS L.

Humulus lupulus L.   Hop. Hopysen.
Perhaps native, but often a deliberate introduction or escape from cultivation; it is most frequent near farms in the lowlands. Map 68.

Humulus japonicus Siebold & Zucc. has once been recorded as a dock alien (27A, Cardiff, JS 1876*).

CANNABIS L.

Cannabis sativa L.   Hemp. Cywarch.
Introduced, usually with bird-seed; a scarce casual weed of rubbish tips, waste ground and gardens, recorded mainly from the Swansea and Cardiff areas. Map 69.

URTICACEAE
URTICA L.

Urtica dioica L.   Common Nettle. Danhadlen.
Native. Very common in woods and scrub on rich soil, in hedgebanks and on roadsides and waste ground, near farmyards and in overgrazed pastures, and near walls and buildings; isolated patches in open country often mark the sites of former dwellings or farm buildings. Recorded from all squares. Map 70.

Urtica urens L.   Small Nettle. Danhadlen Leiaf.
Native. Locally common as a weed of farmland, gardens and waste ground in Gower and the Vale, especially near the sea and on sandy soils; very rare in the Uplands. Map 71.

Urtica pilulifera L., Roman Nettle, Danhadlen Belaidd, was recorded as a casual from three localities in the last century (17B, Cardiff, JS *c*.1876*; 17D, Penarth, JS *c*.1876; 88C, Kenfig, J.E. Bicheno *c*.1840).

PARIETARIA L.

Parietaria judaica L.   Pellitory-of-the-wall. Murlys.
Probably native. Common on old walls and limestone sea-cliffs in Gower and the Vale, scarce in the Uplands. Map 72.

SOLEIROLIA Gaud.-Beaup.

Soleirolia soleirolii (Req.) Dandy   Mind-your-own-business. Mam Miloedd.
Introduced as a rockery and greenhouse plant; now increasingly common as a virtually ineradicable weed of gardens, lawns, damp walls, waste ground and paths, especially in towns. Map 73.

SANTALACEAE

Thesium humifusum DC., Bastard-toadflax, Geulin y Forwyn, was once established on waste ground at

Barry Docks, (16A, RLS 1923!*, EV 1930*), and abundant on newly broken ground at Cwrt-y-fil (17D, JS 1886).

LORANTHACEAE
VISCUM L.

Viscum album L.   Mistletoe. Uchelwydd.
Perhaps native in some sites, but introduced in most localities. Recorded on **Crataegus, Malus, Poplus, Robinia, Tilia**, and **Ulmus**. Scattered throughout the county.
09, near Mountain Ash, T. Evans, c.1958; 16B, Sully, Mr Hodge c.1900; 17A, St. Fagans, JS c.1876, Coedrhiglan Woods, HOB 1907; 17B, Crock-herbtown, JS c.1876, Llandaff, 1977!, Roath Park, V.G. Ellis 1988!; 28A, Ruperra, J. Delhenty c.1925; 69A, Penllergaer, G. Traherne c.1900; 69C, Singleton, H.R. Wakefield 1910, C. Marks 1926, near Swansea, H & CH 1974; 79, near Neath, EV 1935*; 99B, Treorchy Cemetery, JTa 1972.

ARISTOLOCHIACEAE
ARISTOLOCHIA L.

Aristolochia clematitis L.   Birthwort. Afal Daear.
Introduced. Well established on dunes at Merthyr Mawr, casual elsewhere on waste ground and roadsides.
07B, Peterson-super-Ely, D.M. Harrison 1945*; 08, BSBI maps scheme, c.1956; 16A, Barry Dock, Mrs C.I. Sandwith c.1936; 87B, Merthyr Mawr, Mr Bryant c.1936, J.F. Rigby 1949*, BAM 1953*, MG 1975!.

Asarum europaeum L., Asarabacca, Carn Ebol y Gerddi, has once been recorded as an alien from Cardiff (17B, Cardiff Castle, JS c.1880).

POLYGONACEAE
POLYGONUM L.

Polygonum oxyspermum Meyer & Bunge ex Ledeb. subsp. **raii** (Bab.) D.A. Webb & Chater Ray's Knotgrass. Canclwm Ray.
Native. Sandy shingle on or near the beach drift-line, usually with **Atriplex** spp.; now rarer than it was in the past, but with recent records from Gower (where it still occurs regularly) and Porthcawl. Map 74.

Polygonum aviculare L.   Knotgrass. Canclwm.
Native. Very common as a weed of farmland and waste ground, sometimes found on streamsides and on or near the drift-line on beaches. **P. arenastrum**, which is abundant on roadsides and trampled ground, is often confused with **P. aviculare**, and some records of **P. aviculare** may refer to that species, especially in the Uplands. Fl. 6–12, fr. 6–11. Map 75.

Polygonum rurivagum Jordan ex Boreau, which is also similar to **P. aviculare**, has been recorded from three localities (49D, near Fairyhill, 1976!; 60C,

Felindre, 1974!; 79C, Jersey Marine, 1974!**); it is probably more frequent than these records indicate and its status in the county needs further investigation.

Polygonum arenastrum Boreau   Equal-leaved Knotgrass, Roadside Knotgrass. Clymog â Dail Bach.
Native. Very common on the edges of roads and tracks and on trampled ground, often associated with **Matricaria matricarioides** and **Poa annua**. Fl. 8–12, fr. 8–11. Map 76.

Polygonum minus Hudson   Small Water-pepper. Clymog Bychan.
Native. On the margins of ponds and lakes; rare, with recent records from only two localities in the Vale.
07B, Warren Mill Pond, 1926!*, St.-y-Nyll Ponds, MG 1965; 69D, near the Ferry, Swansea, JWGG c.1842; 78D, Port Talbot, HJR 1907; 87B, Ogmore by Sea, MG 1973*.

Polygonum mite Schrank   Tasteless Water-pepper. Penboeth Diflas.
Native. In ditches and by ponds and rivers; rare, with only one recent record.
17A, Ely, EV 1944*; 69D, Swansea, TBF & E. Lees 1842; 79C, Neath, TBF & E. Lees 1842; 97B, Llangan, 1973!*. The record for 07 in Perring & Walters (1962) probably refers to 17A.

Polygonum hydropiper L.   Water-pepper. Tinboeth.
Native. Very common in ditches and damp trampled pastures, in marshy places, as a weed of damp cultivated soil, and at the edges of streams and ponds. Fl. 7–9, fr. 8–10. Map 77.

Polygonum persicaria L.   Redshank. Coesgoch, Dail y Groes.
Native. Very common as a weed of farmland and disturbed roadsides; also found on streamsides and in marshy habitats similar to those in which **P. hydropiper** grows. Fl. 6–11, fr. 8–11; Hb (n), E.ten (n), Ly.phl (n). Map 78.

Polygonum lapathifolium L.   Pale Persicaria. Costog y Domen.
Native. Common as a weed of cultivated land, disturbed roadsides and waste ground; sometimes found on streamsides and in marshy places. A pink-flowered form with densely glandular peduncles (*P. nodosum* Pers.) is locally frequent in western Gower (48A!, 49C!, 58A!) as a weed of arable land, often occurring with typical **P. lapathifolium** from which it appears to be quite distinct. *P. nodosum* has also been recorded from several other sites in western Glamorgan and in the eastern Vale and eastern Uplands. In recent taxonomic treatments **P. lapathifolium** is regarded as a variable species which includes *P. nodosum* and some other variants which have also been given specific rank in the past. (!) Fl. 7–10, fr. 8–11. Map 79.

Polygonum lapathifolium × **P. persicaria** (**P.** ×

lenticulare Hyl.) has been recorded once (18C, Radyr Church, HJR 1912).

Polygonum amphibium L. Amphibious Bistort. Canwraidd Goch.
Native. Locally common in Gower, the Vale and the eastern Uplands in ponds, ditches and shallow streams and in nearby marshy ground, especially near to the sea in Gower and the Vale; sometimes found as a weed of damp or low-lying cultivated land. Absent from the central and western Uplands. Fl. 6. Map 80.

Polygonum bistorta L. Common Bistort. Llys y Neidr.
Native or introduced. Uncommon but widely distributed, growing as scattered patches on roadsides and in hedgebanks or the corners of meadows, usually near houses; decreasing in frequency. Fl. 5. Map 81.

Several species of **Polygonum** have been found on one or a few occasions as casuals or garden escapes. **Polygonum maritimum** L., Sea Knotgrass, (17B, Cardiff, JS *c.*1876; 17D, Penarth, JS *c.*1876; 18C, Radyr, RLS 1921*), **P. patulum** Bieb. (16A, Barry, RLS 1924; 17B, Canton, HJR 1907; 27A, Splott, 1925!*; 78B, Port Talbot, HJR 1905), and **P. arenarium** Waldst. & Kit. subsp. **pulchellum** (Loisel.) D.A. Webb & Chater (16A, Barry Docks, RLS 1924!; 17D, Cardiff, 1933!*) are annual or short-lived species which were once found as aliens, growing near docks or with grain cleanings. In contrast to these, **Polygonum amplexicaule** D. Don, Red Bistort, (59B, Gorseinon, G.M. Reyland 1968*; 79C, Briton Ferry, H & CH 1972), **P. polystachyum** Wallich ex Meissner, Himalayan Knotweed, (17B, Llandaff North, GSW 1983; 69C, Olchfa, JAW 1951; 78B, Margam, JAW 1954), and **P. campanulatum** Hooker fil., Lesser Knotweed, (07D, near Bonvilston, JIL 1972; 17B, Maendy, RLS 1938*; 69C, Clyne Castle, V. Gordon 1956*) are vigorous perennials which may persist and spread as garden escapes.

FALLOPIA Adans.

Fallopia convolvulus (L.) Á. Löve (*Polygonum convolvulus* L.). Black-bindweed. Taglys yr Ŷd.
Native or introduced. A common weed of gardens and arable land, sometimes also found on disturbed roadsides and waste ground. Fl. 6–10, fr. 7–10. Map 82.

Fallopia dumetorum (L.) J. Holub, Copse-bindweed, Taglys y Berth, was recorded from Sully (16B) by H.M. Hallett *c.*1906 (Trow, 1911), possibly in error for a form of **F. convolvulus**, or the following species.

Fallopia aubertii (Louis Henry) J. Holub (*Polygonum baldschuanicum* auct.). Russian-vine. Taglys Tibet.
Introduced as an ornamental plant; often rampant and very persistent in garden hedges, but rarely naturalized.

09A & 09B, Mountain Ash, AMP 1970*; 17B, Cardiff, 1976!; 18C, Whitchurch, Cardiff, GH 1984; 68A, near Mumbles, *c.*1975!; 69C, Blackpill, JAW 1939*; 96B, Llantwit Major, 1983!.

REYNOUTRIA Houtt.

Reynoutria japonica Houtt. (*Polygonum cuspidatum* Siebold & Zucc.). Japanese Knotweed. Pysen Saethwr.
Introduced as an ornamental plant or with ballast; now widespread and common as a vigorous and dominant weed of roadsides, hedgerows, waste ground, rubbish and slag tips, railway land, woodland, scrub and riversides in urban areas. Increasingly common in similar habitats in the countryside, where it is often initially introduced on roadsides with garden waste. R. japonica was recorded by John Storrie (1886) as being abundant on cinder tips near Maesteg, but was not mentioned by H.J. Riddelsdell (1907) or A.H. Trow (1911) in their Floras of the county, perhaps being excluded as a garden escape that appeared unlikely to persist. It is now one of the most abundant plants in urban areas in Glamorgan. Its introduction and spread in the British Isles have been described by Conolly (1977). Fl. 8–10, fr. 10; (Syr) E.ten (p), S.balt (p), (Dip) Call (n), Luc (n). Map 83.

Reynoutria japonica × R. sachalinensis (**R. × bohemica** Chrtek & Chrtkova) has been recorded from Whitchurch, Cardiff (18C) by G. Hutchinson in 1985!*, determined morphologically and cytologically (2n=66) by J.P. Bailey and A. Conolly; also from 00A, Cyfarthfa Castle, Merthyr Tydfil, GH & B.A. Thomas 1990*; 08C, Brynsadler, GH & J. Bevan 1990* and 18D, Rhiwbina, Cardiff, GH 1988!*, determined morphologically by J.P. Bailey.

Reynoutria sachalinensis (F. Schmidt Petrop.) Nakai Giant Knotweed.
Introduced as an ornamental plant, usually by streamsides.
18D, Llanishen, A. Rowlands 1969, GH & RAH 1985; 79D, by Cwm-du Brook, JAW *c.*1950; 88D, Llangewydd, GH 1988*.

FAGOPYRUM Miller

Fagopyrum esculentum Moench Buckwheat. Gwenith yr Hydd.
Occasionally found as a casual on waste or cultivated ground, but with only three recent records. It is sometimes sown near coverts to provide seeds for game birds, and may also occur in commercial bird-seed mixtures. Map 84.

Fagopyrum tataricum (L.) Gaertner has been found twice (16A, Barry Docks, RLS 1923, 1924*).

Rheum rhabarbarum L. (Rhubarb) has recently been discovered as a garden escape or throw-out on a railway cutting near Bridgend (97A, GH 1990).

RUMEX L.

Rumex acetosella L. *s.l.* Sheep's Sorrel. Suran yr Ŷd.
Native. Common in open grassland on well-drained acid soils, on colliery tips and acid smelter slag, in dry heathland, and on acid dunes; sometimes found as a weed of arable land on light acid soils. The common form in Glamorgan appears to be **R. acetosella** L. *s.s.*, but specimens collected in four localities have been identified as **R. tenuifolius** (Wallr.) Á. Löve (18C, Craig-yr-Allt, THT 1902*; 58B, Three Cliffs Bay, AJES 1961**; 88C, near Pen-y-mynydd, 1956!*; 99B, above Ferndale, JTa 1970*), and the status of this segregate species in Glamorgan requires investigation. A specimen collected on the N.M.W. building site (17B!*) *c*.1925, growing with a number of introduced species, appears to be **R. angiocarpus** Murb. Fl. 5–7 (both sexes). Map 85.

Rumex acetosa L. Common Sorrel. Suran y Cŵn.
Native. Very common and widely distributed in grassland on neutral and moderately acid or calcareous soils, especially on roadsides, in hay meadows and long-established pastures, and in natural or semi-natural grassland. Fl. 5–6 (both sexes), fr. 6–7. Map 86.

Rumex frutescens Thouars Argentine Dock. Tafol yr Ariannin.
Introduced. Established on dunes and waste ground in a few sites in the west of the county. When it was first found by Miss Mary Thomas at Kenfig (1932) and by Eleanor Vachell on the foreshore at Swansea (1940), it was abundant at both sites, but it is now hard to find, with only a few plants surviving in each locality.
58A, Oxwich, on drift-line to north of salt-marsh, GTG 1955, 1976!; 60A, Cwmbwrla, JAW 1951; 69C, Cockett, JAW 1951*; 69D, Swansea foreshore and docks, EV 1940, JAW 1951*, CH 1974, waste ground near R. Tawe, 1974!**; 78B, Port Talbot, M. Humphrey 1961*; 78D, Kenfig Burrows, EV 1934, TGE 1980; 99D, Tonypandy riverbank, 1975!.

Rumex hydrolapathum Hudson Water Dock. Tafol y Dŵr.
Native. Growing in **Phragmites** fens, by ditches or streams, and at the edges of ponds or lakes. **R. hydrolapathum** has a surprisingly limited distribution in Glamorgan but is locally abundant in Oxwich marsh (48B, 58A), in Crymlyn Bog (69D, 79C) and intermittently along the marshes (Thaw and Ely valleys) and lakesides (e.g. Hensol Ponds, 07A) in the southern and eastern Vale. Map 87.

Rumex hydrolapathum × R. obtusifolius (**R. × weberi** Fischer-Benzon) has been recorded once (07B, Peterston-super-Ely, HJR 1907).

Rumex cristatus DC. Greek Dock. Tafol Groeg.
Introduced; established on the banks of rivers and streams and by ponds at a few sites in the Cardiff area.

16B, Sully, 1954!; 17B, Roath Park, 1973!, GSW 1986, Pontcanna, GSW 1986; 27A, by Roath Brook, 1950!, GH 1985, by R. Rhymney, 1972!, GSW 1982; 27C, Cardiff Docks, GSW 1986.

Rumex cristatus × R. obtusifolius (**R. × lousleyi** D.H. Kent) has been recorded from three sites near the mouth of the Rhymney River and Cardiff Docks (27A & C) by GSW, 1983*–1986*, conf. D.H. Kent.

Rumex crispus L. Curled Dock. Tafol Crych.
Native. Common as a weed of cultivated land, waste ground and roadsides, and in maritime habitats on or near the drift-line, especially on shingle beaches. Maritime forms have dense panicles and inflated tubercles on the fruiting perianth, sometimes on all three valves (var. **trigranulatus** J.T. Boswell). Fl. 6–11, fr. 6–11. Map 88.

Rumex crispus × R. obtusifolius (**R. × pratensis** Mert. & Koch) has been recorded from several localities throughout the county; it is probably more frequent than the limited number of records indicates and is likely to occur whenever the parents grow together in a disturbed habitat.
09D, near Mountain Ash, 1972!; 16A, Barry, 1974!; 17B, Cardiff, JWD 1940*; 17C, St. Lythans, EV n.d.; 17D, Cogan Hall, AHT 1905–1907, Dinas Powis, AHT 1905, Llandough, AHT 1907; 18D, Cyncoed, E.G. Roberts 1981*; 50D, Pontardulais, HJR 1906; 97D, Cowbridge, HJR *c*.1906 (also recorded from 'Graigafan', an untraced locality, by J. Motley in *c*.1840).

Rumex conglomeratus Murray Clustered Dock. Tafol Blaen, Tafol Mair.
Native. Fairly common in damp grassland and marshes and on the banks of ditches and streams near the coast, but local elsewhere and scarce or absent in the central Uplands. Map 89.

Rumex sanguineus L. Wood Dock. Tafol Coch, Tafol y Coed.
Native. Common in hedgebanks, in woodland rides and clearings, and in similar shaded habitats. The common form is var. **viridis** Sibth. with green or reddish veins; var **sanguineus**, with purple veins was once cultivated as a medicinal herb but is now very rare in the county. Map 90.

Rumex rupestris Le Gall Shore Dock. Tafolen y Traeth.
Native. Marshy dune-slacks near the transition zone to salt-marsh, and in wet flushes on sea-cliffs. Rare, with no recent records.
58A, Three Cliffs Bay, G. Bentham 1859, J.R. Shepherd 1910; 78D, Kenfig Burrows, EV 1932*; 87B, Merthyr Mawr Warren, BAM 1954; 87D, Dunraven Bay, EV 1934*.

Rumex crispus × rupestris has been found twice on Kenfig Burrows (78D, JSCG 1932*, JEL 1948*).

Rumex pulcher L.   Fiddle Dock. Tafol Crwth-ddail.
Native or introduced. Growing in dry grassland and
on waste ground. Rare, with recent records only from
the Cardiff district.
17B, Cardiff, EV 1903, 1969–1976!*; 17C, Dinas Powis,
GMB 1973; 27A, Splott, 1925!*; 59B, Loughor, JM
c.1850; 78B, Port Talbot, HJR c.1907; 87A, Porthcawl,
AHT 1905*, Newton, Porthcawl, EV 1912, JAW 1931,
EMT 1943*; 87D, Southerndown, WWB, 1909.

Rumex obtusifolius L.   Broad-leaved Dock. Dail
Tafol.
Native. Very common as a weed of cultivated land,
farmyards and pastures, and on roadsides, waste
ground and tips. Map 91.

Rumex obtusifolius × R. sanguineus (R. × dufftii
Hausskn.) has been recorded once (17B, N.M.W.
building site, 1923!*).

Rumex maritimus L.   Golden Dock. Tafol Arfor.
Perhaps native in the west of the county on disturbed
ground near water at Crymlyn Bog and formerly at
Morfa Pools; also occasionally found in the past as a
dockside alien at Cardiff and Barry. Rare, with no
recent records.
16A, Barry Docks, AHT 1891–2; 17B, Cardiff, HJR
c.1909; 27A, Splott, 1925!*; 69D, foreshore near
Swansea Pier, C. Marks 1924, in abundance in
Crymlyn Bog near Tir John Power Station, JAW 1940,
EV 1941; 78B, Port Talbot, HJR 1909, apparently
native at Morfa Ponds, EV 1945. The record given by
Turton in Oldisworth (1802), from Port Eynon (48B),
is doubtful.

Five species of **Rumex** have been found as rare
casuals in the Cardiff area, on rubbish tips or
docksides. **Rumex triangulivalvis** (Danser) Rech. fil.,
from four localities, with one recent record (16A,
Barry Docks, JPC 1986; 17B, Cardiff Docks, RLS
1924!*; 27A, Splott, 1925!*, Roath, RLS 1939*), **R.
obovatus** Danser from two (17D Grangetown, RLS
1925*; 27A, Splott, 1925!*), and **R. bucephalophorus**
L. (16A, Barry Docks, RLS 1927*, 1940), **R. dentatus** L.
(27A, Splott, 1925!*, RLS 1928), **R. longifolius** DC.
(18C, Radyr, GCD c.1922), and **R. scutatus** L. (88C,
near Cornelly, J.E. Bicheno c.1800, JWGG c.1840) from
one each.

CHENOPODIACEAE

BETA L.

Beta vulgaris L. subsp. **maritima** (L.) Arcangeli   Sea
Beet. Betys Gwyllt, Melged Arfor, Betys Arfor.
Native. Drift-line and spray-zone, especially on
shingle beaches and low cliffs. Common in suitable
habitats on the coast. Map 92.

Beta vulgaris subsp. **vulgaris** has occasionally been
found as an alien in the Barry and Cardiff areas (16A,
1925!*; 17B, RLS 1938*, JWD 1956*; 17D, JPC 1981*;
18C, GH 1985*; 27A, RLS c.1930*).

CHENOPODIUM L.

Chenopodium bonus-henricus L.   Good-King-
Henry, All-good. Sawdl y Crydd, Llys y Gŵr Da.
Introduced. Grassy places on rich soils near long-
established farms and occasionally on waste ground.
Scattered through the county, but scarce and usually
in small numbers. Map 93.

Chenopodium glaucum L.   Oak-leaved Goosefoot.
Troed yr Ŵydd Dderw-ddeiliog.
Casual, on waste ground. One recent record.
06A, Aberthaw, GMB 1972; 17B, Canton, HJR 1907*;
17D, Penarth, JS c.1886, Ely Estuary, AHT 1907*,
Leckwith, EV 1913*, 1922; 27A, Splott, 1925!*; 69C,
Swansea, JWGG c.1841; 78B, Port Talbot Docks, HJR
1904.

Chenopodium rubrum L.   Red Goosefoot. Troed yr
Ŵydd Ruddog.
Introduced. Farmyards, cultivated land, waste
ground and rubbish tips, mainly near the coast, and
on sea-shores. Map 94.

Chenopodium polyspermum L.   Many-seeded
Goosefoot. Troed yr Ŵydd Luos-hadog.
Native or introduced. Cultivated land and waste
ground, local. Map 95.

Chenopodium urbicum L.   Upright Goosefoot.
Troed yr Ŵydd Syth-ddail.
Casual on waste ground, roadsides and river banks.
06A, Aberthaw, near Burton Bridge, EV 1923*, 1933*;
17B, Cardiff, JS c.1886; 27A, Splott, EV 1932*.

Chenopodium murale L.   Nettle-leaved Goosefoot.
Troed yr Ŵydd Ddynad-ddail.
Native or introduced. Cultivated land, sand-dunes
and waste places. Only three recent records. Map 96.

Chenopodium ficifolium Sm.   Fig-leaved Goosefoot. Troed yr Ŵydd Ffigys-ddail.
Casual on waste ground. Only three recent records.
00A, Merthyr, 1975!*; 00C, Aberdare, HJR c.1907;
16A, Barry Docks, JPC & TGE 1986; 19C, Ystrad
Mynach, M.E. Russell 1972; 27A, Splott, EV 1938*;
78B, Port Talbot Docks, HJR c.1907; 90B, Hirwaun,
HJR c.1907.

Chenopodium opulifolium Schrader ex Koch &
Ziz   Grey Goosefoot. Troed yr Ŵydd Lwyd.
Casual on waste ground. No recent records.
00C, Aberdare, HJR 1908*; 16A, Barry Docks, EV
1926*; 17D, Cardiff, 1932!*; 18C, Radyr, RLS 1920*;
27A, Splott, EV 1926!*; 78B, Port Talbot Docks, HJR
c.1904.

Chenopodium album L. subsp. **album**   Fat-hen.
Tafod yr Oen.
Native. Cultivated land and waste ground. Common
in suitable habitats throughout the county; often a
troublesome weed, especially in potatoes, **Brassica**
spp. and other vegetable crops. Fl. 7–11, fr. 7–10.
Map 97.

Chenopodium album subsp. striatum (Krašan) J. Murr (16A, 1924!*; 17B, 1925!), subsp. album × subsp. striatum (16A, RLS & EV 1926; 17B, GCD 1924), and C. album × C. opulifolium (C. × preissmannii J. Murr) (16A, GCD 1926; 17D, 1922!) have all been recorded as rare aliens at Cardiff or Barry Docks. [C. album × C. suecicum (C. × fursajewii Aellen & Iljin) was once recorded in error.] Several other species of Chenopodium have been found as scarce aliens, in most cases near Cardiff, Barry or Port Talbot Docks. Chenopodium vulvaria L. (16B, JS c.1876; 17B, RLS 1907*; 18C, 1921!*, 27A, 1925!*, EV 1927*; 78B, HJR c.1907) and C. suecicum J. Murr (00C, HJR c.1907; 27A, 1926!*; 69C, JWGG c.1844; 78B, HJR c.1907) were most widespread but have not been seen recently. The American species C. pratericola Rydb. was recorded three times in the 1920s and has been refound recently in one of these localities and in a new locality (16A, ABP 1974!*; 17B, 1924!*, 1973!; 27A, 1925!*; 78B, GCD 1929). C. multifidum L. (16A, RLS 1925; 17B, JS c.1870; 18C, RLS 1921*) and C. hircinum Schrader (00C, HJR 1901; 17B, GCD 1928; 27A, RLS & F. Norton 1937), have been found in three localities. C. hybridum L. (17B, HJR 1907*, GCD 1916, 27A, EV 1927*), C. ambrosioides L. (16A, RLS 1921*; 27A, JS 1876*), C. foliosum Ascherson (17B, Miss Pears 1941*; 27A, RM 1927, 1929!*) and C. berlandieri Moq. subsp. zschackei (J.Murr) Zabel, (17B, GCD c.1930; 78B, HJR 1905) have been found in two localities, while C. botrys L. (17B, JS c.1870, RLS 1923), C. capitatum (L.) Ascherson (27A, J.H. Salter 1927*), C. graveolens Lag. & Rodr. (27A, 1926!*) and C. pumilo R. Br. (17B, GCD 1927) have been found in only one, with no recent records. [C. virgatum Thunb. was recorded from 27A in 1929! but this is now believed to have been in error for C. foliosum Ascherson.]

ATRIPLEX L.

Atriplex laciniata L. Frosted Orache. Llygwyn Arianaidd.
Native. Drift-lines on sand or shingle beaches. Scarce and sporadic in its appearance except at Oxwich. Recent records from 49C, Whiteford Burrows, 1956!*; 58B, Oxwich, 1976!**, 1980; 78D, Kenfig Burrows, TGE 1971; 79C, Crymlyn Burrows, H & CH 1976. Map 98.

Atriplex littoralis L. Grass-leaved Orache. Llygwyn Arfor.
Native. Drift-lines, usually on muddy sand or shingle. Locally abundant around the Burry Inlet but scarce or sporadic elsewhere. Recent records from 17D, Leckwith, SGH, 1962*; 49A, B&C, Whiteford Point and Berges Island, 1972!, 1984!; 59B, Llwchwr, 1974!**, 1984!; 96A, St. Donat's Bay, 1971!. Map 99.

Atriplex patula L. Common Orache. Llygwyn Culddail.

Native. Cultivated land, roadsides and waste ground (occasionally on drift-lines). Common and widespread in suitable habitats, except in parts of the eastern Uplands. Fl. 8, fr. 9. Map 100.

Atriplex prostrata Boucher ex DC. (A. hastata auct., non. L.). Spear-leaved Orache. Llygwyn Tryfal.
Native. Cultivated land, waste ground and drift-lines. Locally fairly common but unevenly distributed; most frequent in the Cardiff and Swansea areas and near the coast; absent from much of the uplands. Fl. 8, fr. 10. Map 101.

Atriplex glabriuscula Edmondston Babington's Orache. Llygwyn Babington, Llygwyn Y Tywod.
Native. Drift-lines. Common in suitable habitats, but variable and scarcely distinguishable from maritime forms of A. prostrata. Map 102.

Atriplex longipes Drejer Long-stalked Orache. Llygwyn Hirgoes.
Native. On or near the drift-line in salt-marshes, usually among Juncus maritimus. Common in the Gower salt-marshes, perhaps overlooked elsewhere. 49A & C, Whiteford Point, 1980!; 49D, Landimore, 1980!; 58A, Three Cliffs Bay, 1980!**; 59A, Crofty, 1980!; 59C, Wern-ffrwd, 1980!**; 78D, Kenfig, O. Stewart, R. Fitzgerald & TGE 1984; 79C, Jersey Marine, 1980!.

Two alien species of Atriplex have been recorded on one occasion each: Atriplex hortensis L. (16B, J. Randall 1974*), and A. palaestina Boiss. (27A, GCD c.1926). [A. tatarica L. has been reported from the Cardiff and Barry areas (16A, 17B & 27A), but specimens in NMW have proved insufficient for determination, P. Taschereau pers. com.]

HALIMIONE Aellen

Halimione portulacoides (L.) Aellen Sea-purslane. Helys Can, Llygwyn Llyswyddaidd.
Native. Salt-marshes, especially along creeks, and occasionally in the spray-zone on sea-cliffs. Abundant in suitable habitats in the western part of the county. (!) Fl. 7–8, fr. 9–10. Map 103.

Halimione pedunculata (L.) Aellen was found at Port Talbot by J. Motley in 1843, presumably as a dock alien (78B**), there are no later records.

Axyris amaranthoides L. was found as a casual in the dock area of Cardiff in 1925 and 1926 (27A!*), and Kochia scoparia (L.) Schrader has twice been found as an alien or garden escape. (17D, HJR 1903, RLS 1937*).

SALICORNIA L.

Salicornia europaea L. Glasswort. Llyrlys.
Native. Sandy salt-marshes, probably rather local, with only two recent records. Map 104.

It is possible that some of the early records may refer to **Salicornia ramosissima**, but since Ingrouille and Pearson (1987) conclude that the pattern of morphological variation 'provides no evidence for the separate recognition of **S. europaea** L. and **S. ramosissima** J. Woods . . . and that it would be better to group them all under **S. europaea** L.', these taxa may not be specifically distinct.

Salicornia ramosissima J. Woods   Bushy Saltwort. Llyrlys Gorweddol.
Native. Common on firm mud in upper salt-marshes, often in shallow pans, mainly in Gower. Map 105.

Salicornia pusilla J. Woods
Native. Locally frequent in the higher parts of salt-marshes, mainly in Gower. Map 106.

Salicornia fragilis P.W. Ball & Tutin   Llyrlys Brau.
Native. Widespread in a variety of sites on mud or muddy sand in salt-marshes, from relatively dry parts of the upper marsh to banks of muddy creeks in the lower marsh. A variable species, with several named forms. Abundant in the north Gower salt-marshes.
The records from 06A and 87B refer to *S. lutescens* P.W. Ball & Tutin, which is probably a variant of this species (Ball in Tutin *et al.* 1964).
06A, Aberthaw, 1963!*; 49C, Llanmadoc, 1980!; 49D, Landimore, 1980!**; 58A, Three Cliffs Bay, 1980!; 59C, Wern-ffrwd, 1980!; 79C, Jersey Marine, 1980!; 87B, Mouth of Ogmore, A. Loosemore 1957, (Ball and Tutin, 1959).

Salicornia dolichostachya Moss   Llyrlys Canghennog.
Native. Locally common on mud and muddy sand in lower salt-marshes, often as a pioneer.
06A, West Aberthaw, EV 1933*, 1963!*; 17D, Taff-Ely estuary, EV 1930*, 1969! GH 1989*; 49B, Berges Island, 1974!; 49D, Landimore Marsh, 1975!, 1980!**; 58A, Oxwich, GTG 1955; 59A, Crofty, 1974!; 59B, Penclawdd, 1974!, Loughor 1984!*; 59C, Llanmorlais, 1974!.

SUAEDA Forskål ex Scop.

Suaeda vera J.F. Gmelin (*Suaeda fruticosa* auct.). Shrubby Sea-blite. Llwynhelys.
Probably introduced. Rare on tidal mud-flats.
17D, Ely estuary, AHT 1899*, 1969!*; 69D, Port Tennant, JWGG *c.*1841. The record from 69D may be an error. Still persisting in the Ely estuary in the late 1980s but threatened by the proposed barrage.

Suaeda maritima (L.) Dumort.   Annual Sea-blite. Helys Unflwydd.
Native. Common on bare or muddy sand in salt-marshes. Map 107.

Suaeda altissima (L.) Pallas was found as an alien at Port Talbot Docks (78B) by H.J. Riddelsdell in 1905.

SALSOLA L.

Salsola kali L. subsp. **kali**   Prickly Saltwort. Helys Ysbigog.
Native. Locally common on the drift-line on sandy beaches. Map 108.

Salsola kali subsp. **ruthenica** (Iljin) Soó was found as an alien at Cardiff Docks (17B*) by R.L. Smith in 1923.

AMARANTHACEAE
AMARANTHUS L.

Amaranthus retroflexus L.   Common Amaranth. Chwyn Moch.
Casual, on waste ground and in cultivated land.
00C, Aberdare, HJR *c.*1907; 16A, Barry Docks, RLS 1920*; 17B, Gorsedd Gardens, Cardiff, ABP 1975*; 27A, Cardiff, JS 1876*, EV 1907*, 1926!, RLS 1937*!; 78B, Port Talbot Docks, HJR *c.*1907; 87B, Merthyr Mawr, A. Jones 1968*.

Several other species of **Amaranthus** have occurred as aliens near Cardiff, Barry or Port Talbot Docks but none have been recorded recently. **Amaranthus hybridus** L. (17B, HJR *c.*1907; 17D, 1932!*, RLS 1967*; 27A, JS 1876, RLS 1927*) and **A. albus** L. (16A, RLS *c.*1924; 17B, 1924!*; 17D, 1933!*; 27A, 1925!*; 78B, HJR *c.*1907) have been found most frequently, **A. graecizans** L. (16A, GCD 1927; 17D, 1933!*; 27A, RLS 1937; 78B, HJR 1905) on four occasions, **A. spinosus** L. twice (17B, 27C, both JS 1876*), and **A. standleyanus** Parodi ex Covas (27A, 1937!*), **A. deflexus** L. (78B, HJR 1905), and **A. acutilobus** Uline & Bray (27A, RLS 1930*) once only. Records of **A. blitum** L. from 16A, 17B, 17D and 78B, in Riddelsdell (1907) may refer to **A. lividus** L. or to one of the other species. A specimen from 27A, collected in 1926 was provisionally determined by G.C. Druce as **A. quitensis** Kunth. It should be noted that most of the identifications of the early records listed above must be regarded as provisional.

PHYTOLACCACEAE

Phytolacca latbenia Maxim. (*P. americana* auct. non L.) has been found as a casual only in waste ground near Cardiff (17B, Cardiff Docks, RLS *c.*1925!, Llandaff, L. Kennard 1931*).

AIZOACEAE

The following three species have all been recorded once in waste places: **Carpobrotus edulis** (L.) N.E. Br., Hottentot-fig, (16A, Barry, RLS, *c.*1925), **Mesembryanthemum crystallinum** L. (*Cryophytum crystallinum* (L.) N.E. Br.), (Gower Coast, Comdr. Chapman, *c.*1955), and **Cleretum bellidiforme** (Burm. fil.) Rowley (*Mesembryanthemum criniflorum* (L.) fil., *Dorotheanthus bellidiformis* (Burm. fil.) N.E. Br.), Livingstone-daisy, (87A, Locks Common, Porthcawl, M.J. Allen 1977).

[Plate 3]

1, *Toninia coeruleonigricans*, a distinctive lichen of sand-dunes and limestone crevices (Whiteford Burrows). **2,** *Hookeria lucens*, frequent near streams in western woods (Pengwern, Gower). **3,** *Ophioglossum vulgatum*, locally common in dune-slacks (Whiteford Burrows). **4,** *Osmunda regalis*, scarce but still widespread on damp heathland in the west (Rhosili Down). **5,** *Ranunculus tripartitus*, a rare and threatened pond crowfoot (Pitton, Gower). **6,** *Matthiola incana*, a scarce plant of limestone cliffs (Nash Point). (Photographs: 1–5, Q. O. N. Kay; 6, M. E. Gillham).

[Plate 4]

1, *Draba aizoides* at the western limit of its European range (Tears Point, Gower). **2,** *Halimione portulacoides* with *Pelvetia canaliculata* in saltmarsh (Loughor). **3,** *Althaea officinalis*, a characteristic plant of the intensively grazed North Gower saltmarshes (Llanrhidian). **4,** *Hypericum elodes*, a western wetland plant (Broad Pool, Gower). **5,** *Glaux maritima*, common in upper saltmarsh (Jersey Marine). **6,** *Nymphoides peltata*, introduced to Broad Pool, where its spread has threatened scarce native species (Broad Pool, Gower). (Photographs: 1–6, Q. O. N. Kay).

PORTULACACEAE
MONTIA L.

Montia fontana L.   Blinks. Gwlyddyn y Ffynnon.
Native. Springs, streamsides and damp grassland, usually on non-calcareous soils. Frequent in Gower and the Uplands but absent from much of the Vale. Map 109.

Montia fontana subsp. **variabilis** Walters
Records confined to the eastern borders of the county.
10A, near Fochriw, 1947!*; 17A, Ely, Cardiff, EV 1940*; 18D, Wenallt, Cardiff, 1921!*; 27A, Pen-y-lan, Cardiff, 1925!*. There is a record for 97 in Perring (1968).

Montia fontana subsp. **amporitana** Sennen (subsp. *intermedia* (Beeby) Walters).
Frequent in Gower and the central Uplands. Map 110.

Montia fontana subsp. **chondrosperma** (Fenzl) Walters
Records mainly from the western parts of the county. Map 111.

Montia perfoliata (Donn ex Willd.) Howell Springbeauty. Trydwll.
Introduced. Occasionally found as a weed of cultivated land and waste places near the sea.
06B, Rhoose Point, MG 1975; 16A, Barry Island, EV 1935*; 17B, Cardiff, AHT 1906*, HJR c.1907; 17D, Penarth, VCB 1972*; 58A, Three Cliffs Bay, MG 1973; 68A, Mumbles, MG 1974; 87A, Porthcawl, VCB 1972; 97C, Monknash, EV 1910.

Montia sibirica (L.) Howell   Pink Purslane. Gwlyddyn Rhudd.
Casual, perhaps naturalized in one locality, (17A), but not seen recently.
17A, St. Fagans, EV 1938*, MG 1960; 17D, Penarth Road, Cardiff, JS c.1886*.

CARYOPHYLLACEAE
ARENARIA L.

Arenaria balearica L.   Mossy Sandwort. Tywodlys Meindwf.
Introduced. Once naturalized on a wall at one locality but not seen recently.
18D, Craig Llanishen, JWD 1957*.

Arenaria serpyllifolia L. agg.
Native. Common throughout the country. Map 112.

Arenaria serpyllifolia L. subsp. **serpyllifolia** Thyme-leaved Sandwort. Tywodlydd Gruwddail.
Native. Shallow soil on cliffs, ant-hills, sand-dunes, cultivated land, colliery spoil-heaps, railway ballast, etc. Common in suitable habitats in Gower and the Uplands but rather local in the central Vale. Fl. 5–9, fr. 6–10; (!) (Lep) Po.ica. Map 113.

Arenaria serpyllifolia subsp. **macrocarpa** (Lloyd) F.H. Perring & P.D. Sell
Native. Shallow soil on cliffs, sand-dunes, grassy drift-lines. Probably common in coastal habitats.
48A, Paviland, 1976!; 48B, Overton Cliff, 1974!; 58A, Pennard Cliff, HJR 1903*; Three Cliffs Bay, 1980!; 58B, High Tor, 1978!; 78D, Kenfig Burrows, MG 1974.

Arenaria serpyllifolia subsp. **leptoclados** (Reichenb.) Nyman   Slender Sandwort. Tywodlys Main.
Native. Similar habitats to **A. serpyllifolia** subsp. **serpyllifolia**, and often associated with it, but less common. (!) Fl. 6–9, fr. 7–10. Map 114.

MOEHRINGIA L.

Moehringia trinervia (L.) Clairv.   Three-nerved Sandwort. Tywodlys Trinerf.
Native. Woods and shaded hedgebanks on the better soils. Common in Gower and the Vale but absent from parts of the Uplands. Fl. 4–7, 10, fr. 5–8, 10. Map 115.

MINUARTIA L.

Minuartia hybrida (Vill.) Schischkin   Fine-leaved Sandwort. Tywodlys Deilfain.
Native. Thin soils near the sea, walls, gravel paths, railway ballast, waste ground, etc. An inconspicuous and easily overlooked plant; always scarce, with only one recent record in the county.
00C, Aberdare, BAW c.1907; 07B, Peterston, HJR c.1907; 17A, St. Fagans, EV 1918*, near St. Brides-super-Ely, 1972!; 17B, Cardiff, JS c.1886; 87A, Porthcawl, J.B. Lloyd c.1907; 87D, Southerndown, WWB c.1907; 90B, Hirwaun, BAW c.1907.

HONKENYA Ehrh.

Honkenya peploides (L.) Ehrh.   Sea Sandwort. Tywodlys Arfor.
Native. Near the drift-line on sandy beaches, in sandy salt-marshes and on low fore-dunes. Common in suitable habitats on the western coasts of Glamorgan. (!) Fl. 6–7, fr. 6–8; ?Dips (n,p) rarely, ?Col rarely. Map 116.

STELLARIA L.

Stellaria media (L.) Vill.   Common Chickweed. Brechlys, Gwlydd y Cywion.
Native. Cultivated ground, farmyards, waste places, etc. and also in or near seabird nesting colonies on the coast. Common everywhere in suitable habitats. Fl. 1–12, fr. 5–11. Map 117.

Stellaria neglecta Weihe   Greater Chickweed. Brechlys Mwyaf.
Native. Woods, hedges and shady places. Rather local in Gower and the Vale, absent from most of the Uplands; very abundant in some woodland localities, for example Cwm Ivy Wood (49C), where it is dominant in the field layer. Fl. 3–5, fr. 4–5. Map 118.

Stellaria pallida (Dumort.) Piré   Lesser Chickweed. Gwlydd y Tywod.
Native. Sand-dunes and shallow soil on sea-cliffs, rather local. Perhaps overlooked in some localities. Recent records from 48B, Port Eynon; 49C, Whiteford Burrows; 87B, Merthyr Mawr; 88C, Kenfig Burrows. Fl. 5, fr. 5–6. Map 119.

Stellaria holostea L.   Greater Stitchwort. Serenllys Mawr.
Native. Hedges and woods, fairly common in most of the county but rather patchy in its distribution and absent from some areas. Fl. 4–6, fr. 5–7; (Syr) Rh.cam, (Dip) Em. tess, Dips (small). Map 120.

Stellaria alsine Grimm   Bog Stitchwort. Serenllys y Gors, Tafod yr Edn y Gors.
Native. Marshes, ditches, flushes and damp grassland. Very common in the Uplands and most of Gower but scarce in parts of the Vale. Fl. 6–8, fr. 7–9. Map 121.

Stellaria graminea L.   Lesser Stitchwort. Serenllys Gwelltog, Tafod yr Edn Lleiaf.
Native. Hedgebanks, grassland and grassy heaths. Common in Gower and most of the Uplands but scarce in the southern Vale. Fl. 5–11, fr. 7–11; (Dip) Luc (n), Sa.car (n), Dips (small). Map 122.

CERASTIUM L.

Cerastium tomentosum L. agg.   Snow-in-summer. Clust Llygoden y Felin.
Introduced; a garden escape. Walls, roadside banks and waste places, now fairly common in eastern Gower and parts of the Vale and northern Uplands. Some records may refer to **C. biebersteinii** DC. Map 123.

Cerastium arvense L., Field Mouse-ear, Clust Llygoden y Caeau, a rare casual of waste ground near docks, has been recorded from two localities (16A, Barry, RLS c.1925, Barry Docks, EV 1935*; 17B, Cardiff, THT 1902*).

Cerastium fontanum Baumg. subsp. **glabrescens** (G.F.W. Meyer) Salman et al. (C. holosteoides Fries). Common Mouse-ear. Clust Llygoden Culddail.
Native. Grassland and waste ground. Very common everywhere. Reported from all squares with the single exception of 49B. Fl. 3–11, fr. 6–11. Map 124.

Cerastium glomeratum Thuill.   Sticky Mouse-ear. Clust Llygoden Llydanddail.
Native. Disturbed grassland, cultivated land, dunes and shallow soils on sea-cliffs. Fairly common in most parts of Gower and the lowlands, but scarce or local in the Uplands. Fl. 4–6, 9–10; fr. 4–7, 9–10. Map 125.

Cerastium semidecandrum L.   Little Mouse-ear. Clust Llygoden Bach.

Native. Sand-dunes and shallow soil over limestone, always associated with other winter annuals. Local and mainly near the coast, but also on limestone at a few inland sites. Fl. 4–5, fr. 5–6, 8. Map 126.

Cerastium diffusum Pers. subsp. **diffusum** (C. atrovirens Bab.). Sea Mouse-ear. Clust Lygoden Arfor.
Native. Sand-dunes and shallow soil on sea-cliffs, occasionally inland on shallow soil over limestone or on railway ballast. Common along the coast. Map 127.

Cerastium dichotomum L. has once been recorded as a casual (27A, RLS 1926*). [The record of **Cerastium pumilum** Curt. by E.F. Linton in 1885 (Bennett, 1905) is now thought to be an error.]

MYOSOTON Moench

Myosoton aquaticum (L.) Moench   Water Chickweed. Llinesg y Dŵr.
Native. Riversides and marshes. Locally frequent along the Ogwr (Ogmore) and Ewenni rivers, occasional, in the past, along the Ely and Taff. Flower & Lees (1842) recorded it as 'Frequent in watery places' (near Swansea). Map 128.

SAGINA L.

Sagina nodosa (L.) Fenzl   Knotted Pearlwort. Corwlyddyn Clymog.
Native. Dune-slacks, and occasionally on roadsides and damp open ground elsewhere, normally growing in open sites that are flooded in the winter but dry in the summer. It is apparently less common now in roadside sites in Gower than it was in the past, when many roads had rutted gravel surfaces. An abundant and characteristic species of dune-slacks in the western part of the county. Fl. 7–8, fr. 8. Map 129.

Sagina procumbens L.   Procumbent Pearlwort. Corwlyddyn Gorweddol.
Native. Open grassland, ant-hills, paths, between paving-stones, cultivated land, lawns, etc. Very common throughout the county. Fl. 5–10, fr. 5–11. Map 130.

Sagina apetala Ard. (including S. ciliata Fries).   Annual Pearlwort. Corwlyddyn Anaf-flodeuog.
Native. Bare soil, paths., walls, etc. A variable species in which a number of autogamous forms with minor differences were recognised as species in the past. **S. apetala** subsp. **apetala** (S. ciliata Fries) is apparently the common form in western Glamorgan, where it grows on paths and between paving-stones as well as in open natural habitats, while **S. apetala** subsp. **erecta** (Hornem.) F. Hermann (S. apetala auct.) has been more commonly recorded in eastern Glamorgan. Fl. 7–8, 10, fr. 7–8, 10. Map 131.

Sagina apetala subsp **apetala** Map 132.

Sagina apetala subsp. **erecta** (Hornem.) F. Hermann Map 133.

Sagina maritima G. Don fil.   Sea Pearlwort. Corwlyddyn Arfor.
Native. Drier sites in upper salt-marshes (especially on ant-hills), on drift-lines and in the spray-zone on cliffs. Locally abundant in suitable habitats; but easily overlooked. Fl. 6, fr. 7. Map 134.

SCLERANTHUS L.

Scleranthus annuus L.   Annual Knawel. Dinodd Blynyddol.
Casual, on waste ground; only one recent record. Map 135.

[**Scleranthus perennis** L., Perennial Knawel, Dinodd Parhaol, recorded by J. Storrie from ballast, Cardiff Docks (17B), c.1876 (Storrie, 1886) is now regarded as dubious.]

Three species of **Herniaria** have been found as aliens at Cardiff or Barry Docks. **Herniaria glabra** L. was found at Barry Docks by R.L. Smith in 1923 (16A) and had been previously found at Cardiff Docks by J. Storrie in c.1870 (27A). The record for Worm's Head (48A) by W. Turton (Oldisworth, 1802) is an error . Smith also found **H. ciliolata** Melderis at Barry Docks in 1925 (16A), this species had been previously found at Cardiff by Storrie (17B). **H. hirsuta** L. has been found only at Cardiff, by Storrie in 1876 and Smith in 1927 (27A).

Corrigiola litoralis L. was found as a casual at Barry Docks (16A) by both E. Vachell and R.L. Smith in 1927*, **Paronychia argentea** Lam. at Cardiff (27A) in 1926!*, **Illecebrum verticillatum** L. at Penarth (17D) by J. Storrie in 1881*, and **Polycarpon tetraphyllum** (L.) L. at Barry Docks (16A) by J.H. Salter in 1929* and at Cardiff Docks (27A) by J. Storrie in c.1876*. [The record for this latter species from 'between Pyle and the sea' by W. Turton (Oldisworth, 1802, Turner & Dillwyn, 1805) is an error.]

SPERGULA L.

Spergula arvensis L.   Corn Spurrey. Troellig yr Ŷd.
Native or introduced. A weed of arable land on acid or sandy soils. Common in Gower and the south-western fringes of the Uplands, but absent from the limestone soils of the central and southern Vale. Fl. 3, 5–11, fr. 6–11. Map 136.

SPERGULARIA (Pers.) J. & C. Presl

Spergularia rupicola Lebel ex Le Jolis   Rock Sea-spurrey. Troellys y Morgreigiau.
Native. In the spray-zone on exposed cliffs. Fairly abundant on Worm's Head but scarce elsewhere in Gower; once recorded from Porthcawl. (!) Fl. 6–8, fr. 7–9.

38B, Worm's Head, EMT 1934, 1979!; 48A, Mewslade, 1970!, 1984; Horse Cliff, 1975!; 48D, Port Eynon Point, 1974!; 49C, Burry Holms, HJR c.1907; 58B, Bacon Hole, JAW 1923*; 68A, Limeslade, MG 1975; 87A, Porthcawl, EV 1929.

Spergularia media (L.) C. Presl   Greater Sea-spurrey. Troellys Mawr.
Native. Salt-marshes. Fairly abundant at all levels in salt-marshes. (!) Fl. 6–8, fr. 6–9; Bb. Map 137.

Spergularia marina (L.) Griseb.   Lesser Sea-spurrey. Troellys Bach.
Native. Upper salt-marshes, in short turf or on bare ground near the drift-line. Much less abundant than S. media, but present in most salt-marshes. Fl. 6–7, 10, fr. 6–7, 10. Map 138.

Spergularia rubra (L.) J. & C. Presl   Sand Spurrey. Troellys Coch.
Native or introduced. Waste ground, colliery spoil-tips, ash-tips near power stations. Locally frequent in suitable habitats in the Uplands. Map 139.

Spergularia bocconii (Scheele) Ascherson & Graebner has twice been recorded as an alien (16A, Barry, EV 1927*; 27A, Cardiff, RLS 1925*).

LYCHNIS L.

Lychnis coronaria (L.) Desr.   Rose Campion.
Garden escape, naturalized in a few localities.
09D, Hopkinstown, on a wall, 1972!; 16B, Lavernock Quarry, MG 1975; 68A, Langland, JAW 1920; 69C, Hendrefoelan Wood, JAW 1920.

Lychnis flos-cuculi L.   Ragged-Robin. Carpiog y Gors.
Native. Marshes and damp grassland. Common in most of the county, but rather scarce in the southern Vale and parts of the Uplands where suitable habitats are infrequent. (!) Fl. 5–6, 10, fr. 6–7, 10. Map 140.

Lychnis chalcedonica L. has once been found as a garden escape in Cardiff (27A, 1953!*).

AGROSTEMMA L.

Agrostemma githago L.   Corncockle. Bulwg yr Ŷd.
Probably extinct. Once a fairly widespread weed of cornfields and disturbed roadsides. Last recorded in the Vale in c.1954 (08, Pontypridd, A. Jones). Map 141.

SILENE L.

Silene vulgaris (Moench) Garcke subsp. **vulgaris**   Bladder Campion. Gludlys Codrwth, Llys y Poer.
Native. Roadsides, rough grassland, edges of cultivated fields and waste ground. Locally common in the Vale, the Swansea district and parts of the Uplands, but not seen recently in western Gower, where Webb (1956b) noted that it had been common

in the 1920s but was much less frequent by the early 1950s. Fl. 6–7. Map 142.

Silene vulgaris (Moench) Garcke subsp. **maritima** (With.) Á. & D. Löve (*S. maritima* With.). Sea Campion. Gludlys Arfor.
Native. Sea-cliffs, shingle beaches, occasionally on dunes; also on copper and zinc slag-heaps in the Lower Swansea Valley, and on mountain cliffs at 450–500m in the Uplands. Rather local along the coast, and apparently extinct in several former sites on the coast of the Vale; most abundant on Worm's Head and other Gower headlands.
Inland records from: 09A, Mountain Ash, M. Davies 1963; 99A, Craig Fawr, *c.*500m, HJR *c.*1907; 99C, Bwlch y Clawdd, J. Williams 1935*. Map 143.

Silene vulgaris subsp. **maritima** × **S. vulgaris** subsp. **vulgaris** has been recorded once (79C, Jersey Marine, MG 1973).

Silene noctiflora L.   Night-flowering Catchfly. Gludlys Nos-flodeuol.
Introduced. Once fairly frequent in cornfields and on waste ground in the south-eastern Vale, but now very rare with only one recent record.
07C, St. Hilary, R. Nicholl *c.*1908; 07D, near Bonvilston, JIL 1971; 16A, Barry, 1925!*; 17A, St Fagans, EV 1935; 17C, Caerau, HAH 1923*; 17D, near Leckwith, 1925!*; 27A, Splott, Cardiff, 1925!*; 68A, Limeslade, JAW 1930s; 69C, Mayals and Blackpill beach, JAW 1930s.

Silene pratensis (Rafn) Godron & Gren. (*S. alba* (Miller) E.H.L. Krause). White Campion. Gludlys Gwyn.
Introduced. Roadsides, cultivated land, waste ground and colliery spoil-tips, common in the south-eastern Vale but scarce or local elsewhere. Always scarce in western Gower (Webb, 1956b) with few recent records there. Fl. 5–11, fr. 6–9. Map 144.

S. pratensis subsp. **divaricata** (Reichenb.) McNeill & Prentice, has twice been recorded as an alien (16A, Barry Docks, 1980!*; 27A, Cardiff, RLS 1927*).

Silene dioica (L.) Clairv.   Red Campion. Gludlys Coch, Blodyn Taranau.
Native. Woods, hedges, roadsides, waste ground, etc. Very common in most of the county, but rather local in parts of the eastern Uplands. Fl. 1, 4–12 (male), 1, 5–10, 12 (female), fr. 6–10; B.hort (n), B.lap (n), B.luc (n–r), B.pasc (n), B.prat (p), B.terr (n–r), (!) Hb (n–r, p) (Syr) Rh.cam (n,p), S.balt (p) (Lep) Pi.bra. Map 145.

Hybrids with **Silene pratensis** occur in places where the species meet (for example, the western end of Crymlyn Burrows, 79C) but most Glamorgan populations of **S. dioica** appear to be unaffected by hybridization. Map 146.

Silene dichotoma Ehrh.   Forked Catchfly. Gludlys Fforchog.

Casual, on waste ground, usually near the sea; no recent records.
00C, Aberdare, HJR *c.*1907; 17B, Cardiff Docks, HJR 1902; 17D, Cardiff, EV 1933*; 27A, Splott, RLS & EV 1936*; 79C, Jersey Marine, RLS & EV 1936*, JAW 1939*; 87A, Porthcawl, HJR *c.*1907.

Silene gallica L.   Small-flowered Catchfly. Gludlys Amryliw.
Native or introduced. Arable land, waste ground, colliery spoil-tips and disturbed sand-dunes. Locally common to the west of Swansea but rare elsewhere. Map 147.

Silene gallica is a polymorphic complex often split into three 'series', the most widespread of which is var. **anglica** (L.) Mert. & Koch, with recent records from 07D, near Bonvilston, JIL 1972*; 49C, fields at Llangennith and Llanmadoc, 1975!; 49D, Landimore, 1975!; 50D, colliery tip at Grovesend, 1974!; 59C, fields south of Blue Anchor, 1975!; 60C, roadside near Pontarddulais, 1974!; 68A, fields at Thistleboon, 1973!; 69C, sand-dunes at Blackpill, 1974!; 79C, Jersey Marine, H & CH 1974!.

Silene gallica var. **sylvestris** (Schott) Ascherson & Graebner (?var. *gallica*), an infrequent casual of waste ground near docks in the past has been recorded only twice recently (16A, EV 1932*; 17B, RLS 1925*; 17D, JS *c.*1876*; 18C, MG 1977; 27A, EV 1926*; 49C, 1971!), and **S. gallica** var. **quinquevulnera** (L.) Boiss., a rare casual, has been recorded only twice (50D, JM *c.*1840; 69C, JAW *c.*1926).

Several species of **Silene** have been found as aliens in two localities: **Silene armeria** L. (70B, Ystalyfera, JAW 1919; 79D Cwmafon, J. Motley *c.*1840), **S. conica** L. (16A, Barry Docks, RLS *c.*1925*; 78B, Port Talbot, HJR 1904 & 1906), **S. csereii** Baumg. (00C, Aberdare, HJR *c.*1910; 78B, Port Talbot Docks, HJR *c.*1910), **S. muscipula** L. (17B, Cardiff, 1920!*; 27A, Cardiff, RLS 1926 & 1936!*). Those found only once at Cardiff or Barry Docks are: **S. antirrhina** L. (27A, Cardiff, RLS 1938*), **S. behen** L. (27A, Cardiff, RLS 1926), **S. conoidea** L. (27A, Cardiff, RLS 1926*), **S. nutans** L. (16A, Barry Docks, RLS *c.*1925), **S. otites** (L.) Wibel (17B, Cardiff, JS 1876), and **S. coeli-rosa** (L.) Godron (27A, Cardiff, RLS 1937!*).

Two species of **Gypsophila** L. have been recorded as aliens in the past.
Gypsophila paniculata L. from two localities (17B, HJR *c.*1906; 69C, JAW *c.*1921) and **G. elegans** Bieb. from one (27A, 1927!*).

SAPONARIA L.

Saponaria officinalis L.   Soapwort. Sebonllys.
Introduced. Roadsides and waste ground. Common in urban areas in the lowlands and in parts of the Uplands. Already abundant on Singleton Burrows (69C) in 1802 (Dillwyn, 1840). Fl. 7–8. Map 148.

Saponaria ocymoides L. has recently been recorded as a spontaneous weed in a garden at Pencoed (98D, H.J. Dawson 1983*).

VACCARIA Medicus

Vaccaria hispanica (Miller) Räuschert (*V. pyramidata* Medicus), Cowherb, has occasionally been recorded as an alien on waste ground, usually near docks, but with no recent records (00C, HJR *c.*1906; 16A, AHT 1890–1892, RLS 1925; 17B, 1921!*; 17D, JS 1876*; 27A, RLS 1925; 78B, HJR *c.*1906; 90C, BAW *c.*1905).

PETRORHAGIA (Ser. ex DC.) Link

Petrorhagia nanteuilii (Burnat) P.W. Ball & Heywood   Childing Pink.
Introduced. Apparently established in Cardiff Docks (27A) where it was first recorded by R.L. Smith in 1926* and rediscovered by J.P. Curtis in 1980; and also on a grassy bank in Cwm Dare, Aberdare (90D), H.J. Dawson 1985*, where it was possibly introduced when the area was re-seeded.

DIANTHUS L.

Dianthus caryophyllus L. Clove Pink. Penigan Rhuddgoch.
Introduced; formerly naturalized on the walls of Cardiff Castle, once recorded as a naturalized garden escape near Swansea, and recently as a garden weed near Bridgend.
17B, Cardiff Castle, WT *c.*1800 (extinct by 1885, JS); 79C, shingle in Crymlyn Bay, JAW 1939*; 88D, Pen-y-fai, Bridgend, T.H. Sheldon 1978*.

Dianthus deltoides L. Maiden Pink. Penigan Gwyryfaidd.
Probably introduced. Recently recorded in only one site, where it maintains a precarious existence.
17B, canal bank, Cardiff, T. Chapman *c.*1907; 18D, railway bank, Llanishen, JS *c.*1886; 69C, Mayals (adventive) JAW 1925–27*; 70C, Mynydd Gellionen, M. Lewis 1953*; 78D, Kenfig Burrows, W. Nelson 1980, 1984!.

Dianthus armeria L. Deptford Pink. Penigan y Porfeydd.
Probably native. Roadsides, dry fields, waste ground. Rare. Occasionally recorded as a probable native in western Glamorgan, with one recent record; once reported as a ballast alien in Cardiff.
17B, Cardiff, THT, 1876*; 59B, disturbed roadside at Kingsbridge (not associated with introduced species) A.J. Lack 1980**; 79C, banks around Briton Ferry, TBF *c.*1843; 79D, Cwmafan, JM 1841**; 87B, dry gravelly field near Merthyr Mawr, HOB 1910*.

Dianthus plumarius L. has been recorded as an alien from Briton Ferry (79C), by H & CH (1973).

NYMPHAEACEAE
NYMPHAEA L.

Nymphaea alba L. White Water-lily. Lili-Ddŵr Wen.
Native in a few localities, but usually introduced. Lakes and canals. Probably native in 58A, Oxwich Marsh, 1973!; 79A & 79C, Crymlyn Bog, 1971! (first recorded there in 1773); perhaps native in 78D, Kenfig Pool. Map 149.

NUPHAR Sm.

Nuphar lutea (L.) Sibth. & Sm. Yellow Water-lily. Lili-Ddŵr Felen.
Probably native in one or two localities, but usually introduced. Marshy pools, lakes, ponds and canals. Probably native in 79A & 79C, Crymlyn Bog, but recently recorded only from the Neath Canal in this area (H & CH 1976, 1980!). Sub-fossil pollen of **N. lutea** has been found at a depth of 130–200cm in peat filling a depression on Cefn Bryn south of Broad Pool in 59C (Q.O.N. Kay, unpublished). Map 150.

CERATOPHYLLACEAE
CERATOPHYLLUM L.

Ceratophyllum demersum L. Rigid Hornwort. Cyrnddail.
Native. Lakes and ponds. Recorded from a few localities in the Vale; perhaps overlooked in other sites.
07A, Hensol, EV 1926*; 07B, St-y-Nyll, BS 1964*; 78B, Kenfig Pool, EV 1936, BS 1964*; Morfa Pools, EV 1935* (habitat now destroyed); 87D, St. Bride's Major, G.G.H. Miller 1964; 88A, Margam Park, MG 1974.

Ceratophyllum submersum L. Soft Hornwort. Cyrnddail Trifforch.
Native. Shallow lakes, ponds and canals. Recently found to be abundant in two localities in western Glamorgan where it was previously unrecorded; perhaps overlooked elsewhere.
58A, Oxwich, 1974!**, 1988!; 79A, Crymlyn Bog, 1972!; 79C, Neath canal, H & CH 1972**.

RANUNCULACEAE
HELLEBORUS L.

Helleborus foetidus L. Stinking Hellebore. Palf yr Arth Ddrewedig, Llewyg y Llyngyr.
Probably native in limestone woodland, cliff scrub and hedges in a few sites in Gower and the Vale; now rare, with recent records from only two localities, one south-west of Parkmill (58A), and the other near Dunraven (87D). It was formerly used as a medicinal plant, and was probably a garden escape in some sites. It was first recorded in the county in *c.*1803 by Dr W. Turton, who found it between Parkmill and Pennard Castle; J.W.G. Gutch found it in great abundance there in *c.*1840, but it is now very scarce in

the Parkmill area although much of the woodland must have changed little since Turton's day. Individual plants of **H. foetidus** are long-lived but seedlings are rare and it may, as J.A. Webb has suggested, have been eradicated or much reduced in this and other localities by over-collecting.
00A, 'on the hill by the tunnel' at Merthyr Tydfil, BAW *c.*1907; 06B, copse E. of Kenstone Farm House, J. Evans 1831, hedge near St. Hilary, EV 1931*, roadside verge near Penmark, J.G. Rees 1952; 16B, Sully House Copse, C.T. Vachell 1876*, Swanbridge, AHT 1892*; 17A, St Fagans, M. Howell 1906; 58A, Parkmill, WT *c.*1803, EV 1910*, JAW 1941*, Crawley Cliff, JAW 1939*, 1982!; 58B, Ilston Cwm, JAW 1948*, Bishopston Valley, D.G. Twomey 1954*; 87D, Witch's Point, Dunraven, JS *c.*1886, MG 1974.

Helleborus viridis L. subsp. **occidentalis** (Reuter) Schiffner   Green Hellebore. Pelydr Gwyrdd, Crafanc yr Arth.
Probably introduced; hedgebanks and woodland on limestone near Oxwich, where it was first recorded in 1925.
48B, near Oxwich, JAW 1925, lane from Penrice Castle to Oxwich Bay, C.M. Griffiths 1929, V.M. Jenkins 1935*; 58A, Oxwich Wood, E. Fenton 1951, JBe 1983!.

Eranthis hyemalis (L.) Salisb., Winter Aconite, Bleidd-dag y Gaeaf, sometimes persists for a while in abandoned gardens or as a naturalized garden escape; it was recorded from a site near Widegate in Gower (58B) by J.A. Webb from 1941 to 1949; Caerau near Cardiff (17A) by HJR in *c.*1900 and Llansannor (97B) in 1980!

Nigella damascena L., Love-in-a-mist, used to grow as a scarce dockside alien at Cardiff and it is sometimes found as an impersistent garden escape; it was found at Cardiff (27A) and Sully (16B) by John Storrie in *c.*1876, and has more recently been found at Dinas Powis (17C, GMB 1971), Llandaff North, Cardiff (17B, R.J. Driscoll 1970) and Langland Cliff (58B, JAW *c.*1956).

TROLLIUS L.

Trollius europaeus L.   Globeflower. Cronnell.
Native. Streamsides and marshy grassland; now rare, with recent records only from a few remote sites in the northern Uplands, but more widespread in the past. In the mid-nineteenth century, according to John Storrie (1886), it was frequent in the Ely, Taff and Rhymney valleys. **T. europaeus** reaches the southern limit of its British range in Glamorgan. Its decline in the county may result partly from the abandoning of many small upland farms and the consequent neglect or disappearance of the small streamside pastures and hay-meadows in which **T. europaeus** grew, and partly from the contamination of streamside habitats by colliery waste. (!) Fl. 6, fr. 7. Map 151.

CALTHA L.

Caltha palustris L.   Marsh-marigold, Kingcup. Melyn y Gors.
Native. Marshes, streamsides and damp meadows. Widely distributed but rather local, and usually absent from base-poor soils. Fl. 3–6, fr. 5–6; ?Syr. Map 152.

ACONITUM L.

Aconitum napellus L.   Monk's-hood. Cwcwll y Mynach.
Native in eastern Glamorgan, where it is locally abundant on streamsides in the Ely, Taff and Rhymney valleys. It also grows at a few sites in the south and west of the county, but usually either as an obvious garden escape or as a probable introduction. Fl. 7. Map 153.

Three species of **Consolida** (DC.) S.F. Gray, have been recorded as rare casuals on waste ground or as garden weeds, in the past probably introduced with ballast or grain cleanings, more recently introduced with bird-seed or as garden escapes. **Consolida ambigua** (L.) P.W. Ball & Heywood, Larkspur, Llyshedydd, has been found most frequently, with records in the past from Cardiff (17B, HJR *c.*1906, R. Davies 1948*; 27A, RLS 1927*), Radyr, (18C, HJR *c.*1906), Merthyr (00A, G.R. Willan 1910*), Port Talbot Docks (78B, HJR *c.*1906) and Porthcawl (87A, EMT 1936), but with only one recent record, from a rubbish tip near Blue Anchor (59B, 1974!). **C. regalis** S.F.Gray has been found four times, most recently in 1974 (17B, Cardiff, JS *c.*1876; 27A, waste ground in Splott, RLS 1949*; 69C, sandy foreshore at Swansea, E. Lees *c.*1849; 79D, waste ground near Briton Ferry, H & CH 1974!), and **C. pubescens** (DC.) Soó has been recorded once (27A, Splott, 1926!*).

ANEMONE L.

Anemone nemorosa L.   Wood Anemone. Blodyn y Gwynt.
Native. Common in woods and hedgebanks in Gower and the Vale, but rather local in the Uplands where it is found only on the better soils. Sometimes found on high cliff-ledges in the hills. Fl. 3–5. Map 154.

Four species of **Anemone** have been found growing as garden escapes or introductions, but only one, **Anemome apennina** L., Blue Anemone, has persisted, growing at Fairy Hill (49D) for at least fifty years after its introduction (JAW *c.*1956); it has also been recorded as an introduction at Penllergaer (69A, JM *c.*1840**) and Sketty (69C, C. Begg pre 1930). **A. ranunculoides** L. is another of the plants that was introduced to L.W. Dillwyn's estate at Penllergaer, but apparently has not persisted (69A, JM *c.*1840**). **A. coronaria** L. was found at Wenvoe by C.T. Vachell in *c.*1890 (17C*), and **A.** × **hybrida** Paxton (*A. japonica* auct.) at Wenvoe Woods (17C), HJR *c.*1900; West Cross (68A), JAW 1940 and Mayals (69C), JAW 1924.

CLEMATIS L.

Clematis vitalba L. Traveller's-joy, Old Man's Beard. Barf yr Hen Ŵr, Cudd y Coed.
Native. Locally common in hedgerows and at the edges of woods on calcareous soils in Gower and the Vale; sometimes found on semi-fixed calcareous dunes, for example at Whiteford Burrows (49A, 49C) where it is locally abundant. Very scarce in the Uplands. Fl. 7–9, fr. 8–11. Map 155.

Clematis montana Buch.-Ham. ex DC. has been recorded once as established on trees, Brynau, Gower (59D) by JAW in c.1956.

Two species of **Adonis** have been found as rare weeds of cultivated land or waste ground in the past. **Adonis annua** L., Pheasant's-eye, was found in five places (17A, Coedrhiglan, Anon c.1890; 17B, Roath Park, Cardiff, EV 1902*; 17D, Penarth, JS c.1886; 78C, near Port Talbot, WAS 1901*; 88D, near Tondu, JS c.1886), **A. aestivalis** L. was found only once (27A, Splott, RLS 1927!*).

RANUNCULUS L.

Ranunculus repens L. Creeping Buttercup. Crafanc y Frân
Native. Very common throughout the county in damp grassland, in damp woodland, and as a weed of arable land, gardens and waste ground on poorly drained or heavy soils; also found in dune-slacks. Fl. 5–12, fr. 5–10; Hb, (Syr) E.ten (n), Rh.cam (n), S.balt (n), S.rib, Sy.pip (p), (Dip) Musca (2 spp)., (Lep) Ma.jur (many). Map 156.

Ranunculus acris L. Meadow Buttercup. Blodyn Ymenyn, Crafanc yr Frân Syth.
Native. Very common in meadows and grassland throughout the county, on all but the poorest or most waterlogged soils. Fl. 5–11, fr. 7–10; (Syr) E.ten, S.rib (Dip) Luc, ?Dips (small) (Lep) Pi.nap, Po.c-al. Map 157.

Ranunculus bulbosus L. Bulbous Buttercup. Chwys Mair.
Native. Common in limestone grassland and on calcareous fixed dunes in Gower and the Vale, but only locally frequent on neutral and mildly acid soils inland and in the Uplands, where it is confined to well-drained pastures and grassy roadsides. Fl. 4–6, fr. 6–7; (Syr) Rh.cam, (Dip) Musca sp., ?Dips. Map 158.

Ranunculus sardous Crantz Hairy Buttercup. Crafanc y Frân Blewog.
Native. Very local in marshy grassland near the sea, and as a rare casual on waste ground. It has been seen recently in only two native localities, near Aberthaw (06A) and East Cardiff (27A) but may still grow in other sites.
06A, banks of R. Thaw near Flemingston, AHT 1905*, marsh by R. Kenson near Aberthaw, 1970!; 16A, ballast at Barry, AHT c.1890; 17B, Cardiff Docks, HJR

c.1907; 17D, marsh at Llandough, EV 1912*, 1920!; 27A, Splott, 1931!*, wet pastures near R. Rhymney, Cardiff, MG 1982, damp waste ground, Howardian Nature Reserve, Cardiff, GH & JPC 1986; 58A, Oxwich Marsh, EV 1918; 69C, marshy field between Swansea and St. Helen's, LWD c.1805.

Ranunculus muricatus L. has once been recorded as a non-persistent alien on sand ballast at Radyr (18C), RLS c.1924!.

Ranunculus arvensis L. Corn Buttercup. Crafanc yr Ŷd.
Introduced. A scarce weed of cornfields and waste ground, with no recent records; the last records in the county appear to be from Barry (16B, abundant in a cornfield, EV 1936) and Swansea (69C, JAW c.1950). It seems to have been most frequent as a cornfield weed in south-eastern Gower, where T.B. Flower and E. Lees (1842) found it in cornfields at Caswell Bay (58B) and Mumbles (68A) in 1843, and in the south-eastern Vale near Barry and Dinas Powis (17C, EV 1928). Map 159.

Ranunculus parviflorus L. Small-flowered Buttercup. Crafanc y Frân Manflodeuog.
Native. Very local, growing on shallow soils over limestone near the sea in Gower and the Vale. It fluctuates in abundance from year to year but is becoming increasingly scarce as a consequence of increased disturbance and trampling of its sea-cliff habitats, especially in Gower. Map 160.

Ranunculus auricomus L. Goldilocks Buttercup. Peneuraidd, Crafanc y Frân Peneuraidd.
Native. Locally frequent in woodlands in the Vale; rare in Gower, where it was first found in 1981 (58A, Green Cwm, A.J. Lack 1981**) and in the Uplands (80B, near Pontneddfechan, 1969!). Map 161.

Ranunculus sceleratus L. Celery-leaved Buttercup. Crafanc yr Eryr.
Native. Locally frequent on brackish mud in the transition zone above estuarine salt-marshes in western and south-eastern Glamorgan, and occasionally found in muddy streamside and pondside sites inland. (!) Fl. 5–6, fr. 6–7. Map 162.

Ranunculus ficaria L. Lesser Celandine. Llygad Ebrill.
Native. Very common in woodland and hedgebanks, under bracken and on grassy heaths, and on streamside banks and hill-slopes. The distributions of the diploid subsp. **ficaria** (2n=16) and the tetraploid subsp. **bulbifer** Lawalrée (2n=32) in the west of the county were investigated by J.C. Read (1978). All 38 native populations that she sampled (from twelve 5km squares in Gower and 10 in the Uplands) were diploid; she found tetraploids only at Llanmadoc (49C) and Sketty (69C), growing as garden weeds in both cases. Tetraploids have also been found at Pontarddulais (50B), and bulbil-producing plants which may be tetraploid have been

reported from 17B, 60, 87, 88D, 96A and 97. Fl. (12, 1–)2–5, fr. 4–5; Hb (n,p), ?Sb, (Syr) E.ten (n,p), S.balt (p), S.rib, (Dip) Luc. caesar, Sc.ste, (Lep) Ag.urt, In.io. Map 163.

Ranunculus flammula L. Lesser Spearwort. Llafnlys Bach.
Native. Very common by streams and in marshes and damp meadows; unrecorded only in a few limestone areas near the sea. Fl. 5–10, fr. 7–10; (Dip) Sc.ste, Tabanus sp., ?Dip. Map 164.

Ranunculus lingua L. Greater Spearwort. Llafnlys Mawr.
Native. Locally abundant in Crymlyn Bog and in the Neath and Tennant Canals (69C, 79C), where it has been known since the early nineteenth century. It may be native in some other coastal sites but is probably introduced or of garden origin elsewhere. Fl. 5–8, fr. 6–8. Map 165.

Ranunculus hederaceus L. Ivy-leaved Crowfoot. Crafanc y Frân Eiddweddail.
Native. On mud and in shallow water in Gower and in northern Glamorgan, where it is the common small-flowered crowfoot of the lowlands and valleys; apparently scarce in the Vale. Fl. 5–8. Map 166.

Ranunculus omiophyllus Ten. Round-leaved Crowfoot. Egyllt y Rhosdir.
Native. In similar sites to **R. hederaceus**; most abundant in the hills of the Uplands, where it replaces **R. hederaceus**, but absent from western Gower and much of the Vale. Fl. 6–8. Map 167.

The putative hybrid **Ranunculus omiophyllus** × **tripartitus** was reported from Rhosili (48A) by HJR in c.1906.

Ranunculus tripartitus DC. Three-lobed Crowfoot. Crafanc Trillob.
Native. Ditches and ponds near Rhosili, where it is abundant in a few suitable sites during wet summers, but scarce in dry years. It is a south-western species which has been lost from many sites in Britain as a result of drainage of its habitats, and is at the limit of its British range in Gower, where it is decreasing as a result of changes in agriculture and the disappearance of the ponds and ditches in which it grows. First recorded by H.J. Riddelsdell before 1907, on Rhosili Down 'in some quantity'.
48A, Rhosili, HJR c.1907, 1910*; Pitton, Pilton Green and Rhosili Down, 1972–1976!**; 49D, Tankeylake Pond, JAW c.1940 (extinct by 1956?); 59D, Clyne Common, JAW 1944*. There is an unconfirmed 1960s record from marshes near Port Talbot (78B).

Ranunculus baudotii Godron Brackish Watercrowfoot. Egyllt y Mordir.
Native. Rather scarce, growing in brackish water in pools and ditches on the fringes of salt-marshes on the coast of Gower and the Vale. Map 168.

Ranunculus aquatilis L. s.l. (incl. **R. aquatilis** L. s.s., **R. peltatus** Schrank and **R. penicillatus** (Dumort.) Bab.). The map of the aggregate is included as not all records have been split into the segregate species. Map 169.

Ranunculus peltatus Schrank Pond Water-crowfoot. Crafanc y Llyn.
Native. Locally frequent in ponds in western Gower and the southern Vale; probably scarce or absent elsewhere, and apparently replaced by **R. penicillatus**, with which it has been confused in the past, in the Uplands. Fl. 5–7, fr. 6–7, (!) ?Dips. Map 170.

Ranunculus penicillatus (Dumort.) Bab. subsp. **penicillatus** Stream Water-crowfoot. Crafanc y Nant.
Native. Locally frequent in rivers and streams, especially in the Uplands where the other species of large-flowered water-crowfoots are scarce or absent. In lowland sites it is usually found in or near rivers flowing from the Uplands, for example along the Ely (07B, Peterston, 1970!; 17A, St. Fagan's, EV 1930*, 1972!) but at Llanrhidian in Gower there is an isolated population growing in the mill-stream and millpond, which were constructed in the late seventeenth century and are fed by a spring flowing from the Gower limestone (49D, 1974!, B.J. Ackers 1978**). Map 171.

Ranunculus penicillatus subsp. **pseudofluitans** (Syme) S. Webster (R. aquatilis subsp. peltatus var. pseudofluitans Syme; R. calcareus Butcher; R. penicillatus var. calcareus Cook).
Native. Rare. Recorded from only one stream in Cardiff.
17B, Whitchurch, GH 1982*; 18D, Whitchurch, GH 1986.

Ranunculus aquatilis L. s.s. Common Water-crowfoot. Crafanc y Dŵr.
Native. Probably rather scarce, growing mainly or exclusively in the Vale. It has been recorded from streams and ponds in sites scattered through the county, more frequently in the Vale than elsewhere, but it has often been recorded in error as a result of confusion with **R. peltatus**, which appears to be the commonest species of the **R. aquatilis** aggregate in ponds in the lowlands, and with **R. penicillatus**, which is apparently the commonest species of the aggregate in flowing water, especially in the Uplands, (Ackers, 1978). Map 172.

Ranunculus trichophyllus Chaix Thread-leaved Water-crowfoot. Egyllt Dail Edafaidd.
Native. Rather scarce in Gower and the southern Vale, growing in shallow ponds and ditches; rare or absent elsewhere. Map 173.

Ranunculus circinatus Sibth. Fan-leaved Watercrowfoot. Egyllt Cylchol-ddail.
Native. Apparently very local, with recent records only from Kenfig Pool, the middle Ely valley and the

Barry area. Past records from the Neath area (79C, TBF & E. Lees 1842) and the upper Neath canal (80B & 80C, HJR c.1907 may have been erroneous identifications of **R. penicillatus**, and no specimen can be traced for J.A. Webb's record from an outlying locality near the Burry Pill, which was probably made before 1930 but not published until 1956 (Webb, 1956b).

07B, St.-y-Nyll, HJR 1909, 1921!*, MG, 1975; 08D, Miskin, MG 1969; 16A, Barry, M. Wade 1981; 17A, St. Fagan's, EV 1941; 27A, ditches, Cardiff, JS c.1876; 49D, ditch near Burry Pill, JAW c.1956, conf. GCD; 78D, Kenfig Pool, HJR c.1907, EV 1929, 1970!; 87D, Southendown, C.T & E. Vachell c.1905*.

Ranunculus fluitans Lam., River Water-crowfoot, Crafanc Hirddail, was recorded from a canal in Cardiff (17B) by JS in 1876. The specimen in **NMW** was examined by R.W. Butcher in 1952 who commented 'I think **R. fluitans** Lam.'

AQUILEGIA L.

Aquilegia vulgaris L. Columbine. Blodau'r Sipsi, Madwysg.
Native in woodland and scrub on calcareous soils in Gower and the Vale, but also occurring in hedgerows and on roadsides and river-banks as a garden escape. Native populations appear to be uniformly blue-flowered, for example the large population growing in cliff-scrub at Seven Slades (58B) on the southern Gower coast, but hedgerow and roadside populations often have a mixture of flower colours. Fl. 5. Map 174.

THALICTRUM L.

Thalictrum minus L. Lesser Meadow-rue. Arianllys Bach.
Native or apparently native populations of this variable species grow in four distinct habitats in Glamorgan; mountain cliffs (subsp. **minus**), crevices of limestone rocks near the sea (subsp. **minus**), fixed sand-dunes (subsp. **arenarium** (Butcher) Clapham) and damp grassland (subsp. **majus** (Crantz) Clapham). It has also been found as a probable garden escape in several places scattered through the county. Map 175.

Native or apparently native plants have been recorded from the following localities:
Thalictrum minus subsp. **minus**, 48A, locally abundant in crevices of limestone pavement east of Ramsgrove, W. Weston 1975!, 1982!, above Paviland Slade, JAW 1932, 1965*, 1982!; 87A, Porthcawl, AHT c.1891*, 1905*, EV 1905, MG 1980; 90C, Craig-y-llyn, AHT 1892, 1905*, ARP 1973!; 99A, above Cwm parc, Graig Fawr, AHT 1898*, PJo 1984, Graig Fach, 1985!.
Thalictrum minus subsp. **arenarium** (Butcher) Clapham, 49C, locally abundant in the southern part of Whiteford Burrows at the foot of Cwm Ivy Tor, EV 1905, 1917, 1983!; 78D, Kenfig Burrows, EV & EMT 1944*, S. Moon 1981, 1983.

Thalictrum minus subsp. **majus** (Crantz) Rouy & Fouc., 48B, damp grassland near Oxwich Marsh, D.O. Elias 1977**.

Thalictrum flavum L. Common Meadow-rue. Arianllys.
Native. Rare, with two recent records which may both be introductions. It was found in the lower Ewenny valley (87B & 97A), where it may still grow, and in a marsh by the Roath Brook (27A) where it is probably extinct following drainage.
07A, Talygarn Lake, MG 1972; 17B, overgrown pond between Roath Park and Heath, Cardiff, ALe 1979; 27A, marsh by Roath Brook, 1950!*; 87B, near Ogmore Castle, HJR 1907*; 97A, Ewenny Bridge, EV c.1906, Southendown, EV & M.A. David 1899*, between Ewenny Bridge and Priory, Miss Rawlins c.1911. A 1958 BSBI mapping scheme record from 09 (Mountain Ash to Pontypridd) is, if correct, almost certainly an introduction.

PAEONIACEAE
PAEONIA L.

Paeonia mascula (L.) Miller Peony. Rhosyn Mynydd.
Introduced. Naturalized on Steep Holm (N. Somerset, v.c. 6) where it was first recorded by F.B. Wright in 1803, and still grows in small numbers. A single plant was recently found on Flat Holm (26, D. Worrall 1982), possibly a planted introduction from Steep Holm.

BERBERIDACEAE
BERBERIS L.

Berberis vulgaris L. Barberry. Eurdrain, Pren Melyn.
Introduced as a planted shrub in hedges, but not naturalized; now mainly confined to gardens. Map 176.

Two species of **Berberis** have been recorded as short-lived escapes from gardens: **Berberis glaucocarpa** Stapf (69C, Blackpill, H.R. Wakefield 1911), and **B. hookeri** Lemaire (87B, Merthyr Mawr, JAW 1943*).

MAHONIA Nutt.

Mahonia aquifolium (Pursh) Nutt. Oregon-grape.
Introduced. Planted in several localities, but hardly naturalized.
Recorded from 09A, MG 1974; 18C, MG 1974; 48B, JAW c.1945; 58A, JAW c.1945; 97A, 1973!.

LAURACEAE
LAURUS L.

Laurus nobilis L. Bay Laurel. Llawrwydden.
Introduced. Planted in several localities and naturalized on limestone slopes at Caswell Bay (58B!) and Mumbles (68A); naturalized at Caswell by 1924

(JAW). Also recorded from 09A, AMP 1975; 18C RAH 1972; and 87D, MG 1974.

PAPAVERACEAE

PAPAVER L.

Papaver somniferum L. Opium Poppy. Cwsglys, Llysiau'r Cwsg.
Introduced. Widespread and fairly frequent as a weed of waste ground and gardens. Map 177.

Papaver rhoeas L. Common Poppy. Pabi Coch.
Introduced as a weed of cultivation. Roadsides, waste ground and arable land; widespread but rather scarce except in parts of the Vale. Fl. 5–11, (!) fr. 7–10. Map 178.

Papaver dubium L. Long-headed Poppy. Pabi Hirben.
Introduced as a weed of cultivation. Cultivated land, waste ground and roadsides; more abundant than P. **rhoeas** in Gower. Fl. 5–11, fr. 7–11; (Syr) S.rib. Map 179.

Papaver lecoqii Lamotte Yellow-juiced Poppy. Pabi Sudd Melyn.
Introduced as a weed. Cultivated land and waste ground; locally frequent in the Vale and occasional to the north and east of Swansea, but apparently absent from most of Gower. Fl. 6–10, fr. 6–10; Hb, B.terr, (Syr) S.balt, S.rib. Map 180.

Papaver argemone L. Prickly Poppy. Pabi Gwry-chog.
Introduced. Cultivated land and waste ground. Formerly found as a cornfield weed near Swansea and in Gower, and as a dockside weed in the Cardiff area, but last recorded in 1927 in Gower (Llangennith, 49C) and in 1949 in Cardiff, 27A. Map 181.

Two species of **Papaver** have been recorded as rare casuals, one from the Cardiff area, **Papaver hybridum** L. (17B, JS 1876; 17D JS 1886; 27A, EV 1937) and the other from near Swansea, P. **orientale** L. (69C, JAW 1923).

MECONOPSIS Vig.

Meconopsis cambrica (L.) Vig. Welsh Poppy. Pabi Cymreig.
Native on stream banks, on cliffs and near waterfalls in the north of the county; an occasional garden escape elsewhere. Not seen recently in several former native localities, and scarce in others in which it was common in the nineteenth century. Fl. 4–11, fr. 6–11; Hb (p), B.luc, B.prat, ?Sb. (All introduced plants) Native or formerly native in 00A, Morlais Castle, G. Fleming c.1907; 61D, Cwm Aman, 1974!; 70D, Cwm Dulais, LWD c.1800, G. Traherne c.1907; 71C, Cwm Aman, 1974!; 80B, Afon Pyrddin, LWD, c.1800, 1969!*, 1978!; 90A, Nedd valley near Hirwaun, JS 1882. Probably a garden escape in 07B, 17A, 69A, and 80A (roadside in Seven Sisters, 1973!). Map 182.

Roemeria hybrida (L.) DC. has been found three times as a casual in the Cardiff area (17D, Penarth, JS c.1870; 27A, Splott, 1926!, T.J. Foggitt 1930, and **Argemone mexicana** L. twice (17B, Cardiff, JS c.1876, HJR 1909).

GLAUCIUM Miller

Glaucium flavum Crantz Yellow Horned-poppy. Pabi'r Môr, Pabi Corniog Melyn.
Native. Shingle beaches and similar seaside habitats, now scarce and surviving only in a few scattered localities, but formerly frequent along most of the coast. Fl. 6. Recent records from: 06A, Aberthaw; 16B, Sully; 48A, Ramsgrove; 48D, Port Eynon; 49B, Whiteford Point; 58A, Three Cliffs Bay; 58B, Pwll-du, Caswell Bay (undercliff); 59C, Llanmorlais (on cock-leshell tip); 77B, Sker. Map 183.

Glaucium corniculatum (L.) J.H. Rudolph, Red Horned-poppy, Pabi Corniog Coch, has been found on four occasions as a casual in Cardiff and Barry (16A, Barry Docks, 1925!; 17B, Canton, HJR 1907; 27A, Splott, 1925!, EV 1935*).

CHELIDONIUM L.

Chelidonium majus L. Greater Celandine. Dil-wydd.
Native or introduced. Roadsides and hedgebanks near houses; once grown as a medicinal plant. Fairly common in villages and near farmhouses in Gower and the Vale, but absent from the central Uplands. Fl. 5–11, fr. 6–9; (Syr) Rh.cam, Leu.luc. Map 184.

Eschscholtzia californica Cham., Californian Poppy, Pabi Califfornia, has been found as a casual on three occasions (17B, Maindy, Cardiff, RLS 1938*; 69B, Llansamlet, T. Davies 1985; 79C, Britton Ferry, H & CH 1973), as has **Hypecoum procumbens** L. (17B, Cardiff, JS c.1876, 1923!; 78B, Port Talbot Docks, HJR c.1906). **Hypecoum pendulum** L. (27A, Splott, Cardiff, RLS 1927*) and **Dicentra formosa** Walpers (17B, Cardiff, R. Davies 1951*) have been found once only.

CORYDALIS Vent.

Corydalis claviculata (L.) DC. Climbing Corydalis. Mwg y Ddaear Gafaelgar.
Native. Woods and rocky places on acid soils. Scarce and mainly confined to the eastern part of the county; old records from the western Vale and Gower may be erroneous. Recent records from 08B, Graig-fach near Pontypridd, JTa 1973*; 09B, above Ynysboeth, AMP 1971*; 19A, Nelson, 1973!*; 70A, Rhyd-y-fro, 1972!. Map 185.

Corydalis lutea (L.) DC. Yellow Corydalis. Mwg y Ddaear Melyn.
Introduced. Naturalized on old walls; frequent in the Vale. Fl. 5–11. Map 186.

Corydalis bulbosa (L.) DC. Hollow-root. Gwagwraidd.
Introduced. Naturalized in mixed woodland in one locality on Gower.
48B, Stouthall Woods, Reynoldston, JBe 1981*. (The specimen in **NMW** resembles subsp. **marschalliana** (Pallas) Chater.)

Corydalis solida (L.) Swartz, Bird-in-a-bush, Caledwraidd, has twice been recorded in the past (07D, Cottrell, JS 1892*; 69A, Penllergaer, JM c.1840).

FUMARIA L.

Fumaria capreolata L. White Ramping-fumitory. Mwg y Ddaear Afreolus.
Native. Hedgebanks and cultivated land. Frequent in southern and western Gower, apparently very local in the Vale. Many old records have been omitted from the map as they may refer to related species. (!) Fl. 5–11, fr. 6–11. Map 187.

Fumaria purpurea Pugsley Purple Rampingfumitory. Mwg y Ddaear Glasgoch.
Native. Cultivated land and hedgebanks near the coast. Scarce.
06A, Gileston, MG 1975; 07C, St. Hilary, RLS 1937*; 17B, Llandaff, EV 1930*; 17C, Dinas Powis, 1930!*; 49C, Whitford Burrows, 1971!; 58B, Caswell Bay, MG 1973; 78B, Port Talbot, HJR c.1906; 79D, Aberafan, JM c.1840.

Fumaria bastardii Boreau Tall Ramping-fumitory. Mwg y Ddaear Grymus.
Native. Cultivated land and waste ground. Locally frequent in western Glamorgan, rare in the east and north. Map 188.

Fumaria muralis Sonder ex Koch subsp. **boraei** (Jordan) Pugsley Common Ramping-fumitory. Mwg y Ddaear Amrywiol.
Native. Cultivated land, waste ground and hedgebanks. Scattered through the lowlands, mainly near the coast. Map 189.

Fumaria officinalis L. Common Fumitory. Mwg y Ddaear.
Native. Cultivated land and disturbed roadsides. Relatively scarce in Gower where it is less frequent than the other fumitories, but fairly common in the Vale. Fl. 3–12, fr. 6–11. Map 190.

Fumaria parviflora Lam., Fine-leaved Fumitory, Mwg y Ddaear Mânflodeuog, has twice been recorded as a dock casual (16A, Barry Docks, RLS c.1925; 17B, Cardiff, JS c.1886).

CRUCIFERAE (BRASSICACEAE)
SISYMBRIUM L.

Sisymbrium irio L. London-rocket. Berwr Caersalem.
Casual, usually near docks and with only one recent record.
00C, Aberdare, BAW c.1907; 16A, Barry Docks, RLS 1924!*; 17A, Coedrhiglan, G. Traherne & H. Evans c.1907; 17B, Cardiff, JS c.1886; Bute Dock, C.T. Vachell 1909*; 17C, Penarth, AHT c.1907; 27A, Splott, EV 1926!*, Roath, 1974!*; 69C, Swansea, HJR c.1907.

Sisymbrium altissimum L. Tall Rocket. Berwr Treigledigol.
Introduced, occasionally locally persistent on waste ground in the Cardiff and Neath districts. Map 191.

Sisymbrium orientale L. Eastern Rocket. Berwr Dwyreiniol.
Introduced, locally established on waste ground and roadsides. Most frequent in urban habitats in the Swansea districts. Map 192.

Sisymbrium officinale (L.) Scop. Hedge Mustard. Cedw'r Berth, Arfog Meddygol.
Native. Common on roadsides and waste places and as a weed of cultivated land and trampled pastures. Fl. 5–10, fr. 7–10; ?Sb. Map 193.

Five other species of **Sisymbrium** have been recorded as casuals or dock aliens in the Cardiff and Barry areas. **Sisymbrium loeselii** L. was found at Barry Docks (16A) by RM in 1923 and at Cardiff (17B) by GCD in 1924, **S. runcinatum** Lag. ex DC. at Splott (27A) in c.1925!, **S. austriacum** Jacq. at Roath Park (17B) in 1920!*, and **S. volgense** Bieb. ex E. Fourn. at Barry Docks (16A) by RM in 1930. [**S. polyceratium** L. has been recorded from 17B and 27A but these records are now regarded as doubtful.]

DESCURAINIA Webb & Berth.

Descurainia sophia (L.) Webb ex Prantl Flixweed. Berwr y Fam, Piblys.
Casual. Waste ground near docks and railways, once fairly frequent but now very rare, with only two recent records. Map 194.

ALLIARIA Scop.

Alliaria petiolata (Bieb.) Cavara & Grande Garlic Mustard. Garlleg y Berth.
Native. Roadsides, hedgebanks, and open woods, most frequent in disturbed sites on the better soils. Common in Gower and the Vale but absent from parts of the Uplands. Fl. 4–6 (8), fr. 6–7; (Syr) Rh.cam (Lep) (!) An.car. Map 195.

ARABIDOPSIS (DC.) Heynh.

Arabidopsis thaliana (L.) Heynh. Thale Cress. Berw'r Fagwyr.
Introduced. An increasingly common weed of cultivated ground, walls and waste land, locally established in open natural habitats. Fl. 4–6 (9–12,1), fr. 4–6 (9–12). Map 196.

Myagrum perfoliatum L. has twice been recorded as a dock alien in Cardiff (17B, Cardiff, HJR c.1907, RLS 1925!). A specimen of **Isatis tinctoria** L., Woad, Llysiau'r Lliw, is in **NMW**; it was collected from a cornfield (without locality) by W. Curtis in 1944. **Bunias orientalis** L., Warty-cabbage, has been recorded as a casual on four occasions in two localities (17D, Penarth, JS 1876*, 1932!; 27A, Cardiff, RLS c.1925, J.H. Salter 1928*). [The record of **Bunias erucago** L. from 17B is now believed to be an error.]

ERYSIMUM L.

Erysimum cheiranthoides L.   Treacle Mustard. Tri-agl-arfog.
Introduced. A rare but persistent weed of cultivated land and waste places, usually in urban habitats; most frequent in industrial West Glamorgan. Map 197.

Erysimum repandum L. has once been recorded as an alien at Cardiff Docks (17B, HJR c.1907).

HESPERIS L.

Hesperis matronalis L.   Dame's-violet. Disawr.
Introduced. Widespread as a garden escape on roadsides and waste ground, mainly in Gower and the Vale. Map 198.

Two species of **Malcolmia** R. Br. (*Wilckia* Scop.) have been recorded as rare aliens; **Malcolmia maritima** (L.) R. Br., Virginia Stock, Murwyll Arfor, on five occasions (17D, 1972!; 27A, 1926!*; 58B, JAW c.1940; 69C, JAW c.1940; 69D, JAW c.1940), and **M. africana** (L.) R. Br. once (27A, RLS 1927*).

CHEIRANTHUS L.

Cheiranthus cheiri L.   Wallflower. Blodyn y Fagwyr, Melyn y Gaeaf, Murwyll.
Introduced. Naturalized and long established on the walls of many old buildings in Gower and the Vale, for example Cardiff Castle, Ogmore Castle, Ewenny Priory and Weobley Castle. (!) Fl. 4–5, fr. 5–6. Map 199.

MATTHIOLA R. Br.

Matthiola incana (L.) R. Br.   Hoary Stock. Murwyll Coesbren.
Native or introduced. Limestone sea-cliffs in two localities, and quarry spoil in a third (06B).
06B, Rhoose, JPC 1989; 68A, Mumbles Head, 1972!; 96A & 97C, near Nash Point, known since before 1900 (Trow, 1911) and still present in the 1980s, although threatened by repeated cliff-falls.

Matthiola sinuata (L.) R. Br.   Sea Stock. Murwyll Tewbannog Arfor.
Native. Sand-dunes and sandy drift-lines. Now locally abundant and spreading. First recorded by Lightfoot in 1773 at Briton Ferry. Dillwyn (1840)

stated that in 1802 it was plentiful on the sand hills between Swansea and the Mumbles, but had become rare there by 1840, although it still grew 'sparingly' between the eastern pier and Crymlyn Burrows. No records have been traced between c.1848 (Woods, 1850) and the early 1960s; Riddelsdell (1907) thought it had probably become extinct. In 1964 it was rediscovered at Witford Point near Baglan, (R. Garrett-Jones) and in 1965 on the western side of the Neath estuary (G. Bates). The rediscovery in the 1970s of several other less conspicuous rare species in the same area, also unrecorded since the 1840s (for example **Equisetum hyemale**) suggests that **M. sinuata** may also have survived unrecorded until post-war disturbance of the dunes enabled it to spread. It is known to have been deliberately introduced at Broughton Burrows (49C) and possibly also at Port Eynon (48B). (!) Fl. 5–8, fr. 6–8; (Syr) E.ten, (Lep) Go.rha, Pi.bra, Pi.nap, noctuid moths.
48B, Port Eynon Burrows (introduced?) JPC 1981, 1982!; 49C, Broughton Burrows, (introduced), P.H. Dunn 1974; 58A, Oxwich Burrows, D.O. Elias 1979; 69C, Singleton Burrows, J. Woods c.1848; 69D, Crymlyn Burrows, 1972!; 78D, Kenfig Burrows, 1973!*, 1986!; 79C, Witford Point, R. Garrett-Jones 1964*, Baglan Bay, H & CH 1972, Crymlyn Burrows, 1972!**, 1986!.

Matthiola tricuspidata (L.) R. Br. has once been recorded as an alien at Cardiff Docks (17B, RLS 1924*).

BARBAREA R. Br.

Barbarea vulgaris R. Br.   Winter-cress. Berwr y Gaeaf.
Native. Damp roadsides, ditches and stream-banks. Fairly common in Gower and the Vale but local in the Uplands. Fl. 5–6. Map 200.

Barbarea verna (Miller) Ascherson American Winter-cress. Berwr Tir.
Introduced. Roadsides, cultivated and waste ground, rare and scattered throughout the county. Map 201.

Barbarea intermedia Boreau Medium-flowered Winter-cress.
Introduced. Roadsides and stream banks, mainly in the eastern part of the county. Map 202.

RORIPPA Scop.

Rorippa austriaca (Crantz) Besser Austrian Yellow-cress. Berwr Melyn Awstria.
Introduced. Roadsides, waste ground and stream banks, rare but locally established in western Glamorgan and Cardiff.
17B, Leckwith, JEL 1960, JPC 1978, Cardiff, MG, 1974; 17D, Cardiff, MG 1974 (first record 1938, RLS!*); 50B, Afon Llwchwr, 1974!; 50D, near Pontarddulais, 1974!; 59B, Dunvant, JAW 1955*, 1989!; 60A, Afon Llwchwr, 1974!; 61C, Garnswllt, 1981!*; 69D, Swansea, JAW 1939*.

Rorippa amphibia (L.) Besser   Great Yellow-cress. Berwr Melyn Mwyaf y Dŵr.
Probably native or formerly native in streamside sites in south-eastern Glamorgan, but it has not been recorded there since 1905, and its habitats may have been destroyed by drainage.
00C, Aberdare, HJR 1897; 07C, Flemingston Moors, AHT 1905; 17B, canal at Llandaff, WWB 1905; 17B, Penarth Road, J. McCulloch 1893; 69B, Llansamlet, JAW 1927.

Rorippa sylvestris (L.) Besser   Creeping Yellow-cress. Berwr Melyn Ymlusgol y Dŵr.
Native. Streamsides, ditches and damp ground. A characteristic plant of the sides of larger rivers in Glamorgan, locally dominant on sloping riverside banks that are submerged during the winter. Map 203.

Rorippa palustris (L.) Besser (*R. islandica* auct.). Marsh Yellow-cress. Berwr Melyn y Gors.
Native. Fairly common on damp disturbed ground where water stands in winter. Map 204.

ARMORACIA Gilib.

Armoracia rusticana P. Gaertner, B. Meyer & Scherb.   Horse-radish. Rhuddygl Poeth.
Introduced. Frequent on roadsides and waste ground in or near towns and villages. Map 205.

NASTURTIUM R. Br.

Nasturtium officinale R. Br.   Water-cress. Berwr y Dŵr.
Native. Common and widespread in streams, ditches and ponds and near springs. (!) Fl. 5–10, fr. 6–10. Map 206.

Nasturtium microphyllum (Boenn.) Reichenb. Narrow-fruited Water-cress. Berwr Dŵr Lleiaf.
Probably native, in similar habitats to **N. officinale**, with which it is often confused; apparently scarce or absent in the western part of the county, and frequent only in the southern and eastern Vale. Map 207.

Nasturtium microphyllum × N. officinale (N. × sterile (Airy Shaw) Oefelein). Hybrid Water-cress. Berwr Dŵr Croesryw.
Probably introduced. Streams, ditches, and ponds, local. Map 208.

CARDAMINE L.

Cardamine pratensis L. Cuckooflower, Lady's-smock. Blodyn y Gog.
Native, in damp grassland and near streams. Common and generally distributed. Fl. 4–7(–8), fr. 6–8; B.pasc (n), (Syr) E.ten, E.per, H.pen, Leu.luc, Rh.cam, Sy.pip, (Lep) An.car, Pi.rap (all n). Map 209.

Cardamine impatiens L.   Narrow-leaved Bitter-cress. Berw Chwerw Culddail.

Native. In shaded sites on limestone north of Merthyr Tydfil. No recent records.
00A, Morlais Castle, HJR 1904 ('very rare'). First recorded in this square by Matthew Moggridge (*c.*1840) at 'Coed Cymmer', perhaps in Breconshire. (J. Storrie's record from 19A, Quaker's Yard, *c.*1886 is possibly an error).

Cardamine flexuosa With.   Wavy Bitter-cress, Wood Bitter-cress. Berwr Cam, Chwerw'r Coed.
Native. Common throughout the county in a variety of shaded or moderately disturbed habitats especially in woods. Fl. 3–12, fr. 5–11. Map 210.

Cardamine flexuosa × C. pratensis (C. × haussknechtiana O.E. Schulz) has been recorded once (17A, St. Fagans, JWD 1972*).

Cardamine hirsuta L.   Hairy Bitter-cress. Berwr Blewog, Chwerw Blewog.
Native. Usually very common as a garden weed and on sand dunes, walls, shallow soil over rock etc., but scarce or absent in parts of the Uplands. Fl. 1–12, fr. 4–12. Map 211.

Cardamine amara L. (Large Bitter-cress, Berwr Chwerw, Chwerw Mawr) was recorded from the Pontypridd area (08/09) by JS in 1877* but its status was uncertain, possibly introduced; it has not been recorded since. **C. trifolia** L. (Trefoil Cress, Berwr Tribys) was recorded 'in great quantity' by Lady Ingleby from 69C, Singleton in 1919 but did not persist.

ARABIS L.

Arabis hirsuta (L.) Scop.   Hairy Rock-cress. Berwr y Graig.
Native. Locally common on sand dunes and on shallow soils over limestone. Changes in farming practice appear to be responsible for its disappearance from many inland sites. (!) Fl. 4–6, fr. 5–7; (Syr) E.ten (n), (Lep) An.car, Pi.nap (n). Map 212.

Arabis caucasica Schlecht.   Garden Arabis. Berwr yr Ardd.
Garden escape, locally naturalized on walls.
07A, Welsh St. Donat's, 1969!; 16A, Cold Knap, 1974! (on shingle); 17D, near Cardiff, MG 1974; 19C, between Nelson and Ystrad Mynach, A. Morgan 1973; 87B, Merthyr Mawr Quarry ('quite naturalized'), JAW *c.*1918.

Arabis verna (L.) R. Br. has once been recorded as a casual (17A, Coedrhiglan, 1882 (Riddelsdell, 1907)).

Aubrieta deltoidea (L.) DC. has twice been found as a garden escape naturalized on walls (09A, Mountain Ash, AMP 1972; 28A, Draethen, 1972!).

LUNARIA L.

Lunaria annua L.   Honesty. Swllt Dyn Tlawd.

Garden escape, occasionally naturalized near houses and on waste ground but usually not persistent. Map 213.

Lunaria rediviva L. has once been recorded as a casual (69A, Llangyfelach, JAW 1922).

Alyssum alyssoides (L.) L., Small Alison, Cyddlin Bach, has twice been recorded as a casual but with no recent records (00C, Aberdare, HJR *c*.1907; 17D, Penarth, JS *c*.1886).

BERTEROA DC.

Berteroa incana (L.) DC. Hoary Alison. Cyddlin Llwyd.
Casual on waste ground usually near docks.
00C, Aberdare, HJR 1905; 16A, Barry Docks, F. Druce 1925, DMcC 1974; 17B, Cardiff Docks, HJR, 1905; 17D, Penarth Dock, THT 1876*; near Leckwith, RLS 1922!*; 69D, Crymlyn, EMT 1937*; 78B, Port Talbot Docks, HJR 1910*; 80B, Glyn Neath, HJR 1911.

LOBULARIA Desv.

Lobularia maritima (L.) Desv. Sweet Alison. Cyddlin.
Introduced. Increasingly common on waste ground and disturbed roadsides, especially near the sea. Map 214.

DRABA L.

Draba aizoides L. Yellow Whitlowgrass. Llysiau Melyn y Bystwn.
Native. Locally frequent in crevices of Carboniferous Limestone cliffs on the southern coast of Gower from Pwll-du Head to Pennard Castle, and Overton to Tears Point; most abundant between Overton and Mewslade, and apparently absent from Oxwich Point and Worm's Head. First recorded 1796 by John Lucas (of Stout Hall) 'near Worm's Head' (probably in the Overton-Mewslade area); known in its present range since the early nineteenth century. The celebrated site at Pennard Castle overlooking Pennard Valley, where it grows both on the walls of the thirteenth century castle and on the limestone cliff below, is an unusual site; most sites are on steep and relatively inaccessible limestone cliffs facing the sea but separated from it by an undercliff of steeply sloping rocky grassland or cliff-heath (Kay & Harrison, 1970). (!) Fl. 2–4(–5), fr. 4–5(–6); B.luc, B.ruderarius, (Sb) Andrena haemorrhoa, Andrena sp., Colletes sp., (Syr) E.aeneus, E.ten, (Dip) Call.erythrocephala, Luc. (Kay & Harrison, 1970).
Recorded from 48A!; 48B!; 48D!; 58A!; 58B!.

Draba muralis L. Wall Whitlowgrass. Bystwn y Fagwyr, Magwyrlys y Bystwn.
Introduced. Old wall by road at Kilvrough Manor near Parkmill (58B); first recorded in 1968!** but probably long-established. Still present in 1986!.

EROPHILA DC.

Erophila verna (L.) Chevall. *s.l.* (incl. **E. verna** (L.) Chevall. *s.s.*; **E. majuscula** Jordan; **E. glabrescens** Jordan). Common Whitlowgrass. Llys y Bystwn.
Native. Locally abundant in Gower and along the coast of the Vale on thin soil on coastal cliffs, sand dunes and wall-tops; scarce inland.
Recent research has indicated the presence of three species in the British flora: **E. verna** (L.) Chevall *s.s.* (incl. *E. spathulata* A.F. Lang, which is at most varietally distinct (var. **praecox** (Steven) Diklic)); **E. majuscula** Jordan and **E. glabrescens** Jordan. All three segregates have been recorded from Glamorgan (Rich & Rich, 1988) but their distribution within the county has not yet been worked out.
(!) Fl. 2–5, fr. 3–5. Map 215.

COCHLEARIA L.

Cochlearia danica L.   Danish Scurvygrass. Llwylys Denmarc.
Native. Thin soil on coastal cliffs, sand dunes, old walls, and waste places near the sea. Locally abundant on the coast. Hybrids with **C. officinalis** occur on Worm's Head! (Fearn, 1977). (!) Fl. 2–5, fr. 3–6. Map 216.

Cochlearia officinalis L.   Common Scurvygrass. Llwylys Cyffredin.
Native. Fissures and ledges on coastal cliffs, and occasionally on or near the drift-line. The Glamorgan populations of this species are robust, with fleshy leaves (subsp. **officinalis**) and are restricted to maritime habitats with the exception of one old record from 09. (!) Fl. 5–7, fr. 6–8. Map 217.

Cochlearia anglica L.   English Scurvygrass. Llwylys Lloegr.
Native. Abundant in coastal and estuarine salt-marshes. (!) Fl. 4–6, fr. 5–7. Map 218.

Kernera saxatilis (L.) Reichenb. has been recorded once as a casual (00C, Aberdare, HJR *c*.1907).

CAMELINA Crantz

Camelina sativa (L.) Crantz Gold-of-pleasure. Cydllin.
Casual. Once fairly frequent on waste ground near docks, but with only one recent record.
00C, Aberdare, HJR *c*.1907; 16A, Barry Docks, HJR *c*.1907, EV 1932; 17B, Bute Dock, JS 1876*, The Hayes, JPC 1983; 18D, Llanishen, JS 1877*; 27A, Splott, 1925!*; 59B, Loughor, HJR *c*.1907; 78B, Port Talbot Docks, HJR *c*.1907.

Camelina alyssum (Miller) Thell. has once been recorded as a dock alien (27A, JS *c*.1876).

Neslia paniculata (L.) Desv., Ball Mustard, was once fairly frequent as a casual on waste ground but not recorded recently (00C, HJR 1902; 17B, A. Langley 1877*; 27A, 1925!*; 79C, HJR 1908; 90D, HJR 1903).

CAPSELLA Medicus

Capsella bursa-pastoris (L.) Medicus Shepherd's-purse. Llys y Tryfal, Pwrs y Bugail.
Native. A common weed of gardens, arable land, and farm tracks, also found in coastal seabird colonies (for example on Worm's Head). Fl. 1–12, fr. 5–12 (1–2). Map 219.

HORNUNGIA Reichenb.

Hornungia petraea (L.) Reichenb. Hutchinsia. Beryn Creigiog.
Native. An inconspicuous winter annual of sand-dunes and thin soils on coastal limestone cliffs. Fl. 2–4, fr. 3–5.
48A, Mewslade cliffs, MG 1968; 49A, Whiteford Burrows, AHT 1908*, 1971!; 49C, Whiteford Burrows 1971!, Llangennith Burrows 1970!; 58A, Nicholaston Burrows 1982!, Pennard Burrows 1976!, Pennard Castle, AHT 1892*; 68A, Langland cliffs, H & CH 1974; 78D, Kenfig Burrows, EV 1924*, JWD 1962*; 79C, Crymlyn Burrows, H & CH 1972.

THLASPI L.

Thlaspi arvense L. Field Penny-cress. Codywasg.
Introduced. A rather scarce and decreasing weed of arable land and waste ground, most frequent in the Vale. Fl. 5(–7), fr. (6–)8. Map 220.

L.W. Dillwyn's records (Turner & Dillwyn, 1805; Dillwyn, 1840) of **Thlaspi caerulescens** J. & C. Presl (*T. alpestre*) L. 'about Pontneddfechan and Aberpergwm' and **Teesdalia nudicaulis** (L.) R. Br. 'not infrequent about Swansea' were probably errors, although **T. nudicaulis** has recently been found as a non-persistent adventive on a newly sown grass bank at Whitchurch in Cardiff (18C, GH 1985).

Three species of **Iberis** L. have been found as garden escapes or casuals. The most frequent of these are **Iberis umbellata** L., Garden Candytuft, Beryn yr Ardd, (09A, Mountain Ash, AMP 1973*; 17D, Penarth, JS c.1876*; 58B, Bishopston, Mr Hillman 1960; 79C, Briton Ferry, H & CH 1974; 79D, Port Talbot, H & CH 1976; 98A, Blackmill, JAW c.1918), and **I. amara** L., Wild Candytuft, Beryn Chwerw, (00A, Cefn-coed-y-Cymmer, MG 1974; 08B/09D, near Pontypridd, JS 1877, AJ c.1955; 16B, Swanbridge, AHT 1891*, 1892*; 18C, Whitchurch, Anon. 1917*; 88C, Kenfig, JM 1841**). **I. sempervirens** L., Perennial Candytuft, has been found once (09A, Mountain Ash, AMP 1971).

LEPIDIUM L.

Lepidium campestre (L.) R. Br. Field Pepperwort. Codywasg y Maes.
Native. Dry grassland, roadside banks and waste ground. Scattered through the county, but apparently less frequent than in the past. Recent records from: 18C, Taff's Well, RAH 1971; 26 A/C,

Flat Holm, 1967!; 50B, near Pontarddulais, 1974!; 59A, Llanmorlais, 1985!**; 60A, Garnswllt, 1974!; 69B, near Llansamlet, 1974!; 69D, Ty-draw, 1972!. Map 221.

Lepidium heterophyllum Bentham Smith's Pepperwort. Pupurlys.
Native. Dry grassland, roadsides and disturbed ground. Once widespread in the county, but with only three recent records (48D, Overton Cliff, 1984!**; 61D, Garnant, 1974!; 87B, Craig y Parcau, H & CH 1974). Map 222.

Lepidium sativum L. Garden Cress. Berwr Gardd.
Introduced. Roadsides and waste ground; four recent records in western Glamorgan. Map 223.

Lepidium ruderale L. Narrow-leaved Pepperwort. Pupurllys Culddail.
Introduced. Roadsides and waste ground. Recent records from: 18A, Llanbradach Quarry, MG 1974; 27A, Pengam Moors, 1969!*. More widespread and frequent in the past. Map 224.

Lepidium latifolium L. Dittander. Berwr Gwyllt, Pupurllys Llydanddail.
Introduced. Waste ground and salt-marshes near docks; very persistent.
16A, Barry Docks, D.G. Holland 1972; 17B, banks of River Taff, EV 1922, Canton, HJR 1907, JWD 1956*, Roath, 1974!*; 17D, Grangetown, EV 1913*, MG 1974, Cardiff Docks, JPC & TGE 1986; 27A, Pengam Moors, 1969!*; 79C, near Neath, T. Westcombe c.1843, bank of tidal creek by Neath Canal, Briton Ferry, EMT 1930*, H & CH 1972–80.

Lepidium graminifolium L. Tall Pepperwort. Pupurlys Tal.
Casual on waste ground near docks.
16A, Barry Docks, RLS 1920*, 1923!*; 17B, Cardiff Docks, EP 1943*; 17D, Penarth Road, Cardiff, EV c.1920*; 27C, Cardiff Docks, 1973!, 1986!; 69C, Swansea, JAW 1952*; 78B, Port Talbot Docks, HJR c.1907.

Five other species of **Lepidium** have been found as rare casuals or dock aliens. **Lepidium densiflorum** Schrader has been found most frequently but with no recent records (16A, Barry Docks, RLS c.1925!*; 17B, Cardiff, GCD c.1927; 18C, Radyr, RLS 1925; 27A, Splott, GCD c.1927; 69C, Mayals, JAW c.1925 (det. GCD)). **L. neglectum** Thell. has been found on four occasions, one recent (06B, Porthkerry, 1921!*; 16A, Barry Docks, 1924!*; 27A, Splott, 1925!*, Pengam, 1969!*). **L. virginicum** L. has been recorded three times (17A, Caerau, HAH 1923*; 18C, Radyr, RLS 1925*; 27A Splott, 1925!*). **L. perfoliatum** L. has been recorded several times but from only two localities (17B, Cardiff Docks, JS c.1876, HJR 1908, 1922!*; 27A, Splott, RLS c.1925!, 1929!*). **L. ramosissimum** A. Nelson has been found only once (27A, Splott, GCD 1927).

CARDARIA Desv.

Cardaria draba (L.) Desv.   Hoary Cress. Pupurllys Llwyd.
Introduced. Almost confined to urban habitats; most frequent in or near Cardiff and Swansea. Roadsides, railway embankments and waste ground. No recent records in the Uplands. All the Glamorgan specimens belong to subsp. **draba**. The first British record of this species was from Hafod Wharfs, Swansea (69D) where L.W. Dillwyn found it in 1802 (Dillwyn, 1848; Scurfield, 1962). Fl. 5–6, 10. Map 225.

CORONOPUS Haller

Coronopus squamatus (Forskål) Ascherson   Swinecress. Olbrain, Berwr y Moch.
Native. Farm tracks, waste ground and cultivated land near the coast. Map 226.

Coronopus didymus (L.) Sm.   Lesser Swine-cress. Olbrain Lleiaf.
Introduced. An increasing weed of cultivated land and waste ground. First recorded in S. Wales at Dale in Pembrokeshire (Turner & Dillwyn, 1805), but abundant in the Swansea area by 1840 (Dillwyn, 1840). Now common in most of Glamorgan. Fl. 4–11, fr. 5–11. Map 227.

CONRINGIA Adanson

Conringia orientalis (L.) Dumort.   Hare's-ear Mustard.
Casual. Once fairly frequent on waste ground near docks, and on rubbish-tips; most recent record in the 1950s (08B). Map 228.

Conringia austriaca (Jacq.) Sweet has once been recorded as a casual (00C, Aberdare, HJR c.1907).

DIPLOTAXIS DC.

Diplotaxis tenuifolia (L.) DC.   Perennial Wall-rocket. Cedw Meindwf y Tywod.
Introduced. Locally common on waste ground and old walls in towns and occasionally found on sand dunes. (!) Fl. 6–11. Map 229.

Diplotaxis muralis (L.) DC.   Annual Wall-rocket. Cedw y Tywod
Introduced. Waste ground, roadsides, walls and cultivated land, mainly in towns. Fairly common near the coast in western Glamorgan. Fl. 6–10, fr. 7–10; ?Sb. Map 230.

Diplotaxis erucoides (L.) DC., White Rocket, Cedw Gwyn yr Ar, has been recorded as a casual at two sites (16A, Barry Docks, RLS 1924!*, J.H. Salter 1927*; 17B, Maindy, RM & RLS 1924*, RLS 1938*, all det. EJC).

BRASSICA L.

Brassica oleracea L.   Wild Cabbage. Bresych Gwyllt.

Native. Locally abundant on the Lias Limestone cliffs of the Vale from Southerndown to Barry, but absent from Gower. Trow (1911) stated that it was regularly collected in hard winters for use as a vegetable. Cultivated forms occasionally occur as garden escapes inland, but the coastal populations appear to be truly native. Map 231.

Brassica napus L.   Rape, Swede. Rêp, Bresych yr Ŷd.
Introduced. Roadsides, cultivated land, waste ground and occasionally on the banks of ditches and streams. Increasingly frequent on roadsides. Most records probably refer to subsp. **oleifera** DC., Oil-seed Rape, which is being increasingly grown as a cash crop in the Vale. (!) Fl. 5–10, fr. 6–11, Hb (n), Bb, (Lep) Pi. Map 232.

Brassica rapa L.   Wild Turnip. Erfinen Wyllt, Meipen.
Probably introduced. A common weed of cultivated land and roadsides.
(!) Fl. 5–11, fr. 6–11, Hb, Bb, (Syr) E., (Lep) Pi. Map 233.

Brassica nigra (L.) Koch   Black Mustard. Cedw Du.
Possibly native on cliffs, but also introduced as a weed and as a relic of cultivation. Locally very abundant in cultivated land and on roadsides in Gower and the south-eastern Vale, and also on the Lias Limestone cliffs of the southern Vale, where Trow (1911) and Riddelsdell (1907) regarded it as native. (!) Fl. 6–11, fr. 7–11, Hb, Bb, (Syr) E., (Lep) Pi. Map 234.

Four species of **Brassica** have been found as casuals or dock aliens. **Brassica elongata** Ehrh. has been recorded from four localities (00C, Aberdare, HJR 1903; 78B, Port Talbot, HJR c.1907; 90B, Hirwaun, HJR 1903**; 90D, Llwydcoed, HJR 1901, 1903), as has **B. juncea** (L.) Czern. (16A, Barry Docks, J.H. Salter 1930*; 17D, Penarth, JS c.1876; 27A, Splott, GCD c.1928; 69D, Swansea foreshore, JAW 1947*). The other two species have been found once only, as dock aliens: **B. fruticulosa** Cyr. (17B, 1924!*) and **B. tournefortii** Gouan (16A, RLS 1924!).

SINAPIS L.

Sinapis arvensis L.   Charlock. Cedw Gwyllt.
Probably introduced as a weed of cultivation. Abundant in arable land on the better soils, and often introduced elsewhere with imported soil, as a seed contaminant, or during road construction. (!) Fl. (1–3–)4–12, fr. 5–12; Hb (n), B.luc (n,p), B.terr (n,p) (Syr) E.arb, E.hor, E.ten, H.pend (all n,p), Leu.luc, S.balt (p), S.?lun (p), S.rib (p), (Lep) (!) Ag.urt, Pi.bra, Pi.nap, Pi.rap, Pl.gam (all n). Map 235.

Sinapis alba L.   White Mustard. Cedw Gwyn.
Introduced. Cultivated land and waste ground, rare; in most cases either introduced with bird-seed or an impersistent relic of recent cultivation. Trow (1911)

[Plate 5]

1, *Helianthemum canum*, a limestone rock-rose limited to refuge sites in the west (Ramsgrove, Gower). 2, *Geranium sanguineum*, locally common on western dunes and limestone cliffs (Broughton Burrows, Gower). 3, *Rosa pimpinellifolia*, abundant in duneland (Nicholaston Burrows). 4, *Potentilla tabernaemontani*, a scarce early-flowering limestone cinquefoil (Seven Slades, Bishopston). 5, *Hippocrepis comosa*, the diploid form found in Gower is, like *Helianthemum canum*, limited to refuge sites in the west (Overton). 6, *Trifolium occidentale*, a rare clover known only from the Atlantic fringe of Europe, and at the northern limit of its known range in Britain in Gower (Fall Bay). (Photographs: 1–6, Q. O. N. Kay).

[Plate 6]

1, *Sorbus porrigentiformis*, a whitebeam known only from western Britain (Horton Slade).
2, *Buglossoides purpurocaerulea*, scarce in limestone scrub near the coast (Nicholaston Wood).
3, *Orobanche hederae*, a local root-parasite of ivy on coastal cliffs (Crawley Cliff). 4, *Mimulus luteus* × *M. guttatus*, conspicuous on streambanks in the uplands (Tonpentre, Rhondda).
5, *Campanula glomerata*, at its western limit in the county (Nash Point). 6, *Cirsium acaule*, frequent in limestone grassland in the Vale but at its western limit (Pant St Brides).
(Photographs: 1–6, Q. O. N. Kay).

stated that it was a common cornfield weed in the Vale and south-western Glamorgan but this is certainly no longer so and he may have confused **S. alba** with **S. arvensis.** Map 236.

Sinapis alba subsp. **dissecta** (Lag.) Bonnier has been recorded three times as a casual in the west of the County (69D, Swansea, H & CH 1974; 78B, Port Talbot Docks, HJR 1910*; 80A, near Seven Sisters, H & CH 1974).

Eruca vesicaria (L.) Cav. subsp. **sativa** (Miller) Thell., Garden Rocket, Berwr yr Ardd, has once been found as a casual at Splott (27A, 1925!*, RLS 1926*), and **Erucastrum gallicum** (Willd.) O.E. Schulz, Hairy Rocket, has been found as a casual on four occasions (16A, Barry Docks, JPC & TGE 1986; 27A, Splott, RLS 1924!, 1932; 69D, Tir John, JAW 1939*).

RHYNCHOSINAPIS Hayek

Rhynchosinapis cheiranthos (Vill.) Dandy (*Hutera cheiranthos* (Vill.) Gómez-Campo). Wallflower Cabbage. Berwr Murwyll y Mor, Bresych Murwyll.
Introduced. Locally abundant and spreading on waste ground, roadsides and sand-dunes from Swansea to Porthcawl; occasional on roadsides and waste ground elsewhere. An increasing species, now a conspicuous part of the flora of urban waste ground in the Swansea district, and also abundant on sand-dunes at Crymlyn Burrows. Map 237.

Rhynchosinapis monensis (L.) Dandy ex Clapham (*Hutera monensis* (L.) Gómez-Campo). Isle of Man Cabbage. Berwr Môn a Manaw, Bresych Môn a Manaw.
Probably native. Sand-dunes at Pennard! and Nicholaston!, locally abundant on the eastern side of Three Cliffs Bay!, (all in 58A). First recorded in this locality in 1838 by M. Moggridge (Dillwyn, 1840). Records from other localities have not been confirmed and are possibly errors for **R. cheiranthos.** **R. monensis** is an endemic British species which reaches its southern limit in Glamorgan. Plants from Three Cliffs Bay have the chromosome number of 2n=24 found in other British populations (Q.O.N. Kay, unpublished).

Rhynchosinapis hispida (Cav.) Heywood (*Hutera hispida* (Cav.) Gómez-Campo) has once been recorded as a casual (27A, Splott, 1926!*).

HIRSCHFELDIA Moench

Hirschfeldia incana (L.) Lagrèze-Fossat Hoary Mustard. Cedw Penllwyd.
Introduced. Roadsides, waste ground, and sand-dunes. Abundant in urban habitats in the Swansea and Cardiff areas, where it is sometimes dominant on waste ground, and increasing elsewhere. Map 238.

Carrichtera annua (L.) DC. has twice been recorded as a casual (27A, Splott, RLS, 1926!, 1927*), as has

Erucaria hispanica (L.) Druce (17D, Penarth, RLS 1937; 27A, Splott, RLS 1931*).

CAKILE Miller

Cakile maritima Scop.   Sea Rocket. Hegydd Arfor.
Native. Fairly common on or near the drift-line on sandy beaches in the western part of the county, and occasionally found in the spray-zone on cliffs and on the drift-line in salt-marshes. (!) Fl. 6–8 (–10), fr. 7–9; B.lap, B.pasc, B.terr (all n). Map 239.

RAPISTRUM Crantz

Rapistrum perenne (L.) All., Steppe Cabbage, has been recorded as a casual on waste ground near docks and rubbish tips (17B, Cardiff Docks, HJR c.1907; 48A, Rhosili, JAW c.1950; 78B, Port Talbot Docks, HJR c.1907). **R. rugosum** (L.) All., Bastard Cabbage, has also been found as a casual in similar habitats. Three subspecies have been recorded: subsp. **rugosum** on four occasions (16A, Barry Docks, RLS c.1925; 27A, Splott, RLS c.1924!; 68A, Langland, JAW c.1950; 78B, Port Talbot Docks, HJR 1910*); subsp. **orientale** (L.) Arcangeli twice (16A, Barry, GCD & RLS c.1927; 27A, Splott, 1925!*); and subsp. **linnaeanum** Rouy & Fouc. once only (27A, Splott, RLS 1925!*).

CRAMBE L.

Crambe maritima L.   Sea-kale. Ysgedd, Bresych y Môr.
Native. Rare and sporadic on shingle beaches, sandy shores and limestone cliffs. No recent records. May still occur in western Gower, where it grew in at least one locality in the 1950s. Map 240.

RAPHANUS L.

Raphanus raphanistrum L.   Wild Radish. Rhuddygl Gwyllt.
Introduced. Cultivated land and disturbed soil on roadsides. Locally abundant, usually on acid soils, but absent or overlooked in some areas, notably the central and eastern Uplands and central Vale. Yellow-flowered plants predominate at Sker (88C, 87D) and Llandarcy (79A) and white-flowered plants in western Gower, but in eastern Gower and the Lliw district most populations are polymorphic and consist of a mixture of yellow and white-flowered plants. (!) Fl. 5–11 (–12), fr. 6–11; Hb (n,p), B.hort, B.lap, B.luc, B.pasc, B.terr (all n,p), (Syr) E.aen, E.arb, E.hort, E.int, E.pert, E.ten, H.pend, Sc.pyr (all n, some p), S.balt (p), S.rib (p) small syrphids (p), (Lep) Ag.urt, Go.rha, Pi.bra, Pi.nap, Pi.rap, Pl.gam. Map 241.

Raphanus maritimus Sm.   Sea Radish. Rhuddygl Glan y Môr, Bysedd yr Îar Arforol.
Native. Drift-lines and disturbed ground on or near shingle beaches. Locally well-established in the Burry Estuary (Penclawdd and Llwchwr) and on the Vale

coast between Barry and Penarth; sporadic elsewhere. (!) Fl. 5–9 (–10), fr. 6–10. Map 242.

Raphanus sativus L.   Radish. Rhuddygl, Radys.
Casual or garden outcast, on waste ground and river-banks, probably of more frequent occurrence than the records indicate.
00C, Aberdare, HJR c.1907; 16A, Barry Docks, 1925!*; 17B, Cardiff, RLS c.1925!; 27A, Splott, GCD c.1927; 50D, banks of Afon Llwchwr, 1974!; 61C, between Garnswllt and Pantyffynnon, 1981!; 69C, Swansea, JAW 1954; 78B, Port Talbot, HJR c.1907.

RESEDACEAE
RESEDA L.
Reseda luteola L.   Weld. Melengu.
Native. Roadsides, paths and waste ground. Common in urban areas, but rather local in the countryside, and absent from parts of the Uplands. Fl. 5–11, fr. 7–11; Hb. Map 243.

Reseda alba L.   White Mignonette. Melengu Wen Unionsyth.
Casual or introduced. Waste ground near the sea, and disturbed sand-dunes; usually an impersistent casual but perhaps naturalized on sand-dunes between Horton and Port Eynon and in one or two localities near Cardiff. Fl. (4–)5–8(–10), fr. 7–10; Hb (n,p), B.lap, B.luc, B.prat (all n,p) ?Sb, (Syr) S.balt, S.rib.
06B, Rhoose, JPC 1989; 07C, St. Hilary, JS c.1876; 16A, Barry, RLS 1920*, 1980!; 17B, Cardiff Docks, 1926!*, Leckwith, JPC 1978; 17D, Cogan Pill, EV 1926*; 27A, East Moors, AL 1877*; 48B, Port Eynon, JAW 1944*, Horton 1970!; 69C, Swansea, JWGG c.1841, Blackpill, JAW c.1956; 87A, Porthcawl, R.E. Cundall c.1898, HJR c.1907.

Reseda lutea L.   Wild Mignonette. Melengu Wyllt Ddisawr.
Native. Roadsides, dry grassland near the sea, cultivated land and waste ground. Common on urban waste ground in the Swansea and Cardiff districts, and fairly frequent on roadsides, field banks, etc. near the coast in Gower and the Vale; scarce elsewhere. Map 244.

Two species of **Reseda** have been recorded as rare casuals, **Reseda odorata** L. from Mayals and Blackpill (69C) by JAW in c.1950 and **R. phyteuma** L. from Cardiff Docks (17B) by JS in c.1876* and RLS in c.1922. [Old records of **R. stricta** Pers. are now believed to be errors for forms of **R. lutea**.]

Caylusea hexagyna (Forskal) M.L. Green has once been recorded as a casual (27A, Splott, RLS 1927).

DROSERACEAE
DROSERA L.
Drosera rotundifolia L.   Round-leaved Sundew. Gwlithlys.

Native. Bogs. Frequent in suitable habitats throughout the north and west of the county, absent from much of the Vale. Map 245.

Drosera anglica Hudson   Great Sundew. Gwlithlys Mawr.
Native. This species was recorded from Crymlyn Bog (69B/D) at the beginning of the 19th century by L.W. Dillwyn '. . . where, before the drainage, it was often found mixed with the other two more common species' (Dillwyn, 1848) and from Town Hill, Swansea (69C) by J.W.G.Gutch in c.1840 '. . . in the boggy ground.' (Gutch, 1844a). It has not been seen since in either locality and must therefore be presumed to be extinct.

Drosera intermedia Hayne   Oblong-leaved Sundew. Gwlithlys Hirddail.
Native. Bogs, and peaty places. Locally frequent in Gower in suitable habitats, for example on Harding's Down (49C). There are several old records in the northern Vale and the Uplands, but only one recent record. Map 246.

CRASSULACEAE
CRASSULA L.
Crassula helmsii (T. Kirk) Cockayne   New Zealand Pigmyweed.
Introduced. Ponds and lakes. An invasive aquatic weed, rapidly spreading in Britain. First seen in Glamorgan at Broad Pool in Gower (59C) in 1984, and soon afterwards found in Dinas Powis (17D).
59C, Broad Pool, R.S. Cropper 1984; 17D, Dinas Powis, J. Kilpatrick 1984.

UMBILICUS DC.
Umbilicus rupestris (Salisb.) Dandy Navelwort, Pennywort. Deilen Gron.
Native. Rocky places both on sea cliffs and inland, walls and stony hedgebanks. Frequent throughout the county. Map 247.

SEMPERVIVUM L.
Sempervivum tectorum L.   House-leek. Llys Pen Tai, Bywfyth.
Introduced. Once fairly frequent on old walls and roofs, but not recorded during the present survey. Map 248.

SEDUM L.
Sedum telephium L.   Orpine. Berwr Taliesin.
Native. Hedgebanks and rocky places. Locally frequent in Gower but with only a few scattered localities in the Uplands and Vale. Map 249.

The following two subspecies have been recorded from Glamorgan but their distribution is imperfectly known.

Sedum telephium subsp. **telephium.** (*S. telephium* subsp. *purpurascens* (Koch) Syme). Recorded only from Gower.
48A, Mewslade, JAW 1928, BAM 1957; 58A, Crawley Cliff, 1985!**; 58B, High Pennard, AHT *c.*1920, Pwll-du, EV 1934*; 59D, Ilston, EV 1923*.

Sedum telephium subsp. **fabaria** (Koch) Kirschleger Many old records scattered throughout the county, but only one recent record. Map 250.

Sedum spurium Bieb. Caucasian Stonecrop. Briweg Rwsieg.
Introduced. Walls, roadsides and, rarely, on shingle. Scarce usually originating as a garden outcast, but often persisting for many years.
06A, Aberthaw, 1955!*; 18A, Pen-y-groes, 1972!; 48B, Port Eynon, 1970!; 49C, Cwm Ivy, JAW *c.*1930; 69C, Tir Hester, JAW *c.*1930; 88C, Kenfig, 1974!; 99C, Ogmore Vale, 1975!.

Sedum reflexum L. Reflexed Stonecrop. Llwynau'r Fagwyr, Bywydog.
Introduced. Well established on walls in several localities, and also on coal tips. Map 251.

Sedum forsteranum L. Rock Stonecrop. Briweg Gymreig.
Probably native on Craig-y-llyn (90C), where it was last seen in 1890, and on limestone in the Taf-fechan Gorge (00A) and at Castell Coch (18C), but introduced, usually on walls, elsewhere. Rare, with recent records only from 00A (MG 1975), 18C (MG 1972), 70D! and 88D!. Subsp. **forsteranum** has been identified from 69A, Llangyfelach (JAW 1941*) and subsp. **elegans** (Lej.) E.F. Warburg from 68A, Langland (G.R. Willan 1909*).

Sedum acre L. Biting Stonecrop. Pupur y Fagwyr, Blywydog Boeth.
Native. Rocky places, dunes, shingle, walls, roadside banks, etc. Frequent throughout the county. Fl. 6–7, fr. 7–8; Hb (n), B.?prat, B.terr (n). Map 252.

Sedum sexangulare L. Tasteless Stonecrop. Briweg Diflas.
Introduced. A garden escape that sometimes becomes established on walls, waste ground and rocky places. Recorded in four scattered localities.
59D, Ilston, JAW 1925*, EV 1936*; 70C, near Pontardawe, 1974!; 79C, Briton Ferry, H & CH 1973; 80D, near source of R. Afan, JM *c.*1840.

Sedum album L. White Stonecrop. Gwenith y Gwylanod.
Introduced. Naturalized on walls, cliffs, quarries etc. mainly in the south and west. Map 253.

Plants resembling **Sedum album** subsp. **micranthum** (Bast.) Syme have been recorded on limestone rocks in two localities (48B, Penrice, JAW 1930; 97A, Cwm Alun, AHT 1907*, EV 1937*) but the validity of this

taxon is now considered doubtful (Clapham, Tutin & Moore, 1987).

Sedum anglicum Hudson English Stonecrop. Briweg y Cerrig.
Native. Shallow acid soil on sandstone outcrops in Gower and at two inland localities. Has apparently decreased in numbers as it has not been refound in many of its former localities. Map 254.

Sedum dasyphyllum L., Thick-leaved Stonecrop, Briweg Praffddail, has only two certain records in the county, as introductions at a disused quarry, Font-y-gary (06B, JPC 1986) and at Neath (79B, EMT 1935*). Riddelsdell reported finding this species at Oxwich, Horton, Port Eynon and Llanmadoc (48B & 49C, HJR *c.*1905) but may have confused **S. dasyphyllum** with **S. album** which is common at Horton and Port Eynon but was not recorded there by him. His records of **S. hybridum** L. as a casual at Bonvilston (07D, HJR *c.*1909) and Llanmadoc (49C, HJR *c.*1905) are also doubtful.

RHODIOLA L.

Rhodiola rosea L. (*Sedum rosea* (L.) Scop.). Roseroot. Pren y Ddannoedd.
Native. Wet cliff ledges. An arctic-alpine species which reaches its southern limit in Glamorgan; known only from the high cliffs of Craig-y-llyn.
90C, Craig-y-llyn, AHT *c.*1890*, HJR *c.*1907, 1920!*, 1979!.

SAXIFRAGACEAE

SAXIFRAGA L.

Saxifraga spathularis Brot. × **S. umbrosa** L. (**S. × urbium** D.A. Webb) (*S. umbrosa* auct.). Londonpride. Balchder Llundain.
Introduced. An occasional garden escape or garden throwout of roadsides, railways and waste places. Scattered throughout the county. Map 255.

Saxifraga tridactylites L. Rue-leaved Saxifrage. Tormaen Tribys.
Native. Sand-dunes, rocks, walls, railways etc. mainly on basic soils. Frequent in Gower and the Vale but almost absent from the Uplands, except on limestone in the extreme north. Fl. 4–5, fr. 5–6. Map 256.

Saxifraga hypnoides L. Mossy Saxifrage. Tormaen Llydandroed.
Native. Rocky places and mountain cliffs in the north of the county; only one recent record as a native.
00A, Morlais Castle Hill, HJR 1904*, EV 1936*; 00B, Morlais Castle Hill, ABP & BS 1973!; 16B, Cadoxton Station, L. Peddle 1983! (introduced); 90C, Craig-y-llyn, AHT *c.*1905*, Padell-y-bwlch, HJR *c.*1905; 99A, Craig Fawr, Treorchy, HJR *c.*1905.

Saxifraga granulata L. Meadow Saxifrage. Tormaen Gwyn y Gweunydd.

Native. Grassy places on base-rich soils. Rare with recent records from a few widely separated sites. Map 257.

Three species or hybrids of **Saxifraga** have been recorded as garden escapes. **Saxifraga cuneifolia** L. twice (70C, near Pontardawe, 1974!; 79C, Briton Ferry, H & CH 1973), **S. cymbalaria** L. (97A, Ewenny Priory, M. Humphrey 1963*) and **S. hirsuta** L. × **S. spathularis** Brot. (S. × polita (Haw.) Link) (79, Neath, EMT 1944*) once each. [**S. aizoides** L. was recorded from the Maesteg area (89) in *c*.1840 but the record is very doubtful (Riddelsdell, 1907).]

CHRYSOSPLENIUM L.

Chrysosplenium alternifolium L.   Alternate-leaved Golden-saxifrage. Eglyn Cylchddail.
Native. Damp woods and shady river banks. Recorded from several localities in the past but not seen recently.
17A, Ely, HJR 1911*, EV 1921; 18D, Heath, PWR 1920*, EMT 1921*; 28A, Draethen, HAH & EV 1926*; 80B, River Pyrddin, N.F. Shepherd, *c*.1900; 87D, Heol-y-mynydd, JS *c*.1876, Southerndown, HOB *c*.1900.

Chrysosplenium oppositifolium L. Opposite-leaved Golden-saxifrage. Eglyn Cyferbynddail.
Native. Streamsides, ditches, damp woods and wet places generally. Abundant in suitable habitats throughout the county. Fl. 3–6. Map 258.

Two other members of the Saxifragaceae have been recorded as garden escapes, both on only one occasion, **Tolmiea menziesii** (Pursh) Torrey & A. Gray, Pick-a-back-plant (49D, near Llanrhidian, MG 1974) and **Bergenia cordifolia** (Haw.) Sternb. (06B, Rhoose Point, MG 1975).

PARNASSIACEAE
PARNASSIA L.

Parnassia palustris L.   Grass-of-Parnassus. Brial y Gors.
Native. Marshy places. Occurred formerly in two localities near Cardiff and a third unconfirmed site in the Rhondda, but now probably extinct in the county.
18A, Pwllypant, JS *c*.1876; 18D, Llanishen, JS *c*.1876; 99D, Ystrad, H. Harris *c*.1900.

HYDRANGEACEAE
PHILADELPHUS L.

Philadelphus coronarius L.   Mock-Orange.
Introduced. Formerly established in a roadside hedge in one locality, but not seen recently.
58A, Parc le Breos, JAW *c*.1930.

ESCALLONIACEAE
ESCALLONIA Mutis ex L. fil.

Escallonia macrantha Hooker & Arnott   Escallonia. Esgalonia.
Introduced. Established on limestone cliffs in three localities in Gower.
48A, Rhosili, D.W. Bloodworth 1958; 48B, near Horton, JAW *c*.1940; 68A, Langland, JAW 1941*, 1947!*.

GROSSULARIACEAE
RIBES L.

Ribes rubrum L.   (*R. sylvestre* (Lam.) Mert. & Koch)   Red Currant. Rhyfon Coch.
Native. Limestone woods, shady streamsides and hedges. Locally frequent in the Vale and Gower, less so in the Uplands. Map 259.

Ribes nigrum L.   Black Currant. Rhyfon Duon.
Doubtful native or introduction. Borders of woods, hedges, riversides etc. Locally frequent in the east of the county, much less so in the west where it has decreased in numbers. Map 260.

Ribes sanguineum Pursh   Flowering Currant. Rhyfon Blodeuog.
Introduced. Established on roadsides especially near houses in several scattered localities throughout the county. Map 261.

Ribes uva-crispa L.   Gooseberry. Gwsberis, Eirin Mair.
Doubtful native. Woods and hedges mainly in the south and west. Map 262.

ROSACEAE
SPIRAEA L.

Spiraea salicifolia L. agg. Bridewort. Erwain Helygddail.
Introduced. More or less naturalized along roadsides in several places. Map 263.

Glamorgan plants belong to the following two taxa (**Spiraea salicifolia** L. *s.s.* has yet to be reliably recorded).

Spiraea douglasii Hooker   An introduction from N. America that occasionally escapes from gardens. 17B, Cardiff, ABP 1959*; 79A, Skewen, JAW 1919.

Spiraea douglasii × S. salicifolia (S. × billardii Hérincq.).
Grown in gardens, and often escaping and becoming more or less naturalized. Most records of the aggregate probably refer to this hybrid but only the following have been confirmed.
70B, Glan-rhyd, 1973!; 78D, Kenfig Castle, 1975!; 90D, Llwydcoed, 1974!.

Spiraea japonica L. fil., a garden plant introduced from Japan, is established on sandy ground at Merthyr Mawr (87B, 1982!*).

FILIPENDULA Miller

Filipendula ulmaria (L.) Maxim. Meadowsweet. Erwain.
Native. Marshy places, damp roadsides, riverbanks and other damp habitats. Abundant throughout the county. Fl. 6–8 (–10), fr. 7–8; B.luc, B.pasc (p), (Syr) E.ten (p). Map 264.

[All records of **Filipendula vulgaris** from the county are errors.]

RUBUS L.

Rubus saxatilis L.   Stone Bramble. Corfwyaren.
Native. Rocky banks in the northern Uplands. Very rare.
00A, Morlais Castle Hill, 1923!*, EV 1929*; 80B, Pyrddin Gorge, HJR c.1900, 1973!; 90A, Rhigos, BAW c.1900; 90C, Craig-y-llyn, MG 1971.

Rubus idaeus L.   Raspberry. Afanen, Mafonen.
Native. Woods, scrub, hedges, waste ground etc. Common in the east and north of the county, becoming less so in the west. Fl. 5–7, fr. 6–8; B.pasc (n), B.prat (n). Map 265.

Rubus phoenicolasius Maxim., Japanese Wineberry, Mafonen Flewgoch, has been found once as a garden escape (08B, Church Village, C.M. Davies 1938*).

Rubus fruticosus L. agg.   Bramble. Mwyaren Ddu
Native. Woods, hedges, scrub, waste ground etc. Very common throughout the county. Map 266.

The following account of the microspecies of this taxonomically difficult genus is based on specimens in **NMW** and elsewhere that have been identified or confirmed by A. Newton, B.A. Miles or E. Edees, and on fieldwork carried out in the county by A. Newton and B.A. Miles. The authors would like to express their thanks to A. Newton for all his helpful advice in compiling this list.

A large number of other records occur in the literature (especially in Riddelsdell 1906, 1907, 1909); those which cannot be substantiated either in the field or from exsiccata have been omitted. Names in square brackets following the accepted name or synonym are those used by Riddelsdell in his lists; they are **not** necessarily nomenclatural or taxonomic synonyms.

The sequence and nomenclature follows Edees & Newton (1988).

Rubus bertramii G. Braun [*R. plicatus* Weihe & Nees].

Native. Damp woodland and scrub.
00C, Aberdare, HJR 1905*; 08C, Llanharan, AHT 1909*, Pontyclun, AN 1973!; 18C, Whitchurch, HJR 1908*, 1909*; 49C, Rhosili Down, HJR c.1887; 59D, Clyne Common, HJR c.1887; 90D, Cwmdare, HJR 1905; 98D, Pencoed, AHT 1909*, BAM 1962.

Rubus fissus Lindley [*R. rogersii* Linton].
Native. Heathy woodland.
00C, Aberdare, HJR c.1905; 07B, Peterston, AHT 1907*; 08C, Pontyclun, HJR c.1908.

Rubus nessensis W. Hall (*R. suberectus* G. Anderson ex Sm.).
Native. Woodland.
08C, Pontyclun, BAM 1963, AN 1973!*; 18C, Castell Coch, AHT 1908*, Forest Ganol, 1976!; 80B, Glyn Neath, AL c.1900, AN 1973; 80C, Resolven, AL 1905, AN 1973.

Rubus scissus W.C.R. Watson (*R. fissus* sensu Focke).
Native. Borders of woodland, moors and railway lines.
00C, Aberdare, HJR 1905; 08C, Pontyclun, AN 1973!; 70C, Pontardawe, AL c.1900; 80C, Resolven, AN 1973!; 90D, Cwmdare, HJR c.1905.

Rubus vigorosus Mueller & Wirtgen (*R. affinis* Weihe & Nees *p.p.*).
Native. Moorland.
07B, Peterston, HJR 1907*.

Rubus albionis W.C.R. Watson [*R. macrophyllus* var. *schlechtendalii* Weihe.]
Native. Hedges and thickets.
07B, Peterston Moors, AL 1905; 18B, Caerphilly & Rudry, AN 1973; 70C, Pontardawe, HJR 1905; 80B, Pont-nedd-fechan, HJR 1905; 98D, Pencoed, BAM 1962.

Rubus godronii var. **foliolatus** Rogers & Ley
Native. Woods and hedges in the Neath valley. First discovered in, and described from, the Pontardawe district (70C). It still persists there and in neighbouring areas.

Rubus gratus Focke
Native. Open woodland.
70B, Ystalyfera, HJR c.1900; 80B, Glyn Neath, AL & WAS c.1890*.

Rubus laciniatus Willd.
Introduced garden escape, becoming naturalized in waste places.
17B Birchgrove, J. Lord 1980!; 17D, Grangetown, C.H. Woodridge 1967*.

Rubus lindleianus Lees
Native. Woodland hedges and scrub. Map 267.

Rubus perdigitatus Newton
Native. Wood margins and hedges.

Only certainly recorded from 80B, Resolven, AN 1973! (recorded, with doubt, from 00, 48, 68, 49, 69).

Rubus polyoplus W.C.R. Watson [*R. salteri* Bab.]
Native. Woodland and hedges.
07B, Peterston, HJR 1907*; 08C, Llantrisant, HJR 1907*, Pontyclun, HJR 1922, BAM 1963; 17B, Penylan, 1969!*; 18B, Caerphilly, HJR 1907*; 18C, Taffs Well, HJR 1905, Radyr, HJR 1909, Castell Coch, AN 1973!; 98D, Llanharan, AHT 1908*.

Rubus pyramidalis Kaltenb.
Native. Wood borders and hedges.
00C, Aberdare, HJR 1905; 07A, Ystradowen, HJR 1905, 1932; 08C, Pontyclun, HJR 1905, AN 1973!*; 18C, Taffs Well, AHT 1908*.

Rubus silurum (Ley) Ley (*R. nemoralis* var. *silurum* Ley).
Native. Hedges and margins of woods and moors. A characteristic upland species in the north of the county. Map 268.

Rubus altiarcuatus W.C. Barton & Riddelsd. (*R. cariensis* sensu Rogers).
Native. Margins of heaths, woods and moors. A widespread and characteristic species, often found in large quantity. Map 269.

Rubus amplificatus Lees has been recorded from hedges in the Pontardawe area (70).

Rubus cardiophyllus Lef. & Mueller (*R. rhamnifolius* auct.).
Native. Woodland and scrub, scarce. Map 270.

Rubus dumnoniensis Bab.
Native. Woodland and scrub.
00C, Aberdare, HJR *c.*1905; 07A, Cwrt Newydd, Welsh St. Donats, BAM 1964; 08C, Llantrisant Common, AN 1973!*; 59D/69C, Clyne Common, HJR *c.*1905; 70A, near Rhyd-y-fro, AN 1973!; 79A, Dyffryn Clydach, AL *c.*1905, Pencaerau, Neath, HJR *c.*1905; 79B, Gilfach, Neath, HJR *c.*1905; 90A, Rhigos, 1969!*

Rubus incurvatus Bab.
Native. Open woods and hedges.
09B, Abercynon, AN 1973!; 70A, near Rhydyfro, AN 1973!; 70B, Ystalyfera, HJR 1906*.

Rubus nemoralis Mueller [*R. villicaulis* subsp. *selmeri* (Lindeb.)]
Native. Woodland, hedges and scrub. Map 271.

Rubus polyanthemus Lindeb. (*R. pulcherrimus* Neuman).
Native. Hedges and scrub. Widespread, but as isolated bushes, mostly on higher ground. Map 272.

Rubus prolongatus Boulay & Letendre ex Corbière [*R. hypoleucus* Lef. & Mueller.]
Native. Common in similar areas and habitats to **R. rubritinctus**. Map 273.

Rubus rhombifolius Weihe ex Boenn.
Native. Scrub.
96A, Nash Point, HJR 1909.

Rubus riddelsdellii Rilstone [*R. godroni* Lec. & Lam. *p.p.*; *R. argentatus* Mueller. *p.p.*]
Native. Hedges and scrub.
08C, Llantrisant Common, AN 1973!, Pontyclun, AN 1973!; 17C, The Tumble, HJR 1909*; 18B, Rudry, AN 1973!; 48B, Horton, AHT 1908*; 97B, Penllyn Castle, AHT 1907*.

Rubus rubritinctus W.C.R. Watson [*R. argenteus* Weihe & Nees.]
Native. Wood borders and hedges. One of the commonest species west of the Lower Wye Valley and south of the Brecon Beacons, predominantly on lower ground. Map 274.

Rubus subinermoides Druce [*R. macrophyllus* Weihe & Nees.]
Native. Woodland and scrub.
08C, Llantrisant, HJR 1905, Pontyclun, AN 1973!; 18B, Rudry, AN 1973!*; 69, near Swansea, EFL 1905.

Rubus sprengelii Weihe
Native. Woodland.
07B, Peterston, HJR 1907*; 70C, Pontardawe, AN 1973!.

Rubus procerus Mueller ex Boulay Himalayan-giant Bramble.
Introduced. Grown widely in gardens and allotments and is frequently found naturalized in hedges and waste places, presumably bird sown (Newton, 1976), but with few localised records.
06B, Porthkerry, *c.*1973!; 08C, Llanharan, 1973!; 16A, Barry, *c.*1973!; 17B, Cardiff, GH 1986*; 27A, Roath, Cardiff, 1972!*; 27C, Cardiff Docks, 1986!; 69A, Mynydd Garn-goch, AN 1973; 99D, Clydach Vale, 1972!.

Rubus rossensis Newton
Native. Woodland.
00C, Aberdare, HJR 1905; 59B, Gorseinon, HJR 1906; 79B, Neath, HJR 1905.

Rubus ulmifolius Schott (*R. rusticanus* Merc.).
Native. A characteristic species of hedgebanks and waste places on lower ground but isolated bushes are occasionally found in the Uplands. Hybrids between this and other lowland species, particularly **R. caesius**, **R. rubritinctus** and **R. prolongatus** are common in lowland areas of the county. Fl. 6–11, fr. 8–11; Hb, B.hort, B.lap, B.luc, B.pasc, B.prat, B.terr (most n,p), ?Vespula, (Syr) E.ten, S.rib, Leu.luc, (Dip) Call.ery, Pol.lar, Poll.rud, (Lep) Ag.urt, Ce.arg, In.io, Ma.jur, Pi.nap, Pi.rap. Map 275.

Rubus longus (Rogers & Ley) Newton [*R. lasiocladus* vars *angustifolius* Rogers and *longus* Rogers & Ley.]
Native. Woodland margins and hedgebanks. Widespread but scattered. Map 276.

Rubus orbus W.C.R. Watson (*R. iricus* var. *minor* Rogers & Riddelsd.).
Native. Wood borders.
08C, Pontyclun, AN 1973!.

Rubus vestitus Weihe (*R. leucostachys* auct., non Sm.).
Native. Woodland and hedges. Widespread and frequent. Map 277.

Rubus melanocladus (Sudre) Riddelsd. [*R. cinerosus* Rogers; *R. podophyllus* Mueller.]
Native. Woodland and hedges.
00C, Aberdare, HJR 1905*; 18B, Caerphilly, AN 1973!.

Rubus aequalidens Newton
Native. Found in only one locality in the county.
87B, Merthyr Mawr, BAM 1961.

Rubus hastiformis W.C.R. Watson (*R. thyrsiger* Bab.).
Native. Woodland and scrub.
07A, Ystradowen, HJR 1922; 08C, Llantrisant, HJR 1905, AHT 1909*, Pontyclun, HJR 1922*, AN 1973!.

Rubus melanodermis Focke (*R. melanoxylon* sensu Bab.).
Native. Woodland and scrub.
18B, Caerphilly and Rudry, AN, 1973!*; 18C, Radyr, HJR 1909; 28A, Hill opposite Machen, HJR 1909, AN 1973!.

Rubus moylei W.C. Barton & Riddelsd. [*R. ericetorum* var. *cuneatus* Rogers & Ley.]
Native. Woodland and hedges. The most widespread glandular species in the county. Map 278.

Rubus raduloides (Rogers) Sudre [*R. scaber* Weihe & Nees *p.p.*]
Native. Woodland and scrub.
07A, Ystradowen, HJR 1904; 08C, Pontyclun and Llantrisant, HJR 1907–1909.

Rubus dentatifolius (Briggs) W.C.R. Watson (*R. vectensis* W.C.R. Watson; *R. borreri* sensu Sudre).
Native. Widespread but local in woods, and hedges. Map 279.

Rubus hibernicus (Rogers) Rogers
Native. Frequent in woods and hedges in the east of the county. Map 280.

Rubus leyanus Rogers [*R. drejeri* subsp. *leyanus* Rogers.]
Native. Common and often abundant in woods and hedges. Map 281.

Rubus morganwgensis W.C. Barton & Riddelsd. [*R. horridicaulis* Mueller.]
Native. Woods and hedges. Map 282.

Rubus echinatus Lindley.
Native. Recorded from only one locality in the county.
18C, Radyr, HJR 1907*.

Rubus flexuosus Mueller & Lef. [*R. foliosus* Weihe & Nees.]
Native. Woodland.
00C, Aberdare, HJR 1906*; 70C, Pontardawe, AN 1973!.

Rubus fuscicaulis Edees (*R. fuscus* auct., non Weihe).
Native. Woods, hedges and scrub.
00C, Aberdare, HJR 1905; 79A, Gilfach near Neath, HJR, 1905; 79B, Neath, HJR 1905; 80B, Glyn Neath, AN 1973!; 80C, Vale of Neath, HJR 1929.

Rubus gallofuscus Newton & Porter.
Native. Woods and hedges.
70B, near Ystalyfera, HJR 1906*, M. Porter 1988; 80C, Resolven, M. Porter 1984.

Rubus insectifolius Lef. & Mueller
Native. Recorded from only one locality.
08C, near Llantrisant, HJR 1907.

Rubus pallidus Weihe
Native. Recorded with certainty only from the north Cardiff area (18) but without precise locality.

Rubus rufescens Lef. & Mueller (*R. rosaceus* var. *infecundus* Rogers).
Native. Woodland, scrub and hedges. Map 283.

Rubus troiensis Newton [*R. scaber* Weihe & Nees *p.p.*]
Native. Woods and hedges.
08C, Pontyclun, BAM 1963, AN 1973!; 09B, Abercynon, AN 1973!; 18C, Radyr, HJR 1907*, Taffs Well, AHT 1908*; 18D, Llanishen, HJR 1908*, Cyncoed, 1972!*.

Rubus angusticuspis Sudre (*R. anglosaxonicus* var. *setulosus* Rogers).
Native. Woods, hedges and scrub.
00C, Aberdare, HJR 1906; 07A, Ystradowen, HJR *c.*1906; 07B, St-y-Nyl, HJR *c.*1908; 17A, near Whitchurch, HJR *c.*1908; 17B, Cardiff, 1941!; 18B, Rudry, HJR *c.*1908; 18D, Lisvane, HJR 1911; 80C, Resolven, HJR 1906, AN 1973!*.

R. breconensis W.C.R. Watson [*R. mutabilis* Genev.].
Native. Woods and hedgebanks.
00C, Aberdare, HJR 1906*, AN 1973!; 69C, Blackpill, JAW 1941*, AN 1973!.

Rubus dasyphyllus (Rogers) E.S. Marshall (*R. koehleri* var. *dasyphyllus* Rogers).
Native. Woods and hedges, remarkably scarce in the county considering its abundance in Breconshire and elsewhere in upland Wales. Map 284.

Rubus hylocharis W.C.R. Watson [*R. rosaceus* var. *hystrix* Weihe & Nees].
Native. Woodland.
08C, Llanharan, AHT 1909*; 68A, Langland Bay, HJR *c.*1905, Norton, JAW 1941*; 70B, Ystalyfera, HJR *c.*1906; 79B, Neath, HJR 1905*.

Rubus pruinosus Arrh. (*R. sublustris* Lees) [*R. corylifolius* Sm.]
Native. Woodland and scrub. Map 285.

Section Caesii Lef. & Courtois
Rubus caesius L. Dewberry. Mwyaren Fair.
Native. Sand dunes, hedges and scrub. Common in Gower and the Vale, far less so in the Uplands. Fl. 5–10, fr. 7–10; Hb (p), B.luc (n,p), Sb (2 spp., p), (Syr) E.ten. Map 286.

Rubus caesius × R. fruticosus (*Rubi corylifolii*)
Hybrids between **R. caesius** and other sections of subgenus **Rubus** are of frequent occurrence in lowland Glamorgan (A. Newton, 1976), but localized records are few.
07C, Cowbridge, V.J. Russell 1970*; 17B, Cardiff, GH 1986*; 49C, Whiteford Burrows, HJR, *c.*1905; 78B, Port Talbot, HJR, *c.*1905.

ROSA L.
Rosa arvensis Hudson   Field Rose. Rhosyn Gwyn Gwyllt
Native. Hedges, woods and scrub. Common in the Vale and Gower, becoming less frequent in the Uplands through the absence of suitable habitats. Fl. 6–8, fr. 8–11. Map 287.

Two hybrids between **Rosa arvensis** and members of **R. canina** agg. have been reported from the county on only one occasion each; they are likely to be of more frequent occurrence: **R. arvensis × R. canina** (**R. × wheldonii** Wolley-Dod) 17D, Llandough, 1920!*, and **R. arvensis × R. dumetorum** 07B, St-y-Nyll, HJR 1908*.

Rosa pimpinellifolia L. Burnet Rose. Mwcog, Rhosyn Draenllwyn.
Native. Sand dunes and limestone cliffs. Frequent in the west of the county, rare in the east and absent from the north. Fl. 6–8, fr. 8–10; (!) B.prat (p). Map 288.

Rosa rugosa Thunb., Japanese Rose, has been found as a garden escape on four occasions (18C, Radyr, P.N. Lewis 1961*; 48B, Penrice, JAW 1923*; 69C, Swansea, H & CH 1974; 99D, Gilfach Goch, 1979!), **Rosa blanda** Aiton, Smooth Rose, twice (48B, Penrice, JAW *c.*1955; 69C, Mayals, Swansea, JAW 1955*) and **Rosa virginiana** J. Herrmann, Virginian Rose, once (17B, Cardiff, 1956!*).

Rosa stylosa Desv.   Short-styled Field-rose. Rhosyn Ungolofn.

Native. A rare plant of hedges in the south of the county. Apparently more frequent in the past, although it may have been overlooked in the present survey. Map 289.

Rosa canina L.   Dog-rose. Rhosyn Coch Gwyllt.
Native. Hedges, scrub, waste ground etc. Very common throughout the county. Fl. 6–7, fr. 9–11; Hb (p), B.prat (p), (Syr) S.lun, (Dip) Call.sp. Map 290.

Rosa canina × R. coriifolia has been recorded once from 07C, near St. Mary Church, ABP 1969!*, det. A.L. Primavesi.

Rosa canina × R. obtusifolia (**R. × concinnoides** W.-Dod) has been recorded once from 28C, near Llanedeyrn, 1969!*, det. A.L. Primavesi.

Rosa canina × R. rubiginosa (**R. x latens** W.-Dod) has been recorded on one occasion only and that specimen was probably back-crossed with **R. canina.** 18C, Rhiwbina, Cardiff, GH 1989*, det. A.L. Primavesi.

Rosa canina × R. stylosa has been recorded on three occasions in woodland margins and hedgerows. 06B, Porthkerry, 1969!*; 16B, Lavernock, 1969!*; 98C, Coity, 1970!*; all det. A.L. Primavesi.

Rosa canina × R. tomentosa (**R. × curvispina** W.-Dod) has been recorded once from 00C, Mountain Ash, 1975!*, det. A.L. Primavesi.

The following species and hybrids (to **R. coriifolia**) have not always been separated from **R. canina** L.; they are probably more frequent than recent records indicate. Much work remains to be done in Glamorgan on members of **Rosa** Sect. Caninae DC.

Rosa afzeliana Fries (*R. dumalis* Bechst. *p.p.*) Glaucous Dog-rose. Rhosyn Llwydwyrdd.
Native. A rare plant of hedges, scattered throughout the county. Map 291.

Rosa afzeliana × R. canina has been recorded on three occasions from hedgerows in the Vale and once in the Uplands: 00C, Abernant, JWD & A. Newton 1973!*, 06B, Porthkerry Bay, 1969!*; 07C, near Llantrithyd, W.H. Nicholls 1970*; 98D, near Coychurch, 1970!*; all det. A.L. Primavesi.

Rosa obtusifolia Desv.   Round-leaved Dog-rose. Rhosyn Deilen Llawban.
Native. A rare plant of hedges, recorded from only six localities with no recent records.
00C, near Aberdare, HJR 1902*; 58B, Bishopston Valley, JAW 1928; 59B, near Loughor, HJR 1906*; 78B, Port Talbot, HJR *c.*1906; 80B, Glyn Neath, HJR *c.*1906; 90D, Llwydcoed, HJR *c.*1906.

Rosa dumetorum Thuill.   Thicket Dog-rose. Rhosyn Dryslwyn.

Native. Hedges. Locally frequent in the east of the county, apparently absent or nearly so from Gower and the Uplands, although this may be due to confusion with **R. canina**. Map 292.

Rosa coriifolia Fr. Hairy Dog-rose. Rhosyn Blewog.
Native. Hedges.
00C, Abernant, JWD 1973!*, det A.L. Primavesi.

Rosa tomentosa Sm. Harsh Downy-rose. Rhosyn Lledwlanog.
Native. A local plant of hedgerows scattered around the county. Very much under-recorded in the present survey. Map 293.

Rosa sherardii Davies Sherard's Downy-rose. Rhosyn Sherard.
Native. A local plant of hedgerows scattered around the county. Probably under-recorded. Map 294.

Rosa mollis Sm. (*R. villosa* auct.). Soft Downy-rose. Rhosyn Gwlanog.
Native. A rare plant of hedgerows in the north of the county.
09A & 09B, Mountain Ash, AMP 1973; 10C, Bargoed, 1958!*; 80B, Glyn Neath and Aberpergwm, HJR 1911*.

Rosa rubiginosa L. Sweet-briar. Drysen Bêr.
Native. A local plant of hedgerows and scrub in scattered localities around the county. Map 295.

Rosa micrantha Borrer ex Sm. Small-flowered Sweet-briar. Rhoslwyn Pêr.
Native. A local plant of scrub, cliffs and hedges, mainly in the south of the county. Map 296.

AGRIMONIA L.

Agrimonia eupatoria L. Agrimony. Llys y Dryw.
Native. Fields, hedges, dune grassland, roadsides and railway embankments usually on basic soils. Common in Gower and the Vale, less so in the Uplands. Fl. 7–8, fr. 8–10. Map 297.

Agrimonia procera Wallr. (*A. odorata* auct., non Miller). Fragrant Agrimony. Llys y Dryw Peraroglus.
Native. In similar habitats to the last but usually on acid soils and much less common. In scattered localities throughout the county. Fl. 7–8, fr. 8–10. Map 298.

SANGUISORBA L.

Sanguisorba officinalis L. Great Burnet. Llysyrlys.
Native. Damp grassland river banks and cliff ledges. Common in the Uplands but very local in the Vale and with only two records in Gower. Map 299.

Sanguisorba minor Scop. subsp. **minor** (*Poterium sanguisorba* L.). Salad Burnet. Gwyddlwdn Cyffredin.

Native. Grassland, sand dunes, cliffs, roadsides and railway embankments, especially on limestone. Common in the Vale and Gower, rare in the Uplands. Fl. 5–6, 9, fr. 6–7. Map 300.

Sanguisorba minor subsp. **muricata** Briq. (*Poterium polygamum* Waldst. & Kit.). Fodder Burnet. Gwyddlwdn Tramor.
Introduced. Waste places by roadsides and railways in several scattered localities. Map 301.

Acaena anserinifolia (J.R. & G. Forster) Druce, Pirri-pirri-bur, a native of New Zealand has once been recorded as an escape (17B, Llandaff, M. Craster 1964*).

GEUM L.

Geum rivale L. Water Avens. Mapgoll Glan y Dŵr.
Native. Moist woods, shaded river sides and damp cliff ledges. Mainly in the north and east of the county. Apparently more frequent in the past. Map 302.

Geum rivale × G. urbanum (**G. × intermedium** Ehrh.) is relatively frequent along the banks of the Afon Taf north of Merthyr (00A! & 01C!). It has also been recorded from a damp wood at Coedrhiglan (17A, H. Evans & G. Traherne c.1900), and the lakeside at Talygarn (07A, H.J. Dawson 1982*).

Geum urbanum L. Wood Avens. Mapgoll.
Native. Woods and hedges. Common in the Vale and Gower, less so in the Uplands. Fl. 5–10, fr. 6–10. Map 303.

POTENTILLA L.

Potentilla palustris (L.) Scop. Marsh Cinquefoil. Pumdalen y Gors.
Native. Marshes, damp grassland and bogs. Local, mainly in the south and west of the county. Decreasing due to loss of suitable habitats through drainage. Fl 5–7, fr. 7–8; Hb (n,p), B.pasc. Map 304.

Potentilla anserina L. Silverweed. Dail Arian, Tinllwyd.
Native. Roadsides, railways and waste places, also on dunes and shingle. Abundant throughout the county. Fl. 5–9. Map 305.

Potentilla argentea L. Hoary Cinquefoil. Pumbys Arian-ddail.
Introduced. On a railway embankment, where it has persisted for over 50 years (16A, Cadoxton, RM 1924*, 1974!), also as a non-persistent alien in Cardiff (17, 1924).

Potentilla norvegica L. Ternate-leaved Cinquefoil. Tribys Tramor.
Introduced. Waste places, railways and roadsides, in a few localities around Cardiff and Swansea. Only two recent records.

00C, Aberdare, J.R. Shepherd 1913; 09A, Mountain Ash, JPC 1987*; 16B, Sully, HJR c.1906; 17B, Cardiff, HJR 1907*, 1925!*; 17D, Grangetown, Cardiff, 1920!*; 18C, Radyr, RLS c.1924, RAH 1972; 18D, Cefn Onn, PWR 1920*; 27A, Splott, 1925!*; 69C, Swansea, JAW 1927*.

Potentilla tabernaemontani Ascherson Spring Cinquefoil. Pumdalen y Gwanwyn.
Native. Limestone sea cliffs in Gower, near outcrops and in crevices; rather local.
48A, Mewslade Bay, HJR 1910*, JAW 1948*, Mewslade to Common Cliff, 1978!; 48B, Overton, 1976!; 48D, Port Eynon, EV 1912*, 1970!; 58A, Oxwich Point, 1975!; 58B, Pwlldu, HJR 1906, 1970!; 58C, Oxwich Point, 1974!.

Potentilla erecta (L.) Räuschel Tormentil. Tresgl y Moch.
Native. Heaths and moors. Abundant throughout the county. Fl. 5–10, fr. 7–10. Map 306.

Potentilla anglica Laicharding Trailing Tormentil. Tresgl Ymlusgol.
Native. Woods, hedgebanks, and along railways. Abundant throughout the county except the extreme south. Map 307.

Potentilla anglica × P. erecta (**P. × suberecta** Zimmeter).
Although only recorded twice during the present survey (18A and 88B) this hybrid, like the following, is possibly of fairly frequent occurrence where the two parent species meet. Map 308.

Potentilla anglica × P. reptans (**P. × mixta** Nolte ex Reichenb. (incl. *P. erecta × P. reptans* (*P. × italica* Lehm.)).
Several old records but not recorded during the present survey (see note above). Map 309.

Potentilla reptans L. Creeping Cinquefoil. Pumdalen Ymlusgol.
Native. Roadsides, hedgerows, duneland, railway embankments, waste places etc. Frequent throughout the county. Fl. 5–11, fr. 7, 10. Map 310.

Potentilla sterilis (L.) Garcke Barren Strawberry. Coegfefusen.
Native. Roadsides, hedgerows, railway embankments, woodland etc. Frequent throughout the county. Fl. 3–5. Map 311.

Potentilla fruticosa L., Shrubby Cinquefoil, Llwyn Pumbys, has once been found as an alien near Cardiff (18C, Castell Coch, JS c.1876), **P. intermedia** L., Russian Cinquefoil, on three occasions (17B, Canton, RLS c.1924; 17D, Leckwith, RLS 1923* (originally determined as *P. inclinata* Vill.), Grangetown, RLS 1938!*), and **P. recta** L., Sulphur Cinquefoil, Pumdalen Talsyth, twice (16A, Pymbylu Moors, J.B. Miller 1972*; 87A, near Merthyr Mawr, VCB 1971).

FRAGARIA L.

Fragaria vesca L. Wild Strawberry. Mefusen y Goedwig.
Native. Woodland banks, hedgebanks, railway lines, duneland etc. Frequent throughout the county. Fl. 4–6, fr. 6–7. Map 312.

Fragaria moschata Duchesne Hautbois Strawberry. Mefusen Fawr.
Introduced. A rare plant of hedgebanks with only one recent record, one old record and two doubtful records, which may have been errors for **F. × ananassa** (16A, Barry, and 18C, Radyr, both RLS c.1925).
69C, Blackpill, C. Marks 1927; 97D, Llanblethian, 1974!*.

Fragaria chiloensis (L.) Duchesne × **F. virginiana** Duchesne (**F. × ananassa** Duchesne). Garden Strawberry. Mefusen.
Introduced. Occasionally found on roadsides, railway embankments and waste ground.
16A, Sully, ABP 1974!*; 17B, Cardiff, 1920!*; 17D, Penarth Dock, HJR c.1900; 18C, Radyr, EV 1933*; 69C, near Mumbles, JAW 1913.

Duchesnea indica (Andrews) Focke, Yellow-flowered strawberry, a rare alien of waste places, has been found near Cardiff on three occasions (16B, Penarth, RAH 1989; 17B, Cardiff, D.B. Sanday 1983; 18D, Birchgrove, R.H. Gwyther 1947*).

ALCHEMILLA L.

Alchemilla vulgaris L. agg. Lady's-mantle. Mantell Fair.
Native. Pastures. Locally frequent in the east of the county but with only two isolated records from the west. Fl. 5–6. Map 313.

The following segregates have been recorded from the county.
Alchemilla xanthochlora Rothm. Intermediate Lady's-mantle. Mantell Fair Cyfryngol.
Native. Pastures, roadsides, hedgerows, railway embankments and woodland. Locally frequent in the east of the county. Map 314.

Alchemilla filicaulis subsp. **vestita** (Buser) M.E. Bradshaw. (*A. vestita* (Buser) Raunk.). Hairy Lady's-mantle. Mantell Fair Flewog.
Native. Pastures, roadsides and railway embankments. Occasional in the east of the county. Fl. 5. Map 315.

Alchemilla glabra Neygenf. Smooth Lady's-Mantle. Mantell Fair y Nant.
Native. Damp grassy places, heaths and woods. Occasional in the north-east of the county but rare elsewhere. Map 316.

Alchemilla mollis (Buser) Rothm., Robust Lady's-mantle, a naturalized garden outcast, has been found in three widely separated localities (08D, Llantrisant, 1975!; 10A, Bute Town, 1978!; 90B, Hirwaun, 1974!).

APHANES L.

Aphanes arvensis L. Parsley-piert. Troed y Dryw.
Native. Dry pastures and cultivated ground, mainly on basic soils. Frequent in the south, almost absent in the north. Map 317.

Aphanes microcarpa (Boiss. & Reuter) Rothm. Slender Parsley-piert. Troed y Dryw Fain.
Native. Dry grassland. Apparently scarce and found mainly in the south-west of the county, but probably more frequent than records suggest due to confusion with the former species. Map 318.

PYRUS L.

Pyrus pyraster Burgsd. (*Pyrus communis* auct., non L.). Wild Pear. Gellygen, Rhwnen.
Introduced. A rare plant of woods and hedges, recorded in the past from sites scattered throughout the county but with only one recent record. Map 319.

Cydonia oblonga Miller (*Pyrus cydonia* L.), Common Quince, has been recorded once as an introduction near Cardiff (18D, near Cardiff, L. Pierce 1937).

MALUS Miller

Malus sylvestris Miller *s.l.*
Native in woods and hedges, but also widespread as an introduction. Fl. 4–6, fr. 8–10. Map 320.

Malus sylvestris Miller *s.s.* Crab Apple. Pren Afal Sur.
Native. Woods and hedges. Frequent in the south and east, less so in the north and west. Map 321.

Malus domestica Borkh. (*Malus sylvestris* subsp. *mitis* (Wallr.) Mansfeld). Apple. Pren Afalau.
Introduced. Widespread in hedges, mainly as isolated trees. Absent from the north-west. Map 322.

SORBUS L.

Sorbus aucuparia L. Rowan. Cerddinen, Criafolen.
Native. Rocky places, woods, hedges. Frequent throughout most of the county. Fl. 5–6, fr. 9–10. Map 323.

Sorbus torminalis (L.) Crantz Wild Service-tree. Cerddinen Folwst.
Native. A rare tree of mainly coastal woodland, with only one recent record.
17C, Wenvoe, C.E. Ollivant *c*.1870; 48B/58A, Penrice, WT *c*.1800; 58A, Nicholaston, EV 1917*, Crawley Cliff, JAW 1956*, M.C.F. Proctor 1989!; 58B, Pwll Du, EV 1934*; 69C, Clyne Wood, JAW 1924–6; 78B, Craig Fawr, Margam, JM 1843; 79C, Briton Ferry, J.M.

Traherne 1808*, JWGG *c*.1840, Crymlyn Burrows, CRH 1990.

Sorbus porrigentiformis E.F. Warburg Green-leaved Whitebeam. Cerddinen Ymledol.
Native. Rocky places and woodland on limestone cliffs and also on Pennant Sandstone cliffs near Aberdare. A rare plant in the county, growing in small numbers in a few localities. Some old records of **S. rupicola** apparently refer to this species. Fl. 5–6, fr. 9–10(!).
48A, near Paviland, 1970!; 48B, near Port Eynon, 1974!; 58A, Nicholaston, 1970+!, Crawley Cliff, 1976!** (det. P.F. Yeo), M.C.F. Proctor 1989!; 90C, Craig-y-llyn, 1969!*; 90C/D, Aberdare, HJR 1902*.

Sorbus rupicola (Syme) Hedl. Rock Whitebeam. Cerddinen y Graig.
Native. Woodland on limestone cliffs. A rare plant in the county, known only from one site in Gower, where a survey by M.C.F. Proctor in 1989 confirmed the presence of both **S. rupicola** and **S. porrigentiformis**.
58A, Crawley Cliff, V.M. Peel 1932*, GTG 1955, 1976!, M.C.F. Proctor 1989!.

Records of other species of **Sorbus** require confirmation or refer to the above species and those of **S. rupicola × S. torminalis** refer to a locality in v.c. 42.

Sorbus intermedia (Ehrh.) Pers., Swedish Whitebeam, Cerddinen Dramor, is an introduced species known from only two localities near Cardiff (16B, Cosmeston Cement works, Penarth, MG 1980*; 18D, Thornhill, JPC 1972*).

COTONEASTER Medicus

Cotoneaster horizontalis Decne Wall Cotoneaster. Cotoneaster y Mur.
Introduced. Often well established on old walls and limestone rocks. Locally frequent in the eastern Vale, less so elsewhere. Map 324.

Cotoneaster simonsii Baker Himalayan Cotoneaster. Cotoneaster y Graig.
Introduced. Often well established in woods and hedges. Scattered localities throughout the south of the county. Map 325.

Cotoneaster dielsianus E. Pritzel ex Diels
Introduced. Becoming established on walls and in rocky places. Found in only four localities. Very similar to **C. simonsii** from which it differs mainly in having leaves densely tomentose beneath; it has in the past been confused with that species.
06B, Rhoose, ABP 1971!*; 18D, Cardiff, H. Morrey Salmon 1980*; 69B, near Morriston, GH 1989*; 98C, Coity, ABP 1974!*.

Cotoneaster microphyllus Wallich ex Lindley Small-leaved Cotoneaster. Cotoneaster Ddeilos.

Introduced. Walls and rocky places. Well established in many localities mainly in the south of the county; spreading on limestone cliffs in Gower. Map 326.

Cotoneaster frigidus Wallich ex Lindley
Introduced. Woods, walls and waste places.
00C, Abernant, KA 1973; 09A, Mountain Ash, AMP 1970*, 1973*; 68A, Lilliput, JAW 1923*, Coltshill Quarries, JAW, 1941!*; 69C, Clyne Valley, Sketty, Cockett, JAW c.1923.

Cotoneaster salicifolius Franchet has been recorded once from waste ground at Llandough near Cardiff (17D, 1974!*), **C. bullatus** Boiss. has been recorded once as a garden weed in Cardiff (18D, HMS 1980*), and **C. × watereri** Exell once as two well established bushes on waste ground at Cefn Onn (18D, GH 1986*), probably planted originally.

Pyracantha coccinea M.J. Roemer has been recorded once from a cliff near Penarth (17D, P. Morgan 1953*) and **Mespilus germanica** L. from four localities (18C, Radyr, JS c.1875; 48B, Stouthall Woods, JBe, 1982; 69C, Singleton, JAW 1939*; 97A, Ewenni, HJR c.1900).

CRATAEGUS L.

Crataegus laevigata (Poiret) DC. (*C. oxyacanthoides* Thuill.). Midland Hawthorn. Draenen Ysbyddaden. Introduced. A rare plant of hedgerows in several scattered localities. Some records may refer to **C. × media** Bechst.
00C, Aberdare, HJR 1906*; 09A, Mountain Ash, AMP 1974*; 58A, Penmaen Hill, JAW 1955*; 58B, Caswell Corner, JAW 1921*; 87B, Merthyr Mawr, JAW 1939.

Crataegus laevigata × C. monogyna (**C. × media** Bechst.) has been recorded once (00C, Abernant Park, Aberdare, HJR 1905 & 1906) but only as a planted tree.

Crataegus monogyna Jacq. Hawthorn. Draenen Wen.
Native. Woods, scrub, hedges, cliffs and rocky places etc. Very abundant throughout the county. Fl. 5–6, fr. 9–11; Hb (n,p), Sb (n,p), (Syr) E.pert (p), E.ten (p). Map 327.

Three other species of **Crataegus** have occurred as rare introductions. **Crataegus intricata** Lange has been recorded from three localities (58B, Caswell Bay, GCD c.1920; 69C, Sketty, JAW c.1930; 78B, near Groeswen, JAW 1941*), **C. crus-galli** L. from four localities around Swansea (59D, Tai-bach, JAW 1918; 69C, Lower Sketty, JAW 1918; 69D, Gors Road, JAW c.1920; 79A, Dyffryn, Neath, JAW, 1953*) and **C. laciniata** Ucria (*C. orientalis* Pallas ex Bieb.) from one locality (69C, Clyne Valley, E.M. Wakefield c.1918).

PRUNUS L.

Prunus cerasifera Ehrh. Cherry Plum. Gaugeiriosen.

Introduced. Established in a hedgerow in one locality.
97A, Colwinston, ABP 1978*.

Prunus spinosa L. Blackthorn. Draenen Ddu.
Native. Woods, thickets, scrub, hedges, cliffs etc. Abundant throughout the county. Fl. 4–5, fr. 9–11; (Syr) E.ten. Map 328.

Prunus domestica L. subsp. **domestica** Wild Plum. Eirinen.
Introduced. Hedges and less frequently in woodland. Locally frequent in Gower and the Vale but absent from much of the Uplands due to the scarcity of suitable habitats. Map 329.

Prunus domestica L. subsp. **institia** (L.) C.K. Schneider (*P. institia* L., incl. *P. domestica* subsp. *italica* (Borkh.) Hegi). Bullace. Eirinen Fwlas.
Introduced. Hedges. Perhaps more frequent than the small number of records suggests. This subspecies is often planted in hedges in other parts of Wales and the relatively small number of records from Glamorgan may be a consequence of its reluctance to flower and the difficulty of separating it vegetatively from subsp. **domestica**.
07D, Bonvilston, JIL 1970; 17A, between Radyr and Llandaff, HJR 1909*, Fairwater, HJR 1909*; 18D, Lisvane, H.H. Thomas 1986*; 69C, Mayals, JAW 1956*; 87A, Newton, JAW 1932*, EMT 1934*, Nottage, EV 1938*; 88C, North Cornelly, 1974!.

Prunus avium (L.) L. Wild Cherry. Ceiriosen Ddu.
Native. Woods, hedges, railway embankments. Locally frequent in west Gower and parts of the Vale; rare in the Uplands. Fl. 4–5. Map 330.

Prunus cerasus L. Dwarf Cherry. Ceiriosen.
Introduced. Hedges. Fairly frequent in Gower but very scarce elsewhere. Map 331.

Prunus padus L. Bird Cherry. Ceiriosen yr Adar.
Native. Hedges and woods. Scarce and local in the county with recent records only from the north and east. **P. padus** reaches the southern limit of its native British range in Glamorgan. Fl. 5. Map 332.

Prunus laurocerasus L. Cherry Laurel. Llawrsirianen.
Introduced. Plantations. Hedges and waste places. Found in several scattered localities. Fl. 2–5, fr. 8–9; Hb, Bn, (Dip) Call, Luc. Map 333.

Prunus serotina Ehrh. has been recorded once as a rare alien (59D, near Ilston, JPB 1942), as has **P. lusitanica** L. (19C, near Llanbradach, 1978!).

LEGUMINOSAE (FABACEAE)
LABURNUM Fabr.

Laburnum anagyroides Medicus Laburnum. Tresi Aur.

Introduced. Occasionally more or less naturalized. 00C, Mountain Ash, AMP 1975; 17B, Cardiff, 1974!; 18C, Radyr, P.N. Lewis 1961*; 49D, Weobley Castle, JAW *c.*1956; 58B, Caswell, JAW *c.*1956; 68A, Langland, JAW, *c.*1956; 89D, near Pencoed, 1976!; 90D, Llwydcoed, HJR *c.*1907.

CYTISUS L.

Cytisus scoparius (L.) Link   Broom. Banadl.
Native. Heaths, dry hillsides, roadsides, railway embankments and waste ground. Locally abundant in suitable habitats, but scarce or absent in most of Gower, the southern Vale and parts of the central and eastern Uplands. Fl. 4–6. Map 334.

GENISTA L.

Genista tinctoria L.   Dyer's Greenweed. Melynog y Waun.
Native. Rough grassland and roadside banks. Frequent in the south-eastern Vale but scarce or local elsewhere. Map 335.

Genista anglica L.   Petty Whin. Cracheithin.
Native. Heathland, particularly in the lowlands and on the lower hills, but absent from maritime heath on sea-cliffs. Frequent in eastern Gower and the northern Vale. Fl. 5–6, fr. 7–9; B.lap (p). Map 336.

SPARTIUM L.

Spartium junceum L.   Spanish Broom. Banadl Sbaeneg.
Rare casual and occasional garden escape. Often planted as an ornamental shrub in gardens and on roadsides as at St. Athan (06A), and used with some success in reclamation of derelict land, for example in the Lower Swansea Valley (69B & D). Once recorded as a dock alien in Cardiff (17B, 1926!).

ULEX L.

Ulex europaeus L.   Gorse. Eithin.
Native. Heathland, rough pastures, open woods and hedges. Particularly abundant in maritime heath, but common throughout the county. Fl. 1–7, 10–12, fr. 6–8; Hb (p), Bb, (Syr) S.rib all (p). Map 337.

Ulex gallii Planchon   Western Gorse. Eithin y Mynydd.
Native. Heathland on acid soils; often associated with **Calluna vulgaris**, **Erica tetralix** and **Molinia caerulea**. Common in suitable habitats throughout the county, and more abundant than **Ulex europaeus** in many areas. Locally abundant in sea-cliff heaths in southern Gower, where it grows on patches of acid drift soil. Fl. 7–10, fr. 4–6; Bb(p). Map 338.

LUPINUS L.

Lupinus arboreus Sims   Tree Lupin. Coeden Fys y Blaidd.

Introduced. Sand-dunes and waste ground in western Glamorgan.
48B, Horton Dunes, JAW 1945*, 1975!; 58B, Pennard Burrows, JAW 1956*, 1974!; 69B, Swansea, 1974!, Llansamlet, T. Davies 1985; 69D, Swansea, 1974!; 79C, Crymlyn Burrows, 1974!; 96B, Llantwit Major, JPC 1981.

Lupinus angustifolius L. has once been found as a casual on sand ballast in Radyr (18C, RLS *c.*1920–1925*), **L. micranthus** Guss. similarly in an allotment at Splott, Cardiff (27A, RLS 1926*–1927*), and **L. nootkatensis** Donn ex Sims on a tip at Blackpill near Swansea (69C, JAW 1924). **L. polyphyllus** Lindley has twice occurred as a garden escape (79C, H & CH 1973; 18C, GH 1980–87).

Robinia pseudacacia L., False-acacia, Ffug-acesia, has been recorded three times as an introduction (17A, St Fagans, 1976!; 17D, Michaelston-le-Pit, 1934*!; 69, Swansea, M.H. Sykes *c.*1942).

GALEGA L.

Galega officinalis L.   Goat's-rue. Llys Llaeth.
Casual on waste ground, usually near docks; possibly naturalized in some localities.
16A, Barry Docks, RLS 1921*, JPC 1980*; 17A, St. Fagans, G. Charteris 1938, EV 1941*; 17B, Canton, RLS 1923*; 17D, Penarth, JS 1876, Penarth Ferry, THT 1902*, Leckwith Bridge, SGH 1967* (perhaps naturalized); 27A, Splott, 1937!*; 27C, Cardiff Docks, 1973!; 48B, Horton Dunes, B.E. Jones 1972 (perhaps naturalized).

ASTRAGALUS L.

Astragalus glycyphyllos L.   Wild Liquorice. Llaeth-wyg.
Native or introduced. Roadsides, rough grassland and waste ground. Well-established and increasing in one locality.
16A, Barry, JS *c.*1886, AHT 1899*, 1907, Cold Knap, HJR 1905, EV 1936*; 16B, Sully, HJR *c.*1908; 17D, Penarth, AHT *c.*1870, RLS *c.*1920*; 18C, Castell Coch 1957!; 78D, Kenfig, B.G. Harris 1970; 79C, Briton Ferry Docks, H & CH, 1972*, 1980! (increasing).

Astragalus boeticus L., has twice been found as an alien in Cardiff (27A, Splott, 1926!; Pengam, RLS 1940*), **A. hamosus** L. on three occasions (18C, Radyr, RLS 1922!*; 27A, East Moors, JS *c.*1876*, Pengam, RLS 1940*), and **A. stella** Gouan once (17B, Cardiff Docks, JS *c.*1870).

Cicer arietinum L. has been found as a dockside alien at three places (16A, Barry Docks, RLS 1923*; 17D, Penarth Ferry, JS *c.*1876*; 27A, East Moors, JS *c.*1876*). **Colutea arborescens** L. has once been found as a spontaneous garden weed (17B, Llandaff, W.H. Dymond 1944*). **Psoralea bituminosa** L. has been found twice as a sand ballast alien (17B, Cardiff Docks 1924!*; 18C, Radyr, RLS 1922*).

VICIA L.

Vicia orobus DC. Wood Bitter-vetch. Pysen y Coed.
Native. Upland woods and mountain cliffs. Scarce, with only two recent records.
00C, Aberdare, HJR c.1907; 00D, near Fochrhiw, 1947!*; 10D, Brithdir, 1974!*; 69A, Penllergaer, JWGG c.1841**, M. Cahn & B.R. Pawson 1990!; 80B, Sgwd Einion Gam, J.D. Hooker c.1841; 90A, Glyn Neath to Hirwaun, HJR c.1907; 99A, Cwm Selsig, HJR c.1907, JAW 1916.

Vicia cracca L. Tufted Vetch. Tagwyg Bysen.
Native. Roadsides, hedges and scrub. Frequent to common throughout the county. Map 339.

Vicia sylvatica L. Wood Vetch. Ffugbysen y Wig.
Native. Woods and roadside banks. Scarce except in the south-eastern Vale, and absent from Gower and most of the Uplands. Map 340.

Vicia villosa Roth Fodder Vetch. Ffugbysen yr Ar.
Casual, on waste ground near docks or railways. Recently recorded only in Cardiff and the Neath district. Map 341.

Four subspecies have been recorded: subsp. **villosa** (16A, RLS 1922*; 17B, JPC 1981*; 17D, c.1923!, JPC 1981; 18C, RLS c.1922; 69A, JAW c.1924; 78B, HJR c.1904), subsp. **varia** (Host) Corb. (00C, HJR 1900*; 16A, HJR c.1904, P.H. Holland 1919*; 17B, HJR c.1904, JWD 1976*; 17D, JS c.1876; 27A, EV 1934*, RLS & F. Norton 1938*; 78B, HJR 1904, EV & RLS 1936*; 79C, 1974!), subsp. **eriocarpa** (Hausskn.) P.W. Ball (17B, RLS c.1938!) and subsp. **pseudocracca** (Bertol.) P.W. Ball (18C, RLS c.1925).

Vicia hirsuta (L.) S.F. Gray Hairy Tare. Corbysen Flewog.
Native. Cultivated land, hedges, rough grassland and waste ground. Locally frequent in the lowlands and in Upland valleys, but absent from many areas. Map 342.

Vicia tetrasperma (L.) Schreber Smooth Tare. Corbysen Lefn Ronynnog.
Native. Rough grassland and waste ground. Locally fairly frequent in the south-eastern Vale and to the north-west of Swansea, but rare elsewhere. Map 343.

Vicia sepium L. Bush Vetch. Ffugbysen y Cloddiau.
Native. Hedges, scrub and rough grassland. Common in the lowlands, but rather local in the lower hills and absent from the central Uplands. Map 344.

Vicia sativa L. *s.l.* (excluding subsp. **nigra**) Common Vetch. Ffugbysen Faethol.
Native, but widely introduced as a weed or as an escape from cultivation. Roadsides, hedges, cultivated land, dune grassland, waste ground and rough grassland. Fl. 4–7, fr. 5–9. Map 345.

Vicia sativa L. subsp. **nigra** (L.) Ehrh. (*Vicia angustifolia* L.). Narrow-leaved Vetch. Ffugbysen Gulddail Ruddog.
Native. Dry grassland, especially on fixed sand-dunes, roadsides and hedgebanks. Apparently restricted to sand-dunes in western Glamorgan but occurring in a wider range of habitats in the eastern part of the county. Most sand-dune populations are polymorphic in flower-colour, with pink-flowered and purple-flowered plants intermingled. (!) Fl. 5–6, fr. 6–7; B.lap, B.pasc, B.terr. Map 346.

Vicia lathyroides L. Spring Vetch. Ffugbysen y Gwanwyn.
Native. Dry grassland on fixed sand-dunes, and occasionally on shallow soil over limestone. Usually associated with other winter annuals. Very local, and apparently restricted to the coast; recently recorded only from Llangennith Burrows, Pennard Burrows and Kenfig Burrows.
16A, Barry, M.A. David c.1907; 49C, Llangennith Burrows, 1964!, 1976!; 58A & 58B, Pennard Burrows, V.M. Peel 1928*, 1955!*, 1979!; 78D, Kenfig Burrows, 1975!; 87A, Porthcawl, PWR 1916*, EV 1924; 87B, near Wig-fach, PJo 1984; 87D, Southerndown, HOB c.1907; 88C, Kenfig Burrows, 1975!. (!) Fl. 4–5, fr. 5–6.

Vicia lutea L. Yellow-vetch. Ffugbysen Felen Arw-godog.
Casual on waste ground, usually near docks. Recently recorded only from Barry Docks (16A, RLS 1939*, DMcC 1974) and Swansea (69D, H & CH 1976). Recorded in the past from 00C, HJR 1897; 07C, J. Rees 1923; 09C, H. Harris c.1906; 17B, JS c.1876, RLS 1928; 17D, JS c.1876, RLS 1925; and 27A (subsp. **lutea** and subsp. **vestita** (Boiss.) Rouy, Splott, RLS 1926!*).

Vicia bithynica (L.) L. Bithynian Vetch. Ffugbysen Ruddlas Arw-godog.
Possibly native, but also occurring as a casual on waste ground. It was listed as a Glamorgan native by Watson (1847), on the basis of a specimen in the British Museum labelled 'Wood near Cardiff' (Trow, 1911). It occurs as an apparent native in several localities on the southern side of the Bristol Channel, in Somerset and Gloucestershire. The status of Eleanor Vachell's Cogan population (see below) is not clear; it grew on a roadside bank now destroyed by road widening. Many of the Somerset and Gloucestershire populations grow or grew (it has disappeared from many sites) on roadside banks.
16A, Barry, M. Cubitt 1853; 17B, Cardiff Docks, RLS 1925*; 17D, Cogan, EV 1936*; 18C, Radyr (adventive), RLS 1922*.

Several other species of **Vicia** have been found as aliens, growing on dockside ballast or from grain cleanings or bird-seed. **Vicia faba** L. is most frequent and has been found in six squares (00C, HJR c.1906; 09C, HJR c.1906; 17B, pre 1930!*; 69C, JAW pre 1950; 78B HJR c.1906; 79A 1974!), the similar **V. narbonensis** L. has been found on two occasions (16A, Barry Docks, RLS 1922*; 27A, Splott, RLS 1926*), **V.**

pannonica Crantz was found at Aberdare by Riddelsdell (00C, HJR 1902*), **V. peregrina** L. was found in Cardiff by R.L. Smith (17D, Grangetown, RLS 1926*; 27A, Splott, RLS 1926), **V. tenuifolia** Roth was found to be 'thoroughly established and spreading at Jersey Marine' (69D, F. Norton & RLS *c*.1936) but now appears to be extinct there, **V. cassubica** L. was recorded from Radyr (18C, RLS 1922*), **V. benghalensis** L. from two localities near Cardiff (16A, Barry Docks, C.I. & N.J. Sandwith 1930*; 18C, Radyr, RLS 1922*), and **V. hybrida** L. from Barry Docks (16A, EV 1935*). [John Storrie's late 19th Century record of **V. tenuissima** (Bieb.) Schinz & Thell. (*V. gracilis* Lois.) from ballast at Cardiff (17B) is now believed to be an error]

Lens culinaris Medicus, Lentil, was found at Barry Docks (16A) by R.L. Smith in 1925.

LATHYRUS L.

Lathyrus japonicus Willd. subsp. **maritimus** (L.) P.W. Ball   Sea Pea. Ytbysen y Môr.
Native. Shingle beach at The Leys, Aberthaw. Probably extinct; the habitat was destroyed during construction work in the late 1950s.
06A, Aberthaw, EV 1933* (non-flowering).

Lathyrus montanus Bernh.   Bitter-vetch. Pysen y Coed Gnapwreiddiog.
Native. Woods, hedges and heaths on acid soils. Very local in the lowlands but more frequent in the Uplands. (!) Fl. 5–6, fr. 6–7; Bb (n). Map 347.

Lathyrus pratensis L.   Meadow Vetchling. Ytbysen y Ddôl.
Native. Meadows, hedges and roadsides. Common in most of the county but local in the central Uplands. Fl. 6–9, fr. 8–9; B.pasc (n,p). Map 348.

Lathyrus palustris L.   Marsh Pea. Ytbysen Las y Morfa.
Native. Marshy meadows, rare. No recent records; discovered at Llanrhidian in 1891, but not seen by Trow or Riddelsdell. Although recent searches of suitable habitats near Llanrhidian have been unsuccessful, the discovery during the 1970s of **L. palustris** at two sites near Pembrey, Carmarthenshire, about 12km north-west of Llanrhidian, suggests that it may still grow in northern Gower.
49D/59C, Llanrhidian, Mr Eve, 1891 (Crouch, 1891).

Lathyrus tuberosus L.   Tuberous Pea. Ytbysen Gnapiog.
Introduced. Naturalized on sand-dunes at Crymlyn Burrows and perhaps also on waste ground in the south-eastern Vale.
08C, Llanharry Common, J. & W. Lovell 1968; 16A, Barry, P.H. Holland 1920*, DMcC 1974, 1980!; 17B, Cardiff Docks, HJR *c*.1902, RLS, 1956!*; 17D, Penarth, ABP 1963*; 79C, Jersey Marine, V.M. Peel 1932, R. Garrett Jones 1963*, H & CH 1972, 1980!.

Lathyrus sylvestris L.   Narrow-leaved Everlasting-pea. Ytbysen Barhaus Gulddail.
Native, but occasionally introduced from gardens. Woods, cliff scrub and hedges. Very local, and mainly on the coast; recently recorded only from southern Gower and the south-eastern Vale. Often confused with **L. latifolius**, which is frequently found as a naturalized garden escape in similar habitats. Map 349.

Lathyrus latifolius L.   Broad-leaved Everlasting-pea. Ytbysen Barhaus Lydanddail.
Introduced. Cliff scrub, hedges and waste ground. Fairly frequent in coastal sites, where it is now more widespread than the similar **L. sylvestris**. Map 350.

Lathyrus nissolia L.   Grass Vetchling. Ytbysen Goch.
Native. Rough grassland and scrub. Very local; found only in the south-eastern Vale and single sites at Cwmavon (79D) and Merthyr Tydfil (00A). Map 351.

Lathyrus aphaca L.   Yellow Vetchling. Ytbysen Felen.
Casual. Once fairly frequent on waste land and disturbed ground, but now rare, with only two recent records (07C, St. Nicholas, MSP 1980; 17C, Dinas Powis, GMB 1971). Not recorded in any other locality since the 1920s. Map 352.

Six southern European species of **Lathyrus** have occurred as aliens. **Lathyrus cicera** L. has been found on four occasions (16A, Barry, RLS *c*.1925; 18C, Radyr, RLS 1922*; 27A, Splott, 1926!*, Pengam, RLS 1940*), **L. ochrus** (L.) DC. has been found three times (16A, Barry, RLS *c*.1925; 17B, Canton, RLS *c*.1925; 27A, Splott, RLS 1927*), **L. clymenum** L. has been found twice (18C, Radyr, RLS *c*.1925; 27A, Splott, RLS 1925!*), as has **L. grandiflorus** Sibth. & Sm. (09A & 09B, Mountain Ash, AMP 1973*; 18D, Llanishen Reservoir, JPC 1972*). Both **L. sphaericus** Retz. (16B, Barry, RLS 1924*) and **L. odoratus** L. (07A, Welsh St. Donats, GCD 1927) have been found once only.

The cultivated pea, **Pisum sativum** L., sometimes occurs on waste ground as an escape from cultivation or as a bird-seed alien, records from: 00C, HJR *c*.1906; 08C, GH & J. Bevan, 1990*; 09A, AMP 1970; 16A, RLS 1925!; 17B, RLS 1925!; 17D, 1920!*; 27A, RLS 1925!; 68A, JAW 1941*; 78B, HJR *c*.1906; 79C HJR *c*.1906, CH 1973).

ONONIS L.

Ononis reclinata L.   Small Restharrow. Tagaradr Bach.
Native. Shallow soil on limestone cliffs. Rare, and probably confined to two areas of cliffland west of Port Eynon. First recorded by D. Sharp in 1828 'on rocks at Port Eynon' (H. & J. Groves, 1907); known in this area by H.R. Wakefield until about 1912, and rediscovered in 1955. The number of plants in the

populations and the periods during which they are in flower differ markedly from year to year.
48, west of Port Eynon, D. Sharp 1828, H.R. Wakefield, *c*.1912, M.H. Bigwood, 1955, AJES & GTG 1961**, 1969!–1986!.

Ononis spinosa L.   Spiny Restharrow. Tagaradr Pigog.
Native or introduced. Roadsides and field banks. Scarce; recently recorded only in the Swansea district and the south-eastern Vale. Often confused with robust forms of **O. repens,** to which some of the older records may refer. Map 353.

Ononis repens L.   Common Restharrow. Tagaradr.
Native. Sand-dunes, cliff grassland and limestone grassland, roadside banks, and occasionally in old meadows. Common near the coast where it is particularly abundant on sand-dunes; locally frequent inland. A few white-flowered plants occur in most sand-dune populations (records from Kenfig, Crymlyn Burrows, Nicholaston Burrows, Whiteford). Fl. 7–9, fr. 8–10; B.luc (n,p). Map 354.

Ononis natrix L. was found as a dock alien by Storrie (27A, East Moors, JS *c*.1876*).

MELILOTUS Miller

Melilotus altissima Thuill.   Tall Melilot. Meillionen y Ceirw.
Introduced. Roadsides, waste ground, disturbed sand-dunes and railway ballast. Common in urban areas in the lowlands, where it is more abundant than **M. officinalis.** Map 355.

Melilotus alba Medicus   White Melilot. Meillion Tair Dalen Gwyn.
Introduced. Roadsides, urban waste ground, construction sites, etc. Locally frequent in western Glamorgan, where it is spreading, and in the Cardiff area; occasional elsewhere. Map 356.

Melilotus officinalis (L.) Pallas   Ribbed Melilot. Gwydro Rhesog.
Introduced. Roadsides, waste ground, railway ballast and similar ruderal habitats. Locally frequent; more widely distributed than **M. altissima** in the Uplands. Fl. 7–8. Map 357.

Melilotus indica (L.) All.   Small Melilot. Gwydro Blodau Bach.
Casual. Fairly widely recorded in the past on waste ground and rubbish tips, especially near docks, but with fewer recent records. Map 358.

Melilotus messanensis (L.) All. has once been recorded as a Cardiff dock alien (17B, JS *c*.1876*).

Several species of **Trigonella** L. used to occur as dock aliens or with grain cleanings. **Trigonella caerulea** (L.) Ser. was most frequent (00C, Abernant, HJR 1901; 17B, Cardiff Docks, JS *c*.1876*, Canton, HJR 1907*;

27A, East Moors, JS *c*.1876*, Splott, 1926!*; 78B, Port Talbot Docks, HJR *c*.1907), **T. hamosa** L. has been found on four occasions (17D, Penarth, JS 1876*; 27A, Splott, RLS *c*.1925!, 1928!, Pengam, RLS 1937!*), **T. corniculata** (L.) L. has been found twice (16A, Barry Docks, RLS 1922*; 27A, Splott, RLS *c*.1925!). Three other species have each been recorded on one occasion: **T. laciniata** L. at East Moors (27A, JS *c*.1876*), **T. monspeliaca** L. at Cardiff Docks (17B, JS *c*.1876*), and **T. polyceratia** L. at Splott (27A, 1926!*).

MEDICAGO L.

Medicago lupulina L.   Black Medick. Maglys, Maglys Gwinenddu.
Native. Dry grassland, thin soil on cliffs, sand-dunes, roadsides, cultivated land, waste ground and similar open habitats. Very common in most of the county, but rather local in the central Uplands. Fl. 5–11, fr. 7–11; Hb (rarely). Map 359.

Medicago sativa L. subsp. **sativa**   Lucerne. Maglys Rhuddlas.
Introduced. Locally naturalized on roadsides, field banks and waste ground near the coast; casual elsewhere. Map 360.

Medicago sativa L. subsp. **falcata** (L.) Arcangeli   Sickle Medick. Meillionen Gorniog.
Introduced. Naturalized in sandy grassland at Crymlyn Burrows and Porthcawl; casual elsewhere. Map 361.

Medicago sativa L. subsp. **varia** Martyn (**M. sativa** subsp. **sativa** × **M. sativa** subsp. **falcata**).
17B, Cardiff Docks, P.H. Holland 1926*; 17D, Grangetown, EV 1929, Penarth Ferry Road, EV 1933*; 27C, Roath Dock, Cardiff, E. Curtis 1969–1981!; 48, near Port Eynon, EF *c*.1800; 69D, Swansea Docks, JAW *c*.1956; 87A, Porthcawl, VCB 1973*.

Medicago arabica (L.) Hudson   Spotted Medick. Maglys Amrywedd.
Perhaps native at Mumbles (no recent records) and Porthcawl, but probably introduced elsewhere. Waste ground and disturbed grassland, usually near the sea. Map 362.

Medicago polymorpha L.   Toothed Medick. Maglys Eiddiog.
Casual. Waste ground, usually near docks. No recent records.
00C, Aberdare, HJR *c*.1906; 16A, Barry, RLS 1925*; 17B Cardiff Docks, AHT *c*.1900; 17D, Penarth Dock, JS 1876*; 18C, Taffs Well, HJR *c*.1906; 18D, near Rhiwbina, W.M. Smith 1955; 27A, East Moors JS 1876*, Splott, EV 1936*; 78B, Port Talbot, HJR 1901; 90A, Hirwaun, HJR *c*.1906.

Several other species of **Medicago** used to occur as aliens on dockside ballast or with grain cleanings. Those found most frequently were **Medicago minima** (L.) Bartal. (16A, RLS 1922*; 17B, EV 1935*; 17D,

[Plate 7]

1, *Inula crithmoides*, frequent on coastal limestone cliffs (Mewslade). 2, *Anaphalis margaritacea*, an attractive and characteristic plant of coal-tips (Godre'r-graig). 3, *Paris quadrifolia*, a scarce plant of ancient limestone woodlands (Parkmill). 4, *Epipactis palustris*, locally common in dune-slacks (Whiteford Burrows). 5, *Sparganium minimum*, at its south-western limit in Britain (Crymlyn Bog). 6, *Scirpus cespitosus*, common in peaty moorland and heathland (Clyne Common). (Photographs: 1, D. Parish; 2–5, Q. O. N. Kay).

[Plate 8]

1, *Aconitum napellus*, a local plant of streamsides in the eastern Vale (St George's). 2, *Pyrola rotundifolia*, now well-established in most dune-land sites (Kenfig Burrows). 3, *Liparis loeselii*, Fen Orchid, in Britain now virtually confined to a few dune-systems around the Bristol Channel, with its major surviving populations in Glamorgan (Whiteford Burrows). 4, *Orchis morio*, still fairly frequent in limestone cliff grassland in the west (Paviland, Gower).
(Photographs: 1: A. R. Perry; 2, D. Parish; 3–4, Q. O. N. Kay).

JS *c*.1886; 18C, RLS *c*.1923; 27A, RLS 1876*) and **M. tornata** (L.) Miller (16A, RLS 1925*; 17B, JS *c*.1876*; 17D, THT 1902*; 18C, 1922*!; 27A, RLS 1937!*). [**M. minima** was also reported from Oxwich Bay (58A) by E. Lees in 1839, but this was probably an error]. **M. ciliaris** (L.) All. has been recorded on five occasions (16A, RLS 1924*; 17B, JS *c*.1876*, GCD 1922; 17D, EV *c*.1925; 27A, 1926!*) as has **M. truncatula** Gaertner (16A, RLS 1924*, RM 1926; 17B, RLS 1923; 27A, JS 1876*, RLS 1940*). **M. doliata** Carmign. (16A, 1922!*; 17B, RLS 1923; 27A, RLS 1927*), **M. murex** Willd. (16A, RLS 1922!*; 18C, RLS *c*.1925; 27A, 1927!*), and **M. littoralis** Rohde ex Loisel. (16A, RLS 1922*; 17B, JS 1876*; 27A, JS 1876*) have been recorded three times, and **M. marina** L. (27A, JS 1876*), **M. turbinata** (L.) All. (17B, RM 1926), **M. rigidula** (L.) All. (27A, 1926!*), and **M. soleirolii** Duby (16A, RLS *c*.1925) once.

TRIFOLIUM L.

Trifolium ornithopodioides L.   Bird's-foot Clover. Corfeillionen Wen.
Native. Dry open grassland, usually near the sea, but also locally on colliery spoil-tips in the Uplands. Recently recorded only from Gower and one site in the Uplands; perhaps overlooked elsewhere. (!)Fl. 5–7, fr. 7–8. Map 363.

Trifolium repens L.   White Clover. Meillionen Wen.
Native, but frequently introduced. Pastures, roadsides, sand-dunes, upper salt-marsh and waste ground. Abundant throughout the county. Fl. (5–)6–9(–11), fr. 7–10; Hb (n,p), B.lap, B.luc, B.luc, B.pasc, B.terr (all n,p) (Lep) Ma.jur, Po.c-al. Map 364.

Trifolium occidentale D.E. Coombe   Western Clover. Meillionen y Gorllewin.
Native. Recently discovered in cliff-top **Festuca rubra** turf at Tears Point and Mewslade (48A, J. Dunn 1987!**) where it is locally frequent on *c*.400m of cliffs (Kay & Ab-Shukor, 1988). (!) Fl. 4–6, fr. 6–7.

Trifolium hybridum L.   Alsike Clover. Meillionen Alsike.
Introduced. Sown pastures, roadsides and waste ground. Widely distributed, but scare in the eastern Uplands. Subsp. **elegans** (Savi) Ascherson & Graebner has been recorded as a casual from two localities (16A, Barry, HJR *c*.1907; 78B, Port Talbot, HJR *c*.1907). Fl. 6–10, fr. 8–10. Map 365.

Trifolium fragiferum L.   Strawberry Clover. Meillionen Fefusaidd.
Native. Dune slacks and the upper fringes of grassy salt-marshes, and occasionally inland on roadside banks. Frequent in suitable habitats in western Glamorgan, occurring at Porthcawl, Kenfig Burrows, Crymlyn Burrows, Three Cliffs Bay and Oxwich, Whiteford Point and Llanmorlais, and inland on roadsides at Pengwern Common. (!) Fl. 7–8, fr. 8–9. Map 366.

Trifolium resupinatum L.   Reversed Clover. Meillionen Dinlan.
Casual or introduced; last reported in 1928 (27A, Splott, EV 1928*). Map 367.

Trifolium campestre Schreber   Hop Trefoil. Meillionen Hopys.
Native. Dry grassland on sea-cliffs, sand-dunes, roadsides, field banks and railway embankments, occasionally on colliery spoil-tips and waste ground. Fairly common in Gower and the Vale but local in the Uplands. Fl. 6–8, fr. 7–9. Map 368.

Trifolium dubium Sibth.   Lesser Trefoil. Meillionen Felen Fechan.
Native. Short grassland, roadsides, lawns, etc. Common in a variety of habitats throughout the county. Fl. 5–10, fr. 6–10. Map 369.

Trifolium micranthum Viv.   Slander Trefoil. Meillionen Felen Eiddil.
Native. Lawns, short grassland near the sea, roadside banks, and occasionally on colliery spoil tips. Fairly common along the coast but rather scarce inland. (!) Fl. 5–7. Map 370.

Trifolium striatum L.   Knotted Clover. Meillionen Rychog.
Native. Shallow soil on sea-cliffs, dry grassland on sand-dunes and roadside banks, and occasionally on colliery spoil-tips in the Uplands, but mainly coastal in its distribution. Locally common in suitable habitats in western Glamorgan but with a few recent records elsewhere. (!) Fl. 6–8, fr. 7–9: B.pasc (n), B.terr (n), Sb. Map 371.

Trifolium arvense L.   Hare's-foot Clover. Meillionen Gedenog. Troed yr Ysgyfarnog.
Native. Sand-dunes and sandy waste ground near the sea, and occasionally on spoil-tips and in dry grassland inland. Common in suitable habitats near the sea in western Glamorgan but local elsewhere. Fl. 7–8. Map 372.

Trifolium scabrum L.   Rough Clover. Meillionen Ger y Mor.
Native. Shallow soil on sea-cliffs, dry grassland on fixed sand-dunes, and occasionally on shallow soils inland. Frequent in Gower, often growing in association with **T. striatum** and other annual clovers. (!) Fl. 5–9, fr. 6–10. Map 373.

Trifolium incarnatum L.   Crimson Clover. Meillionen Ysgarlad.
Casual or introduced; recently recorded only from 78B, Margam. Recorded in the past from several places, mainly near the coast. Map 374.

Trifolium pratense L.   Red Clover. Meillionen Goch.
Native, but often sown for fodder. Old pastures and hay-meadows, roadsides and field banks, grassy sand-dunes, etc. Common in most of the county but

rather local in the central Uplands. Fl. 4–11, fr. 6–11; B.lap (n), B.prat (n), B.pasc (n), (lep) Ma.jur, Pi.bra, Po.c-al, Pl.gam. Map 375.

Trifolium medium L.   Zigzag Clover. Meillionen Wyrgam.
Native. Roadside banks, hedges and rough grassland. Locally common on neutral or acid soils, especially among the lower hills and in Upland valleys. Fl. 5–9, fr. 6–9; B.hort (n,p), B.lap (p), B.pasc (n). Map 376.

Trifolium squamosum L.   Sea Clover. Meillionen y Morfa.
Native in salt-marsh turf, and occasionally found as a casual on waste ground near the sea. Apparently extinct in some localities, recently recorded only at Aberthaw and Swansea.
06A, Aberthaw, 1958!*, MG 1975; 16A, Barry Island and Barry Docks, AHT *c*.1907; 17B, Roath Park, 1927!*; 17D, Penarth, W.G. Cross 1899*; 27A, Pengam Moors, HJR 1905* ('in great quantity'); 69C, Blackpill, JAW, 1939*; 69D, waste ground near Marina, Swansea, 1986!; 79C, Crymlyn Burrows, J. Evans 1832?.

Trifolium subterraneum L.   Subterranean Clover. Meillionen Wen Ymgudd.
Native in Gower, and probably also in the southeastern Vale, but it is occasionally found as an introduced casual on waste ground. Recently recorded only from cliffs between Port Eynon and Horton in Gower, where it grows on shallow soil, disturbed ground and near footpaths; it was discovered in this locality by Phoebe Simons in 1910.
16A, Barry, RLS 1924*; 17B, Cardiff, JS *c*.1886, Pontcanna, EV 1904; 17C, Dinas Powis Common, 1929!*; 48A, Rhosili, JAW *c*.1956; 48B, Port Eynon, AHT 1910*, 1973!**, 1986!, Horton, AHT 1913*. (!) Fl. 6–7, fr. 8–9.

Many species of **Trifolium** used to occur as dockside aliens or with grain cleanings. They have each been found on one or a few occasions and are listed in alphabetical order. **Trifolium angustifolium** L. (16A, Barry Docks, RLS 1922*; 17B, Cardiff Docks, JS *c*.1876*; 18C, Radyr, RLS 1922*), **T. aureum** Pollich (18B, Cefn Onn, EV 1896* (possible native?); 17B, Cardiff Docks, 1928!*; 18C, West of Radyr, RLS *c*.1925!, Whitchurch, H.T. George 1938*), **T. cherleri** L. (18C Radyr, 1922!*), **T. constantinopolitanum** Ser. (27A, Splott, RLS 1926*), **T. diffusum** Ehrh. (17B, Canton, HJR 1907*), **T. glomeratum** L. (00C, Aberdare, HJR *c*.1907; 17B, Cardiff Docks, JS *c*.1876; 69C?, Sketty, Swansea, TBF *c*.1839), **T. lappaceum** L. (00C, Abernant, Aberdare, HJR 1901; Barry Docks, 1924!*; 17B, Cardiff Docks, JS *c*.1876*; 27A, Splott, RLS 1925!; **T. michelianum** Savi (16A, Barry, RLS *c*.1925), **T. nigrescens** Viv. (17B Cardiff Docks, JS *c*.1876*; 18C, Radyr, 1922!*), **T. ochroleucon** Hudson (27A, Splott, JS *c*.1876, RLS 1926; 79C, Port Talbot Dunes, EV *c*.1920*), **T. pannonicum** Jacq. (17D, Penarth, JS 1876–1881), **T. stellatum** L. (16B, Sully, N.D. Wood 1937*; 17B, Cardiff, JS *c*.1876; 17D, Penarth Ferry,

THT 1902*; 69A, Fforest Fach, Mr Rowlands 1921; 78B, Port Talbot Docks, HJR 1904), **T. suffocatum** L. (16A, Barry, JS *c*.1876; 17D, Penarth, JS 1876*; 48A?, Rhossili, E.M. Wakefield 1910), **T. tomentosum** L. (27A, Splott, RLS 1926!*; 16A, Barry Docks, RLS 1925*).

Three species of **Dorycnium** have been found as aliens. **Dorycnium hirsutum** (L.) Ser. has been found at three sites (16A, Barry Docks, RLS 1924; 17B, Cardiff, RLS 1924 & 1926; 18C, Radyr, RLS 1921*). **D. pentaphyllum** Scop. (16A, Barry Docks, RLS 1921*–1924), and **D. rectum** (l.) Ser. (16A, Barry Docks, 1923!*, GCD 1930) have each been found in one locality.

LOTUS L

Lotus tenuis Waldst. & Kit. ex Willd.   Narrow-leaved Bird's-foot-trefoil. Troed Aderyn Culddail.
Probably native. Rough grassland in the southeastern Vale; last recorded in 1954.
16A, Barry Island, near the Pond, AHT 1907*; 16B, Sully, 1954!*; 17D, Cogan Hall, AHT 1907*, Stoneylands Farm, AHT *c*.1907, Grangetown, 1920!*; 27A, Cardiff Marshes, HJR *c*.1907.

Lotus corniculatus L.   Common Bird's-foot-trefoil. Pysen y Ceirw.
Native. Natural and semi-natural grassland on cliffs, dunes and hillsides, old pastures, field banks, roadsides, railway embankments and other grassy places. Very common throughout the county. Fl. (4–)5–7(–10), fr. 6–10; Hb, B.lap, B.luc, B.pasc. (Sb) Anthidium manicatum, (!) (Lep) Ly.phl, Po.ica. Map 377.

Lotus uliginosus Schkuhr   Greater Bird's-foot-trefoil. Pysen y Ceirw Fwyaf.
Native. Marshes and damp grassland. Very common in suitable habitats in most of the county but rather local in Gower and the Vale, and absent from the Gower headlands and from some squares in the southern Vale. Fl. 6–8(–9), fr. 7–9; B.luc. Map 378.

Six Mediterranean species of **Lotus** used to occur as dockside aliens or with grain cleanings. **Lotus edulis** L. has been found on two occasions (17B, Cardiff, RLS *c*.1925; 18C, Radyr, RLS 1922!), **L. conimbricensis** Brot. (27A, Splott, RLS 1926*), **L. ornithopodioides** L. (18C, Radyr, 1922!*), **L. creticus** L. (16A, Barry Docks, 1923!*), **L. subbiflorus** Lag. (16A, Barry Docks, RLS 1927), and **L. cytisoides** L. (16A, Barry Docks, RLS 1926*) have each been recorded once, as have the related species **Tetragonolobus purpureus** Moench (17B, Cardiff Docks, JS *c*.1876*), and **Securigera securidaca** (L.) Degen & Dörfler (17D, Penarth, JS *c*.1876*).

ANTHYLLIS L.

Anthyllis vulneraria L.   Kidney Vetch. Plucen Felen.

Native. Sand-dunes, sea-cliffs and cliff grassland, especially in the spray-zone, and occasionally on roadsides, railway embankments and waste ground inland. Common on the coast but scarce inland. Red-flowered forms (var. **coccinea** L) occur at Crymlyn Burrows (79C) and Porthcawl (87A). Fl. 5–8, 10, fr. 6–9; (!) B.lap, B.luc, B.pasc, B.prat, (Lep) Po.ica. Map 379.

Anthyllis tetraphylla L. from the Mediterranean region has been recorded twice as a casual (16A, Barry, RLS 1921*; 18C, Radyr, RLS 1921*).

ORNITHOPUS L.

Ornithopus perpusillus L. Bird's foot. Troed yr Aderyn.
Native. Short turf on sand-dunes, and on dry soils inland; occasionally on waste ground. Local and with few recent records, but still fairly frequent in the southern part of Kenfig Burrows towards Porthcawl. Map 380.

Ornithopus compressus L. and **O. pinnatus** (Miller) Druce were found growing as aliens at Barry (16A) by R.L. Smith in 1927* and the latter also at Cardiff (17B) by J. Storrie in *c*.1876. **O. sativus** Brot., which is often cultivated as a fodder plant in Europe, has also been found as an alien on one occasion (17B, rubbish tip at Maendy, RLS 1938*).

CORONILLA L.

Coronilla varia L. Crown Vetch. Ffugbysen Goronog.
Introduced. Naturalized on waste ground and roadsides in a few localities in the Vale. Map 381.

Coronilla scorpioides (L.) Koch has been found as a dockside alien on four occasions (17D, Penarth, JS *c*.1876*; 27A, East Moors, JS *c*.1876*, Splott, RLS 1926*; 78B, Port Talbot, HJR *c*.1906) and once on waste ground inland (00C, Aberdare, HJR *c*.1906).

HIPPOCREPIS L.

Hippocrepis comosa L. Horseshoe Vetch. Pedol y March.
Native. Limestone sea-cliffs and limestone grassland in Gower and the western Vale. Abundant in a few localities, but very restricted in its distribution. There is one unconfirmed record from the east of the county (17C) The Gower sea-cliff plants are diploid, with a chromosome number of 2n=14 (Fearn, 1972) and are thus probably a Late-glacial relict population (the majority of the English populations of the species are tetraploid, and probably reached the British Isles from continental Europe later than the western diploids). Map 382.

The Mediterranean species **Hippocrepis unisiliquosa** L. has been found as an alien on one occasion (18C, Radyr, 1922!*).

Scorpiurus muricatus L. used to grow as a dockside alien in Cardiff (17B, JS *c*.1876*; 17D, JS 1892*; 18C, 1922!*; 27A RLS 1925*) where it was last recorded in 1925 at Splott (27A, RLS 1925*), it has recently been recorded again in the Cardiff area, as a garden weed and ruderal, perhaps from birdseed, (07D, Bonvilston, JIL 1974*; 17B, Llandaff, P. Jarvis 1981, Cathays, R.J. Tidswell 1985*). **Hedysarum coronarium** L., which is cultivated as a fodder plant in the Mediterranean region, also used to occur as a dockside alien and was last recorded in 1927 at Barry (16A, J. Waite 1927*), it had been found in 17D, THT 1902*; 18C, RLS *c*.1925!; 27A JS 1876*. The related **Hedysarum glomeratum** F.G. Dietrich was once recorded in Cardiff Docks (17B) by J. Storrie in *c*.1876*.

ONOBRYCHIS Miller

Onobrychis viciifolia Scop. Sainfoin. Codog.
Probably introduced. Once widely cultivated for forage, and naturalized in several localities at one time, but recently recorded at only four sites: (06A, Aberthaw, MG 1980, 1983; 17C, Cwm George, MG 1971; 17D, Dinas Powis, 1971!*; 87A, Porthcawl, P. Bewley 1965). Map 383.

Arachis hypogaea L., Peanut, Cneuen Fys, has twice been recorded as a rare casual on rubbish tips (17D, Penarth Road, Cardiff, RLS 1937*; 27A, Splott, Cardiff, RLS *c*.1925!).

Cassia nictitans L., an alien from N. America, has been recorded on one occasion (16A, Barry, RLS *c*.1927).

LIMNANTHACEAE

Limnanthes douglasii R. Br., Poached-egg-flower, has been recorded three times as a dock alien or casual of waste places, but only once recently (17D, Penarth, JS 1870*, Penarth Ferry, J.H. Thomas 1876*, Penarth Marina, GH 1990*; 71C, near Gwaun-cae-Gurwen, 1974!).

OXALIDACEAE
OXALIS L.

Oxalis corniculata L. Procumbent Yellow-sorrel. Suran Felen Orweddol.
Introduced. An increasingly frequent weed of gardens, lawns and waste places, probably underrecorded. Map 384.

Oxalis europaea Jordan Upright Yellow-sorrel. Suran Felen Unionsyth.
Introduced. An increasing garden weed.
00A, Cyfarthfa Park, Merthyr, J. Davies 1987*; 07D, Dyffryn, EV 1941; 17B, Llandaff, M. Craster 1964*; 18D, Rhiwbina, HAH 1933*, 1941*, 49C, Cwm Ivy, 1980!; 69C, Swansea, 1983!; 88C, near Pyle, G. Campbell 1958.

Oxalis acetosella L.   Wood-sorrel. Suran y Coed, Bara'r Gog.
Native. Woods, hedgebanks, among rocks and on cliff ledges in the hills. Common in suitable habitats throughout the county. Fl. 4–5, fr. 5,8. Map 385.

Two other species of **Oxalis** have recently been found as weeds or garden escapes and may increase. **Oxalis articulata** Savigny has been found at five localities (17B, Cardiff Docks, GSW 1983; 69C, Swansea, H & CH 1975; 69D Landore, 1976!; 87B, Ogmore-by-Sea, AMP 1973; 88C Kenfig Burrows, 1972!*) and **O. corymbosa** DC. at three (00A, Cyfarthfa Park, Merthyr, J. Davies 1987*; 48B Horton, 1974!; 59B, Blue Anchor, 1974!).

GERANIACEAE
GERANIUM L.

Geranium sanguineum L.   Bloody Crane's-bill. Pig yr Aran Rhuddgoch.
Native. Sand-dunes, limestone sea-cliffs, and occasionally in limestone grassland inland. Locally abundant in Gower; recently recorded in the Vale only from Corntown (97A), where it was last seen in 1970! and may be extinct; formerly grew in Cwm Mawr, Dunraven (87D) in small numbers (JS c.1886, HOB c.1907). (!) Fl. 5–7(–8), fr. 7–8; B.terr (n,p). Map 386.

Geranium pratense L.   Meadow Crane's-bill. Pig Garan y Weirglodd.
Native in some localities, particularly in the northern Uplands and parts of the Vale, but often introduced as a garden escape. Meadows, roadsides and waste ground. Local. Map 387.

Geranium endressii Gay has once been found as a garden escape (18D, Lisvane, JPC 1972), **G. nodosum** L. twice (17C, Cwrt-yr-ala, EV 1931; 79B, Baglan, JAW c.1933), **G. versicolor** L. several times, mainly in the Cardiff district (17A, AHT c.1900, HJR 1912*, K. Collins 1966*; 17B, AHT c.1900; 17C, K. Richards 1907*; 17D AHT c.1900), but also in Gower (58A, Nicholaston, 1978!) and at Baglan (79D, Mrs. Llewellyn c.1907), and **G. ibericum** Cav. × **G. platypetalum** Fischer & C.A. Meyer (**G. × magnificum** Hyl.) also several times in widely scattered localities (00C, Abernant, AMP 1974*; 18C, Rhydlafr, 1974!*; 61C, near Garn-swllt, 1980!*; 68A, West Cross, JAW 1942*; 79C, Jersey Marine, JAW 1939*; 90D, Llwydcoed, ABP 1974!; 97D, Cowbridge, ABP 1975!).

Geranium phaeum L.   Dusky Crane's-bill. Gweddw Galarus, Pig yr Aran Dulwyd.
Introduced as a garden escape, rather rare and with only two recent records.
17A, Coedrhiglan, JS c.1876; 17B, Coopers Field, Cardiff, JS c.1876; 17C, Cwrt-yr-Ala, EV 1907; 48B, Stouthall Woods, JBe 1982; 58B, Bishopston, MG 1971; 80B, Glyn Neath, J. Ballinger 1902; 87B, Merthyr Mawr, JS c.1876*; 90B, near Hirwaun, HJR 1903**.

Geranium pyrenaicum Burm. fil. Hedgerow Crane's-bill. Pig Garan y Gwrych.
Native. Roadsides, hedgebanks and waste ground. Very local; most frequent in the Port Talbot area. Map 388.

Geranium rotundifolium L.   Round-leaved Crane's-bill. Pig yr Aran Crynddail.
Native. Roadsides, hedgebanks and waste ground. Locally frequent in the Swansea and Cardiff areas and at Porthcawl. Map 389.

Geranium molle L.   Dove's-foot Crane's-bill. Troed y Golomen, Pig yr Aran Cyffredin.
Native. Shallow soil on sea-cliffs, sand-dunes, dry grassland, paths, roadsides, walls, cultivated land and waste ground. Usually very common on the better soils but scarce or absent in parts of the Uplands and the northern Vale. Fl. 5–10, fr. 6–10; (!) (Lep) Ar.agr. Map 390.

Geranium pusillum L.   Small-flowered Crane's-bill. Pig yr Aran Manflodeuog.
Native. Shallow soil on sea-cliffs, often on headlands where sheep and cattle gather, sand-dunes, occasionally on disturbed ground inland. Local and rather infrequent, usually associated with abundant **G. molle**. Map 391.

Geranium columbinum L.   Long-stalked Crane's-bill. Pig yr Aran Hirgoesog.
Native. Shallow soils, path-sides, open scrub and roadsides, usually on limestone, grassy sand-dunes. Locally frequent in Gower and the Vale. Fl. 6. Map 392.

Geranium dissectum L.   Cut-leaved Crane's-bill. Pig yr Aran Llarpiog.
Native. Path-sides, disturbed grassland, roadsides, cultivated land and waste ground. Common in suitable habitats throughout the county. Fl. 5–11, fr. 6–11. Map 393.

Geranium lucidum L.   Shining Crane's-bill. Pig yr Aran Disglair.
Native. Walls, stony hedgebanks and shaded limestone rocks. Locally common in Gower and parts of the Vale, and growing in the Uplands on Carboniferous Limestone near Merthyr Tydfil. (!) Fl. 5–9, fr. 6–9. Map 394.

Geranium robertianum L.   Herb-Robert. Llys y Llwynog.
Native. Woods, hedges, walls, shaded places in gardens, ruins, cliff scrub and limestone screes and shingle beaches. Common throughout the county. Populations growing in natural habitats on limestone cliffs in southern Gower from Mumbles to Worm's Head show various combinations of the characters of **G. robertianum** subsp. **maritimum** (Bab.) H.G. Baker and **G. robertianum** subsp. **celticum** Ostenf., but appear to intergrade with typical **G. robertianum** subsp. **robertianum**. Fl. 4–12 (1,2), fr. 6–11; Hb (n),

B.pasc, B.terr (n,p), B.prat (n), (Syr) Rh.cam (n), S.balt (p), S.rib (p), Sy.pip (p) (Lep) An.car, Pi.nap, Pi.rap (all n). Map 395.

Geranium robertianum subsp. **celticum** Ostenf. has been recorded from 09B, Mountain Ash, AMP 1970*; 48B, near Oxwich, E.F. Warburg 1961; 58A, Three Cliffs Bay, H.G. Baker 1946, R.J Pankhurst 1985,

G. robertianum subsp. **maritimum** (Bab.) H.G. Baker has been recorded from a number of coastal localities in the past but with no recent records. Map 396.

The record of **Geranium purpureum** in E.F. Linton, 1892, is now believed to be an error for **G. robertianum** subsp. **maritimum**.

ERODIUM L'Hér.

Erodium maritimum (L.) L'Hér. Sea Stork's-bill. Pig y Crëyr Arfor.
Native. Shallow soils on sea-cliffs and walls overlooking the sea, occasionally on dry soils in other maritime habitats. Frequent in suitable sites on the coast of Gower but scarce on the Vale coast, with few recent records there. Map 397.

Erodium cicutarium (L.) L'Hér. (excluding subsp. **bipinnatum** Tourlet). Common Stork's-bill. Pig y Crëyr .
Native. Shallow soils on sea-cliffs, sand dunes, dry grassland and sandy waste ground near the sea, and occasionally on dry soils inland (railway ballast, old slag-tips, etc.) Common along the coast but scarce inland. Most sand-dune populations have some of the characters of **E. cicutarium** subsp. **dunense** Andreas, but do not appear to be sharply distinct from more typical **E. cicutarium** subsp. **cicutarium**, and have not been cytologically studied. Fl. 5–8, 10, fr. 7–8, 10. Map 398.

Erodium cicutarium subsp. **bipinnatum** Tourlet (*E. glutinosum* Dum.). Sticky Stork's-bill. Pig y Crëyr Glydog.
Native. Sand-dunes. Fairly common on major areas of sand-dunes from Merthyr Mawr to Whiteford Point, often growing with **E. cicutarium**. Map 399.

Erodium cicutarium subsp. **anaristatum** Andreas (subsp. **bipinnatum** × subsp. **cicutarium**) has been recorded from one locality (87A, Newton Burrows, HJR 1904, 1925!*).

Erodium moschatum (L.) L'Hér. Musk Stork's-bill. Pig y Crëyr Mwsgaidd. Native. Shallow soils on sea-cliffs, especially on headlands where sheep and cattle gather, sand-dunes and occasionally in limestone grassland or as a dockside alien. Recently recorded only from maritime sites in Gower. Map 400.

Five Mediterranean species of **Erodium** have occured as aliens in Splott, Cardiff (27A) or Barry (16A); each

on one occasion only: **Erodium acaule** (L.) Becherer & Thell. (16A, 1922!*), **E. botrys** (Cav.) Bertol. (16A, RLS *c*.1925!), **E. chium** (L.) Willd. (27A, RLS 1927*), **E. laciniatum** (Cav.) Willd. (27A, RLS 1926!), and **E. malacoides** (L.) L'Hér. (27A, 1926!*).

TROPAEOLACEAE

Two species of **Tropaeolum** L. have been recorded as casuals of waste places: **Tropaeolum majus** L., Nasturtium, (69C, Mayals, Swansea, JAW 1921; 17B, Cardiff, GCD 1916), records of **T. minus** L. are now believed to be errors.

ZYGOPHYLLACEAE

Tribulus terrestris L., Puncture-vine, has once been recorded as a casual on ballast (27A, East Moors, Cardiff, JS *c*.1876).

LINACEAE
LINUM L.

Linum bienne Miller Pale Flax. Llin Culddail.
Native. Limestone grassland, usually near the sea. Frequent in the south-eastern Vale and scattered along the coast to western Gower. Inconspicuous and easily overlooked. Sometimes confused with **L. usitatissimum**. Map 401.

Linum usitatissimum L., Flax Llin Amaeth, is occasionally found, usually as a birdseed alien on waste ground in towns. Map 402.

Linum catharticum L. Fairy Flax. Llin y Tylwyth Teg.
Native. Short open grassland on sea-cliffs, sand-dunes, field banks, roadsides, open hedgebanks, etc., base-rich soils in the hills, railway embankments, and colliery spoil-tips. Common or frequent in most of the county but surprisingly absent from the central Vale and an area of the Uplands north of Swansea. Fl. 5–7 (–10), fr. 7–10. Map 403.

RADIOLA Hill

Radiola linoides Roth Allseed. Gorhilig.
Native. Damp open sites on heaths and acid dunes. Now apparently rare with no recent records, but this inconspicuous plant may still be present at some of the many sites on heathland in Gower from which it was recorded in the 1930s. Map 404.

EUPHORBIACEAE
MERCURIALIS L.

Mercurialis annua L. Annual Mercury. Bresych y Cŵn Blynyddol. Clais yr Hydd Blynyddol.
Introduced. A scarce weed of roadsides, waste ground and gardens, well-established only in the Cardiff area. It has been observed to persist for many years in some sites, as for example on the roadside

near Barry Railway Station where it was first recorded about 1900. Fl. 5–9, fr. 6–9. Map 405.

Mercurialis perennis L.   Dog's Mercury. Bresych y Cŵn, Cwmlwm yr Asgwrn.
Native. Woods and hedgebanks, mainly on base-rich soils, and on limestone scree-slopes on the clifflands of southern Gower. Very common in Gower and the Vale but rather local in the Uplands. Fl. 2–5, fr. 4–6. Map 406.

Ricinus communis L., Castor-oil-plant, Trogenllys, has twice been recorded from the Cardiff area (16A, Barry Docks, RLS 1921*; 17B, Cardiff Docks, GCD *c.*1922).

EUPHORBIA L.

Euphorbia peplis L.   Purple Spurge. Fflamgoed Ruddlas.
Extinct?. James Motley reported finding this species on sand-hills 'growing abundantly at Porth Cawl.' (87A) in 1834, when he was a boy; it had gone by 1841 (Motley, 1847).

Euphorbia platyphyllos L.   Broad-leaved Spurge. Fflamgoed Lydanddail.
A rare weed of cultivated or disturbed ground, with one recent record (16B, Lavernock, JPC 1978*) where it appeared for one year only in a cornfield. Recorded in the past from 16B, cliffs near Lavernock, JS 1886, Upper Cosmeston 1949!*; 17B, Ely, RLS 1947*; 17D, Llandough, AHT 1899*, 1909, Penarth, 1940!*.

Euphorbia helioscopia L.   Sun Spurge. Llaeth Ysgyfarnog.
Native. Very common as a weed of arable land, gardens and waste ground. Fl. 6–11, fr. 7–10; (Syr) E.ten, S.rib. Map 407.

Euphorbia lathyris L.   Caper Spurge. Llysiau y Cyfog, Fflamgoed Gaperol.
Native or introduced in open woods and scrub at a few sites, and as a casual on waste ground and in gardens. Occasional in the south-east, rare elsewhere. Map 408.

Euphorbia exigua L.   Dwarf Spurge. Fflamgoed Fach yr Yd, Fflamgoed Eiddil Flaenfain.
Apparently native in cliff-heath on the limestone cliffs of southern Gower, where it appears with **Anagallis arvensis**, **Kickxia elatine** and other annuals after heath fires; decreasing in frequency as a weed of arable land on limestone soils in Gower and the Vale. Map 409.

Euphorbia peplus L.   Petty Spurge. Llaeth y Cythraul.
Native. Very common as a garden weed; less frequent on farmland and waste ground. Fl. 1–12, fr. 5–11. Map 410.

Euphorbia portlandica L.   Portland Spurge. Llaethlys Portland.

Native in or near the spray-zone on limestone cliffs, and on sand-dunes. Common in suitable habitats on the Gower coast, but less frequent on the coast of the Vale. (!) Fl. 5–8, fr. 7–9; (Syr) Rh.cam, S.sp (Dip) Call, ?Dip, (Lep) Ar.agr, Py.mal. Map 411.

Euphorbia paralias L.   Sea Spurge. Llaethlys y Môr.
Native. Sand-dunes, mainly on young dunes not far from the beach, and sometimes on sandy shingle. Fairly common in suitable habitats along the coast. Map 412.

Euphorbia paralias × **E. portlandica** was recorded by H.J. Riddelsdell from Margam Moors in *c.*1906; it has not been seen since.

Euphorbia × **pseudovirgata** (Schur) Soó   Hungarian Spurge.
Introduced often as a garden escape. Occasionally established on waste ground.
06A, Aberthaw, 1979!*, JPC 1981; 16A, Barry Docks, RLS 1922*, EV 1927*, 1939, 1974!*; 17B, Cardiff Docks, P.H. Holland 1926*, Roath Park, AHT 1912*; 18C, Radyr, RLS 1923; 27C, Cardiff Docks, JPC 1980, GSW 1983*, 1986; 79C, Aberavan, HJR 1909.
There is some confusion over the identity of this taxon. It has been called *Euphorbia esula* L., *E. esula* subsp. *tommasiniana* (Bertol.) Nyman and *E. uralensis* Fischer ex Link.

Euphorbia cyparissias L.   Cypress Spurge. Fflamgoed Gedrol, Fflamgoed Gyprysol.
Introduced. Occasionally established on waste ground, roadsides, and fixed sand-dunes, and as a garden weed. Recent records from 07B, Peterston-super-Ely, 1976!; 49C, Llangennith, MG 1970; 79C, Briton Ferry, H & CH 1973. Map 413.

Euphorbia amygdaloides L.   Wood Spurge. Llaethlys y Coed.
Native in woodland and old hedgerows on base-rich soils; locally frequent in Gower and the Vale, but absent from most of the Uplands. Fl. 4–6. Map 414.

Four species of **Euphorbia** have been found as rare aliens in Cardiff or the southern Vale. They each occurred in only one locality and all are perennials so may still persist at the sites where they were originally found. **Euphorbia villosa** Waldst. & Kit. ex Willd. (16B, Ely, garden weed, RLS 1947*), **E. ceratocarpa** Ten. (16A, Barry Docks, 1923!*, RLS 1927*), **E. corallioides** L. (87A, near Porthcawl, EMT 1932*), and **E. falcata** L. (17B, Cardiff, JS *c.*1876, RLS 1923).

POLYGALACEAE

POLYGALA L.

Polygala vulgaris L.   Common Milkwort. Llysiau Crist, Llaethlys Cyffredin.
Native. Grassland and grassy scrub on basic soils, usually on limestone or on sand-dunes, fairly com-

mon near the coast and in the southern Vale. Many populations have blue-flowered plants, red-flowered plants and sometimes white-flowered plants intermingled; blue-flowered plants are most abundant, but red-flowered plants are sometimes in the majority on limestone cliffland. (!) Fl. 5–8 (–10), fr. 6–10; Hb, B.lap, B.luc, B.pasc, B.terr, (Sb) Colletes sp., (Lep) Ar.agr, Er.tag, Po.ica (none common, all n-s; Lack & Kay, 1987). Map 415.

Polygala serpyllifolia J.A.C. Hose   Heath Milkwort. Llysiau'r Groes, Llaethlys y Rhos.
Native. Heathland and heathy grassland on acid soils. Common in the Uplands and on suitable soils in Gower, but absent from the southern Vale. Flowers normally slaty blue, but red-flowered and white-flowered plants also occur on the Gower coast. (!) Fl. 4–7 (–10), fr. 6–10; Bb (very rarely). Map 416.

ANACARDIACEAE

Rhus typhina L. Stag's-horn Sumach has twice been been recorded recently; once on a railway bank (27A, Pengam, Cardiff, 1973!), and once on a cycle trackway (69D, Lower Swansea Valley, GH 1989*).

ACERACEAE
ACER L.

Acer campestre L.   Field Maple. Masarnen Leiaf.
Native. Woods and hedges. Frequent in Gower and the Vale where it is an indicator of old woodland, but very local in the Uplands where it is found mainly in hedges in the valleys. Fl. 6, fr. 10–11. Map 417.

Acer pseudoplatanus L.   Sycamore. Masarnen.
Introduced. Woods and hedges; often planted near farms in the Uplands. Generally common, but still absent from a few areas of native woodland; an invasive alien tree, which forms species-poor woods with an impoverished field layer and few associates. Fl. 5–6, fr.10; Hb (n). Map 418.

Acer platanoides L.   Norway Maple, Masarnen Norwy, has been recorded several times, usually as a planted tree (08D, MG 1973; 09D, MG 1970+; 18D, MG 1973; 90D, H.J. Dawson 1980+; 96A, MG 1975) but on Crawley Cliff, Oxwich, (58A) T.A.W. Davis recorded it in 1961 with the comment 'several small trees, much regeneration'. **A. negundo** L., Box-elder, has been recorded from only one locality, and then as a planted tree (18D, RAH 1971, MG 1973).

HIPPOCASTANACEAE
AESCULUS L.

Aesculus hippocastanum L.   Horse-chestnut. Castanwydden y Meirch.
Introduced. Often planted in gardens and parks, and sometimes self-sown, but rarely naturalized except in some parts of the Vale. Fl. 5–6, fr. 9–10. Map 419.

BALSAMINACEAE
IMPATIENS L.

Impatiens capensis Meerb.   Orange Balsam. Ffromlys Oren. Ffromlys Melyngoch.
Introduced. Locally frequent in the valley of the Thaw in the central Vale. First recorded by J. Storrie, at Llandough south of Cowbridge, in or before 1886. It was apparently very rare when Storrie first discovered it but had become 'Thoroughly naturalized and abundant ... from Llansanwr to Gigman Bridge' by 1908 (Trow, 1911). Since then it has made relatively little further progress but has spread to the valley of the Ely.
07B, Peterston-super-Ely, R.J.H. Lloyd 1959*, ABP 1981*, St. George's, 1979!; 07C, Gigman Bridge, EV 1908*, 1969!; 97B, Llansanwyr, AHT 1907*, G.R. Willan 1931*; 97D, Llandough-juxta-Cowbridge, JS c.1886, AHT c.1908, ABP 1969!

Impatiens parviflora DC.   Small Balsam. Ffromlys Bach, Ffromlys Lleiaf.
Introduced. River-banks, rarely on waste ground. Locally frequent on the banks of the Afon Llwchwr (Loughor) between Garn-swllt and Pontarddulais, once recorded elsewhere. First recorded by H.J. Riddelsdell in or before 1907.
50D, banks of Afon Llwchwr, HJR c.1907, 1974!; 60A, banks of Afon Llwchwr, 1974!; 61C, banks of Afon Llwchwr, 1981!*; 79B, Neath, H & CH 1974.

Impatiens glandulifera Royle Indian Balsam, Policeman's Helmet. Jac y Neidiwr, Ffromlys Chwarennog.
Introduced. River-banks, canal banks, ditches, occasionally on waste ground. Now well naturalized and abundant in the valleys of the Tawe, Nedd (Neath), Ogwr (Ogmore), Llynfi and Ewenni in western Glamorgan and along the Taf, Ely (Elai), lower Rhymni and some minor rivers in the east. The date of its original introduction is uncertain, but it was recorded only on the Ely at Peterston and on the Ogwr at Merthyr Mawr by Riddelsdell (1907). Trow (1911) did not record it at all, perhaps because it was a common cottage garden plant and naturalized plants, if any were seen, were regarded as garden escapes that would not persist. Nevertheless, it evidently established itself and spread, and by 1920 was 'increasing rapidly' along the banks of several Glamorgan rivers (Vachell, 1920). It is still steadily spreading today, aided by occasional cultivation as a garden flower. Fl. 7–9 (–11), fr. 8–10. Map 420.

AQUIFOLIACEAE
ILEX L.

Ilex aquifolium L.   Holly. Celynnen.
Native. Hedges and woods, often planted. Common in old mixed hedges in many areas, but less frequent in woods; scarce in the central Uplands. Fl. 5–6 (both sexes), fr. (9–) 10–12; Hb (n,p), B.prat, (Syr) E.ten, (Lep) Ce.arg. Map 421.

CELASTRACEAE
EUONYMUS L.

Euonymus europaeus L.   Spindle. Piswydden.
Native. Mixed woods and old hedges on base-rich soils, cliff-scrub on limestone sea-cliffs, sand-dunes. Frequent and locally common in Gower and the Vale but very scarce in the Uplands. Fl. 5–6, fr. 9–11. Map 422.

Euonymus japonicus L. fil., an evergreen shrub often used for hedges especially in coastal areas, sometimes escapes and persists for a time.
09A, Mountain Ash, AMP 1972; 17D, Cardiff Docks, H. McKenzie 1975*; 58B, SE Gower Cliffs, MG 1973; 68A, Mumbles, MG 1973; 69D, Swansea, M.H. Sykes c.1942; 87B, Merthyr Mawr, 1973!.

STAPHYLEACEAE
Staphylea pinnata L.   Bladder-nut, Dagrau Addaf, was once recorded as a rare alien on a hillside (18C, Castell Coch, C.E. Ollivant 1873).

BUXACEAE
BUXUS L.

Buxus sempervirens L.   Box. Pren Bocs, Bocyswydden.
Introduced. Scrub, hedges, roadsides. Occasional in several scattered localities, probably always planted originally but showing regeneration in some sites.
07A, Tal-y-garn, MG 1972; 07D, Llancarfan, AHT 1890*; 17D, Penarth, E. Cawood & K. Baker 1970; 18C, Coed y Bedw, MG 1971; 48B, Stouthall Wood, JBe 1982; 60D, Rhyd-y-pandy, JAW 1941*; 87D, Dunraven, MG 1974; 96A, St. Donats, MG 1975; 98B, Gilfach Goch, 1978!.

RHAMNACEAE
RHAMNUS L.

Rhamnus catharticus L.   Buckthorn. Rhafnwydden.
Native. Woods and old hedges on limestone, scrub on limestone sea-cliffs, occasionally on sand-dunes. Scarce and usually in small numbers, mainly confined to coastal areas of Gower and the Vale, but with one locality to the north of Cardiff. (!) Fl. 5–6, fr. 9–10; Hb (n,p), B.luc, B.prat (n), (Sb) Andrena spp., Halictus spp. (n, ?p), Sphecodes sp. (n) (Syr) E.ten, H.pen, Sy.pip, (Dip) Call, Luc spp., Sa.car, (Lep) Ca.rub. Map 423.

Rhamnus alaternus L. has been found on cliff scree near Penarth Head, presumably as a garden escape (17D, 1943!*).

FRANGULA Miller

Frangula alnus Miller Alder Buckthorn. Breuwydd.

Native. Damp woodland and hedges on acid soils. Locally frequent in the Uplands and in the northern parts of Gower and the Vale. Map 424.

VITACEAE
PARTHENOCISSUS Planchon

Parthenocissus inserta (A. Kerner) Fritsch Common Virginia Creeper. Dringwr Fflamgoch Cyffredin.
Introduced. Occasionally naturalized on dunes, walls, roadside banks, and similar habitats.
09A & 09B, Mountain Ash, AMP 1972; 09C, Wattstown, 1972!; 16A, Cadoxton, 1974!*; 17B, Cardiff, GH 1987*; 48B, Port Eynon, J.P.M. Brenan 1958*, 1975!; 58A, Oxwich, JEL 1948*, JWD 1957*, 1978!; 68A, Mumbles, 1974!; 69D, Jersey Marine, H & CH 1974; 99A, Blaen-y-Cwm, 1975!.

Parthenocissus tricuspidata (Siebold & Zucc.) Planchon and **P. quinqueflora** (L.) Planchon were both recorded by J.A. Webb and M.H. Sykes from bombed sites in Swansea (69D) in c.1945. The latter species may be a misidentification for **P. inserta**.

Vitis vinifera L.   Grape-vine, Gwinwydden, has been recorded as a casual on waste ground in four localities (09A, Mountain Ash, AMP 1972, 1984*; 16A, Barry, RLS c.1925!; 17B, Cardiff, 1989!*; 27A, Splott, RLS c.1925!).

TILIACEAE
TILIA L.

Tilia cordata Miller   Small-leaved Lime. Pisgwydden Deilen Fach.
Native, but sometimes planted. Old woodlands on limestone, and occasionally in riverside woodland in deep valleys and gorges in the Uplands. Scarce and usually in small numbers. Probably native in the sites listed below, and perhaps native in some others.
00A, Taf Fechan Gorge, MG 1975, 1978!; 17C, Cwm George, ARP 1984*; 58A, Crawley Cliff, GTG 1955, 1980!, North-hill Wood, 1979!; 58B, Bishopston Valley, 1978!, Bishop's Wood, Caswell, 1980!; 80B, Pyrddin Valley, 1969!, 1975!; 80C, Melincourt, 1972!, 1978!; 80D, Tonplanwydd, 1974!; 90A, Pont-nedd-fechan, 1974!. Map 425.

Tilia cordata × T. platyphyllos (**T. × vulgaris** Hayne). Lime. Pisgwydden Cyffredin.
Introduced. Often planted and occasionally more or less naturalized. Fl. 6–7, fr. 8; Hb. Map 426.

Tilia platyphyllos Scop. Large-leaved Lime, Pisgwydden Deilen Fawr, has been recorded twice (88A, Margam, JAW 1954*; 96A, St. Donats, MG 1975) but was probably originally planted in both cases.

MALVACEAE
MALVA L.

Malva moschata L.   Musk Mallow. Hocysen Fws.

Native. Rough grassland, roadsides, etc., on the better soils, sometimes with white flowers. Frequent in the Vale but scarce in Gower and rare in the Uplands. Fl. 6–11 (white-fld var. 6–9), fr. 7–11; Hb (n), B.hort, B.luc (n), B.prat (n,p). Map 427.

Malva sylvestris L.   Common Mallow. Hocysen Gyffredin.
Native. Roadsides, cultivated land, waste places, etc., and also on cliff ledges and near seabird colonies on sea-cliffs. Common near the coast but local or scarce inland, and absent from parts of the Uplands. Fl. 6–11, fr. 7–11; Hb (n), B.luc, B.prat, B.pasc, B.terr (all n; B.terr seen shedding pollen). Map 428.

Malva neglecta Wallr.   Dwarf Mallow. Hocys Bychan.
Native. Roadsides, waste ground, occasionally on drift-lines. Locally frequent but sporadic in its appearance near the sea, rare elsewhere. Map 429.

Four species of **Malva** have been recorded in the past as aliens of waste ground and docksides but none have been seen recently. **Malva pusilla** Sm., Small Mallow, Hocys Blodau Bychan, was the most frequent being recorded from 6 localities (00C, Aberdare, HJR 1901*; 07D, St. Hilary, F. Nicholl 1908*; 17D, Cardiff, RLS 1925*; 27A, Splott, RLS 1926*, 1927*, EV 1929*, Newport Road, Cardiff, 1937!*, Pengam, RLS 1940*). **M. parviflora** L., Least Mallow, Hocysen Leiaf, has been recorded from 3 localities (16A, Barry, GCD c.1927; 17D, Cardiff, RLS 1935*; 27A, Splott, 1925!*, EV 1927*, 1929*), as has **M. nicaeensis** All. (16A, Barry, RLS 1925; 27A, East Moors, JS c.1876, Splott, EV 1929*). **M. hispanica** L. has been recorded once only (27A, Splott, RLS 1927*).

LAVATERA L.

Lavatera arborea L.   Tree-mallow. Hocyswydden.
Native, but sometimes cultivated near the sea. Sea-cliffs, especially near seabird colonies, and waste ground near the sea. Very local along the coast, but certainly native at a few sites on the south Gower cliffs, for example at Mewslade (48A), and on the southern Vale cliffs. Map 430.

Lavatera cretica L.   Smaller Tree-mallow, Môr-hocysen Fychan, has been recorded as an introduction on four occasions (17B, Canton, Cardiff, HJR 1907, Maendy Pool, RLS 1938*; 27A, Splott, RLS 1929!; 96B, Llantwit Major, MG 1974). **L. punctata** All. has been recorded once (27A, East Moors, JS 1876*).

ALTHAEA L.

Althaea officinalis L.   Marsh-mallow. Hocys y Morfa.
Native. Upper salt-marshes, near the extreme limit of the highest tides; occasionally found as a garden escape on roadsides or waste ground. Common and locally abundant in the Gower salt-marshes from Whiteford Point to Pontarddulais and at Oxwich; very local on the Vale coast, with recent records only from Aberthaw. (!) Fl. 7–9, fr. 8–10; Hb, B.pasc, B.terr. Map 431.

Althaea hirsuta L.,   Rough Marsh-mallow, Hocysen Flewog, has been recorded as a rare casual on waste ground with no recent records, (17A, Ely, RLS 1922*; 17B, Canton, HJR 1907*; 27A, East Moors, JS c.1876, Splott, EV 1937*; 90B, Hirwaun, HJR 1903).

Four other members of the Malvaceae have occurred as dock aliens or garden escapes: **Alcea rosea** L., Hollyhock, Hocysen Fendigaid, is the most frequent of these, having been recorded on six occasions (16A, Barry Docks, 1978!; 17A, Fairwater, 1921!*; 17B, Cardiff, L. Peddle 1977; 68A, Langland, JAW c.1945; 69D, Swansea 'ruins', JAW 1942), ?, Bayswater [Swansea area], JAW c.1945; **Hibiscus trionum** L. has been recorded on four occasions (17B, Cardiff Docks, JS c.1876*, GSW 1982; 27A, Splott, RLS 1926*, 1930*), **Abutilon theophrasti** Medicus twice (08B, near Church Village, W. Curtis, 1941*; 17B, Cardiff Docks, GSW 1983), and **Sida spinosa** L. (17B, Cardiff, JS c.1876) and **Sidalcea malvaeflora** (DC.) A. Gray ex Bentham (08D, Creigiau, C.P. Perman 1940*) once only.

THYMELEACEAE
DAPHNE L.

Daphne laureola L.   Spurge-laurel. Clust yr Ewig.
Native. Woods on limestone soils, occasionally in scrub. Rare in Gower and very local in the Vale. Map 432.

Daphne mezereum L., Mezereon, Bliwlys, was found near Draethen (28A) in 1908, presumably as an introduction (Riddelsdell, 1909); there are no recent records from the site.

ELAEAGNACEAE
HIPPOPHAE L.

Hippophae rhamnoides L.   Sea-buckthorn. Rhafnwydd y Môr, Môr-rhafnwydd.
Introduced. Sand-dunes, occasionally on sea-cliffs. Native in eastern Britain but introduced in the west (Groves, 1958). In Glamorgan it is a recent introduction which can spread rapidly on sand-dunes and form deep, impenetrable scrub which overwhelms the native vegetation. It was reported only from Penarth, where it may no longer grow, in Riddelsdell (1907), but had become established on sand-dunes in several places by 1922. It had become very abundant in parts of the Whiteford National Nature Reserve (49) in the 1960s and was removed by hand to save the dune-slack flora; it still persists there in small numbers, but may have been successfully eradicated at Oxwich NNR. Fr. 8–10. Map 433.

GUTTIFERAE (HYPERICACEAE)
HYPERICUM L.

Hypericum calycinum L.   Rose-of-Sharon. Rhosyn Saron.
Introduced. Roadside banks, plantations, parks, etc., persistent but scarcely naturalized and no longer present in several sites where it was once established. Fl. 6–11, fr. 9–10; B.luc (p). Map 434.

Hypericum hircinum L.   Stinking Tutsan, Eurinllys Drewllyd, was recorded from Penmaen (58A) by J.A. Webb in 1927 but apparently did not persist.

Hypericum androsaemum L.   Tutsan. Dail y Beiblau, Dail y Fendigaid.
Native. Woods, hedges and roadside banks. Widespread but never abundant; frequent among the lower hills where it characteristically grows on shaded hedgebanks in lanes, and also in limestone woods in the lowlands. Fl. 7–8, fr. 8–9. Map 435.

Hypericum androsaemum × H. hircinum (H. × inodorum Miller) (*H. elatum* Aiton), Tall Tutsan, Eurinllys Tal, an introduced hybrid that is similar to **H. androsaemum** but rather larger, has been recorded from three localities as a garden escape (16B, Sully, JS c.1876; 17B, Cardiff, 1969–73!*; 69C, Sketty Cwm, JAW 1951–2*).

Hypericum hirsutum L.   Hairy St John's-wort. Eurinllys Blewog.
Native. Open woods, hedges and rough grassland on limestone soils. Frequent in the southern Vale but scarce and very local in Gower, where there are no recent records. Fl. 7, fr. 8. Map 436.

Hypericum pulchrum L.   Slender St John's-wort. Eurinllys Tlws, Eurinllys Mân Syth.
Native. Rough grassland, grassy heaths, roadside banks and hedges on acid soils. Fairly common in most of the county but scarce or absent in the central Uplands and southern Vale. Map 437.

Hypericum montanum L.   Pale St John's-wort. Eurinllys Gwelw, Eurinllys Mynyddig.
Native. Open woods, cliff-scrub and hedgebanks on limestone. Scarce and in small numbers; it has been seen recently only in the western Vale and on Oxwich Point in Gower. Fl. 8–9, fr. 9–10; (Syr) S.rib (p).
Recent records from 58A, Oxwich Point, 1980!**; 58C, Oxwich Point, 1980!; 87A, near South Cornelly, H & CH 1974!; 97A, near Ewenni, H & CH 1974**; 97C, railway cutting in Cwm Alun, 1974! Map 438.

Hypericum elodes L.   Marsh St John's-wort. Eurinllys y Gors.
Native. Marshes, bogs and shallow ponds on acid soils. Rather local; most frequent in Gower and the western Uplands, but scattered throughout the county in suitable habitats. Map 439.

Hypericum humifusum L.   Trailing St John's-wort. Eurinllys Mân Ymdaenol.
Native. Open grassland and heaths, roadsides, colliery tips and waste ground, usually on acid soils. Frequent and locally common in the Uplands, but rather scarce elsewhere; absent from parts of Gower and most of the southern Vale. Map 440.

Hypericum tetrapterum Fries   Square-stalked St John's-wort. Eurinllys Pedrongl.
Native. Marshes, damp grassland and ditches. Widely distributed but rather local; absent from parts of the Vale and eastern Uplands. Map 441.

Hypericum maculatum Crantz subsp. **obtusiusculum** (Tourlet) Hayek   Imperforate St John's-wort. Eurinllys Mawr.
Native. Roadsides, hedgebanks, railway embankments and waste ground. Widely distributed and locally abundant in suitable habitats; usually more abundant than **H. perforatum** on non-calcareous soils in the western and northern Uplands and eastern Gower, but absent from western Gower and parts of the eastern Uplands and Vale. Map 442.

Hypericum maculatum × H. perforatum (H. × desetangsii Lamotte) was recorded from Little Garth (18C) by E. Vachell in 1944*.

Hypericum perforatum L.   Perforate St John's-wort. Eurinllys Trydwll.
Native. Rough grassland, roadsides, hedgebanks and railway embankments. Widely distributed and frequent in Gower and the Vale, especially on calcareous soils; fairly frequent in the Uplands, but absent from parts of the centre and north. Fl. 7–8, 10, fr. 8–11; B.prat (p), B.terr (p), (Syr) S.rib. Map 443.

VIOLACEAE
VIOLA L.

Viola odorata L.   Sweet Violet. Fioled Bêr, Millyn.
Native, but often planted or as a garden escape in villages. Fairly common in limestone woodland in the Vale, scarce and usually introduced elsewhere, but perhaps native in limestone woodland in 48B, 58A and 58B (southern Gower). Map 444.

Viola hirta L.   Hairy Violet. Gwiolydd Flewog.
Native. Limestone grassland and scrub, especially on sea-cliffs. Locally common near the coast in Gower and the Vale. **V. hirta** subsp. **calcarea** (Bab.) E.F. Warburg has been recorded from 87B (Ogmore Down). Map 445.

Viola hirta × V. odorata (V. × permixta Jordan) was recorded by G. Traherne (Trow, 1907) without locality but possibly from Coedrhiglan (17A)

[E. Vachell's record of **Viola rupestris** Schmidt from Ogmore Down (87B) is now believed to be an error for a form of **V. riviniana**.]

Viola reichenbachiana Jordan ex Boreau Early Dog-violet, Pale Wood-violet. Gwiolydd y Goedwig. Native. Woodland and hedges on limestone. Locally frequent in Gower and the Vale; sometimes abundant in native limestone woodland, but usually less frequent than **V. riviniana**. Map 446.

Viola reichenbachiana × V. riviniana was twice recorded by H.J. Riddelsdell in the early 1900s (00A, Morlais Castle; 18C, Taffs Well) but has not been seen since.

Viola riviniana Reichenb. (incl. subsp. **minor** (Murb. ex E.S. Gregory) Valentine). Common Dog-violet, Common Violet. Gwiolydd Cyffredin. Native. Woods, hedges, roadside banks, cliff-scrub, heathland, streamsides and cliff ledges in the hills. Common throughout the county. Map 447.

Viola canina L.   Heath Dog-violet. Fioled y Cŵn. Native. Sand-dunes, and occasionally in heathy grassland. Locally very abundant on the major dune-systems from Whiteford Point to the Ogwr (Ogmore). All records refer to subsp. **canina**. Map 448.

Viola canina × V. riviniana has twice been recorded in the past (48A, Rhossili, HJR 1904*; 'Glamorgan Coast', EV *c.*1933) but with no recent records.

Viola canina × V. lactea (**V. × militaris** Savouré) has been reliably recorded only once (48D, Overton Cliff, with the parents, 1984!).

Viola lactea Sm.   Pale Dog-violet. Millyn Welw Grugog. Native. Grassy heathland. Rare; recorded from several localities in the past but recently seen at only four sites, three near Port Eynon in western Gower and one at Pant St. Brides (87B). Map 449.

Viola lactea × riviniana has been recorded twice (18D, near Llanishen, with the parents, PWR, 1922*; 97A, Ewenni Down, EV 1937).

Viola palustris L.   Marsh Violet. Fioled y Gors. Native. Marshes and bogs, usually on acid soils. Common in suitable habitats in the Uplands but scarce or absent in southern Gower and the southern Vale. Map 450.

Viola palustris subsp. **juressi** (Link ex K. Wein) Coutinho was recorded from Aberdulais (79B) by E. Vachell in 1938*, and a specimen intermediate between this and subsp. **palustris**, from Caerphilly Mountain (18B) by E. Vachell in *c.*1930.

Viola lutea Hudson Mountain Pansy. Fioled y Mynydd. Possibly native. Grassy moorland? **V. lutea** is widespread and frequent in central Wales but apparently reaches the southern limit of its present British range in northern Carmarthenshire. It has been reported from four localities in Glamorgan, but none of the

records is satisfactory. Storrie (1886) reported it from 90C, near Llyn Fach, and 99B, Ystrad; there is an unsatisfactory specimen in **NMW** labelled Ystrad, July 1877, A. Langley. In addition there is an undated specimen of **V. lutea** in Storrie's collection (in **NMW**) labelled 'near Pontypridd Rocking Stone' (09D) a rather unlikely locality not referred to in Storrie (1886). Trow (1911) stated that he had entirely failed to verify Storrie's localities, although the three had been searched, in the case of Llyn Fach, repeatedly. There is however, a later field record of **V. lutea** from 'Craig Koynoch' (an uncertain locality) *c.*1926 (Drabble, 1927). The absence of any records of **V. lutea** from the Brecon Beacons and Carmarthenshire Vans, where there are extensive upland habitats apparently far more suitable for the species than any that occur on the Glamorgan hills, suggest that **V. lutea** does not occur as a native species in Glamorgan, and that all the records of the species in the county result from mis-identification or confusion of localities.

Viola tricolor L. subsp. **tricolor**. Wild Pansy, Heartsease. Trilliw. Introduced; a rare weed of acid soils, and also a scarce garden escape on waste ground. Map 451.

Viola tricolor L. subsp. **curtisii** (E. Forster) Syme (*Viola curtisii* E. Forster).   Dune Pansy. Trilliw y Tywyn. Native. Sand-dunes; locally common, but apparently absent from Oxwich Burrows. (!) Fl. 5–9 (–10), fr. 6–10; Hb, B.lap, B.pasc, B.terr, (Syr) Rh.cam, (Dip) ?Dip, (Lep) Po.ica. Map 452.

Viola arvensis Murray   Field Pansy. Ofergaru, Trilliw yr Ar. Introduced. Cultivated land and waste ground. A frequent cornfield weed in the Gower and the Vale, but rather local in the Uplands. Map 453.

Viola × wittrockiana Gams (*V. hortensis* auct.),   Garden Pansy, has been recorded on three occasions as a garden escape (07B, St. Georges, RLS 1927*; 17B, Cardiff Castle grounds, E.G. Roberts 1984*; 69C, Clyne Valley, GH 1985*), and **V. cornuta** L. once, by J.A. Webb at Mayals, Swansea (69C) where it occurred for a year or two only.

CISTACEAE

HELIANTHEMUM Miller

Helianthemum nummularium (L.) Miller Common Rock-rose. Cor-rosyn Cyffredin. Native. Limestone grassland and low limestone scrub, occasionally on sand-dunes. Common in suitable habitats on or near the coast in Gower and the Vale, especially on coastal cliffland; also on limestone near Merthyr Tydfil (00A, GH & J. Bevan 1989*). (!) Fl. (5–)6–7(–9), fr. 7–9; Bb, Sb (p). Map 454.

Helianthemum apenninum (L.) Miller White Rock-rose. Cor-rosyn Gwyn y Mynydd.
Introduced. Planted among **H. nummularium** and **H. canum** on West Cliff, Southgate (58A) in the early 1960s; it had apparently disappeared by 1976.

Helianthemum canum (L.) Baumg. Hoary Rock-rose. Cor-rosyn Lledlwyd.
Native. Limestone cliffs overlooking the sea, on shallow soil near outcrops and in crevices of rock-faces. Locally abundant on the Carboniferous Lime-stone cliffs of Gower from Limeslade to Caswell Bay and Pwll-du Head to Worm's Head. Absent from north Gower and the Vale. (!) Fl. 5–6, fr. 6–7; Sb (p). Map 455.

TAMARICACEAE
TAMARIX L.
Tamarix gallica L.   Tamarisk. Grugbren.
Introduced. Planted near the sea, and more or less naturalized on sand-dunes at three localities, but in small numbers (49C, Whiteford; 79C, Crymlyn Bur-rows; 87A, Porthcawl). Map 456.

FRANKENIACEAE
FRANKENIA L.
Frankenia laevis L.   Sea-heath. Grugeilyn Llyfn.
Native or introduced. Salt-marshes. Known from only one locality in the county where it was first discovered by S. Waldren in 1981. It is present in quite large numbers in two areas of salt-marsh and appears to be native or at least a natural introduction. It similarly appeared spontaneously in a salt-marsh in Anglesey in the 1960s.
87B, Merthyr Mawr, S. Waldren 1981!*.

ELATINACEAE
ELATINE L.
Elatine hexandra (Lapierre) DC.   Six-stamened Waterwort. Gwybybyr Chwe Brigerog.
Native (perhaps extinct). In shallow water or on mud at the edges of lakes on acid soils. Known only from Pysgodlyn Mawr and Hensol Lake in the central Vale; last certainly recorded in 1928.
07A, Hensol Lake, EV 1921–26*, Pysgodlyn Mawr, 1926!*, EV 1928*. **Pilularia globulifera**, also perhaps extinct in the county, used to grow at the latter site.

CUCURBITACEAE
BRYONIA L.
Bryonia cretica L. subsp. **dioica** (Jacq.) Tutin (*B. dioica* Jacq.)   White Bryony. Bloneg y Ddaear, Eirin Gwion.
Native. A rare plant of hedges in the east of the county. The record from west Gower in Perring & Walters (1962) is almost certainly an error. Map 457.

The following four aliens have been recorded in the past from the Cardiff area: **Ecballium elaterium** (L.) A. Richard, Squirting Cucumber (16A, 1925; 27A, 1878*), **Citrullus lanatus** (Thunb.) Mansfeld (*C. vulgaris* Schrader), Water-melon (27A, 1926*), **Lagenaria siceraria** (Molina) Standley (*L. vulgaris* Ser.), Bottle-gourd (27A, 1926*), and **Cucurbita pepo** L., Marrow (17D 1933!*).

LYTHRACEAE
LYTHRUM L.
Lythrum salicaria L.   Purple-loosestrife. Llys y Milwr.
Native. Streamsides, marshes, dune slacks etc. Fre-quent in Gower and the Vale but absent from the central Uplands. Fl. 8–10, fr. 9–10; B.pasc (n), (Lep) Go.rha, Pa.aeg, Pi.nap. Map 458.

Lythrum portula (L.) D.A. Webb (*Peplis portula* L.)   Water-purslane. Troed y Gywen.
Native. Marshy places, wet trampled ground, wet margins of ponds etc. Locally frequent in Gower, less so elsewhere. Var. **longidentata** Gay (subsp. *longidentata* (Gay) P.D. Sell) has been recorded from three localities (07B, Welsh St. Donats, 1969*!; 18D, Llanishen, JPC 1973*; 58A, Cefn Bryn, J. Groves 1903). Map 459.

Two other species of **Lythrum** have been recorded as aliens in the past: **Lythrum junceum** Banks & Solander (*L. graefferi* Ten.) (27A, RLS 1938*), and **L. hyssopifolia** L., Grass-poly, Gwyarllys Isopddail, (16A, 1924!*; 17D, JS 1876*).

ONAGRACEAE
FUCHSIA L.
Fuchsia magellanica Lam., Fuchsia Ffwsia, a widely-grown garden shrub, is perhaps naturalized near Mumbles (68A, MG 1970) and was once recorded from Cardiff (17B, RLS 1938!).

CIRCAEA L.
Circaea lutetiana L.   Enchanter's-nightshade. Llysiau Steffan.
Native. Woods and hedgebanks, and as a persistent weed in gardens. Abundant throughout the county. Fl. 6–9, fr. 7–10; (!) (Syr) S.rib, Sp.scr (n). Map 460.

Circaea alpina L.   Alpine Enchanter's-nightshade. Llysiau Stefan Mynyddig.
Native. Once recorded by a waterfall at the head of the Rhondda Valley (99A, Rhondda, AL 1890) but not seen since.

OENOTHERA L.
Oenothera fallax Renner emend. Rostański (*O. lamarkiana* auct.).
Introduced. Well established on dunes at Whiteford Burrows (49C, K. Rostański 1977!*) and on waste

ground in Swansea (69C, K. Rostanski; 1977!*). This species arose from the hybrid **O. biennis** × **O. erythrosepala**, and is often grown in gardens under the name *O. lamarkiana.*

Oenothera erythrosepala Borbás Large-flowered Evening-primrose. Melyn yr Hwyr Mwyaf, Briallu yr Hwyr Mwyaf.
Introduced. Sand dunes, roadsides, railway lines and waste ground etc. Locally frequent in the south and west. Map 461.

Oenothera novae-scotiae Gates (*O. cambrica* Rostański). Small-flowered Evening-primrose, Welsh Evening-primrose. Melyn yr Hwyr Cymreig, Briallu yr Hwyr Cymreig.
Introduced. Sand dunes, roadsides, railway lines, waste ground etc. Frequent in the west of the county, more scattered in the east. Map 462.

Oenothera novae-scotiae var. **impunctata** (Rostański) Rostański, which lacks much of the red coloration of the type, especially in the bulbous-based hairs, was first described from specimens collected on Jersey Marine, Swansea (79C, DMcC, 1975); it has also been recorded from three other localities in the county, (16A, Barry Docks, 1986! 17B, Cardiff Castle grounds, JWD 1954*; 58A, Oxwich Burrows, 1977!*).

Oenothera cambrica was first described by K. Rostański, a Polish botanist, in 1977. Prior to this, plants belonging to this taxon were variously and erroneously referred to *O. ammophila, O. biennis, O. muricata* and *O. parviflora.* Only the last species has been confirmed as having occurred in the county, and then only twice. It is now accepted that *O. cambrica* is conspecific with the Canadian species *O. novae-scotiae* Gates and this latter name has priority.

Oenothera novae-scotiae × O. erythrosepala (O. × britannica Rostański). Recorded from four localities on sand dunes and waste ground but is probably more widespread. It should be looked for wherever the two parent species grow together.
17B, Cardiff, JWD 1956*; 58A, Oxwich Burrows, K. Rostański 1977!; 69D, Swansea, K. Rostanski 1977!; 78D, Kenfig Burrows, R.M. Speed 1963.

Oenothera stricta Ledeb. ex Link Fragrant Evening-primrose. Melyn yr Hwyr Peraroglys, Briallu yr Hwyr Peraroglys.
Introduced. Once established on dunes near Porthcawl but not seen since 1941. (87A, Porthcawl, RE & FC 1901, PWR 1916*, EMT & EV 1941*). It has also been recorded as a casual from four other localities (17B, Cardiff Docks, HJR c.1907; 27A, East Moors, Cardiff, JS 1876*; 69C, Derwen Fawr Tip, Swansea, JAW 1932; 78B, Port Talbot, HJR c.1907).

Six other species of **Oenothera** have been recorded as rare introductions or casuals: **Oenothera perangusta** Gates from three localities (09B, Abercynon, K. Chamberlain, 1961; 16A, Barry Docks, 1980!*; 18A,

Nantgarw, EV 1935*), **O. parviflora** L. from two localities (00C, Aberdare, HJR c.1905; 78B, Port Talbot Docks, HJR 1905), and the other four from one locality each, **O. salicifolia** Desf. ex G. Don (27A, Splott, Cardiff, RLS 1927*!), **O. rubricuspis** Renner ex Rostański (17D, Grangetown, Cardiff, RLS 1921*–23*), **O. laciniata** Hill (17B, Cardiff Docks, RLS & CS 1925*), **O. rosea** L. (17B, Maindy, Cardiff, RLS 1938*).

The following four alien species of **Clarkia** have all been recorded in the past from the Cardiff area on ballast or rubbish tips: **Clarkia unguiculata** Lindley (*C. elegans* Douglas, non Poiret), (16A, J.N. David 1954*), **C. pulchella** Pursh (17D, JS 1876*), **C. purpurea** (Curtis) A. Nelson & Macbride subsp. **viminea** (Douglas) F.H. & M.E. Lewis (*Godetia viminea* Spach.), (17B, RLS 1938*), and **C. tenella** (Cav.) F.H. & M.E. Lewis (*Oenothera tenella* Cav., *Godetia tenella* (Cav.) Spach ex Steudel), (17B, RLS c.1922; 17D, JS c.1876, THT 1902*, RLS c.1922).

EPILOBIUM L.

Epilobium angustifolium L. (*Chamaenerion angustifolium* (L.) Scop., *Chamerion angustifolium* (L.) J. Holub). Rosebay Willowherb. Helyglys Hardd.
Native in rocky places but widespread as an introduction in woodland clearings, on sand-dunes, along railway lines, and on roadsides. Abundant throughout the county. Fl. 6–10, fr. 8–10; Hb (n), B.lap (n), B.luc, B.pasc, B.terr (n), Paravespula germanica. Map 463.

Epilobium hirsutum L. Great Willowherb. Helyglys Pêr.
Native. Ditches, streamsides and other wet places. Frequent throughout the county. Fl. 7–9, fr. 8–10; Hb (n,p), B.prat (n), B.terr (n), (Syr) S.balt (p), S.rib (p) (lep) Pi.nap, Pi.rap. Map 464.

Epilobium hirsutum × E. montanum (E. × erroneum Hausskn.) and **E. hirsutum × E. parviflorum** have both been recorded several times in the past but not during the present survey; the former from 00C, Aberdare, HJR 1903; 17B, Cardiff, 1929!*; 18C, Castell Coch, EV 1935, and the latter from 17A, Fairwater, HJR 1913; 17B, 17C, 17D, all Cardiff, AHT c.1905.

Epilobium parviflorum Schreber Hoary Willowherb. Helyglys Lledlwyd.
Native. Streamsides, marshes and similar damp habitats, and as a weed of waste ground and gardens. Frequent throughout the county. (!) Fl. 7–10, fr. 8–10. Map 465.

Epilobium parviflorum × E. roseum (E. × persicinum Reichenb.) has been recorded once in the past (00C, Aberdare, HJR c.1903).

Epilobium montanum L. Broad-leaved Willowherb. Helyglys Llydanddail.

Native. Woods, hedges, sand-dunes, waste ground and gardens. Abundant throughout the county. (!) Fl. 5–11, fr. 7–11; B.pasc (n), B.terr (n), (Sb) Colletes sp. (p), ?Sb (n,p), (Syr) E.int, H.pen, S.balt (p), Sc.pyr (n), (Dip) Em.tess (n), ?Dips. Map 466.

Three hybrids of **Epilobium montanum** with other members of the genus have been recorded in the past: **E. montanum × E. obscurum** (E. × **aggregatum** Čelak.), (00C, Aberdare, HJR 1903; 69C, Swansea, EFL 1892), **E. montanum ×E. parviflorum** (E. × **limosum** Schur), (00C, Aberdare, HJR 1903; 16A, Cadoxton, 1921*!; 17B, Cardiff, HJR 1912, 1951!*), **E. montanum × E. roseum** (E. × **mutabile** Boiss. & Reuter), (00C, Aberdare, HJR 1903; 07D, Dyffryn Gardens, EV 1942; 69C, Swansea, EFL 1892).

Epilobium lanceolatum Sebastiani & Mauri   Spear-leaved Willowherb. Helyglys Gwayw-ddail.
Native. Roadsides, dunes, banks and walls. Apparently spreading, but still uncommon; most frequently recorded in the east of the county. (!) Fl. 6–7, fr. 7–8. Map 467.

Epilobium tetragonum L. subsp. **tetragonum** (*E. adnatum* Griseb.) Square-stalked Willowherb. Helyglys Pedrongl.
Native. Damp woods, hedges, roadsides, etc. Locally frequent in the Uplands and Vale, less frequent in Gower. Fl. 6–10, fr. 7–11. Map 468.

Epilobium tetragonum L. subsp. **lamyi** (F.W. Schultz) Nyman (*E. lamyi* F.W. Schultz). Glaucous Willowherb. Helyglys Prin, Helyglys Llwydwyrdd.
Native. Damp woods. Recorded only from two localities near Cardiff.
17, near Cardiff, EV 1935*; 18C, near Whitchurch canal, MG 1971.

Epilobium obscurum Schreber   Short-fruited Willowherb. Helyglys Rhedegydd Tenau.
Native. Wet heaths and marshes, and as a weed of waste ground. Frequent throughout most of the county but absent from the extreme south of the Vale. (!) Fl. 6–9, fr. 7–10. Map 469.

Epilobium obscurum × E. palustre (E. × **schmidtianum** Rostk.).
Although recorded recently only once, this hybrid may, like many others in the genus, be of more frequent occurrence.
58A, Oxwich, HJR c.1903; 98C, near Heol-y-cyw, ABP 1978!*.

The following two hybrids have been recorded from the county in the past: **Epilobium obscurum × E. parviflorum** (E. × **dacicum** Borbás), (17A, St. Fagans, EV 1941*), **E. obscurum × E. tetragonum** (E. × **thuringiacum** Hausskn.), (00C, Aberdare, HJR c.1903; 17B, Cardiff, HJR c.1908; 78B, Port Talbot, HJR c.1904).

Epilobium roseum Schreber   Pale Willowherb. Helyglys Coesig.
Native or introduced. Old walls, roadsides and hedges. A local species scattered throughout the county. Map 470.

Epilobium palustre L.   Marsh Willowherb. Helyglys Culddail.
Native. Marshes, bogs, dune slacks, etc. Frequent throughout the county but less so in the south east. (!) Fl. 7–8, fr. 7–9; B.pasc, (Lep) Oc.ven. Map 471.

Epilobium ciliatum Rafin. (*E. adenocaulon* Haisskn.). American Willowherb. Helyglys Americanaidd.
Introduced. Now abundant in waste places and roadsides throughout the county. First recorded by Eleanor Vachell in the grounds of Rookwood Hospital, Cardiff (17A or B) in 1943. Fl. 6–10, fr. 7–11. Map 472.

Of the following four hybrids, three have been recorded during the present survey: **Epilobium ciliatum × E. hirsutum** (17B, Cardiff, JWD 1975*, provisional determination), **Epilobium ciliatum × E. montanum** (99A, Treherbert, 1970!), **Epilobium ciliatum × E. parviflorum** (18C, Radyr, MSP 1951*), **Epilobium ciliatum × E. roseum** (09A, Mountain Ash, AMP 1970, provisional determination).

Epilobium brunnescens (Cockayne) P.H. Raven & Engelhorn (*E. pedunculare* auct., non A. Cunn., *E. nerteroides* auct., non A. Cunn.). New Zealand Willowherb. Helyglys Gorweddol.
Introduced in the 1940s. Now widely naturalized in damp rocky places and on walls in the Uplands, but almost absent from Gower and the Vale. Map 473.

HALORAGACEAE
MYRIOPHYLLUM L.

Myriophyllum spicatum L.   Spiked Water-milfoil. Myrddail Tywysennaidd.
Native. Ponds, canals, lakes and ditches. Recorded from scattered localities throughout the county, most frequent in the Vale. Early records from western Gower may have been errors for the next species. Map 474.

Myriophyllum alterniflorum DC.   Alternate Water-milfoil. Myrddail Bob yn Ail, Myrddail Cylchynol.
Native. Ponds, ditches and lakes with acid water. Scarce except in western Gower. Map 475.

Myriophyllum aquaticum (Velloso) Verdc. (*M. brasiliense* Camb.).   Parrot's- feathers.
Introduced. A South American species often grown in aquaria and garden pools, and occasionally naturalised.
16B, Cosmeston Pond, Penarth, V.P. Robins 1975*, the first Welsh record. (Originally recorded as **M. verticillatum** L. which is not known to occur in the county).

Gunnera tinctoria (Molina) Mirbel, Giant Rhubarb, Rheonllys Mawr, a native of South America, has twice been recorded as an introduction (17B, Cardiff, RLS 1938!; 69C, Fairy Hill, JAW 1931).

HIPPURIDACEAE
HIPPURIS L.

Hippuris vulgaris L. Mare's-tail. Rhawn y Gaseg. Native. Ponds, marshes, canals etc. A very local plant of the Vale and Gower, only once recorded in the Uplands. Map 476.

CORNACEAE
CORNUS L.

Cornus sanguinea L. (Thelycrania sanguinea (L.) Fourr., Swida sanguinea (L.) Opiz). Dogwood. Cwyros.
Native. Woods and hedges on basic soils. Frequent in Gower and the Vale but almost completely absent from the Uplands. Map 477.

Cornus sericea L. (C. stolonifera Michx, Thelycrania sericea (L.) Dandy, Swida sericea (L.) J. Holub). Red-osier Dogwood.
Introduced. Woodland and scrub. Recorded several times in the past, mainly in Gower.
48B, Penrice Woods, JAW 1944*, 1974!; 69C, Mayals, JAW 1923*, c.1945; 69C, Green Cwm, JAW 1942*; 69C, Blackpill, JAW 1951; 78B, Margam, JAW c.1945.

ARALIACEAE
HEDERA L.

Hedera helix L. s.l. Ivy. Iorwg.
Native. Woods, hedges, cliffs, walls etc. Abundant throughout the county. Fl. 9–12, fr. 3–5; Hb (n,p), Pv.germ, (Syr) E.ten (n), H.pen, S.rib, ?Syrs, (Dip) Call.ery, Luc. spp., Sc.ste, (Lep) Ag.urt, Va.ata. Map 478.

It has recently been found that Hedera helix L. s.l. can be split into two distinct species on morphological and cytological grounds : H. helix L. s.s. a diploid with 2n=48, and H. hibernica (Kirchner) Bean, a tetraploid with 2n=96 (McAllister, 1980). H. hibernica has a western distribution in the British Isles, occurring in Ireland, western Wales and south-western England, while H. helix s.s. has been recorded from Scotland, central, eastern and northern England, and parts of eastern Wales. Although only H. hibernica has been recorded from Glamorgan so far, very few determinations of the segregates have been made (78B, Port Talbot, H. McAllister 1980, 49C!, 58A!, 99A! 1985). Diploid H. helix extends to Brecon and the Wye Valley (H. McAllister, personal communication) and may occur in eastern Glamorgan and perhaps elsewhere.

Hedera colchica (C. Koch) C. Koch, an introduction from the Caucasus, is well established on a wall in Aberdare, the only known locality in the county. 00C, Aberdare, AMP 1981*.

UMBELLIFERAE (APIACEAE)
HYDROCOTYLE L.

Hydrocotyle vulgaris L. Marsh Pennywort. Ceiniog y Gors, Cron y Gweunydd.
Native. Marshes, bogs, dune slacks etc. Locally frequent throughout the county except in the extreme south of the Vale. Map 479.

SANICULA L.

Sanicula europaea L. Sanicle. Clust yr Arth.
Native. Woods, usually on base-rich soils. Locally frequent in Gower and the Vale but less so in the Uplands mainly through the lack of suitable habitats. Fl. 5–6, fr. 7–8. Map 480.

ASTRANTIA L.

Astrantia major L. Astrantia.
Introduced. Established along roadsides in one locality (08B), and recorded once as a casual (17B). 08B, Ton-teg, D.M. Howarth 1974*; 17B, Cardiff, H.E. Salmon 1926*.

ERYNGIUM L.

Eryngium maritimum L. Sea-holly. Celyn y Môr.
Native. Sand-dunes and fore-shores. Frequent in suitable habitats in Gower and the western Vale. Formerly occurred in several localities in the south and east of the Vale, but not recorded there during the present survey. Map 481.

Eryngium campestre L. Field Eryngo. Ysgallen Ganpen, Boglynnon.
Introduced. Waste places. Established near docks in one locality in the Vale; a few old records from similar habitats.
16A, Barry Docks, RLS 1926*, DMcC 1974!; 17B, near Cardiff, F. Brent, 1848; 27A, Roath Dock, Cardiff, JS 1876; 78B, Port Talbot, HJR c.1904, R.L. Burges 1933.

Eryngium amethystinum L., has been recorded as a spontaneous garden weed in Penarth (17D, L. Evans 1979*).

CHAEROPHYLLUM L.

Chaerophyllum temulentum L. Rough Chervil. Perllys y Perthi.
Native. Hedgebanks and roadsides. Common in Gower and the Vale, where it is often the commonest Umbellifer to flower after Anthriscus sylvestris, but almost completely absent from the Uplands. Map 482.

ANTHRISCUS Pers.

Anthriscus sylvestris (L.) Hoffm. Cow Parsley. Gorthyfail.

Native. Hedges, woodland, scrub. Abundant along roadsides in the Vale and Gower, where it is usually the first Umbellifer to flower in the spring. Less frequent in the Uplands where it is absent from large stretches of the coal measures. Fl. 4–6, fr. 6–7; B.luc (n), Pv.germ, (Syr) E.ten, S.rib, Sy.pip, (Dip) Luc.caesar, Sc.ste, Sepsis punctum. Map 483.

Anthriscus caucalis Bieb. (*A. neglecta* Boiss & Reuter). Bur Chervil. Gorthyfail Cyffredin.
Native. Sandy places and sea-cliffs. A rare plant with few recent records. Grows in similar habitats to the more abundant **Torilis nodosa,** with which it may have been confused in the past. Map 484.

Anthriscus cerefolium (L.) Hoffm., Garden Chervil, Gorthyfail y Gerddi, has been recorded once, as a casual (87A, Newton Nottage, EV pre 1930).

SCANDIX L.

Scandix pecten-veneris L. Shepherd's-needle. Crib Gwener, Nodwydd y Bugail.
Introduced. Cultivated ground and waste places. Formerly a quite widespread, although never common, agricultural weed in the south of the county. Now found in only three localities. This decline is probably due, in part at least, to improved agricultural techniques. Map 485.

MYRRHIS Miller

Myrrhis odorata (L.) Scop. Sweet Cicely. Cegiden Bêr.
Doubtfully native. Roadsides, hedges and railway banks. A rare plant found in a few scattered localities, mainly in the eastern Uplands. Map 486.

CORIANDRUM L.

Coriandrum sativum L. Coriander. Brwysgedlys.
Introduced. A rare casual of waste ground and ballast, with only one recent record – in a shrub border.
16A, Barry Docks, EV 1925*; 17B, Cardiff, JPC 1980!; 17D, Cardiff, JS c.1876, EV 1933; 18C, Radyr, HJR 1911; 27A, East Moors, Cardiff, JS 1876*, RLS 1925*; 90D, Llwydcoed, Aberdare, HJR 1906.

SMYRNIUM L.

Smyrnium olusatrum L. Alexanders. Dulys
Introduced. Well naturalized on sea-cliffs and roadsides near the coast. Frequent in the south of the county, usually near the sea. Fl. 4–5 (–6), fr. 5–6. Map 487.

CONOPODIUM Koch

Conopodium majus (Gouan) Loret Pignut. Cneuen Ddaear
Native. Woods and pastures, especially those close to woods or which were formerly wooded. Abundant

throughout most of the county. Fl. 5–6 (–7, 10), fr. 6–7, 10; (Syr) E.ten (n), (Dip) Em.tess (n). Map 488.

PIMPINELLA L.

Pimpinella saxifraga L. Burnet-saxifrage. Gwreiddiriog, Tormaen Burnet.
Native. Grassland, usually on base-rich soils. Widespread throughout the county but less frequent in the north. Fl. 7–9, fr. 8–10. Map 489.

Pimpinella major (L.) Hudson was once recorded as a casual (87A, Porthcawl, J.B. Lloyd pre 1900).

AEGOPODIUM L.

Aegopodium podagraria L. Ground-elder. Llys y Gymalwst.
Introduced. Roadsides and waste places. Frequent throughout the county except for a few areas of the Uplands. Fl. 5–7, fr. 7–8. Map 490.

BERULA Koch

Berula erecta (Hudson) Coville Lesser Waterparsnip. Pannas y Dŵr, Dyfrforonen Gulddail.
Native. Streamsides, ponds and ditches. Frequent in the Vale and Gower, less so in the Uplands. Fl. 7. Map 491.

CRITHMUM L

Crithmum maritimum L. Rock Samphire. Corn Carw'r Môr.
Native. In the spray zone on sea-cliffs and sometimes on shingle beaches. Frequent in suitable habitats all along the coast. (!) Fl. 6–8, fr. 7–9; ?Dips. Map 492.

OENANTHE L.

Oenanthe fistulosa L. Tubular Water-dropwort. Dibynlor Pibellaidd.
Native. Streamsides and marshes. Scarce in the Vale and recently recorded from only one site in Gower (49C, Llangennith Moors, 1987!); some old Gower records may have resulted from confusion with **O. lachenalii.** Apart from a few old records, it is unrecorded from the Uplands. Fl. 7–9, fr. 8–10. Map 493.

Oenanthe lachenalii C.C. Gmelin Parsley Waterdropwort. Dibynlor Perllysddail.
Native. Upper salt-marsh, especially in the **Juncus maritimus** zone. Locally frequent in suitable habitats along the western coast, but with few recent records in the east. Map 494.

Oenanthe crocata L. Hemlock Water-dropwort. Cegid y Dŵr.
Native. Ditches, stream banks and marshes. Abundant throughout the county. Fl. 6–7, fr. 7–8. Map 495.

Oenanthe aquatica (L.) Poiret   Fine-leaved Water-dropwort. Cegid Manddail y Dŵr.
Native. Marshy ground. Very rare; known from only one site, where it was first found in 1979.
17C, Culverhouse Cross, Cardiff, ALe 1979*, P. Randerson 1990.

AETHUSA L.

Aethusa cynapium L.   Fool's Parsley. Gauberllys, Persli'r Ffŵl.
Introduced. A common weed of cultivated ground and waste places. Frequent in Gower, the southern Uplands and the Vale. Fl. 4–11, fr. 6–11; ?Sb (n), (Syr) S.balt, (Dip) Luc. Map 496.

FOENICULUM Miller

Foeniculum vulgare Miller   Fennel. Ffenigl.
Introduced. Well naturalized on wasteland, especially near the sea, and sometimes on sand-dunes and sea-cliffs. Locally frequent in parts of Gower, the southeastern Uplands and southern parts of the Vale, occasional elsewhere. (!) Fl. 7–10, fr. 8–10; Pv.germ, (Syr) E.arb, E.ten, (Dip) Luc.spp, ?Dips. Map 497.

Anethum graveolens L. (*Peucedanum graveolens* (L.) Bentham & Hooker fil.), Dill, Llys y Gwewyr, has once been recorded as a casual (27A, Splott, RLS 1926!*).

SILAUM Miller

Silaum silaus (L.) Schinz & Thell.   Pepper-saxifrage. Ffenigl yr Hwch.
Native. Limestone grassland. Confined to a few scattered sites in the south of the Vale with only one old record from the Uplands, absent from Gower. Map 498.

CONIUM L.

Conium maculatum L.   Hemlock. Cegiden.
Native or introduced. Roadsides and waste ground, especially near farms. Locally frequent in parts of Gower and the Vale much less so in the Uplands. Fl. 6–7, fr. 7–9. Map 499.

Prangos uechtritzii Boiss. & Hausskn. was once recorded as a dock alien (16A, Barry Docks, RLS & RM 1931).

BUPLEURUM L.

Bupleurum rotundifolium L.   Thorow-wax. Paladr Trwyddo.
Casual. Waste places. Recorded several times in the past, but all recent records have proved to be the very similar *B. subovatum*.
00C, Aberdare, HJR 1897; 17B, Cardiff Docks, HJR c.1906; 17D, Temperance Town, Cardiff, JS c.1876, Penarth Ferry, 1920!*; 69C, Blackpill, H.R. Wakefield c.1920; 69D, Hafod, W.G. Williams c.1920.

Bupleurum subovatum Link ex Sprengel (*B. lancifolium* auct., non Hornem.; *B. intermedium* (Loisel. ex DC.) Steudel). False Thorow-wax. Ffugbaladr Trwyddo.
Introduced; often found in bird-seed. Recorded from waste places and roadsides in the south eastern Vale; scarce.
06B, Penmark, D.J. Sully 1965*, K. Hoddle 1971*, 1976!; 16A, Barry, JS c.1876; 17B, Cardiff Docks, JS c.1876*, Roath, Cardiff, 1972!*; 17D, Penarth Rd, Cardiff, EV 1935*; 27A, Splott, RLS 1927*; 98D, Pencoed, H.J. Dawson 1985.

Bupleurum tenuissimum L.   Slender Hare's-ear. Paladr Trwyddo Eiddilddail.
Native. Salt-marshes, very rare and confined to the south-eastern Vale; no recent records.
06A, Aberthaw, EV c.1935; 17D, Ely Estuary, Cardiff, EV 1903, HJR 1908*, 1956!*; 27A, near mouth of River Rhymney, HJR c.1909, EV 1920!, 1926!*.

Bupleurum falcatum L., Sickle-leaved Hare's-ear, Paladr Trwyddo Crymanddail, was once recorded as a casual (27A, Splott, Cardiff, RLS 1926*).

APIUM L.

Apium graveolens L.   Wild Celery. Perllys y Morfa, Mers.
Native. Upper salt-marshes especially in the brackish transition zone, and sometimes on streamsides inland. Fairly frequent in Gower and the southern Vale. Map 500.

Apium nodiflorum (L.) Lag.   Fool's Water-cress. Dyfrforonen Sypflodeuog.
Native. Streamsides and riversides, ditches, etc. Common throughout Gower and the Vale but absent from parts of the Uplands. Fl. 7–8, fr. 8–9. Map 501.

Apium inundatum (L.) Reichenb. fil. Lesser Marshwort. Dyfrforonen Leiaf, Dyfrforonen Nofiadwy.
Native. In shallow water in ponds and canals, and sometimes in dune- slacks. Locally frequent in Gower but increasingly scarce in the Vale. Map 502.

Apium leptophyllum (Pers.) F. Mueller ex Bentham (*A. tenuifolium* Thell.) was once recorded as a casual (17B, Cardiff, EV 1923).

PETROSELINUM Hill

Petroselinum crispum (Miller) A.W. Hill Garden Parsley. Persli.
Introduced. Walls, hedgebanks and waste places. Always very scarce and recently recorded only at Horton in Gower (48B) and at Barry (16A).
06A, Gileston, EV 1936*; 16A, Barry, H. Thomas 1983; 16B, Swanbridge, JS c.1876, AHT 1899*; 48B Horton, JAW, c.1925, 1974!; near Port Eynon, JAW 1923*.

Petroselinum segetum (L.) Koch Corn Parsley. Eilunberllys.
Native. Rocky places, cultivated ground and roadsides. Rare, confined to a few sites in the southern Vale.
06A, Aberthaw, EV 1936*, 1958!*, Fonmon, ABP 1975!, 1977!; 06B, Fontygary, 1972!*; 16A, Hayes Farm, Anon. 1899, Barry and Merthyr Dyfan, HJR c.1904; 16B, Swanbridge, AHT 1899*, HJR c.1904, Cosmeston, 1949!*, Barry Link Road, JPC 1982; 17C, Dinas Powis, 1931!*, GMB 1973!; 87A, Nottage, HJR c.1904, EMT 1931*; 88C, Mawdlam, HJR c.1904.

SISON L.

Sison amomum L.   Stone Parsley. Githran.
Native. Hedgebanks and roadsides. A local plant, confined to the southern Vale. Map 503.

Ammi majus L.   Bullwort, Esgoblys, a rare casual of waste places has been recorded on four occasions (00A, Merthyr Tydfil, R. FitzGerald 1982*; 00C, Aberdare, HJR 1900; 27A, Splott, 1925!*, RLS 1927*).

Falcaria vulgaris Bernh.   Longleaf, a casual of waste places, was twice recorded in the past (16A, Barry Docks, EV 1931*; 78B, Port Talbot, HJR 1909*).

CARUM L.

Carum carvi L.   Caraway. Carwas.
Casual. Recorded in the past from docksides or waste ground in Cardiff, Swansea and Aberdare; only one recent record.
00C, Aberdare, HJR 1904; 16A Barry Link Road, H. Thomas 1983; 17B, Cardiff Docks, HJR 1906; 17D, Penarth, JS c.1876, Grangetown, Cardiff, JS c.1876; 27A, Cardiff, 1943!*; 69C, Swansea, J.R. Shepherd 1912; 69D, Crymlyn Bog, HJR c.1907; 87A, Porthcawl, J.B. Lloyd c.1906.

Carum verticillatum (L.) Koch   Whorled Caraway. Carwas Troellog.
Native. Damp acid pastures and marshy areas. Locally frequent in the north-western part of the county, but almost completely absent from the southern Uplands and the Vale. Map 504.

ANGELICA L.

Angelica sylvestris L. Wild Angelica. Llys yr Angel.
Native. Marshes, streamsides, damp roadside verges and damp woods. Frequent throughout the county. Fl. 7–9, fr. 8–9; (Dip) Call.sp, Luc.caesar. Map 505.

PASTINACA L.

Pastinaca sativa L.   Wild Parsnip. Panasen Wyllt, Moronen y Moch.
Native or introduced. Roadsides, railway lines and waste places. Locally common in the east and west of the county but absent from large areas of the central

Uplands and Vale and south Gower. Fl. 7–10, fr. 8–10. Map 506.

HERACLEUM L.

Heracleum sphondylium L. Hogweed. Efwr, Panasen y Cawr.
Native. Hedgebanks, rough grassland, roadsides, railway lines, waste places, etc. Abundant through most of the county but local or absent in a few areas in the Uplands. Fl. 7–8 (–12), fr. 8–10; Hb, B.luc, B.terr, Paravespula germanica, (Syr) E.ten, H.pen, Leu.luc, S.balt, S.rib, (Dip) Call.ery, Chloromyia formosa, Em.tess, Luc.sp, Mesembrina meridiana, Musca sp., Sa.car, Sepsis punctum, Sc.ste. Map 507.

Heracleum mantegazzianum Sommier & Levier Giant Hogweed. Efwr Enfawr.
Introduced. River banks and roadsides. Recorded from five widely separated areas.
17B, Bank of R. Taff, Cardiff, ABP 1973, GH 1986; 18D Thornhill, JPC 1973*; 79D Aberavon, H & CH 1979; 96B, Llantwit Major, ABP 1973! (A specimen recorded by J.A. Webb in 1923 from a railway bank beyond Skewen (79A) as *H. villosum* Fisch. may be **H. mantegazzianum**).
H. mantegazzianum can cause severe dermatitis and it is now a criminal offence to knowingly introduce this species into the wild.

TORILIS Adanson

Torilis nodosa (L.) Gaertner Knotted Hedge-parsley. Troed-y-cyw Clymog.
Native. Shallow soils on sea-cliffs, especially on headlands where sheep and cattle gather, dry banks along roadsides and in grassland. Frequent in Gower in suitable habitats but increasingly local in the Vale and absent from the Uplands. Map 508.

Torilis japonica (Houtt.) DC. Upright Hedge-parsley. Troed-y-cyw Syth.
Native. Hedges and scrub. Abundant throughout the county. Fl. 6–8(–11), fr. 7–11. Map 509.

Two other species of **Torilis** were recorded as casuals in the past: **Torilis arvensis** (Hudson) Link (00C, HJR c.1907; 69D, TBF & E. Lees c.1840) and **T. leptophylla** (L.) Reichenb. fil. (27A, RLS & EV 1926).

Caucalis platycarpos L.   Small Bur-parsley, Eilunberllys Bychan, was recorded several times in the past as a dock alien or casual weed (00C, HJR 1903; 16A, RLS 1924*, EV 1932*; 17B, HJR 1902, RLS 1936; 68A, V.M. Reel 1929*) as was **Turgenia latifolia** (L.) Hoffm., Greater Bur-parsley (00C, HJR 1903; 16A, RLS 1927*; 17B, HJR 1897, EV & RLS 1936*; 78B, HJR 1906).

DAUCUS L.

Daucus carota L.   Wild Carrot. Moronen y Maes.
Native. Common on grassy sea-cliffs and field banks near the sea (usually subsp. **gummifer** Hooker fil.,

Sea Carrot, Moronen y Mor, on the cliffs of southern Gower and the southern Vale, intermediate with subsp. **carota** elsewhere), locally frequent on roadsides and waste ground and in limestone grassland elsewhere, especially in the Vale (subsp. **carota**). Map 510.

PYROLACEAE
PYROLA L.

Pyrola minor L. Common Wintergreen. Coedwyrdd Bychan, Glesyn y Gaeaf.
Native. Very rare, and possibly extinct in the only locality in which it grew, the upper Pyrddin valley (80B) where it was found by Joseph Hooker in the 1840s. 'Found by Dr. Jos. Hooker, near Uscoed Eynon Gam, above the fall' (Dillwyn, 1848). Hooker's locality may have been in Breconshire (the Pyrddin forms the county boundary) but it was credited to Glamorgan in Watson (1883). The plant has not been seen recently despite careful searches of the Pyrddin valley, but may appear elsewhere in the county; it has recently been found on Tywyn Burrows on the coast of Carmarthenshire, growing with introduced conifers in Pembrey Forest (Vaughan, 1977), and appears to be spreading there.

Pyrola rotundifolia L. subsp. **maritima** (Keynon) E.F. Warburg Round-leaved Wintergreen. Coedwyrdd Crynddail.
Native. Locally common on sand-dunes on the coast of Gower and the western Vale, usually growing in dune-slacks with **Salix repens**. It appears to have spread to Glamorgan during the 1930s, and was first found in the county at Kenfig Burrows by John Lord in 1938; by 1940 about 300 shoots were present (Kay, Roberts & Vaughan, 1974), and it has subsequently spread to all the other major dune systems in Glamorgan. Its flowering tends to be sporadic and irregular, extending over a season from June to November with much variation from clone to clone. 49C, Whiteford Burrows, 1976!** (first record), 1983!; 58A, Oxwich Burrows, S. Herbert 1968 (first record), 1982!; 69D, Crymlyn Fen, P. Mock 1981* (first record); 78D, Kenfig Burrows, J. Lord 1938 (first record), 1939*, 1982!; 79C, Crymlyn Burrows, R.D. Tweed 1953** (first record), 1982!; 87B, Merthyr Mawr Burrows, G. Campbell 1965 (first record), 1973.

ORTHILIA Rafin.

Orthilia secunda (L.) House Serrated Wintergreen. Coedwyrdd Bylchog.
Native. Very rare; found by A.H. Trow in two sites on Craig-y-llyn (90C*); Trow found the first site in 1892 and the second, in 1905, with J.H. Salter. It has not been seen recently but may still grow on the precarious and often virtually inaccessible ledges of the north-facing cliffs of Craig-y-llyn above Llyn Fawr; Trow (1911) observed that the 1892 site was 'almost inaccessible now'. No other localities are known in Glamorgan, but **O. secunda** grows on Craig Cerrig-gleisiad about 10km to the north.

MONOTROPA L.

Monotropa hypopitys L. *s.l.* Yellow Bird's-nest. Cytwf.
Native. Dune-slacks, where it is associated with **Salix repens**, beech-woods, and conifer plantations. Rare, and sporadic in its appearance, with recent records only from dune-slack sites. The dune-slack plants from Whiteford Burrows (49A), Oxwich Burrows (58A), Crymlyn Burrows (79C) and Kenfig Burrows (78D) are glabrous, as were R.C. McLean's plants from Garth Wood (18C), and would thus be attributable to **M. hypophegea** Wallr., if this is regarded as a separate species. T.H. Thomas's specimens collected in 1899 in Castell Coch woods (also 18C) are pubescent and would thus be regarded as **M. hypopitys** *s.s.*
17A, Great Wood, St. Fagans, JS 1874, 1885; 18C, Castell Coch, JS & THT 1899*, RAH 1989, Garth Wood, R.C. McLean 1927, Fforest Ganol, JPC 1978, 1982; 49A, Whiteford Burrows, 1972!, 1982!; 58A, Oxwich Burrows, D.O. Elias 1978, 1985!; 78D, Kenfig Burrows, H.M. Hallett 1906, V. Insole 1927, EV 1935*, G. Campbell 1965*, JPC 1978; 79C, Crymlyn Burrows, CH 1986**; 87A, Merthyr Mawr dunes, S. Waldren 1981; 87B, Merthyr Mawr dunes, PJo 1981, JPC 1983; 88A, Margam Woods, WT 1802.

ERICACEAE
ERICA L.

Erica tetralix L. Cross-leaved Heath. Grug Deilgroes.
Native. Very common on damp acid heaths and moorland, and in peaty bogs. White or pale-flowered plants are common in some populations, for example on Cefn Bryn near Broad Pool (59A!). Scarce or absent in the southern Vale, where there are few suitable habitats. (!) Fl. 6–9 (–11), fr. 8–11; Bb. Map 511.

Erica cinerea L. Bell Heather. Clychau'r Grug, Grug Lledlwyd.
Native. Dry heaths, grassy banks and moorland on acid soils, and on sea-cliffs where it may grow close to the spray-zone. Common in Gower and the southern Uplands but rather local or even rare elsewhere. Map 512.

CALLUNA Salisb.

Calluna vulgaris (L.) Hull Heather. Grug.
Native. Very common on acid heaths and moorland, in peaty bogs, on mountain cliff-ledges and boulder-screes, in acid grassland generally in the Uplands and in Gower, on old colliery spoil-tips, on acid dunes, and on sea-cliffs, where it may sometimes, like **Erica cinerea**, grow in sites that occasionally receive substantial quantities of salt spray. Fl. 6–9, fr. 7–10; Hb, B.luc, B.mont, B.terr. Map 513.

RHODODENDRON L.

Rhododendron ponticum L. Rhododendron.
Introduced as an ornamental shrub or to provide cover; now widely naturalized and spreading in

woodland on acid soils, and to a lesser extent on moorland and in bogs. Map 514.

Arbutus unedo L., Strawberry-tree, Mefusbren, was recorded as naturalised on limestone scree at Caswell (58B) by J.A. Webb in *c*.1927; it was still there in 1952* but has not been reported since; and **Gaultheria shallon** Pursh has recently been reported as established at 79B, near Neath, J.C. Watkins 1988*.

ANDROMEDA L.

Andromeda polifolia L. Bog-rosemary. Rhosmari Gwyllt.
Native, probably extinct. Recorded from Crymlyn Bog (probably at the northern end, in 69B) by E. Forster *c*.1805 (Turner & Dillwyn, 1805), where until drainage took place it was rather plentiful 'towards the northern extremity' (Dillwyn, 1848). It has not been seen there since the early nineteenth century despite recent careful searches of likely habitats stimulated by the rediscovery of **A. polifolia** at Llanllwch Mire west of Carmarthen in 1971.

VACCINIUM L.

Vaccinium oxycoccos L. Cranberry. Llygaeron, Ceirios y Waun.
Native. Scarce and very limited in its distribution, growing in wet heaths and bogs, usually among Sphagnum, **Aulacomnium palustre** or **Polytrichum commune**; most frequent in the northern Vale and southern Uplands, for example on Llantrisant Common (08C!) and Gelli-gaer Common (19A). In Gower it has been found only on Clyne Common and Fairwood Common (59D!). Map 515.

Vaccinium macrocarpum Aiton American Cranberry. Llygaeron America.
Naturalized alien? Vegetative material resembling this species was found on the western side of Crymlyn Bog (69D) in 1981, and had spread to cover about 4 square metres by 1984 (R. Meade, pers. com.).

Vaccinium vitis-idaea L. Cowberry. Llus Coch.
Native. Very scarce, growing on high cliff-ledges in a few localities in the northern Uplands; recently recorded only from the Craig-y-llyn escarpment. 80D, Craig y Pant, 1974!; 90C, Craig-y-llyn, AHT 1892*, above Llyn Fach, EV 1924, 1983!, Padell-y-Bwlch, HJR *c*.1907; 99A, Nant-y-gwyddon, JS *c*.1886, Blaenafan, THT *c*.1911; 99C, near Braich-y-cymer, JS *c*.1886.

Vaccinium myrtillus L. Bilberry. Llus.
Native. Common and widespread on acid soils in the Uplands, growing on dry heaths, roadside banks, steep-sided valleys and hill slopes and in rocky places; rather local in Gower and absent from parts of the Vale where suitable habitats do not occur. It is locally abundant on acid sand-dunes at Penmaen and Pennard Burrows (58A) in Gower, and in similar habitats on the western coast of the Vale. Fl. 5–6, fr. 7–8; B.lap. Map 516.

EMPETRACEAE
EMPETRUM L.

Empetrum nigrum L.   Crowberry. Creiglys.
Native. Very scarce, growing on high moorland and cliff-ledges in a few places in the northern Uplands. 00C, Mynydd Merthyr, THT 1890, HJR *c*.1907, HAH 1927*; 09B, Abercynon, D.P.M. Guile 1941; 10C, Cefn Brithdir, HAH 1926*, KA 1954*, A. Burn 1982/3; 80(B), hills around Pont-nedd-fechan, LWD *c*.1840; 90C, above Llyn Fach, 1983!; 99B, Taren Maerdy, JTa 1970. The record for St. Lythan's Down (see Trow, 1911) is now thought to be an error.

PRIMULACEAE
PRIMULA L.

Primula vulgaris Hudson   Primrose. Briallu.
Native. Common in woods and on hedgebanks in Gower and the Vale, but scarce in the Uplands. The pink-flowered form is much rarer than the yellow and is usually found only near old gardens. Fl. (1–)3–5, fr. 5–6. Map 517.

Primula veris L.   Cowslip. Briallu Mair.
Native. Still fairly common in old pastures and meadows, on field banks, and on roadsides and railway embankments on base-rich soils in the Vale, and locally common in similar habitats in Gower; sometimes abundant in dune grassland, for example on Cwm Ivy Tor (49C); rare in the Uplands. Fl. 4–5, fr. 7–8; B.hort (n,p). Map 518.

Primula veris × P. vulgaris (**P. × tommasinii** Gren. & Godron), False Oxlip, Briallu Tal Ffug, can usually be found where populations of the parent species adjoin one another and an intermediate habitat is available; a range of back-cross and intermediate plants sometimes occurs, for example at the edge of Parkmill Wood near Kilvrough Manor in Gower (58B). Map 519.

Primula × polyantha Hort. has been found as a garden escape or throw-out in one locality (87B, between Merthyr Mawr and Bridgend, H.T. Davies 1983).

Hottonia palustris L., Water-violet, Pluddalen, Fioled y Dŵr, has been recorded once (16B, Cosmeston Ponds, MG 1978) where it is probably an aquarium throw-out or deliberate introduction.

Cyclamen hederifolium Aiton, Cyclamen, Bara'r Hwch, was found in Coedrhiglan Woods (17A) by John Storrie in 1886, 'no doubt planted', and does not appear to have been seen there subsequently. Planted populations have persisted, and have been recorded, at Fairyhill (49D, JAW 1931) and near Radyr (18C, M. Voyce, 1979!).

LYSIMACHIA L.

Lysimachia nemorum L. Yellow Pimpernel. Gwlydd Melyn Mair.
Native. Widely distributed and often common, especially in the Uplands, in woods, on shaded hedgebanks and roadsides, and sometimes in damp heathland and grassland. Fl. 5–7, fr. 7–8. Map 520.

Lysimachia vulgaris L. Yellow Loosestrife. Trewynyn.
Native. Locally common in marshes and fens, by the margins of ponds and lakes, and in ditches and by streams, usually on base-rich or neutral soil. Map 521.

Lysimachia nummularia L. Creeping-Jenny. Siani Lusg.
Native, but sometimes naturalized as a garden escape. Scarce in Gower and the Uplands but fairly frequent in the Vale, growing in marshes and on stream-banks, and less often in damp woods or on hedgebanks. Fl. 7. Map 522.

Lysimachia punctata L. Dotted Loosestrife. Trewynyn Brych.
A naturalized garden escape, usually growing as isolated clumps on roadsides and waste ground; not common, but widespread and perhaps increasing. Map 523.

GLAUX L.

Glaux maritima L. Sea-milkwort. Glas yr Heli.
Native. Abundant and ubiquitous in upper salt-marshes, growing in the upper marsh with **Festuca rubra** and **Juncus gerardi** in the **Juncus maritimus** zone; also found in the transition zone between salt-marsh and sand-dunes, for example at Crymlyn Burrows (79C). (!) Fl. 6–8, fr. 7–9; B.terr (n), Sb. Map 524.

ANAGALLIS L.

Anagallis minima (L.) E.H.L. Krause Chaffweed. Corfrilys, Bril-lys Coraidd.
Native. Damp sandy places and the gravelly margins of pools. Very scarce, with only one recent record, from Kenfig Burrows; it probably still grows in some of the sites in which it has been found in the past in Gower.
07A, Pysgodlyn Mawr, AHT 1905*, 1921! (probably extinct), Pendoylan, AHT *c.*1911; 48B, Penrice, PS 1910*; 58A, below Pennard Castle, J. Ball 1849, Oxwich Bay, HJR *c.*1907; 59D, Fairwood Common, HJR *c.*1907; 77B, Sker, EV *c.*1935; 78B, Kenfig Burrows, AHT 1905*, EMT 1944*, MG 1974; 79C, Jersey Marine, B.W. Bryant 1957; 96A St Donats, AHT 1920*.

Anagallis tenella (L.) L. Bog Pimpernel. Gwlyddyn Mair y Gors.
Native. Widespread and sometimes abundant in grazed marshes, damp grassland (including, at least

in the west, damp lawns), dune-slacks in Gower, and the Uplands and the northern Vale, increasing in frequency in the western part of the county. Map 525.

Anagallis arvensis L. Scarlet Pimpernel. Llys y Cryman, Gwlydd Mair.
Native. Locally abundant with other annuals on sand-dunes, on shallow soil on sea-cliffs, in cliff-heath after fires, and on disturbed ground in long-established grassland on base-rich soils inland; widespread and common as a weed of cultivation. Pink-flowered plants are frequent in some cliff populations, for example at Overton (48B) and Penmaen (58A). Blue-flowered plants appear to be most frequent in the Vale, but have not always been distinguished from the blue-flowered **A. foemina**. They have been recorded in 08A, 16A, 17B, 27A, 69C, 87A, 88C and 97A. Fl. 4–11, fr. 6–11. Map 526.

Anagallis foemina Miller Blue Pimpernell. Gwlyddyn Mair Benyw.
Introduced; occasional on waste ground, roadsides and as a garden weed in a few localities in the neighbourhood of Cardiff and Barry.
16A, Barry, M.T. Hamilton 1929*; 17B, Canton, HJR 1907*; 17C, Dinas Powis, GMB 1953*; 27A, Splott, 1927!, EV 1935, Pengam, L. Browning 1962*. Old unconfirmed records from 16B, 17D, 69A, 78B, 87A, 87D, 88C, probably refer to blue-flowered forms of **A. arvensis**.

SAMOLUS L.

Samolus valerandi L. Brookweed. Claerlys.
Native. Locally common in brackish marshes, on the banks of drainage ditches and streams near estuaries or the sea, and in damp places on sea-cliffs on the coast of Gower and the Vale. Recorded in the past from several inland sites in the Vale by J. Storrie, A.H. Trow and others, perhaps in error; there are no recent records of this species from any inland site in Glamorgan, and it is noteworthy that H.J. Riddelsdell did not himself observe the species at any inland locality although he included Storrie's records in his *Flora* (Riddelsdell, 1907). Map 527.

PLUMBAGINACEAE

ARMERIA Willd.

Armeria maritima (Miller) Willd. Thrift. Clustog Fair.
Native. Very common in upper salt-marshes, in the spray-zone at the foot of sea-cliffs, and in grassland and on walls on cliff-tops in exposed places, on the coast of Gower and the Vale. Fl. 5–8, fr. 7–9; Hb (n), B.terr, ?Sb. Map 528.

LIMONIUM Miller

Limonium vulgare Miller Common Sea-lavender. Lafant y Môr.
Native. Locally common in salt-marshes on the coast of Gower and the Vale, growing in the upper marsh

below the **Juncus maritimus** zone; scarce or absent if the marsh is heavily grazed. Map 529.

Limonium procerum (C.E. Salmon) Ingrouille subsp. **procerum** (*L. binervosum* (G.E.Sm.) C.E. Salmon *p.p.*).   Rock Sea-lavender. Lafant y Morgreigiau.
Native. Fairly common in the spray-zone on limestone sea-cliffs and in sandy upper salt-marshes, where it may grow with **L. vulgare**, as for example at Berges Island (49B), Nicholaston Pill (58A) and Crymlyn Burrows (79C) (Ingrouille, 1986). Fl. 7–8, 10, fr. 8–9; Hb, (Lep) Ma.jur, Zygaena filipendulae. Map 530.

OLEACEAE
FRAXINUS L.
Fraxinus excelsior L.   Ash. Onnen.
Native. Very common and widespread in woods, hedges and scrub, ranging from sheltered sea-cliffs and sand-dunes on the coast to rocky slopes and ravines in the Uplands, but most abundant in deciduous woodland and old hedges on relatively well-drained and base-rich soils, especially on limestone. Map 531.

SYRINGA L.
Syringa vulgaris L.   Lilac. Leilac.
Introduced; sometimes more or less naturalized in hedgerows, but usually not spreading far from the site where it was planted. Recorded from sites scattered through the Vale and the Uplands, but with no recent records from Gower. Map 532.

LIGUSTRUM L.
Ligustrum vulgare L.   Wild Privet. Pryfet, Yswydden.
Native. Common in Gower and the Vale in hedges and scrub and at the edges of woods on limestone soils, and also on limestone sea-cliffs and calcareous sand-dunes; rare elsewhere, sometimes growing as a relic of cultivation in the Uplands. Map 533.

Ligustrum ovalifolium Hassk.   Garden Privet. Pryfet yr Ardd, Yswydden yr Ardd.
Introduced. Sometimes more or less naturalized, growing as a relic of cultivation in hedges, and on waste ground, river-banks and rubbish tips where soil and rubble from the clearance of former gardens and hedgebanks has been tipped. Map 534.

Forsythia × intermedia Zabel has twice been recorded as a probable garden escape (27A, GH 1985; 97A, MG 1972) and **Jasminum officinale** L. once (09A, AMP 1975).

GENTIANACEAE
BLACKSTONIA Hudson
Blackstonia perfoliata (L.) Hudson   Yellow-wort. Canrhi Felen.

Native. Locally common in coastal grassland on limestone soils and calcareous sand-dunes in Gower and the Vale; rare inland. Fl. 7–10, fr. 8–10. Map 535.

CENTAURIUM Hill
Centaurium erythraea Rafn.   Common Century. Canrhi Goch.
Native. Common in limestone grassland and on calcareous sand-dunes, and fairly common and widespread on roadsides, by woodland rides and paths, on railway ballast, and on slag-tips and waste ground. White-flowered plants are frequent in some populations, for example at Crymlyn Burrows (79C!). **C. erythraea** is a variable species in which local habitat-forms show varying degrees of genetic differentiation from one another. Some of these forms have been given specific rank, for example **Centaurium capitatum** (Willd.) Borbás, which is now regarded as a variety of **C. erythraea**. **C. capitatum** has been found in coastal limestone grassland between Overton and Mewslade in Gower (48A, above Mewslade Bay, 1974!; 48B, near Longhole, 1974!**; 48D, Overton, R.C. McLean *c.*1930), along the Vale coastal cliffs (16B, Sully Island, AHT 1905*; 87D, Southerndown, AHT 1907*), and on sand-dunes at Whiteford Burrows (49C, EFL *c.*1892), Sker Point (77B, 1978!*), Kenfig Burrows (78D, WAS 1901*, AHT 1908*) and Porthcawl (E.S. Marshall & WAS 1901*); it appears to be clearly distinct from typical **C. erythraea**, certainly in the Mewslade and Longhole sites where **C. capitatum** occupies a well-defined microhabitat in which the two forms grow side by side. There are less satisfactory records of **C. capitatum** from limestone north of Cardiff (08B, B. Pearce 1949*; 18B, EV 1944*). Fl. (6–) 7–10, fr. 8–10. Map 536.

Centaurium littorale (D. Turner) Gilmour   Seaside Centaury, Canrhi Goch Arfor, grows on dunes at Pembrey Burrows in Carmarthen, on the northern side of the Burry Estuary which separates Carmarthen from Glamorgan; **C. littorale** has not been found at Whiteford Burrows on the southern side of the Burry Estuary (49A, 49C) despite repeated searches there (1976!, 1981!). It was reported from Mumbles (68A) by J.W.G Gutch in *c.*1840 (Gutch, 1844a) and from Kenfig Burrows near Kenfig Pool (87A) by E.S. Marshall and W.A. Shoolbred in 1902 (Marshall & Shoolbred, 1902), but subsequent examination of their specimens and of living plants in the same locality, by Eleanor Vachell and A.E. Wade in 1930 showed that Marshall and Shoolbred's plant was a form of **C. erythraea**, and it is highly probable that Gutch's plant was also.

Centaurium pulchellum (Swartz) Druce   Lesser Centaury. Canrhi Leiaf.
Native. Locally frequent in sandy salt-marshes and brackish dune-slacks on the coast of Gower and the Vale; particularly characteristic of the transition zone between sand-dune and salt-marsh. Map 537.

GENTIANELLA Moench

Gentianella campestris (L.) Börner   Field Gentian. Crwynllys y Maes.
Native. Apparently once fairly frequent in pastures and sand-dune grassland, but with no recent records; last recorded in 1953 (58A). **G. campestris** has shown a similar decline in most of southern Britain, perhaps as a consequence of the ploughing and reseeding of old meadows. Map 538.

Gentianella amarella (L.) Börner   Autumn Gentian. Crwynllys Chwerw.
Native. Locally common in limestone grassland near the sea and on calcareous sand-dunes in the western part of the county; once fairly frequent in the southern Vale, but now rare east of Porthcawl. Fl. (7–) 8–9, fr. 9. Map 539.

Gentianella uliginosa (Willd.) Börner   Dune Gentian. Crwynllys Cymreig.
Native. Known only from dune-slacks at Whiteford Burrows and Oxwich Burrows, where it is fairly abundant in a few slacks in some years, but varies greatly in abundance from year to year.
49A, Whiteford Burrows (northern part), 1971!, 1981!; 49C, Whiteford Burrows (southern part), JEL 1952*, 1981!; 58A, Oxwich Burrows, JEL 1952*, N.M. Pritchard 1959, D. & S. Adams 1971!, 1981!. Hybrids with **G. amarella** can sometimes be found where dune-slacks with **G. uliginosa** adjoin dune grassland with the later-flowering **G. amarella**, as at Oxwich Burrows (58A), and Whiteford Burrows (49) (N.M. Pritchard, 1959). (!) Fl. (5), 7–8, fr. (6), 8–9.

MENYANTHACEAE

MENYANTHES L.

Menyanthes trifoliata L.   Bogbean. Ffa'r Gors.
Native. Locally frequent in marshes, at the edges of shallow lakes and ponds, and in the wetter parts of bogs; most frequent in Gower, but growing in sites scattered through the county. Fl. 4–6, fr. 5–7. Map 540.

NYMPHOIDES Séguier

Nymphoides peltata (S.G. Gmelin) O. Kuntze   Fringed Water-lily. Ffaen Gors Eddiog.
Introduced to ponds, lakes and streams in a few localities; at some sites, for example in Broad Pool (59C) and in Fairwood Corner Pool (59D), it has overgrown and choked formerly open water, smothering or seriously threatening a native plant and animal community of an increasingly rare type.
07A, Pysgodlyn Mawr, MG 1972, GH & B.A. Thomas 1986 (still limited to eastern end of pool); 09A, Blaenllechau near Ferndale, JTa 1975; 17B, Roath Brook, 1942!; 18C, canal near Melingriffith Works, D.M. Howarth 1942; 59C, Broad Pool, GTG 1963, 1983!; 59D, Fairwood Common Pool, 1983!; 69C, Clyne, JAW 1926, University College, 1964 – 1986!, Cwmbwrla, GH 1986*; 79C, Crymlyn East, 1972!. Fl. 7–8.

APOCYNACEAE

VINCA L.

Vinca minor L.   Lesser Periwinkle. Perfagl.
Introduced; occasionally naturalized as a garden escape in hedges and woods, as for example at Stouthall near Reynoldston (48B). Map 541.

Vinca major L.   Greater Periwinkle. Perfagl Mwyaf.
Introduced; occasionally naturalized as a garden escape in hedges or scrub, especially near the sea. (The specimen of **V. major** subsp. **hirsuta** (Boiss.) Stearn from Dunraven Castle Estate (87D) collected by B.A. Miles in 1955 is now thought to be **V. major** cv. 'Oxyloba'). Map 542.

RUBIACEAE

SHERARDIA L.

Sherardia arvensis L.   Field Madder. Mandon Las yr Ŷd.
Native. Common in Gower and the Vale with other winter annuals on shallow soils near limestone outcrops, on anthills in limestone grassland, and on fixed dunes; less frequent on field banks, roadsides and old walls and in cultivated land. Rare in the Uplands. Map 543.

Phuopsis stylosa (Trin.) B.D. Jackson, Caucasian Crosswort, a garden annual, has been recorded as a garden escape on two occasions and appears to be established at St. Donat's (08D, between Creigiau and Pentyrch, M. Davies 1952*; 96A, St. Donat's, JAW 1940, J. Rees 1954*, J.M. Thomas 1970*).

ASPERULA L.

Asperula cynanchica L. *s.l.* (including *A. occidentalis* Rouy). Squinancywort. Mandon Fechan.
Native. Locally common in Gower in short limestone grassland, especially near outcrops; less frequent in calcareous fixed dune grassland. Now very local in the Vale in similar habitats where it has declined in abundance as a consequence of the destruction of much of the limestone grassland in which it grew. Absent from the Uplands. Populations growing in dune grassland in Gower and the Vale (for example in 48A, Rhosili dunes, HJR 1904, V.M. Peel 1929*; 49C, Whiteford Burrows, 1981!; 58B, Pennard Castle, EMT 1920*; 87A, Porthcawl, WAS 1901, 1931!*, E.K. Horwood 1953) appear to be intermediate between **A cynanchica** and **A. occidentalis** and require further investigation. Map 544.

Asperula arvensis L., Blue Woodruff, used to occur as a rare casual on waste ground but has not been seen recently (00C, Aberdare, HJR *c.*1907; 17B, Cardiff Docks, EV 1936*, N.M.W. site, 1937!*; 78B, Port Talbot Docks, HJR *c.*1907). The closely related **A. orientalis** Boiss. & Hohen., sometimes grown as a garden flower, has been found once as a casual (27A, rubbish tip at Pengam, RLS 1940*).

GALIUM L.

Galium boreale L.   Northern Bedstraw. Briwydden Fynyddig.
Native. Very rare in Glamorgan, with no recent records. **G. boreale** was found on Craig-y-llyn (90C) by Augustine Ley before 1907, probably on ledges of the cliffs above Llyn Fawr where **Rhodiola rosea** still grows. Both species are abundant on the north-facing cliffs of Bannau Sir Gaer (the Carmarthen Van) about 10km northwest of Craig-y-llyn, where **G. boreale** may still grow on the higher inaccessible cliffs.

Galium odoratum (L.) Scop. (*Asperula odorata* L.) Woodruff. Briwydden Bêr.
Native. Locally frequent in woods and old hedge-rows in Gower and parts of the Vale, usually on limestone soils; scarce in the Uplands. Map 545.

Galium uliginosum L.   Fen Bedstraw. Briwydden y Fign.
Native. Marshes and fens. Rare in Glamorgan, with no confirmed recent records; in the past it was probably confused with **G. palustre** L., which is very common in similar habitats. Most or all of the records of **G. uliginosum** given in Trow (1911) appear to have been erroneous and Riddelsdell's records (Riddelsdell, 1907) are also rather doubtful. **G. uliginosum** may still grow in a few base-rich marshes in the northern Uplands and the eastern Vale, and possisbly elsewhere. Map 546.

Galium palustre L. *s.l.* (including *G. elongatum* C. Presl). Common Marsh-bedstraw. Briwydden y Gors.
Native. Very common in marshes and fens, in dune-slacks and damp grassland, in ditches and by streams and ponds. Fl. 6–8, (–9), fr. 7–9 (–10). Map 547.

Galium verum L.   Lady's Bedstraw. Briwydden Felen.
Native. Common in limestone and sand-dune grass-land and on field banks, railway embankments and roadsides on base-rich soils in Gower and the Vale; scarce in the Uplands. Fl. 7–9, fr. 8–10. Map 548.

Galium mollugo L. *s.l.*   Hedge bedstraw. Briwyd-den y Clawdd.
Native. Common in the Vale in hedgerows and scrub, in old pastures and hay meadows, and on roadsides and railway embankments, usually on limestone soils; very local in Gower and rare in the Uplands. Fl. 6–7. Map 549.

Galium album Miller (*G. mollugo* subsp. *erectum* Syme), Upright Hedge-bedstraw, Briwydden Syth, has been recorded from only three localities (06A, Aberthaw, MSP 1960; 18D, Llanishen, EV 1932*; 90C, Craig-y-llyn, 1970+!) but has probably been included with **G. mollugo** by many recorders.

Galium mollugo × **G. verum** (**G. × pomeranicum** Retz.) (*G. ochroleucum* Wolf.).

Native. Apparently rare, with records from only two localities, but probably under-recorded and perhaps not infrequent at sites where the parents grow near to one another.
49C, Cwm Ivy Tor, 1979!; 78D, Kenfig Burrows, 1956!*.

Galium sterneri Ehrend.   Limestone Bedstraw. Briwydden y Garreg Galch.
Native. Very local, known only from limestone grassland on Morlais Castle Hill in the northern Uplands; it is locally frequent in similar habitats in southern Breconshire.
00A, Morlais Hill, HJR 1904*, EV 1929*, MG 1973*.

Galium saxatile L.   Heath Bedstraw. Briwydden y Rhosdir, Briwydden Wen.
Native. Very common on well-drained acid soils, growing on grassy heaths, in open woodland, and in unimproved grassland generally; absent or scarce only in the southern part of the Vale. Map 550.

Galium aparine L.   Cleavers. Llau'r Offeiriad, Cyn-ghafan.
Native. Very common in hedges and scrub, at the edges of woods, as a weed of gardens and cultivated land, and on roadsides and waste ground; oc-casionally found on shingle beaches. Fl. 5–10, fr. 6–10. Map 551.

Galium tricornutum Dandy, Corn Cleavers, used to grow as a weed of docksides and rubbish tips, but was last recorded in 1932, at Barry Docks (16A, EV 1932*); it has been recorded from 00C, HJR c.1906; 06A, EV 1904*; 07A, HJR c.1906; 16A, HJR c.1906, EV 1032*; 17D, 1920!*; 69D, HJR c.1906; 78D, HJR c.1906. **G. spurium** L., False Cleavers, has been found once, at Splott (27A, EV 1931*).

CRUCIATA Miller

Cruciata laevipes Opiz (*Galium cruciata* (L.) Scop., *Cruciata chersonensis* auct.).   Crosswort. Croeslys, Briwydden Groes.
Native. In grassy hedgebanks and on roadsides, railway embankments and field banks, especially on limestone soils; sometimes on fixed sand-dunes. Common in Gower and parts of the Vale but scarce in the Uplands. Map 552.

RUBIA L.

Rubia peregrina L.   Wild Madder. Cochwraidd Gwyllt.
Native. Locally common in Gower and along the southern Vale coast in cliff scrub and in woods and hedgerows near the sea, usually on limestone soils; very rare inland. Fl. (5–) 6–8, fr. 7–9. Map 553.

POLEMONIACEAE

Polemonium caeruleum L., Jacob's-ladder, Ysgol Jacob, has been found as a naturalized garden escape or garden weed on three occasions (00C, Aberdare, HJR 1907; 18D, near Great Garth, JS *c*.1886; 89D, Maesteg, R. Cotton 1978*), and **Gilia capitata** Sims as a dock alien in two localities (16A, Barry, 1922!*, GCD 1927; 17B, Cardiff, JS *c*.1876, RLS 1926*).

CONVOLVULACEAE
CUSCUTA L.

Cuscuta epithymum (L.) L. Dodder. Llindag Lleiaf.
Native. Once fairly frequent on the southern coast of Gower, but always scarce in the Vale; now apparently extinct in Gower and very rare in the Vale, with only one recent record there, at Llandough (17D). In coastal sites it grew in local abundance on **Ulex** (for example at Pwlldu (58B) in 1892), but it also grew on a variety of herbaceous plants (**Centaurea nigra**, **Hypochoeris radicata** and **Lotus corniculatus** are recorded as hosts), so its disappearance cannot be entirely attributed to an increase in the frequency of gorse fires on the cliffs. It was occasionally found as a parasite of clover crops inland, but has presumably been eliminated from this habitat by better seed-cleaning and other agricultural changes.
06A, Gileston, T.W. Proger *c*.1908*; 17B, Pontcanna, JS *c*.1886; 17D, near Cogan, Miss Corbett 1897*, Llandough near Cardiff, 1922!*, EV 1934*, GMB 1970; 48B, Horton, AHT 1908; 58A, near Oxwich, AHT 1908; 58B, Pwlldu, AHT 1892; 68A, Langland Bay, M. Howell 1897*; 87D, Dunraven, AHT *c*.1908.

Five other species of **Cuscuta** have been found as rare casuals, parasitic on cultivated plants and presumably introduced with them; **Cuscuta australis** R. Br. subsp. **tinei** (Insenga) Feinbrun was found in experimental clover plots at St. Fagan's (17A) in 1906 (H. Wallace) (Hemsley, 1908) and in Mayals (69C) in 1949 (B. Viney*); **C. campestris** Yuncker has been recorded on only one occasion, parasitic on various plants in W. Palmer & Son's nursery at Wenvoe (17C, D.H. Lewis 1958*). **C. racemosa** C.F.P. Mart. has also been recorded on only one occasion, parasitic on **Nepeta mussinii** in a Cardiff garden (17B, RLS *c*.1925). **C. suaveolens** Ser. has been found in central Cardiff (17B) on four occasions (garden, J. Grimes 1921*, Victoria Park, EV 1924*, N.M.W. site, R. Wells 1932*, Canton, R. Wells 1940*) a record of this species from St. Fagan's (17A) refers to **C. australis** (see above), and **C. epilinum** Weihe has been found in the same area on one occasion (17B, Bute Park, EV 1929*) and also at Llandough Quarry (17D, RLS 1933*).

CALYSTEGIA R. Br.

Calystegia soldanella (L.) R. Br. Sea Bindweed. Taglys Arfor, Cynghafog Arfor.
Native. Still locally abundant on coastal sand-dunes from Whiteford Point (49A) to Porthcawl (87A); no recent records from the southern coast of the Vale, where it was always very local. Map 554.

Calystegia sepium (L.) R. Br. subsp. **sepium** Hedge Bindweed. Taglys Mawr, Clych y Perthi.
Native. Common and widely distributed in hedgerows, in **Phragmites** fens and in river and streambank vegetation, and as a garden weed. Fl. 6–8 (–11), fr. 9–10; B.luc (n), B.pasc (n), (Syr) Rh.cam (n,p), S.balt (p), S.lun, S.rib (p), Sy.pip, (Dip) Em.tess (n). Map 555.

Pink-flowered forms of **Calystegia sepium**, probably subsp. **roseata** Brummit, were recorded between Swansea and Crymlyn Burrows by C.C. Babington in 1840, from several places in Gower by A.H. Trow in 1908, and have been seen recently at Rhosili (48A, 1976!) and Jersey Marine (79C, 1975!). A specimen resembling **Calystegia sepium** × **C. silvatica** (**C. × lucana** (Ten.) G. Don) was collected from Llanmadoc (49C) by J.A. Webb in 1952*; expert opinion differs on its identity.

Calystegia silvatica (Kit.) Griseb. Large Bindweed. Taglys Estron, Cynghafog Fawr.
Introduced. Locally common in hedgerows and as a weed of gardens and waste ground, especially in urban and suburban areas; less common than **C. sepium** in the countryside, where it is usually found near houses or rubbish-tips. Fl. 6–8 (–10), fr. 9–10; Hb (n), B.luc (n). B. pasc (n), (Syr) H.pen (n), S.balt, S.rib (p). Map 556.

Calystegia pulchra Brummit & Heywood Hairy Bindweed. Taglys Blewog.
Introduced. Hedges and waste ground, usually in urban areas; apparently scarce, but probably sometimes overlooked or confused with **C. sepium** subsp. **roseata**. Recorded more frequently in the Uplands than in Gower or the Vale. Map 557.

CONVOLVULUS L.

Convolvulus arvensis L. Field Bindweed. Cwlwm y Cythraul, Cynghafog Fechan, Taglys.
Native. Very common in a variety of open habitats; found as a weed of cultivated land and waste ground, in open grassland especially on slopes and banks, in hedges, on roadsides and by paths, on fixed dunes, on stream-banks and cliffs, and sometimes in shingle above the drift-line. The colour of the flowers varies; white-flowered plants are most frequent but plants with pink-striped flowers are fairly common, especially in maritime habitats, and plants with a purple central marking in a white or pink-striped corolla occur in some populations (for example at Broughton Burrows, 49C). Fl. 6–8 (–10), fr. 7–10 (but seeds often replaced by a smut fungus, *Thecaphora seminis-convolvuli* (Duby) Liro); Hb (n), (Syr) Rh.cam, S.balt, S.rib, Sy.pip. Map 558.

Convolvulus tricolor L. has been found on one occasion, as a weed of waste ground near docks in Cardiff (27A, Splott, RLS 1926).

HYDROPHYLLACEAE

Phacelia ciliata Bentham has twice been recorded as a casual of waste places (17B, Maendy, Cardiff, RLS 1938*; 87A, Porthcawl, RLS 1938*).

BORAGINACEAE

LITHOSPERMUM L.

Lithospermum officinale L. Common Gromwell. Maenhad Meddygol.
Native. Locally frequent on basic soils near the coast, on sand-dunes, in open scrub and on roadsides. Fl. 6–7, fr. 7–8; (!) B.prat. Map 559.

BUGLOSSOIDES Moench

Buglossoides purpurocaerulea (L.) I.M. Johnston (*Lithospermum purpurocaeruleum* L.). Purple Gromwell. Maenhad Gwyrddlas, Y Gromandi Gwyrddlas.
Native. Locally frequent in open woodland or scrub on sea-cliffs or near the sea in the Vale, and rare in similar habitats in southern Gower. First recorded 'at the west end and on top of a cliff called Nine-acre Cliff about ½ a mile from Porth Kerrig Church, July 2, 1793' by Joseph Banks (specimen in Herb. **BM**). It still survives in this locality. Fl. 5–6, fr. 7–8. Map 560.

Buglossoides arvensis (L.) I.M. Johnston (*Lithospermum arvensis* L.), Field Gromwell, Maenhad yr Âr, used to occur as a dock alien and may have persisted for a time as a weed of cornfields and sandy waste ground; it was last recorded in 1962 by E.A. Jenkins at Dyffryn (07D*). There are earlier records from 16A, Barry Docks, EV 1932*; 17B, Cardiff, THT 1902*; 17D, Penarth, AHT 1892*; 27A, East Moors, JS *c*.1876, Splott 1926!*; 87A, Newton, EMT pre 1930*, other records are now thought to be errors.

Heliotropium europaeum L., Common Heliotrope, used to grow as a casual on waste ground near docks, but has not been found since 1926; it was recorded in 16A*, Barry Docks, RM 1924*; 18C, sand-ballast, Radyr, 1921!*; 27A, East Moors, Cardiff, JS 1876*, Splott, 1926!*; 87A, Porthcawl, RE & FC 1898–1899.

ECHIUM L.

Echium vulgare L. Viper's-bugloss. Glas y Graean, Bwglos y Wiber.
Native. Locally common near the coast in calcareous sand-dune grassland and on shallow soils over limestone; occasionally found inland in similar habitats, especially near quarries, and as a weed of waste ground. Fl. 5–8 (–11), fr. 7–11; Hb (n,p), B.lap, B.luc, B.pasc, B.prat, B.terr (all n,p), Sb, (Syr) E.ten, S.balt, S.rib (Lep) Pl.gam. Map 561.

Three introduced species of **Echium** have been found in the past as dock aliens, and one, **Echium rosulatum** Lange, still grows at Barry Docks (16A, RLS 1927*, DMcC 1974, JPC 1983) it was also recorded from Cardiff (17B) by J.H. Salter in 1930*. **E. italicum** L. has been found once (78B, Port Talbot

Docks, HJR 1909*) and **E. plantagineum** L. three times (16A, Barry Docks, 1922!*; 17B, Cardiff Docks, JS *c*.1876*; 78B, Port Talbot Docks, HJR 1910).

PULMONARIA L.

Pulmonaria officinalis L. Lungwort. Llys yr Ysgyfaint.
Introduced. Naturalized in open woods and in hedgerows in several places, but rare and usually impersistent. Map 562.

Pulmonaria longifolia (Bast.) Boreau, Narrow-leaved Lungwort, Llys yr Ysgyfaint Culddail, was found naturalized in woods near Ilston (59D) by J.A. Webb in 1929, (an 1891 record from Swansea is probably from the same locality).

SYMPHYTUM L.

Symphytum officinale L. Common Comfrey. Llysiau'r Cwlwm, Dail Cwmffri.
Native. Occasional by streams and on damp roadsides in the Vale, scarce in Gower and very scarce or absent in most of the Uplands. Some of the early records may refer to **S.** × **uplandicum**. Fl. 5–9; B.pasc (n). Map 563.

Symphytum asperum × **S. officinale** (**S.** × **uplandicum** Nyman). Russian Comfrey. Cyfardwf Glas. Introduced and naturalized (Wade, 1958). Widespread and fairly frequent on roadsides and hedgebanks, especially in Gower and the Vale. Map 564.

Symphytum asperum Lepechin, Rough Comfrey, Cyfardwf Garw, has been recorded twice, from roadsides between Trecynon and Penywain (90D, 1974!*) and near St. Fagans (17A, H & CH 1974). **S. tuberosum** L., Tuberous Comfrey, Cyfardwf Oddfynog, has been recorded once, from the side of a woodland track in Penrice Woods (48B, B.E. Jones 1972*). Both species were probably naturalized introductions in these localities.

ANCHUSA L.

Anchusa arvensis (L.) Bieb. Bugloss. Bwglos, Bleidd-drem.
Native. Widely distributed, but scarce, on sand-dunes; rare as a weed of cultivated or waste ground. Map 565.

Anchusa officinalis L., Alkanet, Alcanet, was found as an alien in Penarth (17D) by A.H. Trow in *c*.1890, near Port Talbot Docks (78B) by H.J. Riddelsdell in 1908, and on sand-dunes near Porthcawl (87A) by R.E. & F. Cundall in 1902; **A. azurea** Miller, Garden Anchusa, has been found as a casual on waste ground in Cardiff (17B, HJR 1907*, 1923!*), at Llwydcoed near Aberdare (90D, HJR 1903*), and at Aberdare (00C, HJR 1900).

PENTAGLOTTIS Tausch

Pentaglottis sempervirens (L.) Tausch ex L.H. Bailey Green Alkanet. Llys y Gwrid.
Introduced. Naturalized and locally common on roadside verges, hedgebanks and railway banks in the Vale and on the eastern borders of the county, scarce elsewhere. Fl. all year, (1–)4–5(–12), fr. 6–12; Hb (n,p), B.hort (n), B.lap (n), B.luc (n), B.pasc (n,p), B.prat (n,p), (Sb) Anthidium manicatum (N) (Syr) Rh.cam. Map 566.

BORAGO L.

Borago officinalis L. Borage. Tafod y Fuwch, Tafod yr Ych.
Introduced. An escape from cultivation which may persist for a time as a garden weed or on sandy waste ground; recently recorded only from Horton dunes (48B) and as a garden weed in Gwaelod y Garth (18C), Pitton (48A) and Llanmadoc (49C). Map 567.

The Borage-like perennial **Trachystemon orientalis** (L.) G. Don fil., Abraham-Isaac-Jacob, has recently been found established on waste ground at Porthkerry Wood (06B, R.J. Tidswell 1983*).

The yellow-flowered American annual **Amsinckia calycina** (Moris) Chater used to grow as a rare dock alien or with grain cleanings in the Cardiff district, and has been recorded twice, most recently in 1938 (17B, RLS 1938*; 18D, W. Evans 1927*); **A. intermedia** Fischer & C.A. Meyer has been recorded from similar habitats, once from Maendy Cardiff (17B, RLS 1938*) and twice from Barry Docks (16A, RM 1925*, JPC 1986*) where it may have established itself. **A. lycopsoides** (Lehm.) Lehm. has been recorded on four occasions as a casual (00C, HJR 1901; 07D, A. David 1927*; 78B, HJR c.1906; 90D, HJR 1902) and twice from Flat Holm (26, A.J. Willmott 1932, MG 1974) where it may have established itself. Specimens in **NMW** of Amsinckia collected from Maendy, Cardiff (17B) by R.L. Smith in 1938 have been provisionally determined as **A. menziesii** (Lehm.) A. Nelson & Macbride and **A. tessellata** A. Gray; the latter species has also been recorded from Barry Docks (16A, RM 1924).

Asperugo procumbens L., Madwort, Cynghafan Mwyaf, has also been found as a rare casual, most recently as a garden weed at Llangeinor (98A, W. Watkins 1952); it has also been found in 00C, Aberdare, HJR c.1906; 17B, Cardiff, JS c.1876; 27A, Splott, T.J. Foggitt c.1920.

MYOSOTIS L.

Myosotis arvensis (L.) Hill Field Forget-me-not. Ysgorpionllys y Meysydd, Llys-Coffa'r Maes.
Native. Common throughout the county as a weed of gardens and arable land, on railway banks, roadsides and sand-dunes, on wall-tops, and sometimes in open scrub and woods. Fl. 4–11, fr. 5–11. Map 568.

Myosotis ramosissima Rochel Early Forget-me-not. Ysgorpionllys Cynnar, Coffa'r Llys-Coffa'r Cynnar.
Native. Frequent in the Vale and Gower, chiefly near the coast on shallow calcareous soils and on sand-dunes, less frequently inland in dry grassland, on roadsides or on walls; often grows on ant-hills or near rabbit burrows. Fl. 4–7, fr. 5–7. Map 569.

Myosotis discolor Pers. Changing Forget-me-not. Ysgorpionllys Amryliw, Llys-Coffa'r Amryliw.
Native. Grassland on basic and light neutral soils, often in similar sites to **M. ramosissima**, with which it frequently grows near the coast. Formerly locally common in the Vale, but now rather scarce there as a result of the conversion of grassland to arable land or the reseeding of pastures. It was common near rabbit burrows on Sully Island and elsewhere before myxomatosis reduced or eliminated the rabbit population. Map 570.

Myosotis sylvatica Hoffm. Wood Forget-me-not. Ysgorpionllys y Coed, Llys-Coffa'r Coed.
Introduced. A frequent garden escape found on hedgebanks, roadsides and waste ground, sometimes naturalized in woods as for example near Candleston Castle, where it has been known for more than 70 years. Map 571.

Myosotis secunda A. Murray Creeping Forget-me-not. Ysgorpionllys Ymlusgaidd.
Native. Common in the Uplands and on the Gower commons in bogs, acid marshes and drainage ditches and by streams, absent from the southern part of the Vale. Map 572.

Myosotis laxa Lehm. subsp. **caespitosa** (C.F. Schultz) Hyl. ex Nordh. (*M. caespitosa* C.F. Schultz). Tufted Forget-me-not. Ysgorpionllys Siobynnog.
Native. Marshy fields and tracks, streambanks, pondsides and ditches. Common in Gower and much of the Vale, locally frequent in the Uplands. Fl. 5–8, 10, fr. 6–8, 10. Map 573.

Myosotis scorpioides L. Water Forget-me-not. Ysgorpionllys y Gors.
Native. Locally common in ditches, on streambanks and at the sides of ponds on basic and neutral soils. Most frequent in the Vale and southern and western Gower and in the Tawe, Neath and Taff valleys. Fl. 5–10, fr. 6–10; B.terr. Map 574.

Lappula squarrosa (Retz.) Dumort., Bur Forget-me-not, used to grow as a casual near docks and on waste ground in the Cardiff and Barry districts (16A, RLS c.1923, N.D. Sandwith 1934; 17B, EV 1907; 17D, JS c.1876, 1920!*; 27A, JS c.1876*, 1925!*) and was once found at Aberdare (00C, HJR 1900, 1902), Port Talbot (78B, HJR 1904) and Hirwaun (90B, HJR 1905); its most recent record was at Barry (16A, 1934).

Omphalodes verna Moench, Blue-eyed-Mary, was recorded as a garden escape from near Penrice (48B) by L.W. Dillwyn in *c*.1840 and near Reynoldston (48B) and Fairyhill (49B) by J.A. Webb in 1924–26.

CYNOGLOSSUM L.

Cynoglossum officinale L.   Hound's-tongue. Tafod y Bytheiad.
Native. Dune grassland, fixed dunes and dry grassland near the sea. Frequent in suitable habitats along the coast from Southerndown westwards, once similarly frequent in the south-east but now scarce there. Fl. 6–8, fr. 7–9. Map 575.

Cynoglossum germanicum Jacq. was once recorded as a casual near Cardiff (17) by J. Storrie in *c*.1876.

VERBENACEAE
VERBENA L.

Verbena officinalis L.   Vervain. Briw'r March.
Native. Roadsides, field banks and waste ground. Locally frequent in Gower and the Vale but scarce in the Uplands. Map 576.

CALLITRICHACEAE
CALLITRICHE L.

Callitriche stagnalis Scop. *s.s.* Common Water-starwort. Brigwlydd y Dŵr.
Native. Ponds, ditches, streams and terrestrial on mud. Common throughout the county. Fl. 5–7, 10; fr. 6–7. Map 577.

Callitriche obtusangula Le Gall Blunt-fruited Water-starwort. Brigwlydd Ffrwyth-aflem.
Native. Slow flowing streams, ponds and ditches. Rare and decreasing. Now found mainly in north Gower, and in the south-western Vale. Map 578.

Callitriche platycarpa Kütz. (*C. verna* auct.). Various-leaved Water-starwort. Brigwlydd y Gwanwyn.
Native. Ponds, ditches and streams. Locally frequent in Gower, the western Uplands and southern Vale, absent from a large area of the central Uplands. Fl. 6–7, fr. 7. Map 579.

Callitriche hamulata Kütz. ex Koch (*C. intermedia* Hoffm., *C. intermedia* subsp. *hamulata* (Kütz. ex Koch) Clapham). Intermediate Water-starwort. Brigwlydd Cyfryngol.
Native. Streams, ponds and ditches. Locally frequent in scattered localities throughout the county. Map 580.

Callitriche brutia Petagna (*C. intermedia* subsp. *pedunculata* (DC.) Clapham). Pedunculate Water-starwort. Brigwlydd Coesog.
Native. Shallow ponds and cart ruts which often dry up in summer. Recently recorded only from a few sites in Gower; perhaps also present elsewhere but not recorded because of confusion with **C. hamulata**. Chromosome counts may be needed to separate these

two species. (Lewis-Jones & Kay, 1977).
00C, Aberdare, HJR, *c*.1904; 07A, Ystradowen, HJR, *c*.1904; 17D, Leckwith Common, EV, 1936; 18B, Caerphilly, 1944*!, 48A, Pitton, 1975!, Kimleymoor, 1975!; 49C, Hardingsdown, 1975!; 59D, Fairwood, 1975!.

LABIATAE (LAMIACEAE)
AJUGA L.

Ajuga reptans L.   Bugle. Glesyn y Coed.
Native. Damp woods, hedgebanks, streamsides and damp grassland. Common in Gower and the Vale but rather local in the Uplands. Fl. 5–7; B.luc (n), B.pasc (n). Map 581.

TEUCRIUM L.

Teucrium scorodonia L.   Wood Sage. Chwerwlys yr Eithin.
Native. Common and widespread in a variety of habitats; open woods, hedgebanks and scrub, heathland and rough grassland, field banks and pathsides, sea-cliffs, old dunes and shingle. Fl. 6–10, fr. 8–10; Hb (n), B.hort (n,p), B.luc (n), B.pasc (n), B.terr (n,p). Map 582.

Teucrium chamaedrys L.   Wall Germander. Chwerwlys y Mur.
Introduced. Naturalized on a limestone outcrop above South Cornelly, where it was first recorded by J.E. Bicheno in 1840 (Dillwyn, 1840) and now grows around the margin of an old quarry in a site that has been isolated by more recent quarries. It was also reported as established on a cottage wall near Merthyr Tydfil (Trow, 1893) but has not been recorded there subsequently.
00A, Pen-yr-heolgerrig, AHT 1891*; 88C, near South Cornelly, J.E. Bicheno 1840, HJR 1904*, EV 1936*, 1966!**, 1974!, 1980!.

SCUTELLARIA L.

Scutellaria galericulata L.   Skullcap. Cycyllog.
Native. Riverside and streamside marshes, fens, and damp dune-slacks. Fairly frequent in suitable habitats in Gower and the Tawe and Neath valleys, but local in the Vale and scarce elsewhere. Map 583.

Scutellaria minor Hudson   Lesser Skullcap. Cycyllog Bach.
Native. Wet heaths, bogs and acid marshes. Fairly common in suitable habitats in Gower, the Uplands and the northern Vale. Map 584.

MARRUBIUM L.

Marrubium vulgare L.   White Horehound. Llwyd y Cŵn.
Probably native. Very local on the coasts of Gower and the Vale, with recent records almost confined to Gower. It grows in scattered small populations on grassy limestone cliff-tops and in disturbed sand-

dune grassland, often associated with **Torilis nodosa** and **Carduus tenuiflorus** in places where cattle or sheep gather. Map 585.

Sideritis montana L. has occasionally been found as a casual on waste ground, usually near docks in the Cardiff area and at Port Talbot; its most recent record was at Llanishen in 1950 (18D, B. Mirfin*); it has also been recorded from Penarth (17D, JS 1876*), Grangetown (17D, RLS 1925*), East Moors (27A, JS c.1876) and Port Talbot (78B, HJR c.1907). **Mellitis melissophyllum** L., Bastard Balm, Gwenynog, has also been recorded as a dockside casual on one occasion, at Bute Docks (17B, JS c.1886) and was claimed to grow in shady hedges near Swansea by W. Turton (Oldisworth, 1823) but was not seen by L.W. Dillwyn, who evidently doubted the record (Dillwyn, 1840). **Phlomis samia** L. was once found as a garden escape at Cefn Onn Halt (18D, L. Sullivan 1950*).

GALEOPSIS L.

Galeopsis angustifolia Ehrh. ex Hoffm. Red Hemp-nettle. Penboeth Culddail.
Native on shingle beaches at Aberthaw (06A, 1971!) and Pwll-du (58B, 1964!, 1978!). Once fairly widespread, though not common, as a cornfield weed in the Vale, but not recorded in this habitat recently. Map 586.

Galeopsis ladanum L., Broad-leaved Hemp-nettle, Penboeth Llydanddail, has recently been recorded, on waste ground near Briton Ferry (79C, H & CH 1972!**), perhaps introduced with other **Galeopsis** spp. and millet in bird-seed.

Galeopsis speciosa Miller Large-flowered Hemp-nettle. Penboeth Amryliw.
Introduced. A rare weed of cultivated land and waste ground, with only one recent record (79C, Briton Ferry, H & CH 1972). Recorded in the past from 00C, near Aberdare, JS c.1880; 09B, Mountain Ash, JS c.1880; 17B, Cardiff, RLS 1936*; 18A, Upper Boat, A. Jones 1950; 59B, Gowerton to Penclawdd, EFL c.1892.

Galeopsis tetrahit L. Common Hemp-nettle. Penboeth.
Native. Hedgerows, gorse and bramble scrub and the edges of woods, among rushes in damp grassland, on river-banks, and as a weed of cultivated land and waste ground. Common and widespread. Fl. 6–9, fr. 7–10. Map 587.

Galeopsis bifida Boenn. Lesser Hemp-nettle. Penboeth Lleiaf.
Native or introduced. In similar habitats to G. tetrahit, with which it sometimes grows, but apparently much less frequent, perhaps because of confusion between these closely similar species. Map 588.

LAMIUM L.

Lamium maculatum L. Spotted Dead-nettle. Marddanhadlen Fraith.
Introduced as a garden escape. Locally and often temporarily established on hedgebanks, roadsides and waste ground. Map 589.

Lamium album L. White Dead-nettle. Marddanhadlen Wen.
Introduced. Roadsides, hedgebanks, and similar habitats, usually in towns or villages. Fairly frequent in the Vale and parts of the eastern Uplands but scarce or absent elsewhere. Probably spreading. Fl. 4–11; B.terr. Map 590.

Lamium purpureum L. Red Dead-nettle. Marddanhadlen Goch.
Native or introduced. Very common as a weed of gardens and cultivated fields, and often found on disturbed roadsides and waste ground. Fl. 1–12, fr. 5–11; B.pasc (n,p). Map 591.

Lamium hybridum Vill. Cut-leaved Dead-nettle. Marddanhadlen Rwygddail.
Introduced. Widespread but not abundant as a weed of cultivated land in parts of the western Vale and the western Uplands, and locally fairly abundant in Gower, especially in vegetable crops; apparently scarce or absent elsewhere. Map 592.

Lamium amplexicaule L. Henbit Dead-nettle. Marddanhadlen Goch Gron.
Native or introduced. A scarce weed of cultivated land and waste ground, especially on sandy soils near the coast; occasionally found on disturbed sand-dunes. Map 593.

LAMIASTRUM Heister ex Fabr.

Lamiastrum galeobdolon (L.) Ehrend. & Polatschek subsp. **montanum** (Pers.) Ehrend. & Polatschek. Yellow Archangel. Marddanhadlen Felen.
Native. Woods and shady hedgebanks. Locally common in the Vale and the western Uplands but scarce or absent on the southern Vale coast and in most of Gower and the central Uplands. Fl. 5–6. Map 594.

LEONURUS L.

Leonurus cardiaca L. Motherwort. Mamlys.
Introduced. Roadsides, hedgebanks and waste ground, probably as an escape from former cultivation as a medicinal herb. Recorded in the past from several localities scattered through the Vale, but seen recently only around Cornelly, Kenfig and Porthcawl; it was first recorded in this area by R.E. & F. Cundall in 1898–99, near Mawdlam Church. 07A, Aberthin, EV 1911*, 1924*, HAH 1927*; 07C, St Hilary, JS c.1880; 17B, Cardiff Docks, 1924!*; 18C, Pentyrch, JS c.1890, Taffs Well, T. Chapman 1905*; 28A, near Machen, Mrs Griffith c.1908, Llanfedw, J. Rees 1941!*; 28C, near Cefn Mably, R.W. Webb 1941*; 78D, Kenfig Dunes, G.O. Rotheray 1978*; 87A, Porthcawl, VCB 1971!; 88C, South Cornelly, RE & FC 1898–1899, VCB 1971!.

BALLOTA L.

Ballota nigra L. subsp. **foetida** Hayek Black Horehound. Marddanhadlen Ddu.
Native or introduced. Hedgerows, roadsides and waste ground. Locally common in the Vale but rare in Gower and absent from most of the Uplands. Fl. 6–11, fr. 8–10; B.hort (n,p), B.terr (n,p). Map 595.

STACHYS L.

Stachys officinalis (L.) Trevisan Betony. Cribau San Ffraid.
Native. Open woods, cliff-heath and scrub, hedge-banks, path-sides and rough grassland. Widespread and fairly common. Fl. 6–12; B.hort (n), B.luc (n,p), B.pasc (n,p), B.terr (n). Map 596.

Stachys sylvatica L. Hedge Woundwort. Briwlys y Gwrych, Briwlys y Goedwig.
Native. Common in hedgerows, in woods and as a weed of cultivated and waste ground. Fl. (3), 6–12, fr. 7–12; Hb (n), B.distinguendus, B.pasc (n), B.prat (n), B.terr (n). Map 597.

Stachys palustris L. Marsh Woundwort. Briwlys y Gors.
Native. Streamsides, marshes and fens, damp road-sides and occasionally as a weed of cultivated land and waste ground. Widespread and fairly common. Fl. 7–11, fr. 8–11; B.hort (n), B.lap (n), B.luc (n), B.pasc (n,p), (Syr) Rh.cam (n,p), S.balt (p), (Lep) Pi.rap. Map 598.

Stachys palustris × S. sylvatica (S. × ambigua Sm.) could occur whenever the parent species meet in quantity; probably overlooked recently. Map 599.

Stachys arvensis (L.) L. Field Woundwort. Briwlys yr Ŷd.
Native or introduced. Locally common as a weed of cultivated farmland, disturbed roadsides and waste ground, especially in the west of the county. Map 600.

Three species of **Stachys** have been recorded as rare weeds of waste ground near docks or with grain cleanings. **Stachys recta** L. appears to have been long established at Barry Docks (16A) where it was first recorded by R.L. Smith in 1923!*, (DMcC 1974, JPC 1983). **S. annua** (L.) L. has been seen several times in the Cardiff area (17B, T.J. Foggitt 1922, EV 1923; 17D, RLS c.1924!, 1933!*; 18C, Radyr, RLS c.1925; 27A, RLS 1926!*, 1931), and **S. cretica** L. once (27A, East Moors, JS 1876*).

NEPETA L.

Nepeta cataria L. Cat-mint. Mintys y Gath.
Probably introduced. Recorded from hedgebanks, roadsides or disturbed sand-dunes at several places in Gower and the Vale in the past, probably as an escape from cultivation. Although it has not been recorded since 1954 (49D, JAW), it may still persist at some sites. Map 601.

GLECHOMA L.

Glechoma hederacea L. Ground-ivy. Eidral.
Native. Hedgebanks, field banks, pathsides, grassy heaths, fixed dunes and rough grassland, and as a garden weed. Common and widespread in Gower and the Vale but less ubiquitous in the Uplands. Fl. 2–6 (gynodioecious), fr. 4 (female), 5–7; Hb, B.hort, B.prat (all visit female fls), B.terr. Map 602.

Dracocephalum parviflorum Nutt. has once been recorded as a casual of waste places (27A, Splott, RLS 1926*).

PRUNELLA L.

Prunella vulgaris L. Selfheal. Craith Unnos.
Native. Very common and widespread in pastures and meadows, in sand-dune grassland, in open woods and scrub, on stream-banks in the hills, in lawns and as a weed of waste ground. Fl. 6–11, fr. 7–11; B.luc, B.pasc, B.prat (all n), B.terr (n,p). Map 603.

MELISSA L.

Melissa officinalis L. Balm. Gwenynddail.
Introduced as a garden escape; scarce and surprisingly impersistent on roadsides, river-banks and rocky slopes. Map 604.

Satureja montana L., Mountain Savory, has recently been found by J.P. Curtis on Roman ruins near Barry (06B, 1988), and **S. hortensis** L., Summer Savory, Safri Fach Flynyddol, has been found on four occasions in the past as a casual on waste ground (17B, Canton, HJR 1907*; 27A, Splott, RLS 1927*; 68A, Mumbles, WAS 1912*; 90D, Llwydcoed, Aberdare, HJR 1897).

ACINOS Miller

Acinos arvensis (Lam.) Dandy. Basil Thyme. Brenhinllys.
Native. Locally frequent on shallow soil near lime-stone outcrops and in rock crevices on the south Gower cliffs from Worm's Head (38A) to Mumbles Head (68A); very scarce in similar habitats in the Vale (recent records from 06A, Aberthaw, JPC 1978, and 87A, Porthcawl, VCB 1970); also found as a rare casual on waste gound. White-flowered plants occur in some Gower populations. Map 605.

CALAMINTHA Miller

Calamintha sylvatica Bromf. subsp. **ascendens** (Jordan) P.W. Ball Common Calamint. Erbin Cyffredin.
Native or introduced. Roadside banks and similar dry habitats on calcareous soils, and on old walls. Still fairly frequent in the Vale but recently recorded in

Gower only from Weobley Castle (49D, 1974–1986!). Map 606.

Calamintha nepeta (L.) Savi, Lesser Calamint, was found as a casual on waste ground at Barry (16A) by R.L. Smith in 1921*.

CLINOPODIUM L.

Clinopodium vulgare L. Wild Basil. Brenhinllys Gwyllt.
Native. Rough grassland, scrub, open woods and hedgebanks on limestone or base-rich soils. Common in southern Gower and the Vale and locally frequent in the eastern Uplands. Fl. 7–9 (–11), fr. 8–11; B.pasc (n), B.prat, (Syr) S.balt, (Lep) Ma.jur, Pi.bra, Pi.rap. Map 607.

ORIGANUM L.

Origanum vulgare L. Marjoram. Penrhudd.
Native. Hedgebanks, open scrub and the edges of woods, field banks, roadsides and rough grassland on limestone. Common in Gower and the Vale. Fl. 7–9 (–10), fr. 9–10; Hb, B.hort, B.lap, B.luc, B.pasc (Syr) E.ten, Rh.cam, S.balt, S.rib (Lep) Ma.jur, Pi.rap, Py.tit. Map 608.

THYMUS L.

Thymus praecox Opiz subsp. **arcticus** (E. Durand) Jalas (*T. drucei* Ronniger). Wild Thyme. Gruw Gwyllt, Teim Gwyllt.
Native. Dry grassland and ant-hills on limestone or base-rich soils, in sand-dune grassland where it is often very abundant, and on colliery spoil-tips and basic smelter slag-tips. Common in Gower and along the Vale coast; locally frequent in the Uplands. Fl. 6–9 (–11), fr. 8–10; Hb (n), B.lap, B.luc, B.prat (Dip) Call, Luc, Sa.car (Lep) Ma.jur. Map 609.

Thymus pulegioides L. Large Thyme. Gruwlys Gwyllt Mwyaf.
Native. Dry calcareous grassland. Apparently very rare, with only one certain record.
17D, Leckwith, GMB 1953*.

LYCOPUS L.

Lycopus europaeus L. Gipsywort. Llys y Sipsiwn.
Native. Streamsides, marshes, fens and dune-slacks. Widespread but rather local, and sometimes absent from apparently suitable habitats. Map 610.

MENTHA L.

Mentha pulegium L. Pennyroyal. Brymlys.
Native. Wet sandy places, damp grassy heaths, village greens and roadsides, often near ponds. Recently recorded only from Kenfig Burrows (78D, M.R. Shanahan 1970) but recorded in the past from several village greens or grassy heaths near villages in Gower and the Vale, for example: Dinas Powys (17D), AHT 1905*; Pitton Green (48A), AHT c.1907;

Reynoldston Common and High Pool (48B), AHT 1908*, EV 1912*, JAW 1941; Oxwich Green (58A), AHT c.1907; Three Crosses (59D), JAW 1924; Nottage (87A), E.S. Marshall c.1907. It may still grow at some of these sites. Map 611.

Mentha arvensis L. Corn Mint. Mintys yr Ŷd.
Native. Locally frequent as a weed of cultivated land, on damp roadsides and in marshy places. Fl. 7–9, fr. 9–10; Hb (n), B.luc, B.prat, (Syr) E.ten, Sy.pip, (Dip) Luc, Musca sp., (Lep) Ly.phl, Py.tit. Map 612.

Mentha aquatica × M. arvensis (**M. × verticillata** L.). Whorled Mint. Mintys Troellaidd.
Native. Widespread and locally common in marshy places and ditches and on damp roadsides. Map 613.

Mentha arvensis × M. spicata (**M. × gentilis** L.). Bushy Mint. Mintys Culddail.
Variable; apparently a hybrid swarm derived from **M. arvensis** and the introduced **M. spicata**. Locally frequent in ditches, on damp village greens and roadsides, and by streamsides in Gower, twice recorded in the Cardiff area where it is probably a garden outcast.
17B, Cardiff, EV 1932*, A.D. Tipper 1973*; 49C, Llangennith to Coity Green, A.L. Still 1934*, R.A. Graham & R.M. Harley 1956; 49D, Fairy Hill, JEL 1948*, near Burry, D. Green 1979*; 58B, Bishopston Valley, ALS 1934, 1937*; 59D, Fairwood Common, JAW 1931.

Mentha aquatica × M. arvensis × M. spicata (**M. × smithiana** R.A. Graham) (*M. rubra* Sm. non Miller). Red Mint. Mintys Coch.
Once locally frequent in parts of Gower and the Vale, in sites similar to those in which **M. × gentilis** grows, but with few recent records. Map 614.

Mentha aquatica L. Water Mint. Mintys y Dŵr.
Native. Very common by streamsides, in ditches, near springs and in marshy places generally, and in dune-slacks. Fl. 7–10, fr. 8–10; Ma.jur. Pi.rap. Map 615.

Mentha aquatica × M. spicata (**M. × piperita** L.). Peppermint. Mintys Poethion, Pupur-fintys.
A garden escape, naturalized on roadsides and in damp places. Locally frequent in the Vale but scarce elsewhere. Map 616.

Mentha × piperita nothomorph **citrata** (Ehrh.) Briq. has been recorded from two localities, Castell Coch Wood (18C, 1972!), and near Nash Point (96A, MG 1972).

Mentha suaveolens Ehrh. (*M. rotundifolia* auct. non (L.) Hudson). Round-leaved Mint. Mintys Deilgrwn.
Native or introduced. Apparently locally frequent on roadsides and waste ground in the Vale and eastern Uplands but rather scarce in the west. Fl. 8–9; Hb,

B.luc, (Syr) E.arb, E.ten, Sy.pip (Dip) Luc.ser, Or.cae, (Lep) Ag.urt, Pa.aeg. Map 617.

Mentha spicata × M. suaveolens (M. × villosa Hudson) (*M. × niliaca* auct. ex Jacq.).   Apple Mint. Mintys Lled Crynddail.
Probably a garden escape. Locally frequent on roadsides and waste ground in the west; perhaps overlooked or confused with **M. suaveolens** and **M. spicata** elsewhere. Map 618.

Mentha spicata L. (incl. *M. longifolia* auct., non (L.) Hudson).   Spear Mint. Mintys Ysbigog.
Probably introduced. Recorded from streamsides, roadsides and waste ground in localities scattered through the county. Fl. 7–10; Hb, B.luc, B.pasc, B.terr, (Syr) E.ten, (Dip) Call, Luc.spp, Sa.car, (Lep) Ma.jur, Pl.gam. Map 619.

Mentha requienii Bentham Corsican Mint, has been found as a garden or roadside weed in three localities (17B, Llandaff, L.S. Pritchard 1965*; 69A, Penllergaer, HJR 1906*; 88A, Margam Park, ARP 1984!) and **M. longifolia** (L.) Hudson × **M. spicata** has been recorded as a garden escape on one occasion (61C, near Garn-swllt, 1981!*).

SALVIA L.

Salvia verbenaca L. (*S. horminoides* Pourret).   Wild Clary. Saets Gwyllt.
Native. Apparently once fairly widespread in dry pastures and on roadsides in the Vale, but now very scarce there with recent records from only one site (97C, Cwm Nash, JPC 1981, PJo 1983). It is locally frequent on limestone sea-cliffs in Gower, at Oxwich Point (48B, 58C, 1982!) and Port Eynon Point (48D, 1981!). It has also been seen recently as a casual on waste ground in western Swansea (69C, 1973!). Map 620.

Salvia verticillata L.   Whorled Clary. Troellennog.
Introduced. Once fairly frequent as a casual on roadsides and waste ground near docks in the Cardiff district, but usually impersistent and with only four recent records (00C, Aberdare, AMP 1980*; 16A, Palmerstown, ABP 1974!*; 17D, Penarth Road, JPC 1982; 18C, Whitchurch, RAH 1970). Map 621.

Five species of **Salvia** have been found as casuals or impersistent introductions on one or a few occasions. **Salvia pratensis** L., Meadow Clary, Saets y Waun, was found at Merthyr Mawr (87B) by H.J. Riddelsdell in c.1907 and persisted for a few years until the field in which it grew was ploughed (W. Cook, in Vachell, n.d.4.) it had previously been recorded from Cardiff (17B) by J. Storrie in *c*.1876 and from Port Eynon 48B by W. Turton (Oldisworth, 1802), a doubtful record. **S. nemorosa** L., Wild Sage, was found as a casual at Cardiff (27A, JS 1876*), Barry Docks (16A, RLS 1926*, 1938*), and Port Talbot (78B, HJR 1906, Wakefield 1919), and also on dunes at Porthcawl (87A, RE & FC 1898–99). **S. aethiopis** L., Woolly Clary, was found by

R.L. Smith at Canton (17B, *c*.1924) and Grangetown (17D, 1926*) and by E. Vachell on Leckwith Common (17D 1926). **S. sclarea** L., Clary, was found on waste ground in Port Talbot (78B) by M. Hamford in 1961* and **S. argentea** L., Silver Sage, in Port Talbot Docks (78B) by H.J. Riddelsdell in 1906.

SOLANACEAE

NICANDRA Adanson

Nicandra physalodes (L.) Gaertner   Apple-of-Peru.
Introduced. Occasionally found as a garden weed or garden escape; still very scarce, but probably increasing.
48B, Horton, R.C. McLean 1924*; 49C, Llanmadoc, 1976!; 69C, University College of Swansea, 1981!; 87A, Newton, S. Oxenham 1969*; 97D, Llysworney, D.J. Thomas 1973*.

LYCIUM L.

Lycium barbarum L.   Duke of Argyll's Teaplant. Ysbeinwydd Hardd.
Occasionally naturalized in hedges and on railway embankments, most frequently near the sea; scarce inland. Fl. 7; B.prat. Map 622.

ATROPA L.

Atropa bella-donna L.   Deadly Nightshade. Ceirios y Gŵr Drwg, Codwarth.
Perhaps once native in a few localities, but more often introduced and naturalized in hedges or scrub, on waste ground or around ruins. Always scarce and only seen recently at Penarth (17D) and Summerhouse Point (96B); it has been deliberately eradicated from some sites, for example at Mumbles (68A).
17B, ruins of White Friars, on rubbish removed during clearance of the site, JS *c*.1890, Llandaff Cathedral, LWD 1848, Llandaff, HAH 1933*; 17D, Penarth, J.W. Zehetmayr 1975; 27A, East Moors, JS *c*.1886; 68A, Mumbles, JWGG 1839, Oystermouth Castle, WT 1802; 96B, Summerhouse Point, EV 1904, E.L. Downing 1907*, JPC 1987.

HYOSCYAMUS L.

Hyoscyamus niger L.   Henbane. Ffa'r Moch, Llewyg yr Iâr.
Native or introduced. Once fairly frequent in Gower and the Vale as a weed of disturbed waste ground and farmyards, and in sandy and gravelly places near the sea; now very scarce. Recently recorded from 16A, Barry, D.G. Holland 1972; 17B, Cardiff, JWD 1976, GSW 1982; 26, Flat Holm, MG 1974; 49C, Broughton Farm, JBe 1983!; 69D, Oystermouth Road, Swansea, H & CH 1979!**; 78D, Kenfig Burrows, H & CH 1978; 87B, Candleston, PJo 1982; 87D, Dunraven, PJo 1983. Map 623.

Hyoscyamus albus L., White Henbane, has been found on two occasions in the past as a casual on dockside waste ground: at Barry Docks (16A, 1923!*), and at Shrimphouse, Cardiff (JS 1882).

Physalis peruviana L., Peruvian Winter-cherry, was found as a casual on a rubbish tip at Penarth Road, Cardiff (17D) by R.L. Smith in 1937*.

SOLANUM L.

Solanum nigrum L.   Black Nightshade. Codwarth Du.
Native or introduced. A common weed of gardens and vegetable crops, farmyards, rubbish tips and waste ground in Gower and the Vale, but surprisingly scarce in the Uplands. Fl. 6–11, fr. 8–11; (!) B.terr (p). Map 624.

Solanum dulcamara L. Bittersweet, Woody Nightshade. Codwarth Caled.
Native. Common in hedges, at the edges of woods and in scrub, on stream banks, on fixed dunes and shingle beaches, and as a weed of waste ground. **S. dulcamara** var. **marinum** Bab., a procumbent form with fleshy leaves, has been recorded from 16B, 48D and 58A. Fl. 5–9, fr. 8–11. Map 625.

Several species of **Solanum** have been found on one or two occasions as casuals or garden weeds at Cardiff or Barry. Four species have been found only once: **Solanum chenopodioides** Lam. at Barry Docks in 1935 (16A, J.P.M. Brenan*), **S. pseudocapsicum** L. on a rubbish tip at Newport Road, Cardiff, in 1937 (27A!*), the similar **S. capsicastrum** Link ex Schauer on waste ground at Splott in 1927 (27A, RLS*), and **S. sisymbrifolium** Lam. as a garden weed at Roath Park, Cardiff in 1974 (17B, M. Davies*). Three species have been found on two occasions: **S. villosum** Miller subsp. **villosum** at a Radyr railway siding in 1921 (18C!*) and **S. villosum** subsp. **puniceum** (Kirschleyer) Edmonds at Barry in 1921 (16A, RLS*), **S. triflorum** Nutt., also in Cardiff Docks in c.1876 (17B, JS*) and at Splott in 1925 (27A, RLS), and **S. cornutum** Lam. at Barry Docks in 1924 (16A, RLS*) and as a garden weed at Penarth in 1982 (17D, M.M. Lennard*). **S. ciliatum** Lam. has been found on three occasions: as a garden weed at Splott in 1927 (27A, RLS*), at Penarth Road in 1932 (17D!*), and at Porthcawl in 1937 (87A, RLS & F. Norton).

Solanum tuberosum L. (Map 626), and **Lycopersicon esculentum** Miller (Map 627), are frequently found on waste ground and rubbish tips, and occasionally as weeds of other crops, but do not persist; they have not been systematically recorded during the present survey and the maps are provisional only.

DATURA L.

Datura stramonium L.   Thorn-apple. Meiwyn.
Introduced. Occasionally found as a weed of gardens, cultivated land and waste ground in Gower and the Vale; rare in the Uplands. Map 628.

Datura ferox L., Angel's-trumpets, has been found on two occasions as a garden weed or throw-out (17B, Cardiff, T.A. Davies 1959*; 18C, Radyr, R.A. Tinker 1970*).

Nicotiana alata Link & Otto, Flowering Tobacco, has been recorded on two occasions (17D, Cardiff, RLS 1934*; 88B, Llynfi Valley, H & CH 1976) and **N. longiflora** Cav. once (17D, Cardiff, 1932!*). **Petunia axillaris** (Lam.) Britton, E.E. Stearns & Poggenb. has been recorded once from a roadside in Radyr (18C, KA 1956*) and **P. × hybrida** Vilm.-Andr. from near Briton Ferry (79C, CH 1973).

BUDDLEJACEAE
BUDDLEJA L.

Buddleja davidii Franchet   Butterfly-bush. Llwyn Iâr Fach.
Introduced. Now very common on waste ground, tips, railway embankments, old walls and ruins in urban areas and increasingly common in similar habitats in the countryside. Map 629.

SCROPHULARIACEAE
LIMOSELLA L.

Limosella aquatica L.   Mudwort. Lleidlys.
Native. Ponds and lake margins. Rare, with no confirmed recent records. Not certainly recorded in the county until 1930, when a single plant was found by C.M.H. Glück in a roadside puddle about 400m from Kenfig Pool. In 1935 Eleanor Vachell found that it was abundant, with **L. australis**, at Morfa Pools (78B), 4–5km north of Kenfig Pool. It has not been seen at Morfa Pools since 1945; they were partly drained and partly converted into a reservoir (Eglwys Nunnydd Lake) for the adjacent steelworks in the following year (Vachell, 1950).
78B, Morfa Pools, EV 1935*, JAW 1941*, EV 1945; 88C, near Kenfig Pool, C.M.H. Gluck 1930.

Limosella aquatica × L. australis.
Found at Morfa Pools (78B) by Eleanor Vachell in 1935* on the muddy cattle-trodden shore of the western pool; chromosome counts by K.B. Blackburn of the University of Durham showed that **L. aquatica** had a chromosome number of 2n=40, **L. australis** 2n=20 and the hybrid 2n=30 (Vachell, 1941b). **L. aquatica × L. australis** was found only at Morfa Pools and was last seen there by Eleanor Vachell in 1945.

Limosella australis R. Br.   Welsh Mudwort. Lleidlys Cymreig.
Native. Ponds and lake margins. Apparently more widespread and less rare than **L. aquatica** in the past, but recently recorded only from Kenfig Pool, where **L. australis** was first found by A.H. Trow in 1897. Both species of **Limosella** are extremely uncertain in their appearance; in 1930, for example, Eleanor Vachell and C.M.H. Glück were unable to find any plants at Kenfig Pool, although they searched the

shores of the pool for three hours; Professor Glück later found a single plant of **L. aquatica** about 400m from the Pool (see above; Vachell, 1941b).
69D, Crymlyn Bog, HJR c.1907; 78B, Morfa Pools, EV 1935, JAW 1941, EV 1945; 78D, Kenfig Pool, AHT 1897*, E.S. Marshall & WAS 1901*, N.Y. Sandwith & EV 1933–1938, DPMG 1943*, M.R. Shanahan 1970.

MIMULUS L.

Mimulus guttatus DC.   Monkeyflower. Blodyn y Mwnci.
Introduced. Now thoroughly naturalized and common in most of the Uplands on streambanks, around springs and in flushes, but still scarce and local in Gower and the Vale. Map 630.

Mimulus guttatus × **M. luteus**.
Introduced. Naturalized in a few places in the western Uplands in similar sites to **M. guttatus**; locally frequent in the Afan and upper Rhondda valleys.
18B, Caerphilly Mountain, Anon. pre 1930*; 79D, Pont-rhyd-y-fen, H & CH 1974; 89B, Glyncorrwg, H & CH 1974!; 89C, Bryn, H & CH 1976; 90D, Aberdare, H.J. Dawson 1985; 99A, Treherbert, 1975!; 99B, Treorchy, CH 1975!; 99D, Tonypandy, CH 1975.

Mimulus (cupreus × M. luteus) × M. guttatus has recently been discovered established in the River Rhymney between Llechryd and Rhymney (10A, TGE & U.T. Evans 1988*, det. A.J. Silverside).

Mimulus moschatus Douglas ex Lindley   Musk. Mwsg.
Introduced. Marshy places and flushed sites, often at the bases of walls. Naturalized in a few places near Cardiff and in the Uplands; scarce.
17B, Cardiff, 1973!; 18B, Caerphilly Common, EV 1930; 19A, Pont-y-Gwaith, MG 1971; 19C, Mynydd Eglwysilan, MG 1973; 90D, Aberdare Country Park, MG 1973!; 99B, Treorchy, G.L. Llewellyn 1969*, Ferndale and Maerdy, JTa 1970, Cwm Aman, AMP 1970.

VERBASCUM L.

Verbascum blattaria L.   Moth Mullein. Gwyfynog, Pannog Gwyfyn.
A rare casual and dock alien, not recorded since 1928 when Eleanor Vachell found it at Splott (27A*); also recorded from Port Talbot (HJR c.1908) and Barry Docks (RLS 1925*) and less certainly from 17A, JS c.1880; 17B, C.E. Ollivant c.1873; 87A, JS c.1880 and 88C, RE & FC 1898–1899. Early field records of this species from the Swansea and Neath districts (69C, 69D, 78B, 79C) by L.W. Dillwyn (1805, 1840) and J.W.G. Gutch (1842) may refer to **V. virgatum**, which is still widespread in this area. Map 631.

Verbascum virgatum Stokes   Twiggy Mullein. Tew-bannog, Pannog Brigog.

Introduced. Scarce but widespread on roadsides, waste ground and rubbish tips in the Swansea, Neath and Cardiff districts; appparently very rare elsewhere. It was first recorded in the county by J.A. Webb at the Derwen Fawr tip (69C) in 1932, but may have been present in western Glamorgan since the early nineteenth century, confused with the similar **V. blattaria**. Map 632.

Verbascum phlomoides L.   Orange Mullein. Pannog Oren.
Casual or introduced; a rare weed of gardens and waste ground, recently seen only in western Swansea but likely to reappear in other localities. 17B, Llandaff, EV 1943*; Roath Park, 1946*!; 69C, Blackpill, JAW 1941*, Singleton, 1977!; 79C, Llansawel, JAW 1945*.

Verbascum thapsus L.   Great Mullein. Pannog Melyn.
Native. Dry grassland and open scrub, roadsides, waste ground and old walls; widespread and frequent. Fl. 6–10, fr. 8–10. Map 633.

Verbascum pulverulentum Vill.   Hoary Mullein. Pannog Blawrwyn, Pannog Lwyd.
Introduced. Locally established on roadsides and waste ground in central Swansea (69D, between Morfa Road and the R. Tawe, 1974!–1987!), Ynystawe (60D, 1975!) and near Jersey Marine (79C, 1956!, H & CH 1976). It was first recorded in the county near Neath (79C*) by E.M. Thomas in 1917. Found elsewhere only at Barry Docks (16A) in 1922 (RLS*) and Cardiff (17B) in 1924 (RM & RLS).

Verbascum nigrum L.   Dark Mullein. Pannog Tywyllddu.
Casual or introduced. A local plant in the county, occurring on waste ground, near railways or on disturbed sand-dunes especially in the west of the country. Map 634.

Verbascum lychnitis L., White Mullein, Hanner Pan, was found as a casual near docks in Cardiff (17B) by Storrie c.1886, at Penarth (17D) by A.H. Trow in 1892, and has recently been found on waste ground at Margam (78B, M.R. Shanahan 1970). Four other species of **Verbascum** have been found in the past as rare casuals or garden escapes: **Verbascum chaixii** Vill. subsp. **chaixii** on four occasions (Aberdare (00C) in 1902 (HJR); Canton, Cardiff (17B) in 1923 (RLS); Port Talbot (78B) in 1905 (HJR); Porthcawl (87A) in 1904 (HJR)), and **V. chaixii** Vill. subsp. **austriacum** (Schott ex Roemer & Schultes) Hayek once (Llwydcoed (90D) in 1902 (HJR**)). The other three species have been found on only one occasion each; **V. sinuatum** L. at Barry Docks (16A) in 1924 (RLS!*), **V. ovalifolium** Donn ex Sims as a garden escape in Cardiff (17B) in 1923 (EV) and **V. phoeniceum** L. at Jersey Marine (79C) in 1936 (RLS).

SCROPHULARIA L.

Scrophularia nodosa L. Common Figwort. Gornerth.
Native. Common throughout the county on hedgebanks, in woods and in damp shady places. Fl. 5–10, fr. 7–11; Paravespula germanica. Map 635.

A form of **Scrophularia nodosa** with the flowers wholly green (var. **bobartii** Pryor) has been recorded from both Sully Island (16B, EV 1906, 1938!*) and Candleston (87B, Miss Leak 1936*). A.H. Trow (1911) observed that this form grew side by side with typical **S. nodosa** on Sully Island, with the forms being linked by a continuous series of intermediates.

Scrophularia auriculata L. (*S. aquatica* auct. non L.). Water Figwort. Gornerth y Dŵr.
Native. Widespread and frequent on streambanks and in ditches, marshes and flushes, but much less common than **S. nodosa**. Map 636.

Scrophularia scorodonia L., Balm-leaved Figwort, Gornerth Gwenynddail, has been found as a rare casual near docks and on sand-dunes, its most recent records are at Cardiff Docks (17B, 1956!*; 17D, JPC 1980–1984; 27C, JPC 1981, H.J. Noltie 1983); it was found at the same sites by J. Storrie c.1876–1886, and it has also been found at Porthcawl (87A, RE & FC, 1898–99, PWR 1920*). **S. canina** L. subsp. **canina** was also found as a casual in the Cardiff district by Storrie in 1876 (17B*, 27A*) and in 1920 at Radyr (18C, RLS*), it was last recorded in 1956! (17B). **S. canina** subsp. **hoppii** (Koch) P. Fourn was found once at Radyr in c.1922 (18C, RLS), and **S. scopolii** Hoppe at Cardiff (17B) in 1923 (GCD).

ANTIRRHINUM L.

Antirrhinum majus L. Snapdragon. Safn y Llew, Trwyn y Llo Mwyaf.
Introduced. A garden escape, often established on old walls. Map 637.

MISOPATES Rafin.

Misopates orontium (L.) Rafin. Lesser Snapdragon, Weasel's Snout. Trwyn y Llo Bychan.
Introduced. A rare weed of cultivated land and waste ground. Recently recorded from only four localities: 16A, Barry, D.G. Holland 1972; 59C, cornfield weed near Grongaer, 1974!; 68A, garden weed in Mumbles, 1970!; 79C, garden weed at Jersey Marine, CH 1981. Map 638.

CHAENORHINUM (DC.) Reichenb.

Chaenorhinum minus (L.) Lange Small Toadflax. Gingroen Bychan.
Introduced. Locally common as a weed of cultivated and waste ground, railway ballast and colliery spoil-tips in the Uplands and parts of the Vale. Map 639.

LINARIA Miller

Linaria purpurea (L.) Miller Purple Toadflax. Gingroen Cochlas.
Introduced. Old walls, waste ground, roadsides and embankments. Fairly frequent and widespread, but not common except in parts of the Vale. Map 640.

Linaria purpurea × **L. repens** (**L.** × **dominii** Druce) is rare; it has been found only at Barry Docks (16A, 1986!*), near Gowerton (59B, 1974!) and at Baglan (79D, H & CH 1974).

Linaria repens (L.) Miller Pale Toadflax. Gingroen Porffor Gwelw.
Introduced. Locally common near docks and on waste ground and tips in urban areas, but most frequent on railway ballast and widely distributed along existing or former railways; sometimes more or less naturalized on disturbed sand-dunes and sandy shingle. Apparently first recorded in the county, c.1837, as a ballast plant near Cardiff by J.C. Collins (Watson, 1837). Map 641.

Linaria repens × **L. vulgaris** (**L.** × **sepium** Allman).
Locally frequent in the Lower Swansea Valley, in the Neath district, and near Barry, sometimes in the absence of one or both parents; formerly recorded from several sites in the Cardiff district, but not seen there since 1931. Recent records from 16A, Barry Docks, JPC 1980; 69D, Lower Swansea Valley, 1974!; 79A, Neath, H & CH 1974; 79C, Jersey Marine, H & CH 1976; 87A, Porthcawl, G. Campbell 1970. Recorded in the past from 17D (Penarth, Grangetown and Llandough).

Linaria vulgaris Miller Common Toadflax. Llin y Llyffant.
Native or introduced. Widespread and locally common as a weed of field banks, neglected grassland, hedgebanks and roadsides, railway embankments and tracks, quarries, smelter slag-tips, colliery spoil-tips and other man-made habitats; occasionally found on shingle beaches and in scrub on sea-cliffs. Fl. 6–11, fr. 8–11; B.hort (n,p), B.lap, B.luc, B.pasc (n,p), B.terr, (Syr) S.rib. Map 642.

Several alien species of **Linaria** have been found as dockside casuals or garden escapes in the Cardiff district and one in Swansea; **Linaria reflexa** (L.) Desf. at Cardiff (17B, GCD 1923), **L. triphylla** (L.) Miller at Splott (27A, RLS 1926!*), **L. genistifolia** (L.) Miller subsp. **dalmatica** (L.) Maire & Petitmengin at Cardiff Docks (17B, RLS 1924*), **L. viscosa** (L.) Dum.-Courset at Radyr (18C, RLS 1920*, 1921), **L. incarnata** (Vent.) Sprengel as a garden relic at Penarth Dock (17D, ABP 1966*) and Cyncoed (17B, SGH 1977*), **L. pelisseriana** (L.) Miller at Cardiff (17B, JS c.1876) and Swansea (69D, S. Kerrow c.1919), **L. supina** (L.) Chaz. on ballast in 17B (JS c.1886), and **L. caesia** (Pers.) DC. ex Chav. at Cardiff and Penarth (17D, JS c.1880).

CYMBALARIA Hill

Cymbalaria muralis P. Gaertner, B. Meyer & Scherb.   Ivy-leaved Toadflax. Llin y Fagwyr.
Introduced. Common and widespread on old walls in towns and villages, sometimes also found on limestone outcrops. A form completely lacking any red pigmentation occurs sporadically as, for example, on walls in the Cardiff area. Fl. 4–12, fr. 6–10. Map 643.

Cymbalaria pallida (Ten.) Wettst., Greater Ivy-leaved Toadflax, Llin y Fagwyr Mwyaf, has recently been recorded, as an introduction, in Gower (58B, Bishopston, L. Wilson 1975) Wilson (1988).

KICKXIA Dumort.

Kickxia elatine (L.) Dumort.   Sharp-leaved Fluellen. Llysiau Llywelyn.
Native or introduced. Locally frequent in cliff-heath and as a weed of arable land and disturbed roadsides on basic soils in Gower; once frequent as a weed of arable land in the Vale, but with few recent records there. In Gower it is apparently native in cliff-heath, where it appears, with **Anagallis arvensis, Euphorbia exigua** and other annuals, after fires. Map 644.

Kickxia spuria (L.) Dumort.   Round-leaved Fluellen. Llysiau Llywelyn Crwnddail.
A weed of arable land, often growing with **K. elatine**; once fairly frequent in the Vale but now rare. Map 645.

Erinus alpinus L., Fairy Foxglove, Clychau'r Tylwyth Teg, is well established as a garden escape on old walls and limestone quarries in a few localities; it has been recorded recently from Vaynor Quarries (00A, G. Morris 1985), Pendoylan (07B, ABP 1974!) and Castle-upon-Alun (97C, ABP 1973!) and in the past from Llandaff (17B, EV 1938).

DIGITALIS L.

Digitalis purpurea L.   Foxglove. Bysedd y Cŵn.
Native. Common on well-drained acid soils, growing in a range of open or partly shaded habitats; in rough grassland and among bracken, on rocky slopes, in open woods and scrub, on hedgebanks and roadsides and on railway embankments. Fl. 5–7 (–11), fr. 7–10; B.hort (n,p), B.lap, B.luc (n,p), B.pasc (n), B.terr. Map 646.

VERONICA L.

Veronica serpyllifolia L.   Thyme-leaved Speedwell. Rhwyddlwyn Gruw-ddail.
Native. Common and widespread, usually on rather moist soils, in short grassland in pastures and on grassy heaths, in lawns, on roadsides and waste ground, and by streams and on cliff ledges in the hills. Fl. 3–9, fr. 5–10. Map 647.

Veronica officinalis L.   Heath Speedwell. Rhwyddlwyn Meddygol.
Native. Common in well-drained sites on grassy heaths, in old pastures on field banks and roadsides, and sometimes in open woods; often grows on ant-hills. Map 648.

Veronica chamaedrys L.   Germander Speedwell. Llygad Doli, Rhwyddlwyn Blewynog.
Native. Very common in open grassy woodland and scrub, on field banks, hedgebanks, roadsides and railway embankments, and in dune grassland. Fl. (2–) 5–7 (–10), fr. 6–8; B.prat, B.terr, Sb, (Syr) Rh.cam, ?Syr. Map 649.

Veronica montana L.   Wood Speedwell. Rhwyddlwyn y Gwrych, Rhwyddlwyn Mynyddol.
Native. Common in woodland on calcareous or neutral soils in most of the county but scarce in the central Uplands. Fl. 4–7, fr. 5–7. Map 650.

Veronica scutellata L.   Marsh Speedwell. Rhwyddlwyn y Gors.
Native. Rather local in wet heaths and marshes on neutral or acid soils; most frequent in Gower and parts of the Vale. Fl. 6–8. Map 651.

Veronica beccabunga L.   Brooklime.   Llysiau Taliesyn, Gorferini, Llychlyn y Dŵr.
Native. Common in shallow streams and ditches, at the edges of lakes and ponds, and in muddy marshes. Fl. 5–7. Map 652.

Veronica anagallis-aquatica L.   Blue Water-speedwell. Graeanllys y Dŵr, Rhwyddlwyn Glas y Dŵr.
Native. Rare, in shallow streams, ditches, dune slacks and small ponds in the Central Vale. Map 653.

Veronica catenata Pennell   Pink Water-speedwell. Graeanllys y Dŵr Rhuddgoch.
Native. Shallow streams and marshes on calcareous soils. Locally frequent in the Vale and western Gower. Fl. 6–8, fr. 7–9. Map 654.

Veronica anagallis-aquatica × **V. catenata** (V. × **lackschewitzii** Keller) has once been recorded from Gower (Marchant, 1970).

Veronica arvensis L.   Wall Speedwell. Rhwyddlwyn y Mur, Mur-rwyddlwyn.
Native. Common and widespread on shallow dry soils, on wall-tops, as a weed of gardens, cultivated land and waste ground, and on ant-hills; often associated with other winter annuals in calcareous grassland on shallow soil over limestone and in dune grassland. Map 655.

Veronica agrestis L.   Green Field-speedwell. Rhwyddlwyn Gorweddol.
Introduced. A weed of cultivated land, waste ground and roadsides. Most frequent in the Uplands, but scattered through Gower and the Vale. Map 656.

Veronica polita Fries Grey Field-speedwell. Rhwyddlwyn Llwyd.
Introduced. A weed of gardens and waste ground, occasionally found in arable land. Locally common in the east but rare or absent in the west of the county. Map 657.

Veronica persica Poiret Common Field-speedwell. Rhwyddlwyn y Gerddi.
Introduced. Very common as a weed of gardens, arable land, waste ground and disturbed roadsides. Fl. 1–12, fr. 4–11; Hb, ?Sb, (Syr) S.rib, Sy.pip. Map 658.

Veronica filiformis Sm. Slender Speedwell. Rhwyddlwyn Crwn-ddail.
Introduced. Rapidly spreading as a weed of lawns, grassy roadsides and river banks, and waste ground. Now locally common in urban areas in the southern half of the county, but still scarce in the countryside and the Uplands. Fl. (3–) 4–7; no seed set. Map 659.

Veronica hederifolia L. Ivy-leaved Speedwell. Rhwyddlwyn Eiddew-ddail.
Probably native. Common in disturbed habitats in the lowlands; found in disturbed places in woods and scrub and under bracken, on hedgebanks, roadsides and tracks, and as a weed of gardens and arable land. Fl. 3–6, fr. 4–6; (Syr) S.rib. Map 660.

Veronica hederifolia subsp. **hederifolia** has been recorded from various habitats at several localities, mainly in the west of the county. Map 661.

Veronica hederifolia subsp. **lucorum** (Klett & Richter) Hartl has been recorded from similar habitats to subsp. **hederifolia** but mainly in the east of the county. Map 662.

Veronica spicata L. Spiked Speedwell. Rhwyddlwyn Pigog, Rhwyddlwyn Tywysennaidd.
Native. On shallow soil near outcrops and in rock crevices on limestone sea-cliffs. Very scarce, with recent records from only three sites between Porteynon and Rhosili (48A!, 48B and 48D!); most records refer to a site near The Knave (48A) where 50–100 plants grow on about 200 metres of the cliff-slopes. V. spicata was probably first found in this area by S.W. Jenkins before 1907 (Riddelsdell, 1907) but was not certainly recorded until 1937, when it was re-found by 'Miss Mitchell and Miss Simpson' (E. Vachell, 1947b).
48A, west of Port Eynon, Miss Mitchell & Miss Simpson 1937, near The Knave, EV 1945*, P.M. Loyn 1963*, 1972!, 1982!; 48B/C, Port Eynon, EV 1945*, JAW 1956*; 48D, Port Eynon, I. Tew 1983!. (!) Fl. (6–) 7–8, fr. 7–9.

Veronica longifolia L. has been found as a garden escape on waste ground in five places (17B, Canton, R. Davies 1948*; 17C, Dinas Powis, 1928!*; 59B, Penclawdd, 1974!; 79C, Briton Ferry, H & CH 1973;

87A, Porthcawl, VCB 1978), and **V. paniculata** L. once (70D, near Crynant, 1974!).

Hebe speciosa (R. Cunn. ex A. Cunn.) Anderson has been recorded once, at Langland (68A) by J.A. Webb in 1920. There is no specimen to support this record; a specimen in **NMW** collected by JAW from Merioneth (v.c. 48) in 1947 as this species has now been identified as the hybrid H. **elliptica** × H. **speciosa** (H. × **franciscana** (Eastwood) Souster, and all subsequent Welsh records of this genus refer to the hybrid; it is probable that the Langland record does also.

SIBTHORPIA L.

Sibthorpia europaea L. Cornish Moneywort. Ceinioglys.
Native. Amongst mosses on damp grassy heaths, damp rocks and walls, and moist shady roadsides. Very local, known only from a small area around Llantrisant, Pontypridd and Caerphilly on the southern fringe of the Uplands. The first record of the species in Glamorgan was made by Sir J. Cullum, who found it near Pontypridd in the latter part of the 18th Century (Turner & Dillwyn, 1805).
08B, near Pontypridd, A. Jones 1951*; 08C, Y Graig, Llantrisant, EV c.1908, Llantrisant Common, MG 1970; 09D, Pontypridd Common, JS c.1885, AHT 1905*; 18B, Caerphilly, H.S. Thompson c.1907; 18A, Pwllypant, JS c.1885; 18C, Garth Hill, EV 1898*, 1975!*, GH 1981; 19C, Carnedd Lwydion, EV 1941.

MELAMPYRUM L.

Melampyrum pratense L. Common Cow-wheat. Gliniogai.
Native. In open oakwoods or mixed deciduous woods and on shady hedgebanks on well-drained acid soils. Locally frequent in the Uplands but very local in Gower and the Vale. Apparently uniformly yellow-flowered in Glamorgan. Map 663.

EUPHRASIA L.

Eyebright. Effros.
Euphrasia rostkoviana Hayne subsp. **rostkoviana** Effros Blodau Bach Gludiog.
Native. In old hay meadows, mainly in the Uplands; now scarce, but probably more frequent and widespread in the past than is indicated by the records. Map 664.

Euphrasia anglica Pugsley English Eyebright. Effros Chwareog Gwalltog.
Native. Locally frequent in northern Gower and the western Uplands in short closely grazed grassland on acid soils, often on roadside banks. Map 665.

Euphrasia arctica Lange ex Rostrup subsp. **borealis** (Townsend) Yeo (E. brevipila auct.) Arctic Eyebright. Effros â Gwallt Byr.

Native. Fairly widespread but not common, growing in hay meadows and old pastures, on field banks and roadsides, and on fixed dunes at Kenfig Burrows and elsewhere. The most frequent large-flowered eyebright in the county. Map 666.

Euphrasia tetraquetra (Bréb.) Arrondeau (*E. occidentalis* Wettst.). Broad-leaved Eyebright. Torfagl ar Graig y Don.
Native. Locally common in limestone grassland on or near the coast. Map 667.

Euphrasia tetraquetra × E. confusa or **E. nemorosa** has been recorded on two occasions: from Crawley Cliff, Oxwich (58A) by T.A.W. Davis in 1961*, and from near Kenfig Pool (78D) by E.K. Horwood in 1953, both det. P.F.Yeo.

Euphrasia nemorosa (Pers.) Wallr. Common Woodland Eyebright, Llygad Effros.
Native. Common and widespread in short grassland. By far the commonest eyebright in the county, growing on most well-drained soils; found in limestone grassland, on fixed dunes, in old pastures and on grassy heaths, and on roadside banks and railway embankments. Map 668.

Euphrasia confusa Pugsley Little Kneeling Eyebright. Effros Bach Gliniog.
Native. Most frequent in hill pastures, but occasionally recorded from fixed dunes; apparently scarce except in some parts of the Uplands. Map 669.

Euphrasia confusa × E. nemorosa has been recorded once, on the short turf of a quarry bank, Pont Alun near Ewenny (87B) by E.K. Horwood in 1953 (det. P.F. Yeo).

Euphrasia confusa × E. tetraquetra was recorded from several sites on dune grassland in the Kenfig – Porthcawl areas by E.K. Horwood in 1953 (78D, Kenfig Burrows; 87A, Porthcawl; 88C, Kenfig golflinks, all det. P.F. Yeo), and by T.G. Evans in 1985* (87B, Merthyr Mawr, det. A.J. Silverside).

ODONTITES Ludwig

Odontites verna (Bellardi) Dumort. Red Bartsia. Gorudd.
Native. Common and widespread in pastures and on grassy roadsides. The late-flowering subsp. **serotina** (Dumort.) Corb. has not been consistently separated from subsp. **verna** during the survey but appears to be more frequent. Fl. 7–10, fr. 8–10. Map 670.

PARENTUCELLIA Viv.

Parentucellia viscosa (L.) Caruel Yellow Bartsia. Gorudd Melyn.
Native. In damp pastures and damp marshy heath or sand-dune grassland, appearing sporadically and usually in small numbers. Scarce, with recent records only from a few sites near Neath and Jersey Marine

and two sites to the west of Swansea, it probably still grows in other sites on Barland, Clyne and Fairwood Commons and in marshy pastures nearby in a small area of eastern Gower; it was also recorded in the past from a site near Bridgend.
58B, Barland, EV & JAW 1929; 59D, Pen-y-bank near Fairwood, Miss Lyons *c*.1910, Wergan-rhos, JAW 1923*, T. Davies 1988!; 68A, near Mumbles, VMP 1927*; 69C, Clyne Valley, JAW 1924*, R.E. Castell 1972; 79, near Neath, J. Ralfs 1840; 79A, Crymlyn Dingle, LWD *c*.1840, Neath Technical College, E.C. Roberts 1966*; 79C, near Jersey Marine, H & CH 1973!**, I. Tew 1987, Baglan Sands, HJR 1906*; 79D, Aberafan, HJR *c*.1907; 98C, north of Bridgend, JS *c*.1886.

PEDICULARIS L.

Pedicularis palustris L. Marsh Lousewort. Melog y Waun.
Native. Base-rich marshes and marshy dune-slacks; local and decreasing. Map 671.

Pedicularis sylvatica L. Lousewort. Melog y Cŵn.
Native. Wet heaths and marshy grassland, usually on acid soils. Common except in the southern Vale. (!) Fl. 5–7, fr. 6–9; Bb. Map 672.

RHINANTHUS L.

Rhinanthus minor L. Yellow-rattle. Cribell Felen.
Native. Old hay-meadows and pastures, field banks and grassy roadsides, limestone grassland and fixed dunes. Still widespread and fairly common. The subspecies into which this species has been split in the past are now regarded as no more than seasonal ecotypes, and of these the vernal type (var. **minor**) is much more common than the autumnal type (var. **stenophyllus** Schur). Fl. 6–8, fr. 7–8; B.hort (n,p), B.lap (n), B.luc (n,p), B.pasc (n), B.prat (n,p). Map 673.

LATHRAEA L.

Lathraea squamaria L. Toothwort. Dantlys.
Native. Parasitic on the roots of hazel or elm in woods on base-rich soils. Very local; now most frequent in limestone woodland in the Vale, for example in Bute Park, Cardiff (17B, JPC 1976–1983), at Cwrt-yr-Ala (17C, GMB 1977), near Cefn Carnau (18A, 1974!), Taff's Well (18C, RAH 1972), Whitchurch (18C, JPC 1976) and near Pencoed (98D, 1970!). Also found in the Parkmill, Ilston and Bishopston Valleys in Gower (58A, 58B, 1981!), and in the north on limestone in the Taf-fechan valley (00A, MG 1975) and on non-limestone soil in the Pyrddin gorge near Scŵd Einion Gam (80B, MG 1973). Map 674.

ACANTHACEAE

ACANTHUS L.

Acanthus mollis L., Bear's-breech, has been recorded from four sites as a garden escape or well-established garden weed (17B, pathside near R. Taff, Bute Park,

JPC 1983, Cathays Park 1977–1987!; 18D, Cyncoed, W.M. Davies 1950*; 79D, Aberafan, L. Thomas 1950*).

OROBANCHACEAE
OROBANCHE L.

Orobanche minor Sm.   Common Broomrape. Gorfanc Lleiaf.
Native. Locally frequent and sometimes abundant on sand-dunes in Gower and the Vale; occasionally found inland in the Vale. On sand-dunes it grows on a range of legumes, composites and other hosts; its commonest host is **Ononis repens**, as for example at Whiteford Burrows (49A), but it is also frequently found on **Eryngium maritimum** and several other hosts at Crymlyn Burrows (79C). Map 675.

Orobanche hederae Duby   Ivy Broomrape. Gorfanc Eiddew.
Native. Parasitic on ivy (**Hedera helix**). Locally frequent on coastal limestone cliffs, especially in Gower, and occasionally found on dunes on which ivy grows, for example at Nicholaston Burrows (58A); rare elsewhere. Map 676.

Orobanche elatior Sutton   Knapweed Broomrape. Gorfanc Hir.
Native. Parasitic on **Centaurea scabiosa**. Rare, with no recent records; **C. scabiosa** often grows on the coastal limestone cliffs on which ivy and **O. hederae** also grow, and some at least of the records of **O. elatior** in Glamorgan may result from confusion with **O. hederae**.
06B, Porthkerry, 1957!, A.W. Crump 1961*; 17C, Cwrt-yr-ala, T. Chapman c.1886; 48B, Penrice, JWGG c.1844; 78B, Port Talbot, HJR c.1907.

Orobanche rapum-genistae Thuill.   Greater Broomrape. Gorfanc Mwyaf.
Native, Parasitic on the roots of leguminous shrubs. Rare, with no recent records; reliably recorded only from the Cardiff district, where it was last seen in 1959.
16A, near Cadoxton Mill and Merthyr Dyfan, JS c.1886; 17A, St. Fagans, T.W.E. David c.1886; 17B, Penylan Hill, 1944!; 18C, Radyr, EV 1907, Taff's Well, C.M. Goodman 1929*; 18D, Thornhill, B. Ogden 1959*; 58A, Penmaen, 1893; 69D, Swansea, JWGG c.1842; 97D, Cowbridge, W.F. Evans c.1900.

LENTIBULARIACEAE
PINGUICULA L.

Pinguicula vulgaris L.   Common Butterwort. Tafod y Gors, Toddaidd Cyffredin.
Native. Near springs and in acid flushes, especially where water flows over cliffs in the hills; on the edges of bogs and on boggy streamsides; and sometimes in dune-slacks, as for example at Jersey Marine. Rather uncommon in the Uplands, where it is most frequent in the north; rare elsewhere. Map 677.

The record of **Pinguicula vulgaris** from Rhosili (48A, HJR c.1907) may refer to **P. lusitanica** L., which should be searched for in suitable habitats on or near Rhosili Down.

UTRICULARIA L.

Utricularia minor L.   Lesser Bladderwort. Swigenddail Lleiaf, Chwysigenddail Lleiaf.
Native. Acid water in ponds on heathland, bog-pools and upland lakes. Rare, with recent records from only two localities.
59C, near Broad Pool on Cefn Bryn, EV 1945*, GTG 1965, 1973!; 60D, Penclun Marsh, EV c.1936; 69B, Llandwr Marsh, JWGG c.1844, Crymlyn Bog, 1971!, 1981!; 79A, Crymlyn Bog, 1971!, 1976!; 90C, Llyn Fach, JS 1886.

Utricularia australis R.   Br.   Bladderwort. Swigenddail, Chwysigenddail Cyffredin.
Native. Eutrophic pools, ditches and canals. Rare, with recent records only from Oxwich marshes, Crymlyn Bog and the Neath Canal. Some older records may refer to **U. vulgaris** L.
17B, Canton Common, JS c.1886; 17D, Leckwith Moors, K. Richards 1894*; 48B, Oxwich marsh below Penrice, EV 1929; 58A, Oxwich marsh, EV 1933*, 1955!*, 1975!(fl.); 68A, Oystermouth, E. Lees & TBF, c.1844; 69B, Crymlyn Bog, HJR c.1907, ALe 1977; 79A, Neath Canal, EV 1936, H & CH 1980.

PLANTAGINACEAE
PLANTAGO L.

Plantago major L.   Greater Plantain. Llwynhidydd Mawr, Llyriad Mwyaf.
Native. Very common on roadsides, pathways and tracks, in trampled grassland and as a weed of cultivated land and waste ground (subsp. **intermedia** (DC.) Arcangeli should be looked for in coastal communities; it has fewer veins to the leaf and more seeds per capsule than the type). Map 678.

Plantago coronopus L.   Buck's-horn Plantain. Llwynhidydd Corn Carw, Llyriad Corn y Carw.
Native. Locally abundant in upper salt-marshes and in the spray-zone on sea-cliffs, often growing with **P. maritima**, **P. coronopus** is particularly characteristic of sandy salt-marshes and of the transition zone between salt-marshes and dry dune-slacks, and also grows to some extent on fixed sand-dunes; it is occasionally found on shallow dry soils inland. Map 679.

Plantago maritima L.   Sea Plantain. Llwynhidydd Arfor.
Native. Locally abundant in salt-marshes and in the spray-zone on sea-cliffs; also observed on saline soil below a large permanent dump of rock-salt (for road maintenance) on Fairwood Common in Gower, at a height of c.90m and 4km from the sea (59D, 1983!). Map 680.

Plantago media L. Hoary Plantain. Llwynhidydd Blewog, Llyriad Llwydion.
Native. Common in limestone grassland in the Vale, rare or absent elsewhere; the records from Gower require confirmation. Map 681.

Plantago lanceolata L. Ribwort Plantain. Llwynhidydd, Llyriad Llwynhidydd.
Native. Very common and widespread in a wide range of grassland habitats; absent only from very acid or waterlogged soils. **P. lanceolata** is a characteristic plant of old pastures and hay-meadows, limestone grassland and grazed roadsides. Map 682.

Three species of **Plantago** used to occur as casuals on waste ground near docks in the Cardiff district. **Plantago lagopus** L. has been recorded at one site (27A, Splott, 1925!*, EV 1926*), **P. afra** L. at two sites (17B, Cardiff Docks, JS *c*.1876*; 27A, Splott, 1926!*) and **P. arenaria** Waldst. & Kit. at six sites (00C, Aberdare, HJR *c*.1906; 17B, Cardiff Docks, EV & RLS 1924*; 17D, Penarth, C.C. Babington *c*.1860, Penarth Road, EV 1929*; 27A, East Moors, JS 1876*, Splott, 1926!*).

LITTORELLA Bergius

Littorella uniflora (L.) Ascherson Shoreweed. Beistonnell.
Native. At the margins of lakes and pools, and sometimes also in small streams and ditches on heathland, as for example on Kennexstone Moor in Gower (49C, 49D). Very local; found only in a few sites in western Gower, in montane lakes and pools near Hirwaun and Merthyr, and in a few lakes and ponds in the Vale. Map 683.

CAPRIFOLIACEAE
SAMBUCUS L.

Sambucus ebulus L. Dwarf Elder. Ysgawen Fair.
Probably an escape from former cultivation as a medicinal plant. Rare, occurring on roadsides and field banks or in hedgerows or similar sites in a few scattered localities, often near old farmsteads or monastic sites, as for example at Monknash on the Vale coast (97C). Map 684.

Sambucus nigra L. Elder. Ysgawen.
Native. Common in hedges and scrub, at the edges of woods, on sea-cliffs and fixed sand-dunes, on derelict buildings, and on railway embankments, old tips and waste ground. Fl. 5–7, fr. (8–) 9–10; Hb (p), Dips. Map 685.

VIBURNUM L.

Viburnum opulus L. Guelder-rose. Corswigen.
Native. Fairly common in hedges and in open woods, especially on acid soils and in damp sites; it sometimes grows with **Salix cinerea** in carr woodland at the edges of fens or marshes, for example on Crymlyn Bog (79C). Fl. 6–7, fr. 9–10. Map 686.

Viburnum lantana L. Wayfaring-tree. Gwifwrnwydd.
Native. Common in hedges and woods on limestone soils in the Vale, but rare in Gower (where the few trees that have been recorded may be introductions) and absent from the Uplands. Fl. 5–6, fr. 7–9. Map 687.

Viburnum tinus L., Laurustinus, was reported by J.A. Webb to be naturalized on screes at East Caswell, Gower (58B) in 1941; it has not been recorded since.

SYMPHORICARPOS Duh.

Symphoricarpos albus (L.) S.F. Blake var. **laevigatus** (Fernald) S.F. Blake Snowberry. Llus Eira.
A well-established introduction. Now fairly common in hedges and former parkland and on waste ground, especially in towns and villages. Fl. 6–11, fr. 7–11; B.prat, Paravespula germanica. Map 688.

LONICERA L.

Lonicera periclymenum L. Honeysuckle. Gwyddfid.
Native. Common in hedges, scrub and woods on most soils. Fl. 6–10, fr. 8–11; B.hort, B.luc, B.prat, B.terr, (Syr) E.ten, Rh.cam, (Lep) Pl.gam, noctuid moths. Map 689.

Four cultivated species of **Lonicera** are occasionally found in hedges; in most cases they do not appear to spread after being planted and scarcely merit the status of introductions.

Lonicera nitida E.H. Wilson, Chinese Honeysuckle, has been recorded most frequently and may sometimes spread spontaneously. Map 690.

Lonicera xylosteum L. Fly Honeysuckle, Gwyddfid Syth, has twice been found in hedges (00A, near Pontsticill, J.H. James 1932*; 18D, between Thornhill and Cefn Onn Ridge, L. Reynolds *c*.1950), and **L. caprifolium** L., Perfoliate Honeysuckle, Gwyddfid Trydwll, has been recorded three times (58A, Parkmill, JAW *c*.1930; 58B, Stonemill, JAW 1956*; 87A, Newton, J.B. Lloyd *c*.1905). **L. japonica** Thunb., Japanese Honeysuckle, has been recorded from western Swansea and two other localities (06B, Rhoose, 1971!*; 68A, Lilliput, JAW 1943*; 69C, Mayals and Derwen fawr, JAW 1957*; 79C, Baglan, JAW 1941*).

Leycesteria formosa Wall., Himalayan Honeysuckle, Bachgen Llwm, is occasionally found as a fairly well-established introduction in hedges, scrub and woods. Map 691.

ADOXACEAE
ADOXA L.

Adoxa moschatellina L. Moschatel. Mwsglys.
Native. In hedgebanks and woods, usually in shady places on a moist loamy soil. Fairly common in

Gower and the Vale, but apparently very local in the Uplands, where it may have been under-recorded to some extent (it is an inconspicuous species which appears very early in the year and has disappeared by early summer). Fl. 3–5. Map 692.

VALERIANACEAE
VALERIANELLA Miller

Valerianella locusta (L.) Betcke Common Cornsalad. Llysiau'r Oen.
Native. Fairly frequent in Gower and the Vale with other annuals on shallow soils near limestone outcrops and on sand-dunes and sandy shingle; occasionally found on roadsides and as a weed of cultivated land. Apparently absent from the Uplands. Fl. 6–7; (Lep) Ma.jur. Map 693.

A taxon from sand-dunes sometimes called **V. locusta** subsp. **dunensis** (D.E. Allen) P.D. Sell has been recorded from Three Cliffs Bay (58A, EMT 1921*); it is probably more frequent than this single record suggests.

Valerianella carinata Loisel. Keeled-fruited Cornsalad.
Native. In similar sites to **V. locusta**, and on old walls. Apparently rare, with few recent records; perhaps overlooked in some sites.
06B, Porthkerry, 1970!; 17A, Ely, EV 1932*; 18D, Llanishen, JPC 1971*; 49C, Llanmadoc, 1971!; 80B, Aberpergwm, HJR c.1907, Glyn-neath, BAW c.1908, EV 1920*; 87A, Newton, EMT c.1930; 97A, Bridgend, BAM 1955*.

Valerianella dentata (L.) Pollich Narrow-fruited Cornsalad. Gwylaeth yr Oen Deintiog.
Native or introduced. A scarce weed of cultivated land and waste ground in Gower and the Vale, last recorded in 1941 from the Porthcawl area (87A) by Miss E.M. Thomas. Map 694.

Valerianella rimosa Bast. Broad-fruited Cornsalad.
Native or introduced. A rare weed of cultivated land and waste ground in Gower and the Vale, with no recent records.
06B, Porthkerry, EV 1920*; 68A, Norton, J. Woods c.1850; 69D, Fabian's Bay, Swansea, JWGG c.1840; 87A, Nottage, EMT 1941*; 97 Bridgend to Cowbridge area, BAM 1954 (BSBI maps scheme).

Valerianella discoidea (L.) Loisel. was found as a casual at Aberdare (00C) by H.J. Riddelsdell in 1907.

VALERIANA L.

Valeriana officinalis L. Common Valerian. Triaglog.
Native. Common in marshes, ditches and wet woods and on wet cliff ledges in the Uplands; less frequent in Gower and rather local in the Vale, where it may have been reduced by drainage. Map 695.

Valeriana dioica L. Marsh Valerian. Triaglog y Gors.
Native. Marshes and bogs. Occasional in the Uplands and the Vale. It was once fairly frequent in parts of the Vale but has, like **V. officinalis** and other marsh plants, been reduced in frequency by drainage of its habitats. Map 696.

CENTRANTHUS DC.

Centranthus ruber (L.) DC. Red Valerian. Triaglog Coch.
Introduced as a garden escape; now common on old walls and frequent on roadside banks and cuttings, railway embankments, quarry faces and sea-cliffs near towns and villages in Gower and the Vale, but still scarce in the Uplands. Fl. 4–11, fr. 7–10; Hb, B.hort (n), (Syr) S.balt (p), S.rib (p), (Lep) Ag.urt, Pi.bra, Pi.nap, Pi.rap, Macroglossa stellatarum. Map 697.

DIPSACACEAE
DIPSACUS L.

Dipsacus fullonum L. Teasel, Crib y Pannwr. Teilau Gwyllt.
Native. Fairly common in Gower and the Vale in hedgerows and scrub and as a weed of roadsides and waste ground; rare in the Uplands. Fl. 7–8, fr. 8–10; B.hort (n), B.luc (n,p), B.pasc (n). Map 698.

Dipsacus pilosus L. Small Teasel. Ffon y Bugail, Teilau Lleiaf.
Probably introduced. Still established on a streamside near Llandough-juxta-Cowbridge (97D, 1969!*), where it was first found by W.F. Evans in or before 1908; Trow (1911) observed that **D. pilosus** was most abundant at this site in a thicket of rhododendrons, and had probably been introduced with them, but was thoroughly naturalized. Recently discovered by J.P. Curtis on a streamside at East Aberthaw (06A, 1987) where it may also become established.

SUCCISA Haller

Succisa pratensis Moench Devil's-bit Scabious. Tamaid y Cythraul, Clafrllys Gwreidd-don.
Native. Very common in pastures and on grassy heaths, especially on waterlogged soils, and in marshes; occasionally found in cliff grassland and dune-slacks. Fl. 7–11, fr. 8–11; Hb (n), B.hort, B.lap, B.luc, B.pasc, B.terr, Psithyrus sp., (Syr) E.arb, E.hort, E.ten, H.pen, Rh.cam, S.balt, S.rib, Sc.pyr, Volucella bombylans (Dip) Luc, (Lep) Ag.urt, Cy.car, In.io, Pi.bra, Pi.nap, Pi.rap. Map 699.

KNAUTIA L.

Knautia arvensis (L.) Coulter Field Scabious. Clafrllys.
Native. Common on basic soils in the Vale, where it occurs in a range of grassland habitats including cliff and dune grassland and old pastures as well as field

banks, hedgebanks, railway embankments and roadsides; less common in Gower where it grows mainly
on hedgebanks, and scarce in the Uplands. Fl. 6–9
(–11), fr. 8–10; B.prat, B.pasc, (Syr) E.ten, H.pen,
Rh.cam, S.balt, S.rib, (Lep) Ly.phl. Map 700.

SCABIOSA L.

Scabiosa columbaria L.   Small Scabious. Claffrllys
Bychan.
Native. Locally frequent in Gower in limestone
grassland near outcrops and on limestone cliffs, and
sometimes on calcareous fixed dunes; rare in similar
habitats in the Vale, where it has been recorded only
from Kenfig Burrows (78D, D.M. Cape 1957), Ogmore Down (87B, MG 1960) and limestone grassland
near the **Teucrium chamaedrys** site at South Cornelly
(88C, 1966!, J. John 1970). Map 701.

CAMPANULACEAE

CAMPANULA L.

Campanula glomerata L.   Clustered Bellflower.
Clychlys Clwstwr.
Native. Limestone grassland at a few sites on the Vale
coast, where it reaches the western limit of its British
range; most frequent on south-facing slopes.
17D Penarth, JS *c.*1876; 80C, Resolven (garden
escape), MG *c.*1970; 87D, Southerndown, JS 1876,
F.H. Perring 1971, Dunraven, EV 1896*, Cwm Mawr,
BAM 1953; 96A, Nash Point, HJR *c.*1906, 1974!, 1983!,
Marcross, JS *c.*1876; 96B, Llantwit Major, JS *c.*1876;
97C, Monk Nash, HAH 1934*, Cwm Nash, EV
pre–1930, 1972!, 1983!.

Campanula portenschlagiana   Schultes   Adria
Bellflower. Clychlys Adria.
A garden escape, naturalised on old walls in several
localities.
08C, Llanharan, 1972!; 09A, Mountain Ash, 1975!;
16A, Barry, 1972!; 17B, Cardiff, RLS 1922*, Roath,
1972!*, GH 1985; 97C, Llandow, 1972!; 97D, Llysworney, 1975!; 98C, Bridgend, 1978!; 99C, Ogmore Vale,
ABP 1975.

Campanula latifolia L.   Giant Bellflower. Clychlys
Mawr.
Probably introduced, persisting after introduction, or
as a garden escape, in several places.
07B, near Peterston-super-Ely, 1969!, St George's,
R.M. Baker 1980; 17A, Coedrhiglan, JS *c.*1880; 17D,
Penarth, JS *c.*1880; 48B, Stouthall Woods, JAW 1946,
JBe 1980!; 58A, Oxwich Point, MG 1973; 69C,
Hendrefoilan, JAW 1947–1948; 87A, Candleston,
AHT 1892*, 1972!; 97C, Cwm Nash, 1972!; 97D,
Cowbridge, JS *c.*1880.

Campanula trachelium L.   Nettle-leaved Bellflower. Clychlys Danadl.
Possibly native in hedgerows and on shaded laneside
banks on limestone soils in Gower and the Vale. Rare,
with recent records only from 07C (Beaupre Castle,
KA 1972), 48B (Pennyhitch Hill west of Penrice Park,

1974, 1982!), and 97D (Llandough-juxta-Cowbridge,
1969!, PJo 1984). First recorded in Flower & Lees 1842
'between Penrice and Port Eynon' (48B). Map 702.

Campanula rotundifolia L.   Harebell.   Cloch   y
Bugail, Clychlys Deilgrwn.
Native. This species has a surprisingly discontinuous
distribution in the county. In southern and western
Gower and parts of the Vale it is locally frequent or
common in calcareous dune grassland and in limestone grassland and rock crevices. It does not occur in
most of the central part of the county, but it is again
frequent or locally common in the north-eastern
Uplands, where it grows in dry grassland and heaths
and on rock ledges. Fl. 7–9, fr. 8–10. Map 703.

Four species of **Campanula** have been found as
garden escapes in a few places: **Campanula
poscharskyana** Degen, Trailing Bellflower, Clychlys
Ymlusgol, naturalized on walls or hedgebanks at
Roath (17B, 1972!), Lisvane (18D, GH 1984),
Llanedeyrn (28C, ABP 1975!) and Bridgend (98C,
ABP 1978!), **C. rapunculoides** L., Creeping
Bellflower, Clychlys Llusg, naturalized at Merthyr
Mawr (87B, RE & FC 1898, EV 1939*), **C. medium** L.,
Canterbury-bells, Clychlys Caergaint, at Cosmeston
(16B, L. Peddle 1979), Splott (27A, RLS 1925*) and
Briton Ferry (79C, H & CH 1972), and **C. latiloba** DC.
at Canton (17B, RLS 1939*).

Legousia hybrida (L.) Delarbre has been found once
as a casual (87A, near Porthcawl, J.B. Lloyd *c.*1907) as
has **L. speculum-veneris** (L.) Chaix (97D?, near
Cowbridge, C.T. Vachell 1881*).

WAHLENBERGIA Schrader ex Roth

Wahlenbergia hederacea (L.) Reichenb.   Ivy-leaved
Bellflower. Clychlys Eiddew.
Native. Locally frequent in the Uplands in wet heaths
and bogs and in marshy grassland by streams, often
with **Polytrichum** spp. and other mosses. Map 704.

JASIONE L.

Jasione montana L.   Sheep's-bit. Clefryn.
Native. Locally common in Gower and the Uplands
on dry heaths, grassy banks and old dry-stone walls
on acid soils, and on acid fixed dunes (for example at
Southgate, 58B); rare in the Vale. Map 705.

LOBELIA L.

Lobelia dortmanna L.   Water Lobelia. Bidoglys y
Dŵr.
Native. Now found only in Llyn Fach above Hirwaun, a montane lake with acid oligotrophic water at
an altitude of 400m below the escarpment of Craig-y-
llyn. It is fairly abundant and grows from the
shallows to a depth of about 2 metres. It has not been
recorded in Llyn Fawr since the conversion of that
lake to a reservoir.

90C, Llyn Fawr, LWD *c*.1800, THT 1890*, AHT *c*.1900, Llyn Fach, LWD *c*.1800, AHT 1897*, EV 1934, 1939*, 1965!, 1983!. Fl. 7–8; B.terr.

Lobelia erinus L., commonly grown in gardens as a bedding plant has twice been recorded as a presumed garden escape in Cardiff (17B, Cathays Park, 1989!*; 27A, Newport Road, 1937!*).

COMPOSITAE (ASTERACEAE)
EUPATORIUM L.

Eupatorium cannabinum L. Hemp-agrimony. Byddon Chwerw.
Native. Streamsides, marshes, ditches, scrub and damp hedges. Common in the lowlands but rather local in the Uplands. Fl. 7–10, fr. 8–11; B.terr, (Syr) E.ten, (Lep) Ag.urt, Ma.jur, Zygaena filipendulae. Map 706.

SOLIDAGO L.

Solidago virgaurea L. Goldenrod. Eurwialen.
Native. Open woods and heathland, and on rocky slopes and cliffs in the hills, usually on acid soil. Fairly common in Gower and the Uplands but absent from most of the Vale. Map 707.

Solidago canadensis L. Canadian Goldenrod. Eurwialen Ganada.
Introduced. Roadsides and waste ground; a naturalized garden escape. Frequent in urban areas. Some of the records may refer to **S. gigantea**. Map 708.

Solidago gigantea Aiton subsp. **serotina** (O. Kuntze) McNeill Early Golden-rod. Eurwialen Gynnar.
Introduced, Roadsides and waste ground; a naturalized garden escape.
16A, Barry, 1969!*; 17B, Cardiff Castle Grounds, A.D. Tipper 1977*; 18D, Cyncoed, E.G. Roberts 1981*; 48B, Scurlage, 1980!; 69A, Swansea, JAW 1952*, 79C, Aberavon, JAW 1943*.

The similar **Solidago graminifolia** (L.) Salisb. has once been recorded as a garden relic (17B, Cardiff, RLS *c*.1925).

BELLIS L.

Bellis perennis L. Daisy. Llygad y Dydd.
Native. Closely grazed pastures, short turf and lawns on most soils. Common everywhere; recorded from all squares. Fl. all year, (1–) 3–7 (–12), fr. 5–8; Hb (n,p) (Syr) E.ten, S.rib, (Dip) Luc (especially 3–4). Map 709.

ASTER L.

Aster novi-belgii L. Michaelmas-daisy. Blodyn Mihangel.
Introduced. Marshes, rough grassland and waste ground. Fairly frequent; the commonest naturalized Michaelmas-daisy in the county. Map 710.

Five other Michaelmas-daisies have been recorded as rare casuals or garden escapes. **Aster lanceolatus** Willd. on seven occasions (09A, Mountain Ash, AMP 1972*; 09B, Mountain Ash, AMP 1972*; 17B, Cardiff, JWD 1952*, RLS 1957*; 69D, Cwmbwrla, JAW 1924, Swansea, A. Lake 1959; 87B, Merthyr Mawr, HJR *c*.1905); **A. lanceolatus × A. novi-belgii** (**A. × salignus** Willd.) five times (00C, Aberdare, HJR *c*.1905; 49D, Cheriton, HJR *c*.1905; 78B, Port Talbot, HJR *c*.1905; 87B, near Merthyr Mawr, MG 1974; 90D, Llwydcoed, HJR 1902*); **A. novae-angliae** L. twice (17B, Cardiff, 1923!*; 78D, Kenfig, 1972!) and on one occasion each **A. laevis × A. novi-belgii** (**A. × versicolor** Willd.) (90D, Llwydcoed, HJR 1901) and **A. adscendens** Lindley (27A, Splott, 1925!*).

Aster tripolium L. Sea Aster. Seren y Morfa.
Native. Salt-marshes, and occasionally in the spray-zone on sea-cliffs and sea-walls. Very common in salt-marshes. Plants with purple ray-florets occur in most populations but are less abundant than rayless plants (var. **discoideus** Reichenb. fil.). Map 711.

Aster linosyris (L.) Bernh. (*Crinitaria linosyris* (L.) Less.) Goldilocks Aster. Gold y Môr.
Native. In grass and amongst gorse on limestone cliff-slopes. Very rare. Eight sites are known on the cliffland between Port Eynon and Mewslade Bay in western Gower. The shoots of **Aster linosyris** are inconspicuous, and although its presence in Gower is geographically and ecologically unsurprising the first site was not discovered until the 1940s and the second and third in 1974 and 1979. The largest population occupies an area of about 40 square metres in low gorse.
48A, near The Knave, EV *c*.1940, DMcC 1948, south of Paviland Farm, M. Page 1974!; 48B/D, west of Port Eynon, 1979!, A. Lack 1985! (five sites). (!) Fl. 8–10, fr. 10–11; (Syr) E.ten, (Dip) Luc, (Lep) Ag.urt.

ERIGERON L.

Erigeron karvinskianus DC. (*E. mucronatus* DC.). Mexican Fleabane. Cedowydd y Clogwyn.
Introduced. Naturalized on old walls in the Cardiff area, and limestone rocks at Caswell Bay; apparently spreading.
17A, St. Fagans, SGH 1966, Radyr to Llandaff, MG 1971; 17B, Llandaff, 1961!, M.P. Morton 1984*; 17C, Dinas Powis, DMcC 1958, 1974!; 18D, Lisvane Reservoir, J.E. Beckerlegge 1947, GH 1984; 58B, Caswell Bay, CH 1978; 87A, Newton Dunes, JAW 1932.

Erigeron acer L. Blue Fleabane. Cedowydd Glas.
Native. Open limestone grassland and sand-dune grassland, and occasionally on base-rich slag-tips and roadside banks. Frequent along the coast, scarce elsewhere. Fl. 6–8 (–10), fr. 7–10. Map 712.

Erigeron philadelphicus L.
Introduced. A naturalized garden escape.
17B, Llandaff, S. Jones 1944*, 1973!*.

CONYZA Less.

Conyza canadensis (L.) Cronq. Canadian Fleabane. Amrhydlwyd Canada.
Introduced. Cultivated land, waste ground, road-sides and sand-dunes. Locally common in urban areas in the lowlands and on some dune-systems; increasing. Fl. 7–9, fr. 9–10. Map 713.

The similar **Conyza bonariensis** (L.) Cronq., Argentine Fleabane, has occurred as an alien at Cardiff Docks (17B, 1924!*, J.H. Salter 1927*).

Grindelia squarrosa (Pursh) Dunal has been recorded as a casual in the Cardiff area on three occasions (17B, RLS *c*.1925!; 17D, JS *c*.1876; 27A, 1925!*).

FILAGO L.

Filago vulgaris Lam. (*Filago germanica* L.). Common Cudweed. Edafeddog.
Native. Sand-dunes, dry grassland and roadsides. Now very scarce, with only three recent records; formerly fairly widespread along the coast. Recent records from: 17D, near Penarth, E. Cawood and K. Baker 1970; 48B, Slade, D.O. Elias 1976; 77B, Sker Point, 1977!. Map 714.

LOGFIA Cass.

Logfia minima (Sm.) Dumort. (*Filago minima* (Sm.) Pers.) Small Cudweed. Edafeddog Lleiaf.
Native. Colliery spoil-tips and restored opencast sites, waste ground and roadsides. Very abundant in suitable habitats in the Uplands but rare in Gower and the Vale. It has apparently become abundant in the Uplands since the introduction of extensive opencast coal mining, and it is now one of the most abundant and characteristic plants of restored opencast sites and colliery tips. Map 715.

OMALOTHECA Cass.

Omalotheca sylvatica (L.) Schultz Bip. & F.W. Schultz (*Gnaphalium sylvaticum* L.). Heath Cudweed. Edafeddog y Rhosdir.
Native. Acid sand-dune grassland and dry grassy heaths. Rare; reliably recorded only from sand-dunes at Kenfig and Porthcawl. No recent records.
00C, Aberdare, JS *c*.1886?; 09B, Cynon Valley, JS *c*.1886?; 70D, Mynydd Marchywel, C. Hawkins *c*.1841?; 78D, Kenfig Burrows, EMT 1947*; 79A, Y Drumau, JWGG *c*.1841?; 81C, Onllwyn, THT *c*.1908?; 87A, Porthcawl Golf Links, EV 1934.

FILAGINELLA Opiz

Filaginella uliginosa (L.) Opiz (*Gnaphalium uliginosum* L.). Marsh Cudweed. Edafeddog Canghennog, Edafeddog y Gors.
Native. Damp open grassland, cultivated land and roadsides. Generally fairly common, but scarce or

absent in the southern Vale. Fl. 7–9, fr. 8–10. Map 716.

Gnaphalium luteo-album L., Jersey Cudweed, Edafeddog Melynwyn, was found as an alien on ballast at Cardiff Docks by J. Storrie (27A, JS *c*.1876*), and Barry Docks by R.L. Smith (16A, RLS 1927). **Helichrysum stoechas** (L.) Moench, was found as a casual at Barry Docks (16A) by R.L. Smith in 1923.

ANTENNARIA Gaertner

Antennaria dioica (L.) Gaertner Mountain Everlasting. Edafeddog y Mynydd.
Native. Dry grassland or grassy heathland, usually on limestone. Rare, with only one recent record. Storrie's records from the Rhondda (99B, 99C) (Storrie, 1886) were probably errors; Dillwyn's record from 'about Pont Nedd Vachn' (Turner & Dillwyn, 1805) probably refers to a site in Breconshire although Dillwyn included it with his Glamorgan records, and the records in Perring and Walters (1962) for 09 and 49 have not been confirmed.
00A, Morlais Castle, THT *c*.1900*; 00C, between Merthyr Tydfil and Aberdare, THT *c*.1908; 18B, Cefn Onn, J. Godbey & L. Reynolds 1944*; 68A, Mumbles Hill, JAW 1921–1931*; 87B, Ogmore Down, D.P.M. Guile 1943*, MG 1976, J. Etherington 1984.

ANAPHALIS DC.

Anaphalis margaritacea (L.) Bentham Pearly Everlasting. Edafeddog Tlysog.
Introduced. Colliery tips, riversides, waste ground, and occasionally on sand-dunes. Locally fairly abundant in the Uplands and northern Vale; on sand-dunes, in small numbers, at Whiteford, Crymlyn and Kenfig Burrows. An early introduction to European gardens from North America, first recorded in Glamorgan by Edward Llwyd before 1709 'on the banks of Rymny river for the space of at least twelve miles, . . .' (Ray, 1690). It is now widely distributed in Britain as a naturalized garden escape, but it is common only in Glamorgan and Monmouthshire. Map 717.

INULA L.

Inula helenium L. Elecampane. Marchalan.
Introduced. Hedgebanks and rough grassland, scarce. Formerly grown as a medicinal plant, and naturalized near villages and old farms. Most frequent in Gower. Map 718.

Inula conyza DC. Ploughman's spikenard. Meddyg Mair, Meddyg y Bugail.
Native. Limestone grassland and scrub, sand-dune grassland, and occasionally on spoil-tips and waste ground. Frequent in coastal districts but scarce inland. Map 719.

Inula crithmoides L. Golden-samphire. Sampier y Geifr, Cedowys Sugnol.

Native. In the spray-zone on limestone cliffs, and occasionally on shingle beaches, sand-dunes or salt-marshes near the drift-line. Rather common in suitable habitats on the limestone cliffs of southern Gower, but absent from most of the Vale coast. Map 720.

Dittrichia viscosa (L.) W. Greuter (*Inula viscosa* (L.) Aiton), Sticky Aster, has been recorded as a dock alien at Cardiff and Barry (16A, RLS 1921*; 17B, RLS 1924; 27A, JS 1876*, EV 1923).

PULICARIA Gaertner

Pulicaria dysenterica (L.) Bernh. Common Fleabane. Cedowydd.
Native. Marshes, damp grassland, ditches and streamsides. Common in Gower and the Vale but scarce or local in the Uplands. Fl. 8–10, fr. 9–11; (Sb) Anthidium manicatum, ?Sb, (Syr) E.ten, S.balt, S.rib, (Lep) Cupido minimus, Ma.jur, Pi.nap, Pi.rap. Map 721.

Pulicaria vulgaris Gaertner, Small Fleabane, Cedowydd Bach, was found growing as a ballast alien at Penarth and Leckwith by J. Storrie (17D, *c.*1876*) and **P. paludosa** Link at Cardiff Docks by G.C. Druce (17B, 1923).

The Mediterranean species **Asteriscus maritimus** (L.) Less., and **A. aquaticus** (L.) Less., were found as aliens at Cardiff Docks, the former by J. Storrie (17B, *c.*1876*) and the latter by G.C. Druce (17B, 1906).

BIDENS L.

Bidens tripartita L. Trifid Bur-marigold. Graban Deiran.
Native. Ponds, lake margins, streamsides and ditches. Rather local, but widely distributed in suitable habitats in Gower and the Vale; scarce in the Uplands. Map 722.

Bidens cernua L. Nodding Bur-marigold. Graban Ogwydd.
Native. Ponds, lake margins, streamsides and ditches. Often associated with **B. tripartita**, but less frequent than that species in the Vale. Fl. 8–10, fr. 9–11; B.terr. Map 723.

Bidens frondosa L. Beggarticks. Llawr Crwydryn.
Introduced from America. Ponds, ditches and marshes, and sometimes on waste ground. Map 724.

GALINSOGA Ruiz & Pavón

Galinsoga parviflora Cav. Gallant Soldier. Llygaid Bach.
Introduced. Cultivated land and waste ground. Found as a garden weed or on waste ground in several localities, but recently recorded from only three places.

08D, Efail-isaf, W. Curtis 1965*; 17D, Grangetown, P.H. Holland 1927*; 27A, Pengam, RLS 1937!*; 69C, Swansea, JAW 1949*, 1975–85!; 79B, Neath, H & CH 1978; 87A, Newton, 1957!*; 88A, Margam Park, D.M. Cape 1957*.

Galinsoga quadriradiata Ruiz & Pavón (*G. ciliata* (Rafin.) S.F. Blake). Shaggy Soldier. Llygaid Bach Blewog.
Introduced. Now grows in several places in the Vale, where it is more frequent than **Galinsoga parviflora**, and also in Swansea. First recorded in 1922 (17D). Map 725.

Several other members of the Heliantheae, mainly from America, have been found as dock aliens or garden escapes. **Rudbeckia laciniata** L., Coneflower, has occurred on waste ground as a garden escape in Cardiff (17B, RLS, 1939*). **Helianthus annuus** L., Sunflower, Blodau'r Haul, is occasionally found as a garden escape or bird-seed alien on waste ground (17A, near Radyr, MG 1973; 17B, Cardiff, JPC 1982; 27A, Cardiff, 1937!; 90D, Llwydcoed, HJR 1902). Cultivated perennial sunflowers, probably **Helianthus rigidus × H. tuberosus** (**H. × laetiflorus** Pers.), sometimes occur as garden escapes on dunes (87A, Newton Burrows, JAW 1953*) or waste ground (17B, Cardiff, P.H. Dunn 1969*, MG 1974; 17D, Cardiff Docks, GSW 1982, JPC 1986; 69C, Blackpill, JAW *c.*1924; 69D, Swansea, M.H. Sykes 1941; 90D, Llwydcoed, HJR 1902). **Helianthus tuberosus** L. has been recorded once (78B, Port Talbot, HJR *c.*1905). **Cosmos bipinnatus** Cav. has been found as a garden escape on waste ground in Cardiff (27A, Roath, RLS 1939*). **Hemizonia pungens** Torrey & A. Gray and **H. kellogii** E.L. Greene were found by Storrie as ballast aliens at Cardiff Docks (17B, JS *c.*1876) and the former also at Barry (16A, RLS 1925!). **Ambrosia artemisiifolia** L., Ragweed, was once frequent as a casual near docks, and has also been found as a garden weed, but it was last recorded in 1947 (17A, Ely, RLS 1947*); it occurred at Barry (16A, RLS *c.*1935), Cardiff (17A, RLS 1947* 17B, HJR 1907, RLS 1927; 18C, Radyr, RLS 1924*) and Port Talbot (78B, HJR 1905). **Ambrosia maritima** L., Sea-ambrosia, has been found in three places (17B, Cardiff Docks, JS *c.*1876*; 17D, Penarth, JS *c.*1876*; 78B, Port Talbot, HJR *c.*1907). **Ambrosia trifida** L., Great Ragweed, has been found three times, all in Cardiff Docks (17B, HJR 1910, RLS *c.*1924, RM 1929) and **Ambrosia coronopifolia** Torrey & A. Gray twice (16A, Barry Docks, RLS 1926*; 27A, Splott, GCD 1926). **Xanthium spinosum** L., Spiny Cocklebur, has occurred several times as a dock alien (16A, Barry, RLS 1926; 17B, Canton, RLS *c.*1925; 17D, Penarth, JS *c.*1876, Leckwith, EV 1924*; 27A, Cardiff, JS *c.*1876*; 78B, Port Talbot, HJR 1904). **Xanthium strumarium** L. subsp. **strumarium**, Rough Cocklebur, Cacamwci Lleiaf, has been found several times (16A, Barry, RLS 1925!, EV & RLS 1930; 17B, Cardiff Docks, JS *c.*1876, RLS 1924*, 1938; 27A, Cardiff Docks, EV 1922; 69D, Tir John tip, JAW 1939*) and **Xanthium strumarium** L. subsp. **italicum** (Moretti) D. Löve, Italian Cocklebur, once

(17B Cardiff Docks, GCD 1923). **Guizotia abyssinica** (L. fil.) Cass., Niger, Olewlys, has been recorded several times in the Cardiff area in the past as a grain alien (17B, JS 1876*, HJR 1913, EV 1922, RLS 1937!*; 18D, W. Evans 1940*; 27A, 1927!*, RLS 1931!) and once at Aberdare (00C, HJR 1904). **Iva xanthifolia** Nutt., Prairie Ragweed, has been recorded twice (08D, Efail Isaf, R.J. Wilks 1935*; 17A, Ely, RLS 1947*), and **Coreopsis tinctoria** Nutt., Tickseed, once (27A, Splott, RLS 1922*), as has **Madia capitata** Nutt., Tarweed, (17B, Cardiff, JS c.1876*).

ANTHEMIS L.

Anthemis arvensis L.   Corn Chamomile. Camri'r Ŷd.

Introduced, but long established as a cornfield weed, and also naturalized on Worm's Head (Kay, 1971b); also occurs as a casual near docks. Rare, and recently recorded as an established plant only in western Gower: 17D, Cardiff, MG 1974; 19B, Hengoed (garden weed), 1974!*; 38B, Inner Head, Worm's Head, 1969!**, 1980!; 49C, field weed at Llangenydd, 1971!, Cheriton, 1977!. Map 726.

Anthemis cotula L.   Stinking Chamomile. Camri'r Cŵn.

Introduced, but long established as a cornfield weed. Cornfields and roadsides, especially on heavy basic soils. Scarce in Gower but locally common in the Vale. Map 727.

Anthemis tinctoria L.   Yellow Chamomile. Camri Melyn

Casual or introduced. Sometimes found as a garden escape. It was recorded in the past as a dock alien at Cardiff and Barry. Recently recorded only from 69B, gravelly bank of Afon Tawe near Ynysforgan, 1977–79!. Past records from: 16A, RM 1924*, EV 1925–1927; 17B, RLS 1925*; 17D, EV 1925, railway embankment by River Ely, 1956!*.

Anthemis tomentosa L. was found as a casual on waste ground in Cardiff (17B) by G.C. Druce in 1923.

Santolina chamaecyparissus L., Lavender-cotton, was found as a casual at Mayals (69C) by J.A. Webb in 1946*.

ACHILLEA L.

Achillea ptarmica L.   Sneezewort. Ystrewlys.
Native. Marshes and damp grassland. Common in suitable sites in Gower and the western Uplands, but rather local in eastern Glamorgan and scarce or absent in the southern Vale. Map 728.

Achillea millefolium L.   Yarrow. Milddail.
Native. Grassland on relatively well-drained soils, especially on roadsides. Very common throughout the county, and recorded from all squares except 49B.

Fl. 6–11, fr. 8–11; (Syr) E.ten, S.balt, (Dip) Call, Luc.caesar. Map 729.

Achillea nobilis L., Graceful Yarrow, (17B, Cardiff, RLS & RM 1923; 78B, Port Talbot Docks, HJR c.1907) and **A. ligustica** All. (16A, Barry Docks, RM 1924; 17B, Cardiff Docks, 1924!*; 18C, Radyr, RLS 1920) have been found as dock aliens or garden escapes (18C), and **A. cretica** L. (17D, Penarth, JS c.1876) as a casual of waste places.

CHAMAEMELUM Miller

Chamaemelum nobile (L.) All.   Chamomile. Camri.
Native. Short grazed turf and roadside banks on acid soils. Recently recorded only from western Gower where it is locally frequent on village greens (48B, Reynoldston; 49C, Coety Green; 49D, Burry Green) and also grows on some grassy commons. Map 730.

Chamaemelum mixtum (L.) All. was found as a dock alien at Barry by R.L. Smith (16A, 1927) and at Cardiff by J. Storrie (17B, c.1876*).

TRIPLEUROSPERMUM Schultz Bip.

Tripleurospermum maritimum (L.) Koch *s.s.* (*Matricaria maritima* L.)   Sea Mayweed. Ffenigl Arfor, Ffenigl y Môr.
Native. Drift-lines on shingle beaches, in the spray-zone on sea-cliffs, near seabird nesting colonies, on toxic slag-tips and locally on railway ballast, road-sides and waste ground inland. Common in the Lower Swansea Valley (69B, D), where Sea Campion, **Silene vulgaris** subsp. **maritima,** also grows on toxic slag-tips. **Tripleurospermum maritimum** hybridizes fairly freely with **T. inodorum** and some roadside and waste ground populations appear to be of hybrid origin. (!) Fl. 5–11, fr. 7–11; Sb, Syr, Dip. Map 731.

Tripleurospermum inodorum Schultz Bip. (*Matricaria perforata* Mérat) Scentless Mayweed. Ffenigl y Cŵn.
Probably introduced, but long established as a weed of farmyards and cornfields. Cultivated land, farmyards, roadsides and waste ground. Locally abundant as a cornfield weed and on roadsides in the lowlands, but less frequent in the Uplands; often associated with disturbed ground around road construction sites. (!) Fl. 6–11, fr. 7–12; Syr, Dip. Map 732.

MATRICARIA L.

Matricaria recutita L. (*Chamomilla recutita* (L.) Raus-chert)   Scented Mayweed. Amranwen.
Probably introduced, but long established as a weed of cornfields. Cultivated land, roadsides and waste ground, usually on acid soils. Scarce or very local in most of the country; common only in parts of the south-eastern Vale where it is locally abundant in arable land and on disturbed roadsides. (!) Fl. 5–10, fr. 7–11; Syr, Dip. Map 733.

Matricaria matricarioides (Less.) Porter (*Chamomilla suaveolens* (Pursh) Rydb.). Pineappleweed. Chwyn Afal Pinwydd.
Introduced in the late nineteenth century. Roadsides, paths, cultivated land and waste ground. Not recorded in Storrie (1886), but very frequent in Cardiff by 1907, when it was present in all regions of Glamorgan except Gower and the Llwchwr valley (Riddelsdell, 1907). Now common in suitable habitats throughout the county. Fl. (3) 5–11, fr. 6–12. Map 734.

Three species of **Anacyclus** have been found as dock aliens in Cardiff and Barry. **Anacyclus clavatus** (Desf.) Pers. has been found in three localities (16A, Barry, RLS c.1924!; 17B, Cardiff, JS c.1876, HJR 1903*, 1907; 27A, Splott, 1925!*), **A. radiatus** Loisel. also in three (16A, Barry, RLS 1923*; 17B, Cardiff, HJR 1903, EV 1922*; 27A, Splott, 1924!*) and **A. homogamos ×radiatus** (**A. × valentinus** L.) in two (17B, Cardiff, RLS 1924; 27A, Splott, 1926!*).

CHRYSANTHEMUM L.

Chrysanthemum segetum L. Corn Marigold. Melyn yr Ŷd, Gold yr Ŷd.
Introduced, but long established as a weed of cornfields, cultivated land and disturbed roadsides, mainly on acid soils. Still locally common in western Gower, but now very scarce elsewhere. Map 735.

Chrysanthemum coronarium L., Crown Marigold, has been found in three places in the eastern Vale (07B, Peterstone, RM 1923*; 17B, Cardiff, JS 1876*, GCD 1922; 27A, Splott, 1926!*, 1927!*) and once at Aberdare (00C, HJR c.1905) as a dock alien or garden escape.

TANACETUM L.

Tanacetum vulgare L. (*Chrysanthemum vulgare* (L.) Bernh.). Tansy. Tanclys.
Introduced. Roadsides, waste ground and occasionally in rough grassland. Widespread and locally frequent, especially in the eastern Vale; once cultivated as a medicinal herb. Map 736.

Tanacetum parthenium (L.) Schultz Bip. (*Chrysanthemum parthenium* (L.) Bernh.). Feverfew. Wermod Wen.
Introduced. Waste ground, old walls, disturbed roadsides and cultivated land. Fairly common in most urban areas; commonly cultivated as a medicinal herb in the past, and still grown to some extent for the same purposes, but more often as an ornamental plant in a double-flowered form that is rather less frequent than the single-flowered form as an escape. Map 737.

Coleostephus myconis (L.) Reichenb. fil. (*Chrysanthemum myconis* L.) was found as a dock alien in Cardiff in 1926 (27A*) and **Balsamita major** Desf.,

Costmary, in Barry (16A) in 1925 and Canton (17B) in 1922 all by R.L. Smith.

LEUCANTHEMUM Miller

Leucanthemum vulgare Lam. (*Chrysanthemum leucanthemum* L.) Oxeye Daisy. Llygad Llo Mawr.
Native. Limestone grassland and cliffs, old meadows, roadsides and waste ground. Widely distributed and locally common in the lowlands and lower hills but scarce or absent in parts of the Uplands. Native populations on limestone cliffland in Gower, and most old meadow populations, have been found to be diploid, with a chromosome number of 2n=18, but most roadside and waste ground populations that have been investigated in western Glamorgan are tetraploid, with 2n=36, although a few are diploid (Atherton, 1975). Fl. 5–9 (–11), fr. 7–10; Hb (n), B.lap (n), B.prat (n), (Syr) E.per, E.ten (n), E.sp, H.pen, S.balt, S.rib, Sy.pip, (Dip) Luc, Sc.ste. Map 738.

Leucanthemum maximum (Ramond) DC. (*Chrysanthemum maximum* Ramond). Shasta Daisy. Llygad Ych Mawr.
Introduced. A garden escape that is occasionally naturalized on roadsides and waste ground. Map 739.

Leucanthemum lacustre (Brot.) Samp. (*Chrysanthemum lacustre* Brot.) was recorded from Langland (68A) by J.A. Webb in 1920.

ARTEMISIA L.

Artemisia vulgaris L. Mugwort. Beidiog Lwyd.
Native. Roadsides, hedges, field banks, riversides, cultivated land and waste ground. Common in Gower and most of the Vale, and frequent in some Upland valleys, but scarce or absent in parts of the Uplands and central Vale. Fl. 8–9, fr. 9–10. Map 740.

Artemisia absinthium L. Wormwood. Wermod Lwyd.
Introduced. Roadsides, waste ground, sand-dunes and pastures on shallow soils. Locally common on waste ground in urban areas; spreading in overgrazed grassland in north-western Gower. Map 741.

Artemisia maritima L. Sea Wormwood. Wermod y Môr.
Native. In the higher parts of salt-marshes, and occasionally in the spray-zone on sea-cliffs. Fairly abundant in suitable habitats in the northern Gower salt-marshes and in the estuary of the Afon Nedd (R. Neath). Map 742.

Artemisia campestris L. Field Wormwood. Llysiau'r Corff, Wermod y Maes.
Introduced. Waste ground and sand-dunes. Locally established and spreading at Crymlyn Burrows (79C), but recorded only once, as a ballast alien, elsewhere in the county. (The 1902 Rhondda record (Harris, 1905) is now thought to be an error).

27A, East Moors, THT, c.1876; 78C, Crymlyn Burrows, 1956!*, DMcC 1973, H & CH 1977!**.

Four other species of **Artemisia** have been found as casuals on waste ground, usually near docks: **Artemisa pontica** L., Roman Wormwood, Wermod Dramor, (69D, Swansea, M.H. Sykes 1949*), **A. biennis** Willd. (17B, Cardiff, RLS 1924*, 1925!*, Maindy, RLS 1938*; 18C, Radyr, HJR 1913*; 18D, Llanishen, W. Evans 1946*; 27A, Penylan to Llanedeyrne, 1927!*), **A. ludoviciana** Nutt., Whitesage, (17B, Cardiff Docks, RLS 1924*, 1926!*; 18C, Radyr, RLS 1922*), and **A. annua** L., Annual Wormwood, (27A, Splott, E.S. Todd 1938).

TUSSILAGO L.

Tussilago farfara L. Colt's-foot. Carn yr Ebol, Alan.
Native. Roadsides, streamsides, dunes, unstable slopes, cultivated land and waste ground. Common in all parts of the country. Map 743.

PETASITES Miller

Petasites hybridus (L.) P. Gaertner, B. Meyer & Scherb. Butterbur. Alan Mawr.
Native. Streamsides and marshes. Locally frequent in the east but scarce in the western part of the country. Fl. 4. Map 744.

Petasites fragrans (Vill.) C. Presl Winter Heliotrope. Alan Mis Bach.
Introduced as a garden escape. Roadsides, hedges and parks. Most frequent in suburban areas in the lowlands; increasing. Fl. 12–2. Map 745.

Petasites japonicus (Siebold & Zucc.) Maxim., Giant Butterbur, Alan Gawr, has been found as an established garden escape especially on the banks of the Rivers Ely (07B) and Taff (17B) in several localities (07B, Peterstone, A.R. Hare 1970, between Pont Tal-y-bont and Pont Sarn, I. Rees 1974*, St. Georges, D.J. Thomas 1983; 17B, Llandaff North, B.R. Savage 1952*, GSW 1983; 88A, near the Pond behind Margam Abbey, D.M. Cape 1966*). **P. albus** (L.) Gaertner, White Butterbur, Alan Bach, has been recorded once, from Stouthall Wood (48B) by J. Berney in 1982.

DORONICUM L.

Doronicum pardalianches L. Leopard's-bane. Llysiau y Llewpard, Llewpard-dag. Introduced; a garden escape on roadsides.
08D, Miskin, MG 1969; 18D, foot of Wenallt hill north of Whitchurch, 1921!*; 97A, Castle-upon-Alun, ARP 1984.

Doronicum plantagineum L. has been recorded once as naturalized in a wood at Merthyr Mawr or Dunraven (87) by E.M. Thomas in 1932*, 1933* (the locality is given as above on the specimen in **NMW**).

D. orientale Hoffm. has also been recorded once (18D, Cyncoed, 1935!*).

SENECIO L.

Senecio jacobaea L. Common Ragwort. Creulys Iago.
Native. Pastures, cliff grassland, sand-dunes, roadsides and waste ground. Common in most of the county but rather local in the higher Uplands. Fl. 6–8 (–12), fr. 8–10; Hb, B.luc, B.terr, Sb, (Syr) E.ten, S.rib, (Dip) Luc, (Lep) Ag.urt, Ly.phl, Ma,jur, Pi.rap, Zygaena filipendulae. Map 746.

Senecio aquaticus Hill Marsh Ragwort. Creulys y Gors.
Native. Marshes, damp grassland and ditches. Common in western Glamorgan but increasingly local towards the east. Map 747.

Senecio erucifolius L. Hoary Ragwort. Creulys Llwyd.
Native. Roadsides and grassland on basic soils, and occasionally on sand-dunes. Fairly common in parts of the southern Vale but absent from most of Gower and the Uplands; a few plants grow in dune grassland at Whiteford Burrows in Gower (49C), perhaps originating from Pembrey Burrows in Carmarthenshire, about 5km to the north-west, where **S. erucifolius** is locally common in dune-slacks. Map 748.

Senecio squalidus L. Oxford Ragwort. Creulys Rhydychen.
Introduced. Waste ground, old walls and ruins, roadsides and sand-dunes. Very abundant in the centres of Swansea, Neath and Cardiff, and locally common in urban areas in the Swansea and Cardiff districts within a few miles of the sea, but still scarce in the Uplands and in the countryside of the Vale and western Gower, although it is steadily spreading. It is abundant on sand-dunes at Crymlyn Burrows (79C) and will probably spread and increase on other dune-systems; it was seen on Whiteford Burrows for the first time in 1977 (49B, Berges Island!) and on Nicholaston Burrows for the first time in 1979 (58A!). In urban areas it is often introduced with builder's rubble (hard-core). It has recently spread along the verges and central reservations of the new motorways and similar roads, as it has done in other parts of Britain. It apparently arrived in Cardiff in the late nineteenth century and was abundant in the central area around the railway stations by c.1909 (Trow, 1911) and was also present at Port Talbot Docks by 1907 (Riddelsdell, 1907), but was still unrecorded in Swansea at that time. Fl. (4–) 6–10, fr. 6–10; B.lap (n), B.terr (n), (Syr) E.ten, Sy.pip. Map 749.

Senecio sylvaticus L. Heath Groundsel. Creulys y Rhosydd, Creulys y Coed.
Native. Dry heathland and open grassland on acid soils. Very local; most frequent in acid cliff-heath in

Gower and on acid dunes in Gower and the Vale. Map 750.

Senecio viscosus L. Sticky Groundsel. Creulys Gludiog.
Native or introduced. Roadsides, railway ballast, colliery tips and waste ground, sometimes on river-banks, and on shingle beaches and in other open maritime habitats. Most abundant in the Uplands; scarce or absent in most of Gower and the Vale, and apparently absent from most of the areas where **Senecio squalidus** is abundant. In the places where the parental species do form mixed populations (for example on the banks of the Afon Tawe south of Ynystawe, 69B), **Senecio squalidus × S. viscosus (S. × londinensis** Lousley) is frequent. Map 751.

Senecio vulgaris L. Groundsel. Creulys Cyffredin.
Native. Very common as a weed of cultivated land and waste ground; also in natural habitats on shallow soils on sea-cliffs, on sand-dunes, on the drift-line and in seabird nesting colonies. **Senecio vulgaris** forma **radiatus** Hegi is frequent in the Cardiff area and locally frequent in the Vale; records of this variety from other areas (10A, Pontlottyn; 68A, Mumbles) are few. Fl. 1–12, fr. 2–12; ?Sb (p). Map 752.

Senecio squalidus × S. vulgaris (S. × baxteri Druce) may be expected to occur wherever the two parents meet; it has been recorded from 16A, Barry, GCD 1927; 17B, Cardiff, HJR c.1906, AHT 1910, 1922!*; 27A, E. Cardiff, 1975!*; 59B, near Gowerton, 1974!; 69C, near Swansea, H & CH 1974.

Senecio cineraria DC., Silver Ragwort, Llys y Lludw, occurs as a garden escape and is more or less naturalized in a few localities (16A, Barry Docks, 1980!*; 17D, Leckwith, RLS 1922*, Dinas Powis and Llandough, JPC 1983; 68A, Langland Bay, J.R. Shepherd 1913; 78A, Aberafon, H & CH 1975, 1981!; 88B, Cwm Felin, ABP 1975!).

Senecio cineraria × S. jacobea (S. × albescens Burbidge & Colgan) also appears to be naturalized in a few places (18D, N. Cardiff, JPC 1970*; 48B, Horton, JAW c.1921; 61C, Garnswllt, 1981!*). **S. leucanthemifolius** Poiret (*S. crassifolius* Willd.) has once been recorded as a dock alien at Cardiff (17B, HJR c.1906). **S. congestus** (R. Br.) DC. (*S. palustris* (L.) Hooker), Marsh Fleawort, was recorded by John Ray from Aberafan Marshes (79C) in 1662 but has not been confirmed. Specimens of **S. nebrodensis** L. collected by H.J. Riddelsdell from Cardiff (17B) in 1908 have proved to be **S. vulgaris** forma **radiatus**.

CALENDULA L.

Calendula officinalis L., Pot Marigold, Melyn Mair, is often found on roadsides and waste ground as a garden escape but rarely maintains itself for more than a year or two. Map 753.

Calendula arvensis L., Field Marigold, Melyn Mair yr Âr, has been recorded three times in the Cardiff area (06B, Rhoose, MG 1975; 17B, Cardiff Docks, JS c.1876*; 27A, Splott, 1926!*).

Arctotheca calendula (L.) Levyns, Cape-dandelion, has been found three times as a dock alien (16A, Barry, C.I. Sandwith 1927*; 17B, Cardiff, JS c.1876*, RLS 1923).

CARLINA L.

Carlina vulgaris L. Carline Thistle. Ysgallen Siarl.
Native. Limestone grassland and calcareous sand-dunes, and also on colliery spoil-tips. Fairly common in cliff grassland and on dunes near the coast, and locally frequent on non-calcareous spoil-tips in the central and eastern Uplands. Map 754.

Echinops sphaerocephalus L., Globe Thistle, Ysgallen Bengrwn, has been found as a garden escape at Leckwith (17D, JPC 1978) and Dinas Powis (17D, JPC 1982).

ARCTIUM L.

Arctium lappa L. Greater Burdock. Cyngaf Mawr.
Native. Rough grassland, near paths, roadsides and hedgebanks. Fairly frequent in the southern and eastern Vale, and also found in a few places in the eastern Uplands. There is only one recent record of **A. lappa** from Gower (49C, Llanmadoc, 1981!**), which is at the edge of the British range of the species. Trow (1911) and Riddelsdell (1907) both recorded it from several sites in Gower, but it is possible that some of these records resulted from confusion with **A. minus**. Map 755.

Arctium minus Bernh. *s.l.* Lesser Burdock. Cyngaf Bychan.
Native. Pathsides, open woodland and scrub, rough grassland, hedgebanks, roadsides and waste ground. Frequent in the lowlands, widespread but scarce in the Uplands. **Arctium minus** *s.l.* (Map 756) has been divided into three ill-defined species, **A. minus** *s.s.*, **A. pubens** Bab., and **A. nemorosum** Lej. **A. minus** *s.s.* appears to be by far the commonest segregate species in Glamorgan and has been reported recently from 23 of the 24 recording squares in which **A. minus** *s.l.* has been critically examined. Map 757. There are recent field records of **A. nemorosum** from only two squares (09B, AMP 1972; 96A, 1971!). Map 758. **A. pubens** does not appear to occur in the county. Fl. 6–9, fr. 8–11; Hb (n,p) B.luc, B.pasc, B.prat (all n), (Lep) Pi.bra.

CARDUUS L.

Carduus nutans L. Musk Thistle. Ysgallen Og-wydd.
Native. Calcareous grassland, colliery spoil-tips and roadsides. Frequent in calcareous grassland near the sea in the Vale, but rather scarce in similar habitats in

Gower; also on non-calcareous colliery tips and roadsides in parts of the Uplands. Fl. 6–8; Psithyrus campestris. Map 759.

Carduus acanthoides L.   Welted Thistle. Ysgallen Grych.
Native. Grassland, roadsides and waste ground. Most frequent in calcareous grassland in the southern Vale; scarce in Gower and absent from the central and western Uplands. Map 760.

Carduus acanthoides × C. nutans (C. × orthocephalus Wallr.) is occasionally found with the parents as at Bishopston Valley (58B, JAW 1943*).

Carduus tenuiflorus Curtis   Slender Thistle. Ysgallen Flodfain.
Native. Grassland on sea-cliffs, especially on headlands where sheep and cattle gather; occasionally on sand-dunes, roadsides and waste ground near the sea. Locally common in suitable habitats in Gower and on parts of the Vale coast. White-flowered plants are fairly abundant in some Gower cliff populations. Map 761.

Carduus pycnocephalus L. has once been recorded as a casual (17B, Cardiff, GCD 1931).

CIRSIUM Miller

Cirsium eriophorum (L.) Scop.   Woolly Thistle. Ysgallen Benwlanog.
Native. Limestone grassland, open scrub, and roadsides. Locally frequent in the southern Vale, usually near the sea; rare and sporadic in Gower (48A, Rhosili, 1973; 58B, Seven Slades, single plant in cliff scrub in 1969!). Map 762.

Cirsium vulgare (Savi) Ten.   Spear Thistle. Marchysgallen.
Native. Pastures, roadsides, cultivated land and waste ground. Common throughout the county; recorded from every square except 19D. Fl. 7–9, fr. 8–10; B.hort (n,p), B.lap, B.luc, B.pasc, B.prat, B.terr, Psithyrus campestris. Map 763.

Cirsium dissectum (L.) Hill   Meadow Thistle. Ysgallen Gorswaun, Ysgallen Mignwern.
Native. Marshes and bogs, on peaty soil, occasionally in marshy dune-slacks. Locally common, but often absent from apparently suitable habitats; most frequent on marshy moorland in the lowlands and lower hills. (!) Fl. 6–7; B.spp, (Lep) Ca.rub, Ly.phl, Oc.ven. Map 764.

Cirsium tuberosum (L.) All.   Tuberous Thistle. Ysgallen Oddfynog.
Native. Limestone grassland and limestone sea-cliffs. Rare, and confined to the limestone coast between Southerndown and St. Donat's in the south-western Vale; most frequent near Nash Point where it grows both in the cliff-top grassland and around seepages on the upper cliff-face, but in small numbers. **Cirsium**

tuberosum × C. acaule (C. × zizianum Koch) has been found here (96A, MG 1976; 97C, J.L. Bruce 1935, EV, 1938*, M.C. Knowles 1954) and **C. tuberosum × C. palustre (C. × semidecurrens** Richter) has also been recorded (96A, J.L. Bruce, 1932). The first record of **C. tuberosum** in this locality was made by T. Westcombe in 1843 'between St. Donat's and Dunraven' (Westcombe, 1844). It has once been found on waste ground at Barry Docks (16A, 1922!*) and at Cardiff Docks (17B, HJR 1909). Fl. 6–8, fr. 7–9; Hb (n), B.hort (n), B.lap, B.pasc, B.prat. B.terr, (Syr) E.ten, S.rib, Syrs.
87D, Southerndown, 1920*, 1968!; between Dunraven Castle and Cwm Mawr, MG 1975; 96A, slopes below multivallate fort on Nash Point, HJR 1908*, 1968!, 1980!; cliffs and cliff-top grassland near lighthouses, 1976!, 1980!; 97C, Cwm Nash, 1972!.

Cirsium acaule Scop.   Dwarf Thistle, Stemless Thistle. Ysgallen Ddigoes.
Native. Limestone grassland. Very local, but abundant in some localities; it is most frequent in cliff-top grassland and on south-facing slopes near the coast of the Vale, but it also grows on or near the limestone of Morlais Castle Hill in the northern Uplands. It is abundant on south-facing slopes in Pant St. Bride's (87B), at the western limit of its range in Glamorgan. The old records from Paviland (48A) (Anon., 1893), Briton Ferry (79C, JWGG c.1841) and the Rhondda (99A) (Harris, 1905) are almost certainly errors, perhaps due to misidentification of **Carlina vulgaris** L. Map 765.

Cirsium palustre (L.) Scop.   Marsh Thistle. Ysgallen y Gors.
Native. Marshes, damp grassland and moorland, roadsides and open woods. Common throughout the county. Many populations show flower-colour polymorphism, with white, pale and dark purple-flowered plants intermingled (Mogford, 1974). Fl. 6–10, fr. 8–11; Hb, B.lap, B.pasc, B.prat, B.terr, (Lep) Ma.jur, Zygaena sp., noctuid moths. Map 766.

Cirsium palustre × C. arvense (C. × celakovskianum Knaf fil.) has been recorded from Penylan, Cardiff (17B, EV 1922*).

Cirsium arvense (L.) Scop.   Creeping Thistle. Ysgallen Gyffredin.
Native. Grassland, roadsides, cultivated land and waste ground. Common throughout the county. Plants with white or near-white flowers are fairly frequent (recorded from 49C, 58A, 69D, 71C, 96A) and probably occur in small numbers in many large populations. **C. arvense** has been recorded from every square in the county. Fl. 7–10, fr. 8–11; Hb, B.lap, B.luc, B.pasc, B.terr, Psithyrus campestris, (Syr) E.ten, (Lep) Ag.urt, Cy.car, Pi.bra, Pi.rap. Map 767.

ONOPORDUM L.

Onopordum acanthium L.   Cotton Thistle. Ysgallen Gotymog.

Introduced. An alien or garden escape which establishes itself for a time but does not appear to persist for more than a few years. Map 768.

SILYBUM Adanson

Silybum marianum (L.) Gaertner Milk Thistle. Ysgallen Fair.
Introduced. Once fairly frequent as a dock alien or garden escape in the Vale, but has been recorded recently only from dunes between Horton and Port Eynon in Gower (49B, B.E. Jones 1972*). Map 769.

The thistle-like Mediterranean species **Galactites tomentosa** Moench was found growing on 'sand ballast' in Radyr by R.L. Smith in 1925 (18C*).

SERRATULA L.

Serratula tinctoria L. Saw-wort. Dant y Pysgodyn.
Native. Rough grassland, grassy heathland and moorland, usually on neutral or moderately acid soils. Abundant in some localities, but absent from many areas of the county; most frequent in cliff-heath in western Gower, on the inland commons of Gower, and on the heathy moorland of the Border Ridges between the Vale and the Uplands. (!) Fl. 7–9, fr. 8–10; B.lap, B.luc, B.pasc, B.terr, (Sb) Halictus spp, (Syr) E.arb, E.ten, S.vit, (Dip) Eriothrix rufomaculatus, Sa.car, (Lep) Ag.urt, Cy.car, In.io, Ma.jur, Pi.bra, Pi.rap, Po.ica, Py.tit. Map 770.

CENTAUREA L.

Centaurea scabiosa L. Greater Knapweed. Pengaled Mawr.
Native. Limestone grassland and cliffs, field banks and roadsides on calcareous soils, calcareous sand-dunes, and occasionally on railway embankments and smelter slag. Common in limestone cliffland in southern Gower and in limestone grassland near the coast of the Vale, and locally frequent in suitable habitats elsewhere in Gower and the Vale. Very scarce in the Uplands, where it occurs mainly on railway embankments. Plants growing on exposed limestone sea-cliffs in southern Gower have short stems and fleshy leaves with few or no basal lobes; they have been described as var. **succisiifolia** E.S. Marshall, and may be a genetically distinct relict population that, like **Helianthemum canum** and **Aster linosyris**, has persisted in open limestone-cliff habitats in Gower since the Late-glacial period (Valentine, 1980). Fl. 6–11, fr. 7–10; Hb (n,p), B.hort, B.lap, B.luc, B.pasc (n,p), B.prat (n,p), B.terr, Psithyrus campestris, (Syr) E.ten, Rh.cam, S.balt, S.rib, (Lep) In.io, Pi.bra, Py.tit. Map 771.

Centaurea nigra L. Common Knapweed. Pengaled.
Native. Common in grassland. Two ill-defined subspecies, sometimes regarded as species, occur in the county. **C. nigra** L. subsp. **nigra** grows in old meadows and damp grassland, and on roadsides and railway embankments on neutral or acid soils. It is common in most of the county but scarce or absent in the central uplands and partly replaced by subsp. **nemoralis** on the limestone soils of the southern Vale. **C. nigra** L. subsp. **nemoralis** (Jordan) Gremli (*C. debeauxii* Gren. & Godron subsp. *nemoralis* (Jordan) Dostal), is found in limestone grassland and sand-dune grassland. It is fairly common in the Vale and may occur in suitable habitats in southern and western Gower, but is absent from most of the Uplands and from acid soils on the lowlands; it is rather variable and not clearly distinguishable from subsp. **nigra**. Fl. 6–12, fr. 7–10; Hb (n.p), B.hort, B.lap, B.luc, B.pasc (all n,p), B.prat (n), B.terr (n), Psithyrus campestris, (Sb) Anthidium manicatum, (Syr) E.ten, Rh.cam, S.balt, S.rib, Sc.pyr, (Lep) Cy.car, Go.rha, In.io, Pa.aeg, Pi.bra, Pi.rap. Map 772.

Centaurea montana L. Perennial Cornflower. Penlas Fythol.
Introduced. Occurring on waste ground and roadsides as a garden escape, and in churchyards. 09A, Mountain Ash, AMP 1973*; 16A, Barry 1974!; 28C, Lisvane, 1972!; 69A, Llangyfelach, JAW 1941*; 70C, Pontardawe 1974!; 79C, Neath, c.1974!; 90D, Aberdare, HJR c.1907; 97D, Llyswyrny 1974!.

Centaurea cyanus L. Cornflower. Penlas yr Ŷd.
Introduced. Formerly a scarce weed of cornfields and a fairly frequent dock alien, but now found only as an occasional garden escape on waste ground. Map 773.

Many species of **Centaurea** have been found as casuals or dockside aliens on one or more occasions in the past: **Centaurea melitensis** L., Maltese Star-thistle, Ysgallen Seraidd Felitaidd, most frequently, in seven squares (00C, Aberdare, HJR 1904; 17B, Cardiff Docks, RLS 1925; 17D, Grangetown, RLS 1926*; 27A, Splott, JS 1876*, RLS 1937!*; 48A, Paviland HJR 1906; 48B, Horton, PS 1944*; 78B, Port Talbot Docks, HJR 1904, EMT 1937*), **C. calcitrapa** L., Red Star-Thistle, Ysgallen Seraidd, in five squares (16A, Barry, RLS 1921*; 17B, Cardiff Docks, HJR c.1907; 17D, Cardiff, T. Clark 1859*, Penarth and Leckwith, JS c.1876; 27A, Splott, RLS 1937!*; 78B, Port Talbot, HJR c.1906), **C. solstitialis** L. subsp. **solstitialis**, Yellow Star-thistle, Ysgallen Seraidd Felen, in six squares (17B, Cardiff, ALe 1978*; 17D, Penarth Ferry, AHT 1891*; 27A, East Moors, JS 1876*, Splott, EV 1923*, 1938; 48B, Port Eynon, JAW 1918; 78B, Port Talbot, JAW 1918; 87A, Newton Nottage, EV 1909*), and **C. solstitialis** subsp. **adamii** (Willd.) Nyman in one only (00C, Aberdare, HJR 1902). **C. aspera** L., Rough Star-thistle, Ysgallen Seraidd Arw, has been found in three squares (16A, Barry Docks, RLS 1921*, DMcC 1974; 17B, Cardiff Docks, JS c.1876; 78B, Port Talbot, R.L. Burges 1933, G. Randall 1987*), **C. jacea** L. Brown Knapweed, Pengaled Llwytgoch, in two (78B, Port Talbot, JAW 1911; 69A, Swansea, A. Lake 1959), and the following species or hybrids in one only **C. jacea × C. nigra** (**C. × drucei** C.E. Britton) (16A, Barry, 1922!*), **C. algeriensis** Cosson & Durieu (27A, Splott, RLS 1926, 1927*), **C. bruguierana** (DC.) Hand.-Mazz.

subsp. **belangerana** (DC.) Bornm. (*C. phyllocephala* Boiss, *Tetramorphaea belangerana* DC.) (27A, Splott, E.S. Todd 1935*), **C. diffusa** Lam. (27A, Splott, EV 1933*), **C. diluta** Aiton (78B, Port Talbot Docks, HJR 1904), **C. iberica** Trev. ex Sprengel (16A, Barry Docks, C.I. Sandwith 1933*), **C. leucophaea** Jordan (27A, Splott, RLS 1939!*), **C. nicaeensis** All. (27A, Splott, RLS 1926*), and **C. orientalis** L. (17B, Cardiff Docks, RLS 1937*).

Mantisalca salmantica (L.) Briq. & Cavillier (*Centaurea salmantica* L.) was found once at Splott (27A, RLS 1926*).

CARTHAMUS L.

Carthamus tinctorius L., Safflower, Cochlys, has been found as a casual on four occasions, probably growing from bird-seed (17B, Cardiff, EV 1938*; 17D, Cardiff, 1932!*, Dinas Powis, RLS 1923; 18D, Rhiwbina, A.M. Webb 1969*).

Carthamus lanatus L., Downy Safflower, Cochlys Gwlanog, has been found on five occasions as a dock or ballast alien (17B, Cardiff, JS *c.*1876*, EV 1922; 18C, Radyr, RLS 1921; 27A, Splott, RLS 1927*; 78B, Port Talbot, HJR 1905).

CICHORIUM L.

Cichorium intybus L. Chicory. Ysgellog.
Native or introduced. Limestone grassland, field banks and roadsides. Rather scarce in Gower and the Vale and rare in the Uplands; most frequent in southern Gower and the south-eastern Vale. Map 774.

Cichorium endivia L., Endive, Ysgallen y Meirch, has been found at two sites as a casual (18C, Radyr, RLS 1922*; 27A, Splott, RLS 1926*).

Tolpis barbata (L.) Gaertner occurred as a dock alien at Barry in 1927 (16A, RLS 1927*). **Rhagadiolus stellatus** (L.) Gaertner and **Hedypnois cretica** (L.) Dum.-Courset were both found as grain aliens at Splott in 1926 (27A, RLS 1926*). **Scolymus maculatus** L., Spotted Golden-thistle, was found as a dock alien at Splott (27A, RLS 1927), and **S. hispanicus** L., Golden-thistle, at four sites (16A, Barry Docks, RLS 1924*; 17B, Cardiff Docks, HJR 1903; 17D, Cardiff, JS *c.*1886; 27A, East Moors, JS *c.*1876*).

HYPOCHOERIS L.

Hypochoeris glabra L. Smooth Cat's-ear. Melynydd Moel.
Native. Sand-dune grassland. Very local; found only at Nicholaston Burrows, Kenfig Burrows and Porthcawl. First recorded by Riddelsdell before 1907, at Porthcawl.
58A, Nicholaston Burrows, 1984!**; 78D, Kenfig Burrows, EMT & EV 1944*, 1975!*; 87A, Porthcawl,

HJR *c.*1907, Porthcawl golf links, EV 1931–32*; 88C, Kenfig Burrows near Kenfig Castle, 1975!*.

Hypochoeris radicata L. Cat's-ear. Melynydd.
Native. Grassland and grassy heaths on all soils, sand-dunes, roadsides and hedgebanks. Common throughout the county. Fl. 5–7 (–12), fr. 6–11; Hb (n,p), B.lap, ?Sbs, (Syr) Criorrhina floccosa, E.ten, Rh.cam, S.balt, S.lun, S.rib, (Dip) Luc caesar, Musca sp, (Lep) Cupido minimus, Ma.jur, Pi.rap. Map 775.

LEONTODON L.

Leontodon autumnalis L. Autumn Hawkbit. Peradyl yr Hydref.
Native. Grassland on all soils, salt-marshes near the upper limit of spring tides, sand-dunes, roadsides, colliery spoil-tips and similar sites. Common throughout the county in a range of habitats, including colliery spoil-tips and opencast reclamation sites where it is often abundant. A variable species with a number of different habitat-forms. Fl. 7–10, fr. 8–11; (Syr) S.balt, S.rib, Sp.scr, Sy.pip, (Dip) Luc. Map 776.

Leontodon hispidus L. Rough Hawkbit. Peradyl Garw.
Native. Meadows, roadsides, sand-dunes and waste ground. Common in the Vale and parts of the Uplands but scarce or absent in western Gower and the central Uplands; most abundant in old haymeadows. Fl. 5–10, fr. 6–10; B.lap, (Syr) S.rib, ?Dips, (Lep) Ma.jur. Map 777.

Leontodon taraxacoides (Vill.) Mérat Lesser Hawkbit. Peradyl Lleiaf.
Native. Dry grassland, sand-dunes, field banks and roadsides, and in the spray-zone on sea-cliffs. Abundant in the lowlands of western Glamorgan but rather local in the Uplands and the central and eastern Vale. Map 778.

PICRIS L.

Picris echioides L. Bristly Oxtongue. Gwylaeth Chwerw.
Native or introduced. Rough grassland, roadsides, hedgebanks and waste ground, usually near the sea. Fairly common near the coast in the Vale but scarce elsewhere; absent from most of Gower but locally frequent at Horton and Port Eynon (48B). Map 779.

Picris hieracioides L. Hawkweed Oxtongue. Gwylaeth yr Hebog.
Native. Rough grassland, roadsides, hedgebanks and waste ground on calcareous soils. Locally common in Gower and the southern Vale but rare in the Uplands. Map 780.

TRAGOPOGON L.

Tragopogon pratensis L. subsp. **minor** (Miller) Wahlenb. Goat's-beard. Barf yr Afr Felen.

Native. Old meadows, sand-dunes, roadsides and waste ground. Fairly frequent in the lowlands and parts of the Uplands, but never abundant. Map 781.

Tragopogon porrifolius L., Salsify, Barf yr Afr Gochlas, has been found in several places in the past as a casual or garden escape but has been recorded recently only once (16A, Barry Island, EV 1898*, Barry Docks, EV *c*.1924; 17B, Cardiff, G. Martin 1921*; 18D, Heath, C. Taylor 1960*; 27A, Pengam, JS 1893, 1969!; 59B, Loughor, JM *c*.1840; 89D, Maesteg, JS *c*.1880), **T. pratensis** subsp. **pratensis** has been found twice as a dock alien (16A, Barry, RM 1924*; 17D, Cardiff, EV 1931*), **T. hybridus** L. has been found once (27A, Pengam, RLS 1940*) as has **T. crocifolius** L. (27A, Splott, RLS 1926).

SONCHUS L.

Sonchus asper (L.) Hill Prickly Sow-thistle. Llaethysgallen Arw.
Native. Cultivated land and waste ground; occasionally on sea-cliffs, drift-lines and sand-dunes. Common in the lowlands and fairly common in the Uplands. Fl. 5–11, fr. 6–10; Luc. Map 782.

Sonchus oleraceus L. Smooth Sow-thistle. Llaethysgallen.
Native. Cultivated land, waste ground, walls, sea-cliffs, sand-dunes, river banks and drift-lines. Common in suitable natural habitats as well as in man-made habitats in the lowlands, fairly common in the Uplands. Fl. 1, 3–12, fr. 5–12; Sb, (Syr) E.ten, H.pen, Rh.cam, S.balt, S.lun, S.rib, Sp.scr, Sy.pip, (Dip) Call, Luc, (Lep) Ag.urt, Ly.phl. Map 783.

Sonchus arvensis L. Perennial Sow-thistle. Llaethysgallen yr Ŷd.
Native. Roadsides, hedgebanks, cultivated land and waste ground; riversides, fens, shingle beaches, drift-lines in salt-marshes and dune-slacks. Fairly common in the lowlands but local in the Uplands. Fl. 7–10, fr. 8–11; Hb, B.lap, B.luc, B.pasc, B.prat (all n), ?Sb, (Syr) E.ten, S.balt, S.rib, (Lep) Po.c-alb. Map 784.

LACTUCA L.

Lactuca serriola L. Prickly Lettuce. Gwylaeth Bigog.
Native or introduced. Sandy shingle, roadsides and waste ground. Locally frequent in the south-eastern Vale, with one old record from Porthcawl (87A, R.E. Cundall 1897). Map 785.

Lactuca virosa L. Great Lettuce. Gwylaeth Chwerwaidd.
Introduced. Grassy slopes, roadsides and waste ground near the sea, usually impersistent. Recently recorded only from Roath Park, Cardiff (17B, P.H. Dunn 1976!) and eastern Cardiff (27A, ABP 1979!); past records from Oystermouth (68A), Swansea (69D), Port Talbot (78B), Barry (16A) and Cardiff Docks (17D).

Lactuca saligna L., Least Lettuce, Gwylaeth Leiaf, has been found in the past as a dock alien (16A, Barry, EV 1910*; 17B, Cardiff Docks, RLS 1924*; 17D, Penarth Dock, AHT 1899*, Llandough, EV 1917–19*), and **L. sativa** L., Garden Lettuce, Gwylaeth, has once been found as a garden escape or bird-seed casual on waste ground (17C, Dinas Powis, JPC 1978).

CICERBITA Wallr.

Cicerbita macrophylla (Willd.) Wallr. Common Blue-sow-thistle. Llaethysgallen Lelog.
Introduced. Roadsides, streamsides, hedge and scrub. Established in several places and probably increasing.
07D, St. Nicholas, MSP, 1970; 09B, Mountain Ash, MG 1986; 16B, near Sully, E. Jones 1982*; 17B, Leckwith Moors, JPC 1978; 58B, Brandy Cove, 1970!; 68A, Mumbles, V.M. Peel 1953; 69B, Morriston, 1974!; 69C, Ty-coch, JAW 1951; 79C, Briton Ferry, H & CH 1972. Fl. 6–8, no good seeds found; Hb, B.luc, B.prat, B.terr, (Sb) Anthidium manicatum, ?Sb, (Syr) S.balt, S.rib, (Dip) Tabanus sp.

MYCELIS Cass.

Mycelis muralis (L.) Dumort. Wall Lettuce. Gwylaeth y Fagwyr.
Native. Limestone woodland and shaded limestone rocks, and on walls. Locally common in woods in Gower and the Vale and on limestone around Morlais Castle; frequent on walls in the eastern part of the county. Rather scarce elsewhere. Fl. 6–11, fr. 7–10; (Syr) S.balt. Map 786.

TARAXACUM Wiggers

Dandelion. Dant y Llew

The **Taraxacum** flora of Glamorgan is relatively well-known as a result of field-work in the county by A.J. Richards and others. Forty-four of the 244 species currently recognised as occurring in the British Isles (Haworth & Rundle, 1986; Haworth, 1988) have been recorded from the county. The genus as a whole has been recorded from almost every square but the distribution of most species within the county is only partly known, because only a few areas have been thoroughly worked. More sites for existing species and further species will undoubtedly be found in the future. The identity of all the records given here has been determined or confirmed by A.J. Richards and/or Chris Haworth and the authors wish to record their grateful thanks to them both. Chris Haworth has also passed his critical eye over the account which has been much improved as a result. The sequence of Sections and species follows Haworth (1988) with amendments.
Fl. all year, (1–3) 4–5 (6–12), fr. 4–12; Hb (n,p), B.lap, B.luc, B.pasc, B.terr, (Sb) Halictus sp., ?Sbs, (Syr) Chrysotoxum festivum, E.ten, H.pen, Leu.luc, Rh.cam, S.balt, S.rib, (Dip) Luc, Sc.ste (Lep) Ag.urt, In.io, Ly.phl, Ma.jur, Pa.aeg, Pi.bra, Pi.nap, Pi.rap.

Taraxacum Section **Erythrosperma** (H. Lindberg fil.) Dahlstedt Lesser Dandelion. Dant y Llew Lleiaf. Map 787.

Taraxacum brachyglossum (Dahlstedt) Raunkiaer
Native. Limestone grassland, sand-dunes and colliery spoil-tips; one of the commonest members of this Section.
06A, Aberthaw, EV 1930*; 10D, Brithdir, 1978!*; 16A, Barry Island, HJR 1912; 77B, Sker, 1981!*; 87D, Southerndown, 1978!*; 88C, Kenfig, ABP 1970!*; 89D, Blaengarw, 1978!*; 90C, Blaenrhondda, 1978!*.

Taraxacum fulviforme Dahlstedt
Native. Colliery spoil-tips and walls.
06B, Porthkerry, HJR 1912; 87B, Merthyr Mawr, 1976!*; 99B, Cwmamman, AMP 1972*.

Taraxacum fulvum Raunkiaer
Native. Grassland and river sides.
01C, Vaynor, 1978!*; 17D, Leckwith, 1922!*.

Taraxacum glauciniforme Dahlstedt
Native. Limestone grassland.
87B, Ogmore Down, 1920!*.

Taraxacum lacistophyllum (Dahlstedt) Raunkiaer
Native. Limestone grassland. One of the commonest British species.
18C, Radyr, GCD 1921*; 87B, by R. Ewenny near Merthyr Mawr, 1984!*; 87D, Ogmore, 1920!*, ABP 1972!*; 88C, Kenfig, ABP 1970!*; 96A, Nash Point, 1978!*.

Taraxacum laetum (Dahlstedt) Raunkiaer
Native. Dune grassland.
58A, Oxwich Burrows, ARP 1981*.

Taraxacum oxoniense Dahlstedt
Native. Limestone grassland, sand-dunes, limestone quarries, walls, railway embankments. The commonest member of the Section in the county and probably in Britain. Map 788.

Taraxacum rubicundum (Dahlstedt) Dahlstedt
Native. Sandy limestone grassland and railway embankments.
16A, Barry Island, HJR 1905, 1912; 17C, St. Lythans, HJR 1912; 18B, Cefn Onn, 1920!*; 48A, Rhossili, M.P. Marsden 1980; 49C, Broughton Burrows, HJR 1905; 58B, Caswell, JAW 1921*; 79C, Crymlyn Burrows, EFL 1885; 90A, Hirwaun to Glyn Neath, HJR 1904.

Taraxacum silesiacum Dahlstedt ex Haglund
Native. Limestone grassland.
18B, Cefn Onn, 1920!*.

[The following members of this Section have been erroneously recorded for the county: T. **disseminatum** Haglund, T. **dunense** van Soest, T. **glaucinum** Dahlstedt, T. **pseudolacistophyllum** van Soest].

Taraxacum Section **Spectabilia** (Dahlstedt) Dahlstedt emend. A.J. Richards. Broad-leaved Marsh-dandelion. Dant y Llew Cochwythien.

Taraxacum faeroense (Dahlstedt) Dahlstedt
Native. Damp grassland especially on acid soils. This is one of the commonest of all British species wherever a suitable habitat occurs. It is morphologically very variable.
00C, Aberdare, HJR 1904; 07A, Mynydd-y-glew, MSP 1961*, Ystradowen, ARP 1981*; 09B, Mountain Ash, HJR 1904; 17A, Pentrebane, EV 1938*; 80B, Glyn Neath, HJR 1904; 90D, Llwydcoed, HJR 1904; 88D, Pen-y-fai, HJR 1902, 1912, Cefn Cribwr, HJR 1912.

[**Taraxacum spectabile** Dahlstedt was recorded in error].

Taraxacum Section **Naevosa** M.P. Christiansen

Surprisingly the only member of this Section found in the county is, as yet, undescribed; it has the manuscript name T. 'richardsianum' and was collected from Brithdir (10D) in 1978!*.

[**Taraxacum maculosum** A.J. Richards was recorded in error].

Taraxacum Section **Celtica** A.J. Richards. Map 789.

Taraxacum bracteatum Dahlstedt
Native. Grassy places.
18D, Llanishen, KA 1970*.

Taraxacum gelertii Raunkiaer
Native. Grassy places along roadsides, woodland stream banks etc.
00C, Abernant Park, HJR 1904; 08A, Llantrisant 1979!*; 09A, Llanwonnow, AMP 1972*; 58B, Caswell Bay, HJR 1904; 88D, Pen-y-fai, HJR 1912; 97A, Wallas Fach, 1973!*.

Taraxacum lancastriense A.J. Richards
Native. Grassy, rocky places.
00A, Taff Fechan, EV 1931*; 07A, Mynydd-y-glew, EV 1945*; 07C, Llantrithyd, HJR 1908; 18A, Watfordfawr, ABP 1972!*; 88D, Pen-y-fai, HJR 1912.

Taraxacum nordstedtii Dahlstedt
Native. Mainly confined to damp or wet habitats; very variable. This is one of Britain's commonest dandelions. It is unique in having six sets of chromosomes, which seems to make for a possibly intractable taxonomy. Map 790.

Taraxacum raunkiaeri Wiinstedt
Native. Limestone grassland and roadsides. In native habitats this is amongst the most widespread and abundant of all British dandelions.
17B, Cardiff, 1973!*; 87D, Dunraven, 1978!*; 97A, Wallas Fach 1973!*.

Taraxacum subbracteatum A.J. Richards
Native. Base of wall.
88A, Margam, 1982!*.

Taraxacum unguilobum Dahlstedt
Native. Wet grassy places. A very distinctive native species, with a markedly western and northern distribution in Britain.
09A, Mountain Ash, AMP 1970*; 11C, Llechryd, 1978!*; 98A, Llangeinor, 1978!*.

An as yet undescribed species, with the manuscript name of **T.** 'hesperium', widespread in the south of Wales, has been recorded from Roath, Cardiff (27A, 1973!*).

Taraxacum Section Hamata H. Øllgaard
Map 791.

Taraxacum boekmanii Borgvall
Probably introduced. Grassy banks.
87D, St. Bride's Major, 1978!*; 99B, Ton Pentre, 1978!*.

Taraxacum hamatiforme Dahlstedt
Probably introduced. Grassy roadsides.
19C, Llanbradach, 1978!*; 27A, Roath, 1973!*; 87D, Pitcot, 1981!*.

Taraxacum hamatum Raunkiaer
The most 'native' member of this Section but widespread in non-native habitats such as damp grassy verges. Map 792.

Taraxacum hamiferum Dahlstedt
Native. Hedgebank.
87D, St. Bride's Major, 1978!*.

Taraxaum lamprophyllum M.P. Christiansen
Probably introduced. Roadsides and grassy places.
00C, Cwmpennar, AMP 1973*; 07A, Ystradowen, 1978!*; 17B, Central Cardiff, ABP 1969*.

Taraxacum marklundii Palmgren
Native. Damp grassy places.
18D, Llanishen Reservoir, J.M. Price 1970*.

Taraxacum pseudohamatum Dahlstedt
Native. Grassy roadsides, hedgebanks and colliery spoil-heaps. Generally the commonest member of this Section. This robust species is widespread in Britain and is often the first species to flower in early Spring. Map 793.

Taraxacum quadrans H. Øllgaard
Introduced. Grassy roadsides and a garden weed.
17B, Cardiff, 1973!*; 99D, Gilfach Goch, 1979!*.

Taraxacum subhamatum M.P. Christiansen
Native. Grassy roadside verge.
97A, Colwinston, ABP 1973*.

Specimens provisionally determined as 'British atactum' have been collected from two localities: 19B, Cardiff, 1973!*; 97A, Colwinston, ABP 1973*.

Taraxacum Section Ruderalia kirschner, Ollgaard & Stepanek (Section *Vulgaria* Dahlstedt).
The majority of the species in this large Section are doubtless introductions. They are weedy in habit and many are casuals. A considerable number are known from only a single site or even plant (Haworth, 1988). Map 794.

Taraxacum aequilobum Dahlstedt
Probably introduced. Roadsides.
09A, Abercwmboi, AMP 1972; 98A, Llangeinor, AMP 1973.

Taraxacum alatum H. Lindberg fil.
Native. Dune grassland.
58A, Oxwich Burrows, A.J. Richards 1968.

Taraxacum ancistrolobum Dahlstedt
Native. Grassy roadsides, margins of woodland, waste ground etc.
17B, Cardiff, 1921!*, 1973!*; 97A, Wallas Fach, 1973!*.

Taraxacum aurosulum H. Lindberg fil.
Probably introduced. Roadsides.
18C, Whitchurch, ABP 1970!*

Taraxacum cordatum Palmgren
Native. Dune grassland and roadsides. One of the commonest members of this Section found in native habitats in Britain.
18C, Whitchurch, ABP 1970!*; 87B, Merthyr Mawr, 1972!*; 88C, Kenfig, ABP 1970!*.

Taraxacum ekmani Dahlstedt
Probably introduced, usually found in ruderal habitats as a weed.
07A, Ystradowen, SGH 1968!*; 17B, Cardiff, 1920!*, 1927!*; 17C, Cwm George, GH 1988*.

Taraxacum expallidiforme Dahlstedt (incl. *T. sub-cyanolepis* M.P. Christiansen).
Probably introduced. A weed of gardens and waste places.
17A, St. Fagans Castle, SGH 1969*; 17B, Central Cardiff, 1925!*; 18D, Cyncoed, KA 1970*; 88C, Kenfig, ABP 1970!*.

Taraxacum linguatum Dahlstedt ex M.P. Christiansen & K. Wiinstedt
Native. Dune grassland.
78D, Kenfig, GCD 1926*.

Taraxacum longisquameum H. Lindberg fil.
Probably introduced. Roadside weed.
27A, Roath, 1973!*.

Taraxacum oblongatum Dahlstedt
Probably introduced. Grassy field entrance.

87D, between St. Bride's Major and Southendown, 1978!*.

Taraxacum ostenfeldii Raunkiaer
Possibly native. Roadsides.
97A , Colwinston, ABP 1973*.

Taraxacum pannucium Dahlstedt
Probably introduced. Roadsides
17B, Roath, 1973!*.

Taraxacum polyodon Dahlstedt
Introduced. Waste places and roadsides. One of the most widespread and common species in Britain.
16A Barry Docks, 1927!; 18C, Radyr, GCD 1923.

Taraxacum rhamphodes Haglund
Introduced. Grassy roadside verge.
97A, Colwinston, ABP 1973*.

Taraxacum sellandii Dahlstedt
Possibly native. Waste places.
00C, Aberdare, HJR 1904.

Taraxacum stenacrum Dahlstedt
Possibly native. Roadsides.
97A, Ewenny Down, 1973!*.

Taraxacum sublucescens Dahlstedt
Probably introduced. Waste places.
69D, Swansea, GCD 1926*.

An as yet undescribed species with the manuscript name of **T.** 'erraticum' was collected from St. Fagans (17A) by A.E. Wade in 1975*.

[The following members of this Section have been erroneously recorded from the county: **Taraxacum adsimile** Dahlstedt, **T. croceiflorum** Dahlstedt, **T. cyanolepis** Dahlstedt, **T. insigne** Ekman ex M.P. Christiansen & K. Wiinstedt, **T. laticordatum** Marklund, **T. lacinosum** Dahlstedt, **T. latisectum** H. Lindberg fil., **T. lingulatum** Marklund, **T. obtusilobum** Dahlstedt, **T. pallescens** Dahlstedt, **T. parvuliceps** H. Lindberg fil, **T. porrectidens** Dahlstedt; **T. privum** Dahlstedt, **T. sagittipotens** Dahlstedt & R. Ohlsen, **T. tenebricans** (Dahlstedt) Dahlstedt].

LAPSANA L.

Lapsana communis L.   Nipplewort. Cartheig.
Native. Roadsides, hedges, open woods, cultivated land and waste ground. Common in the lowlands, fairly frequent but rather local in the Uplands. Fl. 5–12, fr. 7–11; B.lap, (Syr) S.balt, S.rib, Sp.scr, ?Dips. Map 795.

CREPIS L.

Crepis paludosa (L.) Moench Marsh Hawk's-beard. Gwalchlys y Gors.

Native. Streamsides and shaded rocks. Very local; known only from the valleys of the Afon Pyrddin and Nedd-fechan in the northern Uplands, but fairly abundant in a few places on the banks of the Pyrddin. **C. paludosa** reaches the southern limit of its British range in Glamorgan. The earliest record of the species in the county was made by E. Forster in c.1805 at Sgwd Einion Gam (80B) (Turner & Dillwyn, 1805); it still grows below the waterfall there. Glamorgan plants have the expected chromosome number of 2n=12 (Q.O.N. Kay, unpublished). Storrie's records from Llanishen & Caerphilly (Storrie, 1886) are errors.
80B, Pyrddin Valley, E. Forster c.1805, HJR c.1905, 1969!*, 1974!*, 1986!; 90A, bank of Nedd-fechan, 1974!

Crepis biennis L.   Rough Hawk's-beard. Gwalchlys Garw.
Native or introduced. Grassland, roadsides and waste ground. Scarce; most records are from sites on limestone soils in the Vale. Recent records from 06B, Rhoose Point, MG 1975; 17A, St. Fagans, SGH 1967*; 18C, Gwaelod-y-Garth, MG 1971; 88C, South Cornelly quarry, MG 1975. Map 796.

Crepis capillaris (L.) Wallr.   Smooth Hawk's-beard. Gwalchlys Llyfn.
Native. Dry grassland, sand-dunes, cliff-heath, roadsides, walls and waste ground. Common in the lowlands and fairly common in most of the Uplands. Fl. 5–11, fr. 7–11; (Syr) S.balt, S.rib, Sp.scr (Lep) Cupido minimus, Pa.aeg, Pi.bra, Pi.nap. Map 797.

Crepis vesicaria L. subsp. **haenseleri** (Boiss. ex DC.) P.D. Sell (*Crepis taraxacifolia* Thuill.) Beaked Hawk's-beard. Gwalchlys Gylfinhir.
Introduced. Roadsides, railway embankments and waste ground. Locally common in or near urban areas in the lowlands. Trow (1911) stated that it was then a recent immigrant almost confined to railway embankments in the Vale. Fl. 5–8 (–10), fr. 6–9; Hb (n,p), B.lap, B.prat, ?Sb, (Syr) E.ten, Rh.cam. Map 798.

Crepis setosa Haller fil., Bristly Hawk's-beard, Gwalchlys Pigog, was found by Eleanor Vachell as a dock alien in Cardiff (27A*, n.d.) and a small population has recently been found on waste ground near the river in Glyn-neath (80B, 1974!**).

Andryala integrifolia L. has been found as a casual on three occasions, 16A, Barry Docks, GCD 1930; 17B, Maindy, Cardiff, RLS 1938*; 18C, Radyr, RLS 1922*).

PILOSELLA Hill

Pilosella officinarum F.W. Schultz & Schultz Bip. (*Hieracium pilosella* L.). Mouse-ear Hawkweed. Clust y Lygoden, Heboglys Torllwyd.
Native. Dry grassland, roadside banks, railway embankments, walls and sand-dunes. Common in

suitable habitats throughout the county. Fl. 5–11, fr. 6–11; Hb (n,p), B.pasc, ?Sb, (Syr) S.rib, ?Syrs, (Lep) Cupido minimus, Pi.nap, Pi.rap. Map 799.

Pilosella officinarum has been split into several subspecies differing mainly in the indumentum of the phyllaries. The following account is based mainly on specimens in **NMW**, **UCSW** & **CGE** and many of the determinations should be regarded as provisional only.

Pilosella officinarum subsp. **officinarum** (*Hieracium pilosella* subsp. *pilosella*).
Native. Grassy places on cliffs, roadsides etc, also in rocky places; mainly in the east of the county. Map 800.

Pilosella officinarum subsp. **micradenia** (Naegeli & Peter) P.D. Sell & C. West (*Hieracium pilosella* subsp. *micradenium* Naegeli & Peter).
Native. Grassy places on banks, dunes, railways and also in rocky places. Map 801.

Pilosella officinarum subsp. **euronota** (Naegeli & Peter) P.D. Sell & C. West (*Hieracium pilosella* subsp. *euronotum* Naegeli & Peter).
Native. Grassy places on sandy banks, disused railways, woodland margins, banks etc.
17A, St. Brides-super-Ely, 1972!*; 18B, Caerphilly, 1920!*; 58A, Pennard Castle, HAH 1935!*; 77B, Sker, 1981!*; 90C, Craig-y-llyn, 1969!*; 98C, Heol-y-Cyw, KA, 1973*; 99C, Ogmore Vale, ABP 1975!*.

Pilosella officinarum subsp. **trichosoma** (Peter) P.D. Sell & C. West (*Hieracium pilosella* subsp. *trichosoma* Peter).
Native. Grassy places on banks, roadsides, old colliery spoil-tips and on walls, rocky places etc. Map 802.

Pilosella officinarum subsp. **melanops** (Peter) P.D. Sell & C. West (*Hieracium pilosella* subsp. *melanops* Peter).
Native. Dune grassland, disused railway track, old colliery spoil-heaps, waste ground etc.
08D, between Efail Isaf and Rhiwsaeson, 1972!*; 17A, St. Brides-super-Ely, 1980!*; 18B, Rudry, ABP 1972!*; 18C, Radyr, 1922!*; 78B, Margam Burrows, 1972!*.

Pilosella caespitosa (Dumort.) P.D. Sell & C. West subsp. **colliniformis** (Peter) P.D. Sell & C. West (*Hieracium caespitosum* Dumort. subsp. *colliniforme* (Peter) P.D. Sell).
Introduced. Once established on a wall 'in Lord Bute's garden in Cardiff' (17B, EV 1905*) but now extinct. It occurs on railway banks and old walls in a few widely scattered sites in Britain.

Pilosella aurantiaca (L.) F.W. Schultz & Schultz Bip. subsp. **carpathicola** (Naegeli & Peter) Sojók (*Hieracium brunneocroceum* Pugsley). Fox-and-cubs. Heboglys Euraid.

Introduced. A widespread and increasing garden escape. Roadsides, hedgebanks and railway embankments. Most frequent in the Uplands. Map 803.

Pilosella aurantiaca (L.) F.W. Schultz & Schultz Bip. subsp. **aurantiaca** (*Hieracium aurantiacum* L.) was twice recorded by J. Storrie in *c*.1876 (18D, Lisvane; 98C, Coity), it has not been seen in either locality, or any other in the county, since.

HIERACIUM L.

The apomictic microspecies of **Hieracium** are notoriously difficult taxonomically. The **Hieracium** flora of Glamorgan was studied in some detail by Riddelsdell (1907), who had a good knowledge of the British species of the genus as they were then understood. Another assiduous worker was the Rev. A. Ley and their specimens, many of which are in the herbarium of the British Museum (Natural History) or the Botany School, The University, Cambridge have been extremely valuable to subsequent workers, including H.W. Pugsley who used them in his revision of the British species (Pugsley, 1948).
The following account still owes much to the work of Riddelsdell and Ley and is based on records accumulated by P.D. Sell and C. West for their **Hieracium** account in Perring (1968), and specimens in **NMW** and elsewhere that have been verified by C.E.A. Andrews, B.A. Miles, P.D. Sell, C. West or, especially, by J. Bevan. The authors are particularly grateful to J. Bevan who has spent many days examining specimens in **NMW** and sorting out at least some of the problems posed by the Glamorgan **Hieracium** flora, and has checked through the text. At the request of J. Bevan and the BSBI's Hieracium Study Group, the sequence of Sections and species follows that of Sell & West in Perring (1968).

Hieracium murorum L. agg. Hawkweed. Heboglys
Native.
The map, showing the distribution of the genus in Glamorgan, is included so that future workers may see at a glance where the genus has been recorded and where their efforts may best be directed. Map 804.

Hieracium Section **Oreadea** Zahn

Hieracium eustomon (E.F. Linton) Roffey
Native. Rocky places on limestone cliffs by the sea with one record from sand dunes.
48A, Rhossili, HJR 1904; 58A, Pennard, AL 1903*, U.K. Duncan & E.P. Beattie 1964, Pobbles Bay, J. Bevan 1985, Penmaen Cove A. Ley 1903; 58B, Pwlldu, AL 1903*, 1909, Southgate Cliffs, J. Bevan 1985.

Hieracium stenopholidium (Dahlst.) Omang
Native. Cliff ledges on Pennant Sandstone.
90C, Craig-y-llyn, BAM 1954, 1978!*; 90D, Tarren y Bwllfa, HJR 1902. A specimen in **NMW**, from

Craig-y-bwlch (90C) is also believed to be this taxon but was collected too late in the year to allow for positive identification.

Hieracium cambricum (Baker) F.J. Hanb.
Native. Cliff ledges on Pennant Sandstone.
99A, Graig Fawr, Treorchy, HJR 1903.

Hieracium argenteum Fries
Native. Cliff ledges on Pennant Sandstone.
99A, Graig Fawr, Treorchy, HJR 1903*, BAM 1944, PJo & GH 1985*.

Hieracium cacuminum (A. Ley) A. Ley
Native. Cliff ledges on Pennant Sandstone.
90C, Craig-y-llyn, AL 1890, 1896, Top of Rhondda Valley, HJR 1903; 99A, Cwm Selsig, Blaen-y-Cwm, 1970!*, Graig Fawr, Treorchy, GH & PJo 1985*.

Specimens from 90C, Craig-y-llyn, previously referred to **Hieracium eustomum**, **H. subbritannicum** (A. Ley) P.D. Sell & C. West, **H. schmidtii** Tausch, **H. leyi** F.J. Hanb., and **H. argenteum**, and from 90D, Tarren y Bwllfa, previously referred to **H. leyi**, match the taxon identified with the continental species **Hieracium stenopholidium** and research is currently being undertaken to clarify the situation.
Other species from Craig-y-llyn, variously identified as **H. eustomon**, **H. cacuminum** and, by earlier botanists, **H. nitidum** Backhouse var. **siluriense** F.J. Hanb., appear to belong to a separate taxon which requires further study before its relationship with other local taxa can be established.

Hieracium Section **Vulgata** F.N. Williams

Hieracium exotericum Jordan ex Boreau agg. (excluding **H. exotericum** *s.s.*)
H. exotericum agg. is used to denote forms whose precise relationships to the many taxa in Hylander (1943) has still to be worked out.
Probably introduced. Rocky ground, walls, lanesides, railway lines, grassy banks etc.
00, near Merthyr Tydfil, C.C. Babington 1839; 17A, Radyr, HJR 1910; 18B, near Caerphilly, AL 1906; 18C, Castell Coch, BAM 1954, Coryton, GH 1986*; 18D, Rudry, Craig Llanishen, AL 1905, Lisvane, HJR 1905; 80B, Glyn Neath, AL 1897, Pyrddin Valley, HJR 1904; 88D, Court Colman, HJR 1916*, GH 1988*; 90A, Hirwaun to Glyn Neath, HJR 1903; 90D, Llwydcoed, HJR 1907*.

Hieracium exotericum Jordan ex Boreau *s.s.*
Probably introduced. Rocky ground on limestone, walls, and railway banks.
18C, Taff's Well, 1925!*, 1937!*, Castell Coch, JEL 1948*.

Hieracium grandidens Dahlst.
Probably introduced. Shaded roadside banks.
18C, Radyr, EV 1943*, GH 1985*; 28C, Llanedeyrn, EV 1921!*.

Hieracium sublepistoides (Zahn) Druce
Probably introduced. Roadsides, railway banks, woodland edge etc.
00B, Morlais Hill, ABP & B. Scotter 1973!*; 07A, Hensol Castle, HAH 1927*; 07B, Pendoylan, DMcC 1968; 08D, Efail Isaf, 1972!*; 18A, Groes-wen, 1974!*, Graig, 1974!*; 18B, Caerphilly, 1920!*, Rudry, ABP 1972!*; 18C, Garth Wood, HJR 1906*, Radyr, HJR 1909*, 1921!*, Taff's Well, 1925!*; 18D, Cefn Onn, 1921!*, Llanishen, EV 1921!*.

Hieracium cinderella (A. Ley) A. Ley
Native. Rocky streamsides and railway banks.
18B, Caerphilly, AL 1905*, 1906*; 18C, Taff's Well, HJR 1904*; 80B, Glyn Neath, HJR 1911*, Pyrddin Valley, HJR 1904, 1911*, BAM 1955, C. West & P.D. Sell 1957; 90A, Rhigos Hill, HJR 1906.

Hieracium pellucidum Laest.
Native. Rocky places on limestone and Pennant Sandstone in the Uplands and on grassy roadside banks in the Vale.
00A, Morlais Castle, HJR 1909*; 18C, Radyr, J. Bevan 1985, Morganstown, GH 1987*; 90C Craig-y-llyn, BAM 1954.

Hieracium stenstroemii (Dahlst). Johans.
Native. Rocky places, roadsides, railway and rocky river banks usually on basic soils.
00A, Cefn-coed-y-cymmer, 1969!*; 01D, Vaynor, 1978!*; 17A, Radyr Quarry, J. Bevan 1985; 18B, Cefn Onn, ABP 1972!*; 18C, Forest Ganol, RAH, 1971*, Radyr, J. Bevan 1985, Morganstown, GH 1987*; 18D, Cefn Onn, EV 1921!*; 28A, Draethen, 1972!*; 90D, Llwydcoed, HJR 1901.

Hieracium cuneifrons (W.R. Linton) Pugsley
Native. Rocky places on limestone and Pennant Sandstone.
49C, Llanmadoc, R.A. Graham *c*.1955; 90D, Llwydcoed, HJR 1901.

Hieracium radyrense (Pugsley) P.D. Sell & C. West
Native. Rocky oak woodland edge on limestone and grassy roadside bank. Endemic to Glamorgan.
17A, Radyr Quarry, HJR, 1910*, J. Bevan 1985; 18C, Radyr, BAM 1955, C. West 1955, J. Bevan 1985. (The population from Court Colman (88D) is now known to differ from that at Radyr. Its true identity has yet to be established; for convenience it is included with other doubtful records under **H. exotericum** agg.).

Hieracium submutabile (Zahn) Pugsley
Native. Grassy banks and sand dunes.
18C, Radyr, GH 1987*; 87B, Merthyr Mawr, E.K. Horwood 1953, S. Waldren 1982*; 90D, Llwydcoed, HJR 1901, ABP 1974!*.

Hieracium diaphanum Fries
Native in rocky places and sand dunes but colonizing roadsides, railway banks, waste ground and industrial areas. Map 805.

Hieracium diaphanoides Lindeb.
Native in rocky places but colonizing, walls, railway banks and waste ground. Map 806.

Hieracium strumosum (W.R. Linton) A. Ley (incl. *H. acuminatum* auct.).
Native in rocky places but colonizing walls, roadsides, railway embankments and waste places. Map 807.

Hieracium cheriense Jordan ex Borreau
Probably introduced. Roadsides.
18C, Radyr, C. West 1958.

Hieracium subamplifolium (Zahn) Roffey
Native in rocky places, but colonizing walls, railway banks, sand dunes etc.
00C, Abernant, AL 1901, Aberdare, HJR 1902*, 1911*; 06B, Porthkerry, BAM, 1954, Rhoose, 1959!*, ABP 1971!*; 10C, Deri, 1972!*; 17B, Llandaff, EV 1948*; 18D, Cefn Onn, EV 1921*; 49C, Whitford Burrows, 1969!*; 70B, Ystalyfera, AL 1906; 71D, Cwm Twrch, AL 1899; 90D, Llwydcoed, HJR 1901, Trecynon to Hirwaun, HJR 1911*.

Hieracium lepidulum (Stenstrom) Omang
Introduced. Roadsides.
06B, Porthkerry, F.R. Browning 1958.

Hieracium vulgatum Fries
Possibly introduced. Rocky places and railway lines.
08D, Efail Isaf, 1972!*; 17A, near St. Fagans, 1972!*; 28A, near Rudry, ARP 1976!*; 87B, Pont Alun, R.D. Tweed & L.J. Hayward 1950.

Hieracium Section **Tridentata** F.N. Williams

Hieracium sparsifolium Lindeb.
Native. Rocky places on Pennant Sandstone.
00C, Aberdare, HJR c.1905; 90C, Craig-y-llyn, AL 1890, WAS 1896*, BAM 1955, 1968!*, GH & D.E. Evans 1987*.

Hieracium substrigosum (Zahn) Roffey
Native. Rocky riversides.
99A, head of Rhondda Valley, HJR c.1905.

Hieracium scabrisetum (Zahn) Roffey
Native. Rocky places, streamsides, hedgebanks, roadsides, open woods, sand dunes etc. Map 808.

Hieracium eboracense Puglsley
Native. Rough grassland.
01D, Pontsticill, AL 1900; 90B, Tre-gibbon, HJR 1905.

Hieracium calcaricola (F.J. Hanb.) Roffey
Native. Grassy roadsides, heaths and woods.
08B, Rhydyfelin, C. West 1957; 90B, Tre-gibbon, HJR 1905; 99C, Graig Fach, Treorchy, BAM 1954.

[**Hieracium cantianum** F.J. Hanb. was recorded in error in Riddelsdell (1907), and the specimen on which the record of **H. trichocaulon** for the county in

Ellis (1983) was based has recently been re-identified as **H. scabrisetum**].

Hieracium Section **Foliosa** Pugsley

Hieracium subcrocatum (E.F. Linton) Roffey
Native. Cliff ledges on Pennant Sandstone.
90C, Craig-y-llyn, AL 1890 ('probably this' P.D. Sell & C. West 1957).

Hieracium Section **Umbellata** F.N. Williams

Hieracium umbellatum L.
Native. Rocky places, roadsides, railways, sand dunes, scrub, damp heath etc. Map 809.

The following two subspecies of H. umbellatum recognised by Sell & West in Perring, 1968 have not been consistently separated during the present survey and consequently the distribution of each is poorly known. Herbarium material at **NMW** has been separated and provides the following records.

Hieracium umbellatum L. subsp. **umbellatum**
Native. Roadsides, heaths, scrub, railway lines.
00B, Penydarren, BS 1975!*; 08A, Llantrisant, ABP 1971!*; 18A, Nantgarw, EV no date*; 18D, Cefn Onn railway station, GH 1987*; 19A, Llancaiach, ABP 1973!*; 89D, near Maesteg, 1973!*.

Hieracium umbellatum L. subsp. **bichlorophyllum** (Druce & Zahn) P.D. Sell & C. West
Native. Grassy places and scrub.
00B, Penydarren, BS 1975!*; 00C, Aberdare, HJR c.1905; 17A, St. Fagans, EV 1904*; 98B, near Pencoed, KA 1971*.

Hieracium Section **Sabauda** F.N. Williams

Hieracium vagum Jordan
Probably introduced. Grassy places along railway lines.
78D, Kenfig, O. Stewart, R. FitzGerald & TGE, 1984; 90D, Aberdare railway station, J.H. Phipps 1953.

Hieracium perpropinquum (Zahn) Druce
Native. Grassy places in open woods along roadsides and on heathy slopes. Probably the commonest species of **Hieracium** in Glamorgan. Map 810.

MONOCOTYLEDONES
ALISMATACEAE
SAGITTARIA L.

Sagittaria sagittifolia L.   Arrowhead. Saethlys.
Native. Streams, canals, ponds. Now rare, surviving only in a few scattered localities; more frequent in the past, but many habitats have been destroyed. Fl. 7–8
07A, Talygarn, WWB c.1900; 08B, Pontypridd, EMT 1949*, MG 1972; 09B, Penrhiwceiber, AHT 1904; 17B, Cardiff, R. Drane 1859, HAH 1947*; 18C, Whitchurch Canal, BS 1966*, RAH 1972!; 69A, Penllergaer, AHT

*c.*1900 (introduced); 78B, Margam Moors, HJR *c.*1904, 1974!; 79B, near Neath, H & CH 1973; 79C, Briton Ferry, JAW 1928*, 1941*.

BALDELLIA Parl.

Baldellia ranunculoides (L.) Parl.   Lesser Water-plantain. Llyren Fechan, Dŵr-lyriad Bychan.
Native. Marshes, margins of pools, canals, dune slacks etc. Rare and now confined to Gower and the western Vale. Formerly more widespread. Much of this decrease is probably due to the destruction of suitable habitats by drainage and changes in agricultural techniques. Map 811.

LURONIUM Rafin.

Luronium natans (L.) Rafin.   Floating Water-plantain. Dŵr-lyriad Nofiadwy.
Native. Marsh and fen. Extinct. Recorded in the last century by J.W.G. Gutch from two localities near Swansea, but not seen since.
69D, Crymlyn Fen, JWGG *c.*1840; 69C, near Singleton, JWGG *c.*1840.

ALISMA L.

Alisma plantago-aquatica L.   Water-plantain. Dŵr-lyriad.
Native. Streams, pools, marshes. Locally frequent throughout the county. Fl. 6–7, fr. 7–8. Map 812.

Alisma lanceolatum With.   Narrow-leaved Water-plantain. Dŵr-lyriad Culddail.
Probably introduced. Pools. Very rare, recorded only from two sites in the Uplands.
00C, Cefnpennar, Mountain Ash, AMP 1975; 09A, Peace Park, Mountain Ash, AMP 1972*.

BUTOMACEAE
BUTOMUS L.

Butomus umbellatus L.   Flowering-rush. Engraff.
Native. Marshes, fens, streams, ponds. Scarce, growing in scattered localities mainly along the coast. Decreasing in numbers because of improved methods of drainage. Map 813.

HYDROCHARITACEAE
HYDROCHARIS L.

Hydrocharis morsus-ranae L.   Frogbit. Alaw Lleiaf.
Native. Ponds, marshes, canals. Rare; recently recorded from only three localities and possibly extinct in one of these (17B).
17B, Penylan, Cardiff, 1969!; 17D, Grangetown and Leckwith Moors, JS *c.*1876, Grangetown, AHT 1891*, 1892*; 27A, East Moors, Cardiff, JS *c.*1876; 69D, Crymlyn Fen, J. Lightfoot 1773; 78B & 78D, Margam Moors, HJR *c.*1905, ABP 1983!; 79B, near Neath, H & CH, 1973!.

ELODEA Michx

Elodea canadensis Michx   Canadian Waterweed. Alaw Canada.
Introduced. Ponds, streams, canals etc. Locally frequent in the Uplands and the Vale but only once recorded from Gower (in 1956) and not seen since. Map 814.

Elodea nuttallii (Planchon) St. John   South American Waterweed. Alaw Nuttall. Introduced. Canal. Recorded from only one locality, but will probably spread to most parts of the county in the future.
08B, Glyntaff, ARP 1979*, det. D.A. Simpson.

LAGAROSIPHON Harvey

Lagarosiphon major (Ridley) Moss   Curly Waterweed. Pib-flodyn Crych.
Introduced. Canals and ponds. Found in three localities, perhaps spreading.
18C, Whitchurch, Cardiff, ARP 1974*; 69C, University College of Swansea, 1970!; 88A, Margam, ARP 1984!.

APONOGETONACEAE
APONOGETON L. fil.

Aponogeton distachyos L. fil.   Cape Pondweed. Dyfryllys Tramor.
Introduced. Pond. Established in one locality in Cardiff, probably planted originally.
18D, Roath Park, Cardiff, MG 1970.

JUNCAGINACEAE
TRIGLOCHIN L.

Triglochin maritima L.   Sea Arrowgrass. Saethbennig Arfor.
Native. Salt-marshes. Frequent in the west, less so in the east. Map 815.

Triglochin palustris L.   Marsh Arrowgrass. Saethbennig y Gors.
Native. Marshes, including the brackish fringes of salt-marshes where it grows with **T. maritima**. Locally frequent in north Gower and the south western Uplands, occasional in a few scattered localities elsewhere. Like many other wetland plants this species is decreasing in numbers because of habitat destruction. Map 816.

POTAMOGETONACEAE
POTAMOGETON L.

Potamogeton natans L.   Broad-leaved Pondweed. Dyfrllys Llydanddail.
Native. Ponds and canals. Frequent in the south of the county, less so in the north. Map 817.

Potamogeton polygonifolius Pourret   Bog Pondweed. Dyfrllys y Gors.

16A, Barry, RLS *c.*1923; 96A, near Marcross, JS *c.*1876*.

HYACINTHOIDES Medicus

Hyacinthoides non-scripta (L.) Chouard ex Rothm. (*Endymion non-scriptus* (L.) Garcke). Bluebell. Clychau'r Gog.
Native. Woods, hedges, under **Pteridium aquilinum** on heathland, etc. Common throughout most of the county. (!) Fl. 4–6, fr. 6–7; Hb, Bb, (Syr) Rh.cam, (Lep) Ag.urt, An.car, Ly.phl, Pi.bra, Pi.nap, Pi.rap. Map 828.

Hyacinthoides hispanica (Miller) Rothm. (*Endymion hispanicus* (Miller) Chouard). Spanish Bluebell. Clychau Cog Sbaen.
Introduced. Parks, grassy places in woods, grave-yards, roadsides, waste ground etc. Occasional in the south of the county, especially in the Vale. Map 829.

MUSCARI Miller

Muscari comosum (L.) Miller   Tassel Hyacinth.
Introduced. Waste places. Apparently established in one locality (16A), but not seen recently in the other (06A).
06A, The Leys, Aberthaw, RLS 1937*, Mrs. Williams 1950; 16A, Barry Docks, 1923!*, DMcC 1974.

Muscari neglectum Guss. ex Ten. (*M. atlanticum* Boiss. & Reuter), Grape Hyacinth, has been recorded four times as a casual and may become established (16B, Cosmeston, GH 1986; 18D, Llanishen, JPC 1971!; 87D, St. Bride's Major, 1978!; 19D, near Llanbradach, 1978!).

ALLIUM L.

Allium schoenoprasum L.   Chives. Cennin Syfi, Seifys.
Introduced. Recorded three times in the past, appar-ently only as a casual. Meadows, walls and railway lines, originating from garden throwouts.
00C, Aberdare, HJR 1904; 06A, St. Athan, RLS *c.*1920; 17B, Llandaff, EV *c.*1920.

Allium roseum L. subsp. **bulbiferum** (DC.) E.F. Warb.   Rosy Garlic. Garlleg Gwridog.
Introduced. Well established at the base of a wall in one locality (96A).
16A, Barry, RLS 1938*; 96A, Marcross, K. Richards 1920*, 1973!*.

Allium triquetrum L.   Three-cornered Leek. Cen-hinen Drichornel.
Introduced. Woods and hedges. Locally established in a few places in the west of the county. Fl. 4–5; B.prat (Lep) Pi.rap.
48A, Stouthall Woods, JBe 1981!; 48B, Horton, B.E. Jones 1970; 49C, near Cwm Ivy, 1975!; 58B, Caswell Bay, JAW 1921, MG 1973; 87A, Newton, Mrs Bruce 1940, EMT 1942*.

Allium ursinum L.   Ramsons. Craf y Geifr.
Native. Woods and hedges. Very abundant on limestone soils in Gower and the Vale but almost absent from the Uplands. Fl. 4–6, fr. 6–7. Map 830.

Allium oleraceum L.   Field Garlic. Garlleg Rhesog y Maes.
Native. Grassy places on dunes and roadsides. Very scarce, with no recent records.
06B, Rhoose, F. White 1945*; 17B, Penylan, Cardiff, E.L. Downing *c.*1904; 18D, Llanishen, EV 1899, 1911, Lisvane, W.M. Smith 1955; 87A, Porthcawl, EMT 1941*.

Allium carinatum L.   Keeled Garlic.   Garlleg Mynyddig.
Introduced. Sand dune and roadside. Recorded twice in the past, and apparently naturalized in Cardiff.
18B, Roath Park, GSW 1983; 18D, Llanishen, Cardiff, EV 1944*; 78D, Kenfig Burrows, D.T. Price 1962*.

Allium ampeloprasum L.   Wild Leek. Cenhinen Wyllt, Garlleg Mawr Pengrwn.
Native or introduced. Rocky sea cliffs. Very rare, recorded from two localities but now known only from Flat Holm where it was first recorded by Dr J. Newton in 1688 (Ray, 1688).
06B, Porthkerry, HJR *c.*1906; 26, Flat Holm, J. Newton 1688, 1981!.

Allium vineale L.   Wild Onion. Garlleg Gwyllt.
Native. Grassy sea-cliffs and sand-dunes, pastures and grassy roadsides. Frequent on the coast of Gower and the Vale but scarce inland and absent from the Uplands. Bulbils produced 8–9. Map 831.

Nectaroscordum siculum (Ucria) Lindley subsp. **bulgaricum** (Janka) Stearn Established on shady bank of Nant Fawr, Roath Park, Cardiff (17B, V.G. Ellis 1983) and on clifftop at Porthkerry (06B, R.J. Tidswell 1983).

Ipheion uniflorum (R.C. Graham) Rafin. (*Brodiaea uniflora* (Lindley) Engler, *Milla uniflora* R.C. Graham), Spring Starflower, has once been recorded as a rare casual of waste places (87A, Newton, EMT 1948*).

CONVALLARIA L.

Convallaria majalis L.   Lily-of-the-valley. Clych Enid.
Native in limestone woodland, and sometimes as an introduction or garden escape elsewhere. Rare and decreasing.
00C, Aberdare, BAW *c.*1904; 09A, Mountain Ash, AMP *c.*1972 (garden escape); 17A, Coedrhiglan, JS *c.*1879; 18C, Garth Wood, Taff's Well, THT 1898*; 18C, Castell Coch, MG 1972; 58A, near Parkmill, JAW 1959* (probably an introduction); 69A, Penllergaer, JAW 1924–26 (introduction); 88C, near Cornelly, EV n.d.*; 88C, Porthcawl, EMT 1946; 96A, near Marcross, 1973!.

POLYGONATUM Miller

Polygonatum multiflorum (L.) All.   Solomon's-seal. Dagrau Job, Sêl Solomon.
Native. Woodland and hedges. Local in the eastern Vale. Found in a few places elsewhere in the past, but not recorded recently in Gower and the Uplands except for one garden escape (09A). Map 832.

PARIS L.

Paris quadrifolia L.   Herb-Paris. Cwlwm Cariad.
Native. Wooded areas especially on limestone. Local in the eastern Vale and Gower, rare in the western Vale and almost absent from the Uplands. Decreasing in numbers. Map 833.

ASPARAGUS L.

Asparagus officinalis L. subsp. **officinalis**   Garden Asparagus. Merllys.
Introduced. Sand-dunes amd waste places, usually near the sea. Scarce, with recent records from only four places; a disused railway near Castell Coch (18C, MG 1972) and on sand-dunes at Whiteford (49C, 1975!), Oxwich (58A, D.O. Elias 1981) and Merthyr Mawr (87B, PJo 1984). Map 834.

Asparagus officinalis subsp. **prostratus** (Dumort.) Corb. (*A. prostratus* Dumort.).   Wild Asparagus. Merllys Gorweddol.
Native. Sandy cliff slopes. Very rare, with recent records only from two sites in south Gower, where very few plants survive. Recorded in the past from several other localities on the coast of Gower and the southern Vale. Map 835.

RUSCUS L.

Ruscus aculeatus L.   Butcher's-broom. Celynnen Fair.
Native or introduced. Woodland and cliff-scrub; local in south Gower and the southern Vale. Apparently native in Gower, where it is locally abundant near Parkmill and Nicholaston, but probably introduced in several localities in the Vale. Map 836.

AGAVACEAE

Yucca recurvifolia Salisb. has recently been recorded on sand dunes at Crymlyn Burrows (79C, B.M. Sturdy 1982), and **Y. gloriosa** L., Spanish-dagger, was recorded once on sands at Jersey Marine (79C, J. Woods *c.*1845 (Woods, 1850)).

AMARYLLIDACEAE

LEUCOJUM L.

Leucojum aestivum L.   Summer Snowflake. Eiriaidd.
Introduced. Woodland. Established in a few localities in Gower and the western Vale.

06B, Porthkerry Woods, G.B. Hughes *c.*1876; 58B, Caswell Bay, JAW 1925*, 1966–1982!; 68A, Langland, P.M. & R. Davies 1970; 88C, Kenfig, JM 1843; 88D, Cefn Cribwr, I.M. Carey 1970*.

GALANTHUS L.

Galanthus nivalis L.   Snowdrop. Eirlys.
Native or introduced. Woods, roadside verges and hedgerows. Fairly frequent in Gower and the Vale, but rare or absent and with no recent records in the Uplands. Fl. 1–4, fr. 3–5; Hb (n,p). Map 837.

NARCISSUS L.

Narcissus poeticus L. (*N. majalis* Curt.)   Pheasant's-eye Daffodil. Gylfinog Barddol.
Introduced. Apparently once established in woods and fields in several localities in the Vale and Gower in the past but not recorded recently.
07C, St. Hilary, JS *c.*1876; near Howmill, JS *c.*1876; 17B, Cathays Park, Cardiff, W.W. Pettigrew *c.*1900; 18A, near Pwll-y-pant, EV *c.*1900; 18D, Llanishen Reservoir, JS *c.*1876; 19C, Llanbradach, EV 1897*; 48A, Penrice Castle, LWD *c.*1840; 58B, near Murton, JAW 1939*; 69C, Mayals, JAW 1939*; 87A, near Southerndown, WWB *c.*1900.

Narcissus poeticus × **N. pseudonarcissus** (**N.** × **incomparabilis** Miller).   Nonesuch Daffodil.
Introduced. Grassland. Like the last, apparently established in several localities in the past but not recorded recently.
17B, near Pontcanna, Cardiff, EV 1933*; 69D, near Swansea, JAW *c.*1950; 78D, Kenfig Burrows, EMT 1943*; 88A, Margam, LWD *c.*1840, D. Llewellyn *c.*1910; 98C, near Pencoed and Coedymwster, JS *c.*1876.

Narcissus poeticus × **N. tazetta** (**N.** × **medioluteus** Miller) (*N.* × *biflorus* Curtis).   Primrose-peerless. Gylfinog Dauflodeuog.
Introduced. Grassy places in fields and on sand dunes. Was well established in at least one locality (88C) for *c.*100 years until the habitat was destroyed by ploughing in 1976.
06B, Porthkerry, E.L. Downing 1909*; 16A, Barry, E.L. Downing *c.*1900; 17D, Penarth, J. Evans 1848; 48B, Oxwich, JAW 1925*; Penrice, HJR 1905; 87A, near Nottage, EMT 1940; 87B, Merthyr Mawr, AHT 1899*; 88C, near Kenfig, JM 1843, 1929*! 1975!, Mawdlan, HJR 1905.

Narcissus pseudonarcissus L. subsp. **pseudonarcissus** (incl. *N. telamonias* Link).   Wild Daffodil. Cenhinen Bedr.
Native, but widespread as an introduction. Woodland, hedgerows, and pastures. Occasional in the southern Vale and sometimes locally in large numbers, rare elsewhere, decreasing in numbers. Map 838.

Narcissus pseudonarcissus subsp. **major** (Curtis) Baker (*N. hispanicus* Gouan). Spanish Daffodil. Cenhinen Sbaen.
Introduced. Woodland and pastures in the Vale. Rare, with only two recent records.
07D, Bonvilston, JIL 1974*; 18D, Llanishen Reservoir, J.M. Price 1970*; 28C, Llanedeyrn, HJR 1912*; 28C, Coed-y-gores, EV 1892, 1922!*; 87B, Candleston, EV 1895*; 97A, Ewenni, EV 1895.

Narcissus tazetta L. Bunch-flowered Narcissus, Gylfinog Bwysi, a rare casual once recorded on waste ground (17D, Penarth, JS 1876*).

DIOSCOREACEAE
TAMUS L.

Tamus communis L. Black Bryony. Gwinwydden Ddu.
Native. Hedgerows, scrub and woodland. Common in most of the county but absent from part of the Uplands. Fl. 5–7 (male), 6–7 (female), fr. 9–11; (!) Sb (female fls, n), (Dip) Empis aestiva (male fls, n), ?Dips (both sexes, n). Map 839.

IRIDACEAE
SISYRINCHIUM L.

Sisyrinchium montanum E.L. Greene (*S. bermudiana* sensu Coste, non L., *S. angustifolium* Miller). American Blue-eyed-grass. Llygatlas America.
Introduced. Established on dunes at Kenfig Burrows (78D, GCa 1968*, S. Moon 1979; 88C, 1974!), and possibly at Whitchurch, Cardiff, where it has been seen recently after a gap of 90 years (18C, E. Walker 1890, EV 1899, E. Wiley 1989). Recorded in the past as a dockside alien or on waste ground in: 17B, Cardiff Docks, A.D. Lewis 1959*; 18C Radyr, H. Husbands c.1900; 97A, Bridgend, D.P.M. Guile 1944*.

IRIS L.

Iris foetidissima L. Stinking Iris. Iris Ddrewllyd.
Native. Dunes, sea-cliffs, open woods, scrub and hedgebanks. Common on the coast of the Vale and locally frequent in Gower but absent from the Uplands. Map 840.

Iris pseudacorus L. Yellow Iris, Yellow Flag. Iris Felen.
Native. Marshes, pondsides, streambanks, etc. Common in suitable habitats throughout most of the county. Map 841.

Iris germanica L. Flag Iris, Iris Farfog, was once established in grassy places on Flat Holm (26) where it probably escaped from a garden, but is long extinct there. **I. latifolia** (Miller) Voss (*I. xiphioides* Ehrh.) was one of L.W. Dillwyn's introductions at Penllergaer (69A, LWD c.1840) but apparently did not persist.

Crocus vernus (L.) Hill (*C. purpureus* Weston) Spring Crocus, Saffrwm y Gwanwyn, has been recorded twice in the past (49D, Fairy Hill, JAW 1926*; 87B, Laleston, EV 1932*) but not recently, probably disregarded as a deliberate introduction by many recorders, and has apparently not become fully naturalized anywhere; **C. biflorus** Miller, Scotch Crocus, has similarly been recorded as an introduction (48B, Penrice Green, JAW 1923*).

TRITONIA Ker-Gawler

Tritonia aurea Pappe ex Hooker × **T. pottsii** (Baker) Baker (**T. × crocosmiflora** (Lemoine) Nicholson (*Crocosmia × crocosmiflora* (Lemoine) N.E. Br.)). Montbretia. Montbretlys.
Introduced. Hedgerows, roadsides and waste ground. Frequent throughout the county. Map 842.

GLADIOLUS L.

Gladiolus communis L. subsp. **byzantinus** (Miller) A.P. Hamilton, Common Gladiolus, is naturalized on a cliff at Nash Point (96A, BSBI field meeting 1983) and appeared at the time of discovery to have been established for at least 10 years; a probable escape from the lighthouse garden nearby.

JUNCACEAE
JUNCUS L.

Juncus maritimus Lam. Sea Rush. Brwynen Arfor.
Native. Salt-marshes and brackish dune-slacks. Abundant in suitable habitats on the coasts of Gower and the western Vale, less frequent in the south and east. Map 843.

Juncus acutus L. Sharp Rush. Llymfrwynen.
Native. Dune-slacks. Locally frequent on the coast of Gower and eastwards to Kenfig Burrows. Map 844.

Juncus inflexus L. Hard Rush. Brwynen Galed.
Native. Marshes, wet heaths, boggy places etc. Abundant throughout the county. Map 845.

Juncus effusus L. Soft-rush. Brwynen Babwyr.
Native. Boggy pastures and wet places generally. Abundant throughout the county. Map 846.

Juncus effusus × **J. inflexus** (J. × **diffusus** Hoppe), Diffuse rush, Brwynen Dryledol, has been recorded four times in the past and should be looked for where the parents occur together.
07B, Peterstone Moors, HJR c.1906; 49C, Cwm Ivy, J.D. Grose 1936*; 69D, Crymlyn Bog, EFL c.1890; 88C, Kenfig Hill, BAM 1957.

Juncus conglomeratus L. (*J. subuliflorus* Drejer). Compact Rush. Brwynen Bellennaidd
Native. Marshy places on acid soils. Abundant throughout the county except in the southern Vale. Map 847.

Juncus conglomeratus × **J. effusus** (J. × **kernreichgeltii** Jansen & Wachter ex Van Ooststr.). There are plants of this hybrid in the Jodrell Botanic Gardens of Manchester University that are labelled as having originated at Barry (ST16A).

[The record of **Juncus conglomeratus** × **J. inflexus** (**J.** × **ruhmeri** Ascherson & Graebner) from Crymlyn Bog (69D, 1890) is now considered to be an error]

Juncus squarrosus L. Heath Rush. Brwynen Droellgorun.
Native. Acid moorlands and boggy heaths. Abundant in the Uplands, less frequent in Gower and the northern Vale and absent from the southern Vale. Map 848.

Juncus gerardi Loisel. Saltmarsh Rush. Brwynen Gerard.
Native. Salt-marshes. Locally very abundant in coastal salt-marshes, especially on the coast of Gower. Map 849,

Juncus tenuis Willd. Slender Rush. Brwynen Fain.
Introduced. Roadsides and tracks. Common in the Uplands, first recorded in 1903 at Aberdare (00C, HJR*), still spreading and increasing in abundance but rather local in Gower and the Vale and absent from the southern Vale. Map 850.

Juncus bufonius L. s.l. Toad Rush. Brwynen y Llyffant.
Native. Wet grassland, wet roadsides etc. Abundant throughout the county. Map 851.

Juncus bufonius has recently been split into several segregate species, three of which have been recorded from Glamorgan. The treatment of these three species that follows is necessarily provisional, especially that of **J. bufonius** s.s. which is probably abundant throughout the county.

Juncus foliosus Desf.
Native. Wet grassland and roadsides.
49D, Llanrhidian, V. Gordon 1980*; 58A, Cefn Bryn, HAH, 1924*; 61C, Garn-swllt, 1981!*.

Juncus bufonius L. s.s. (incl. *J. minutulus* V. Krecz. & Gontsch.).
Native. Wet places generally. Map 852.

Juncus ambiguus Guss. (*J. ranarius* Song. & Perr.).
Native. Salt-marshes and mud-flats. Probably locally frequent in this habitat.
06A, Aberthaw, 1979!; 17D, Grangetown, Cardiff, HJR c.1908; 49C, The Groose, 1980!*; 49D, Landimore, 1981!; 59A, Llanmorlais, 1981!; 59C, Penclawdd, 1981!; 79C, Crymlyn Burrows, R. FitzGerald 1988.

Juncus subnodulosus Schrank Blunt-flowered Rush. Brwynen Flodbwl.
Native. Base-rich marshes and dune-slacks. Frequent in Gower and the western Vale, rare elsewhere in the Vale and almost absent from the Uplands. Map 853.

Juncus bulbosus L. s.l. Bulbous Rush. Brwynen Oddfog, Brwynen Fwlbaidd.
Native. Wet heaths, marshes and bogs. **Juncus bulbosus** s.l. is sometimes split into two ill-defined segregate species, **J. bulbosus** L. s.s. and **J. kochii** F.W. Schultz. **J. bulbosus** s.s. appears to be very rare in the county with only two certain records (07A, Mynydd y Glew, RLS 1930*; 90C, Llyn Fach, 1969!), but **J. kochii** is common throughout, except for the southern Vale. Map 854.

Juncus acutiflorus Ehrh. ex Hoffm. Sharp-flowered Rush. Brwynen Flodfain.
Native. Marshes, bogs, streamsides etc. Common throughout the county except in parts of the Vale. Map 855.

Juncus articulatus L. Jointed Rush. Brwynen Gymalog.
Native. Marshes, dune slacks, wet roadsides and railway lines. Abundant throughout the county. Map 856.

LUZULA DC.

Luzula campestris (L.) DC. Field Wood-rush. Coedfrwynen y Maes.
Native. Pastures, dune grassland, dry heaths, roadsides etc. Abundant throughout the county. Map 857.

Luzula multiflora (Retz.) Lej. Heath Wood-rush. Coedfrwynen Luosben.
Native. Heaths and pastures on acid soil. Common throughout the county except in the southern Vale. Map 858.

Luzula sylvatica (Hudson) Gaudin Great Wood-rush. Coedfrwynen Fawr.
Native. Woods, rocky places and stream-banks. Frequent in the central Uplands and the eastern Vale, rare elsewhere. Map 859.

Luzula pilosa (L.) Willd. Hairy Wood-rush. Coedfrwynen Flewog.
Native. Woods. Locally frequent in Gower, the eastern Vale and parts of the Uplands. Map 860.

GRAMINEAE (POACEAE)

Several species of bamboo occur in the county. They were often planted in ornamental gardens as at Margam (88A). A few populations remain after the garden has disappeared or have spread vegetatively beyond the confines of the garden. **Sasa palmata** (Burbidge) E.G. Camus is most frequently encountered in these situations (18C, Castell Coch, MG 1972; 87A, Newton, GH 1986!; 87B, Merthyr Mawr Village, 1982!*; 88A Margam Park, 1982!*) and **Arundinaria japonica** Siebold & Zucc. ex Steudel has been seen twice (61C, Garn-swllt, 1981!*; 88A Margam Park 1982!*).

FESTUCA L.

Festuca altissima All. Wood Fescue. Peisgwellt y Gwigoedd.
Native. Wooded ravines in the northern Uplands, usually near waterfalls; very scarce. Known from only two localities and probably extinct in one of these.
70B, Ystalyfera, HJR *c.*1907; 80B, valley of Afon Pyrddin near Sgŵd Einion Gam and Sgŵd Gwladys, 1925!, EV 1948*, 1973!**, 1985!.

Festuca gigantea (L.) Vill. Giant Fescue. Peisgwellt Mawr.
Native. Woods and hedges. Common in Gower and the Vale but local in the Uplands. Map 861.

Festuca pratensis Hudson Meadow Fescue. Peisgwellt y Waun.
Probably native in various types of grassland in the east of the county, but rather local in western Glamorgan where it occurs mainly in sown grassland or as a subspontaneous introduction. Map 862.

Festuca arundinacea Schreber Tall Fescue. Peisgwellt Tal.
Native. Dune slacks, cliff-scrub, grassy hedgebanks and roadsides, and occasionally on waste ground. Locally frequent in cliff-scrub and in suitable dune-slacks in the major dune-systems from Whiteford to Merthyr Mawr, and locally abundant on grassy hedgebanks, especially near the sea. Map 863.

Festuca nigrescens Lam. (*F. rubra* subsp. *commutata* Gaudin, *F. rubra* var. *fallax* (Thuill.) Hayek), Chewings Fescue, has been recorded on a few occasions as an introduction in lawns, grassy roadsides, wall tops and waste ground.
17B, Cardiff 1972!*; 17D, Llandough, 1920!*; 18B, Caerphilly Common, 1974!*; 18D, Castell Coch, EV 1932*; 88C, Pyle, EV *c.*1930*. The recent determinations are provisional only.

Festuca rubra L. Red Fescue. Peisgwellt Coch.
Native. Grassland on limestone and base-rich to neutral soils, sand-dunes, salt-marshes and sea-cliffs, cliff-ledges in the mountains, roadsides, etc.; often sown as an amenity grass, especially in lawns, on roadsides, and on reclaimed tips and waste ground. Common throughout the county. Several subspecies of F. rubra are described in Markgraf-Dannenberg (1980), and at least four occur in Glamorgan. These subspecies, however, appear to be little more than genetically distinct habitat forms which, with the possible exception of some sand-dune forms, intergrade with one another in intermediate habitats and presumably interbreed freely. Forms with blue, glaucous leaves (subsp. **pruinosa** (Hackel) Piper) are common on exposed sea-cliffs in Gower and the Vale. Populations growing on mobile sand-dunes are often robust, with rigid leaves and long stolons (subsp. **arenaria** (Osbeck) Syme). The fine-leaved salt-marsh forms, which are often dominant in the turf of upper salt-marshes, appear to be referable to subsp. **litoralis** (G.F.W. Meyer) Auquier. Most inland populations are referable to subsp. **rubra**. Map 864.

Festuca juncifolia St.-Amans Rush-leaved Fescue. Pesgwellt Brwynddail.
Native. Sand-dunes, in the **Ammophila arenaria** community. Recorded from only two localities, but may occur on other dune-systems (Stace & Cotton, 1974).
58A, Oxwich Bay, H.C. Evans 1959, above Three Cliffs Bay, 1978!**; 78B, Kenfig Burrows, P.L. Thomas *c.*1973

Festuca tenuifolia Sibth. (*F. ovina* L. subsp. *tenuifolia* (Sibth.) Peterm.) Fine-leaved Sheep's-fescue. Pesgwellt Meinddail.
Native. Acid grassland. Recorded from only four localities, but possibly widespread. F. tenuifolia and F. ovina are closely similar to one another and may be conspecific.
07A, Hensol Castle, HAH 1927*; 58A, Pennard, EV 1932*; 79B, Pont-rhyd-y-fen, SGH 1970*; 99A, Mynydd Maendy, SGH 1966.

Festuca ovina L. Sheep's-fescue. Peisgwellt y Defaid.
Native. Limestone, neutral and acid grassland, sand-dune grassland and grassy heathland. Common except in the southern Vale. In limestone grassland, F. ovina is usually less frequent than F. rubra, and occurs mainly on shallow soils and near rock outcrops. On sand-dunes F. ovina is usually confined to the older fixed dunes, but is sometimes abundant on old dune slacks. It is common on well-drained acid soils inland and it is abundant in Upland grassland and grassy heath. Several subspecies have recently been recognised within the F. ovina complex, some of which will undoubtedly be recorded from the county in the future. Map 865.

× FESTULOLIUM Ascherson & Graebner.

× **Festulolium loliaceum** (Hudson) P. Fourn. (**Festuca pratensis × Lolium perenne**). Hybrid Fescue.
A spontaneous hybrid, not uncommon with the parents, and probably more frequent than the small number of records suggests. Meadows and roadsides.
00C, Aberdare, HJR 1902*; 07D, St. Nicholas, MP 1970*; 48B, Penrice, WT *c.*1800; 80B, Pont-neddfechan, JWGG *c.*1842; 87B, near Ogmore Castle, EV 1935*; 90B, Hirwaun, 1974!; 97C Marcross, PJo 1982*. (Also recorded from 17 and 69 (Perring, 1968).

× FESTULPIA Melderis ex Stace & R. Cotton

× **Festulpia hubbardii** Stace & R. Cotton (**Festuca rubra × Vulpia fasciculata**)
Occasionally found with the parents on sand dunes (Stace & Cotton, 1974).

49C, Llangennith Burrows, C.A. Stace & R. Cotton c.1974 (2n=35), 1985!; 78D, Kenfig Burrows, 1969!; 79C, Jersey Marine, P.L. Thomas 1969.

× **Festulpia melderisii** Stace & R. Cotton (**Festuca juncifolia** × **Vulpia fasciculata**)
Closely similar to × **Festulpia hubbardii** but distinguished by its longer (9.5–10.5mm), often conspicuously hairy lemmas.
78B, Kenfig Burrows, P.L. Thomas c.1973 (Stace, 1975).

LOLIUM L.

Lolium perenne L.   Perennial Rye-grass. Rhygwellt Lluosflwydd.
Native, but so commonly sown to form pastures and hay-meadows and as an amenity grass that its native distribution is uncertain. Meadows, roadsides, grassy paths and grazed grassland on the better soils. Very common throughout the county, although it is restricted to roadsides and nutrient enriched grassland in the hills. Recorded from all squares except 49B. Map 866.

Lolium multiflorum Lam.   Italian Rye-grass. Rhygwellt Eidalaidd.
Introduced. Sown grassland, footpaths, roadsides, cultivated and waste land. Most frequent as a naturalized or sub-spontaneous plant in the Vale. Map 867.

Lolium multiflorum × **L. perenne** (**L.** × **hybridum** Hausskn.) has been recorded from the county but without locality (Stace, 1975).

Lolium temulentum L. Darnel. Efrau.
Casual, on waste ground, usually near docks. No recent records.
90D, Llwydcoed, HJR 1902; 17B, Cardiff Ballast, JS c.1876*, Maendy, RLS 1938*; 17D, Grangetown, JS c.1876, 1920!*; 27A, Splott, 1925–26!*; 78B, Port Talbot Docks, HJR 1907.

Lolium rigidum Gaudin was twice recorded as a casual of waste places by H.J. Riddelsdell (00C, Aberdare, 1902; 78B, Port Talbot, 1905).

VULPIA C.C. Gmelin

Vulpia fasciculata (Forskål) Samp. (**V. membranacea** auct.).   Dune Fescue. Peisgwellt Uncib.
Native. Sand-dunes, often associated with **Phleum arenarium** on open but stabilized dunes. Locally or generally abundant in the major dune-systems from Whitford Burrows to the Ogwr (Ogmore); probably introduced at Porthkerry (06B) and on tips near Hirwaun (90). Map 868.

Vulpia bromoides (L.) S.F. Gray Squirreltail Fescue. Peisgwellt Anhiliog.
Native. Dry open grassland, roadsides, wall-tops, waste ground and colliery tips. Fairly common in Gower and in the lower Uplands but rather local in the Vale. Map 869.

Vulpia myuros (L.) C.C. Gmelin   Rat's-tail Fescue. Peisgwellt y Fagwyr.
Native or introduced. Railway ballast, waste ground near collieries and factories, occasionally on disturbed roadsides. Closely associated with railways, especially those on which coal has been transported, and often abundant around railway sidings. Map 870.

Vulpia geniculata (L.) Link and **V. ligustica** (All.)   Link have both been recorded as aliens of waste ground in Splott (27A), the former by R.L. Smith in 1926 and the latter in 1925!*.

DESMAZERIA Dumort.

Desmazeria marina (L.) Druce   Sea Fern-grass. Corwenithwellt y Morfa.
Native. On sea-cliffs in or near the spray-zone, on consolidated shingle-beaches, and occasionally on sand. Locally abundant on the coast, especially in southern and western Gower. Map 871.

Desmazeria rigida (L.) Tutin   Fern-grass. Gwenithwellt Caled.
Native. Shallow soils on limestone, especially on sea-cliffs, sand-dunes, wall-tops, and occasionally on railway ballast. Common near the coast but scarce inland, where it is most frequent on old walls. Map 872.

POA L.

Poa annua L.   Annual Meadow-grass. Gweunwellt Unflwydd.
Native. Open grassland, paths, farmyards, cultivated land, roadsides and waste ground. Very common as a weed and in trampled grassland throughout the county. Recorded from all squares except 49B. Map 873.

Poa trivialis L.   Rough Meadow-grass. Gweunwellt Lledarw.
Native. Grassland, woods and hedgebanks, cultivated and waste ground. Common throughout the county. Map 874.

Poa pratensis L. s.l.
The segregates of **Poa pratensis** (**P. subcaerulea**, **P. pratensis** and **P. angustifolia**) intergrade to some extent and have not been consistently distinguished from one another by recorders. Map 875.

Poa subcaerulea Sm.   Spreading Meadow-grass. Gweunwellt Helaeth.
Native. Dune grassland, limestone grassland, roadsides, railway embankments, wall-tops and occasionally as a weed of gardens. Probably underrecorded. Map 876.

Poa pratensis L. *s.s.* Smooth Meadow-grass. Gweunwellt Llyfn.
Native. Pastures, limestone grassland, dune grassland, roadsides, footpaths, railway embankments, lawns etc. Frequently sown for forage and in amenity grassland. Probably common throughout the county, normally as a minor constituent of mixed grassland but under-recorded in the west. Map 877.

Poa angustifolia L. Narrow-leaved Meadow-grass. Gweunwellt Culddail.
Native. Limestone grassland and roadside banks. Locally abundant in the eastern Vale, occasionally reported elsewhere; perhaps under-recorded. Map 878.

Poa chaixii Vill. Broad-leaved Meadow-grass, Gweunwellt Llydanddail, has once been recorded as a casual in an allotment (00C, Abernant Park, Aberdare, HJR 1905*).

Poa compressa L. Flattened Meadow-grass. Gweunwellt Cywasg.
Native or introduced. Grassy shingle-banks by the sea, roadsides, railway ballast and embankments, wall-tops, basic slag-tips, waste ground. Scarce, but scattered throughout the county; most frequent in the Uplands and the Cardiff district. Map 879.

Poa palustris L. Swamp Meadow-grass. Gweunwellt yr Afon.
Introduced or casual. Damp places near the sea, and waste ground near docks; recently recorded from Aberthaw.
06A, Aberthaw, BSBI field meeting 1979, H & CH 1980; 16A, Barry, GCD 1925; 17B, Cardiff Docks, HJR c.1907; 27A, Splott, 1925–26!, EV 1927*.

Poa nemoralis L. Wood Meadow-grass. Gweunwellt y Coed.
Probably native, but more frequent as an introduction. Woods, shaded hedgebanks and parks, often associated with planted trees. Scarce and local; probably native only on the eastern borders of the county. Map 880.

Poa bulbosa L. var. **vivipara** Koeler Bulbous Meadow-grass. Gweunwellt Oddfog.
Native. Shingle beaches. Rare, limited to one locality on the border of two recording squares.
06B, Bull Cliff, Porthkerry, 1956!, R.J. Tidswell 1983*; 16A, Cold Knap, HJR & EV 1907*, 1974!.

PUCCINELLIA Parl.

Puccinellia distans (L.) Parl. Reflexed Saltmarsh-grass. Gweunwellt Gwrthblygedigaidd.
Native. In the drier parts of salt-marshes. Rather local, and far less abundant than **P. maritima**, with recent records from only five localities; perhaps overlooked in grazed salt-marshes. Map 881.

Puccinellia fasciculata (Torrey) E.P. Bicknell. Borrer's Saltmarsh-grass.
Native or introduced; recorded from only one site that appears to have been destroyed.
27A, Salt-marsh near Cardiff Docks, HJR 1904, 1905*.

Puccinellia maritima (Hudson) Parl. Common Saltmarsh-grass. Gweunwellt Arfor.
Native. Salt-marshes, from the pioneer zone to the **Juncus maritimus** level; occasionally in the spray-zone on cliffs. Very abundant in the Gower salt-marshes and in salt-marshes near Cardiff but scarce elsewhere. Map 882.

Puccinellia rupestris (With.) Fernald & Weatherby Stiff Saltmarsh-grass. Gweunwellt Anhyblyg.
Native. In salt-marshes near the drift-line. Scarce, with confirmed records only from the Pengam marsh near Cardiff.
27A, Pengam, HJR 1905*, RLS 1925*, 1969!*.

DACTYLIS L.

Dactylis glomerata L. Cock's-foot. Byswellt, Troed y Ceiliog.
Native. Cliff grassland, rough grassland in the lowlands and valleys, hedgebanks, roadsides and waste ground. Occasionally sown for forage. Common throughout the county and recorded from every square except 49B. Map 883.

CYNOSURUS L.

Cynosurus cristatus L. Crested Dog's-tail. Rhonwellt y Ci.
Native. Grassland on all but the most acid soils. Often the most abundant grass in old unimproved pastures on neutral soils. Common throughout the county and recorded from all but three squares. Map 884.

Cynosurus echinatus L. Rough Dog's-tail. Rhonwellt y Ci Pigog.
Casual, on waste ground. Only one recent record.
08D, near Llantrisant, A.L. Still 1935*; 16A, Barry, RLS 1925, Barry Docks, JPC 1980!; 17D, Grangetown, RLS 1922*; 27A, Splott, JS c.1876*, RLS 1927*.

Cynosurus elegans Desf. has been recorded as an alien from one site in Cardiff (27A, Splott, RM & RLS 1926, RLS 1927*).

CATABROSA Beauv.

Catabrosa aquatica (L.) Beauv. Whorl-grass. Brigwellt Dyfrdrig.
Probably native, but with no recent records, and fairly certainly extinct in one of the two localities in which it has been found.
27A, East Moors and Pengam Moors, JS c.1886, probably extinct by 1933 (Vachell, 1934); 79C, between Neath and Briton Ferry, EMT 1938*.

APERA Adanson

Apera spica-venti (L.) Beauv. Loose Silky-bent.
Maeswellt Sidanaidd.
Casual, on waste ground, usually near docks.
Recently recorded from 16A, Barry Docks, JPC &
TGE 1986; 27C, Cardiff Docks, JPC & TGE 1986.
Recorded in the past from: 00C, Aberdare, HJR 1904;
17A, Caerau, HAH 1927*; 17D, Grangetown and
Penarth Ferry, JS *c*.1887; 27A, Splott, 1926!*, E.
Cardiff, RLS 1938*; 69D, Kilvey Hill, JWGG *c*.1840;
78B, Port Talbot, HJR 1904.

Apera interrupta (L.) Beauv. Dense Silky-bent,
Maeswellt Sidanaidd Trwchus, has three times been
recorded as a casual on waste ground (00A, Merthyr
Tydfil, G.R. Willan 1912*; 17B, Cardiff Docks, RM
1935*; 17D, Grangetown JS *c*.1886).

MIBORA Adanson

Mibora minima (L.) Desv. Early Sand-grass. Eid-
dil-welltyn Cynnar.
Native. Open grassland on stabilized sand-dunes.
Rare and known only from Whiteford Burrows,
where it was first found in 1964, and is fairly
abundant in an area of about two hectares.
49A & 49C, Whiteford Burrows, E. Duffy & GTG
1964, 1971!**, 1985!.

BRIZA L.

Briza media L. Quaking-grass. Crydwellt.
Native. Grazed grassland on basic, neutral and
mildly acid soils from dunes and limestone sea-cliffs
to hill pastures. Fairly abundant in Gower and the
Vale, but absent from the central and north-western
Uplands and local elsewhere in the hills. Map 885.

Briza maxima L. Great Quaking-grass. Crydwellt
Mwyaf.
Casual, on waste ground; sometimes introduced with
bird-seed or as a garden escape.
07C, Pendoylan, E. Elward 1977*; 18C, Radyr, 1922!*;
27C, Cardiff Docks, RM 1935*; 68A, Mumbles, R.E.
Castell 1984!; 87A, Porthcawl, D.C. Griffith 1982.

The following grain-sifting aliens were found at
Splott (27A): **Briza minor** L. (RLS 1926!); **Lamarckia
aurea** (L.) Moench (RLS 1926!, 1927); **Beckmannia
eruciformis** (L.) Host (1925!*) and **B. syzigachne**
(Steudel) Fernald (RM 1926), this latter species has
also been recorded recently from Barry Docks (16A,
TGE 1981).

MELICA L.

Melica nutans L. Mountain Melick. Meligwellt
Gogwydd.
Native. Limestone woodland in the Uplands. Rare,
occurring only on Morlais Castle Hill.
00A, G.R. Willan 1908*, HAH 1923!*, 1969!.

Melica uniflora Retz. Wood Melick. Meligwellt,
Meligwellt y Goedwig.
Native. Woods and shaded hedgebanks, mainly on
limestone. Locally abundant in limestone woodland
in Gower and the Vale, and on limestone or relatively
base-rich soils in the northern Uplands; scarce or
absent in the rest of the county. Map 886.

GLYCERIA R. Br.

Glyceria maxima (Hartman) Holmberg Reed Sweet-
grass. Perwellt.
Native. Canals, reens and slow rivers; local and
mainly near the coast except along the Swansea and
Neath canals. Most Glamorgan rivers are too fast-
flowing for **G. maxima**, but the canals and deep
drainage ditches (reens) in coastal levels provide it
with excellent man-made habitats into which it
probably spread from natural habitats in a few
coastal marshes and slow-flowing rivers. Map 887.

Glyceria declinata Bréb. Small Sweet-grass. Perwellt Llwydlas.
Native. Muddy ditches, pools and small streams, and
on muddy ground near the drinking-place of cattle
and horses, mainly on neutral or acid soils. Rather
common in Gower and parts of the Uplands (perhaps
under-recorded in the eastern Uplands) but scarce in
the Vale. Map 888.

Glyceria fluitans (L.) R. Br. Floating Sweet-grass.
Glaswellt y Dŵr.
Native. Ditches and ponds on the better soils. Locally
common in Gower and the Vale, but scarce in the
western Uplands. Map 889.

Glyceria plicata Fries Plicate Sweet-grass. Perwellt
Plygedig.
Native. Ditches, pools and small streams. Com-
mon in Gower, the north-western Uplands and the
southern Vale but with rather few records elsewhere.
Map 890.

Glyceria fluitans × G. plicata (G. × pedicellata
Townsend). Hybrid Sweet-grass. Perwellt Croes-
ryw.
Native, in habitats similar to those of the parents.
Probably fairly widespread in Gower and the
southern Vale, but under-recorded. Map 891.

BROMUS L.

Bromus diandrus Roth Great Brome. Pawrwellt
Mawr.
Casual or introduced. It occurs occasionally as an
impersistent casual on waste ground near the coast. It
has been increasing as a weed of roadsides and waste
places in southern England and recent records from
the Cardiff district suggest that it may also establish
itself in Glamorgan. Map 892.

Bromus sterilis L. Barren Brome. Pawrwellt Hysb.

Native. Roadsides, cultivated land and waste ground. Fairly common in Gower and the coastal regions of the Vale, and locally common on the borders of the Uplands, but scarce or absent in the central Vale and the central Uplands. Map 893.

Bromus madritensis L. Compact Brome. Pawrwellt Dwysedig.
Native or introduced. Open habitats on cliffs, roadsides and waste ground near the sea. Possible native on cliffs from Porthkerry to Penarth. Map 894.

Bromus tectorum L. Drooping Brome, Pawrwellt Llipa, a casual on waste ground usually near docks has been recorded several times, but not recently (16A, Barry Docks, RLS & RM 1925*; 17B, Cardiff, HJR *c*.1903, RM 1925, EV 1937*; 27A, Splott, 1925!*, 1926!*; 78B, Port Talbot Docks, HJR 1904, EV 1937*; 90D, Llwydcoed, HJR 1902), **B. rigidus** Roth, Ripgut Brome, has been recorded as an alien of waste places on three occasions (17B, Cardiff, HJR *c*.1905; 78B, Port Talbot, HJR *c*.1905; 87A, Porthcawl, C.W. Banister 1946), and **B. rubens** L. in similar habitats twice (17B, Cardiff, JS *c*.1876*; 27A, Splott, RLS 1926!*).

Bromus ramosus Hudson Hairy-brome. Pawrwellt Blewog.
Native. Woods and shaded hedges, usually on limestone except in the eastern Uplands. Local and rather scarce in Gower, more frequent in the Vale and eastern Uplands. Map 895.

Bromus erectus Hudson Upright Brome. Pawrwellt Unionsyth.
Native. Limestone grassland, usually on sea-cliffs; occasionally on calcareous dunes. Common on the Vale coast, but scarce in Gower where it appears to be a relatively recent introduction still occurring in small numbers but spreading along cliff footpaths. A long-established patch of **B. erectus** sward on sand-dunes near the former Blackpill railway station (69C) was destroyed by landscaping in 1975. Map 896.

Bromus arvensis L. Field Brome. Pawrwellt y Maes.
Casual, usually found on waste ground near docks but occasionally in grassland in the past. Recently recorded from 17D, Penarth, E. Cawood and K. Baker 1973, and 18C, Forest Farm, RAH 1978*. Map 897.

Bromus secalinus L. Rye Brome. Pawrwellt Ller.
Casual, on waste ground near docks and formerly as a rare weed of cornfields. No recent records. Last recorded by E. Vachell in 1937 from Barry Island (16A).
00C, Aberdare, HJR 1903*; 06A, St. Athan's, HJR *c*.1907; 07D, Bonvilston, JS *c*.1880; 16A, Cadoxton, HJR *c*.1907, Barry Docks, 1922!*, Barry Island, EV 1937*; 16B, Cog Moors & Swanbridge, AHT 1913*; 17B, Cardiff, HJR *c*.1907; 18D, Llanishen, HJR *c*.1907; 58A, Pennard Castle, J. Ball *c*.1849; 78B, Port Talbot, HJR *c*.1907; 88C, Pyle, AHT 1913*.

Bromus commutatus Schrader Meadow Brome. Pawrwellt Mwyaf y Maes.
Native or introduced. Meadows, roadsides, cultivated land and waste ground. Scarce, with only five recent records. Map 898.

Bromus racemosus L. Smooth Brome. Pawrwellt Llyfn.
Native. Meadows, damp grassland and roadsides. Rather infrequent, but widely distributed. Map 899.

Bromus hordeaceus L. subsp. **hordeaceus** (*Bromus mollis* L.). Soft-brome. Pawrwellt Masw.
Native. Grassland, sand-dunes, roadsides, cultivated land and waste ground. Very common throughout the county; by far the commonest annual brome. Map 900.

Bromus hordeaceus L. subsp. **ferronii** (Mabille) P.M. Sm. (*Bromus ferronii* Mabille). Least Soft-brome. Pawrwellt Arfor.
Native. Shallow soils on limestone sea-cliffs. Locally abundant in Gower, and also recorded from Flat Holm. Map 901.

Bromus hordeaceus L. subsp. **thominii** (Hard.) Maire & Weiller
Native. Dune grassland. Rare and reliably recorded from only two sites.
49C, Whiteford Burrows, 1984!** (det. P.M. Smith); 79C, Crymlyn Burrows, W.R. Linton 1888* (det. F.H. Perring). (All other records for the county refer to **B.** × **pseudothominii**).

Bromus hordeaceus × **B. lepidus** (**B.** × **pseudothominii** P.M. Sm.) (*B. thominii* sensu Tutin, non Hard.). Lesser Soft-brome. Pawrwellt Minffordd.
Native. Limestone grassland, dune grassland and roadsides. Recorded chiefly from the Vale; common on Kenfig Burrows. Map 902.

Bromus lepidus Holmberg Slender Soft-brome. Pawrwellt Gweirglodd.
Native. Roadsides, waste ground, and occasionally in meadows. Rather infrequent and mainly in the east. Map 903.

Four species of **Bromus** have been recorded as casuals of waste places. **Bromus squarrosus** L. (27A, Splott, 1926!*; 79C, Jersey Marine, EMT *c*.1930*), **B. japonicus** Thunb. (27A, Cardiff Docks, HJR 1902, Splott, 1926!*), **B. lanceolatus** Roth (27A, Splott, 1926!*, Pengam, RLS 1940*), and **B. briziformis** Fischer & C.A. Meyer (90D, Llwydcoed, HJR 1907). Two further species of **Bromus**, sometimes separated as *Ceratochloa* Beauv. have been recorded: **B. willdenowii** Kunth (*Ceratochloa unioloides* (Willd.) Beauv.), Rescue Brome, as a casual of waste ground on several occasions (00C, Aberdare, HJR 1902*; 17B, Cardiff, HJR 1902, RLS 1925; 18C, Radyr, RLS *c*.1922; 27A, Splott, M. Hall 1932; 50D, Pontardulais, HJR *c*.1905), and **B. carinatus** Hooker & Arnott (*Ceratochloa carinata* (Hooker & Arnott) Tutin),

California Brome, which is apparently established at Barry Docks (16A, 1922!*, RLS 1957*, DMcC 1974).

BRACHYPODIUM Beauv.

Brachypodium sylvaticum (Hudson) Beauv.   False Brome. Breichwellt y Coed.
Native. Woods, hedgebanks and scrub, and in rough grassland on limestone cliffland. Common in Gower and the Vale, locally common in most of the Uplands but scarce or absent in the high central uplands. Map 904.

Brachypodium pinnatum (L.) Beauv.   Tor-grass. Breichwellt y Twr.
Native. Limestone grassland near the coast; occasionally found as a casual on waste ground.
00C, Aberdare, HJR 1902*; 06A, near Aberthaw, MG 1975; 06B & 16A, Porthkerry and Cold Knap, 1972!*; 16A, Barry Docks, EV 1935*; 17D, The Kymin, Penarth, JPC 1982; 96A, Tresilian Bay, EV 1932*.

LEYMUS Hochst.

Leymus arenarius (L.) Hochst. (*Elymus arenarius* L.).   Lyme-grass. Amdowellt, Lymwellt.
Native. Sand-dunes, often on low foredunes. Locally common on Whiteford and Llangennith Burrows in western Gower and at Crymlyn Burrows. It has increased in abundance in western Gower since 1965, when it was scarce and confined to part of Whiteford Burrows, and may spread to the other major dune-systems.
48A, between Paviland and Rhosili, 1970!; 49A, Whiteford Burrows, 1956!*, 1972!; 49B, Berges Island, 1974!; 79C, Crymlyn Burrows, B.W. Bryant 1957, 1972!. (Also recorded, probably as a casual (Trow, 1911) from 06B, Porthkerry, JS *c*.1876; 69, Swansea, M. Moggridge *c*.1840; 78B, Port Talbot, HJR 1910*).

ELYMUS L.

Elymus caninus (L.) L. (*Agropyron caninum* (L.) Beauv.).   Bearded Couch. Marchwellt y Coed.
Native. River-banks, occasionally in woods and hedgebanks. Locally frequent in the Uplands where it is a characteristic river-bank plant; rather scarce in the Vale, where it occurs more often in woods and hedgebanks than by rivers, and rare in Gower. Map 905.

Elymus pycnanthus (Godron) Melderis (*Agropyron pungens* auct.).   Sea Couch. Marchwellt Arfor.
Native. Sandy or relatively dry high-level salt-marshes and low dunes near the drift-line, shingle-beaches, sea-walls. Locally frequent in suitable habitats on the coast, especially in northern Gower. Map 906.

Elymus pycnanthus × **E. repens** (E. × **oliveri** (Druce) Melderis & D. McClintock) (*Agropyron* × *oliveri* Druce) was once recorded from Port Talbot Docks (78B, HJR 1904).

Elymus repens (L.) Gould (*Agropyron repens* (L.) Beauv.).   Common Couch. Marchwellt.
Native. Cultivated land, waste ground, roadsides, and rough grassland near the sea. Common in Gower and the Vale and in parts of the Uplands, but remarkably scarce in the central Uplands, where it most frequently occurs, if at all, in allotment gardens. Map 907.

Elymus farctus (Viv.) Runemark ex Melderis subsp. **boreoatlanticus** (Simonet & Guinochet) Melderis (*Agropyron junceiforme* (Á. & D. Löve) Á. & D. Löve).   Sand Couch. Marchwellt Tywyn.
Native. Sand-dunes, especially on fore-dunes. Common in suitable habitats on the major dune-systems from Whiteford Burrows to the Ogwr (Ogmore); fairly frequent above the drift-line on sandy shingle-beaches. Map 908.

Elymus farctus × **E. repens** (E. × **laxus** (Fries) Melderis & D.McClintock) (*Agropyron* × *laxum* (Fries) Tutin).
Native. Recorded from three localities in the Vale.
06A, Aberthaw, 1958!*; 78B, Port Talbot, HJR 1904*; 87A, Porthcawl, HJR *c*.1907.

Elymus farctus × **E. pycnanthus** (E. × **obtusiusculus** (Lange) Melderis & D. McClintock) (*Agropyron acutum* auct, *A.* × *obtusiusculum* Lange).
Native. Fairly frequent in Gower, usually occurring with the parents in intermediate habitats. Recently recorded from 06A!*, 48B, 49B!, 58A!, 68D, 69C!, 79C!.

Five species of **Aegilops** L. have been found in the past in Cardiff or Port Talbot as dockside or grain aliens: **Aegilops geniculata** Roth three times (27A, JS *c*.1876*, 1926!*; 78B, HJR 1906**), **A. triuncialis** L. (17B, JS *c*.1876; 27A, Splott, RLS 1926) and **A. neglecta** Req. ex Bertol. (27A, Splott, 1926!*, T.J. Foggitt 1929) twice, and **A. ventricosa** Tausch (27A, 1926!*) and **A. cylindrica** Host (17D, RLS 1937*) once each.

TRITICUM L.

Triticum aestivum L., Bread Wheat, Gwenith, has been recorded on several occasions as an impersistent relic of cultivation. Map 909.

Triticum spelta L. has once been recorded as a grain alien (27A, Splott, 1926!*).

Secale cereale L., Rye, Rhyg, has been recorded on several occasions as a dock or grain alien or, rarely, as an impersistent relic of cultivation (00A, Aberdare, HJR *c*.1905; 16A, Barry Docks, 1924!*, Barry, H. Thomas 1983; 17B, Cardiff Docks, RLS 1924*; 27A, Splott, 1926!*; 78B, Port Talbot, HJR *c*.1905).

HORDEUM L.

Hordeum murinum L.   Wall Barley. Heiddwellt y Mur.

Native or introduced. Roadsides, footpaths and waste ground. Common in urban areas near the sea but local inland and in the countryside; scarce in the Uplands. Map 910.

Hordeum marinum Hudson Sea Barley. Heiddwellt y Morfa.
Native. Maritime grassland and waste ground in the eastern Vale. Recently recorded only at Aberthaw.
06A, The Leys, Aberthaw, HJR c.1907, H & CH 1980; 16A, Barry Island, P.N. Lewis 1961*; 17D, Ely estuary, AHT 1907; 27A, Rhymney estuary, 1923!*, Splott, 1925!*. [Two 19th Century records from the Swansea area are now considered doubtful (69D, about Swansea, LWD c.1805; 79C, Briton Ferry, WT 1802)]

Hordeum secalinum Schreber Meadow Barley. Heiddwellt y Maes.
Native. Meadows and grassy banks, usually near the sea. Locally frequent in the southern Vale, where it occurs both by the sea and inland; very local in Gower where it grows only on grassy sea-walls and near the upper fringes of salt-marshes. Map 911.

Hordeum hystrix Roth (27A, Splott, RLS & RM 1925, RLS 1926*) and **H. jubatum** L. (17B, Cardiff Docks, RM 1924*; 27A, Splott, 1925!, EV 1933*) have been found in the past as rare grain aliens in Cardiff, and the latter has recently been found again both in Cardiff (18C, disturbed ground GH 1982*) and near Swansea (58B, garden weed, Bishopston, W. Hankinson 1985). **H. distichon** L., Two-rowed Barley, and **H. vulgare** L., Six-rowed Barley, often grow on roadsides or waste ground as impersistent relics of cultivation. They have not been systematically recorded, but **H. distichon** has been found in ten localities (06A, 09A, 16A, 17B, 17C, 18B, 69, 88D, 90D, 98D), and **H. vulgare** in six (00C, 06A, 17D, 78B, 79A, 80C).

AVENA L.

Avena fatua L. Wild-oat. Ceirchwellt Gwyllt y Gwanwyn.
Introduced. Cultivated land, roadsides and waste ground; increasing as a weed of cereals, and now locally frequent in the Vale but still relatively scarce in Gower. Map 912.

Avena sativa L. Oat. Ceirch.
Introduced. Cultivated land, roadsides and waste ground. Frequently occurs in small numbers as a weed of farmland or as a casual on waste ground; the least impersistent of the cultivated cereals. Map 913.

Avena strigosa Schreber, Bristle Oat, Blewgeirch, has been recorded as a casual on waste ground, or as a grain alien, on several occasions (06A, near Fonmon, 1975!; 16A, Barry Docks, 1923!*; 17A, St. Fagans, EV 1943*; 27A, Splott, EV 1930*, 1937, Pengam, ABP 1979!), **A. sterilis** L. subsp. **sterilis**, Animated Oat, has occurred as a grain alien at Cardiff and Barry Docks (16A, Barry, 1922!*, 1980!*, H. Thomas 1983; 27A,

Splott, 1925!*; 27C, Roath Dock, ABP 1979!*), and **A. sterilis** subsp. **ludoviciana** (Durieu) Nyman, Winter Wild-oat, Ceirchwellt Gwyllt yr Hydref, has been recorded as a casual in three sites (27A, Splott, 1926!*; 27C, Cardiff Docks, 1986!*; 61C, Garnswllt, 1981!*).

AVENULA (Dumort.) Dumort.

Avenula pubescens (Hudson) Dumort. (*Helictotrichon pubescens* (Hudson) Pilger). Downy Oat-grass. Ceirchwellt Blewog.
Native. Limestone cliffs, limestone grassland, calcareous fixed dunes, roadsides, field banks and railway embankments. Fairly common in suitable habitats in Gower and the Vale; rare in the Uplands. Map 914.

Avenula pratensis (L.) Dumort. (*Helictotrichon pratense* (L.) Pilger). Meadow Oat-grass. Ceirchwellt Culddail.
Native. Calcareous grassland. Rare, occurring only in a few sites in the southern Vale.
06B, Porthkerry and Swanbridge, HJR 1906; 18B, Caerphilly, A.G. Cadogan 1954; 16B, Lavernock, 1972!*; 87B, Pant Mari Flanders, ARP 1984; 97C, Cwm Nash, P. Grubb c.1975.

ARRHENATHERUM Beauv.

Arrhenatherum elatius (L.) Beauv. ex J. & C. Presl False Oat-grass. Ceirchwellt Tal.
Native. Rough grassland, roadsides, hedgebanks, orchards and gardens. Common in Gower and the Vale and in most Upland valleys; scarce or absent only in the central Uplands. Map 915.

GAUDINIA Beauv.

Gaudinia fragilis (L.) Beauv. French Oat.
Casual or introduced. Waste ground and open grassy places near the sea. May still persist west of Porthkerry (06B).
06B & 16A, shore near Cold Knap, HJR 1906*, RLS c.1925!; 17B, Cardiff, RLS c.1925; 27A, Splott 1926!*.

KOELERIA Pers.

Koeleria macrantha (Ledeb.) Schultes (*Koeleria cristata* (L.) Pers. *p.p.*). Crested Hair-grass. Cribwellt.
Native. Limestone grassland and calcareous sand-dunes. Locally common on limestone cliff-land in southern and western Gower and the western Vale, and less frequent inland in the same areas; few recent records in the eastern Vale. Map 916.

Koeleria glauca (Schrader) DC. Dune Hair-grass. Cribwellt y Tywod.
Native. Sandy soils. Not separated from **K. macrantha** during the survey, possibly locally frequent in suitable habitats along the coast. The following records are from specimens provisionally determined as this species in **NMW**.

16B, Sully Island, 1921!*; 58A, Pennard Castle, EMT 1920*; 97B, Llansannor, 1970!*.

Lophochloa cristata (L.) Hyl. (*Koeleria phleoides* (Vill.) Pers.) has been recorded in the past as a casual on waste ground near docks (16A, Barry Docks, RLS 1924; 17B, Cardiff Docks, JS *c*.1876*; 27A, Splott, RLS 1926!*).

TRISETUM Pers.

Trisetum flavescens (L.) Beauv. Yellow Oat-grass. Ceirchwellt Melyn.
Native. Grassland on cliffs and dunes, old meadows, field banks and roadsides, usually on calcareous soils. Common in Gower and the Vale on limestone, also locally frequent in the north-eastern Uplands. Map 917.

LAGURUS L.

Lagurus ovatus L. Hare's-tail. Cwt Ysgyfarnog.
Casual, on waste ground near the sea as a grain alien or escape from cultivation; one recent record.
16A, Barry Docks, MG 1979*; 17D, Grangetown, JS *c*.1876; 27A, Splott, RLS 1927*; 69C, Blackpill, JAW 1927; 87A, Porthcawl, JAW 1923.

DESCHAMPSIA Beauv.

Deschampsia cespitosa (L.) Beauv. Tufted Hair-grass. Brigwellt Cudynnog.
Native. Damp grassland, also in woods. Fairly common throughout the county, but unrecorded in several coastal squares. Map 918.

Deschampsia flexuosa (L.) Trin. Wavy Hair-grass. Brigwellt Main.
Native. Dry acid soils. Common in the Uplands but rather local in Gower and the northern Vale, where it is to some extent replaced by **Agrostis curtisii**; absent from the limestone soils of the southern Vale. Map 919.

AIRA L.

Aira praecox L. Early Hair-grass. Brigwellt y Gwanwyn.
Native. Shallow dry soils in open grassland, ant hills, sand-dunes; usually on acid soils. Common in most of Gower and the Uplands but absent or overlooked in some areas; local in the Vale where it is frequent only on sand-dunes in the west and acid soils in the north. Map 920.

Aira caryophyllea L. Silver Hair-grass. Brigwellt Arian.
Native. In habitats similar to those of **A. praecox**, but usually more abundant on coastal cliffs, and less frequent in the Uplands. Map 921.

ANTHOXANTHUM L.

Anthoxanthum odoratum L. Sweet Vernal-grass. Perwellt y Gwanwyn, Eurwellt.

Native. Grassland and grassy heaths on all soils from calcareous to acid. Very common throughout the county; recorded from all squares except 49B, 60C and 27C. Map 922.

Anthoxanthum aristatum Boiss. (*A. puelii* Lecoq & Lamotte), Annual Vernal-grass, Perwellt Barfog, has twice been recorded in the past as a casual on waste ground (00C, Aberdare, HJR 1902*; 16A, Barry, EV 1927).

HOLCUS L.

Holcus lanatus L. Yorkshire-fog. Maswellt.
Native. Grassland and grassy heath on all soils, open woodland and waste ground. Often abundant in neglected pastures. Very common throughout the county; recorded from all squares. Map 923.

Holcus mollis L. Creeping Soft-grass. Maswellt Rhedegog.
Native. Woods and hedgerows on neutral and acid soils. Common in Gower and the Uplands but local in the Vale, where it is absent from some areas with limestone soils. Map 924.

CORYNEPHORUS Beauv.

Corynephorus canescens (L.) Beauv. Grey Hair-grass. Brigwellt Llwyd.
Native or introduced. Sand-dunes and sandy waste ground at Aberafan and around Port Talbot Docks. Last certain record 1938; perhaps still present, considered by Riddelsdell to be native (Riddelsdell, 1907); also recorded as a dock alien at Cardiff.
17B, Cardiff Docks, JS *c*.1876; 78B? & 78D, Aberafan, HJR 1904*, 1910*, EV 1938*.

AGROSTIS L.

Agrostis canina L. (*A. canina* subsp. *fascicularis* (Curtis) Hyl.). Velvet Bent. Maeswellt y Cŵn.
Native. Damp acid grassland, especially in marshy dips on heaths and moors. Common in western Glamorgan but rather local in the east and south; old records from the southern Vale may be errors. Map 925.

Agrostis vinealis Schreber (*A. canina* subsp. *montana* (Hartman) Hartman). Brown Bent. Maeswellt y Rhos.
Native. Relatively dry heathland and moorland. Apparently widely distributed but less common than **A. canina**, from which it differs in being rhizomatous and lacking leafy axillary shoots. Probably over-looked in many localities. Map 926.

Agrostis curtisii Kerguélen (*A. setacea* Curtis, non Vill.). Bristle Bent. Maeswellt Gwrychog.
Native. Dry acid heathland. Locally frequent on heaths in Gower and the western Vale and adjacent parts of the Uplands, often locally dominant. **A.**

curtisii reaches the northern limit of its European range in the county. Map 927.

Agrostis capillaris L. (*A. tenuis* Sibth.). Common Bent. Maeswellt Cyffredin.
Native. Grassland and grassy heaths, usually on neutral or acid soils. Common or fairly common throughout the county; recorded from all squares except 10D, 19B, 49A, 49B and 97C. Map 928.

Agrostis gigantea Roth Black Bent. Maeswellt Mawr.
Native or introduced. Cultivated land, roadsides and waste places. Probably widespread as a rhizomatous weed of gardens and farmland but often overlooked. Map 929.

Agrostis stolonifera L. Creeping Bent. Maeswellt Rhedegog.
Native. Grassland on a wide range of soils, especially in unstable or disturbed sites, in the spray-zone on sea-cliffs, upper salt-marshes, cultivated land, roadsides and waste ground. Comon in suitable habitats throughout the county. Map 930.

Agrostis scabra Willd. has been recorded as a casual on railway ballast in Cardiff (17B, GH & G.M. Barter 1986*) and **A. avenacea** J.F. Gmelin (*Deyeuxia retrofracta* Kunth) as a dock alien in Barry and Cardiff (16A, RLS 1925!; 17B, RLS 1924*).

GASTRIDIUM Beauv.

Gastridium ventricosum (Gouan) Schinz & Thell. Nit-grass. Llauwair.
Native. Open grassland on shallow soils on south-facing limestone sea-cliffs. Very local in Gower and the Vale; a rare species that reaches the north-western limit of its native range in Gower. Also recorded as a dock alien at Cardiff and Barry. Map 931.

POLYPOGON Desf.

Polypogon monspeliensis (L.) Desf. Annual Beard-grass. Barfwellt Blynyddol.
Casual, occasionally found in the past on damp ground or near ponds in dockland areas.
16A, Barry Docks, 1923!*; 17B, Cardiff, Anon. 1869*, Cardiff Docks, JS c.1876*, 1924!*, Maendy Pool, RLS 1938; 17D, Penarth Ferry, JS 1902*; 27A, Splott, RM 1927, 1937!; 27C, pond margin near Cardiff Docks, EV 1925*.

Polypogon viridis (Gouan) Breistr. (*Agrostis semiverticillata* (Forskål) C. Chr.). Water Bent. Barfwellt Diffaith.
Introduced. Established on waste ground in the Barry area.
16A, Barry Docks, 1924!*, EV 1925*, TGE & A.L. Grenfell 1981, Barry Link Road, H. Thomas 1983; 27A, East Moors, Cardiff, JS 1876*, GCD c.1924.

Polypogon maritimus Willd. has once been recorded as a grain alien (27A, Splott, RLS 1927!*).

× AGROPOGON P. Fourn.

× **Agropogon littoralis** (Sm.) C.E. Hubbard (**Agrostis stolonifera** x **Polypogon monspeliensis**). Perennial Beard-grass. Barfwellt Bythol.
Once found as an apparently spontaneous hybrid (with both parents) on the muddy edge of a pool in Cardiff Docks, where it persisted for several years (27C, EV, RM & RLS 1923, 1924!*, EV 1929*).

AMMOPHILA Host.

Ammophila arenaria (L.) Link Marram. Morhesg.
Native. Sand-dunes. Abundant on all the dune-systems in Glamorgan and usually dominant on the higher dunes; also common on fore-dunes, with **Elymus farctus**. Map 932.

CALAMAGROSTIS Adanson

Calamagrostis epigejos (L.) Roth Wood Small-reed. Mawnwellt.
Native. Grassy heathland on sea-cliffs and occasionally inland, dune-slacks, open hedgebanks, roadsides, sometimes in woods. Scattered through Gower and the Vale, usually forming dense stands where it does occur. Most frequent near the coast, apparently growing on neutral soils that are periodically flushed with base-rich water in most of its localities. Map 933.

Calamagrostis canescens (Weber) Roth Purple Small-reed. Mawnwellt Blewog.
Native. Fens and shaded pond margins. Rare and known from only two localities. Discovered in the county, in 1977, during a survey by the Nature Conservancy Council of Crymlyn Bog near Swansea; **C. canescens** reaches the south-western limit of its British range in Glamorgan.
69D, Crymlyn Bog, ALe 1977*; 18D, near Llanishen, JPC 1978*.

PHLEUM L.

Phleum pratense L. subsp. **pratense** Timothy, Meadow Cat's-tail. Rhonwellt.
Native, but often sown with other grasses for forage. Meadows, especially on moderately damp soils, roadsides, cultivated land and waste ground. Common in Gower and the Vale and in meadows in many of the Upland valleys, but scarce in the central Uplands. Map 934.

Phleum pratense L. subsp. **bertolonii** (DC.) Bornm. (*Phleum bertolonii* DC., *P. nodosum* L.). Smaller Cat's-tail. Rhonwellt Penfain.
Native. Limestone grassland on dry soils, usually near the sea; occasionally on dry neutral soils. Locally abundant on dry grassy banks in southern and western Gower and the coastal regions of the Vale,

especially on or near sea-cliffs; scattered records inland. **P. pratense** subsp. **bertolonii** intergrades to some extent with small forms of **P. pratense** subsp. **pratense**, but it is both ecologically and morphologically distinct in coastal limestone grassland. Map 935.

Phleum arenarium L.   Sand Cat's-tail. Rhonwellt y Tywyn.
Native. Sand-dunes, usually on open but relatively stable sand. Abundant in the dune-systems of western Glamorgan from Whiteford Burrows to the Ogwr (Ogmore). Map 936.

Phleum phleoides (L.) Karsten was recorded as a dock alien on ballast at Cardiff (17B) by JS in c.1876.

ALOPECURUS L.

Alopecurus pratensis L.   Meadow Foxtail. Cynffon-wellt y Maes.
Native. Meadows and roadsides. Locally fairly common in the eastern Vale and frequent in the eastern Uplands but rather uncommon in the western Vale, scarce in Gower, and probably absent from the western Uplands. Map 937.

Alopecurus geniculatus L.   Marsh Foxtail. Cynffon-wellt Elinog.
Native. Marshy grassland, ponds and ditches. Common in Gower and the western Uplands but rather local elsewhere. Map 938.

Alopecurus bulbosus Gouan   Bulbous Foxtail. Cynffonwellt Oddfog.
Native. Short turf in upper salt-marshes. Locally frequent in eastern Glamorgan and in northern Gower.
06A, The Leys, AHT c.1909; 06B, Porthkerry, AHT 1899*, 1971!; 16B, Cog Moors, AHT 1899*, 1913*, Sully Moors, AHT 1899*; 17D, Ely Estuary, AHT 1907*, near Llandough, 1920!*; 27A, Rumney salt-marshes, HJR 1909*, 1937!*; 49D, Llanrhidian, 1985!**; 59A, Llanmorlais, 1984!; 59B, Gowerton, EFL 1891.

Alopecurus myosuroides Hudson   Black-grass. Cynffonwellt Du.
Native or introduced. Cultivated land and waste ground. A frequent cornfield weed in part of the south-eastern Vale, but rare elsewhere, though recently recorded in western Gower. **A. myosuroides** has become one of the most abundant cornfield weeds on heavy soils in England over the last thirty years, increasing as a result of the use of selective herbicides that have little effect on weedy grasses, and it is likely to become more frequent in Glamorgan if this trend continues. Map 939.

Alopecurus rendlei Eig (*A. utriculatus* auct. non Solander) has been recorded as a dock alien at Cardiff and Barry (16A, RLS c.1925!; 17B, Anon. c.1880*; 27A, JS c.1876).

PARAPHOLIS C.E. Hubbard

Parapholis incurva (L.) C.E. Hubbard   Curved Hard-grass. Corwelltyn Camaidd.
Casual or introduced. Open ground near the sea; established at The Leys, Aberthaw.
06A, The Leys, 1955!, GMB 1972*, MG 1979; 16A, Barry Docks, 1923!*; 17B, East Moors, JS c.1876!; 27A, Splott, M. Hall 1927; 69D, near Swansea Marina, 1986!**.

Parapholis strigosa (Dumort.) C.E. Hubbard   Hard-grass. Corwelltyn y Morfa.
Native. Upper salt-marshes. Locally abundant in the Gower and Cardiff salt-marshes; perhaps overlooked in some sites on the Vale coast. Map 940.

Hainardia cylindrica (Willd.) W. Greuter (*Lepturus cylindricus* (Willd.) Trin.) was recorded as a dock alien by R. Melville at Barry (16A) in 1924*.

PHALARIS L.

Phalaris arundinacea L.   Reed Canary-grass. Pefrwellt, Gwyran.
Native. Marshes, fens, canals and ditches. Common in the southern Vale and locally common in western Glamorgan; more frequent in the Uplands than **Phragmites australis**. Map 941.

Phalaris canariensis L.   Canary-grass. Pefrwellt Amaethol.
Casual or introduced. Roadsides and waste ground. A characteristic plant of rubbish-tips, where it is often more or less naturalized on disturbed ground. Map 942.

Four species of **Phalaris** have been recorded in the past as dock aliens in Cardiff, Barry or Port Talbot; **Phalaris minor** Retz., Lesser Canary-grass, Pefrwellt Lleiaf, in four localities (16A, Barry, 1923!*; 17B, Cardiff, HJR c.1908, GCD 1924, RLS 1938*; 27A, Splott, 1925!*, EV 1927*; 78B, Port Talbot, HJR c.1905), **P. paradoxa** L., Awned Canary-grass, in three (17B, Cardiff, JS 1876*, HJR 1907, GCD 1916; 17D, Penarth, JS c.1876; 27A, Splott, 1926!, EV 1927*), **P. coerulescens** Desf. (*P. aquatica* auct., non L.) in two (27A, East Moors, JS 1876*, Splott, 1926!*) and **P. angusta** Nees ex Trin. in one only (27A, Splott, RLS 1927).

MILIUM L.

Milium effusum L.   Wood Millet. Miledwellt.
Native. Mature woods, usually on limestone. Locally frequent in suitable habitats in Gower and in the eastern Vale, rare elsewhere. Map 943.

Piptatherum miliaceum (L.) Cosson (*Oryzopsis miliacea* (L.) Bentham & Hooker ex Ascherson & Graebner) has twice been recorded as a dock alien (16A, Barry, 1924!*; 17B, Cardiff, 1924!*), and **Arundo donax** L., Giant Reed, Corsen Fawr, was recorded from sands near the West Pier at Swansea (69D) by J.A. Webb in 1951.

PHRAGMITES Adanson

Phragmites australis (Cav.) Trin. ex Steudel (*Phragmites communis* Trin.)   Common Reed. Corsen.
Native. Reed-swamps and fens, drainage ditches near the sea, shallow lakes and canals, damp dune-slacks. Most of the sites where **P. australis** grows in western Glamorgan are on coastal flats or near the mouths of valleys close to sea-level, with the exception of Gors-llwyn (81C & 81D) which is an isolated **P. australis** fen at an altitude of 230m in the northern Uplands. Crymlyn Bog (69B, 69D, 79A, 79D) is one of the largest surviving **P. australis** fens in western Britain. In the Vale **P. australis** is fairly widespread inland, but also grows in coastal sites as in the west. Map 944.

DANTHONIA DC.

Danthonia decumbens (L.) DC. (*Sieglingia decumbens* (L.) Bernh.).   Heath-grass. Glaswellt y Rhos.
Native. Grassland on acid and neutral soils, grassy heaths, limestone cliff-heath, usually, but not always, on moderately damp soils. Common in Gower, the Uplands and the northern Vale but scarce in the southern Vale. Map 945.

MOLINIA Schrank

Molinia caerulea (L.) Moench   Purple Moor-grass. Glaswellt y Gweunydd.
Native. Moorland, damp heaths and marshes, usually on neutral or acid soils. Very common in Gower, the Uplands and the northern Vale where it is often dominant on poorly drained peaty heathland and moorland especially where these are regularly burned during the spring. Unrecorded in most of the southern Vale, where suitable habitats do not exist. Map 946.

NARDUS L.

Nardus stricta L.   Mat-grass. Cawnen Ddu.
Native. Grassy heaths and moorland on acid peaty soils. Common on suitable soils in Gower and the Uplands; often abundant and sometimes dominant on overgrazed moorland in the hills, growing on relatively well-drained soils. Absent from the southern Vale but locally frequent on acid soils in the northern Vale. Map 947.

CYNODON L.C.M. Richard

Cynodon dactylon (L.) Pers.   Bermuda-grass.
Introduced. Naturalized on sandy ground near the sea between Swansea and Mumbles, and perhaps near Cardiff Docks; casual elsewhere. Map 948.

SPARTINA Schreber

Spartina anglica C.E. Hubbard   Common Cord-grass. Cordwellt.
Introduced, but now thoroughly naturalized. Salt-marshes, from the pioneer zone to high-level creeks and pans; often colonises previously unvegetated areas. Introduced to the salt-marsh at the mouth of the Afon Cynffig (River Kenfig) in 1921 (Vachell, 1934) and to the north Gower salt-marshes in 1931. Now abundant in all salt-marshes but usually dominant only in low-level marsh; often fairly heavily grazed by ponies in the Gower salt-marshes. **Spartina x townsendii** H. & J. Groves, the seed-sterile diploid parent of **S. anglica**, apparently does not occur in Glamorgan. Map 949.

PANICUM L.

Panicum miliaceum L.   Common Millet. Miled.
Casual, on waste ground and rubbish-tips; the recent records are probably from bird-seed.
00C, Aberdare, HJR c.1905; 17B, Cardiff, JS c.1876, JPC 1981*; 17D Penarth, JS 1876*, Penarth Road, Cardiff, JPC 1982; 18D, Llanishen, W. Evans 1940*, Whitchurch JPC 1984; 27A, Splott, RLS 1925!*, Pengam 1937!*; 69D, Crymlyn, EMT 1937*; 79C, Briton Ferry, H & CH 1973–74; 97A, Colwinston, ABP 1979*.

Panicum capillare L. occurs as a casual in similar habitats to **P. miliaceum** (00C, Aberdare, HJR c.1905; 16A, Barry, RM 1924*; 17B, Cardiff, EV 1923; 18C, Radyr, RLS 1923; 27A, Splott, 1926!*; 90D, Llwydcoed, HJR c.1905) and **P. maximum** Jacq. has once been recorded as a grain alien (27A, Splott, RLS 1925!).

ECHINOCHLOA Beauv.

Echinochloa crus-galli (L.) Beauv. Cockspur. Cibogwellt Rhydd.
Casual, occurring on waste ground and rubbish-tips, now fairly frequent as a bird-seed alien. Map 950.

Digitaria sanguinalis (L.) Scop., Hairy Finger-grass, has been recorded several times as a casual on waste ground and rubbish-tips with two recent records (08D, Efail-isaf, W. Curtis 1967*; 16A, Barry Docks, AHT 1890*, 1892*; 17B, Cardiff Docks, JS c.1876, 1924!*, EV 1927*; 17D, Penarth Road, Cardiff, 1932!; 18D, Llanishen, JPC 1972*; 27A, Cardiff, RLS 1937!*), and **D. ischaemum** (Schreber) Muhl., Smooth Finger-grass, has twice been recorded as a dock alien (17B, Cardiff, JS c.1876*; 17D, Penarth, JS c.1876).

SETARIA Beauv.

Setaria viridis (L.) Beauv. Green Bristle-grass. Cibogwellt Gwyrddlas.
Casual, on waste ground and rubbish-tips, scattered throughout the county, recent records may be from bird-seed. Map 951.

Setaria pumila (Poiret) Schultes (*S. lutescens* F.T. Hubbard; *S. glauca* auct. non (L.) Beauv.).   Yellow Bristle-grass. Cibogwellt Melyn.
Casual, in similar habitats to the last and recent records probably from the same source.

00C, Aberdare, HJR c.1905; 16A, Barry, RLS c.1925!; 17B, Cardiff, EV 1923*, A.D. Tipper 1987*; 17D, Penarth, JS 1902*; 18C, Radyr, RLS c.1925!*; 27A, Splott, RLS c.1925, EV 1936*; 89C, Pontrhydyfen, H.J. Dawson 1983.

Three other species of **Setaria** have been recorded as casuals of docks, rubbish-tips or waste ground; **Setaria verticillata** (L.) Beauv., Rough Bristle-grass, Cibogwellt Troellog (16A, Barry Dock, AHT 1891*, 1892*; 17B, Cardiff, JS c.1876, EV 1922*; 17D, Penarth, JS c.1876; 27A, Splott, EV 1922*, RLS 1930), **S. italica** (L.) Beauv., Foxtail Bristle-grass, Cibogwellt (17D, Grangetown, RLS 1937*; 18D, Llanishen, W. Evans 1941*; 27A, Newport Road, Cardiff, 1939!*), and **S. geniculata** (Lam.) Beauv. (17B, Maindy, RLS 1938*).

Several other alien grasses have been recorded in the past mainly as dock aliens in the Cardiff area: **Eragrostis cilianensis** (All.) F.T. Hubbard (*E. megastachya* (Koeler) Link) (17B, EV 1923, RLS 1938*; 27A, 1926!*), **E. minor** Host, (27A, RLS 1930*; 78B, Port Talbot, HJR 1905), **E. pectinacea** (Michx) Nees (17B, RLS 1938*), **E. chloromeles** Steudel (27A, RLS 1957*), **Sporobolus pyramidatus** (Lam.) A.S. Hitchc. (*S. argutus* (Nees) Kunth) (17B, RLS & RM 1924), **Muhlenbergia distichophylla** (Presl) Kunth (90D, Llwydcoed, HJR 1903), **Crypsis aculeata** (L.) Aiton (*Pallasia aculeata* (L.) Druce) (27A, JS c.1876), **C. schoenoides** (L.) Lam. (*Heleochloa schoenoides* (L.) Host) (27A, JS 1875, 1884), **Paspalum dilatatum** Poiret (17B, EV et al. 1924, 1926!*, **Sorghum halepense** (L.) Pers. (17B, RLS 1938*, 17D, RLS & F. Norton 1937; 27A, RLS 1937!*), and **S. bicolor** (L.) Moench (17B, RLS 1938*; 17D, RLS 1937*), and **Zea mays** L., Maize, Indraw (00C, Aberdare, HJR c.1905; 17B, HJR 1907, RLS 1925!; 17D, JS c.1876*; 18D, W. Evans 1940*; 79C, Jersey Marine, HJR c.1905).

PALMAE
PHOENIX L.

Phoenix dactylifera L., Date Palm, Palmwydden, has twice been recorded as a casual on waste ground (27A, Splott, Cardiff, RLS 1925!, 1929!*); the seedings do not survive the winter.

ARACEAE
ACORUS L.

Acorus calamus L.   Sweet-flag. Gellesgen Bêr. Introduced. Stream and pond sides. Found in several scattered localities in the county.
07A, Hensol, MG 1974; 17B, Roath Park, Cardiff, 1977!; 27A, Roath Brook, Cardiff, 1950!*; 69A, Penllergaer, HJR c.1906, AJES 1962; 69C, Brynau, JAW 1924*; 79C, Briton Ferry, TBF 1842; 88A, Margam Park, MG 1974, ARP 1984!.

LYSICHITON Schott

Lysichiton americanus Hultén & St. John Yellow Skunk-cabbage.

Introduced. Streamsides and ditches. Well established in four localities in the Vale, obviously originally planted but now spreading alongside ditches and streams.
18D, Cefn On, R.A. Coniber 1969*, 1982!; 78D, Kenfig Burrows, S. Moon & JPC 1983; 88A, Margam Park, MG 1974, 1982!*; 88D, Tondu, G. Grant 1981.

Lysichiton camtschatcense (L.) Schott. (88A, D.M. Cape 1956) and **Zantedeschia aethiopica** (L.) Sprengel (69C, Mr Hillman 1960) have both been reported as non-persistent garden outcasts, the former possibly in error for **Lysichiton americanus**.

ARUM L.

Arum italicum Miller subsp. **neglectum** (Townsend) Prime (*A. neglectum* (Townsend) Ridley. Italian Lords-and-Ladies. Pidyn y Gog Eidalaidd.
Native. Woods. Very rare and recorded only from two adjacent localities near Cardiff.
17C, Cwm George, EV 1930*, 1953!*, Cwrt-yr-Ala, EV 1935*, MG 1968.

Arum italicum × **A. maculatum** has been recorded once from a wood near Cardiff (17D, Cwm George 1955!*) as a non-flowering plant.

Arum maculatum L.   Lords-and-Ladies. Pidyn y Gog.
Native. Hedgerows, woods, roadsides etc. Common throughout Gower and the Vale but less frequent in the Uplands. Fl. 4–5, fr. 8–10; (!) (Dip) ?Psychoda sp., ?Dips. Map 952.

LEMNACEAE
WOLFFIA Horkel ex Schleiden

Wolffia arrhiza (L.) Horkel ex Wimmer   Rootless Duckweed. Llinad Diwraidd.
Native (extinct). Ponds. Once grew in two ponds in Cardiff, both of which have now been built over.
17D, Penarth Rd, Cardiff, AHT 1899; 27C, Roath Dock, Cardiff, JS c.1876.

LEMNA L.

Lemna trisulca L.   Ivy-leaved Duckweed. Llinad Eiddew.
Native. Lakes, ditches, streams, canals etc. Locally abundant in suitable habitats in the south of the county. Map 953.

Lemna gibba L.   Fat Duckweed. Llinad y Dŵr Crythog.
Native. Canals and streams. Very scarce; recently found at only one site (09B, near Mountain Ash, ARP 1984*). Recorded in the past from three other localities (00C, Aberdare, HJR c.1906; 17D, Grangetown, Cardiff, AHT 1891*; 27A, Penylan, Cardiff, 1925!*).

Lemna minor L.   Common Duckweed. Llinad.

Native. Ponds, ditches, streams, canals etc. Common and widespread throughout most of the county. Map 954.

SPIRODELA Schleiden

Spirodela polyrhiza (L.) Schleiden (*Lemna polyrhiza* L.) Greater Duckweed. Llinad Mawr.
Native. Streams, ditches, ponds, etc. Rare and decreasing. Recorded in the past from several sites in the Vale and one in the Uplands but with only two recent records.
17, Cardiff district, 1956!; 27A, East Moors, Cardiff, JS *c.*1876; 28C, Llanedeyrn, 1969!; 69A, Penllergaer, AJES 1962; 78B, Port Talbot and Margam Moors, HJR *c.*1906; 87A, Sker, EV *c.*1930; 97A, Bridgend, D.P.M. Guile 1944*; 99A, Blaenycwm, H. Harris *c.*1900.

SPARGANIACEAE
SPARGANIUM L.

Sparganium erectum L. (*S. ramosum* Hudson). Branched Bur-reed. Cleddlys Canghennog.
Native. Streams, ditches, ponds etc. Frequent throughout most of the county. Map 955.

The distribution of the following three subspecies has not been fully worked out.
Sparganium erectum subsp. **erectum**.
Native. Streams and ditches, ponds etc. Frequent in the south eastern Vale, occasional elsewhere. Map 956.

Sparganium erectum subsp. **microcarpum** (Neuman) Domin
Native. Marshes. Recorded in the past from four localities.
07B, St. y Nyll, HJR 1908*; 07C, Flemingstone Moor, HJR *c.*1906; 17A, Fairwater, HJR 1912; 69D Crymlyn Bog, J.R. Shepherd 1913.

Sparganium erectum subsp. **neglectum** (Beeby) Schinz & Thell.
Native. Ditches, streams, pools, etc. Recorded from scattered localities throughout the county. Map 957.

Sparganium emersum Rehmann Unbranched Bur-reed. Cleddlys Di-gainc.
Native. Streams and ditches. Locally frequent in Gower and the eastern Vale, occasional elsewhere. Map 958.

Sparganium angustifolium Michx Floating Bur-reed. Cleddlys Culddail.
Native. Base-poor lakes. Rare, with recent records from only one locality.
07A, Pysgodlyn Mawr, EV 1920*; 90C, Llyn Fach, AHT 1892, 1899*, 1969!*, 1984!, Llyn Fawr, HJR *c.*1906.

Sparganium minimum Wallr. Least Bur-reed. Cleddlys Bach.

Native. Acid water in lakes, marshes and canals. Rare, recently recorded only from Crymlyn Fen where it is locally frequent, and Hensol Lake; recorded in the past from western Gower and the central Vale.
07A, Hensol Lake, JS *c.*1876, BS & MSP 1962; Pysgodlyn Mawr, JS *c.*1876, HJR *c.*1906; 48A, Pitton Common, HJR *c.*1906; 69C, Singleton Marsh, JWGG, *c.*1840; 69B/79C, Crymlyn Fen, EF *c.*1800, JAW 1925*, AJES 1962**, 1983!, Tennant Canal, JWGG *c.*1840, HJR *c.*1906.

TYPHACEAE
TYPHA L.

Typha angustifolia L. Lesser Bulrush. Cynffon y Gath Gulddail.
Native. Swampy edges of lakes and marshes. Locally abundant at Oxwich Marsh (58A) and Crymlyn Fen (69B & 79C), rare elsewhere. Map 959.

Typha latifolia L. Bulrush. Cynffon y Gath.
Native. Swampy edges of lakes, pools, ditches, marshes etc. Frequent throughout the county except in the central Uplands. Map 960.

CYPERACEAE
SCIRPUS L.

Scirpus sylvaticus L. Wood Club-rush. Clwbfrwynen y Coed.
Native. Marshes, streambanks, etc. Occasional in the eastern Vale, rare elsewhere. Map 961.

Scirpus maritimus L. Sea Club-rush. Clwbfrwynen Arfor.
Native. Salt-marshes. Frequent all along the coast but especially so in Gower. Map 962.

Scirpus lacustris L. subsp. **lacustris** (*Schoenoplectus lacustris* (L.) Palla). Common Club-rush. Llafrwynen.
Native or introduced. Lakes, streams and ponds; sometimes planted as an ornamental. Rare in the Vale, and not recorded recently in Gower and the Uplands. Map 963.

Scirpus lacustris subsp. **tabernaemontani** (C.C. Gmelin) Syme (*Schoenoplectus tabernaemontani* (C.C. Gmelin) Palla). Grey Club-rush. Llafrwynen Arfor.
Native. Ponds, marshes and streams, especially near the sea, and in upper salt-marshes. Frequent in Gower and the western Vale, occasional elsewhere in the Vale but absent from most of the Uplands. Map 964.

Scirpus holoschoenus L. (*Holoschoenus vulgaris* Link). Round-headed Club-rush. Clwbfrwynen Pengrwn.
Probably introduced. Wet places near docks. Recorded in four sites, one near Neath and three in the south-east.

16A, Barry Docks, RLS 1926*, GCD 1930; 17B & 17D, Bute Dock, GSW 1983*; 27A, East Moors, Cardiff, JS 1879; 79C, near Neath, EMT 1937*, T.R. Lovering 1959, Briton Ferry Docks, H & CH 1980!**.

Scirpus setaceus L. (*Isolepis setacea* (L.) R. Br.). Bristle Club-rush. Clwbfrwynen Fach.
Native. Marshy places, wet roadsides etc. Frequent in Gower and the Uplands, less so in the Vale. Map 965.

Scirpus cernuus Vahl (*Isolepis cernua* (Vahl) Roemer & Schultes). Slender Club-rush. Clwbfrwynen Gwychog Eiddilaidd.
Native. Dune slacks and marshy places near the sea. Frequent in Gower and the western Vale, rare or absent elsewhere. Map 966.

Scirpus fluitans L. (*Eleogiton fluitans* (L.) Link). Floating Club-rush. Clwbfrwynen Nawf.
Native. Lakes and pools with acid water. Frequent in Gower, rare in the Uplands and Vale. Map 967.

Scirpus cespitosus L. subsp. **germanicus** (Palla) Broddeson (*Trichophorum cespitosum* subsp. *germanicum* (Palla) Hegi). Deergrass. Clwbfrwynen y Mawn.
Native. Damp heaths and moorlands. Common in Gower and the Uplands, absent from the southern Vale. Map 968.

BLYSMUS Panzer

Blysmus rufus (Hudson) Link (*Scirpus rufus* (Hudson) Schrader). Saltmarsh Flat-sedge. Corsfrwynen y Morfa.
Native. Salt-marshes. Found once in west Gower but not seen since, despite careful searches of the locality.
49C, Llanmadoc, EV 1905*.

ERIOPHORUM L.

Eriophorum angustifolium Honckeny Common Cottongrass. Plu'r Gweunydd.
Native. Bogs and wet peaty moorland. Common in Gower and the Uplands but absent from the southern Vale. Map 969.

Eriophorum latifolium Hoppe Broad-leaved Cottongrass. Plu'r Gweunydd Llydanddail.
Native. Wet, boggy places on basic soils. Recorded from several scattered localities in the past but now known from only one, near Cardiff.
18C, Cwm Nofydd, HAH 1923*, MG 1979; 58A, Cefn Bryn, PS c.1900; 69D, Crymlyn Fen, J. Westcombe 1843; 70, between Pontardawe and Ystalyfera, E.M. Payne 1959; 81C, Mynydd-y-drum, JAW 1927; 81D, Gors Llwyn, E. Lees c.1840; 90C, near Llyn Fach, JS c.1880; 99A, Rhondda Waterfalls, JS c.1880.

Eriophorum gracile Koch ex Roth Slender Cottongrass. Plu'r Gweunydd Eiddil

Native. Acid fens. Recently discovered in two sites near Swansea.
69D, Crymlyn Fen, ALe 1977!*; 79C, Pant-y-Sais Fen, CH 1990.

Eriophorum vaginatum L. Hare's-tail Cottongrass. Plu'r Gweunydd Unben.
Native. **Sphagnum** bogs and wet peaty moorland. Frequent in parts of the Uplands and the central Vale, very local elsewhere. Map 970.

ELEOCHARIS R. Br.

Eleocharis quinqueflora (F.X. Hartmann) O. Schwarz Few-flowered Spike-rush. Sbigfrwynen Goch
Native. Dune slacks and marshes. Locally abundant at Whiteford Burrows, (49A & 49C, 1969–85!), Crymlyn Burrows (79C, H & CH 1980) and Kenfig Burrows (78D, 1977!) but scarce or with no recent records elsewhere; absent from the Uplands. Map 971.

Eleocharis palustris (L.) Roemer & Schultes subsp. **vulgaris** Walters Common Spike-rush. Sbigfrwynen y Gors.
Native. Pool margins and marshy places. Common in Gower, parts of the Uplands and parts of the Vale. Map 972.

Eleocharis uniglumis (Link) Schultes Slender Spike-rush. Sbigfrwynen Un Plisgyn.
Native. Dune slacks, margins of ponds, and brackish marshes. Locally frequent near the coast in the west. Map 973.

Eleocharis multicaulis (Sm.) Desv. Many-stalked Spike-rush. Sbigfrwynen Gadeiriog.
Native. Flushed sites on acid heathland. Frequent in Gower, rare elsewhere. Map 974.

CYPERUS L.

Cyperus longus L. Galingale. Ysnoden Fair.
Introduced. Marshy places. Established in a few localities in the Vale.
17B, Maendy Pool, Cardiff, RLS 1938* (habitat destroyed); 17B, Roath, Cardiff, 1972!*; 18D, Llanishen, Cardiff, JPC 1978*; 88A, Margam, C. Marks 1924, JAW 1942*.

Cyperus eragrostis Lam. (*C. declinatus* Moench) was once recorded as a casual on ballast (17D, Penarth Ferry, JS 1876*), and had recently been found established in the same area (17B, Bute East Dock, JPC & GH 1985), but development of the site may have destroyed the population.

CLADIUM Browne

Cladium mariscus (L.) Pohl Great Fen-sedge. Llemfrwynen.
Native. Dune slacks, fens and lake margins. Rare. Known for many years in Crymlyn Fen near Swan-

sea, where it is locally dominant, and recently recorded from two other sites in the Vale.
07A, Hensol Lake, MSP *c*.1963; 69D, Crymlyn Fen, LWD *c*.1800, JAW 1924*, AJES 1962**, 1983!; 78D, Kenfig Burrows, JPC 1978.

RHYNCHOSPORA Vahl

Rhynchospora alba (L.) Vahl White Beak-sedge. Corsfrwynen Wen.
Native. Wet peaty moorlands. Recently recorded only from one site in Gower and one in the western Uplands, but recorded in the past from a few other sites in the west of the county; probably still present on Clyne Common and perhaps at Cwmllynfell and Crymlyn Bog.
49D, Cefn Bryn above Reynoldston, PS 1911, 1973!**, D. Green 1979; 69C, Singleton, Swansea, JWGG *c*.1840; 59D/69C, Clyne Common, W. Weston 1943, JAW 1944; 60A, Penlanau, I.K. Morgan 1984; 69B/D, Crymlyn Fen, EF *c*.1800; 71C, Cwmllynfell, JAW 1943.

Rhynchospora fusca (L.) Aiton fil. Brown Beak-sedge. Corsfrwynen Rudd.
Native (extinct). Peaty marsh. Recorded in the past from one locality near Swansea.
69B/D, Crymlyn Fen, E. Forster *c*.1800.

SCHOENUS L.

Schoenus nigricans L. Black Bog-rush. Corsfrwynen Ddu.
Native. Peaty fen near Swansea and on flushed ledges on a sea cliff near Barry.
06B, Bull Cliff, Barry, HAH 1956*, 1972!; 69D, Crymlyn Fen, JWGG *c*.1844, EMT 1938, ABP 1970!, 1971!**, 1983!.

CAREX L.

Carex paniculata L. Greater Tussock-sedge. Hesgen Rafunog Fwyaf.
Native. Marshes, dune slacks, streamsides etc. Frequent in east Gower the south west Uplands and the western Vale, occasional elsewhere. Map 975.

Carex paniculata × C. remota (**C. × boenninghausiana** Weihe) has been recorded once in the Uplands (99A, Cwm Parc, C.S. Nicholson *c*.1900).

Carex diandra Schrank Lesser Tussock-sedge. Hesgen Rafunog Leiaf.
Native (extinct). Marsh and fen. Recorded from two localities in the past but not seen this century.
09B, near Quakers Yard, JS *c*.1876; 69C, Sketty Bog, EF *c*.1800.

Carex otrubae Podp. False Fox-sedge. Hesgen Dywysennog.
Native. Ditches and pondsides, base-rich marshes, upper levels of salt-marshes, etc. Frequent in Gower and the Vale, rare in the Uplands. Map 976.

Carex otrubae × C. remota (**C. × pseudaxillaris** K. Richter) was once recorded on a roadside between Penarth and Sully (16B, Cog Farm, EV 1905*); it has not been seen recently.

Carex muricata L. agg.
This complex includes the three following species: **Carex spicata**, **C. muricata** and **C. divulsa**, which are closely similar and form a taxonomically difficult group. Only records that have been confirmed or accepted by A.O. Chater, R.W. David or A.C. Jermy are used in the text or included on the maps of the segregate species. The map of the aggregate shows all records. Map 977.

Carex spicata Hudson (*C. contigua* Hoppe). Spiked Sedge. Hesgen Dywysennog Borffor.
Native. Damp roadsides. Local in the Vale and eastern Uplands, apparently absent from Gower. Map 978.

Carex muricata L. subsp. **lamprocarpa** Čelak. (*C. muricata* subsp. *pairaei* (F.W. Schultz) Čelak., *C. pairaei* F.W. Schultz). Small-fruited Prickly-sedge Hesgen Bigog Hwyr.
Native. Roadsides and hedgebanks. Scarce, mainly in the eastern Vale and Gower.
27A, Roath, 1973!*, 28A, Draethen, 1972!*; 48A, Rhossili, HJR *c*.1905; 88C, Pyle, HJR 1900*, R.W. David 1950+.

Carex divulsa Stokes subsp. **divulsa** Grey Sedge. Hesgen Lwydlas.
Native. Roadsides, hedgebanks, railway embankments etc. Scarce, mainly in the eastern Vale and Gower.
06A & 07C, Llanblethery, R.M. Walls 1987*; 16A, Barry, 1969!*; 17A, Fairwater 1920!*; 17D, Dinas Powys, ARP 1984*; 18B, Caerphilly, JPC 1972; 58B, Penmaen, 1927!*.

Carex divulsa subsp. **leersii** (Kneucker) W. Koch (*C. muricata* subsp. *leersii* (Kneucker) Ascherson & Graebner, *C. polyphylla* Kar. & Kir.). Many-leaved Sedge. Hesgen Lwydlas Digainc.
Native. Grassy places on limestone near the sea. Known with certainty from only two widely separated localities.
26, Flat Holm, HAH 1932*, SGH 1967!*; 68A, Oystermouth Castle, W.R. Linton 1886*, R.W. David 1976.

Carex arenaria L. Sand Sedge. Hesgen Arfor.
Native. Sand dunes and shingle. Locally common along the coasts of Gower and the western Vale, less so in the east. Map 979.

Carex disticha Hudson Brown Sedge. Hesgen Lygliw Benblydd.
Native. Marshes and wet meadows. Rather rare, with scattered localities in the south and east of the county.
00B, Penydarren, J. Davies 1988*; 07A, Ystradowen Bog, EV 1904; 07B, Peterstone, HJR 1904, Pendoylan,

BS 1966*, near St. George's, 1973!; 16B, Cog Moors, AHT 1908*, border of Sully Moor, AHT 1913*; 17A, near Fairwater, HJR 1908*, 1920!*; St. Brides-super-Ely, EV 1943*, 1972*; 78B, Margam Moors, E.S. Marshall & WAS 1901, 1974!; 78B & 78D, near Eglwys Nunydd, MG 1973; 78D, Kenfig, EV 1927, MG 1970; 79C, Baglan Bay, MG 1973*; 99D, Glyn Cornel, MG 1972!.

Carex divisa Hudson Divided Sedge. Hesgen Deiligol.
Native. Tidal fields and saltmarshes. Very rare. Recorded from only two localities in the past, and not seen during the present survey.
17B, Canton Common near Cardiff, EV, 1937*; 69C, Blackpill Burrows, JAW 1941*.

Carex remota L. Remote Sedge. Hesgen Anghyfagos.
Native. Damp places in woods, along roadsides and on the banks of ditches. Common throughout the county except in the extreme south of the Vale. Map 980.

Carex ovalis Good. Oval Sedge. Hesgen Hirgrwn, Hesgen Hirgylchaidd.
Native. Grassland, hedgerows, marshes and peaty places. Common in the north of the county, less so in the south. Map 981.

Carex echinata Murray Star Sedge. Hesgen Seraidd.
Native. Wet heathy places, bogs, marshes and streamsides especially on acid soils. Common in the Uplands, less so elsewhere and absent from parts of Gower and the southern Vale. Map 982.

Carex dioica L. Dioecious Sedge. Hesgen Ysgar.
Native. Very rare. Now known in only one locality although recorded in the past from three others.
00A, between Llwydcoed and Merthyr Tydfil, HJR c.1904; 69C, Sketty Bog, JWGG c.1840; 69D, Crymlyn Fen, JWGG c.1840, ALe 1977*!; 79B, near Aberdulais waterfalls, LWD c.1800.

Carex curta Good. White Sedge. Hesgen Benwen.
Native. Marshy and boggy places. Locally frequent in the eastern Vale, and at Crymlyn Bog, rare or absent elsewhere. Map 983.

Carex hirta L. Hairy Sedge. Hesgen Flewog.
Native. Damp grassy roadsides, damp meadows, dune-slacks, waste ground etc. Common in most parts of the county but absent from areas of the southern Uplands and Vale. Map 984.

Carex acutiformis Ehrh. Lesser Pond-sedge. Hesgen Ganolig-dywysennog.
Native. Marshes, swamps, streamsides etc. Locally frequent in the eastern Vale, very local or absent elsewhere. Map 985.

Carex acutiformis × **C. riparia** has twice been recorded from marshes in the eastern Vale, but in the absence of specimens both records must be regarded as doubtful (07B, Peterston-super-Ely, HJR c.1907; 17A, Fairwater, HJR c.1909).

Carex riparia Curtis Greater Pond-sedge. Hesgen Braff-dywysennog.
Native. Marshes, stream and pond sides etc. Locally frequent in Gower and the Vale, less so in the Uplands. Map 986.

Carex pseudocyperus L. Cyperus Sedge. Hesgen Hopysaidd.
Native. Stream, canal and lake sides, marshes etc. Scarce; known from a few localities scattered throughout the county.
07B, St.-y-Nyll, EV 1904*, BS 1964; 17B, Cardiff, 1949!*, JWD, 1972*!; 17C, Culverhouse Cross, P. Randerson 1990*; 58A, Oxwich, 1974!; 69B, Crymlyn Fen, ALe 1977!; 69C, Singleton Marsh, JWGG c.1840, JAW 1926 (habitat destroyed); 69D, Crymlyn Fen, J. Lightfoot 1773, 1974!**, ALe 1977*; 88D, Cefn Cribwr, D.A. Baxter 1975*.

Carex rostrata Stokes Bottle Sedge. Hesgen Ylfin-fain.
Native. Marshes, bogs, streamsides etc., usually in peaty places. Locally frequent in scattered localities. Map 987.

Carex rostrata × **C. vesicaria** (**C.** × **involuta** (Bab.) Syme) was recorded from one locality in the past, but has not been seen recently (07A, Mynydd-y-glew, RM 1923*, GCD & EV 1927).

Carex vesicaria L. Bladder-sedge. Hesgen Chwysigennaidd.
Native. Marshes and ditches. Recently recorded only in the eastern Vale, where it is scarce; once found in the northern Uplands, but absent from the west of the county.
00C, Aberdare, HJR c.1904*; 07A, Pysgodlyn Mawr, EV 1927*, MG 1972; 07B, St.-y-Nyll, HJR c.1909, Peterston-super-Ely, HJR 1904*, 1949!*; 17A, near Fairwater, HJR 1912*; 17C, Culvershouse Cross, P. Randerson, 1990*; 18D, Llanishen, JPC 1972*; 19A, Llancaiach, HJR 1909; 79C, Crymlyn Burrows, NCC 1970; 90A, Hirwaun ponds, AHT 1907*, 1909*.

Carex pendula Hudson Pendulous Sedge. Hesgen Bendrymus, Hesgen Ddibynnaidd Fwyaf.
Native, but sometimes introduced as an ornamental plant. Damp woodland, stream and pond sides. Occasional in the eastern Vale and the western Uplands, rare in Gower where it has been introduced in some localities. Map 988.

Carex sylvatica Hudson Wood-sedge. Hesgen y Coed, Hesgen Ddibynnaidd y Goedwig.
Native. Woods, especially on limestone soils. Common in Gower and the Vale, but scarce or absent in most of the Uplands. Map 989.

Carex strigosa Hudson   Thin-spiked Wood-sedge. Hesgen Ysbigog Denau.
Native. Woods. Rare, with recent records from only two sites, one in Gower, the other in the western Vale. Gutch's records from Sketty (69C) and Crymlyn (69D) Bogs (Gutch, 1844a) are doubtful.
17C, Cwrt yr Ala, EV 1904*, HJR 1908*; 58A, Parkmill, 1974!; 78B, Margam, 1974!.

Carex flacca Schreber   Glaucous Sedge. Hesgen Oleulas.
Native. Limestone grassland, dune-slacks, grassy roadsides, and marshy places, usually on base-rich soils. Abundant in Gower and the Vale, less so in the Uplands and apparently absent from some areas. Map 990.

Carex panicea L.   Carnation Sedge. Hesgen Benigen-ddail.
Native. Marshes, bogs, and wet heaths on acid soils. Common in Gower, parts of the Uplands and the northern Vale, but absent from much of the southern Vale. Map 991.

Carex laevigata Sm.   Smooth-stalked Sedge. Hesgen Ylfinog Lefn.
Native. Marshy copses and thickets, boggy places etc. Occasional in the Uplands, rare in Gower and the Vale. Map 992.

Carex binervis Sm.   Green-ribbed Sedge. Hesgen Ddeulasnod.
Native. Dry acid moorland. Frequent in Gower and the Uplands, but absent from much of the southern Vale. Map 993.

Carex distans L.   Distant Sedge. Hesgen Bell, Hesgen Anghysbell
Native. Salt and brackish marshes, and sometimes in the spray-zone. Occasional on the coast of Gower and the southern Vale. Map 994.

Carex extensa Good.   Long-bracted Sedge. Hesgen Hiriain.
Native. Salt-marshes and tidal banks of rivers. Locally frequent on the Gower coast, but scarce on the coast of the Vale. Map 995.

Carex hostiana DC.   Tawny Sedge. Hesgen Dywyll-felen.
Native. Marshes. Locally frequent in Gower, rare or absent elsewhere. Map 996.

Carex lepidocarpa Tausch   Long-stalked Yellow-sedge. Hesgen Felen Baladr Hir.
Native. Base-rich flushes and marshes. Rare; found only in the north-eastern Uplands and the eastern Vale.
00A, Taf Fechan, 1969!; 01C, Vaynor, 1978!*; 01D, Twynan Gwynion, ARP 1973!*; 07A, Ystradowen, EV 1904*; 07D, Bonvilston, HJR 1909; 18C, Cwm Nofydd, C. Walker 1969*.

Carex demissa Hornem.   Common Yellow-sedge. Hesgen Felen.
Native. Marshes, streamsides, flushed sites on moorland and wet heaths on acid soils. Common in Gower and the Uplands, less so in the Vale and absent from the south. Map 997.

Carex serotina Mérat   Small-fruited Yellow-sedge. Hesgen Goraidd, Hesgen Oeder.
Native. Dune slacks. Locally frequent in a few sites in Gower and the western Vale. Plants from Oxwich have been identified as subsp. **pulchella** (Lönnr.) Van Ooststr. (PJo 1982, det. A.O. Chater & R.W. David).
49C, Whiteford Burrows, 1956!*, 1972!*; 58A, Oxwich Burrows, HJR 1906*, 1977!; 78D, Kenfig Burrows, WAS 1902*, 1973!; 79C, Crymlyn Burrows, 1972!.

Carex pallescens L.   Pale Sedge. Hesgen Welwlas.
Native. Grassland and grassy places in woods, usually damp. Occasional in the north central Uplands and eastern Vale, rare in Gower. Map 998.

Carex caryophyllea Latourr.   Spring-sedge. Hesgen Gynnar.
Native. Dry grassland, usually on south-facing slopes. Frequent in Gower and the Vale, but apparently absent from large areas of the Uplands. Map 999.

Carex montana L.   Soft-leaved Sedge. Hesgen Feddal, Hesgen Mynydd-dir.
Native. Limestone grassland. Locally frequent in one area in the western Vale, and with one locality in the Uplands.
00A, Morlais Castle Hill, EV 1932*, R.W. David 1980; 87A, Newton, Porthcawl, EMT 1942*, R.W. David 1979; 87B, Ogmore Down, HJR 1905*, R.W. David, 1975, Pant St. Brides, AHT 1907*, 1972!**, ARP 1979*; 97A, Ewenny Down, EV 1907, R.W. David 1975.

Carex pilulifera L.   Pill Sedge. Hesgen Bengron.
Native. Hill pasture and heaths. Occasional in scattered localities throughout the county but decreasing. Map 1000.

Carex limosa L.   Bog-sedge. Hesgen Eurwerdd.
Native. Wet bogs; Very rare; for many years it was thought to be extinct in the county, but it was recently rediscovered in Crymlyn Bog, where it had not been seen since c.1800.
69C, Sketty Bog, EF c.1800 (habitat destroyed); 69B/D, Crymlyn Bog, EF c.1800, ALe 1977*!.

Carex elata All.   Tufted-sedge. Hesgen Oleulas Sythddail.
Native. Pool sides, fens and marshy places. Rare, with recent records only from Crymlyn Fen and one locality in the eastern Vale.
07C, Flemingston Moors, HJR 1904; 18C, near Radyr, MG 1981*; 69D, Crymlyn Fen, HJR 1904, GTG & AJES 1962*, 1983!; 78D, Kenfig Pool, JM c.1840, BS 1964.

Carex elata × C. nigra (C. × turfosa Fries) has been recorded once (07C, Flemingston Moors, HJR 1904).

Carex nigra (L.) Reichard   Common Sedge. Swp-hesgen y Fawnog.
Native. Marshes, bogs, wet moorland etc. Abundant throughout the county except in the south-eastern Vale. Map 1001.

Carex acuta L.   Slender Tufted-sedge. Hesgen Eiddil Dywysennog.
Native. Streamsides and marshy places. Occasional in the Vale, and with two localities in the eastern Uplands; absent from Gower. Map 1002.

[**Carex acuta** × **C. elata** (**C.** × **prolixa** Fries) and **C. acuta** × **C. nigra** (**C.** × **elytroides** Fries) have both been recorded once from the same locality (69D, Crymlyn Fen, HJR 1904) but since **C. acuta** has not been recorded from the area, both must be regarded as doubtful.]

Carex pulicaris L.   Flea Sedge. Chwein-hesgen.
Native. Base-rich wet flushes and boggy places. Locally frequent in Gower and the north-central Uplands, rare in the Vale. Map 1003.

ORCHIDACEAE

EPIPACTIS Zinn

Epipactis palustris (L.) Crantz   Marsh Helleborine. Caldrist y Gors.
Native. Dune-slacks, occasionally in **Phragmites** fen; formerly in marshy calcareous grassland inland. Locally abundant at 49A, Whiteford Burrows, 58A, Oxwich Burrows; 79C, Crymlyn Burrows (also growing in small numbers among **Phragmites australis** in the south part of Crymlyn Bog and at Pant-y-Sais, JPC 1982); 78D & 88C, Kenfig Burrows and 87B, Merthyr Mawr. Recently found on a quarry floor at Rhoose (06A, JPC 1985). There are no recent records from several inland localities in the Vale, where it may have been eliminated by drainage and changes in agriculture, or by suburban spread; some of the inland records are dubious and may refer to **E. helleborine**, but **E. palustris** undoubtedly grew at 07A, Ystradowen Moor; 17A, near St. Fagans; and 18D, Llanishen (Trow, 1911). Fl. (6–) 7–8, fr. 8–9; (!) B.terr (n). Map 1004.

Epipactis helleborine (L.) Crantz   Broad-leaved Helleborine. Caldrist Llydanddail.
Native. Woods, dune grassland and roadside banks. Local in Gower and the Vale, and also in the northern Uplands where it grows on laneside banks; usually occurs in small numbers. Map 1005.

Epipactis phyllanthes G.E. Sm. (including *E. cambrensis* C. Thomas). Green-flowered Helleborine. Caldrist Melynwyrdd.
Native. Sand-dunes in or near **Salix repens** slacks. Very scarce and recorded from only four localities. *E. cambrensis*, which has been described from Kenfig Burrows and Margam Burrows (Thomas, 1950) is

probably a form of this species, and grows with **E. phyllanthes** var. **pendula** D.P. Young at Kenfig. 49A & C, Whiteford Burrows, D. Lang 1990; 58A, Oxwich, D. & S. Adams 1971!, 1975!, D. Lang 1990; 78B, Margam Burrows, C. Thomas, 1941; 78D, Kenfig Burrows, C. Thomas 1940*, S. Moon 1981.

NEOTTIA Ludwig

Neottia nidus-avis (L.) L.C.M. Richard   Bird's-nest Orchid. Tegeirian Nyth Aderyn.
Native. Woods, usually on limestone soils. Very scarce and sporadic in its appearance. Many of the older records are rather doubtful; most of the reliable past records and all recent records are in limestone woodland in southern Gower or the south-eastern Vale, and it is most frequent in the beechwoods around Taff's Well. Map 1006.

LISTERA R. Br.

Listera ovata (L.) R. Br.   Common Twayblade. Ceineirian.
Native. Woods, usually on limestone, and sand-dune grassland. Fairly frequent in suitable habitats in Gower and the Vale, but very scarce in the Uplands. Fl. 6–7, fr. 8. Map 1007.

SPIRANTHES L.C.M. Richard

Spiranthes spiralis (L.) Chevall.   Autumn Lady's-tresses. Ceineirian Troellog.
Native. Short limestone turf on shallow soils, occasionally on sand-dunes and in lawns. Locally frequent in most years on the limestone sea-cliffs of southern Gower and in a few places on the Vale coast, but scarce and decreasing inland. Fl. 8–9. Map 1008.

HERMINIUM Guett.

Herminium monorchis (L.) R. Br.   Musk Orchid. Tegeirian Mwsg.
Native. Dune grassland, and perhaps formerly in limestone pasture inland. Rare, with recent records only from Kenfig Burrows, where it was first found in 1958; there is a single rather doubtful past record from another locality.
07C, Beaupre, Mrs. Traherne c.1882; 78D, Kenfig Burrows, A. Matthews 1958!*, J.W. Lewis, 1968.

PLATANTHERA L.C.M. Richard

Platanthera bifolia (L.) L.C.M. Richard   Lesser Butterfly-orchid. Tegeirian Llydanwyrdd Bach.
Native. Woods and grassland. Rare, with only two recent records. Many of the older records are doubtful and some may refer to **P. chlorantha**. The records from 48A, Rhosili (HJR c.1907) and 48B, Mead Moor (PS c.1911) are probably correct. Map 1009.

Platanthera chlorantha (Custer) Reichenb.   Greater Butterfly-orchid. Tegeirian Llydanwyrdd.

Native. Woods, usually on limestone. Rare except in the south-eastern Vale where it is scarce but fairly regular in its appearance. Map 1010.

GYMNADENIA R. Br.

Gymnadenia conopsea (L.) R. Br. Fragrant Orchid. Tegeirian Pêr.
Native. Damp limestone grassland and calcareous marshes, and in dune-slacks. Increasingly scarce, occurring only in a few localities in the Vale and one in the Uplands. Most frequent in dune-slacks at Kenfig Burrows. Map 1011.

PSEUDORCHIS Séguier

Pseudorchis albida (L.) Á. & D. Löve Small-white Orchid. Tegeirian Broga Gwyn.
Native. Damp pastures in the hills. Rare, with no recent records, perhaps overlooked in some places. It may still grow in the Llangiwg area.
00C?, Merthyr district, J. Evans & G. Fleming c.1909!; 70A, Llangiwg, EV 1917, Cwm-du Falls, EV c.1922; 98C, near Coity, JS c.1876; 99A, Cwm Selsig, JAW c.1925, Cwm Ffrwd?, JS c.1876.

COELOGLOSSUM Hartman

Coeloglossum viride (L.) Hartman Frog Orchid. Llys Ysgyfarnog.
Native. Pastures, limestone grassland and sand-dune grassland. Scarce, with no recent records; it may still grow in Gower and at Kenfig.
07C, near Llantrithyd, JS c.1886; 17D, Llandough, J. Evans c.1831 (abundant). 18B, Caerphilly Common, L. Reynolds 1946*; 48A, Rhosili, F.W. Benison 1904*, PS c.1909; 48B, Port Eynon, PS c.1909; 78B, Kenfig Burrows, C. Thomas, 1942*; 87B, St. Bride's Major, EV 1900, Dunraven, AHT c.1909; 90B, Hirwaun, JS c.1886; 96A, Marcross, K. Richards c.1909; 97A, Bridgend, F.M. Williams c.1909.

DACTYLORHIZA Necker ex Nevski

Dactylorhiza incarnata (L.) Soó Early Marsh-orchid. Tegeirian Rhuddgoch, Tegeirian y Gors Cynnar.
Native. Dune-slacks, damp meadows and marshes. Locally fairly abundant in dune-slacks at Whiteford Burrows, Llangennith Burrows, Oxwich, Crymlyn Burrows, Kenfig and Merthyr Mawr, it also occurs at Ystradowen Moors and in a few other base-rich marshes in the Vale. The crimson-flowered form (subsp. **coccinea** (Pugsley) Soó) is the predominant one in the dune populations (49A & C, 58A, 78D, 87A), but plants with the flesh-pink flowers of the typical subspecies occur with the crimson-flowered ones in the same population. The cream- or pale yellowish-flowered subsp. **ochroleuca** (Boll) P.F. Hunt & Summerhayes has been found at Kenfig Burrows (78D, D.M. Turner Ettlinger 1982), and the purple-flowered subsp. **pulchella** (Druce) Soó at Kenfig Burrows (78D, H.W. Pugsley 1935, D. Lang

1990) and Gors-Llwyn (81D 1978!). (!) Fl. (5–) 6–7, fr. 7–8; B.lap, (Lep) Coenonympha pamphilus (both very rarely). Map 1012.

Dactylorhiza incarnata × D. praetermissa (D. × wintonii (A. Camus) P.F. Hunt) has been recorded from Kenfig Burrows (78D, D. & A. Amos 1952*, D. Lang 1990).

Dactylorhiza purpurella (T. & T.A. Stephenson) Soó Northern Marsh-orchid. Tegeirian y Fign.
Native. Dune-slacks. Recently recorded from one locality in the Vale.
78D, Kenfig Burrows, D. Lang 1990.

Dactylorhiza praetermissa (Druce) Soó Southern Marsh-orchid. Tegeiran y Gors.
Native. Base-rich marshy meadows, dune-slacks, marshy waste ground. Common in the lowlands of western Glamorgan where it often colonizes disturbed waste ground. (!) Fl. 6–7, fr. 7–9; (Dip) Bombylius sp. (Lep) An.car, Pi.rap (all rarely). Map 1013.

Dactylorhiza maculata (L.) Soó Heath Spotted-orchid. Tegeirian Brych.
Native. Heathland and marshy grassland on acid soils. Locally very abundant in Gower, parts of the Uplands and the northern Vale, where some damp meadows and moorland are sometimes pale lilac with its flowers. Map 1014.

Dactylorhiza maculata × D. praetermissa (D. × hallii (Druce) Soó) has been recorded on several occasions in the past (00C, Aberdare, HJR c.1905; 48A, Rhosili, W.C. Barton 1916; 48B, Oxwich, 1927!*; 49C, Whiteford Burrows, H.W. Pugsley 1932; 59C, Pengwern Common, J.D. Grose 1936; 78D, Kenfig Burrows, AHT 1908, 1909).

Dactylorhiza fuchsii (Druce) Soó Common Spotted-orchid. Tegeirian Mannog, Tegeirian Brych Cyffredin.
Native. Base-rich marshes, marshy meadows and dune-slacks. Widely distributed but usually in rather small populations in Gower and the Uplands; widespread and fairly common in the Vale. Map 1015.

Dactylorhiza fuchsii × D. incarnata (D. × kernerorum (Soó) Soó) has been recorded once (78D, Kenfig Burrows, D. Lang 1990).

Dactylorhiza fuchsii × D. praetermissa (D. × grandis (Druce) P.F. Hunt) has been recorded from three localities (10C, Fochriw, 1947!*; 69B, Crymlyn Burrows, D. Lang 1990; 78D, Kenfig Burrows, D. & A. Amos 1952*, 1955!*, D. Lang 1990).

Dactylorhiza fuchsii × D. purpurella (D. × venusta (T. & T.A. Stephenson) Soó) has been recorded once (81D, Gors-llwyn, 1978!). One of the parents (**D. purpurella**) has not been recorded from the Glamorgan part of Gors-llwyn, but does occur in the same area (81D) just over the county boundary in Breconshire.

ORCHIS L.

Orchis morio L.   Green-winged Orchid. Tegeirian y Waun.
Native. Limestone grassland, mainly on sea-cliffs, and dune grassland; occasionally in old meadows inland. Apparently common in meadows on Lias limestone in the Vale eighty years ago (Trow, 1911). Most of these meadow populations have been destroyed by ploughing, but a few fragments survive on roadsides and above quarries. **O. morio** is still fairly frequent in grazed turf above the limestone cliffs of Gower and the Vale, and it is locally abundant in sand-dune grassland at Kenfig. Map 1016.

Orchis mascula (L.) L.   Early-purple Orchid. Tegeirian Coch.
Native. Limestone woodland, grassland and scrub on limestone sea-cliffs, and occasionally in grassland elsewhere. Fairly common in southern and western Gower and the southern and eastern Vale, but rare in the Uplands. Fl 5, fr. 6–7. Map 1017.

ANACAMPTIS L.C.M. Richard

Anacamptis pyramidalis (L.) L.C.M. Richard Pyramidal Orchid. Tegeirian Bera.
Native. Limestone grassland and calcareous sand-dune grassland. Fairly abundant in some years on fixed dunes at Whiteford, Oxwich and Kenfig, but scarce or in rather small numbers at its other localities. Recently recorded only from coastal sites. Map 1018.

OPHRYS L.

Ophrys insectifera L. Fly Orchid. Tegeirian Pryfyn.

Perhaps native but with only two rather doubtful records.
07D, near St. Nicholas, 'said to have been found by Archdeacon Bruce in 1883' (Storrie, 1886); 16A, Barry Island, 'reported', c.1907, (Riddelsdell, 1907).

Ophrys apifera Hudson   Bee Orchid. Tegeirian y Gwenyn.
Native. Limestone grassland, sand-dune grassland, and occasionally on roadside banks and in open woods. Infrequent and usually in small numbers. Now found chiefly but not exclusively in sand-dune grassland at Whiteford, Oxwich, Crymlyn and Kenfig Burrows and on cliff grassland in the southeastern Vale. Map 1019.

LIPARIS L.C.M. Richard

Liparis loeselii (L.) L.C.M. Richard   Fen Orchid. Gefell-llys y Fignen, Tegeirian y Fign.
Native. Damp dune-slacks. An increasingly rare species throughout its range; its chief British localities are now in Glamorgan. It shows great apparent fluctuations in numbers from year to year; in a good year (for example 1968, 1974 and 1980) it is locally frequent in 10–20 dune-slacks at both Whiteford Burrows (49A & C) and Kenfig Burrows (78D), but in a poor year (for example 1978) few if any plants produce leafy shoots in the same localities. **L. loeselii** was not discovered in Glamorgan until 1905, when H. J. Riddelsdell found several hundred plants in an undisclosed locality, usually thought to have been Kenfig Burrows. Riddelsdell did not then know that the species had previously been found at another locality in South Wales, Pembrey Burrows in Carmarthenshire in 1897 and 1899 (Barker, 1905; Bennet, 1918). Riddelsdell's locality remained the only one that was known in Glamorgan until 1927, when more than a hundred plants were found at Margam Burrows (78B) (now the site of Margam Steelworks) by Eleanor Vachell and Miss Insole; it was found at Crymlyn Burrows (79C) in 1931 or 1932 and subsequently at Oxwich (58A) (where it is rare) and Whiteford (49A, 49C). (!) Fl. 7 (–9), fr. 8–9.

Distribution Maps of Flowering Plants and Ferns

on all maps

○ pre-1960

● 1960 or later

× introduction pre-1960

+ introduction 1960 or later

An open circle or cross in the centre of a 10km square indicates records which it has not been possible to assign to a 5km square and which are the only known records for that 10km square.

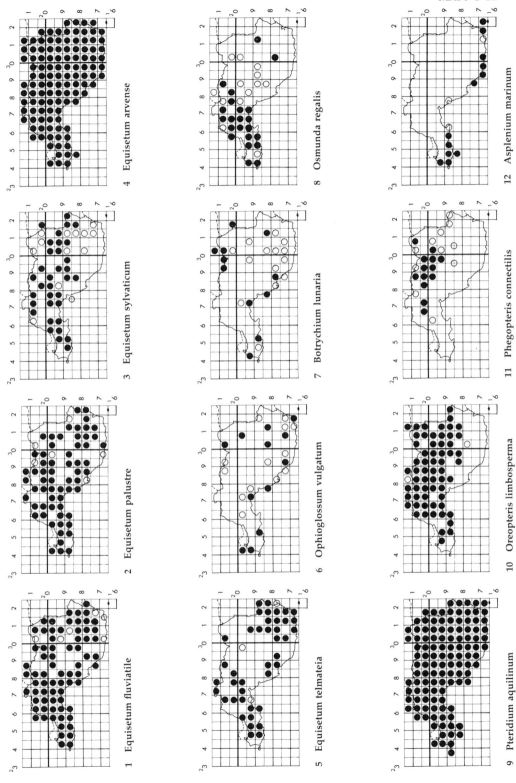

1 Equisetum fluviatile
2 Equisetum palustre
3 Equisetum sylvaticum
4 Equisetum arvense
5 Equisetum telmateia
6 Ophioglossum vulgatum
7 Botrychium lunaria
8 Osmunda regalis
9 Pteridium aquilinum
10 Oreopteris limbosperma
11 Phegopteris connectilis
12 Asplenium marinum

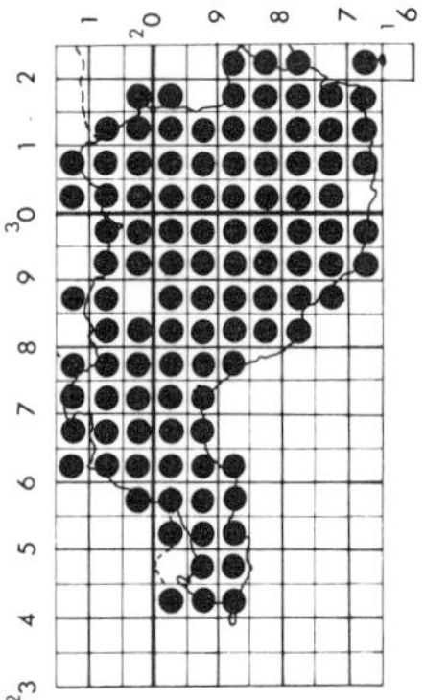

MAPS 13–24

13 Asplenium trichomanes
14 Asplenium adiantum-nigrum 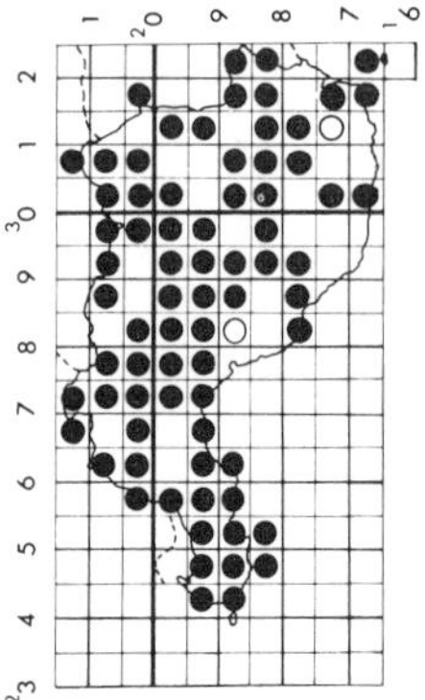
15 Asplenium ruta-muraria 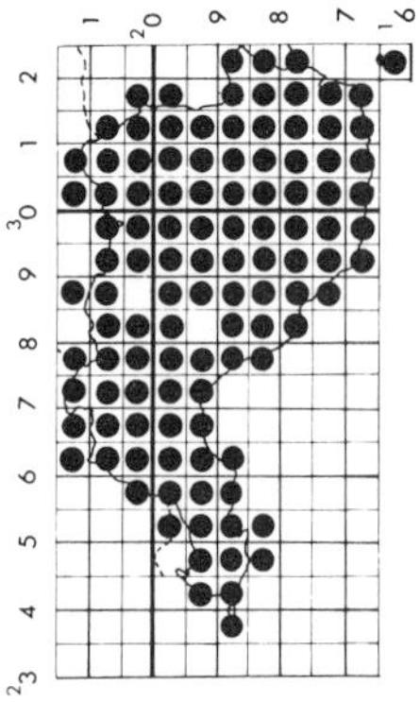
16 Asplenium ceterach 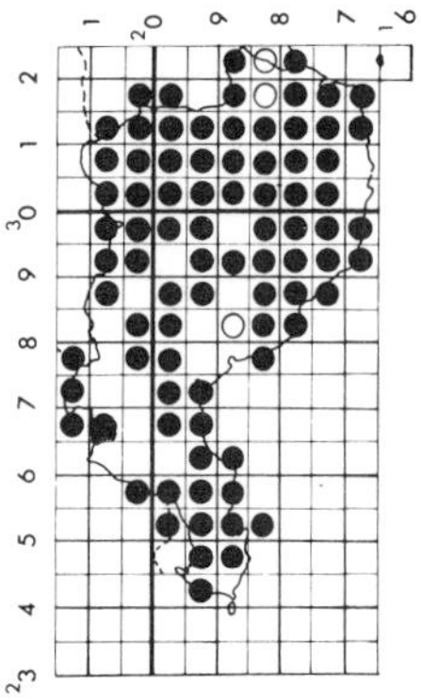
17 Asplenium scolopendrium 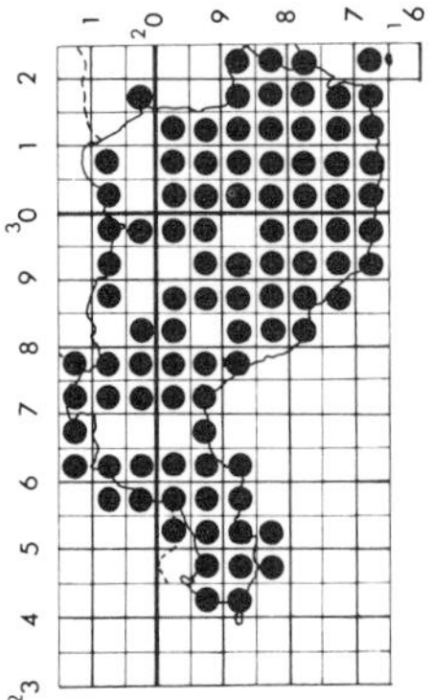
18 Athyrium filix-femina 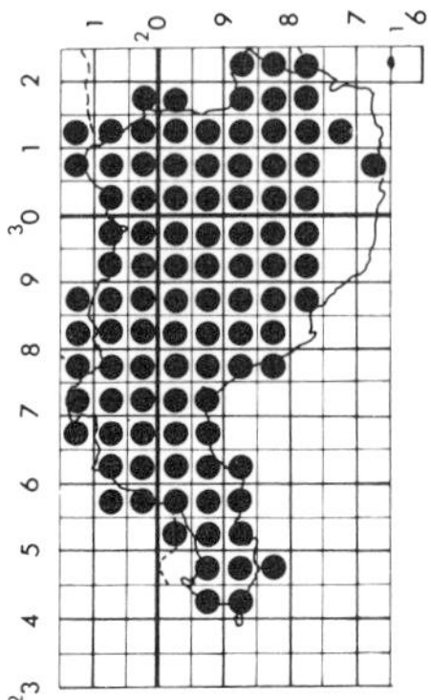
19 Cystopteris fragilis 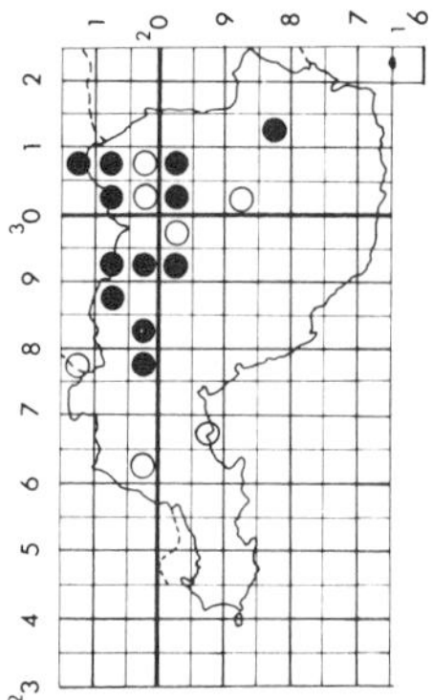
20 Polystichum aculeatum 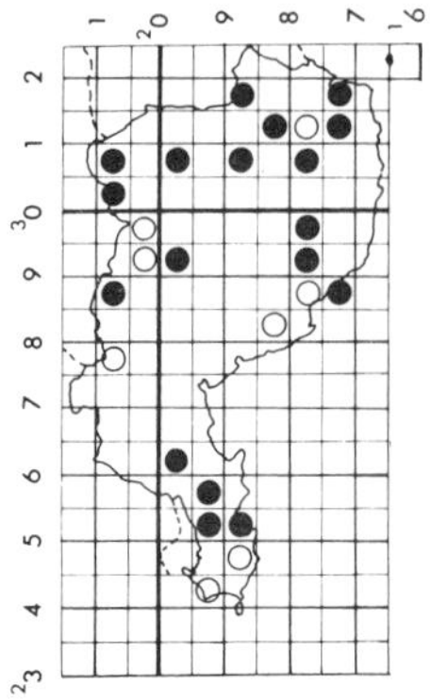
21 Polystichum setiferum 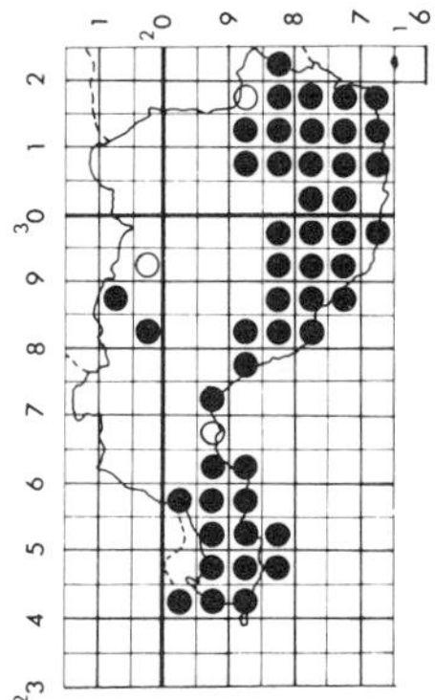
22 Dryopteris filix-mas 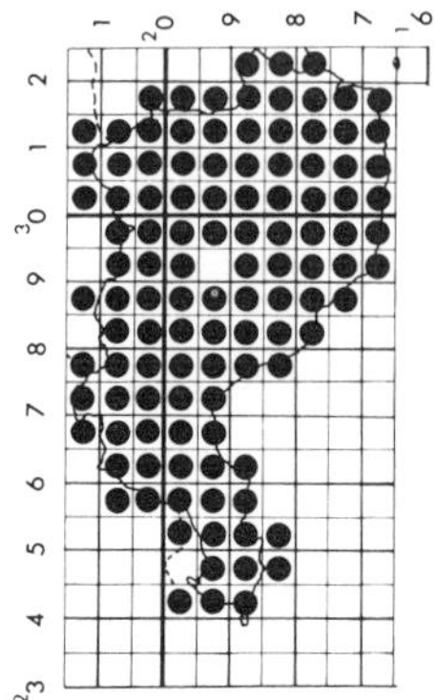
23 Dryopteris affinis 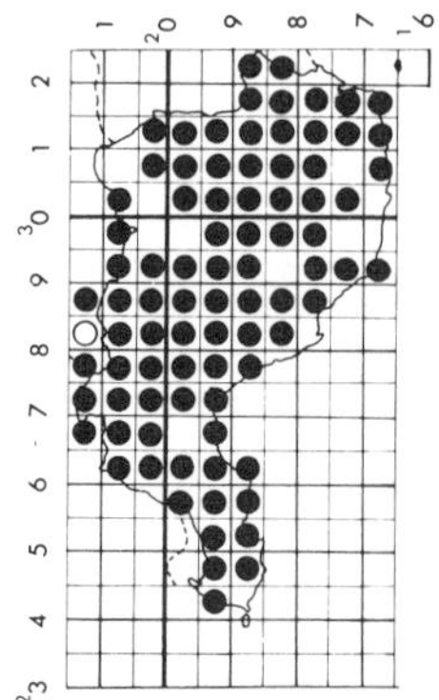
24 Dryopteris affinis subsp. affinis

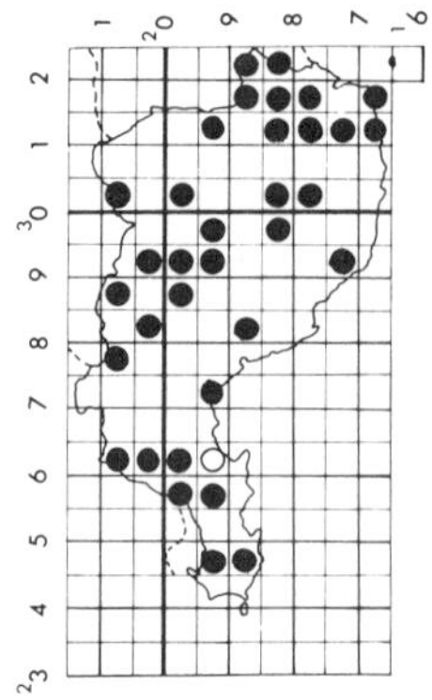

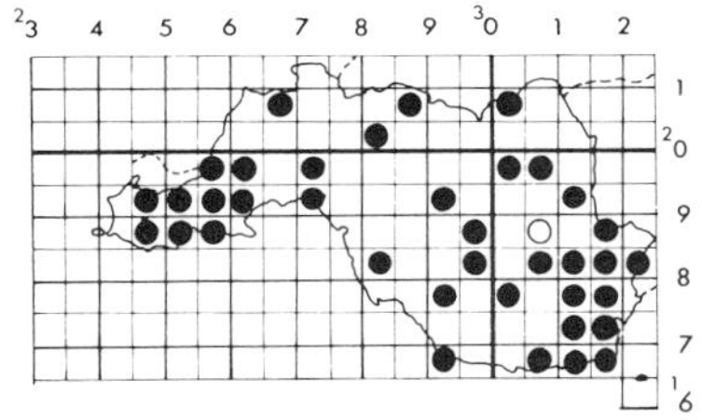

28 Blechnum spicant

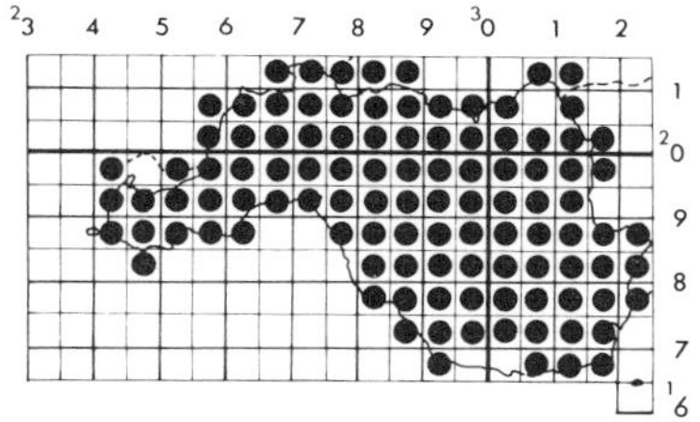

27 Dryopteris dilatata

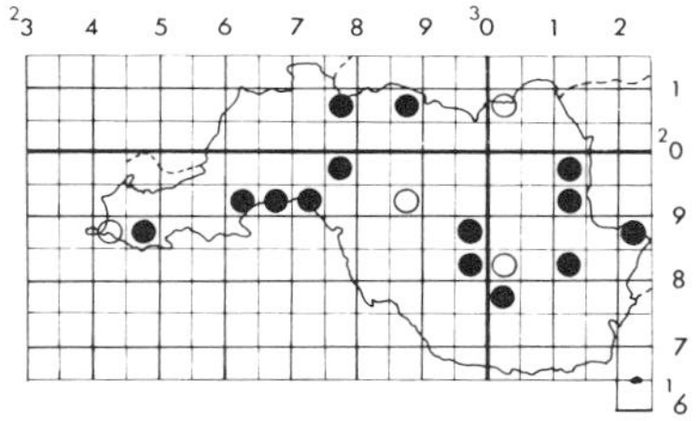

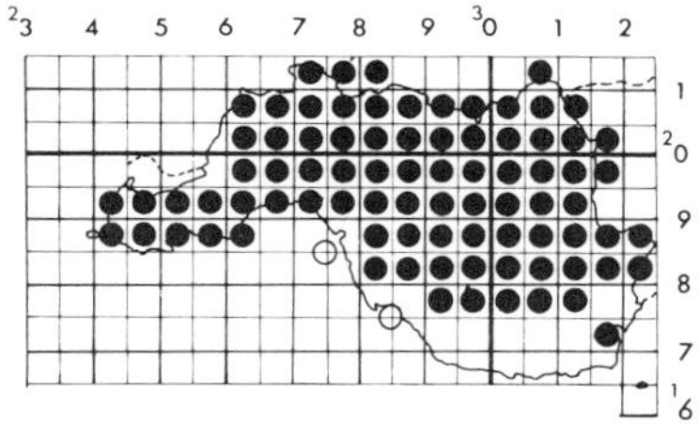

26 Dryopteris carthusiana

25 Dryopteris affinis
subsp. borreri

32 Pinus sylvestris

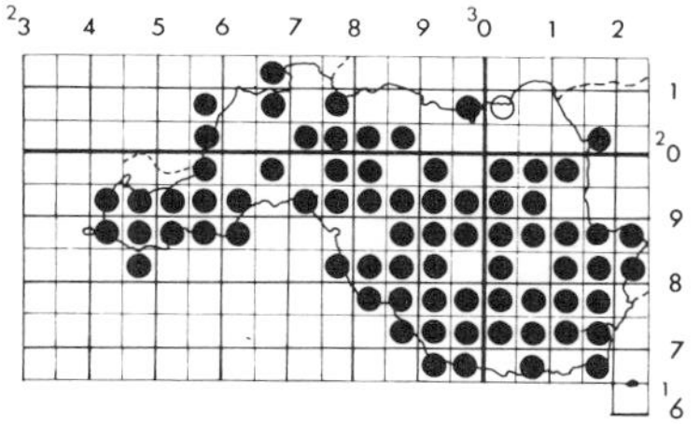

31 Polypodium interjectum

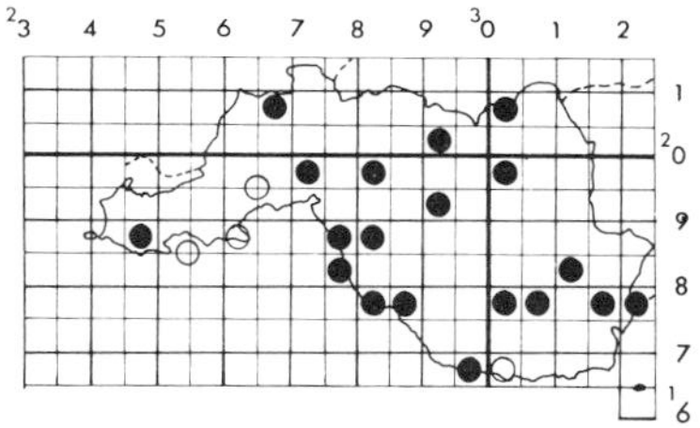

30 Polypodium vulgare

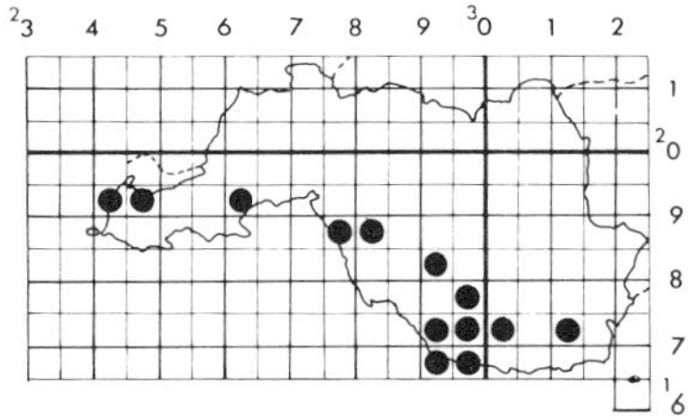

29 Polypodium cambricum

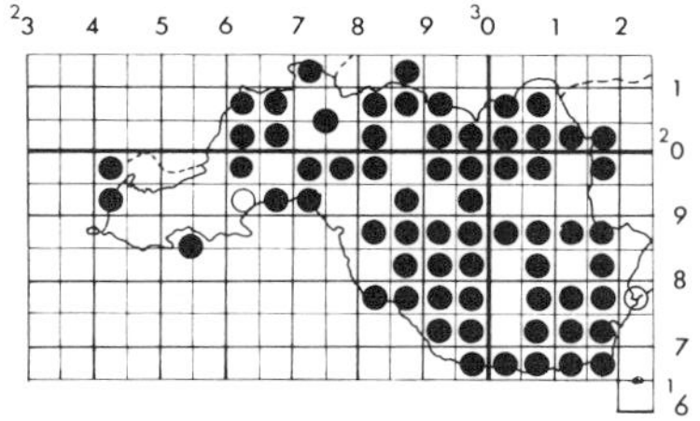

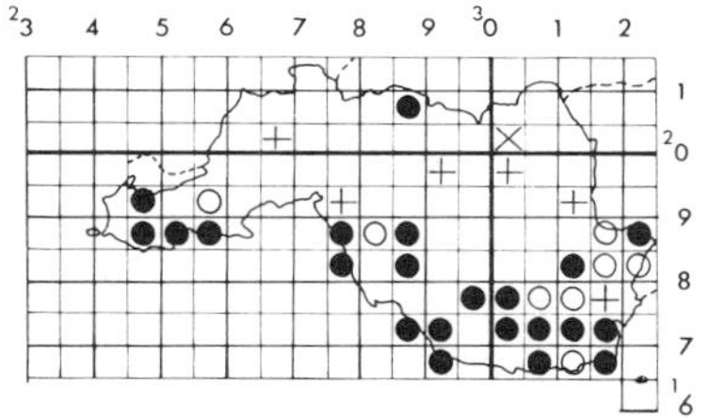

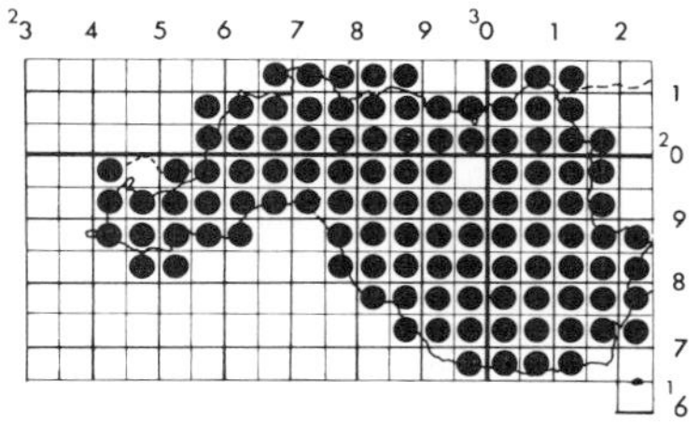

36 Salix cinerea subsp. oleifolia

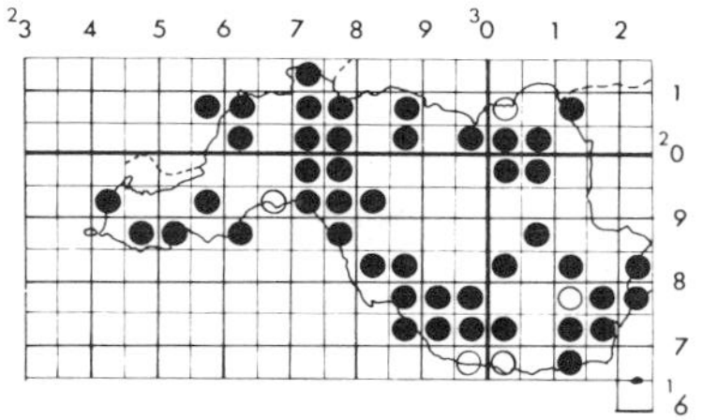

35 Salix alba

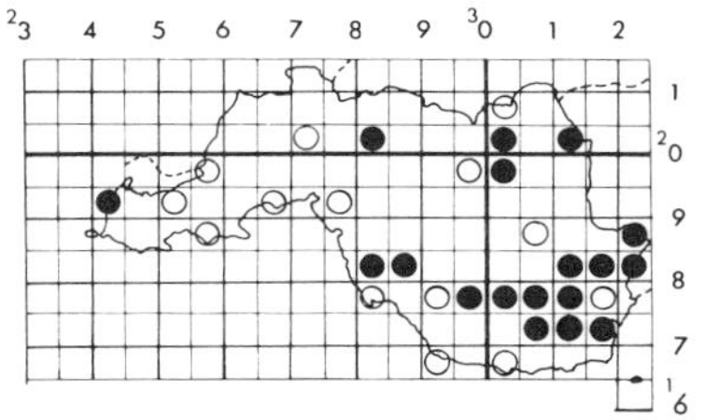

34 Salix fragilis

33 Taxus baccata

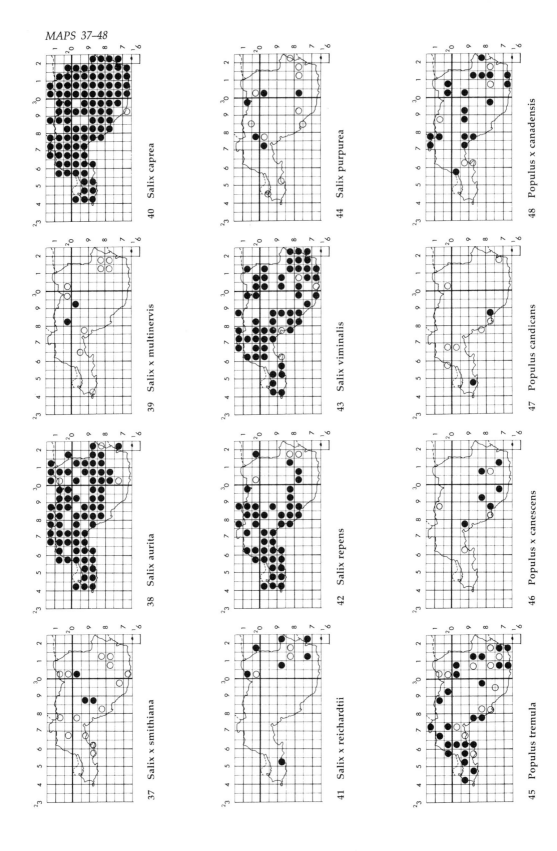

40 Salix caprea

44 Salix purpurea

48 Populus x canadensis

39 Salix x multinervis

43 Salix viminalis

47 Populus candicans

38 Salix aurita

42 Salix repens

46 Populus x canescens

37 Salix x smithiana

41 Salix x reichardtii

45 Populus tremula

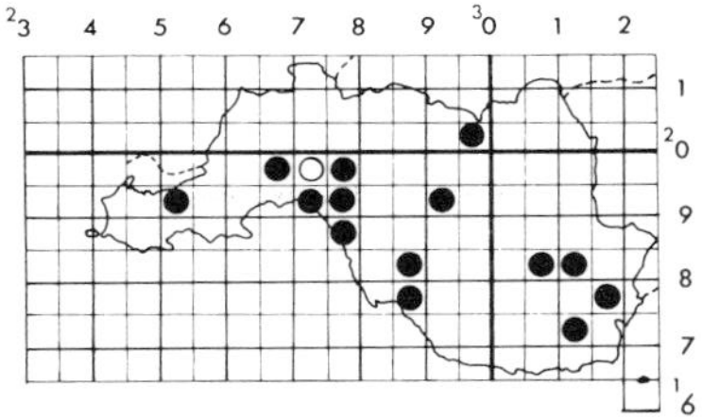

49 Populus nigra s.l.

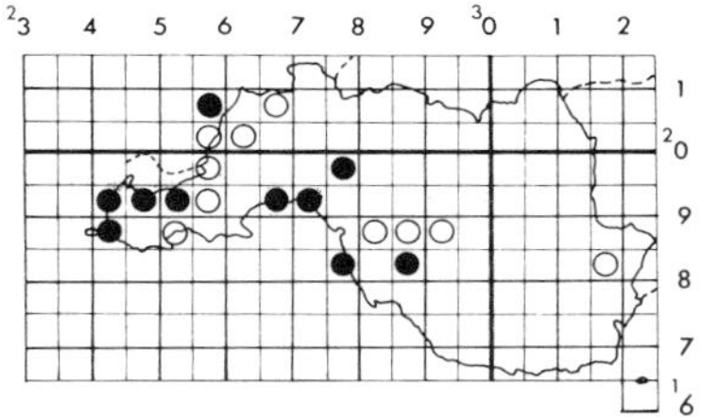

50 Myrica gale

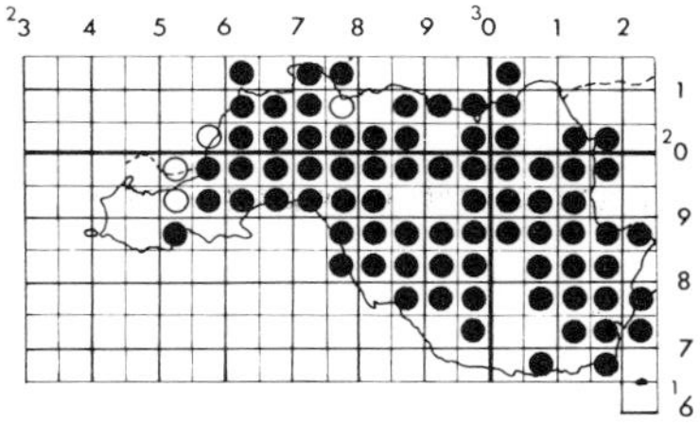

51 Betula pendula

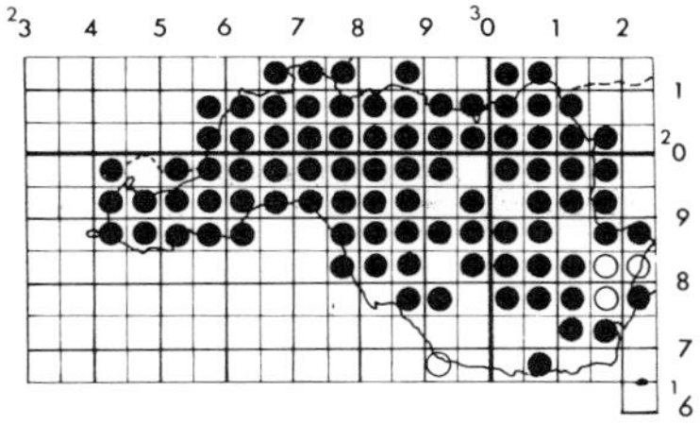

52 Betula pubescens

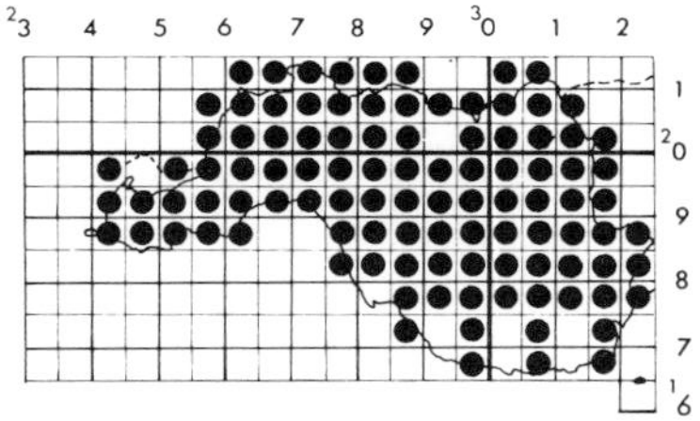

53 Alnus glutinosa

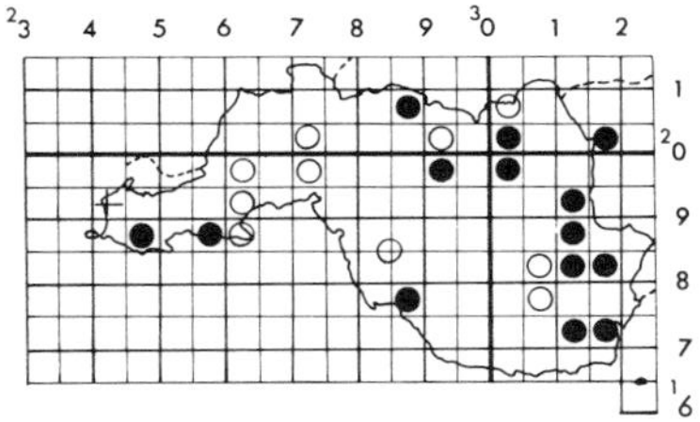

54 Carpinus betulus

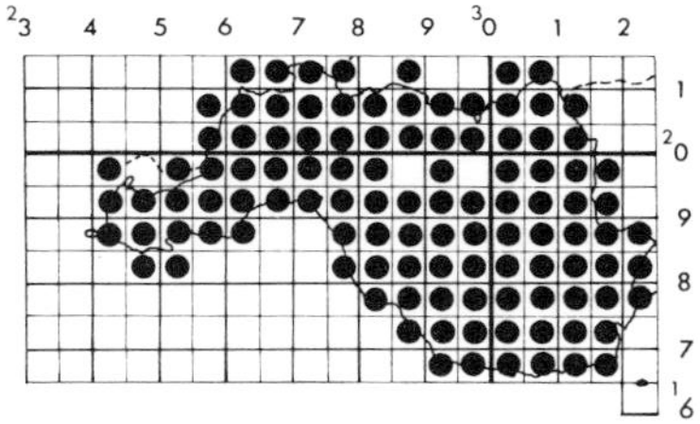

55 Corylus avellana

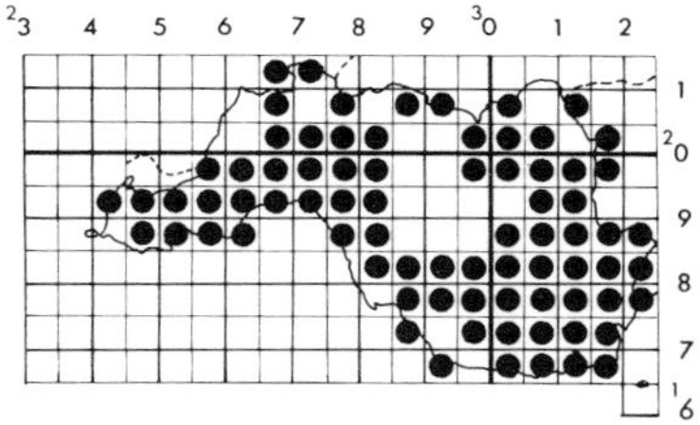

56 Fagus sylvatica

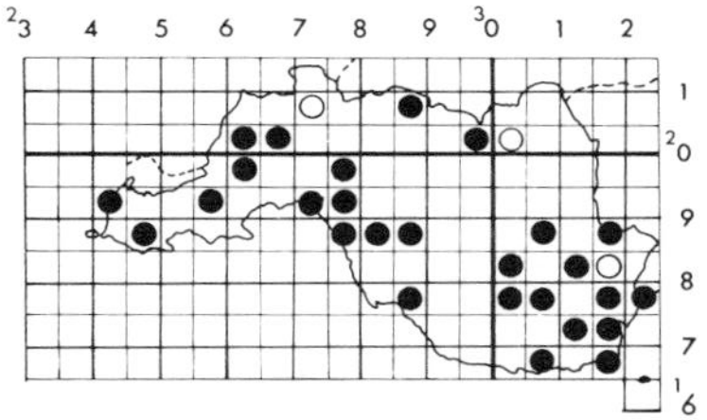

57 Castanea sativa

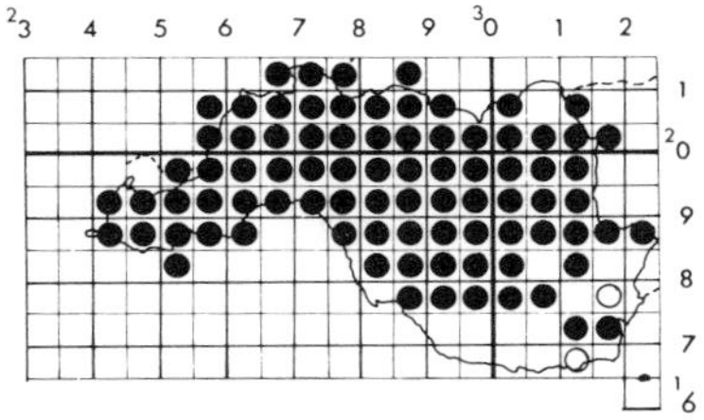

58 Quercus petraea

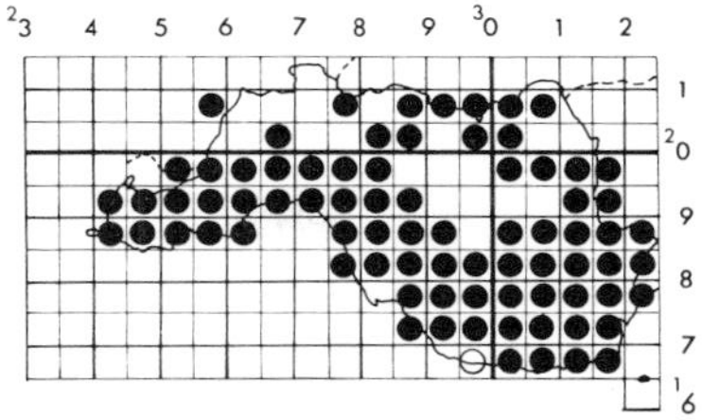

59 Quercus robur

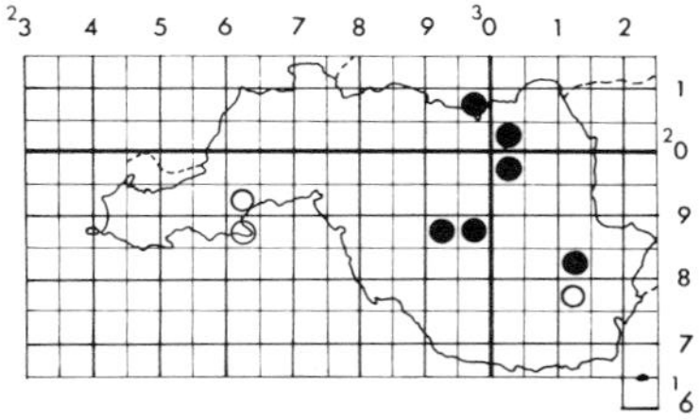

60 Quercus x rosacea

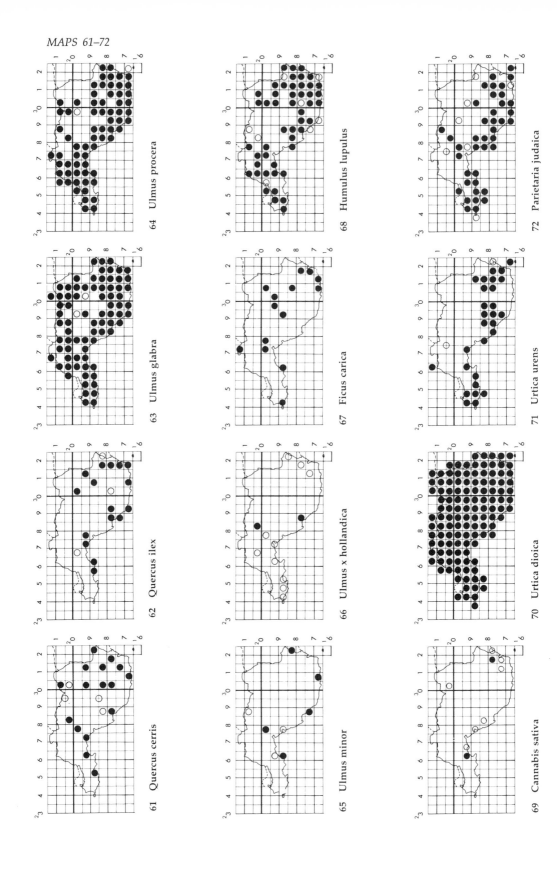

64 Ulmus procera

63 Ulmus glabra

62 Quercus ilex

61 Quercus cerris

68 Humulus lupulus

67 Ficus carica

66 Ulmus x hollandica

65 Ulmus minor

72 Parietaria judaica

71 Urtica urens

70 Urtica dioica

69 Cannabis sativa

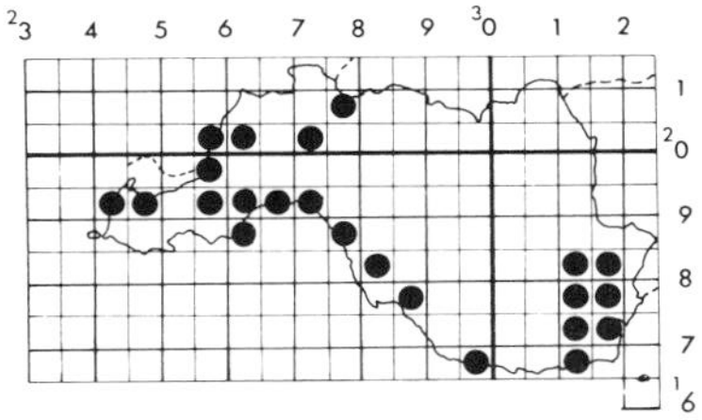

73 Soleirolia soleirolii

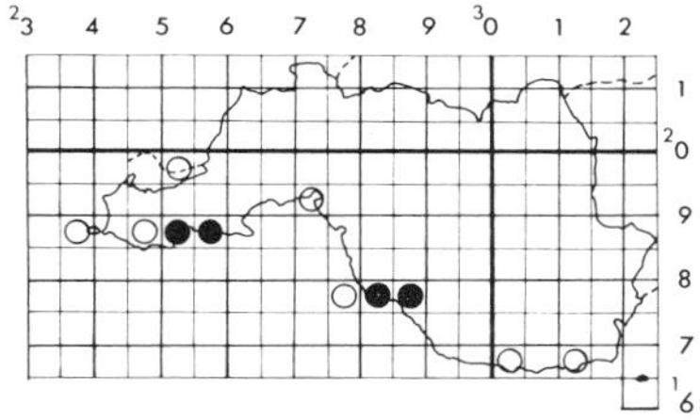

74 Polygonum oxyspermum subsp. raii

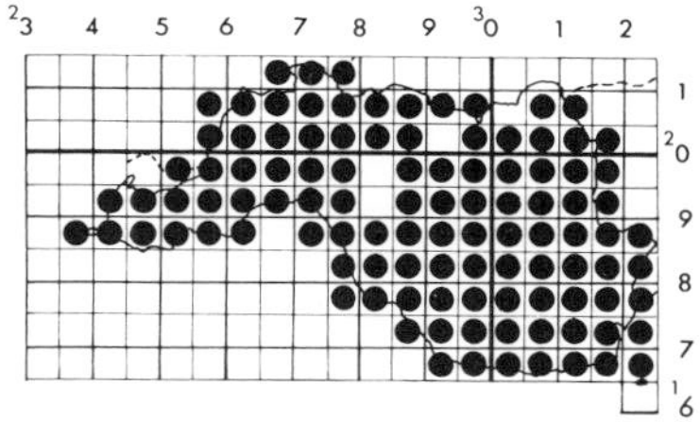

75 Polygonum aviculare

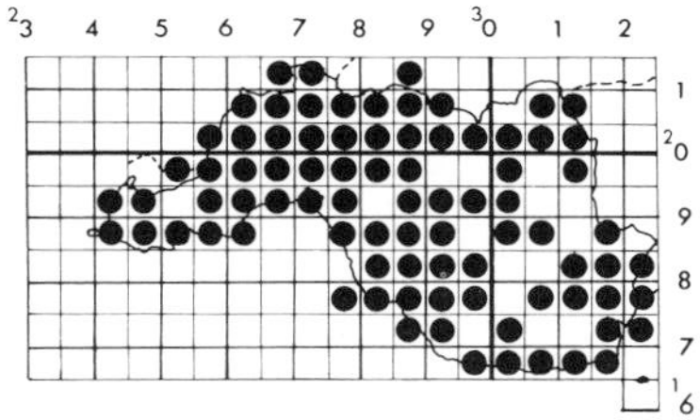

76 Polygonum arenastrum

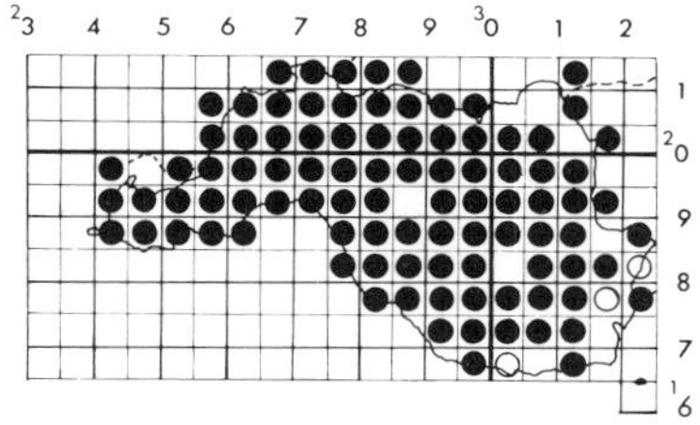

77 Polygonum hydropiper

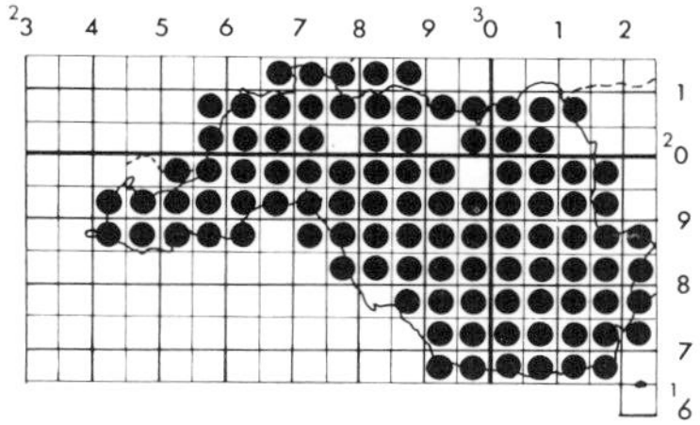

78 Polygonum persicaria

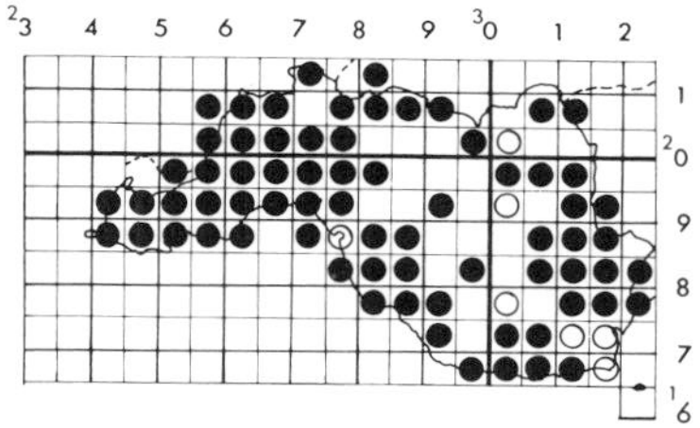

79 Polygonum lapathifolium

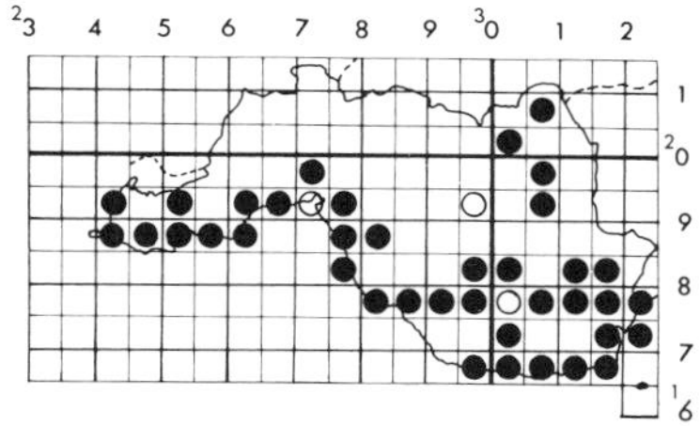

80 Polygonum amphibium

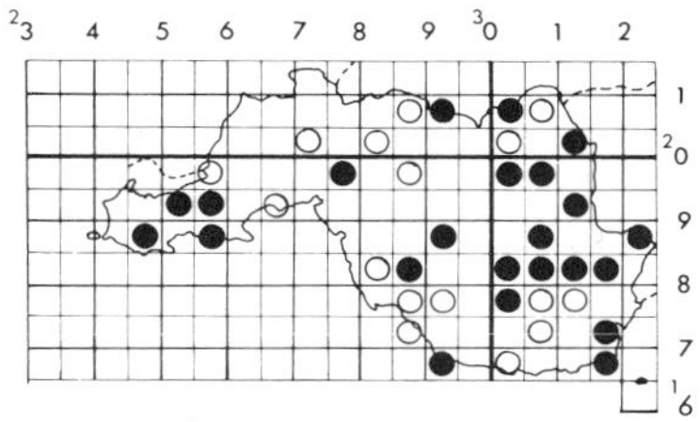

81 Polygonum bistorta

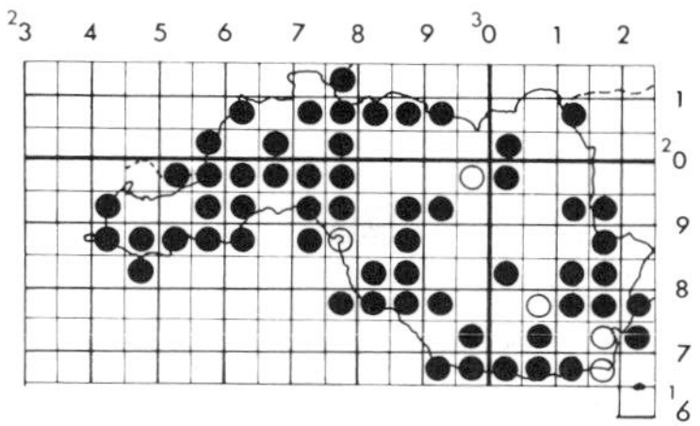

82 Fallopia convolvulus

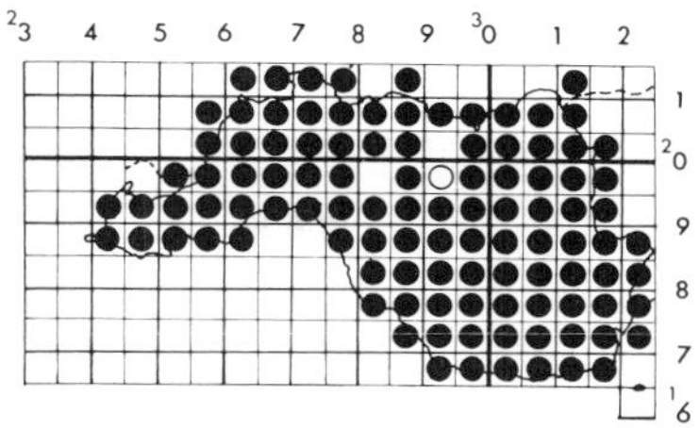

83 Reynoutria japonica

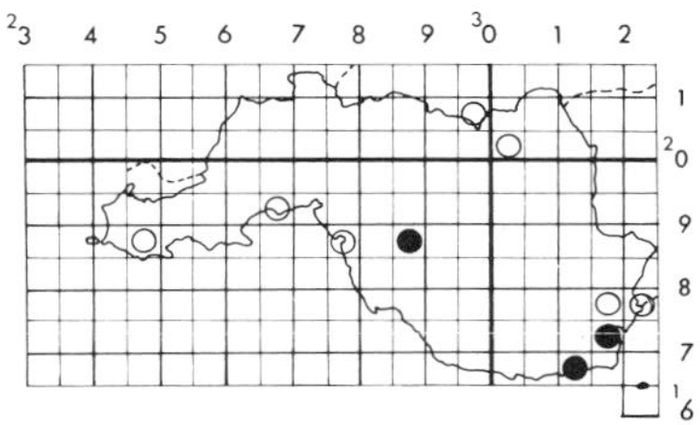

84 Fagopyrum esculentum

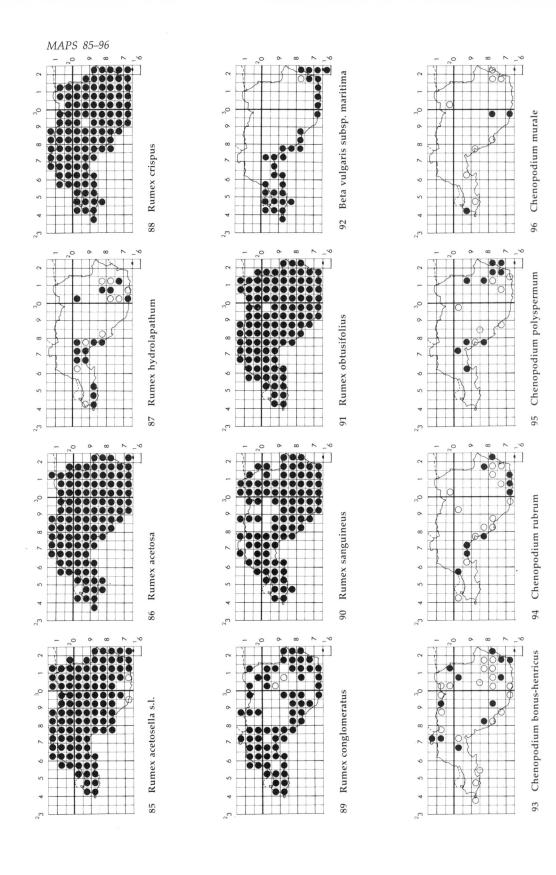

MAPS 85–96

85 Rumex acetosella s.l.

86 Rumex acetosa

87 Rumex hydrolapathum

88 Rumex crispus

89 Rumex conglomeratus

90 Rumex sanguineus

91 Rumex obtusifolius

92 Beta vulgaris subsp. maritima

93 Chenopodium bonus-henricus

94 Chenopodium rubrum

95 Chenopodium polyspermum

96 Chenopodium murale

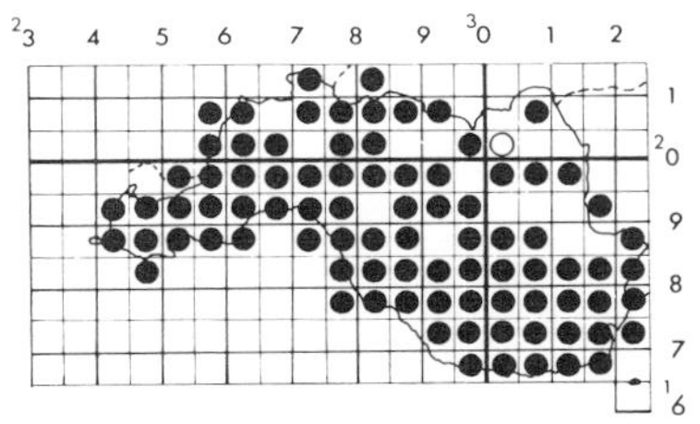

97 Chenopodium album subsp. album

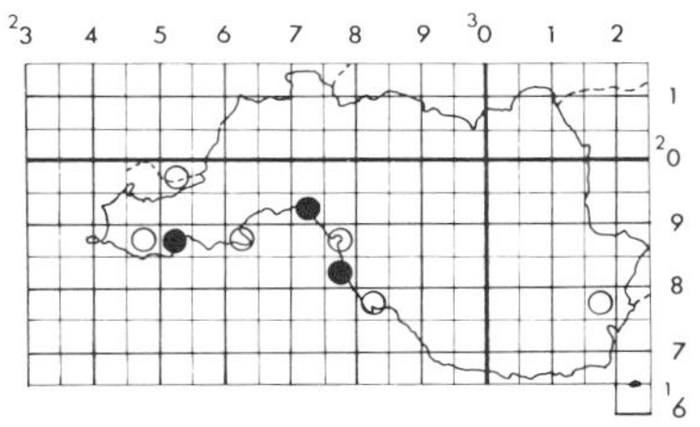

98 Atriplex laciniata

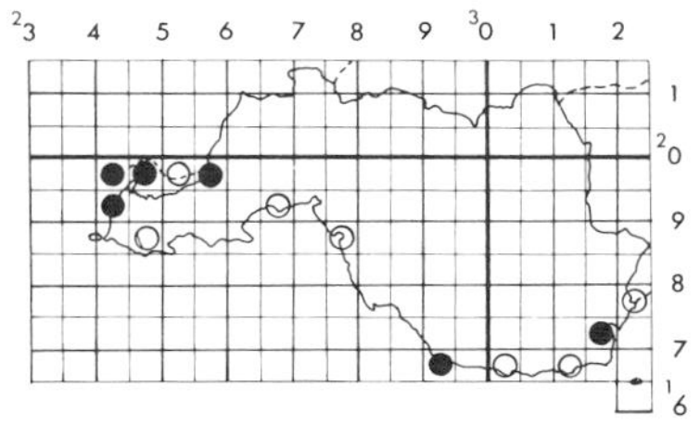

99 Atriplex littoralis

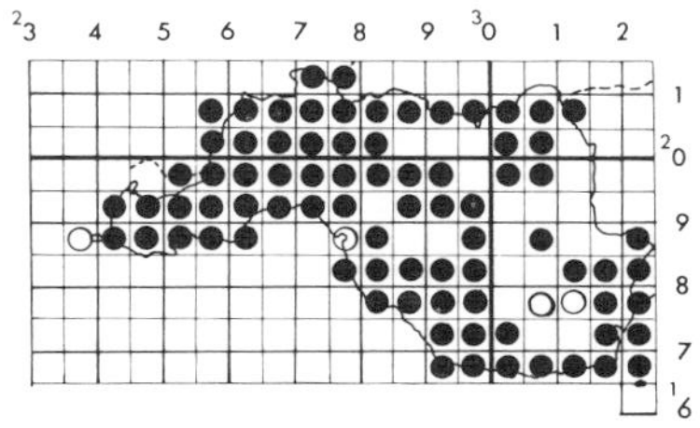

100 Atriplex patula

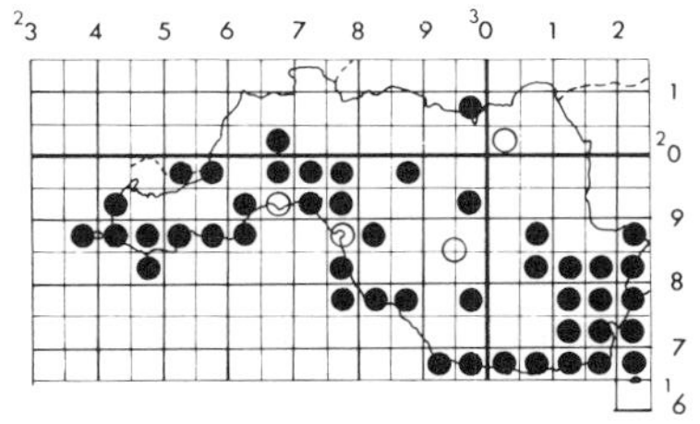

101 Atriplex prostrata

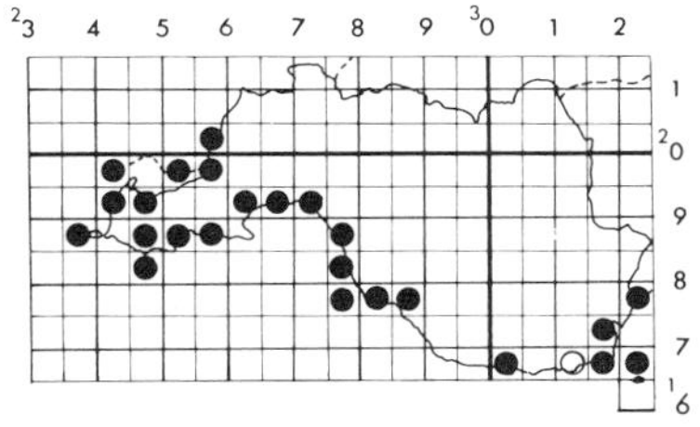

102 Atriplex glabriuscula

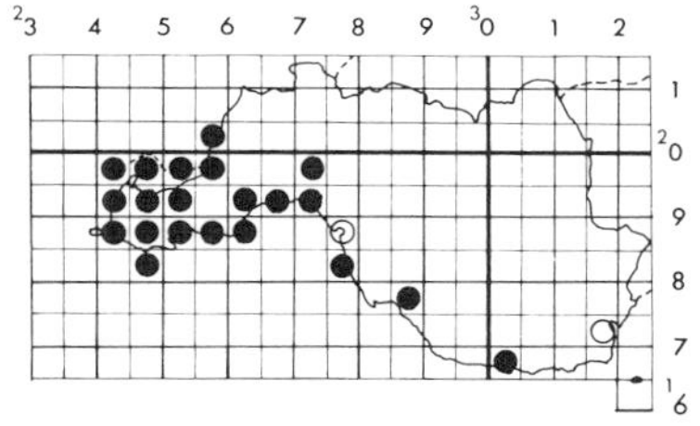

103 Halimione portulacoides

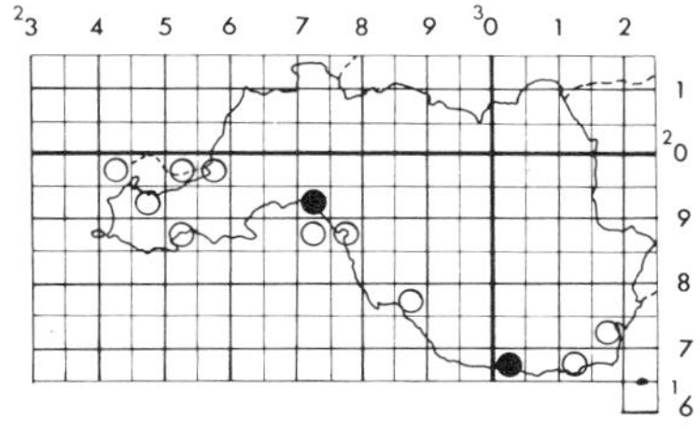

104 Salicornia europaea

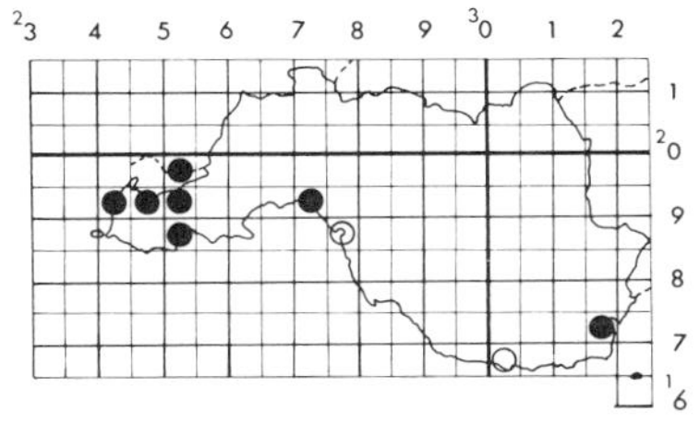

105 Salicornia ramosissima

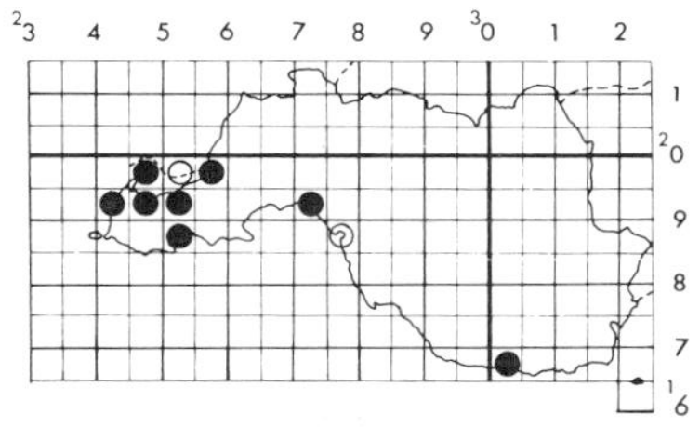

106 Salicornia pusilla

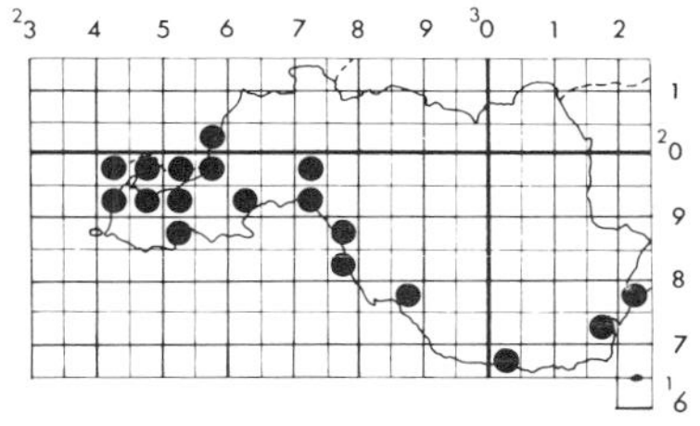

107 Suaeda maritima

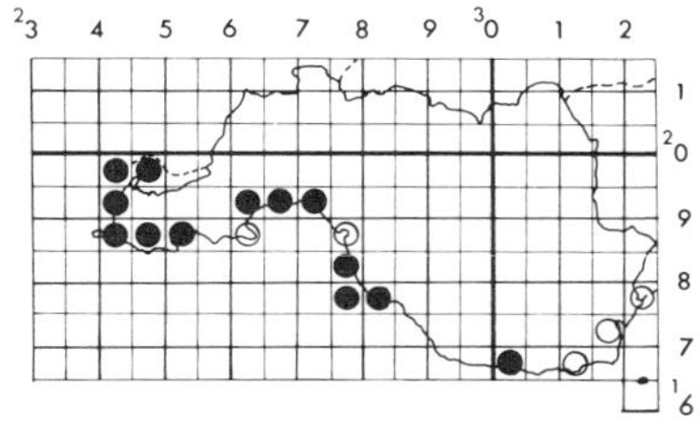

108 Salsola kali subsp. kali

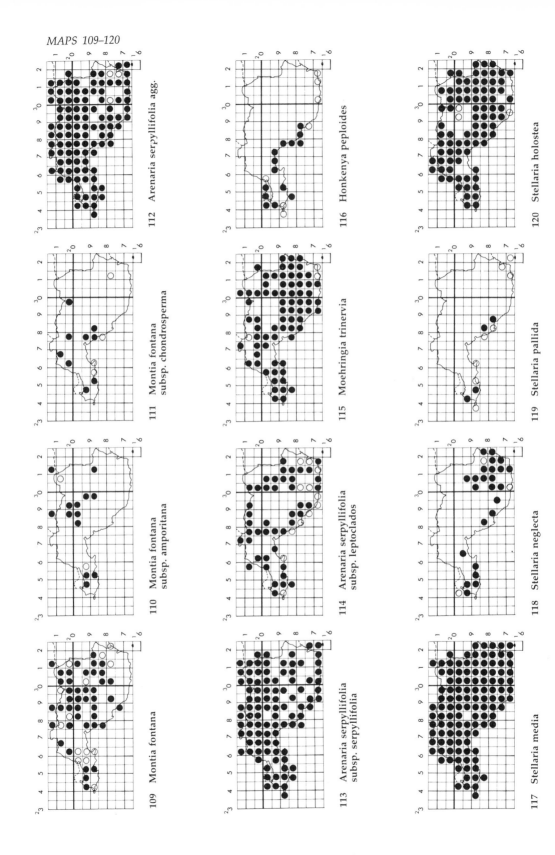

109 Montia fontana

110 Montia fontana
subsp. amporitana

111 Montia fontana
subsp. chondrosperma

112 Arenaria serpyllifolia agg.

113 Arenaria serpyllifolia
subsp. serpyllifolia

114 Arenaria serpyllifolia
subsp. leptoclados

115 Moehringia trinervia

116 Honkenya peploides

117 Stellaria media

118 Stellaria neglecta

119 Stellaria pallida

120 Stellaria holostea

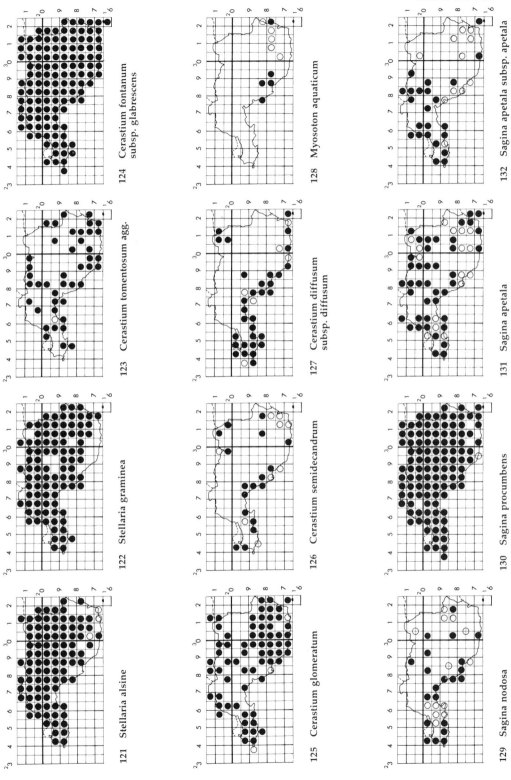

124 Cerastium fontanum subsp. glabrescens

128 Myosoton aquaticum

132 Sagina apetala subsp. apetala

123 Cerastium tomentosum agg.

127 Cerastium diffusum subsp. diffusum

131 Sagina apetala

122 Stellaria graminea

126 Cerastium semidecandrum

130 Sagina procumbens

121 Stellaria alsine

125 Cerastium glomeratum

129 Sagina nodosa

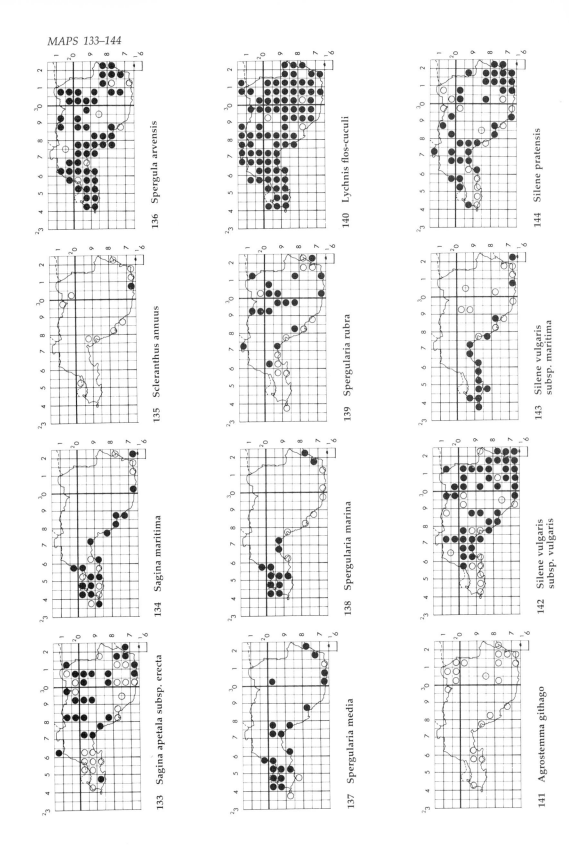

133 Sagina apetala subsp. erecta

134 Sagina maritima

135 Scleranthus annuus

136 Spergula arvensis

137 Spergularia media

138 Spergularia marina

139 Spergularia rubra

140 Lychnis flos-cuculi

141 Agrostemma githago

142 Silene vulgaris subsp. vulgaris

143 Silene vulgaris subsp. maritima

144 Silene pratensis

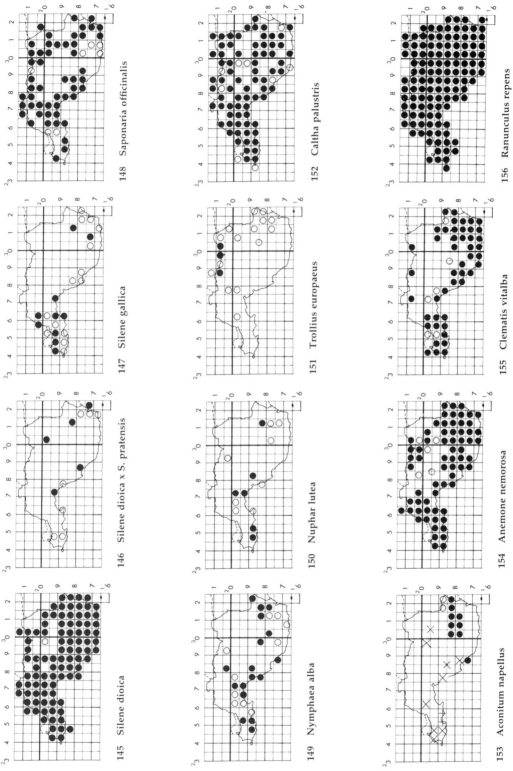

145 Silene dioica
146 Silene dioica x S. pratensis
147 Silene gallica
148 Saponaria officinalis
149 Nymphaea alba
150 Nuphar lutea
151 Trollius europaeus
152 Caltha palustris
153 Aconitum napellus
154 Anemone nemorosa
155 Clematis vitalba
156 Ranunculus repens

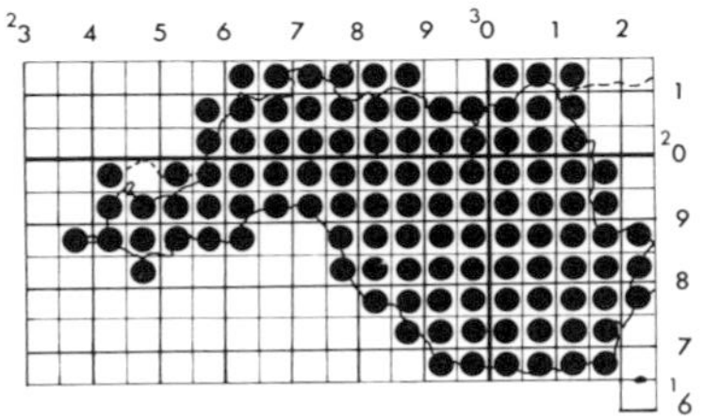

157 Ranunculus acris

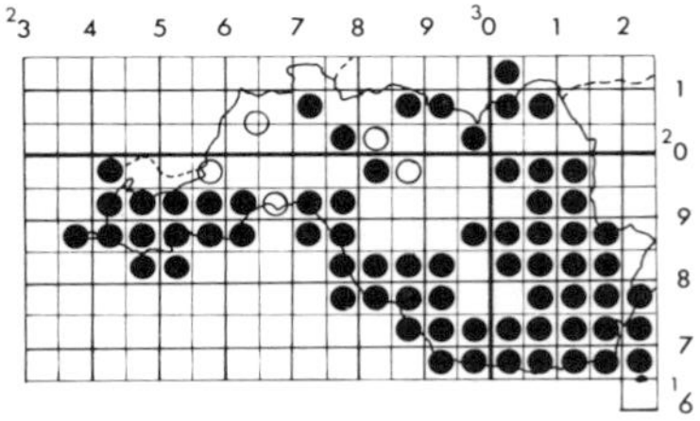

158 Ranunculus bulbosus

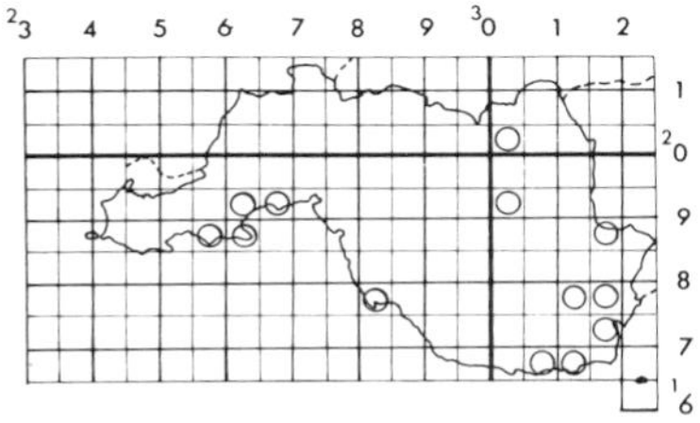

159 Ranunculus arvensis

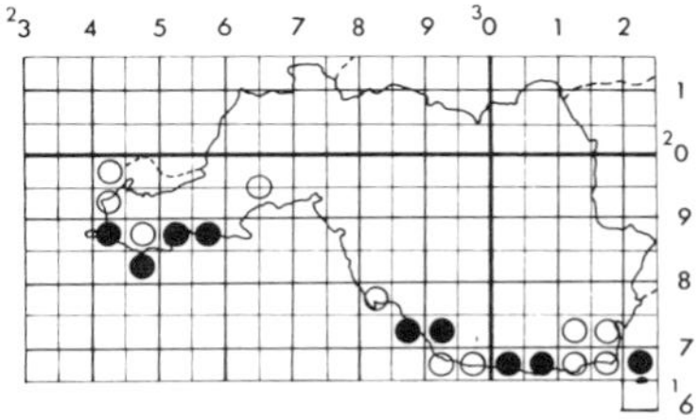

160 Ranunculus parviflorus

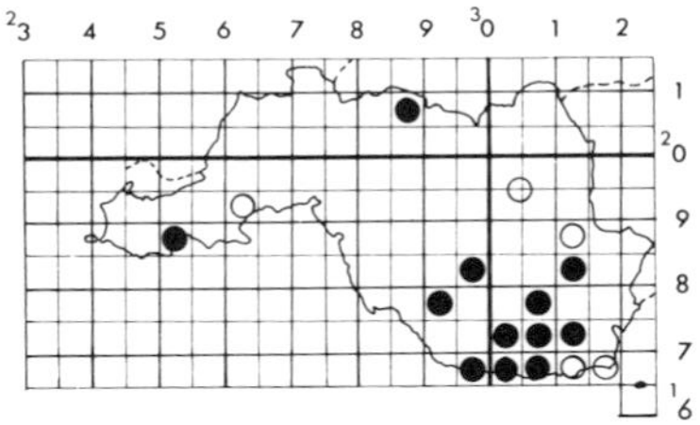

161 Ranunculus auricomus

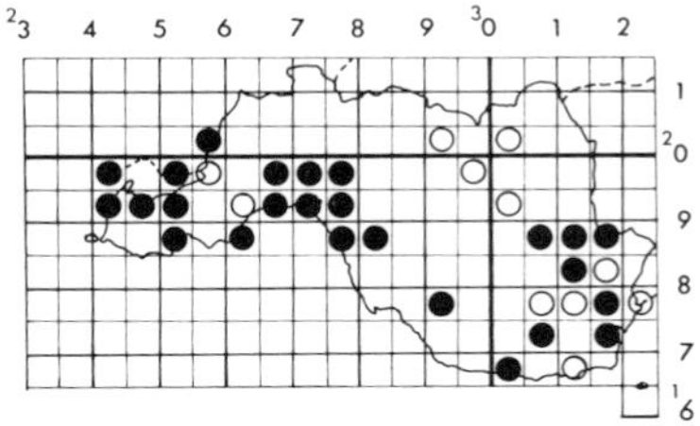

162 Ranunculus sceleratus

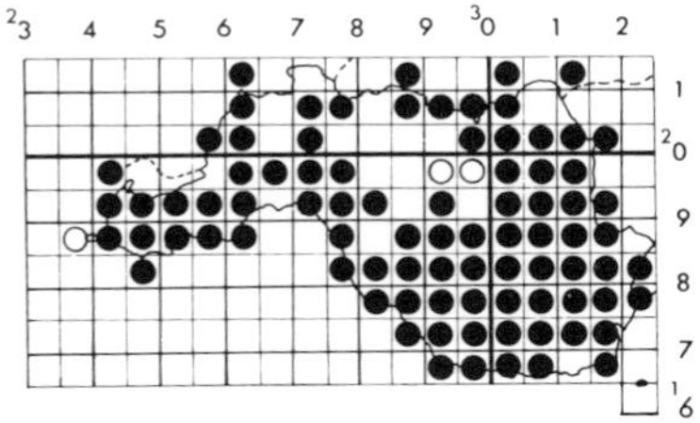

163 Ranunculus ficaria

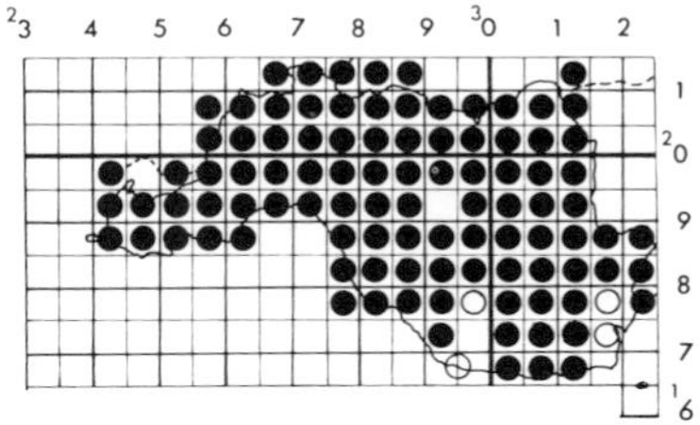

164 Ranunculus flammula

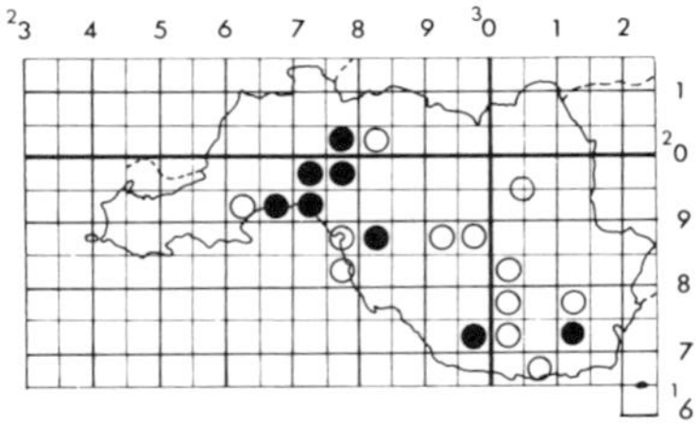

165 Ranunculus lingua

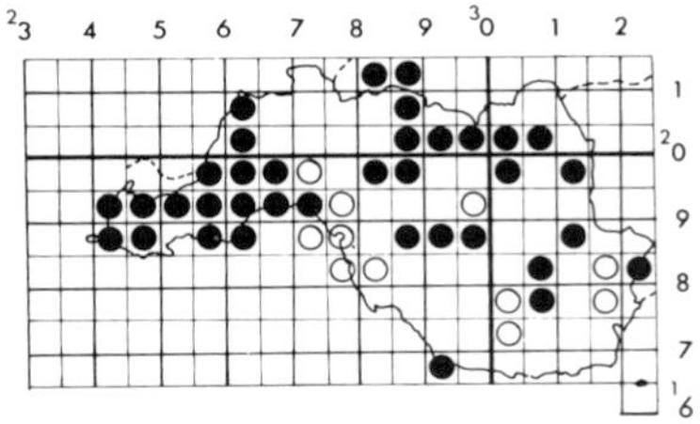

166 Ranunculus hederaceus

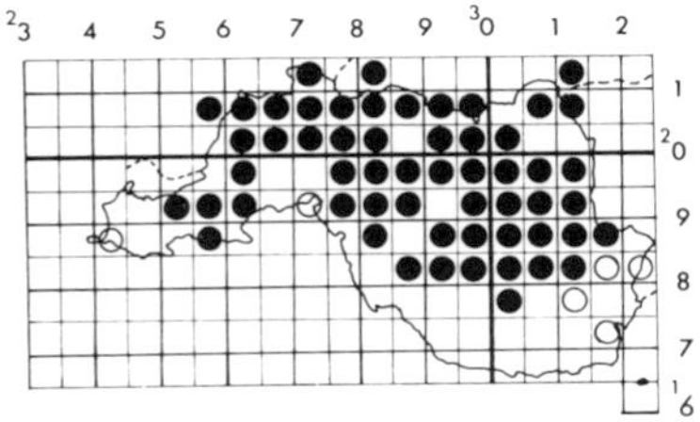

167 Ranunculus omiophyllus

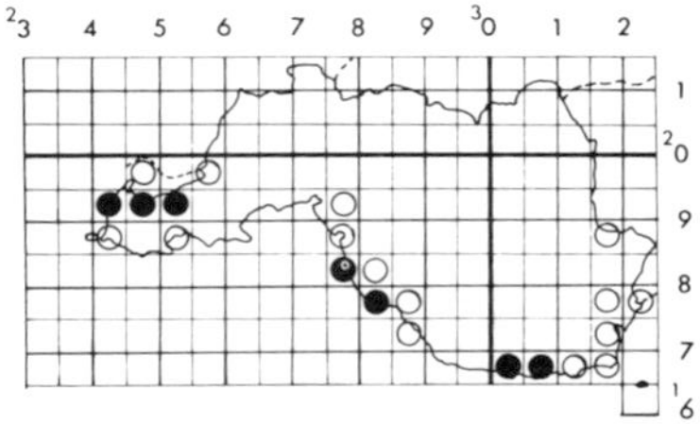

168 Ranunculus baudotii

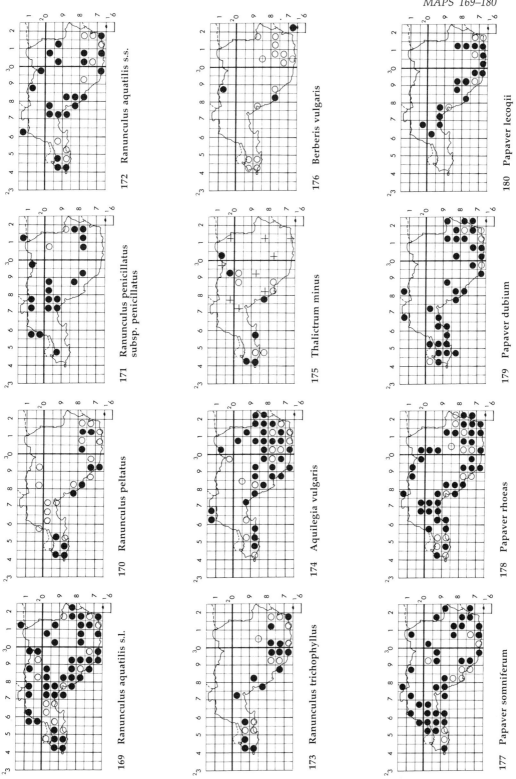

169 Ranunculus aquatilis s.l.

170 Ranunculus peltatus

171 Ranunculus penicillatus subsp. penicillatus

172 Ranunculus aquatilis s.s.

173 Ranunculus trichophyllus

174 Aquilegia vulgaris

175 Thalictrum minus

176 Berberis vulgaris

177 Papaver somniferum

178 Papaver rhoeas

179 Papaver dubium

180 Papaver lecoqii

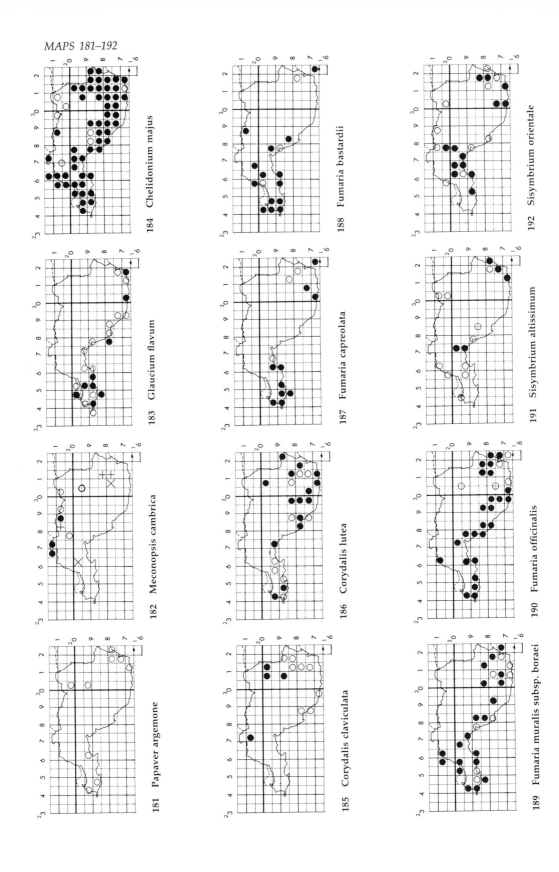

184 Chelidonium majus

188 Fumaria bastardii

192 Sisymbrium orientale

183 Glaucium flavum

187 Fumaria capreolata

191 Sisymbrium altissimum

182 Meconopsis cambrica

186 Corydalis lutea

190 Fumaria officinalis

181 Papaver argemone

185 Corydalis claviculata

189 Fumaria muralis subsp. boraei

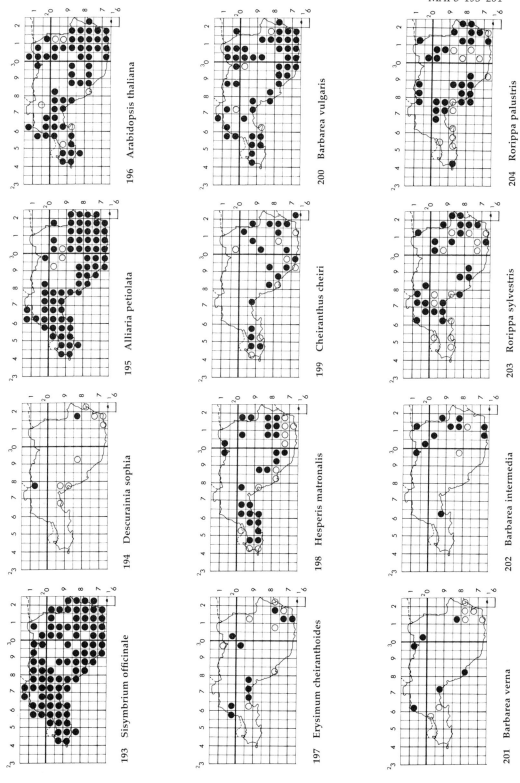

193 Sisymbrium officinale

194 Descurainia sophia

195 Alliaria petiolata

196 Arabidopsis thaliana

197 Erysimum cheiranthoides

198 Hesperis matronalis

199 Cheiranthus cheiri

200 Barbarea vulgaris

201 Barbarea verna

202 Barbarea intermedia

203 Rorippa sylvestris

204 Rorippa palustris

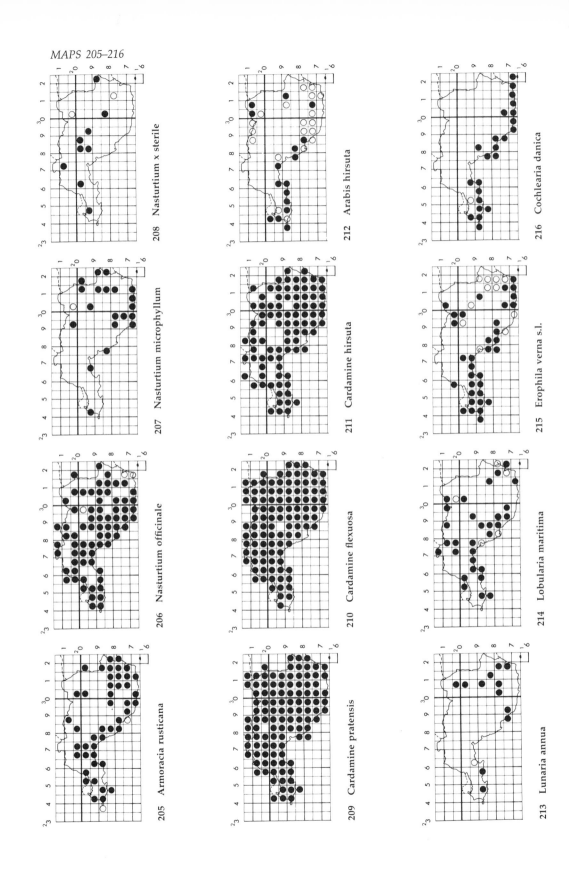

205 Armoracia rusticana

206 Nasturtium officinale

207 Nasturtium microphyllum

208 Nasturtium × sterile

209 Cardamine pratensis

210 Cardamine flexuosa

211 Cardamine hirsuta

212 Arabis hirsuta

213 Lunaria annua

214 Lobularia maritima

215 Erophila verna s.l.

216 Cochlearia danica

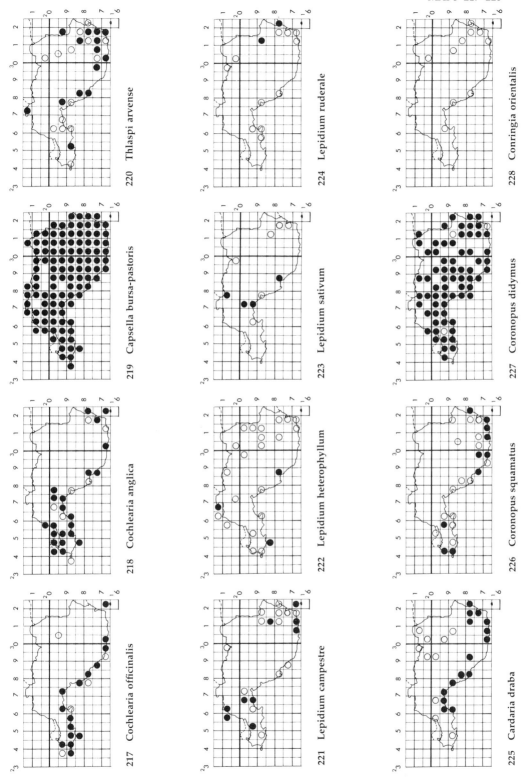

217 Cochlearia officinalis
218 Cochlearia anglica
219 Capsella bursa-pastoris
220 Thlaspi arvense
221 Lepidium campestre
222 Lepidium heterophyllum
223 Lepidium sativum
224 Lepidium ruderale
225 Cardaria draba
226 Coronopus squamatus
227 Coronopus didymus
228 Conringia orientalis

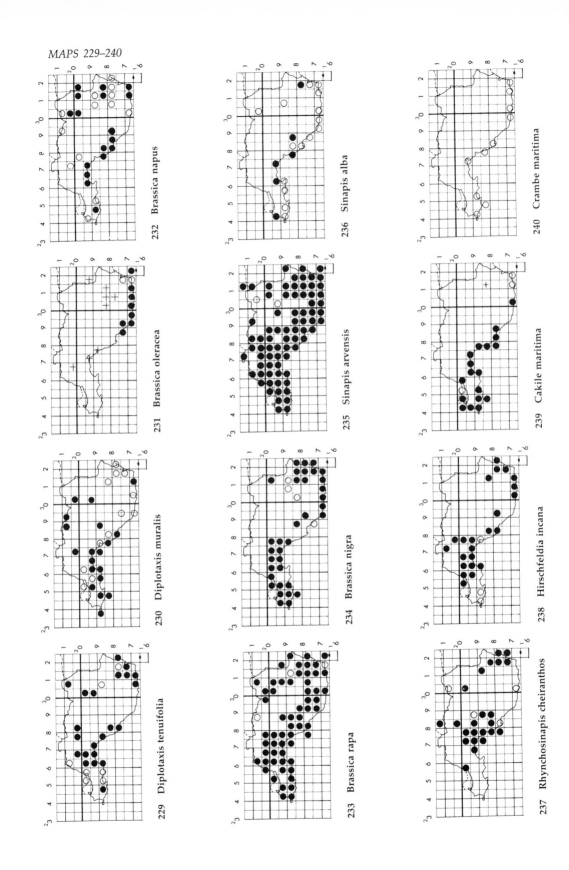

229 Diplotaxis tenuifolia
230 Diplotaxis muralis
231 Brassica oleracea
232 Brassica napus
233 Brassica rapa
234 Brassica nigra
235 Sinapis arvensis
236 Sinapis alba
237 Rhynchosinapis cheiranthos
238 Hirschfeldia incana
239 Cakile maritima
240 Crambe maritima

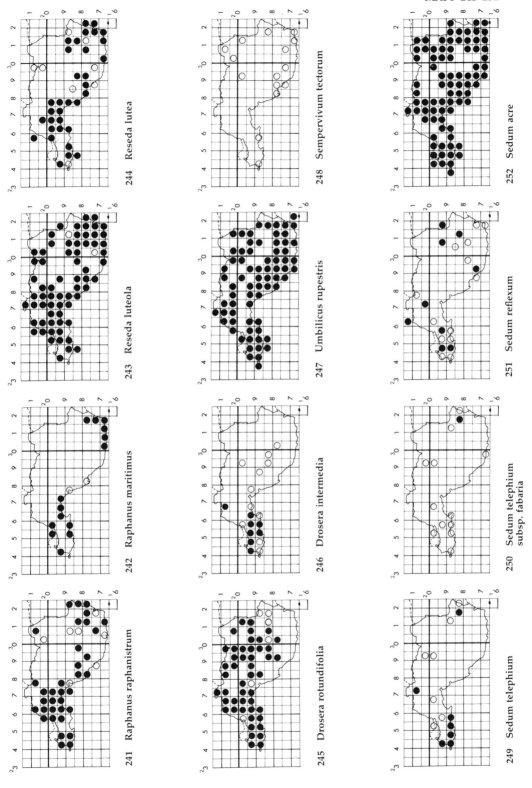

244 Reseda lutea

248 Sempervivum tectorum

252 Sedum acre

243 Reseda luteola

247 Umbilicus rupestris

251 Sedum reflexum

242 Raphanus maritimus

246 Drosera intermedia

250 Sedum telephium
subsp. fabaria

241 Raphanus raphanistrum

245 Drosera rotundifolia

249 Sedum telephium

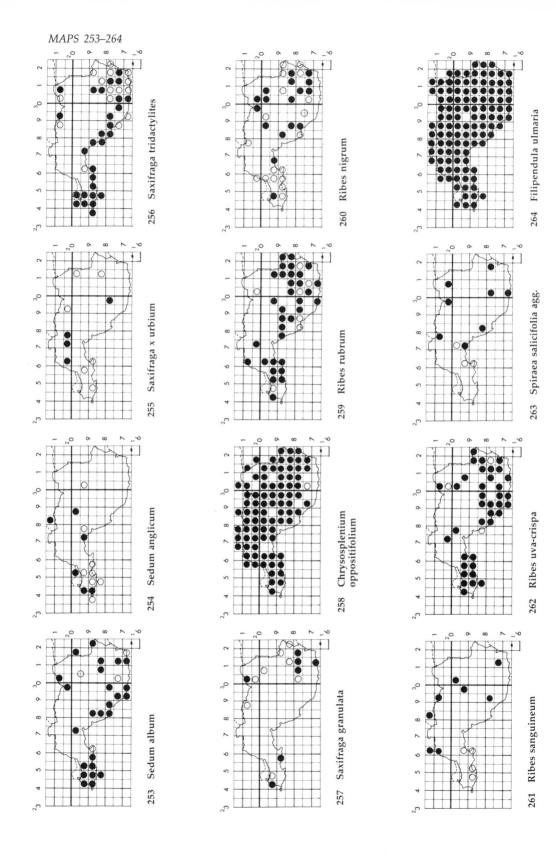

253 Sedum album

254 Sedum anglicum

255 Saxifraga x urbium

256 Saxifraga tridactylites

257 Saxifraga granulata

258 Chrysosplenium oppositifolium

259 Ribes rubrum

260 Ribes nigrum

261 Ribes sanguineum

262 Ribes uva-crispa

263 Spiraea salicifolia agg.

264 Filipendula ulmaria

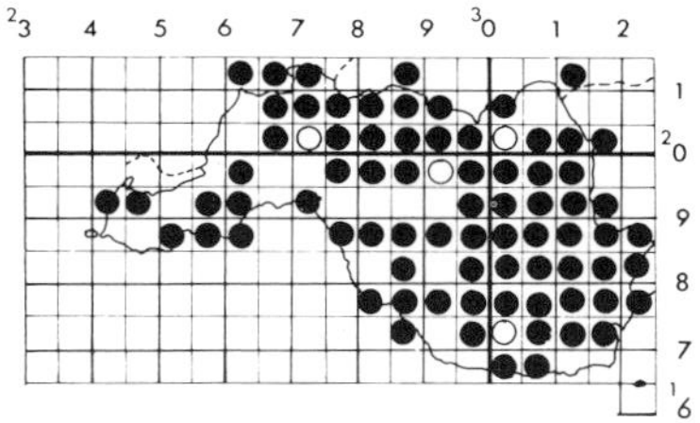

265 Rubus idaeus

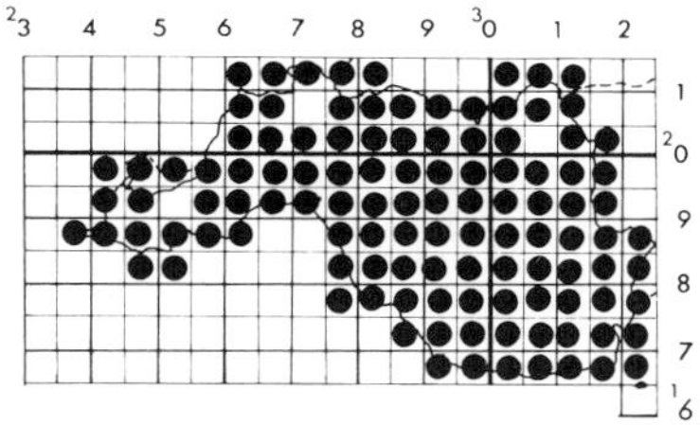

266 Rubus fruticosus agg.

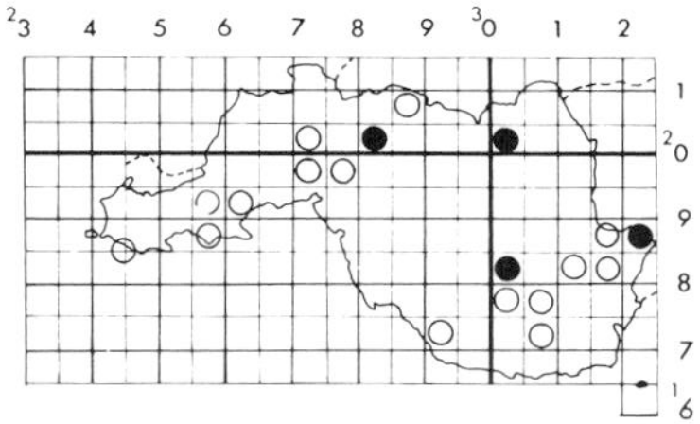

267 Rubus lindleianus

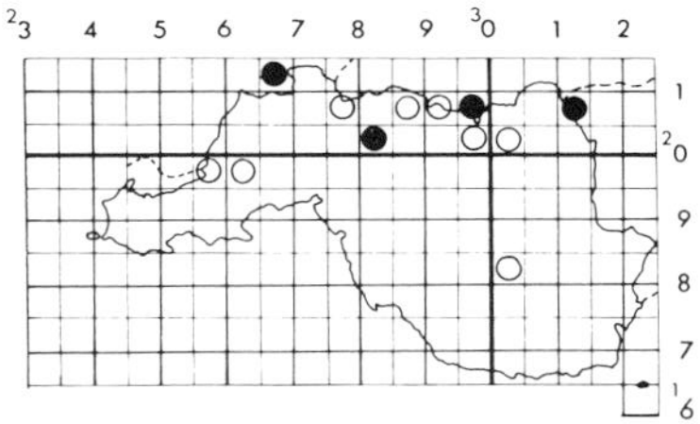

268 Rubus silurum

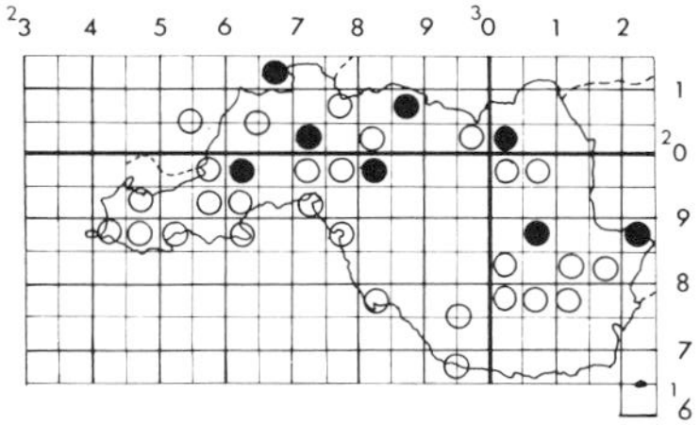

269 Rubus altiarcuatus

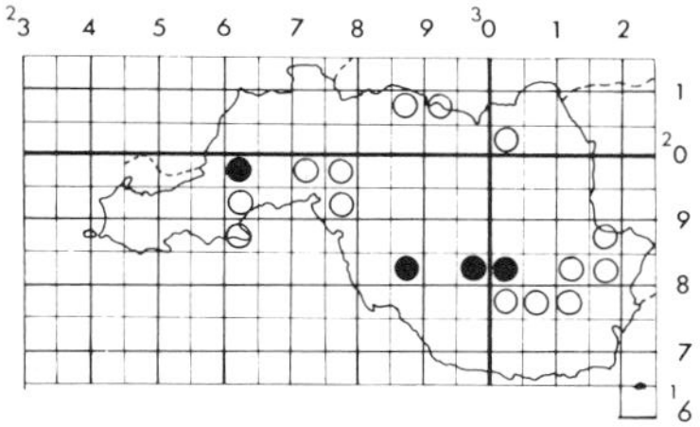

270 Rubus cardiophyllus

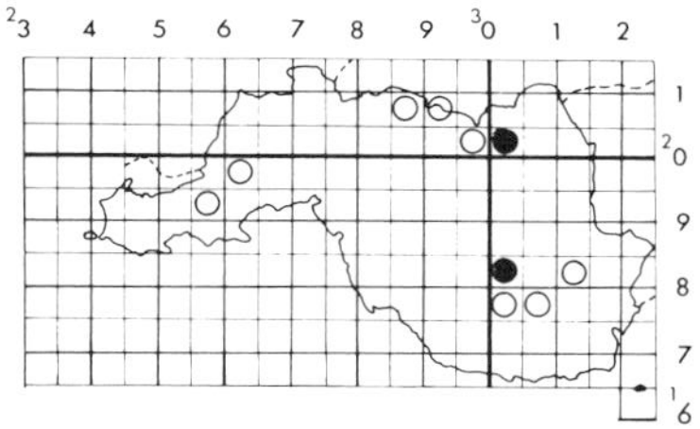

271 Rubus nemoralis

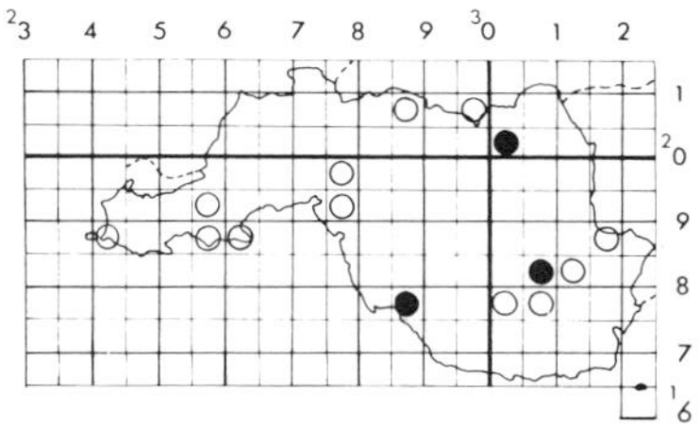

272 Rubus polyanthemus

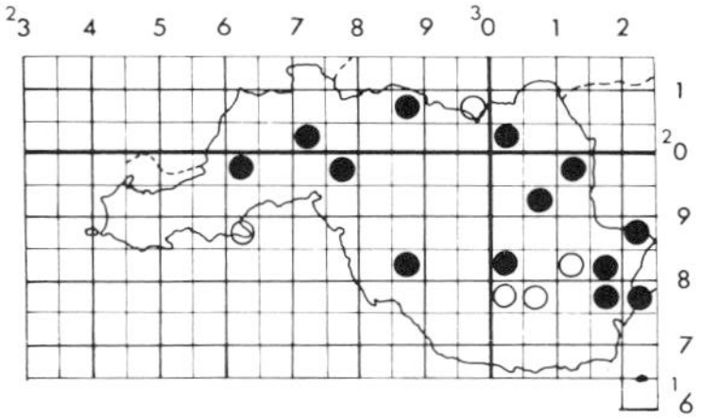

273 Rubus prolongatus

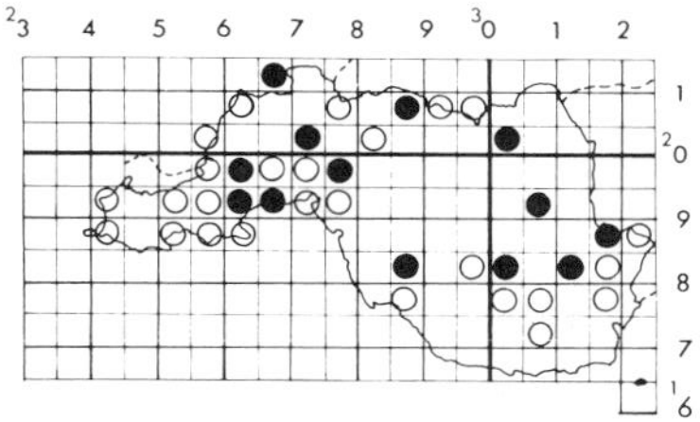

274 Rubus rubritinctus

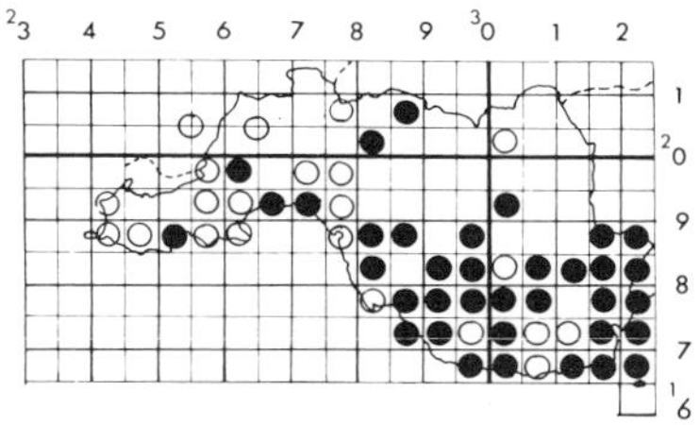

275 Rubus ulmifolius

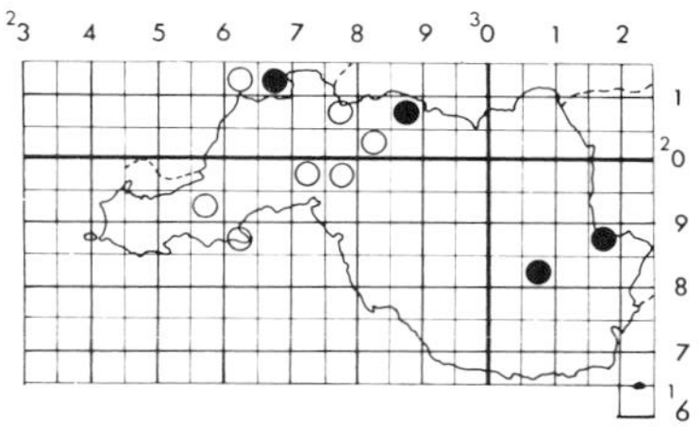

276 Rubus longus

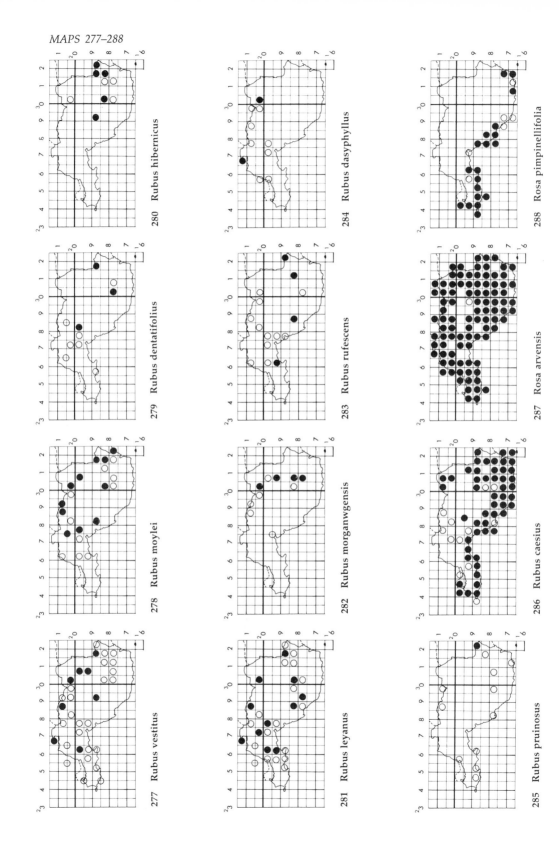

280 **Rubus hibernicus**

279 **Rubus dentatifolius**

278 **Rubus moylei**

277 **Rubus vestitus**

284 **Rubus dasyphyllus**

283 **Rubus rufescens**

282 **Rubus morganwgensis**

281 **Rubus leyanus**

288 **Rosa pimpinellifolia**

287 **Rosa arvensis**

286 **Rubus caesius**

285 **Rubus pruinosus**

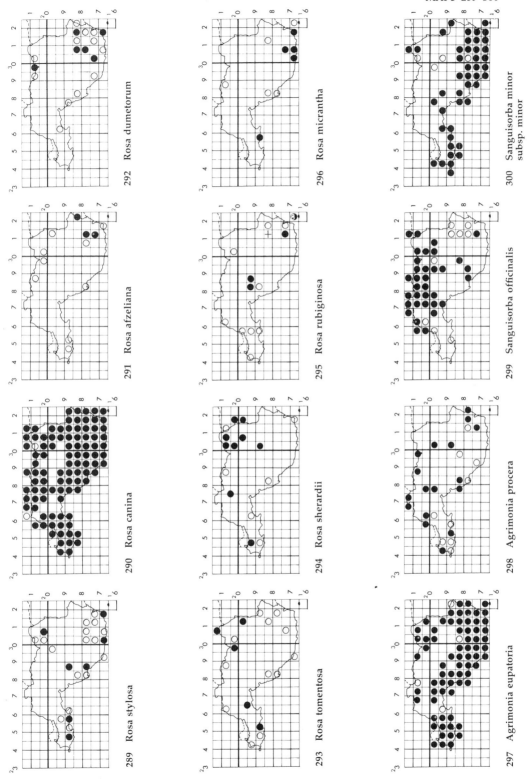

289 Rosa stylosa

290 Rosa canina

291 Rosa afzeliana

292 Rosa dumetorum

293 Rosa tomentosa

294 Rosa sherardii

295 Rosa rubiginosa

296 Rosa micrantha

297 Agrimonia eupatoria

298 Agrimonia procera

299 Sanguisorba officinalis

300 Sanguisorba minor subsp. minor

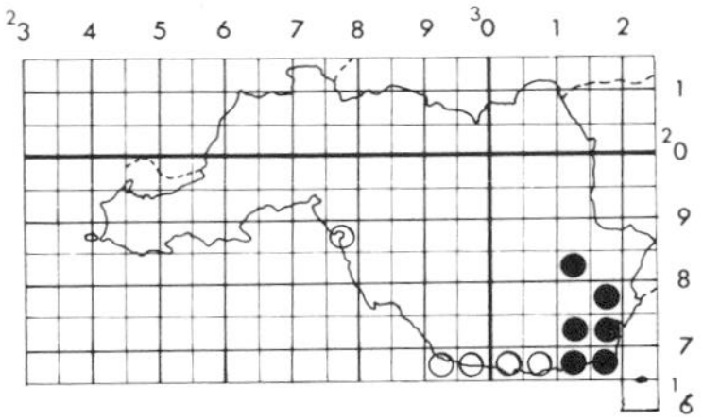

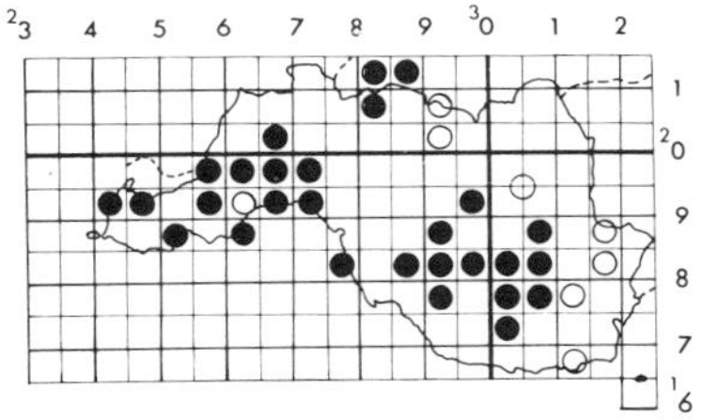

304 Potentilla palustris

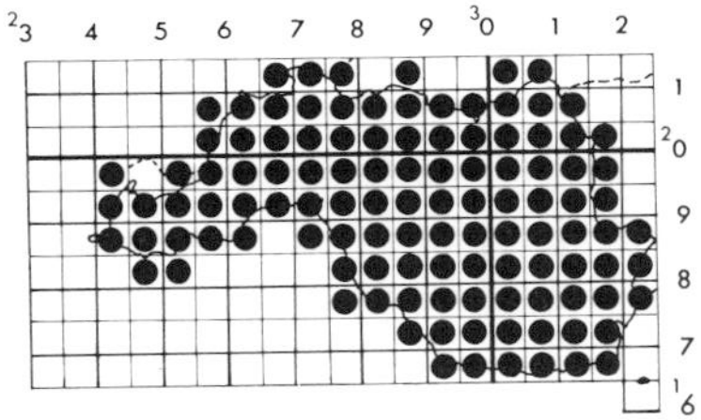

303 Geum urbanum

302 Geum rivale

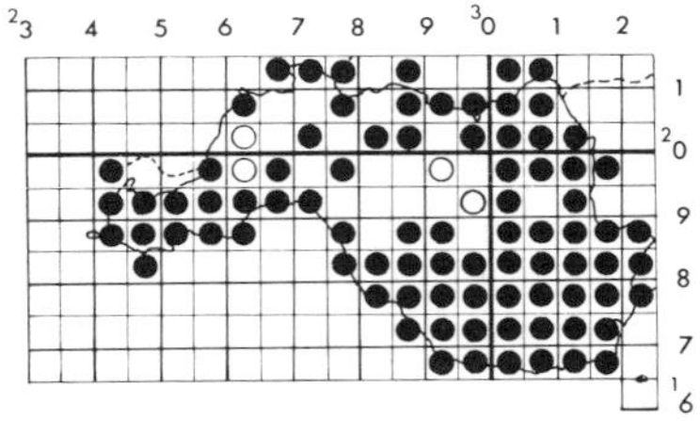

301 Sanguisorba minor
subsp. muricata

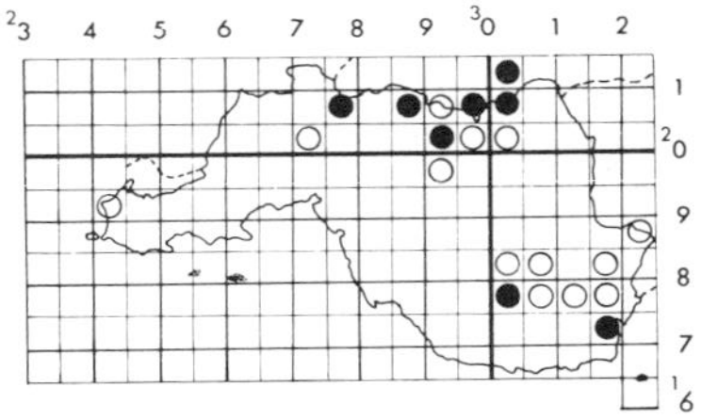

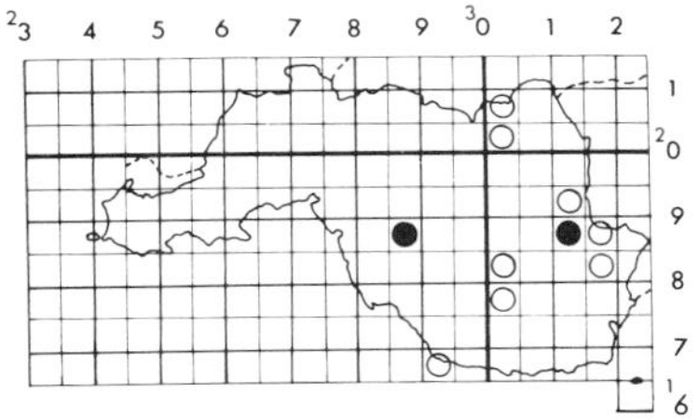

308 Potentilla x suberecta

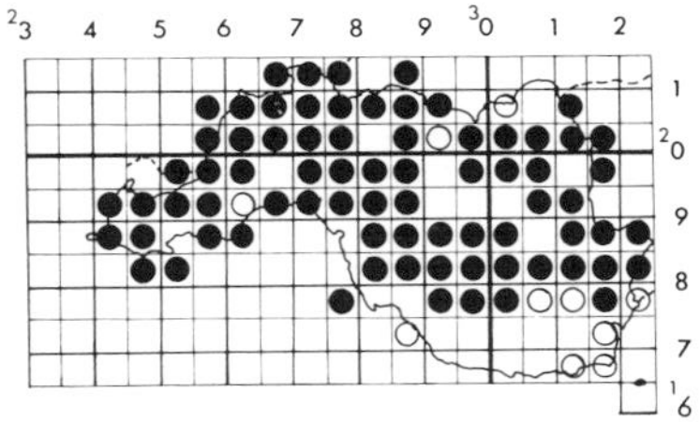

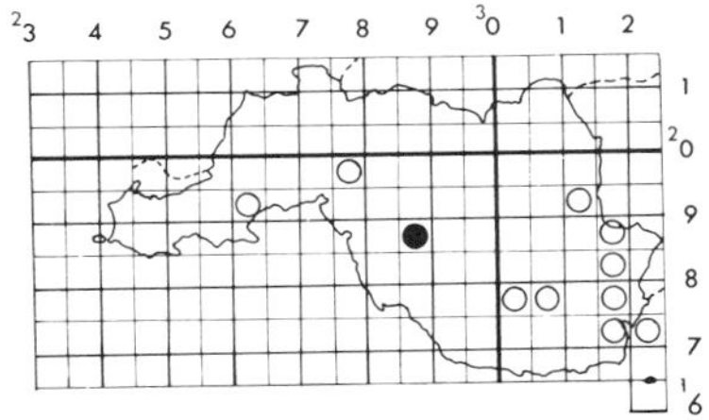

307 Potentilla anglica

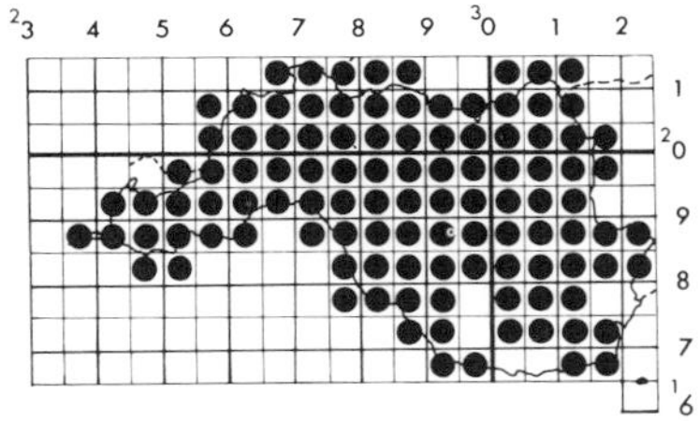

306 Potentilla erecta

305 Potentilla anserina

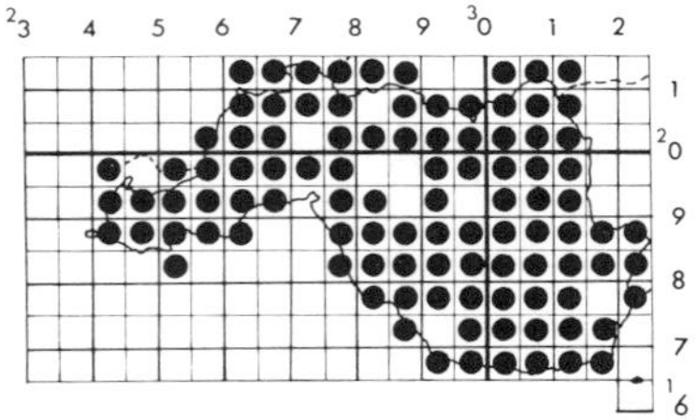

312 Fragaria vesca

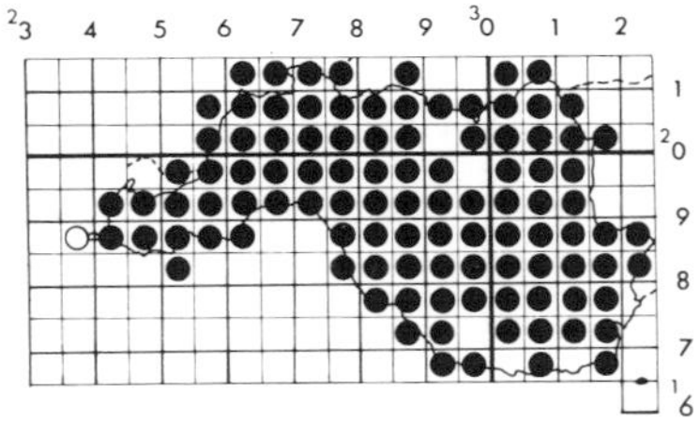

311 Potentilla sterilis

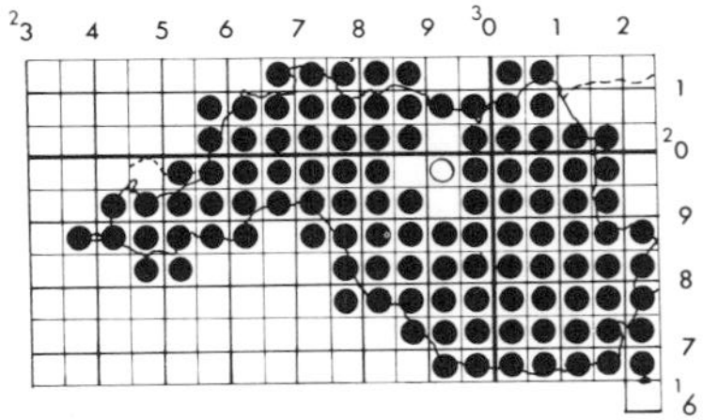

310 Potentilla reptans

309 Potentilla x mixta

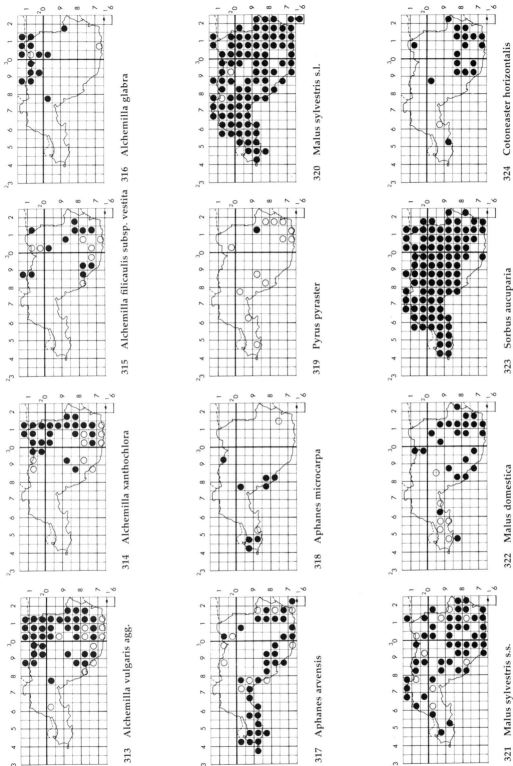

316 Alchemilla glabra

315 Alchemilla filicaulis subsp. vestita

314 Alchemilla xanthochlora

313 Alchemilla vulgaris agg.

320 Malus sylvestris s.l.

319 Pyrus pyraster

318 Aphanes microcarpa

317 Aphanes arvensis

324 Cotoneaster horizontalis

323 Sorbus aucuparia

322 Malus domestica

321 Malus sylvestris s.s.

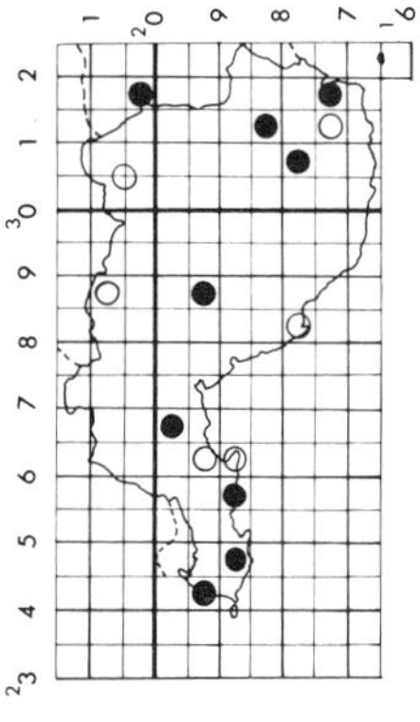

325 Cotoneaster simonsii

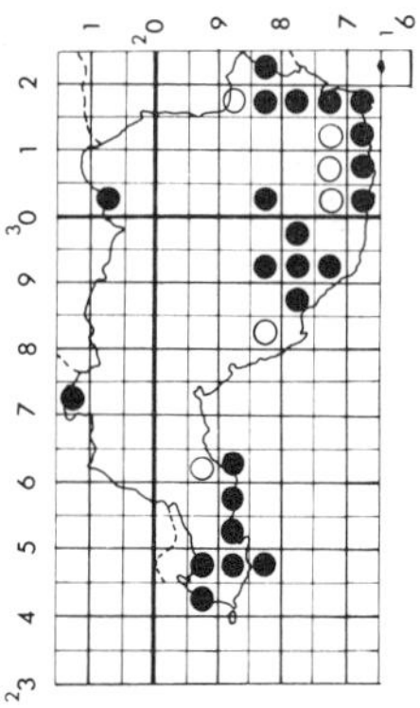

326 Cotoneaster microphyllus

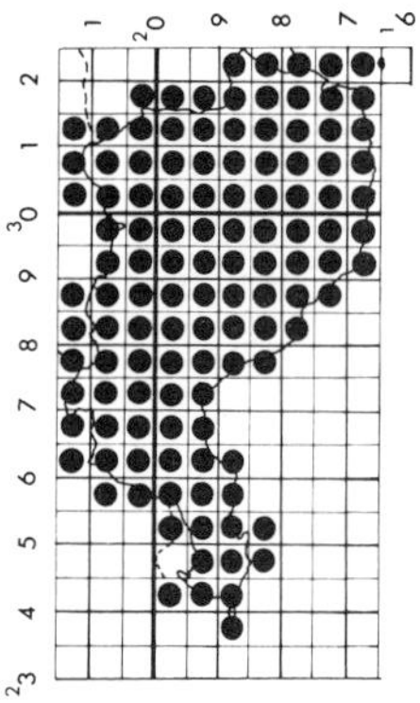

327 Crataegus monogyna

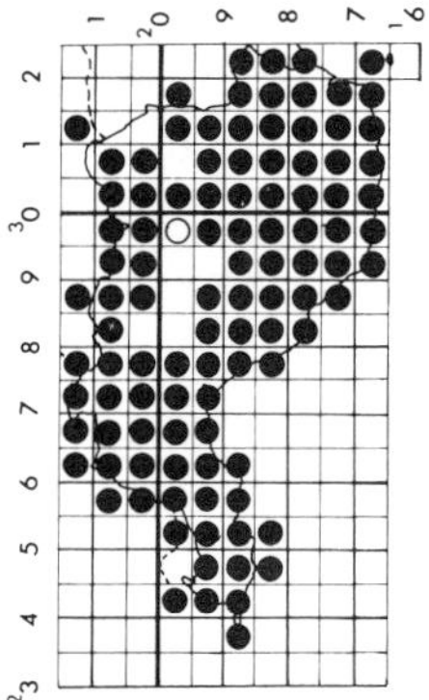

328 Prunus spinosa

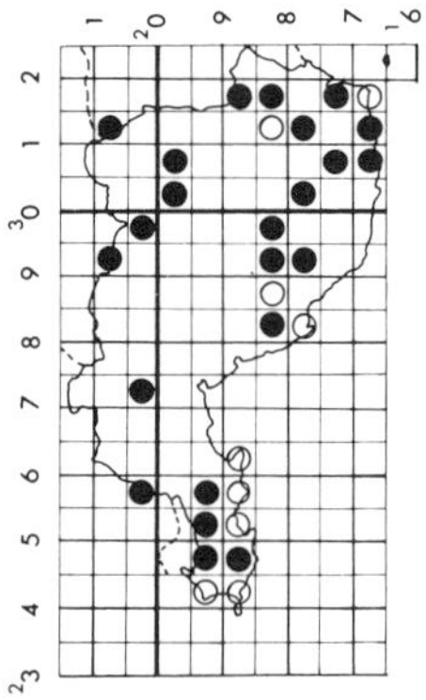

329 Prunus domestica subsp. domestica

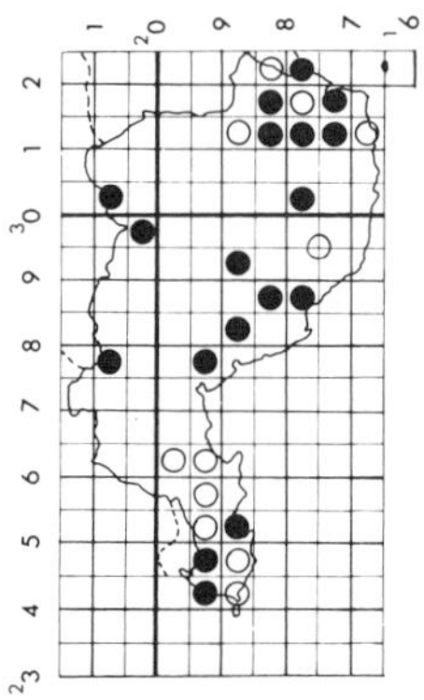

330 Prunus avium

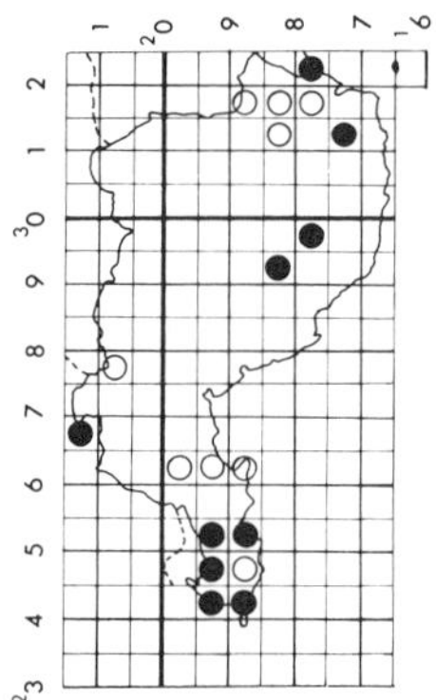

331 Prunus cerasus

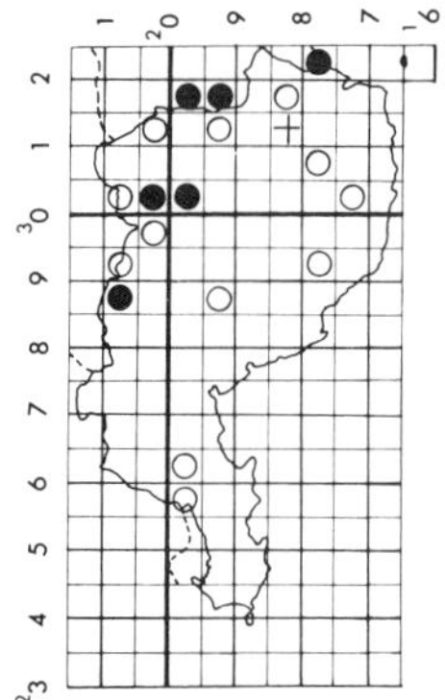

332 Prunus padus

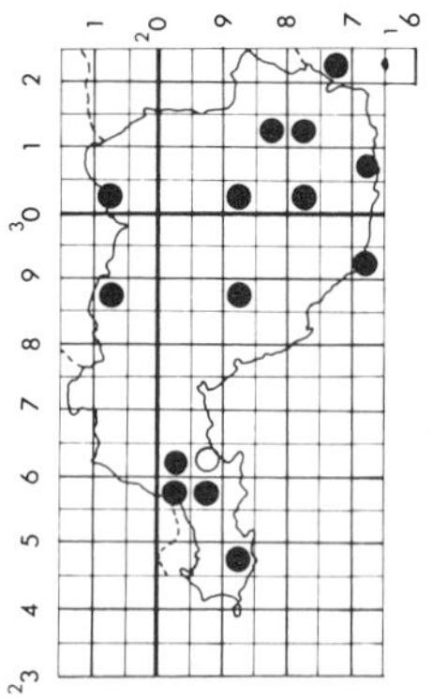

333 Prunus laurocerasus

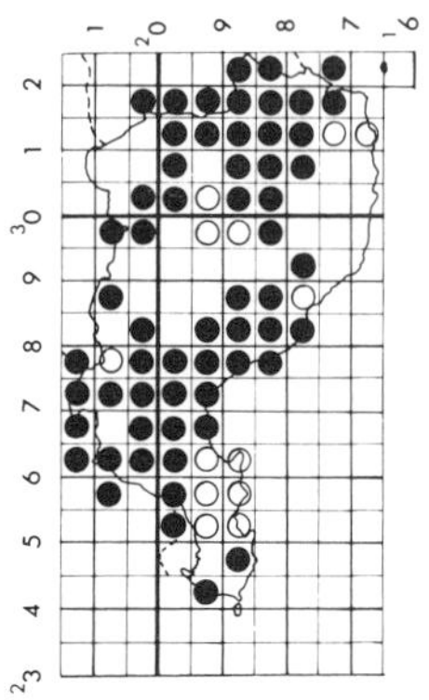

334 Cytisus scoparius

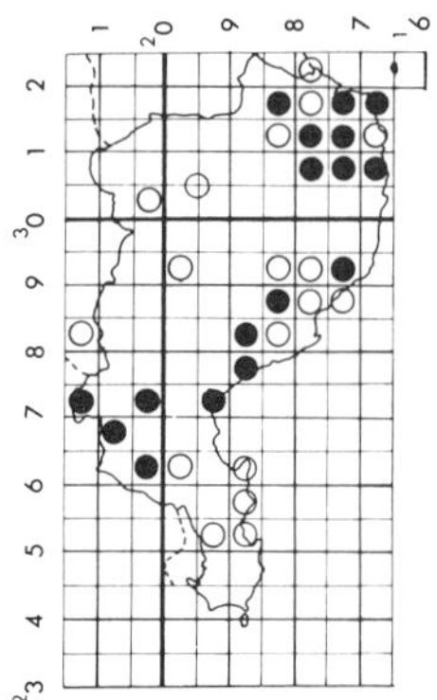

335 Genista tinctoria

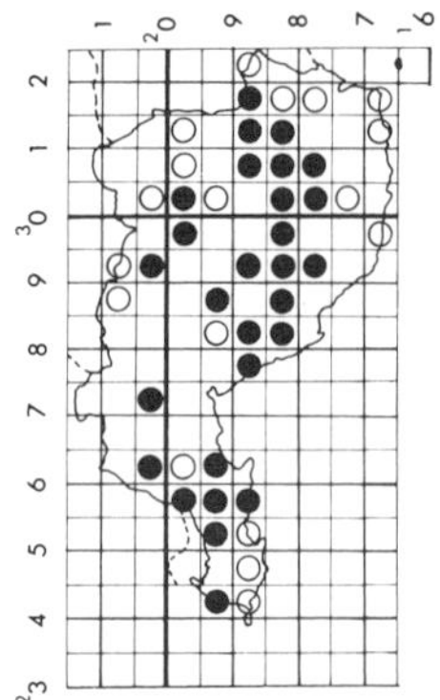

336 Genista anglica

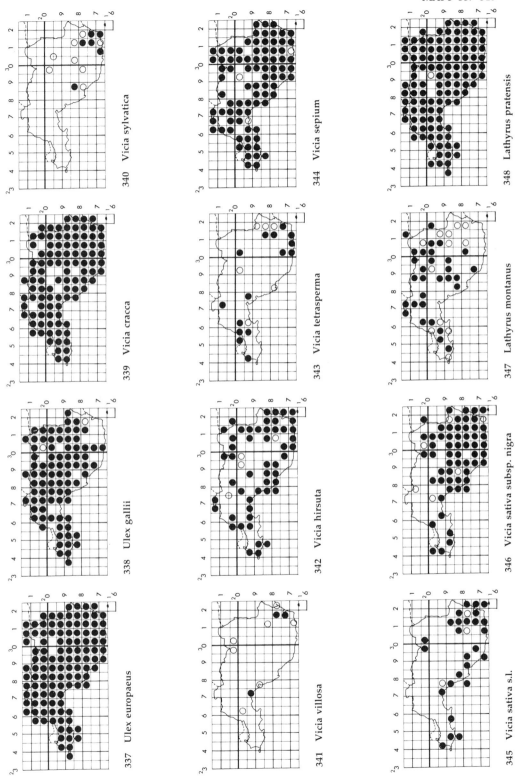

337 Ulex europaeus

338 Ulex gallii

339 Vicia cracca

340 Vicia sylvatica

341 Vicia villosa

342 Vicia hirsuta

343 Vicia tetrasperma

344 Vicia sepium

345 Vicia sativa s.l.

346 Vicia sativa subsp. nigra

347 Lathyrus montanus

348 Lathyrus pratensis

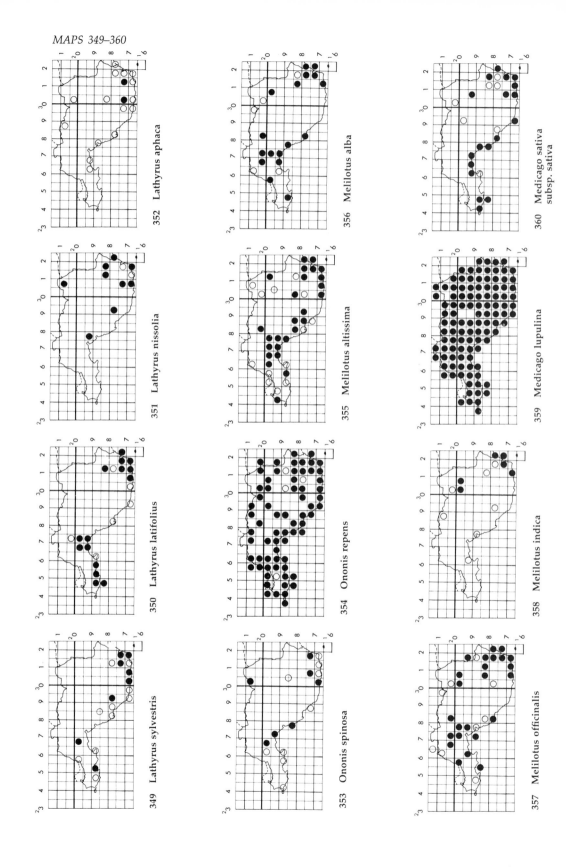

349 Lathyrus sylvestris
350 Lathyrus latifolius
351 Lathyrus nissolia
352 Lathyrus aphaca
353 Ononis spinosa
354 Ononis repens
355 Melilotus altissima
356 Melilotus alba
357 Melilotus officinalis
358 Melilotus indica
359 Medicago lupulina
360 Medicago sativa subsp. sativa

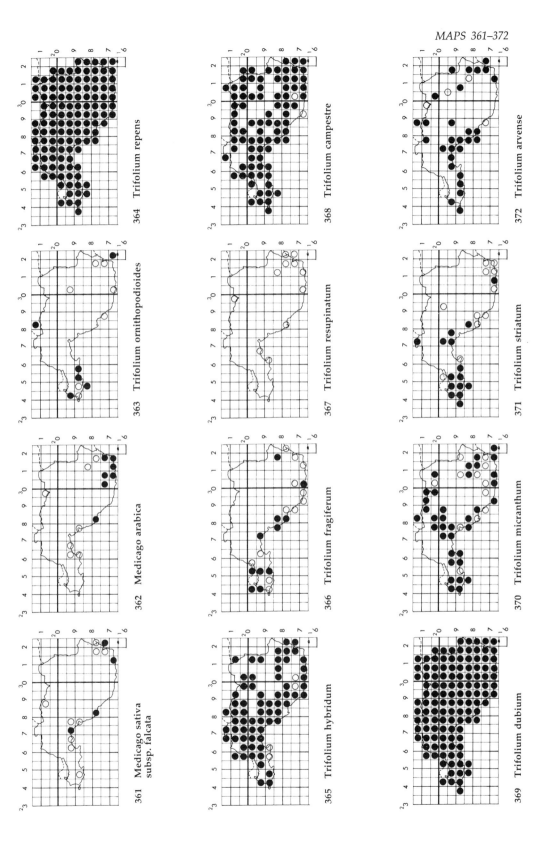

361 Medicago sativa subsp. falcata

362 Medicago arabica

363 Trifolium ornithopodioides

364 Trifolium repens

365 Trifolium hybridum

366 Trifolium fragiferum

367 Trifolium resupinatum

368 Trifolium campestre

369 Trifolium dubium

370 Trifolium micranthum

371 Trifolium striatum

372 Trifolium arvense

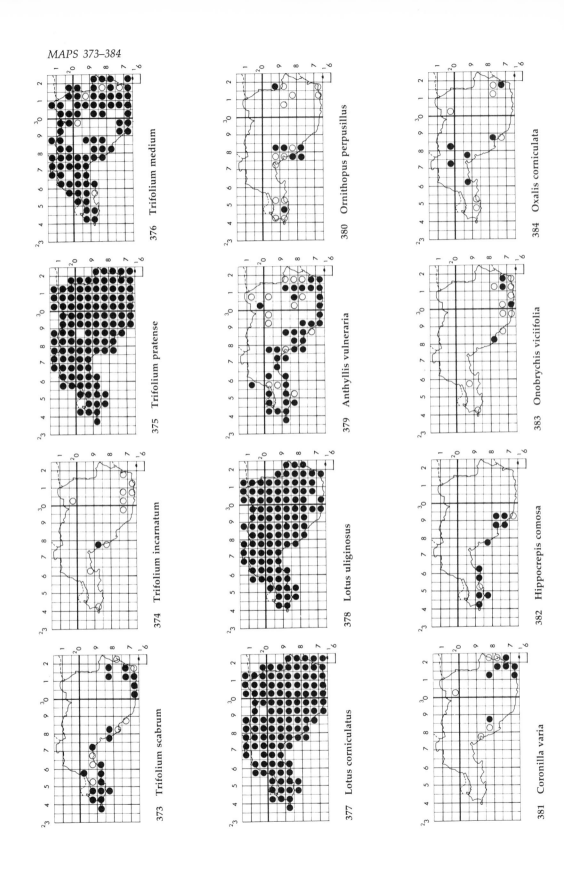

376 Trifolium medium

380 Ornithopus perpusillus

384 Oxalis corniculata

375 Trifolium pratense

379 Anthyllis vulneraria

383 Onobrychis viciifolia

374 Trifolium incarnatum

378 Lotus uliginosus

382 Hippocrepis comosa

373 Trifolium scabrum

377 Lotus corniculatus

381 Coronilla varia

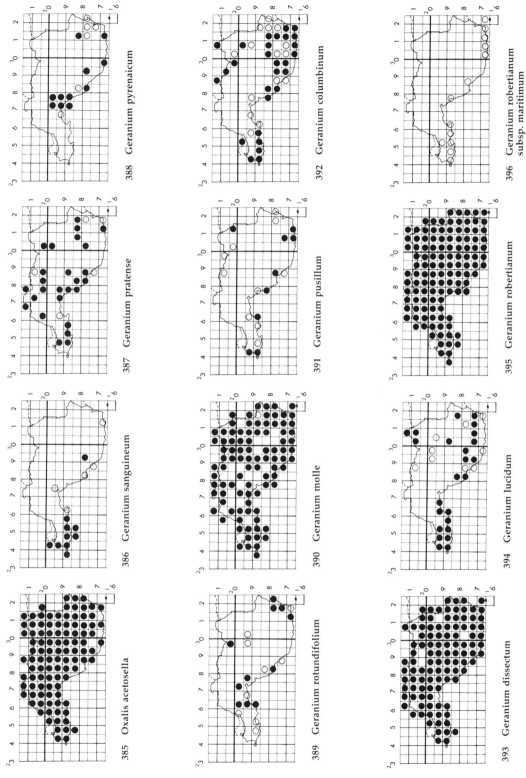

385 Oxalis acetosella

386 Geranium sanguineum

387 Geranium pratense

388 Geranium pyrenaicum

389 Geranium rotundifolium

390 Geranium molle

391 Geranium pusillum

392 Geranium columbinum

393 Geranium dissectum

394 Geranium lucidum

395 Geranium robertianum

396 Geranium robertianum subsp. maritimum

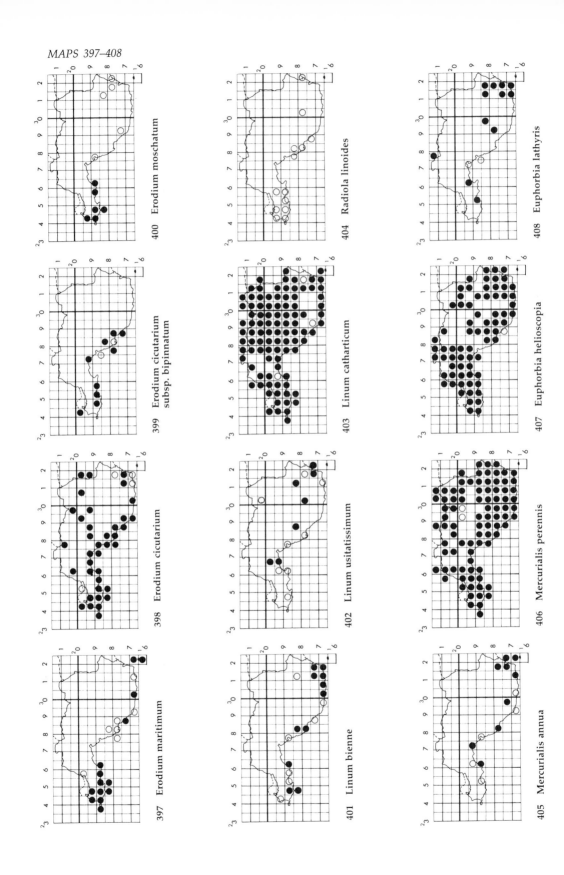

397 Erodium maritimum

398 Erodium cicutarium

399 Erodium cicutarium
     subsp. bipinnatum

400 Erodium moschatum

401 Linum bienne

402 Linum usitatissimum

403 Linum catharticum

404 Radiola linoides

405 Mercurialis annua

406 Mercurialis perennis

407 Euphorbia helioscopia

408 Euphorbia lathyris

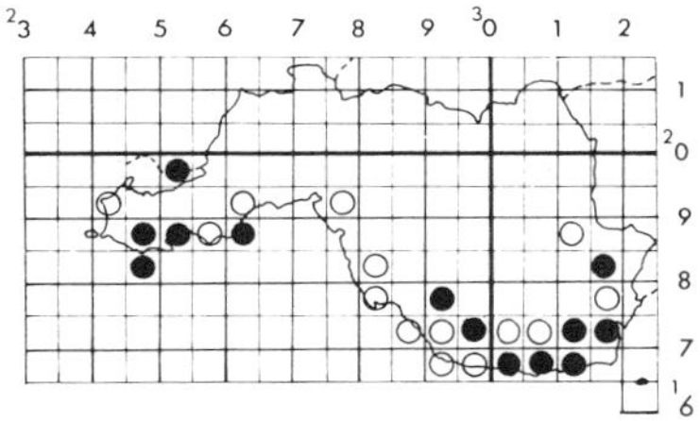

409 Euphorbia exigua

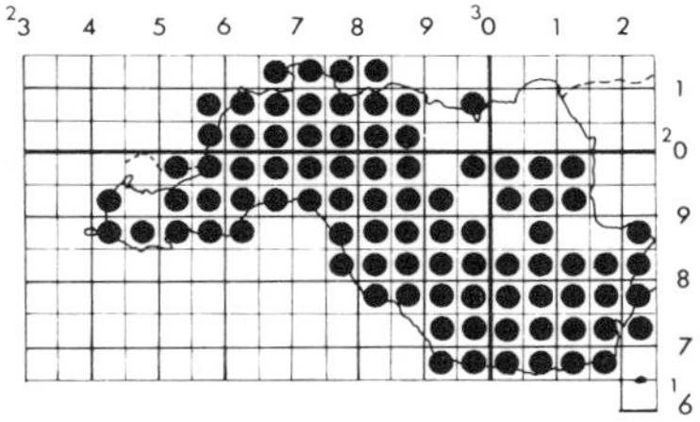

410 Euphorbia peplus

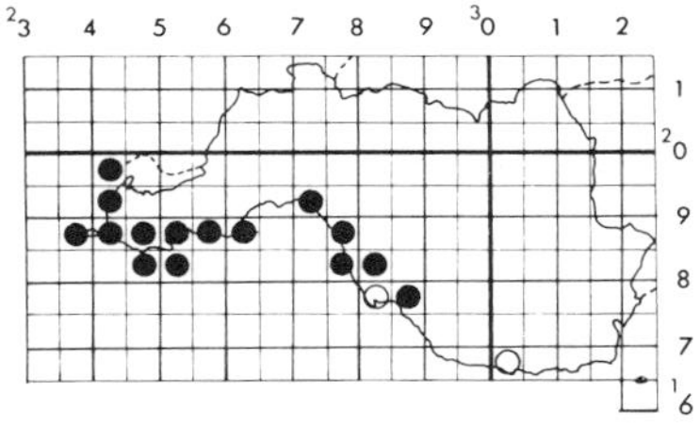

411 Euphorbia portlandica

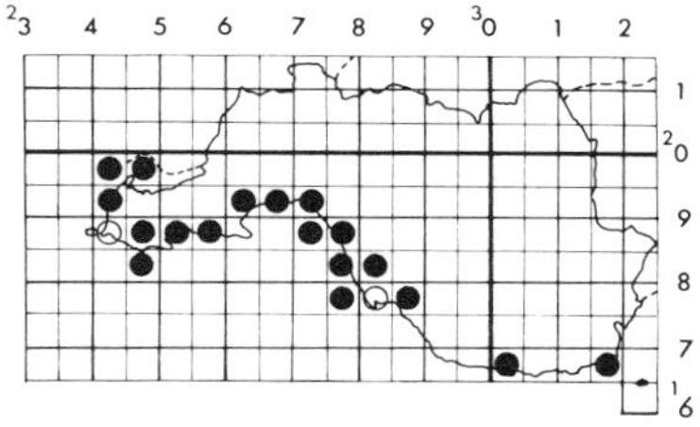

412 Euphorbia paralias

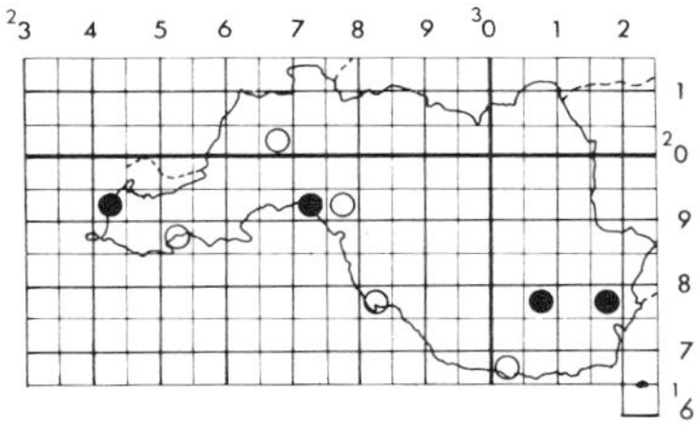

413 Euphorbia cyparissias

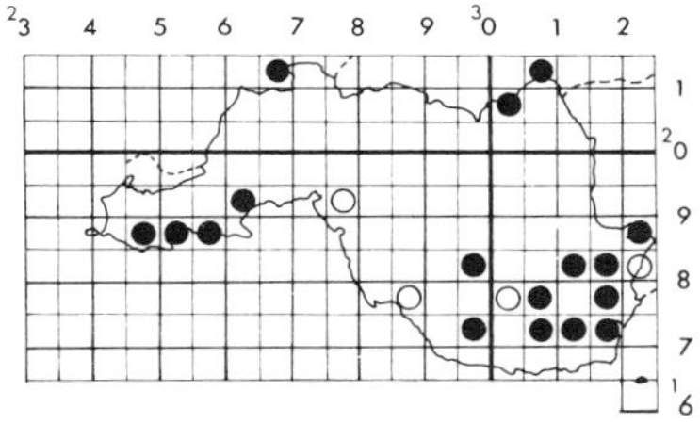

414 Euphorbia amygdaloides

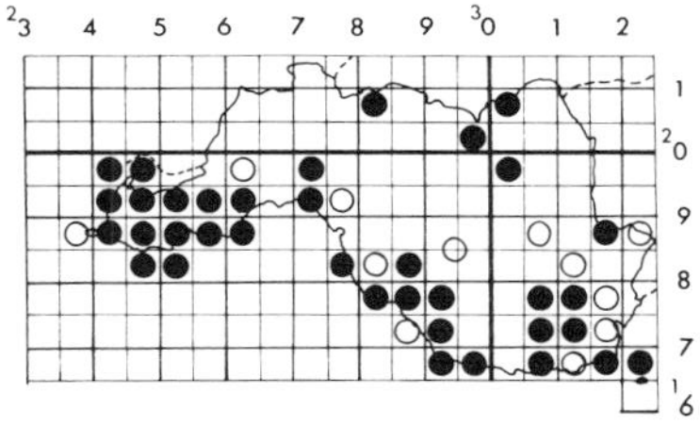

415 Polygala vulgaris

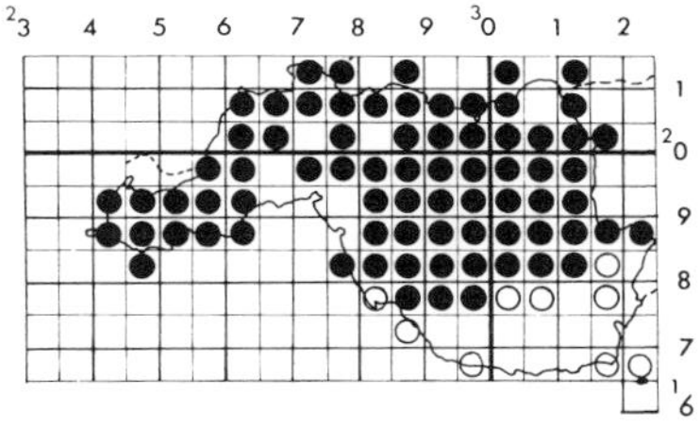

416 Polygala serpyllifolia

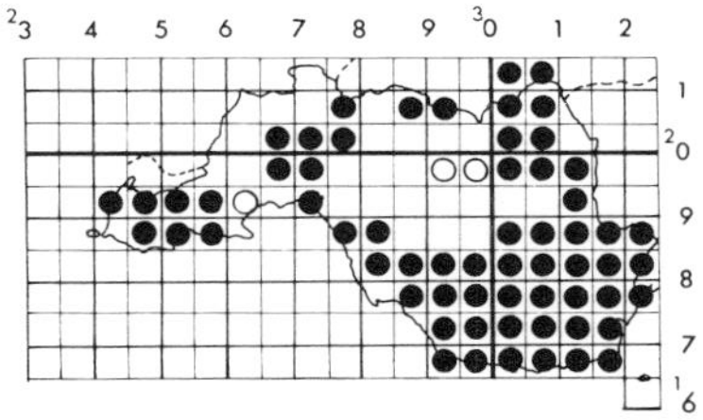

417 Acer campestre

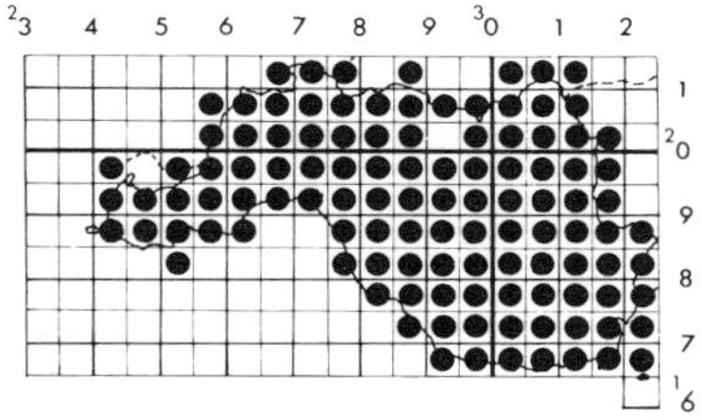

418 Acer pseudoplatanus

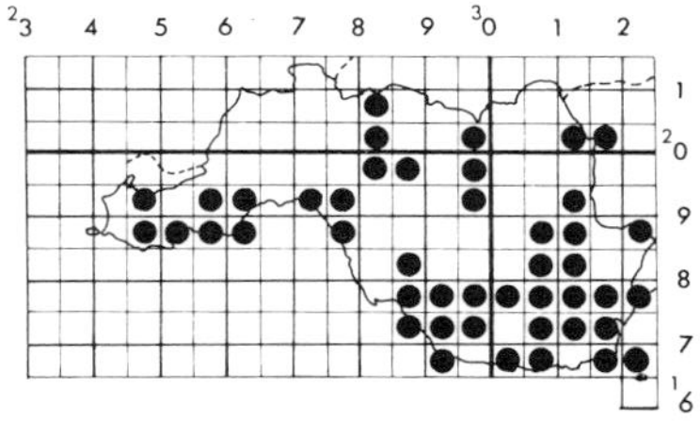

419 Aesculus hippocastanum

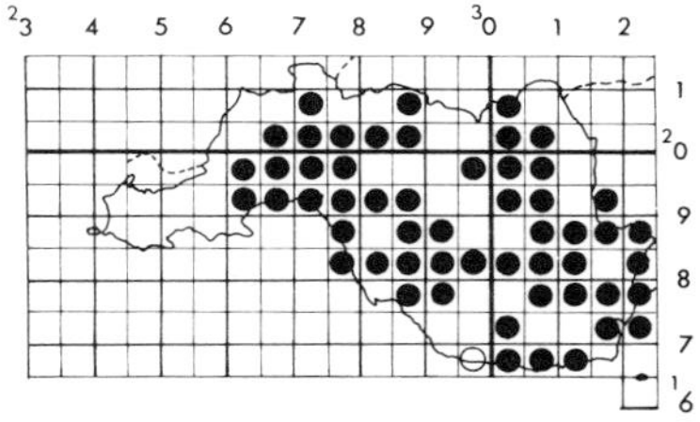

420 Impatiens glandulifera

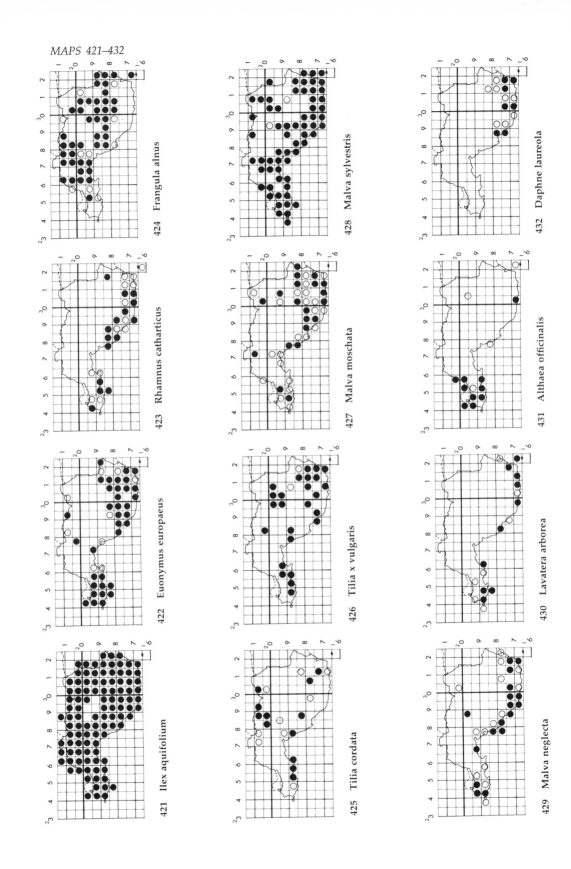

421 Ilex aquifolium
422 Euonymus europaeus
423 Rhamnus catharticus
424 Frangula alnus
425 Tilia cordata
426 Tilia x vulgaris
427 Malva moschata
428 Malva sylvestris
429 Malva neglecta
430 Lavatera arborea
431 Althaea officinalis
432 Daphne laureola

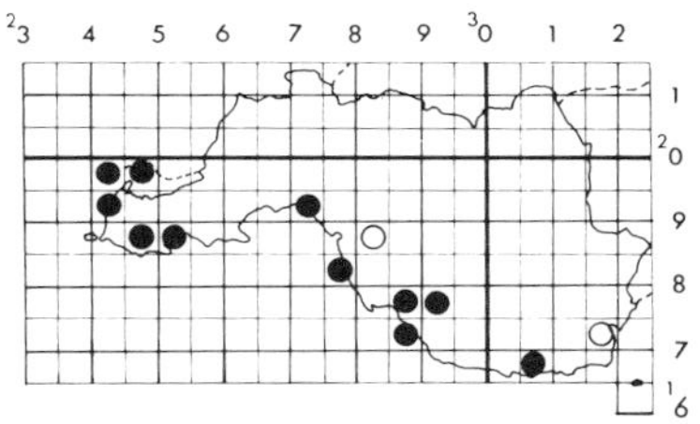

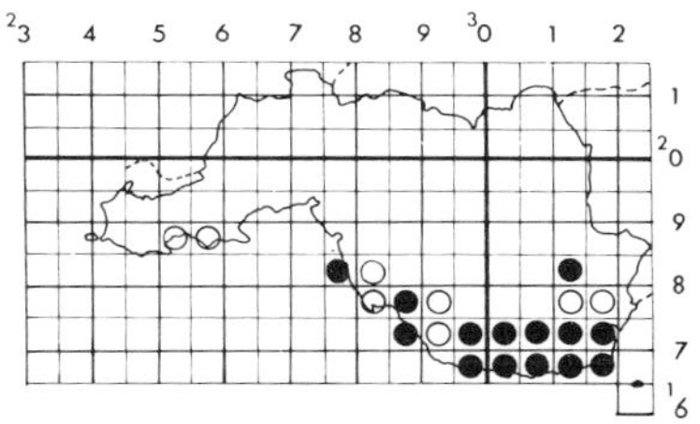

436 Hypericum hirsutum

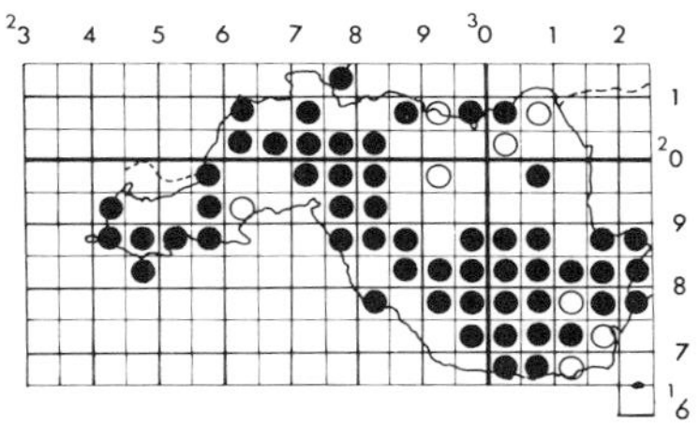

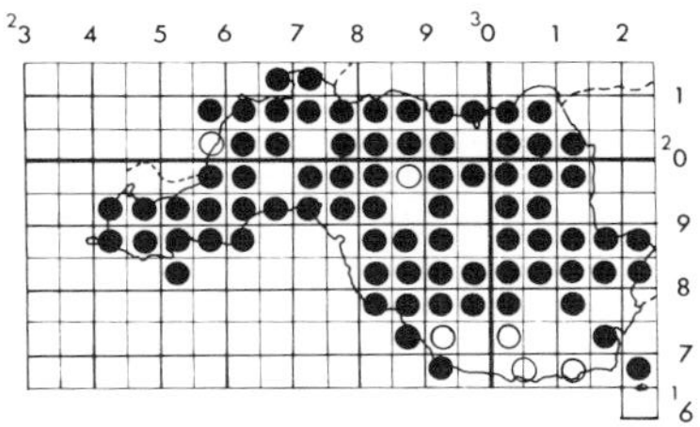

435 Hypericum androsaemum

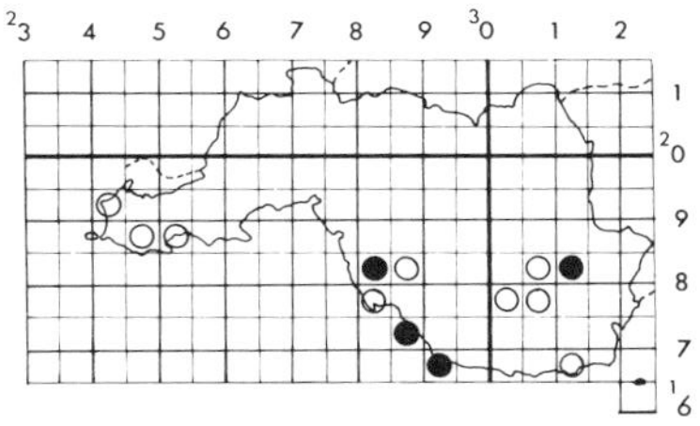

434 Hypericum calycinum

433 Hippophae rhamnoides

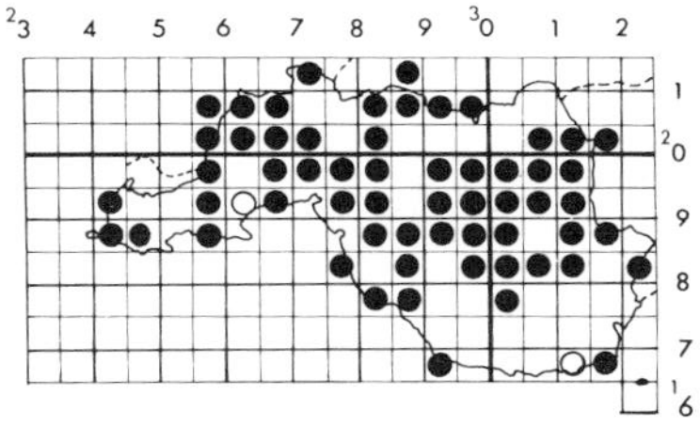

440 Hypericum humifusum

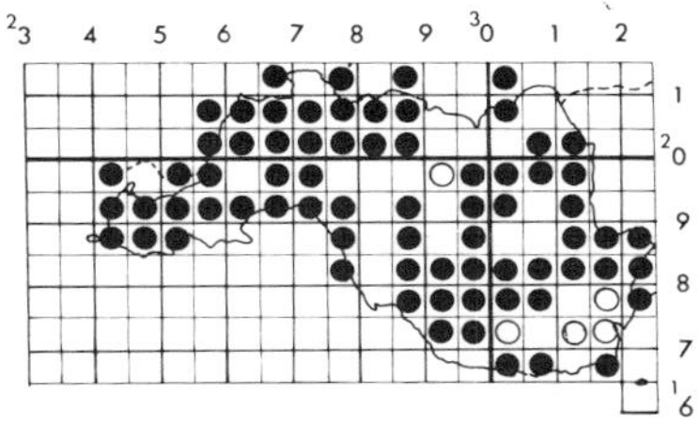

439 Hypericum elodes

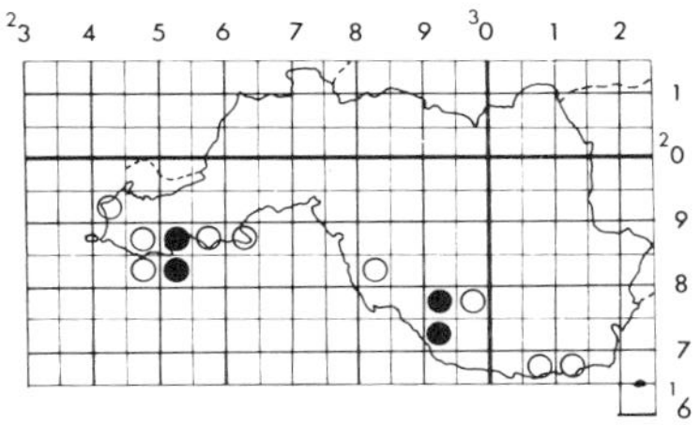

438 Hypericum montanum

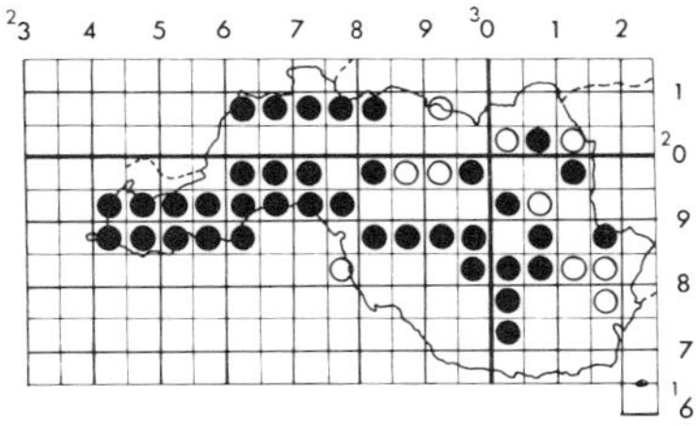

437 Hypericum pulchrum

444 Viola odorata

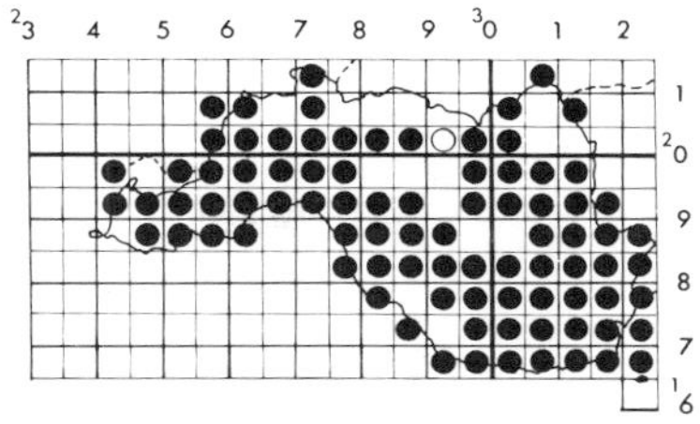

443 Hypericum perforatum

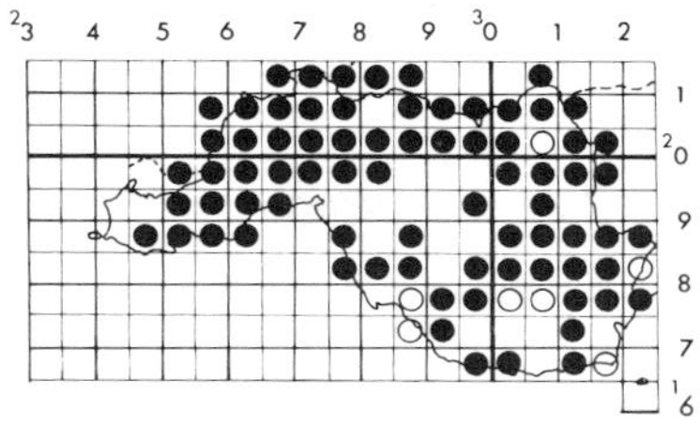

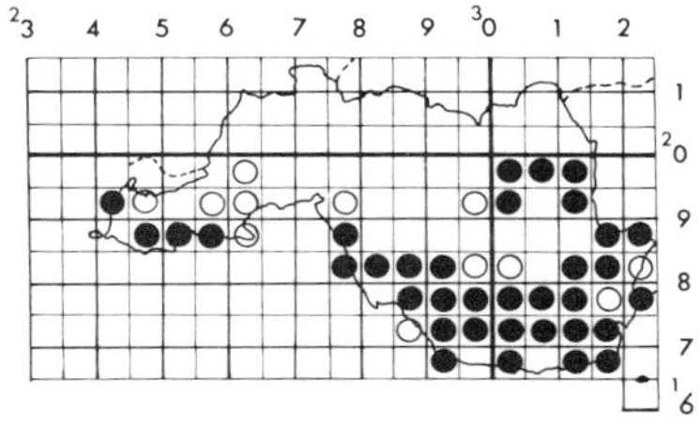

442 Hypericum maculatum subsp. obtusiusculum

441 Hypericum tetrapterum

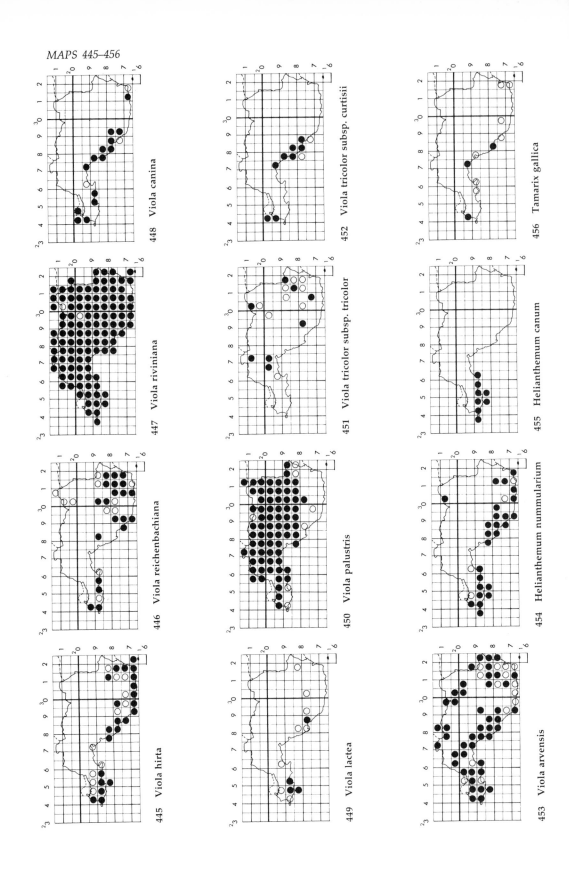

448 Viola canina

452 Viola tricolor subsp. curtisii

456 Tamarix gallica

447 Viola riviniana

451 Viola tricolor subsp. tricolor

455 Helianthemum canum

446 Viola reichenbachiana

450 Viola palustris

454 Helianthemum nummularium

445 Viola hirta

449 Viola lactea

453 Viola arvensis

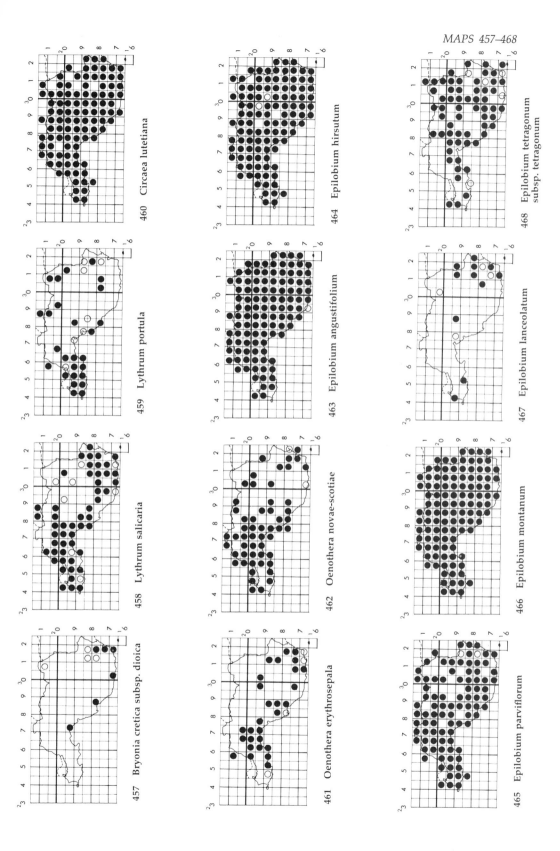

457 Bryonia cretica subsp. dioica

458 Lythrum salicaria

459 Lythrum portula

460 Circaea lutetiana

461 Oenothera erythrosepala

462 Oenothera novae-scotiae

463 Epilobium angustifolium

464 Epilobium hirsutum

465 Epilobium parviflorum

466 Epilobium montanum

467 Epilobium lanceolatum

468 Epilobium tetragonum subsp. tetragonum

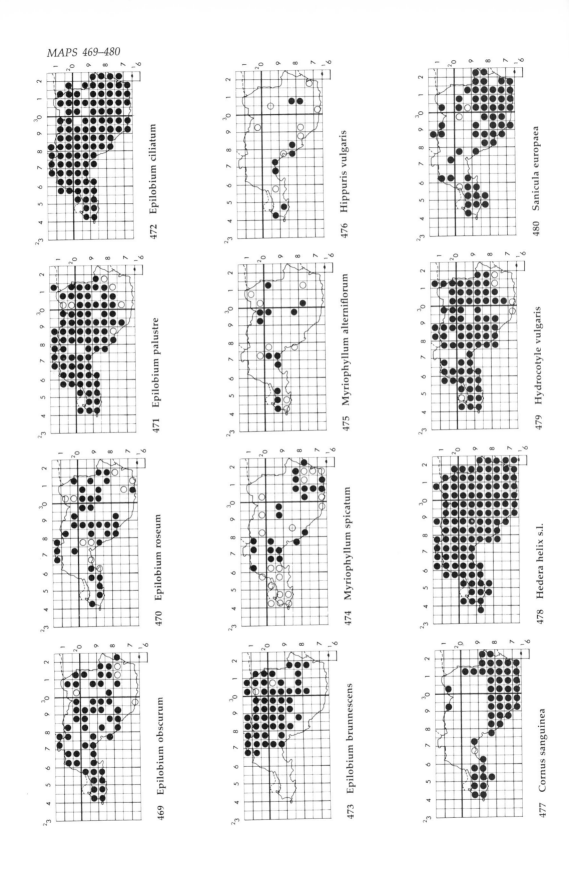

469 Epilobium obscurum

470 Epilobium roseum

471 Epilobium palustre

472 Epilobium ciliatum

473 Epilobium brunnescens

474 Myriophyllum spicatum

475 Myriophyllum alterniflorum

476 Hippuris vulgaris

477 Cornus sanguinea

478 Hedera helix s.l.

479 Hydrocotyle vulgaris

480 Sanicula europaea

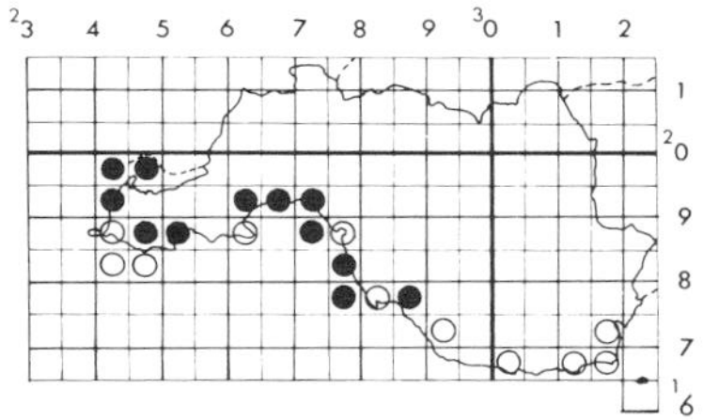

481 Eryngium maritimum

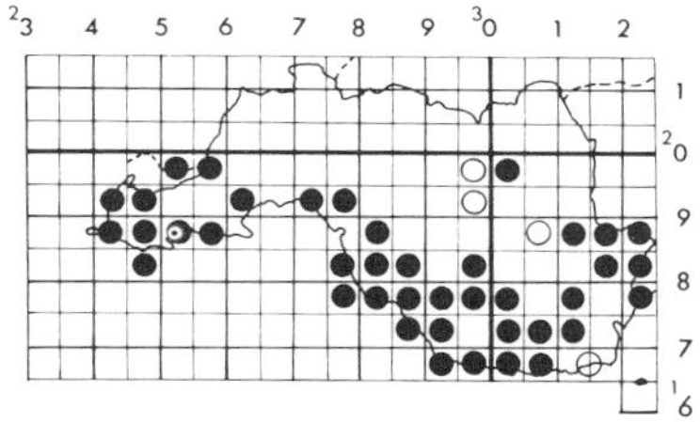

482 Chaerophyllum temulentum

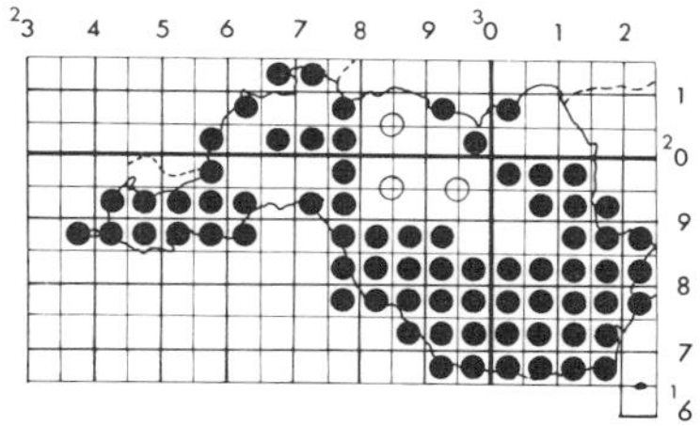

483 Anthriscus sylvestris

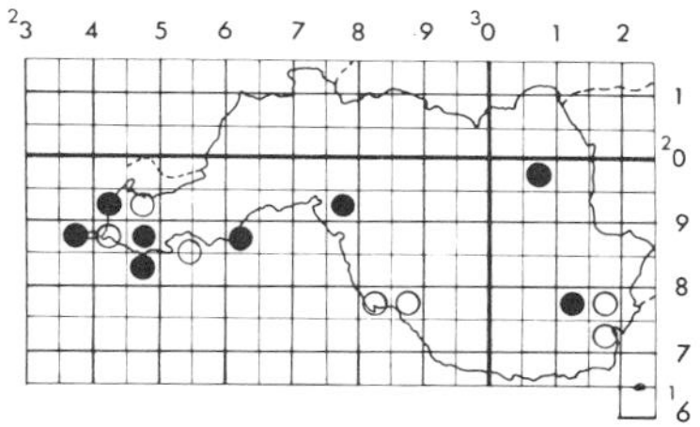

484 Anthriscus caucalis

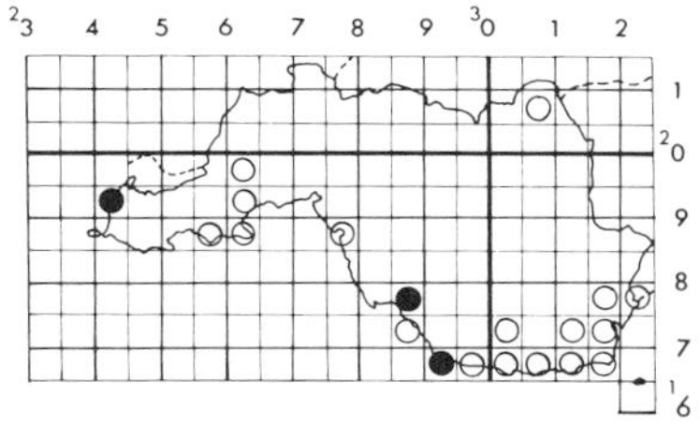

485 Scandix pecten-veneris

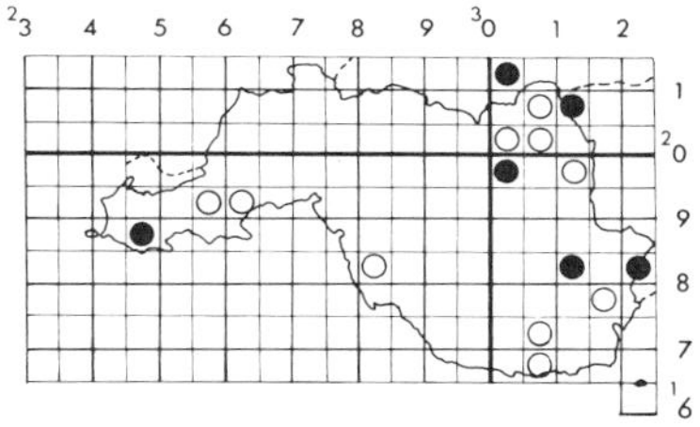

486 Myrrhis odorata

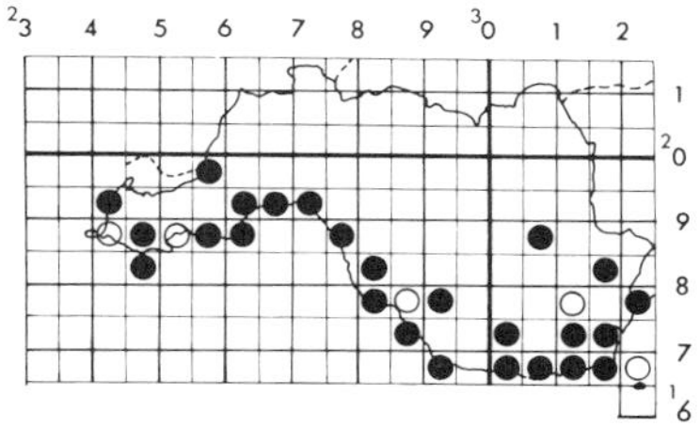

487 Smyrnium olusatrum

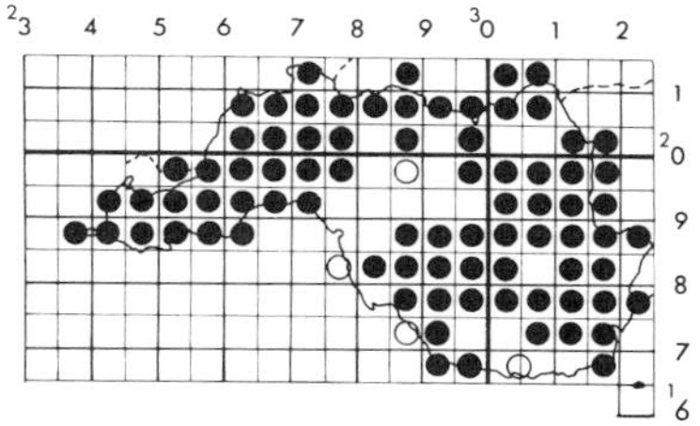

488 Conopodium majus

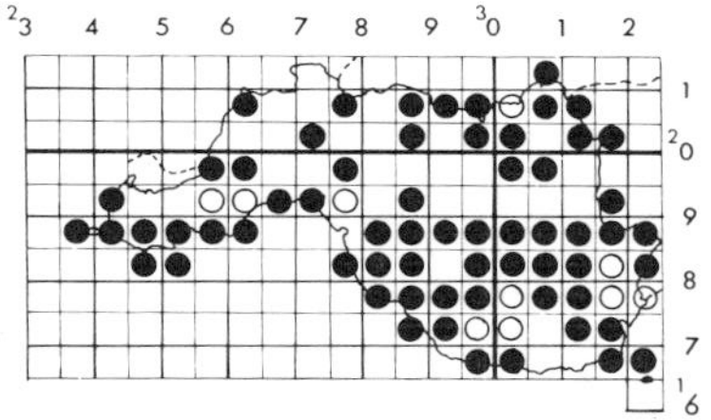

489 Pimpinella saxifraga

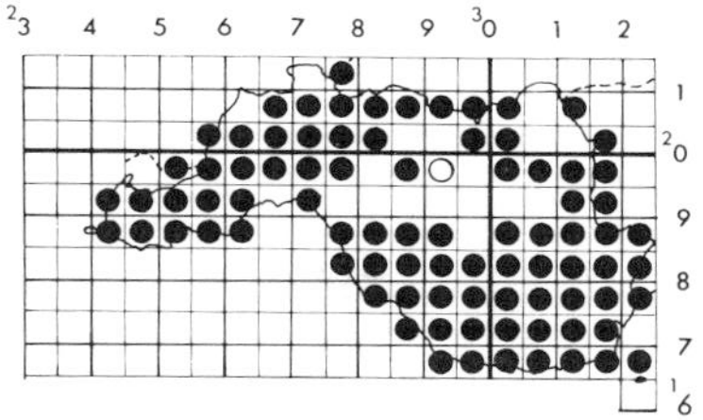

490 Aegopodium podagraria

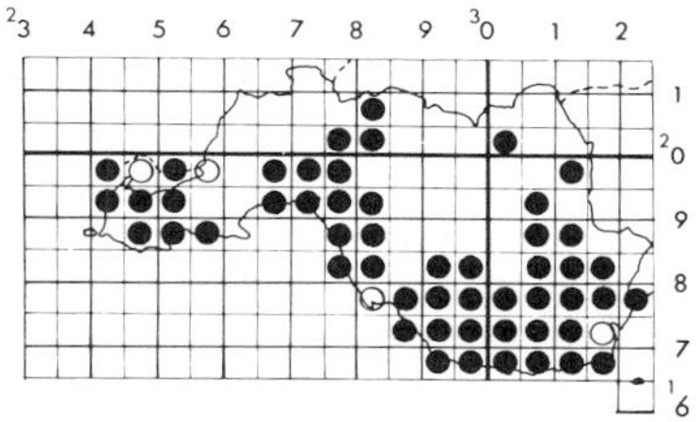

491 Berula erecta

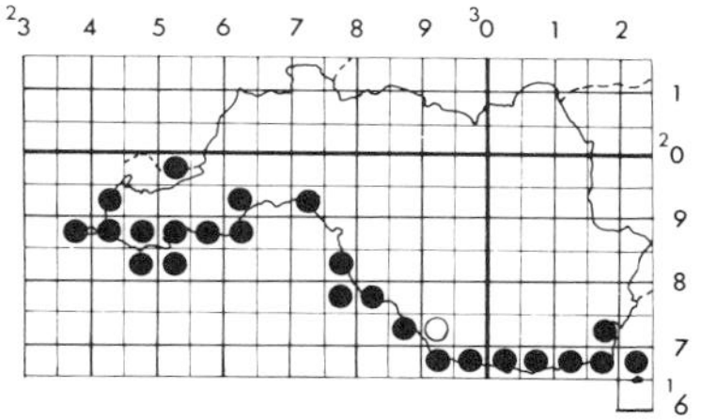

492 Crithmum maritimum

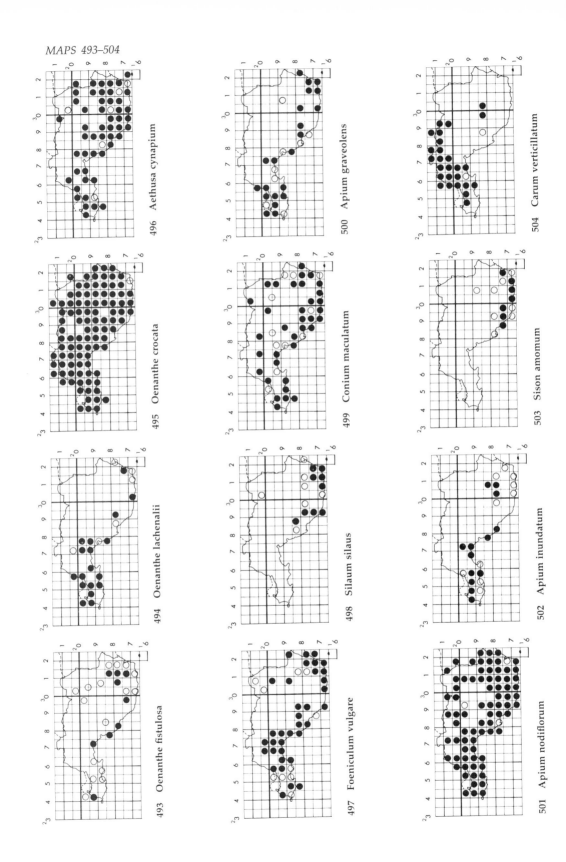

493 Oenanthe fistulosa
494 Oenanthe lachenalii
495 Oenanthe crocata
496 Aethusa cynapium
497 Foeniculum vulgare
498 Silaum silaus
499 Conium maculatum
500 Apium graveolens
501 Apium nodiflorum
502 Apium inundatum
503 Sison amomum
504 Carum verticillatum

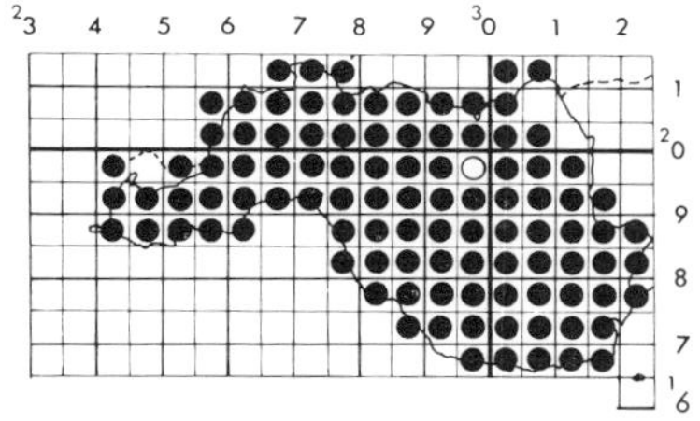

505 Angelica sylvestris

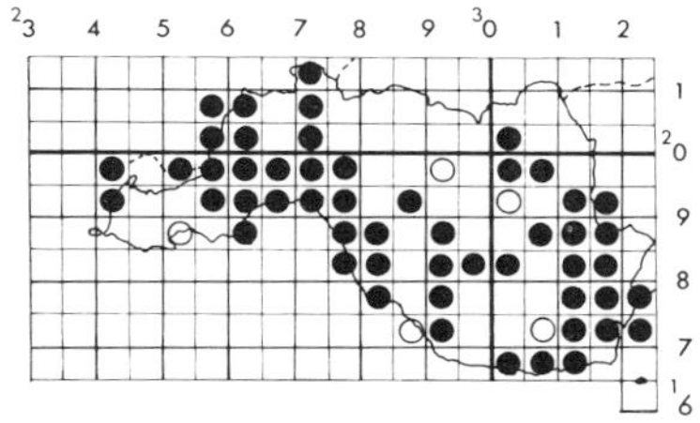

506 Pastinaca sativa

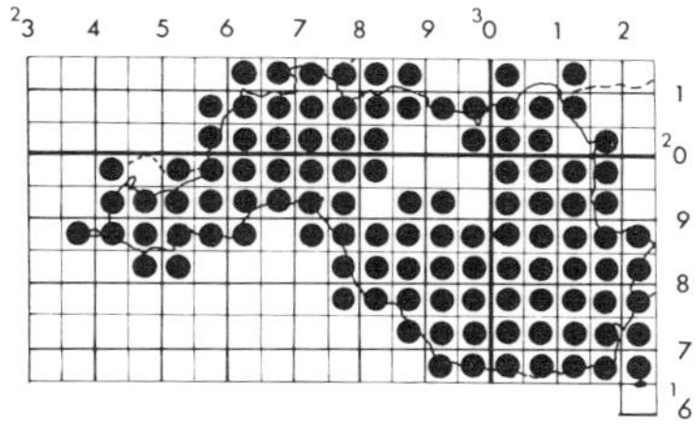

507 Heracleum sphondylium

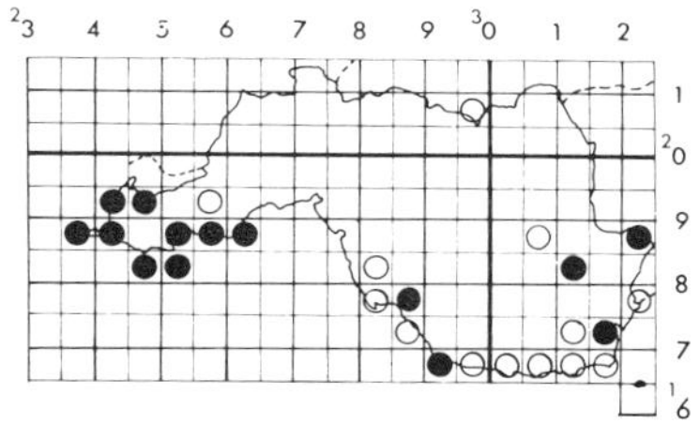

508 Torilis nodosa

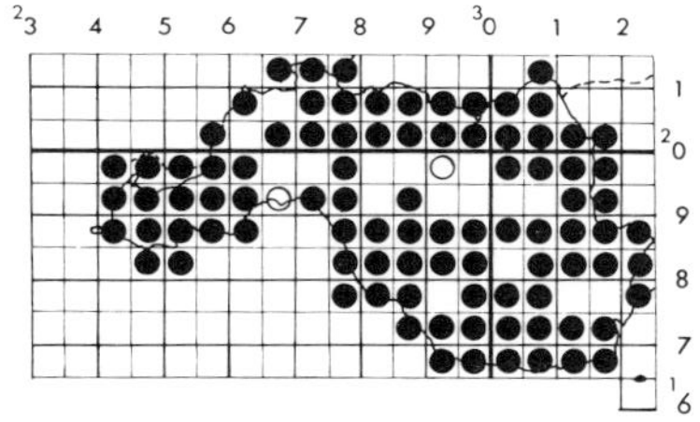

509 Torilis japonica

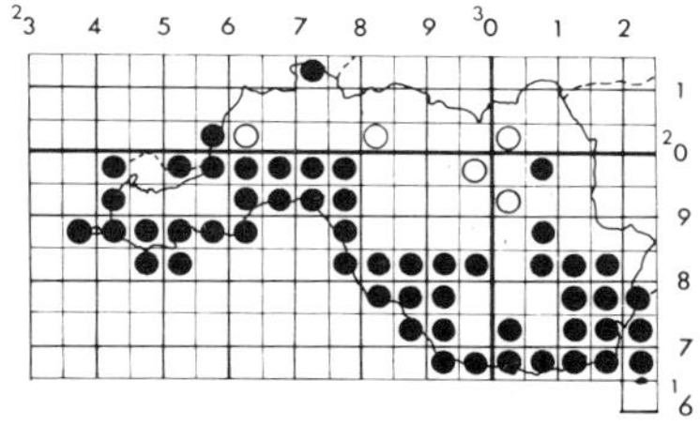

510 Daucus carota

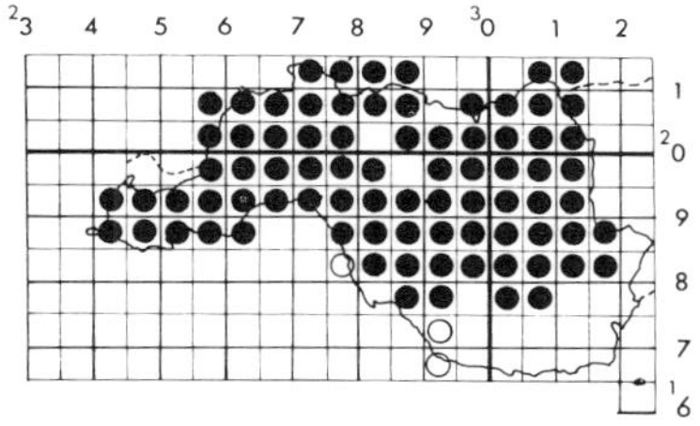

511 Erica tetralix

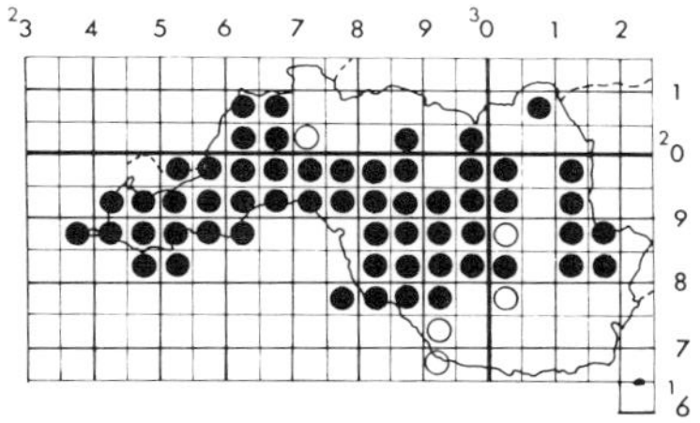

512 Erica cinerea

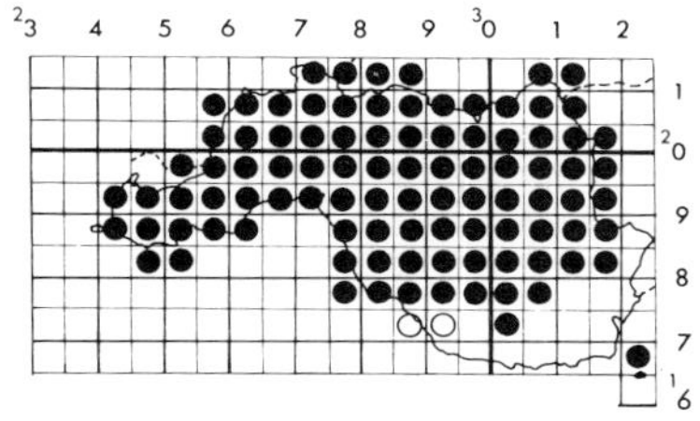

513 Calluna vulgaris

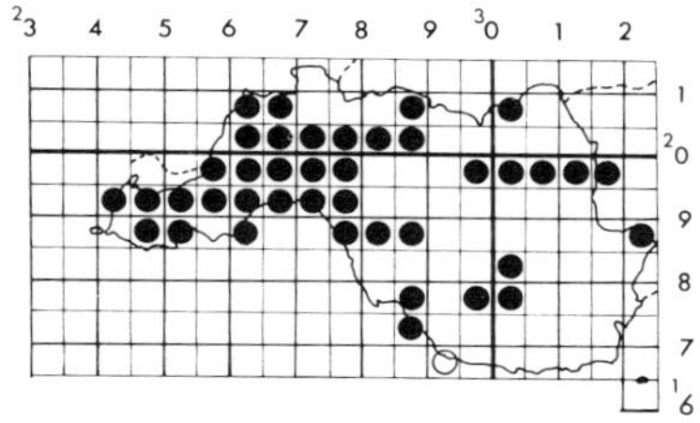

514 Rhododendron ponticum

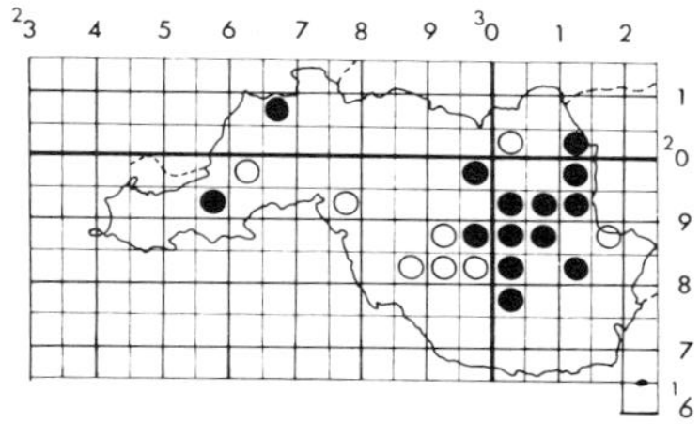

515 Vaccinium oxycoccos

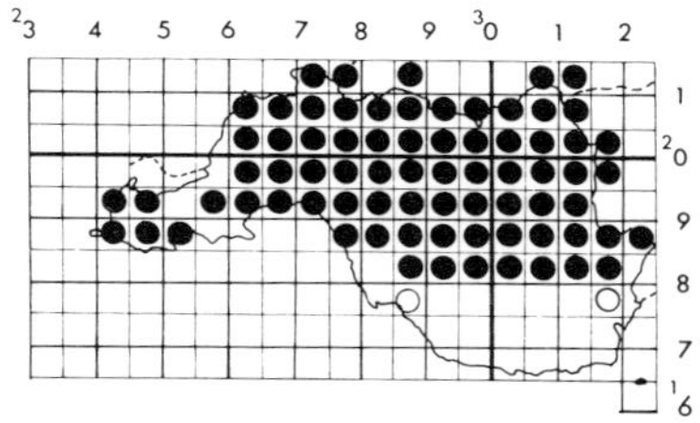

516 Vaccinium myrtillus

517 Primula vulgaris

518 Primula veris

519 Primula x tommasinii

520 Lysimachia nemorum

521 Lysimachia vulgaris

522 Lysimachia nummularia

523 Lysimachia punctata

524 Glaux maritima

525 Anagallis tenella

526 Anagallis arvensis

527 Samolus valerandi

528 Armeria maritima

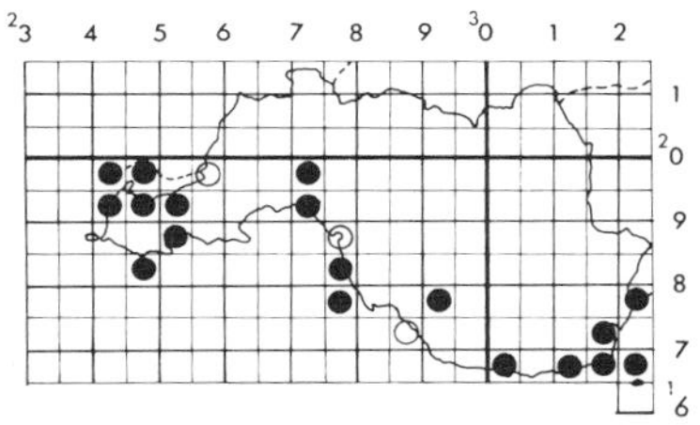

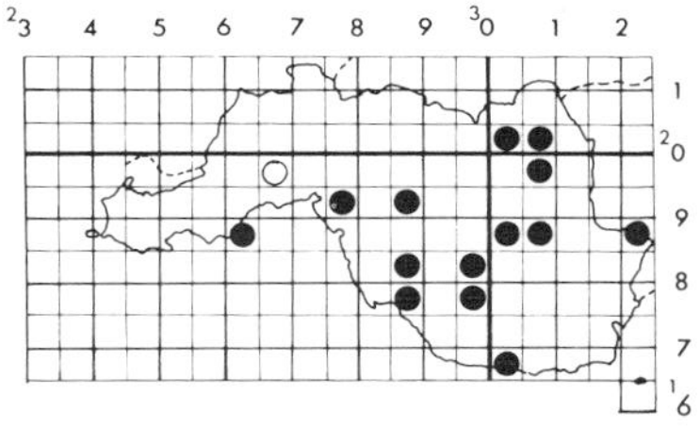

532 Syringa vulgaris

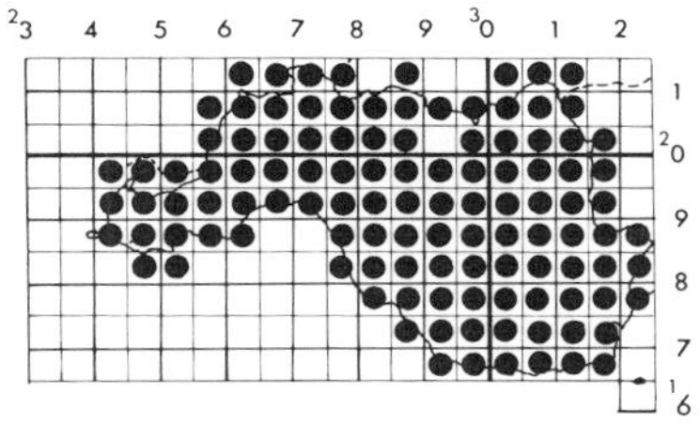

531 Fraxinus excelsior

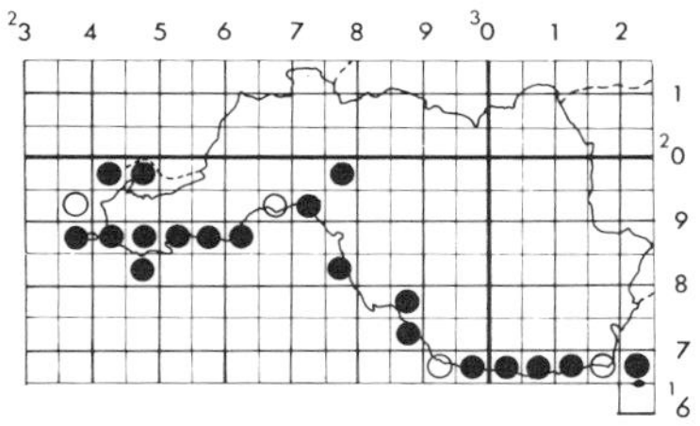

530 Limonium procerum

529 Limonium vulgare

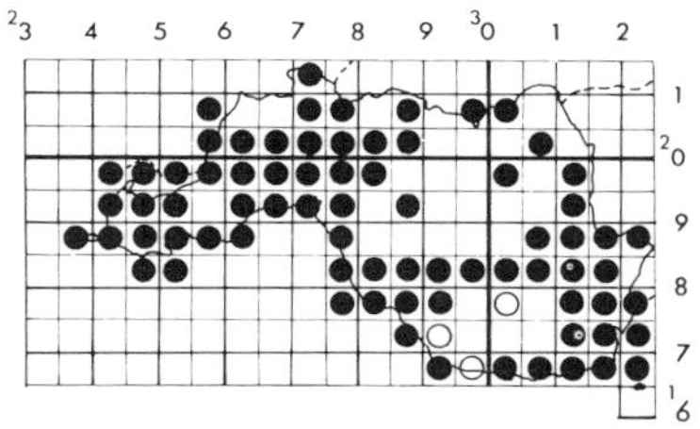

536 Centaurium erythraea

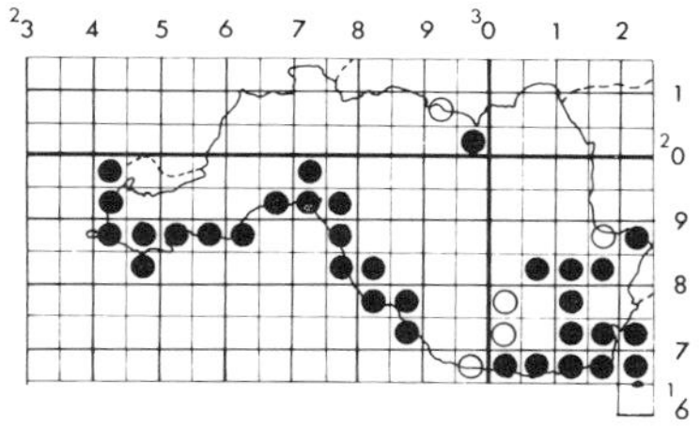

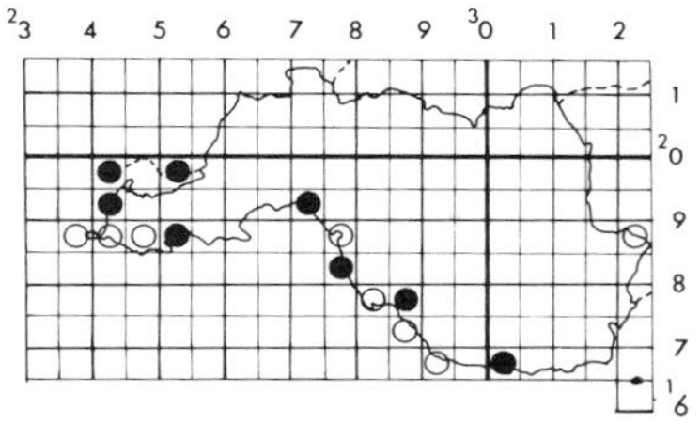

535 Blackstonia perfoliata

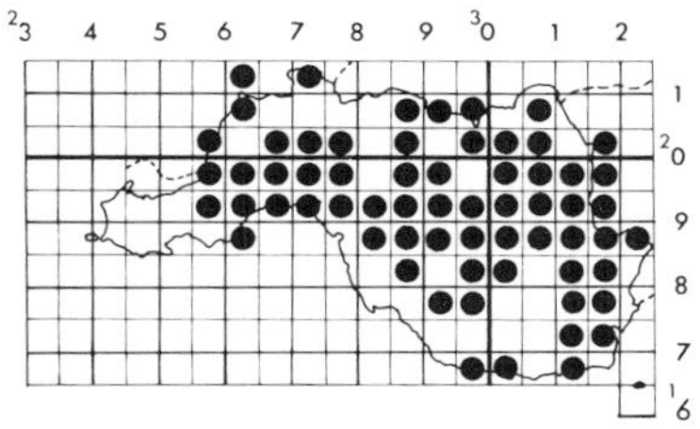

534 Ligustrum ovalifolium

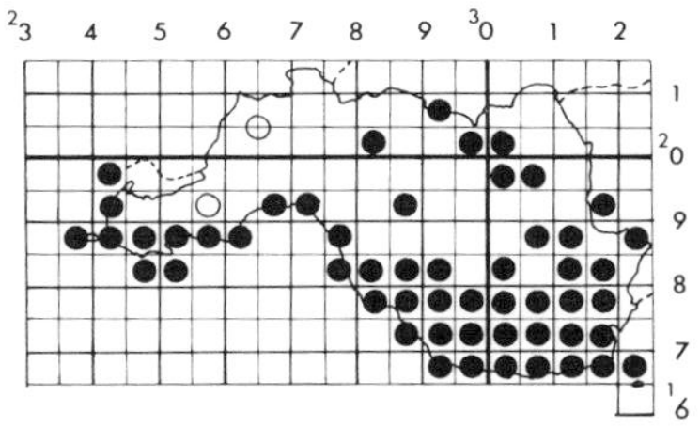

533 Ligustrum vulgare

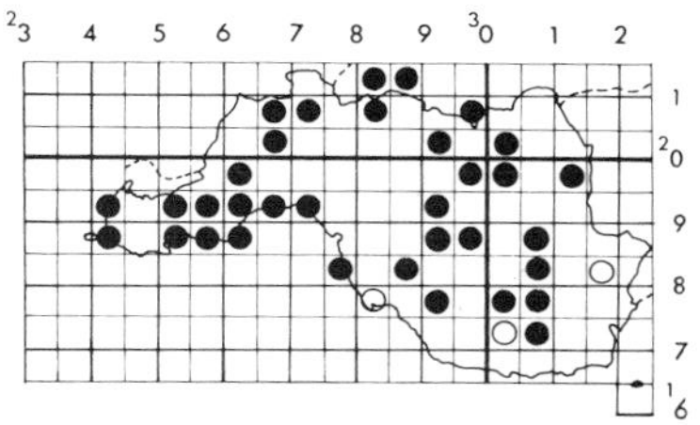

540 Menyanthes trifoliata

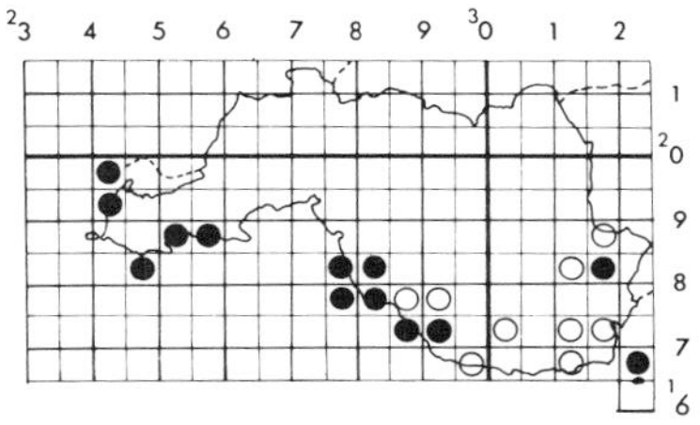

539 Gentianella amarella

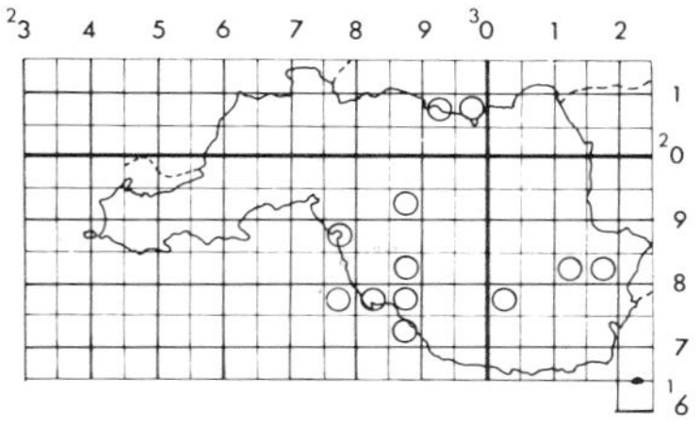

538 Gentianella campestris

537 Centaurium pulchellum

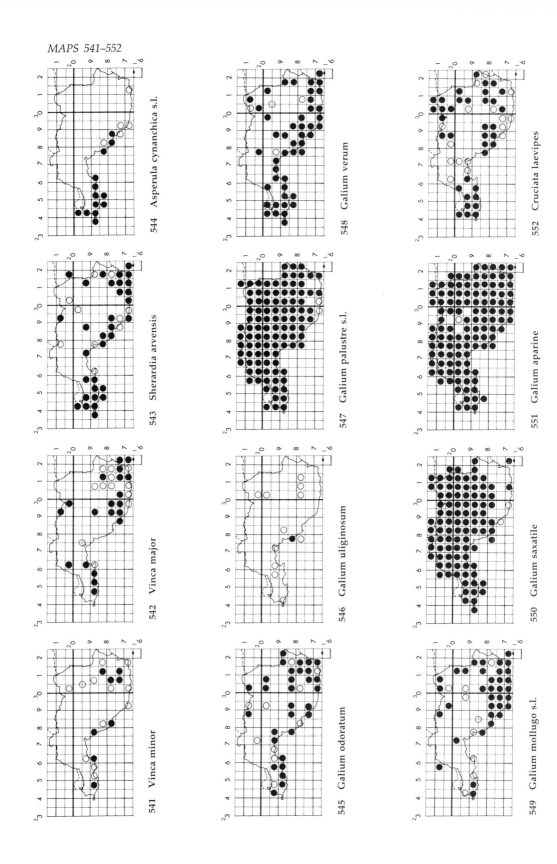

541 Vinca minor

542 Vinca major

543 Sherardia arvensis

544 Asperula cynanchica s.l.

545 Galium odoratum

546 Galium uliginosum

547 Galium palustre s.l.

548 Galium verum

549 Galium mollugo s.l.

550 Galium saxatile

551 Galium aparine

552 Cruciata laevipes

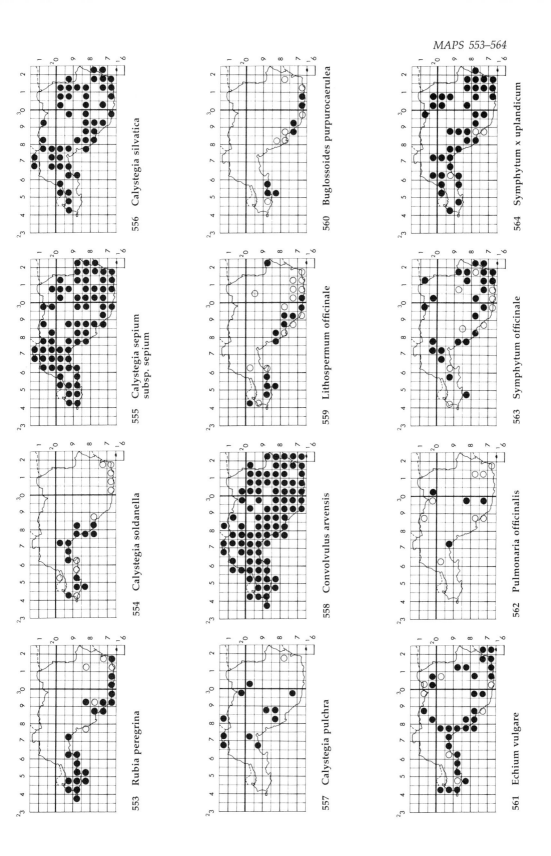

553 Rubia peregrina

554 Calystegia soldanella

555 Calystegia sepium subsp. sepium

556 Calystegia silvatica

557 Calystegia pulchra

558 Convolvulus arvensis

559 Lithospermum officinale

560 Buglossoides purpurocaerulea

561 Echium vulgare

562 Pulmonaria officinalis

563 Symphytum officinale

564 Symphytum x uplandicum

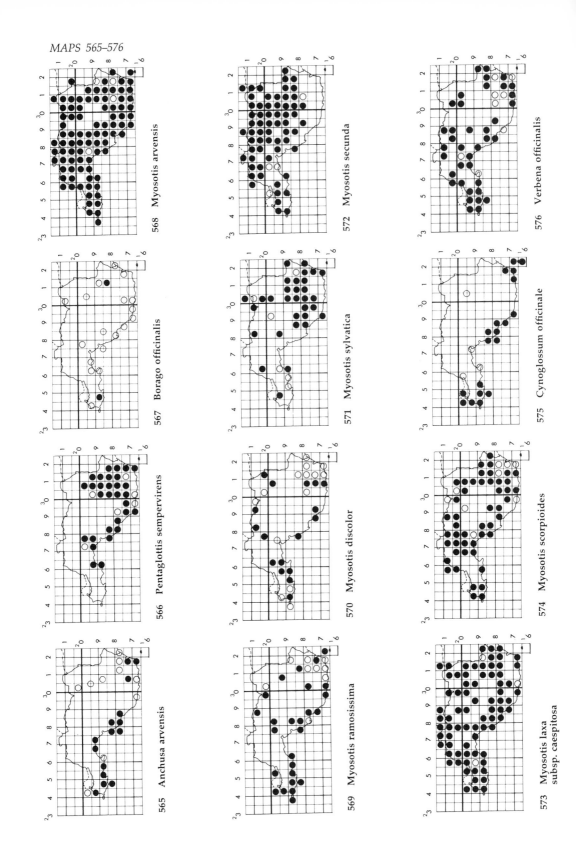

565 Anchusa arvensis

566 Pentaglottis sempervirens

567 Borago officinalis

568 Myosotis arvensis

569 Myosotis ramosissima

570 Myosotis discolor

571 Myosotis sylvatica

572 Myosotis secunda

573 Myosotis laxa subsp. caespitosa

574 Myosotis scorpioides

575 Cynoglossum officinale

576 Verbena officinalis

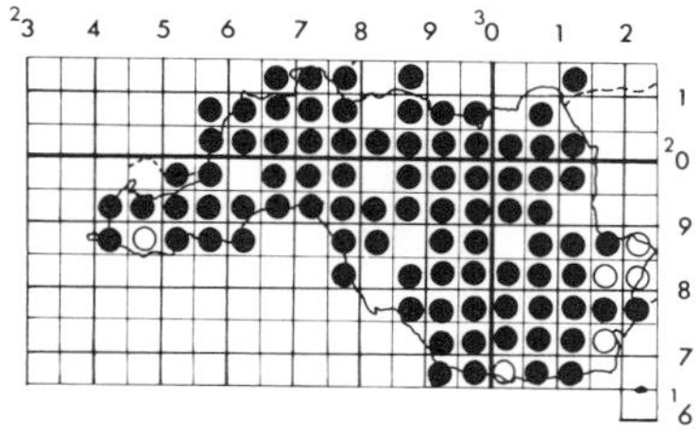

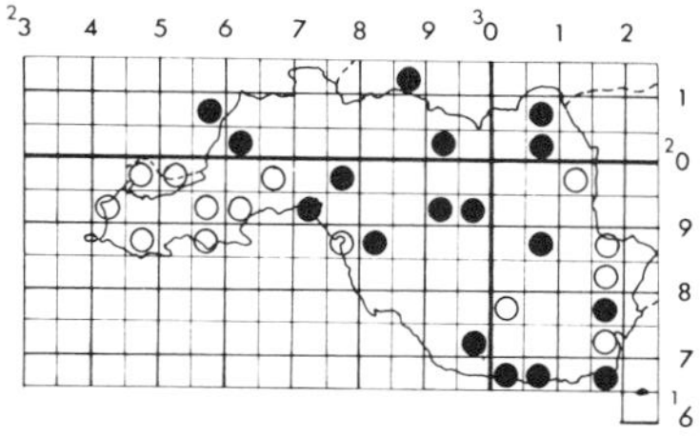

580 Callitriche hamulata

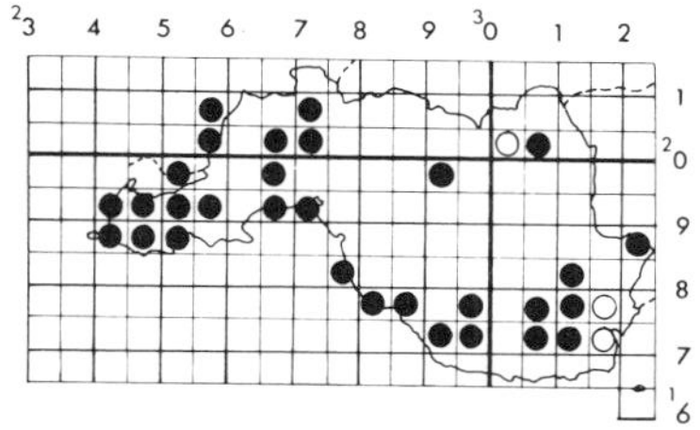

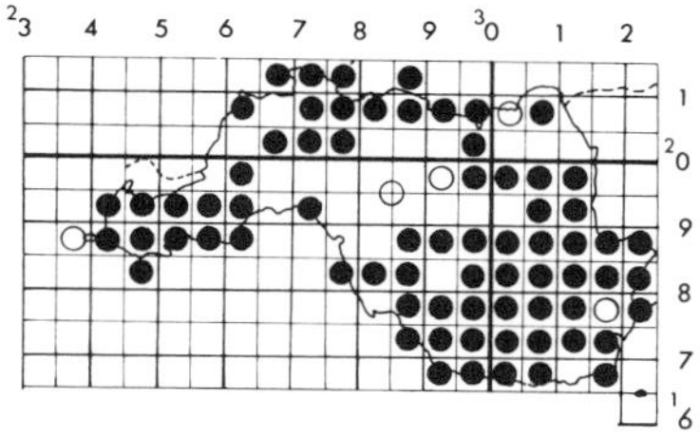

579 Callitriche platycarpa

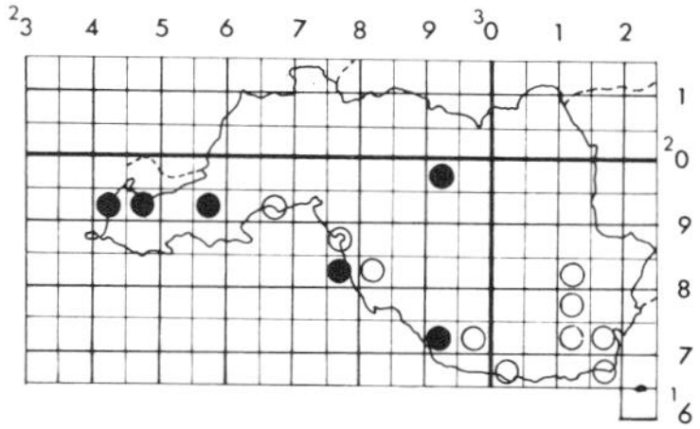

578 Callitriche obtusangula

577 Callitriche stagnalis s.s.

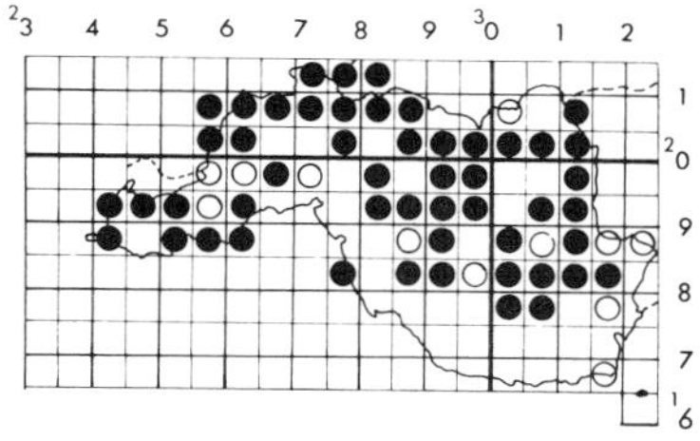

584 Scutellaria minor

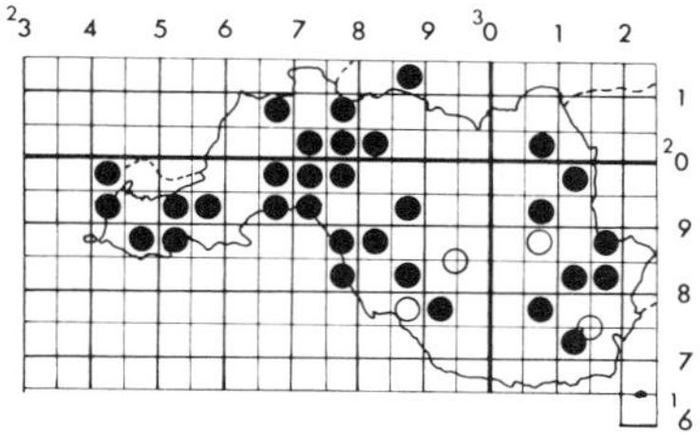

583 Scutellaria galericulata

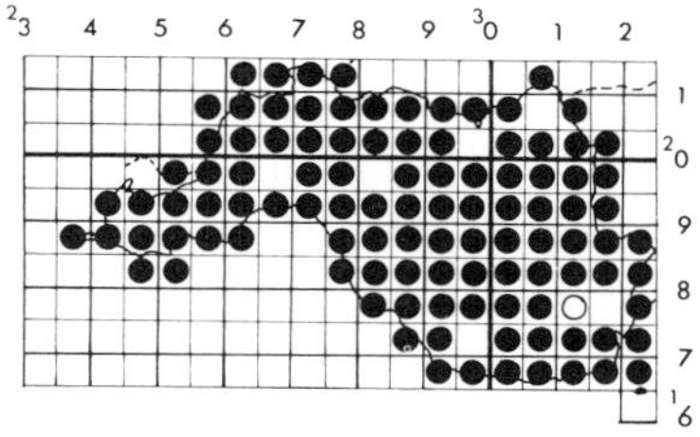

582 Teucrium scorodonia

581 Ajuga reptans

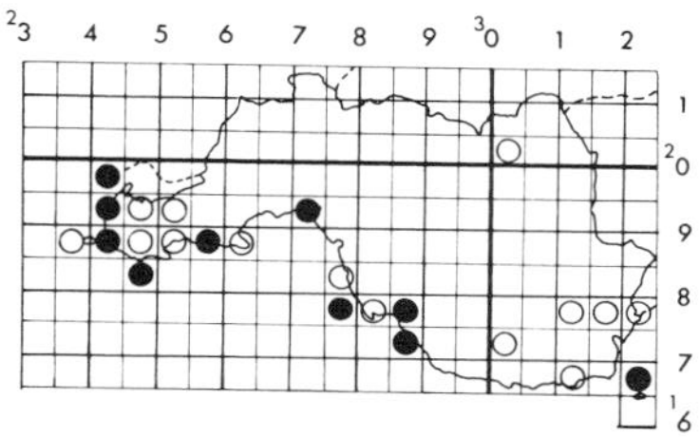

588 Galeopsis bifida

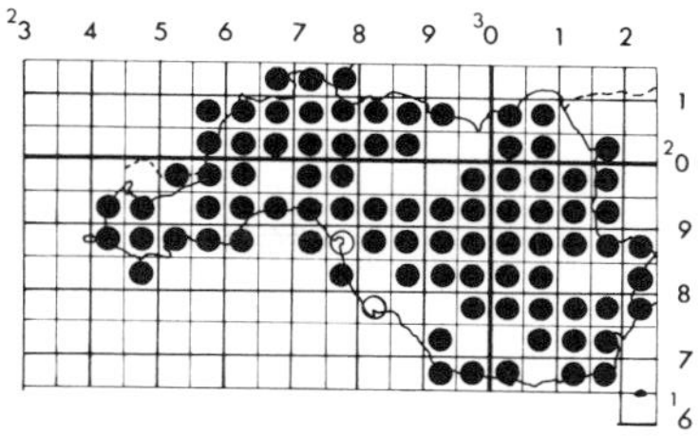

587 Galeopsis tetrahit

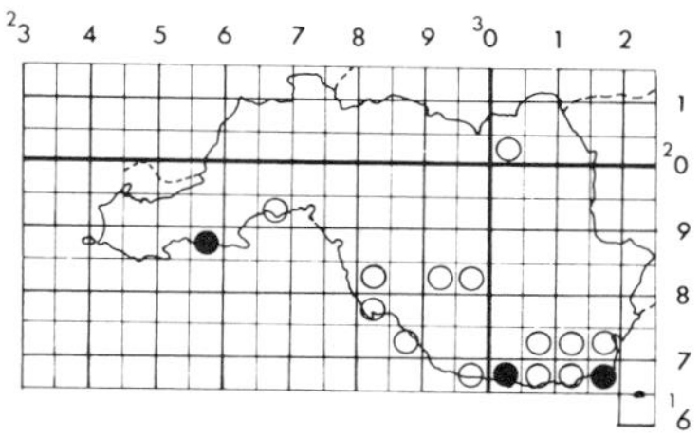

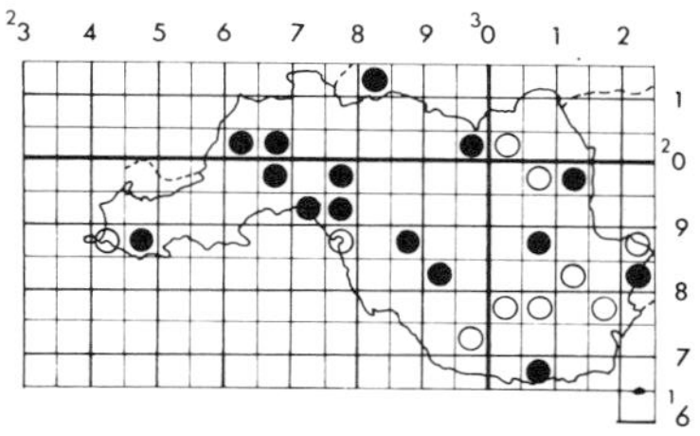

586 Galeopsis angustifolia

585 Marrubium vulgare

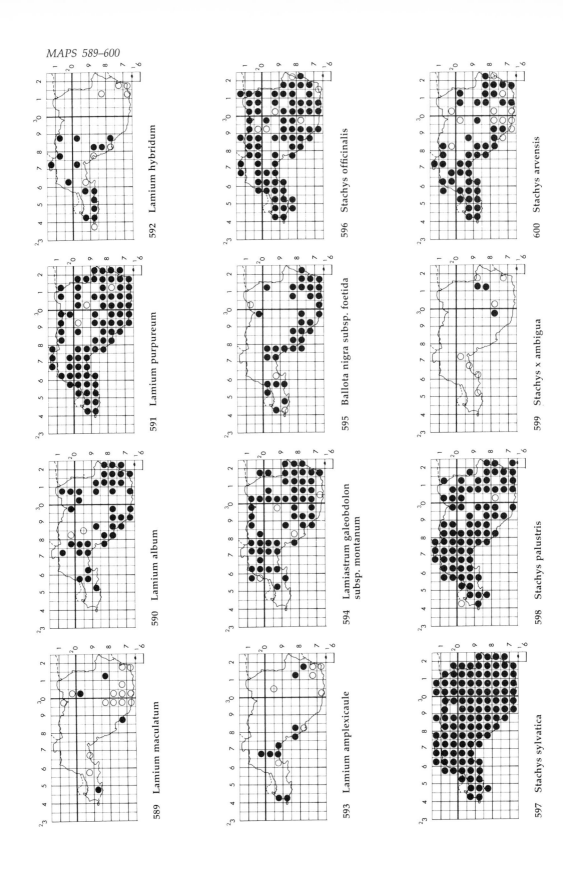

589 Lamium maculatum

590 Lamium album

591 Lamium purpureum

592 Lamium hybridum

593 Lamium amplexicaule

594 Lamiastrum galeobdolon subsp. montanum

595 Ballota nigra subsp. foetida

596 Stachys officinalis

597 Stachys sylvatica

598 Stachys palustris

599 Stachys x ambigua

600 Stachys arvensis

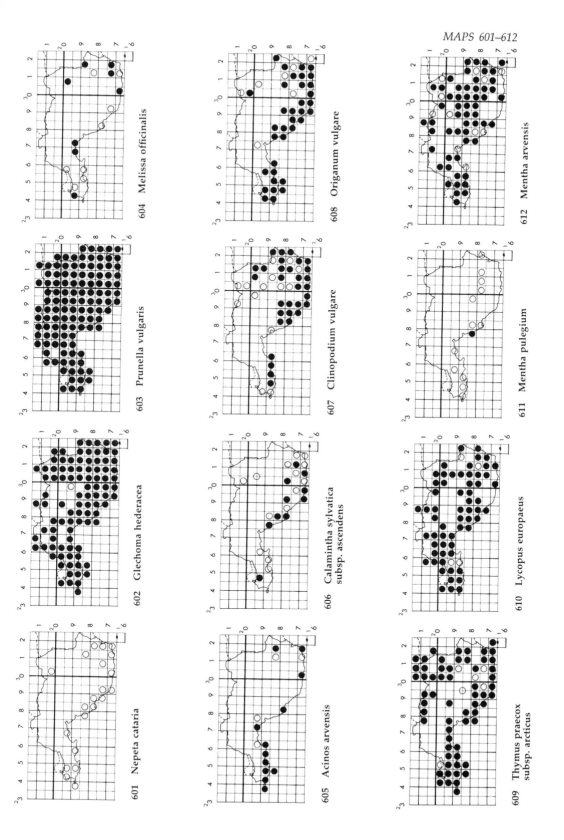

601 Nepeta cataria

602 Glechoma hederacea

603 Prunella vulgaris

604 Melissa officinalis

605 Acinos arvensis

606 Calamintha sylvatica subsp. ascendens

607 Clinopodium vulgare

608 Origanum vulgare

609 Thymus praecox subsp. arcticus

610 Lycopus europaeus

611 Mentha pulegium

612 Mentha arvensis

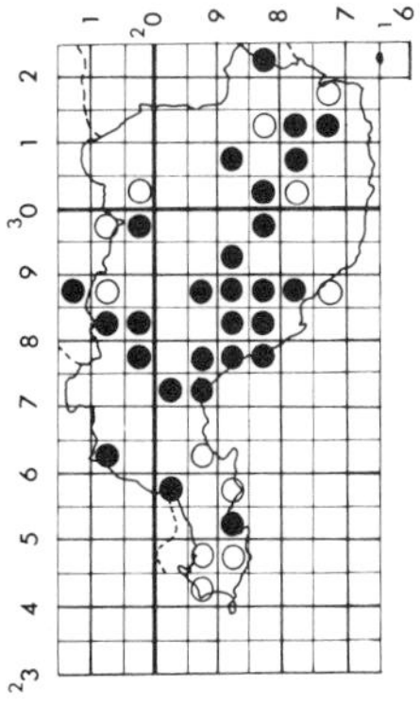

613 Mentha x verticillata

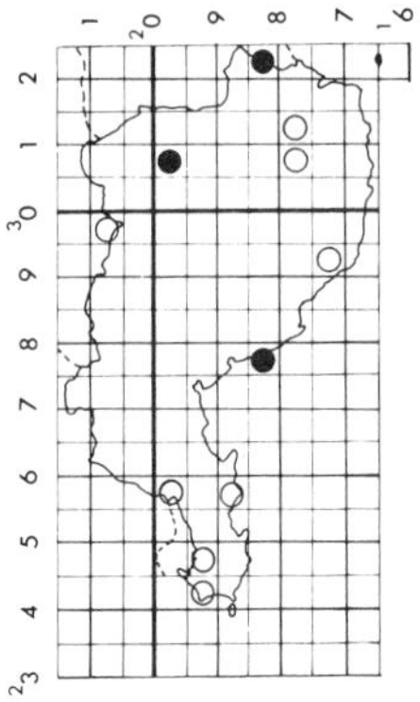

614 Mentha x smithiana

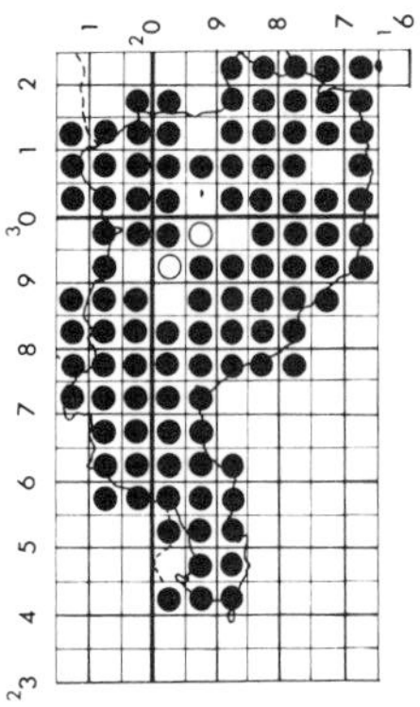

615 Mentha aquatica

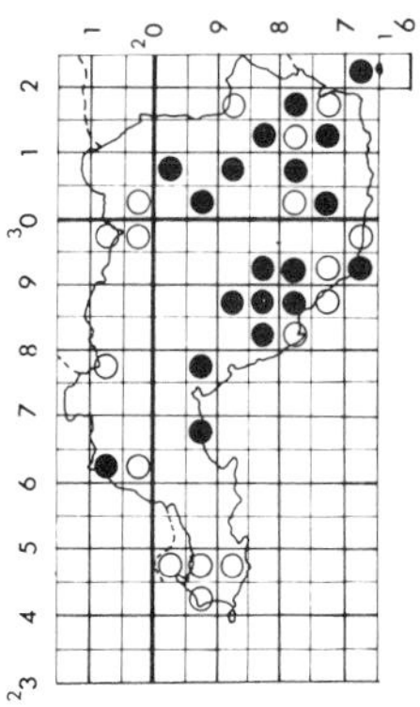

616 Mentha x piperita

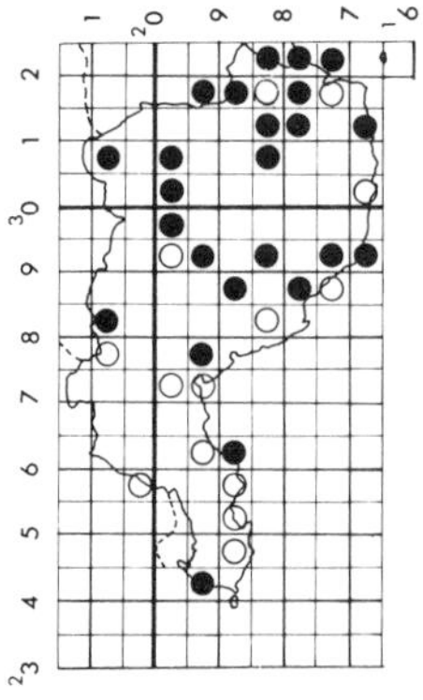

617 Mentha suaveolens

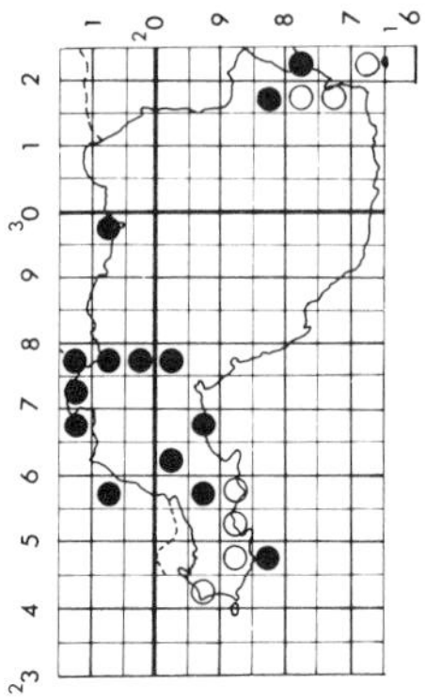

618 Mentha x villosa

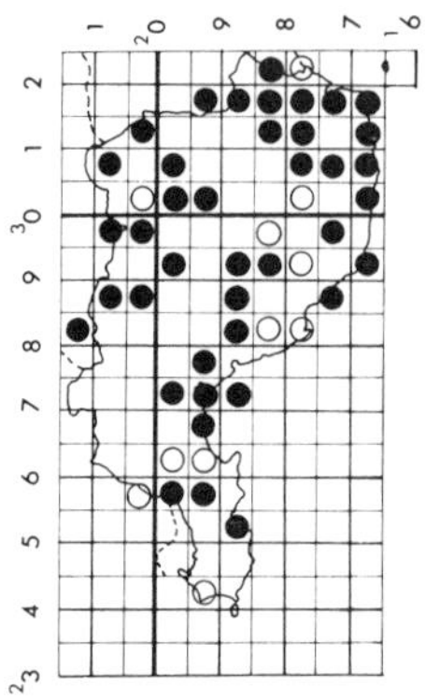

619 Mentha spicata

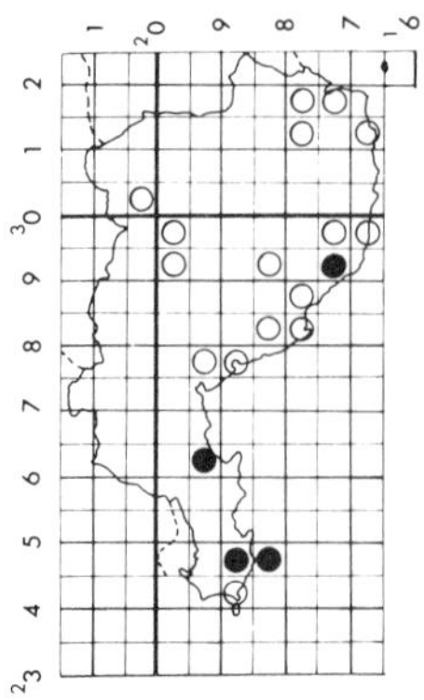

620 Salvia verbenaca

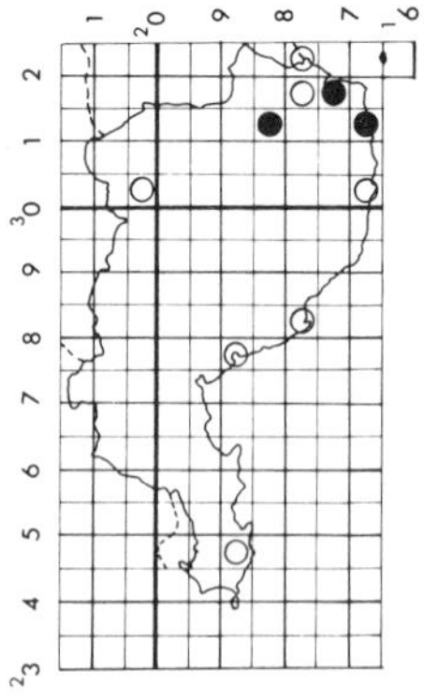

621 Salvia verticillata

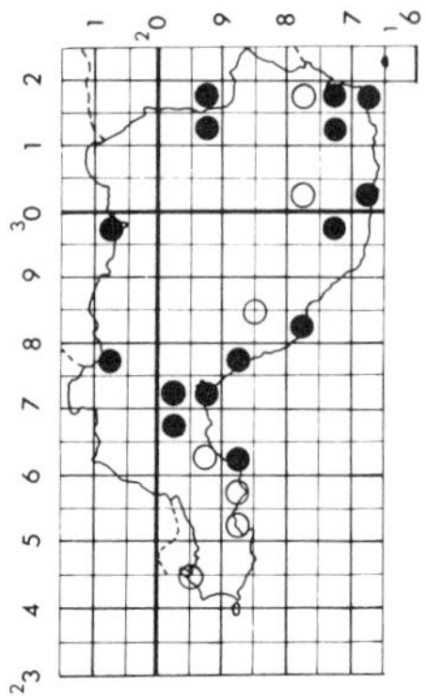

622 Lycium barbarum

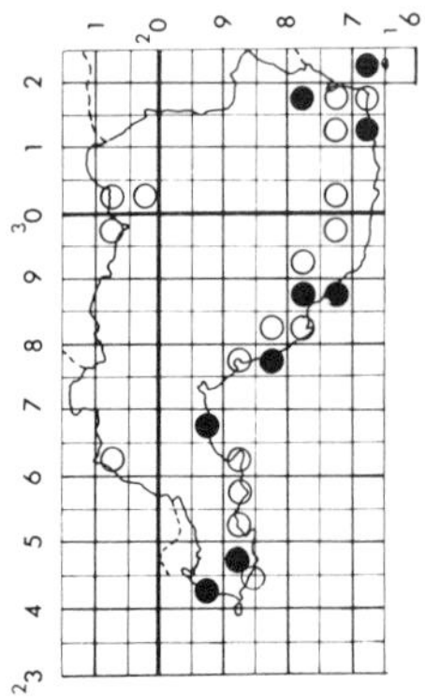

623 Hyoscyamus niger

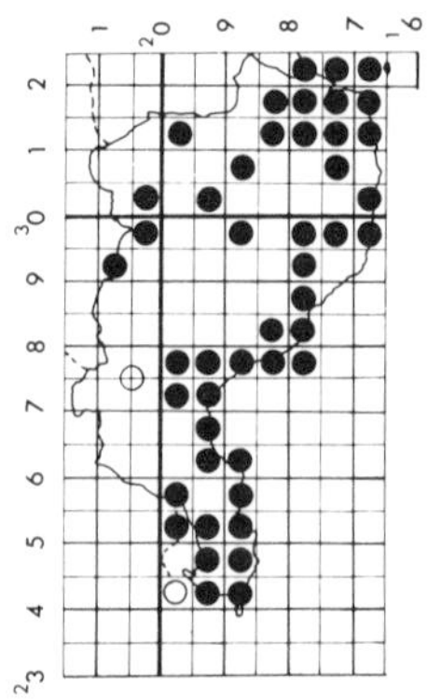

624 Solanum nigrum

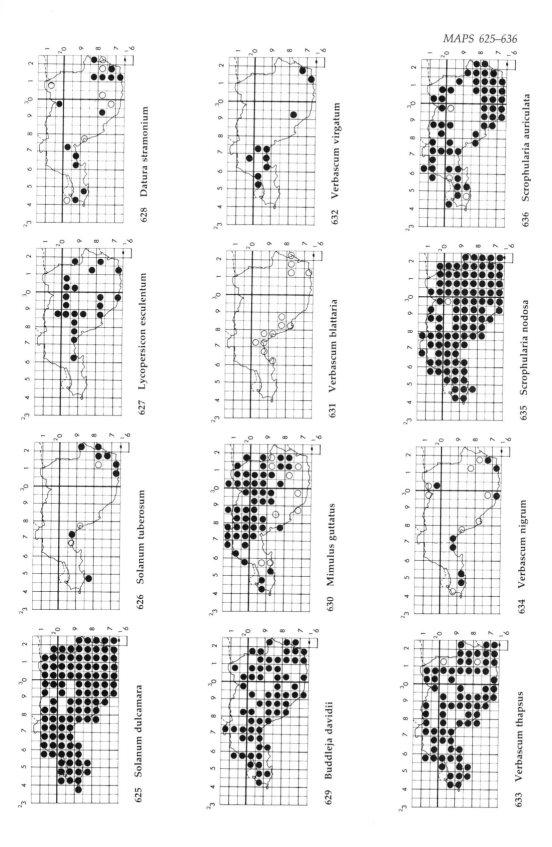

625 Solanum dulcamara

626 Solanum tuberosum

627 Lycopersicon esculentum

628 Datura stramonium

629 Buddleja davidii

630 Mimulus guttatus

631 Verbascum blattaria

632 Verbascum virgatum

633 Verbascum thapsus

634 Verbascum nigrum

635 Scrophularia nodosa

636 Scrophularia auriculata

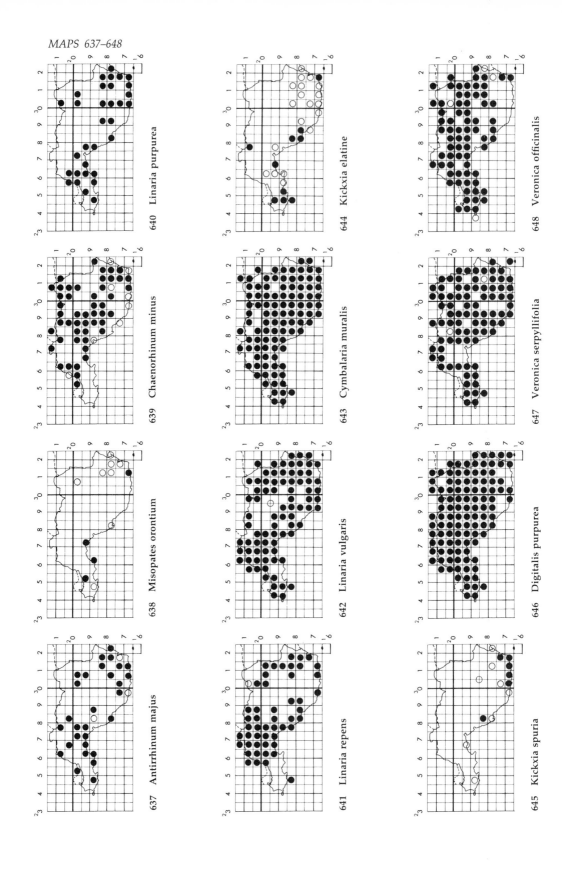

637 Antirrhinum majus

638 Misopates orontium

639 Chaenorhinum minus

640 Linaria purpurea

641 Linaria repens

642 Linaria vulgaris

643 Cymbalaria muralis

644 Kickxia elatine

645 Kickxia spuria

646 Digitalis purpurea

647 Veronica serpyllifolia

648 Veronica officinalis

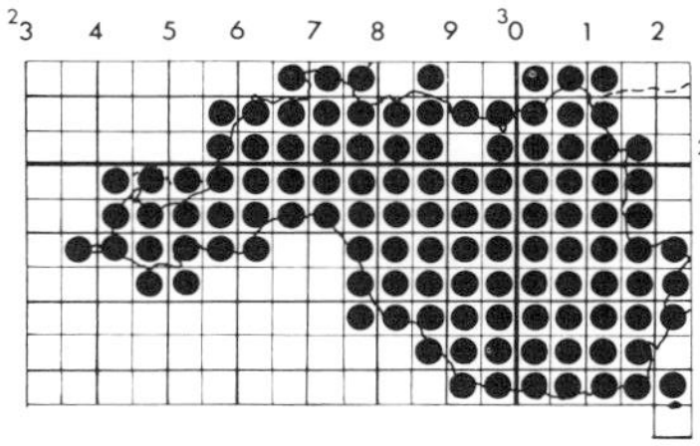

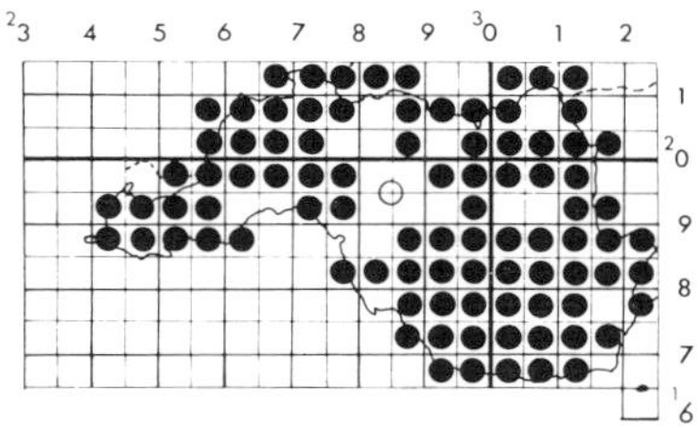

652 Veronica beccabunga

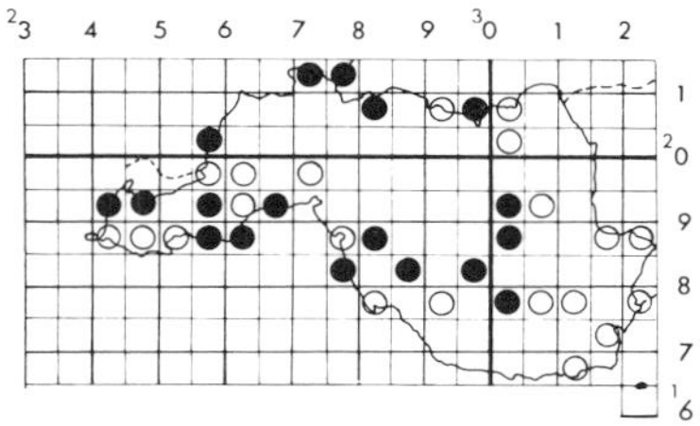

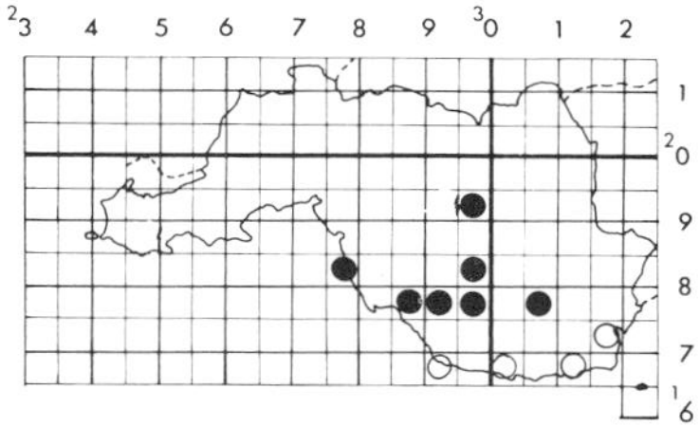

651 Veronica scutellata

650 Veronica montana

649 Veronica chamaedrys

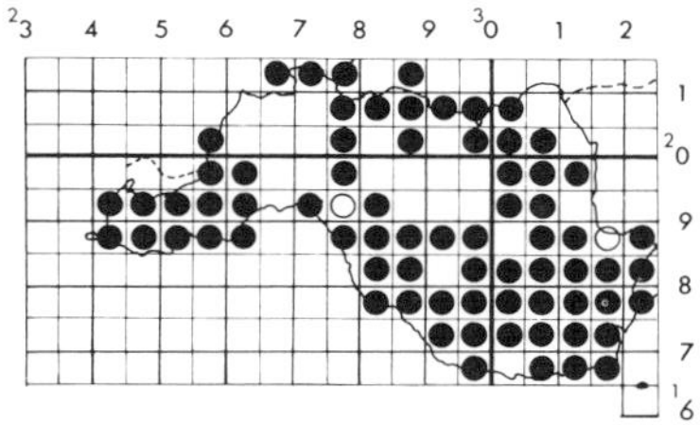

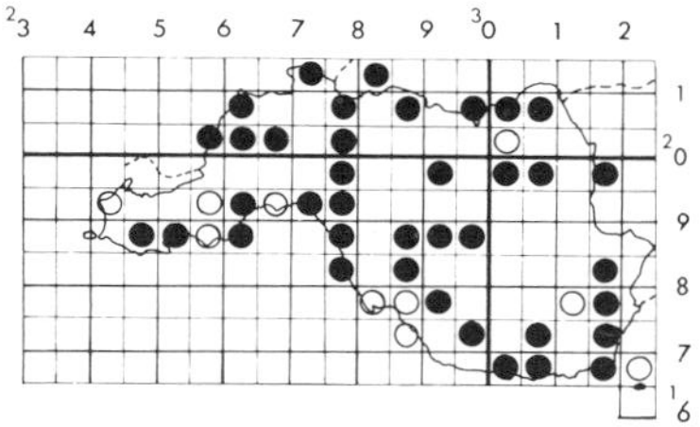

656 Veronica agrestis

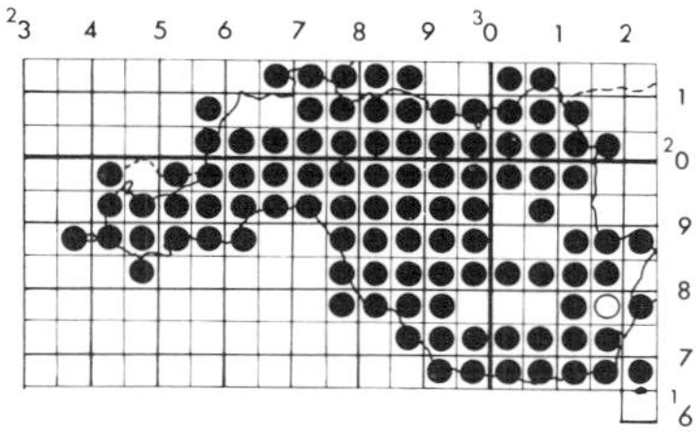

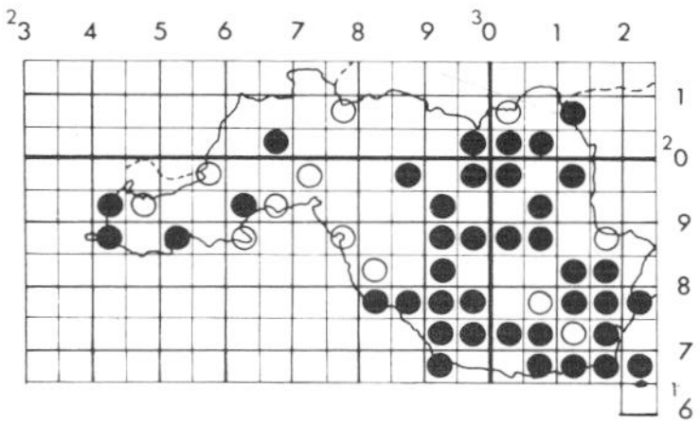

655 Veronica arvensis

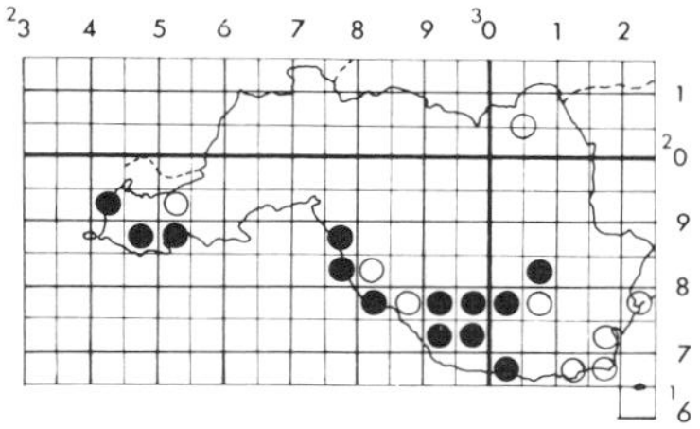

654 Veronica catenata

653 Veronica anagallis-aquatica

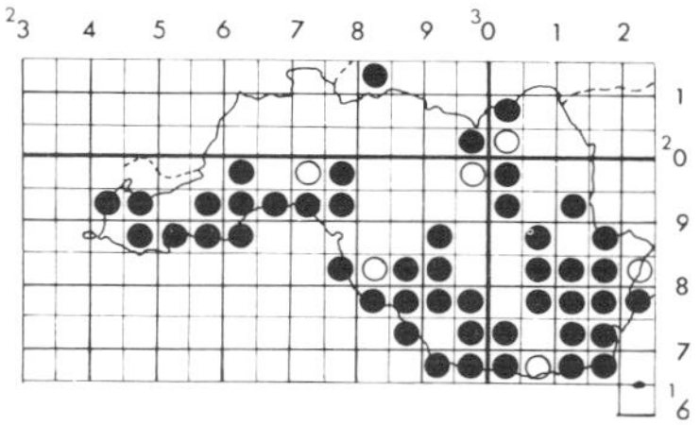

660 Veronica hederifolia

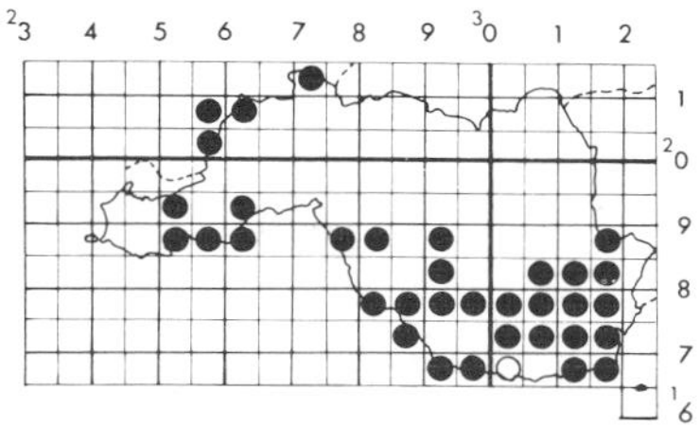

659 Veronica filiformis

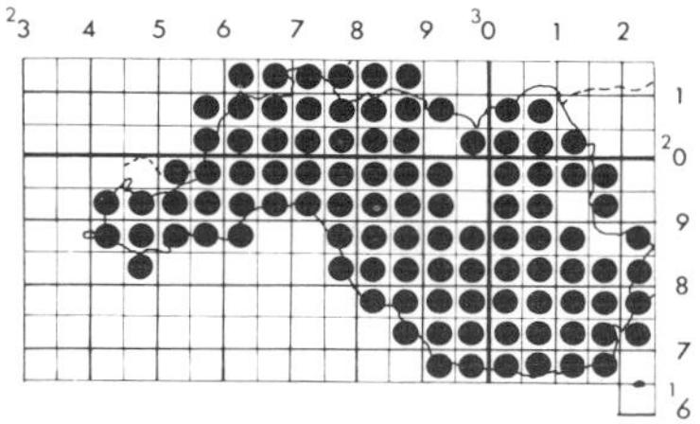

658 Veronica persica

657 Veronica polita

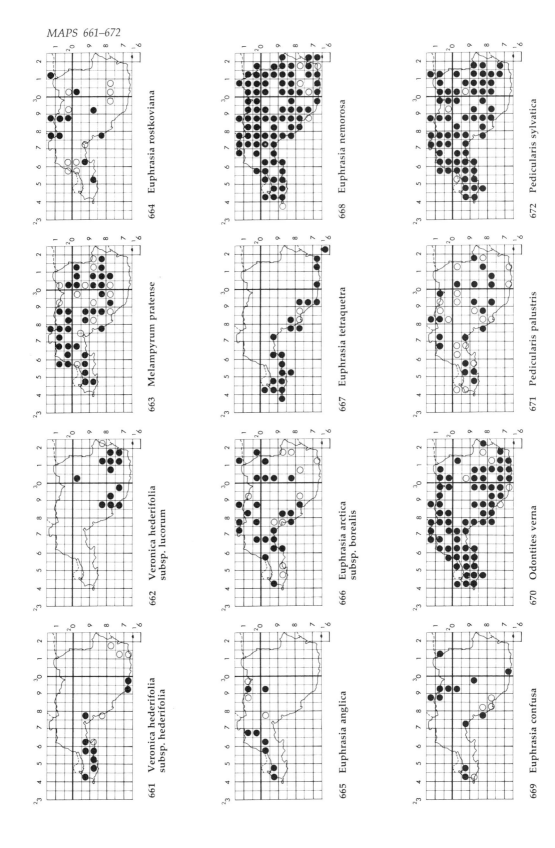

661 Veronica hederifolia subsp. hederifolia

662 Veronica hederifolia subsp. lucorum

663 Melampyrum pratense

664 Euphrasia rostkoviana

665 Euphrasia anglica

666 Euphrasia arctica subsp. borealis

667 Euphrasia tetraquetra

668 Euphrasia nemorosa

669 Euphrasia confusa

670 Odontites verna

671 Pedicularis palustris

672 Pedicularis sylvatica

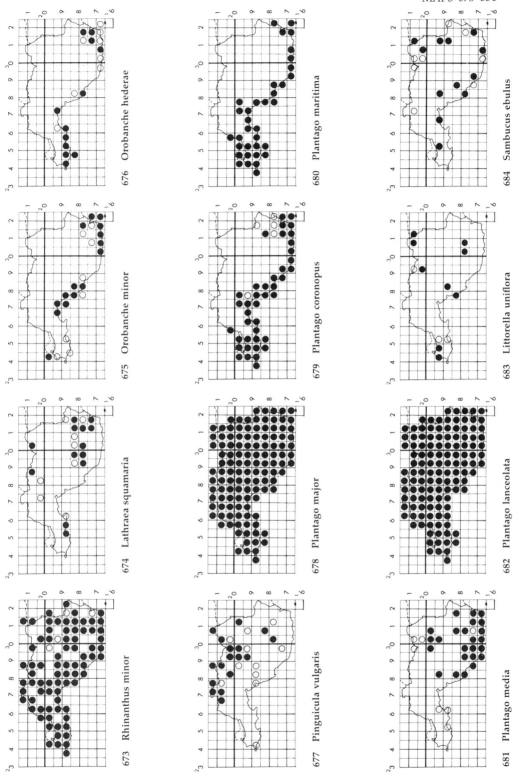

676 Orobanche hederae

675 Orobanche minor

674 Lathraea squamaria

673 Rhinanthus minor

680 Plantago maritima

679 Plantago coronopus

678 Plantago major

677 Pinguicula vulgaris

684 Sambucus ebulus

683 Littorella uniflora

682 Plantago lanceolata

681 Plantago media

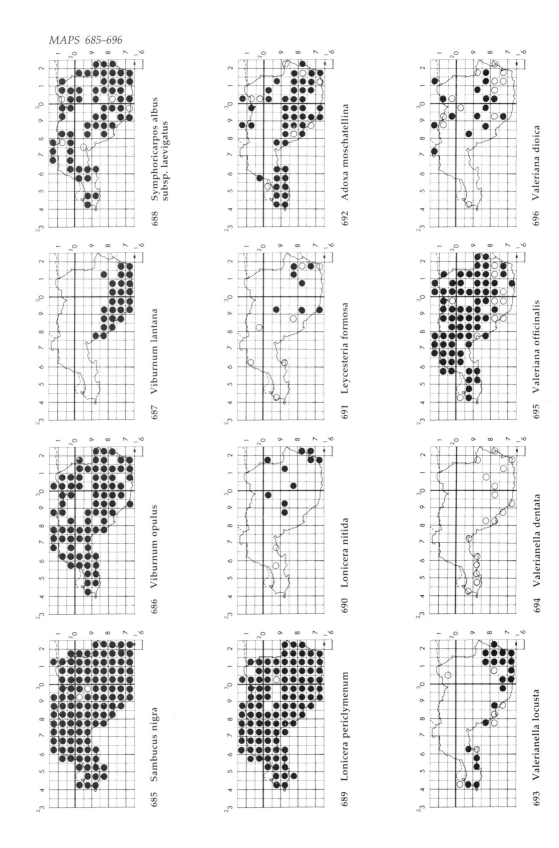

685 Sambucus nigra

686 Viburnum opulus

687 Viburnum lantana

688 Symphoricarpos albus subsp. laevigatus

689 Lonicera periclymenum

690 Lonicera nitida

691 Leycesteria formosa

692 Adoxa moschatellina

693 Valerianella locusta

694 Valerianella dentata

695 Valeriana officinalis

696 Valeriana dioica

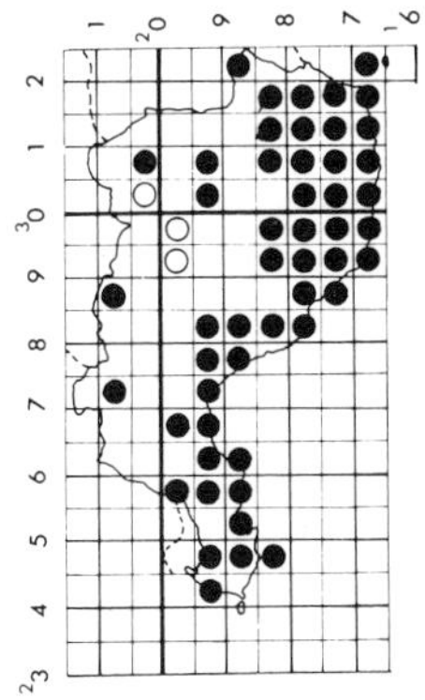

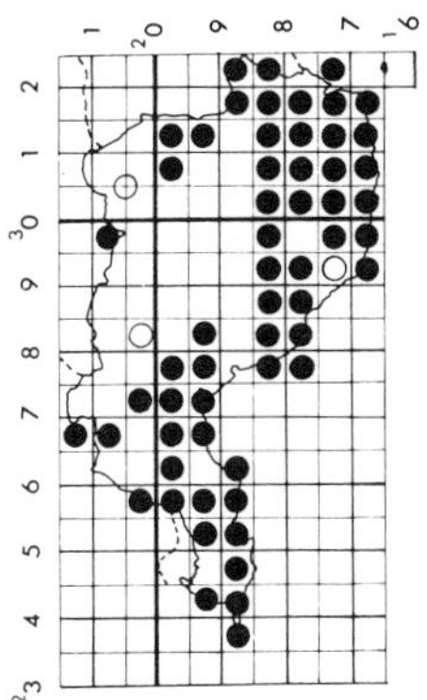

700 Knautia arvensis

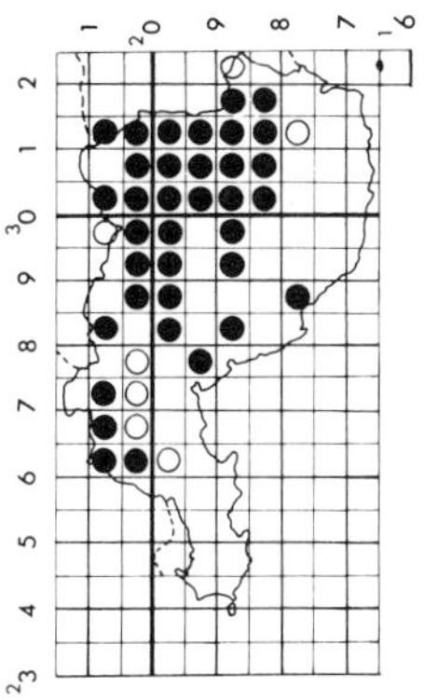

704 Wahlenbergia hederacea

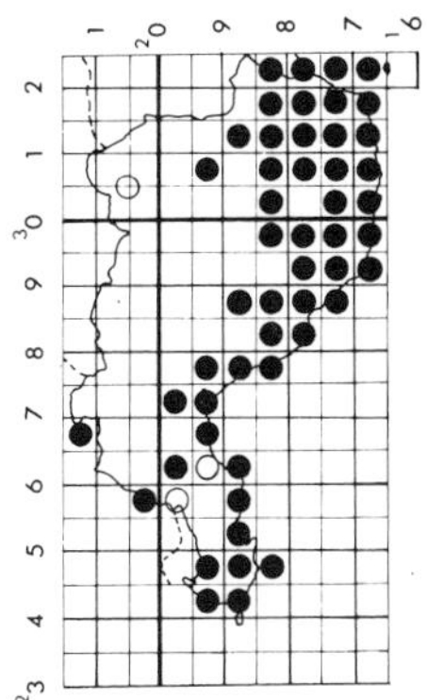

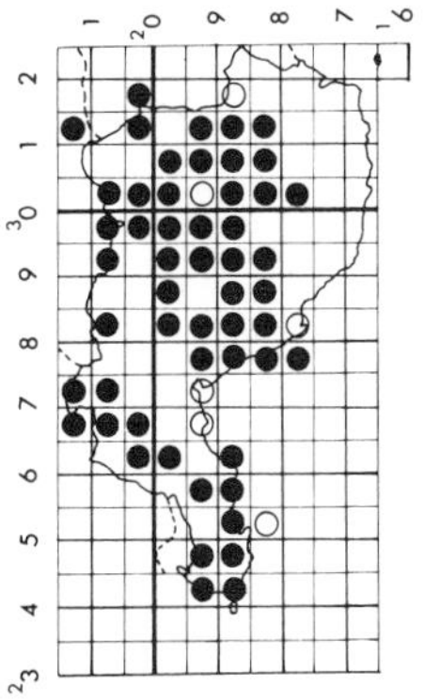

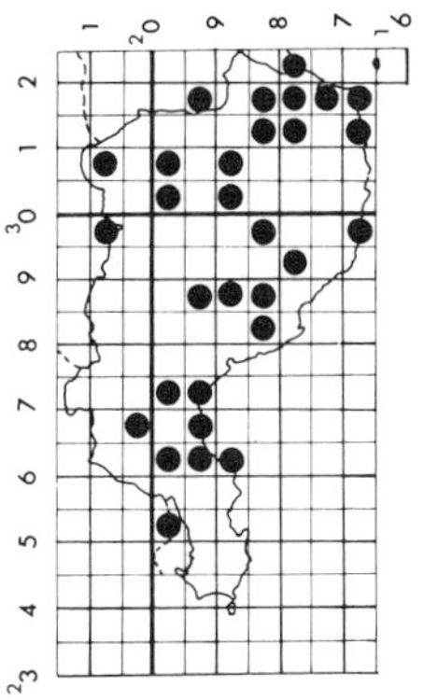

708 Solidago canadensis

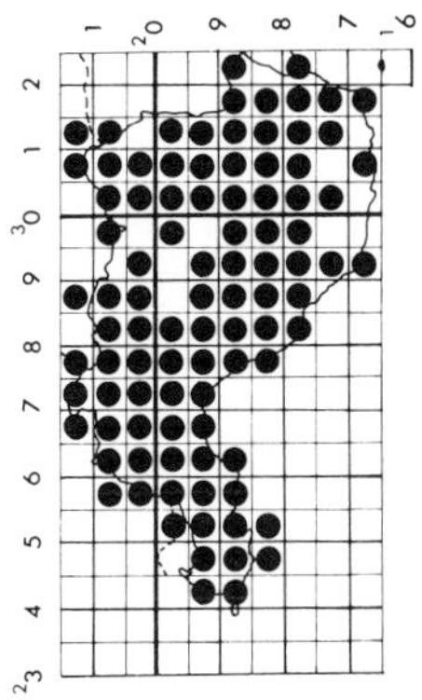

699 Succisa pratensis

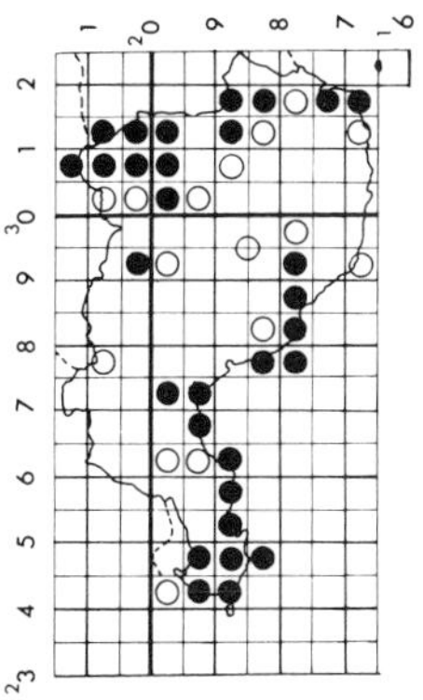

703 Campanula rotundifolia

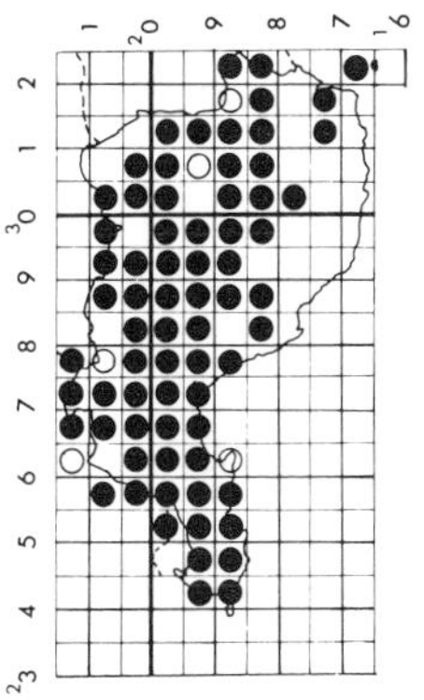

707 Solidago virgaurea

698 Dipsacus fullonum

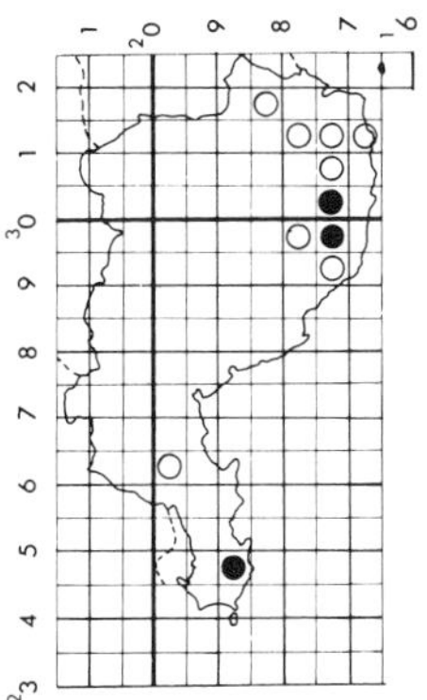

702 Campanula trachelium

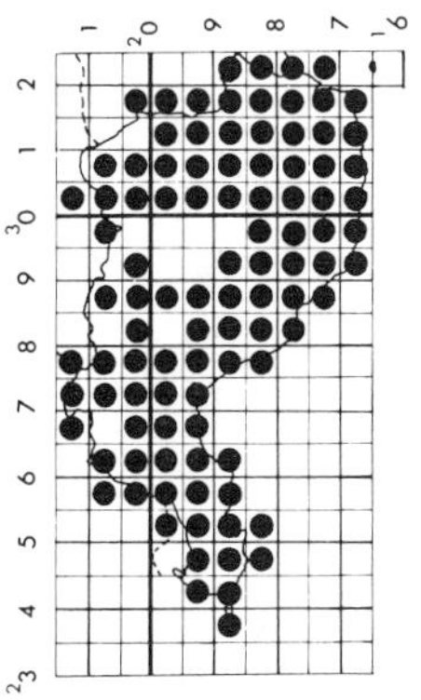

706 Eupatorium cannabinum

697 Centranthus ruber

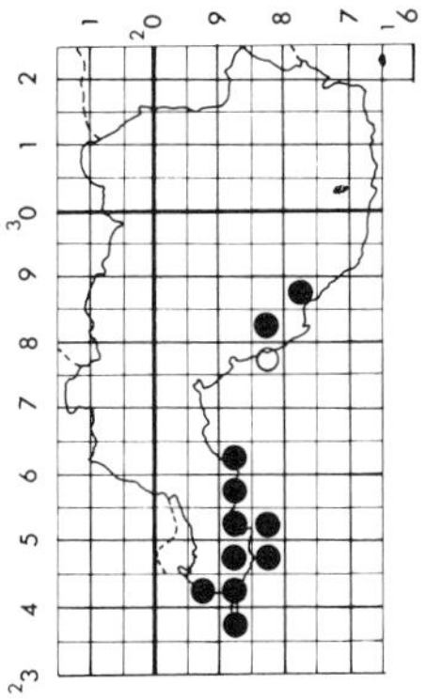

701 Scabiosa columbaria

705 Jasione montana

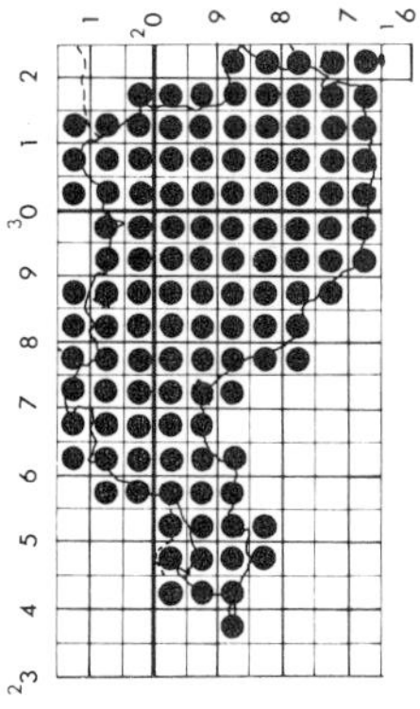

712 Erigeron acer

711 Aster tripolium

710 Aster novi-belgii

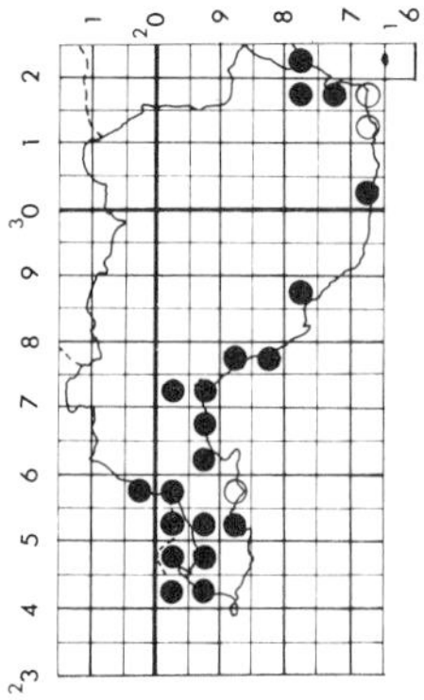

709 Bellis perennis

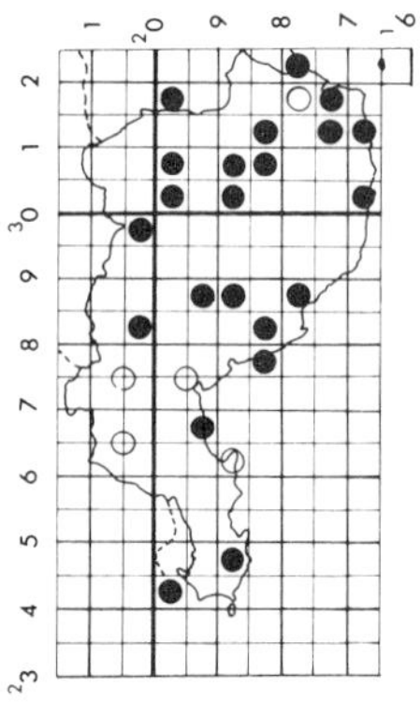

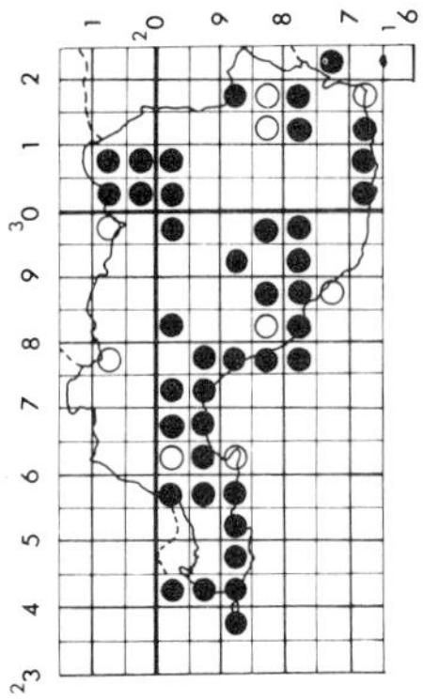

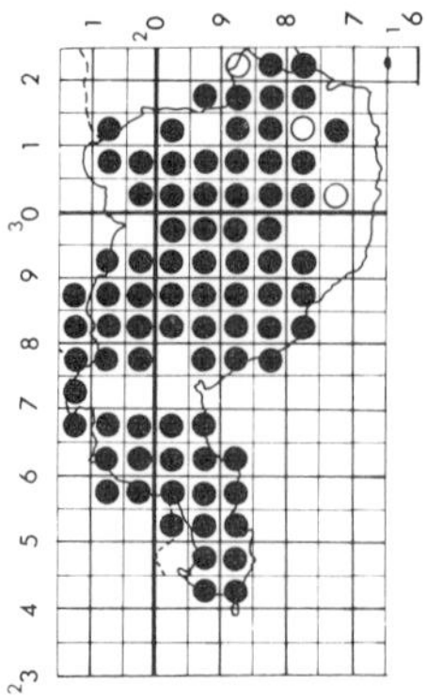

716 Filaginella uliginosa

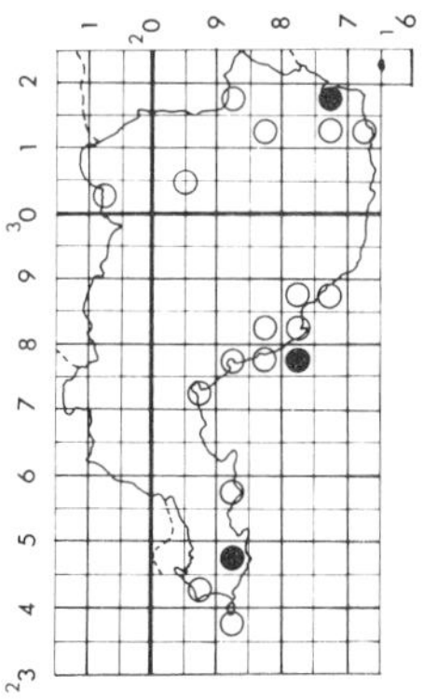

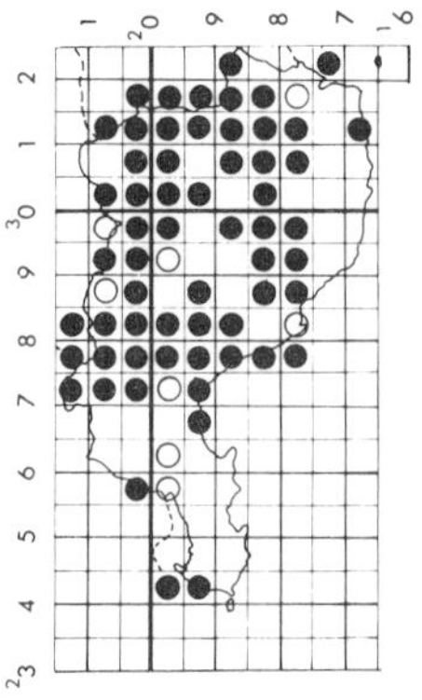

715 Logfia minima

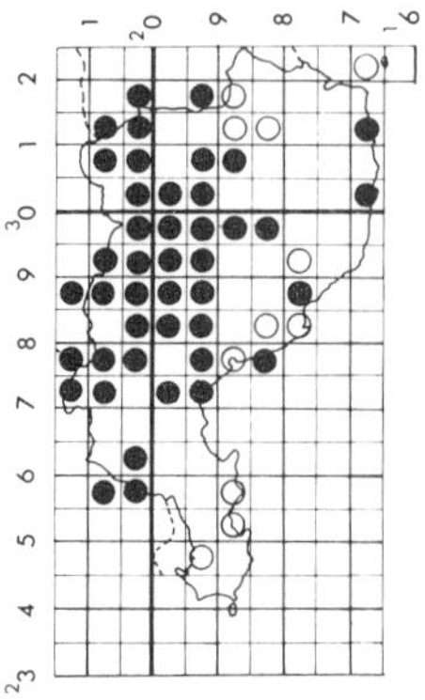

714 Filago vulgaris

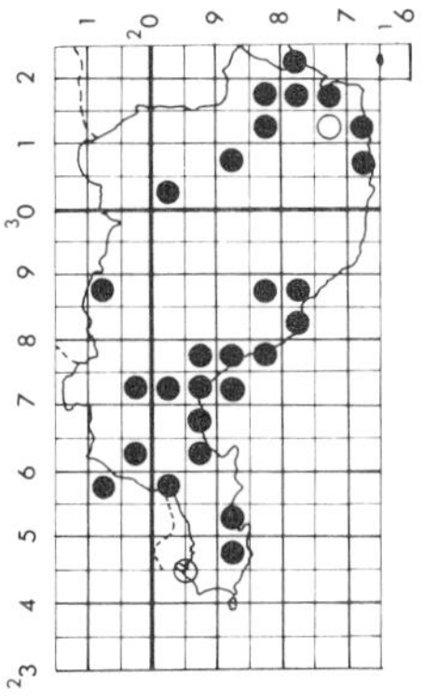

713 Conyza canadensis

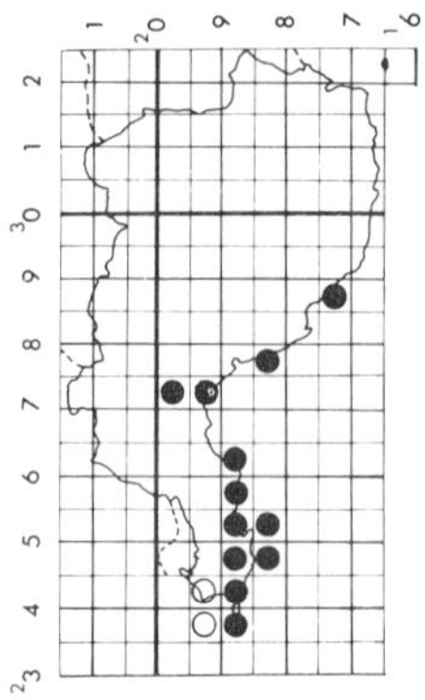

720 Inula crithmoides

719 Inula conyza

718 Inula helenium

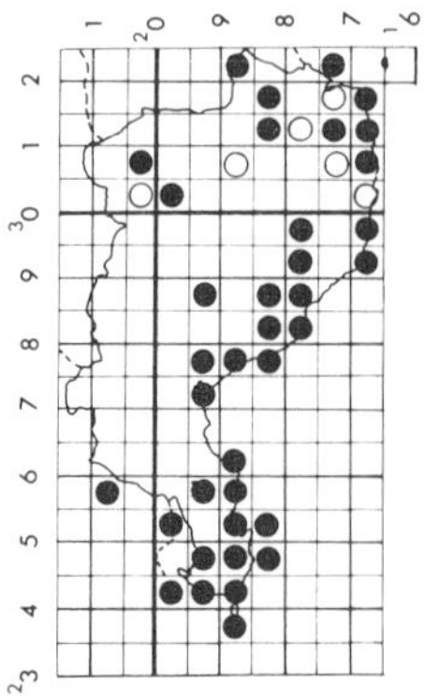

717 Anaphalis margaritacea

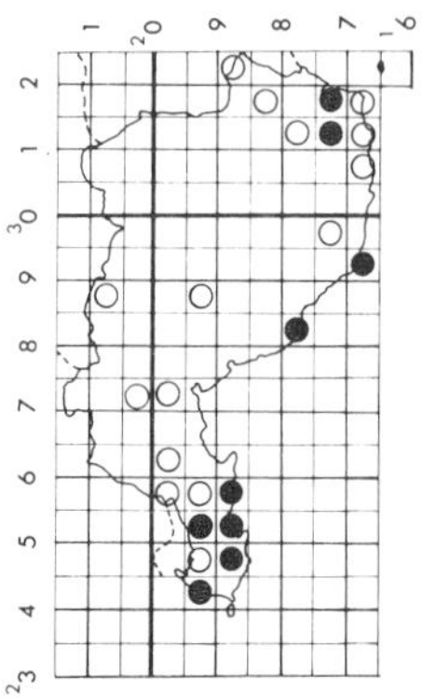

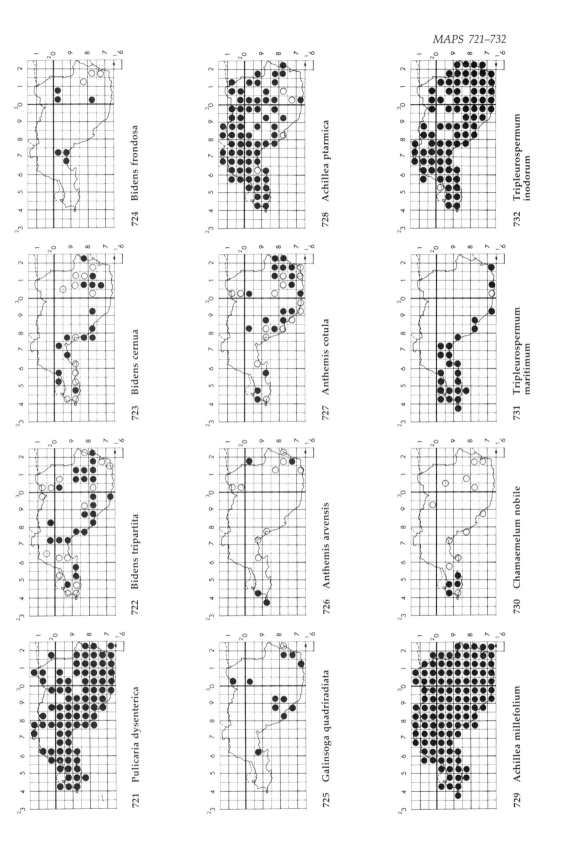

721 Pulicaria dysenterica
722 Bidens tripartita
723 Bidens cernua
724 Bidens frondosa
725 Galinsoga quadriradiata
726 Anthemis arvensis
727 Anthemis cotula
728 Achillea ptarmica
729 Achillea millefolium
730 Chamaemelum nobile
731 Tripleurospermum maritimum
732 Tripleurospermum inodorum

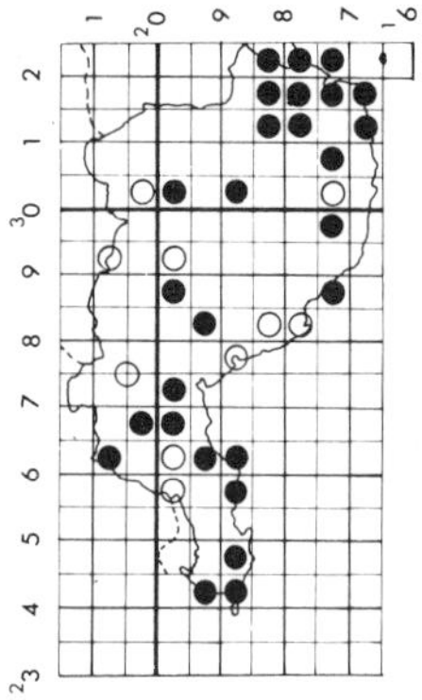

733 Matricaria recutita

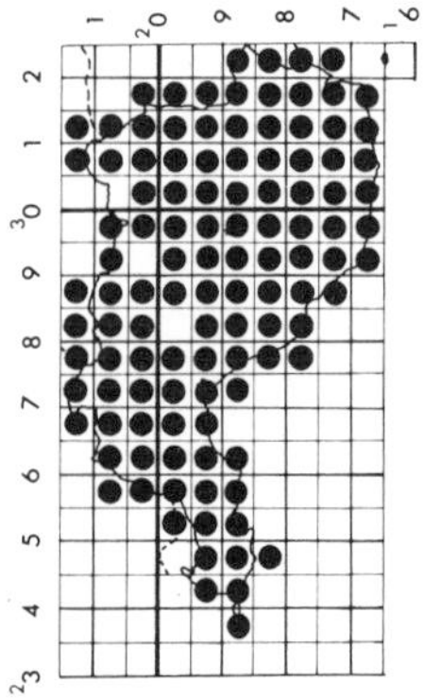

734 Matricaria matricarioides

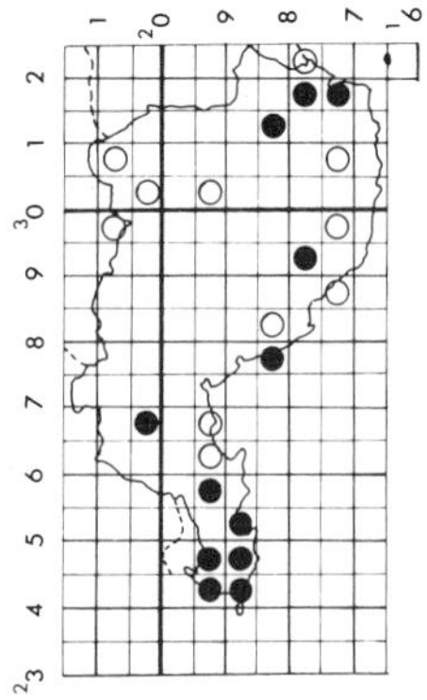

735 Chrysanthemum segetum

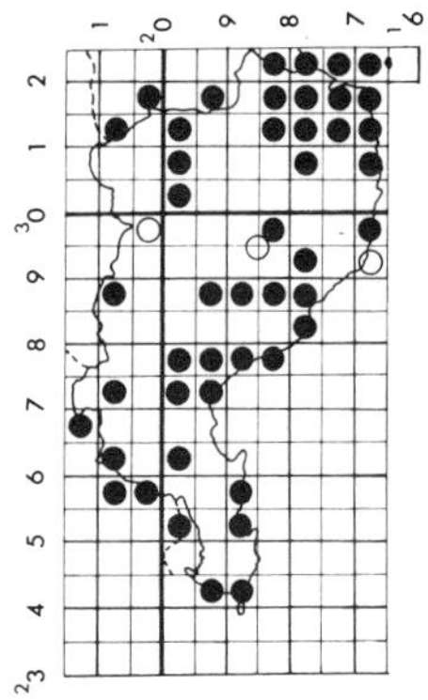

736 Tanacetum vulgare

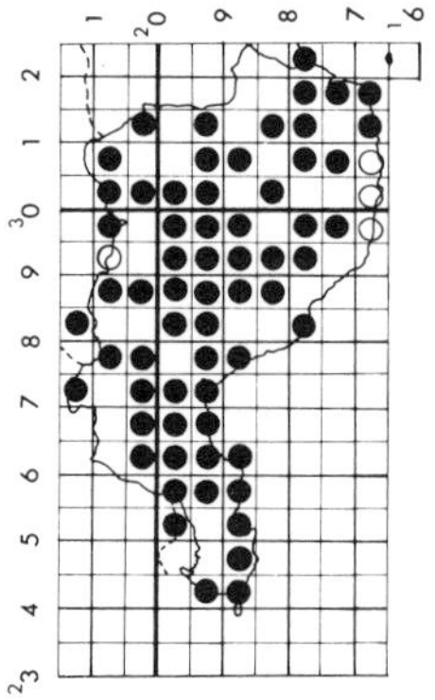

737 Tanacetum parthenium

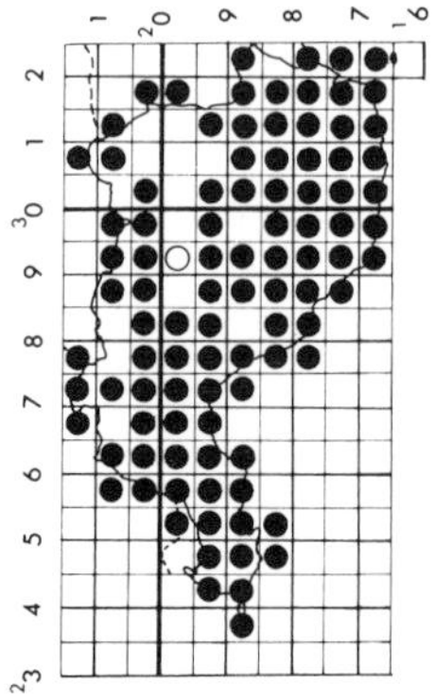

738 Leucanthemum vulgare

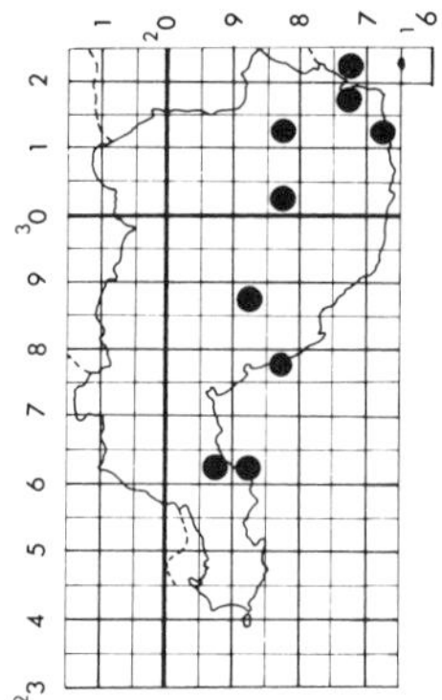

739 Leucanthemum maximum

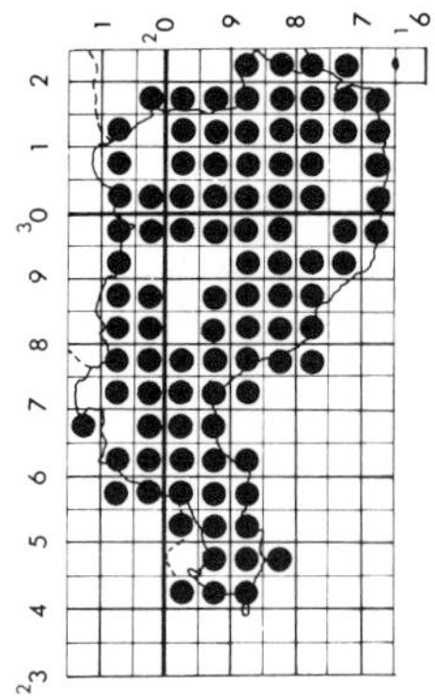

740 Artemisia vulgaris

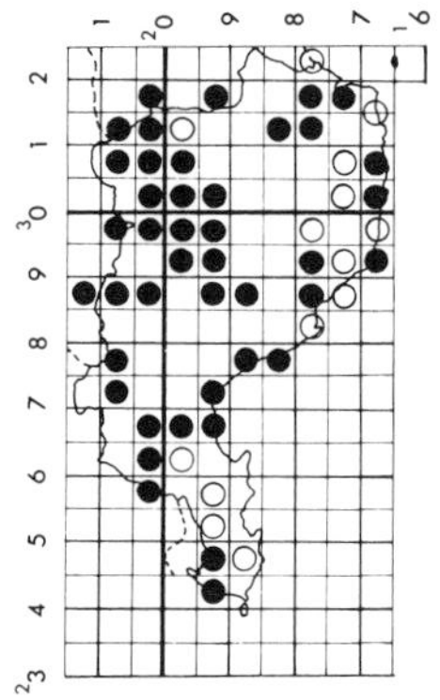

741 Artemisia absinthium

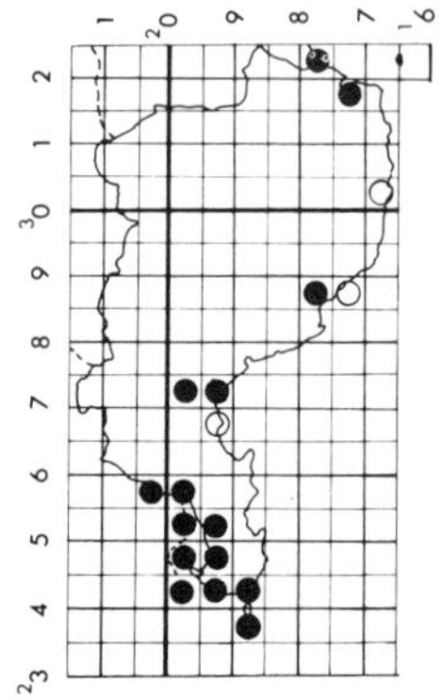

742 Artemisia maritima

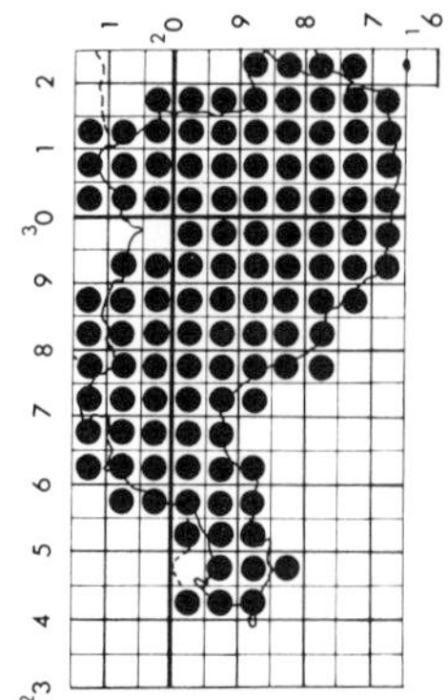

743 Tussilago farfara

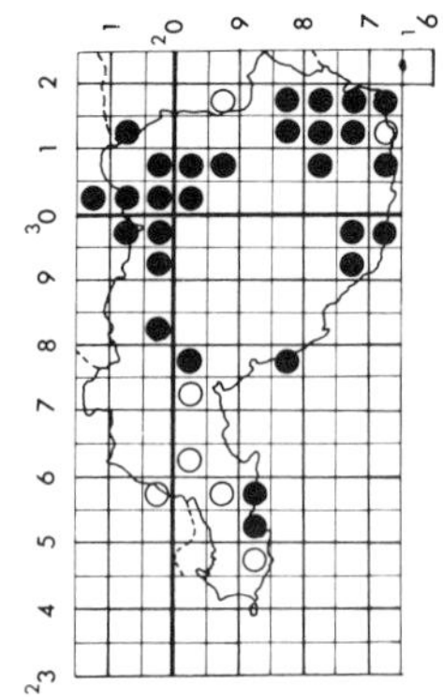

744 Petasites hybridus

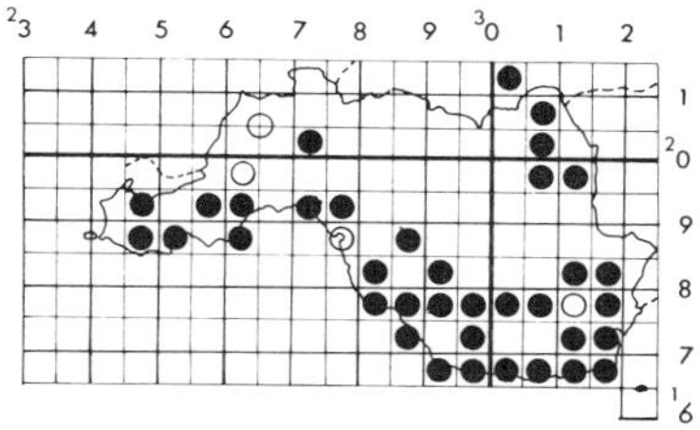

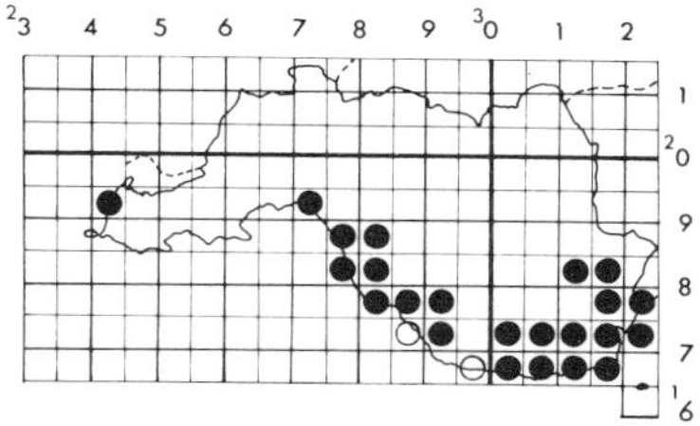

748 Senecio erucifolius

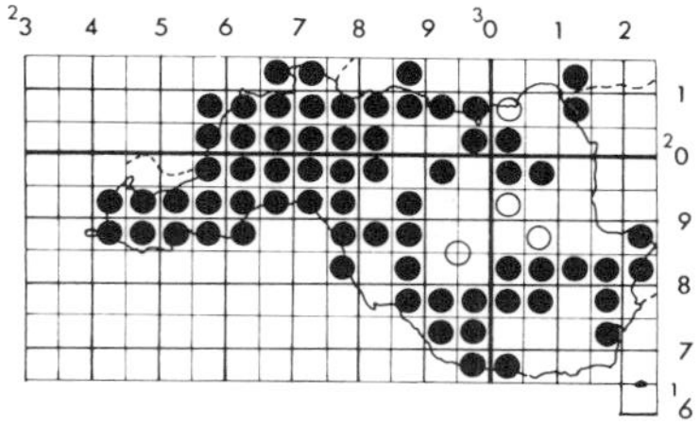

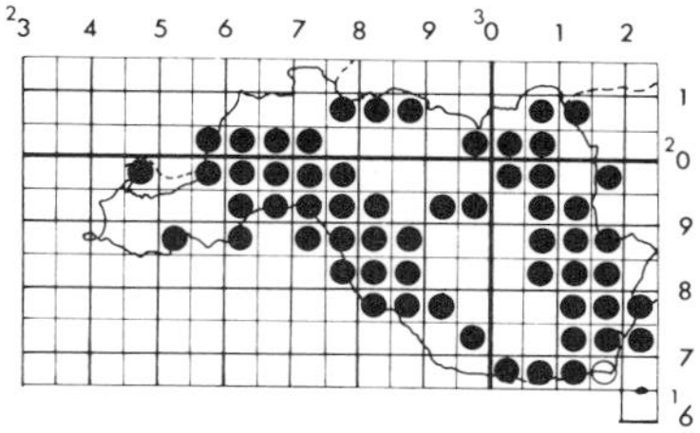

747 Senecio aquaticus

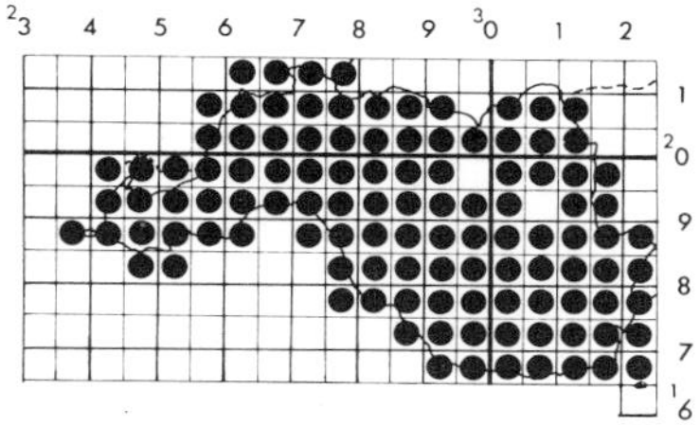

746 Senecio jacobaea

745 Petasites fragrans

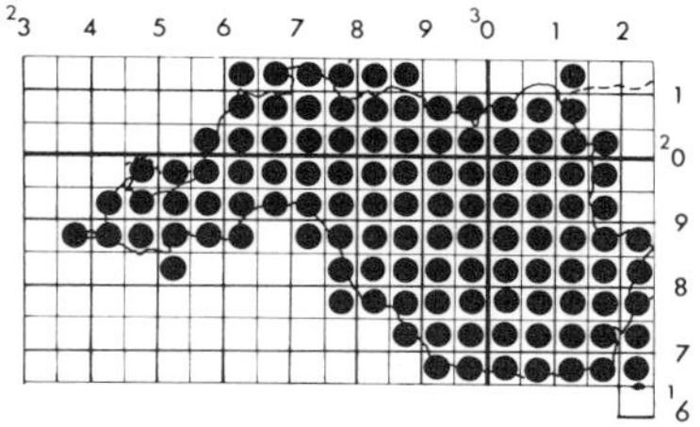

752 Senecio vulgaris

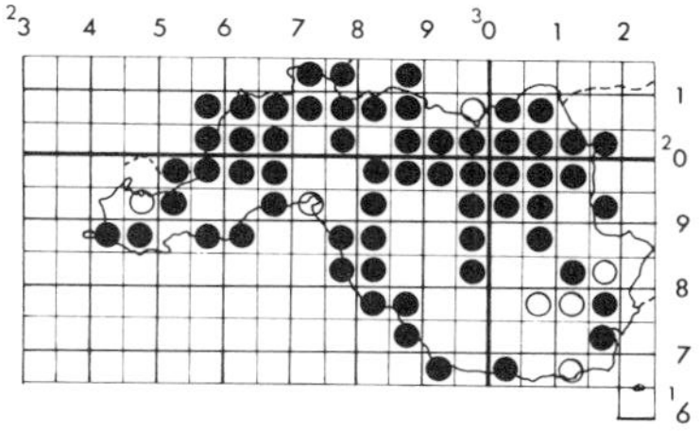

751 Senecio viscosus

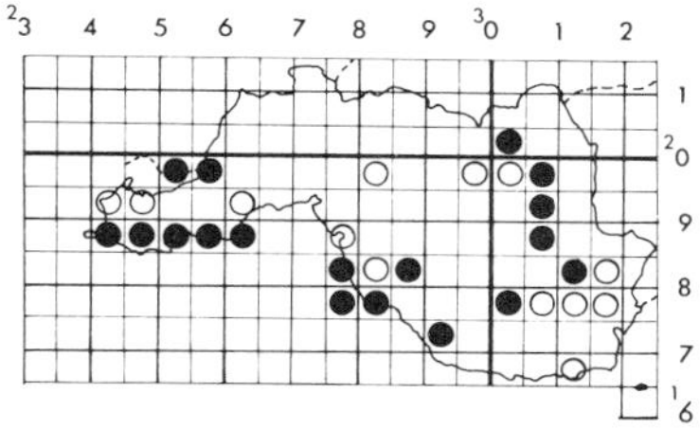

750 Senecio sylvaticus

749 Senecio squalidus

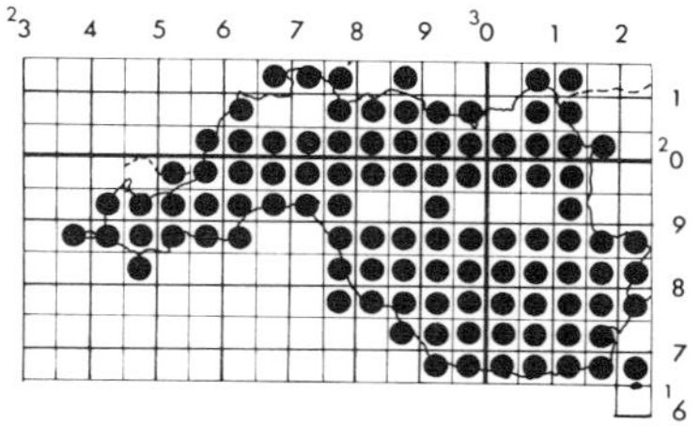

756 Arctium minus s.l.

755 Arctium lappa

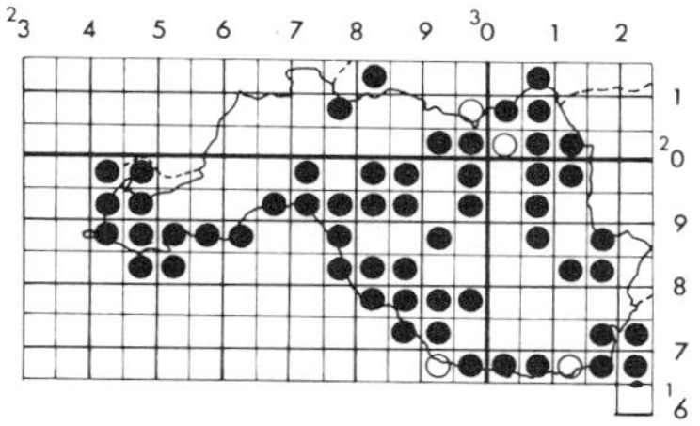

754 Carlina vulgaris

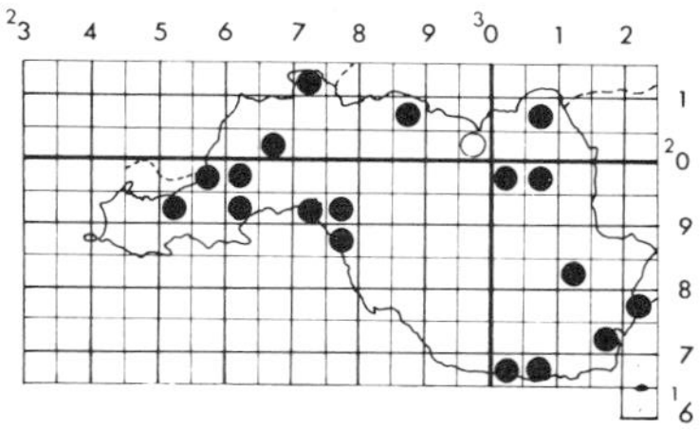

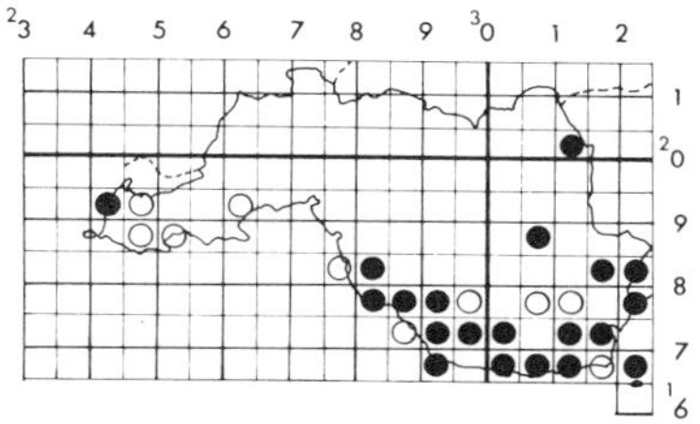

753 Calendula officinalis

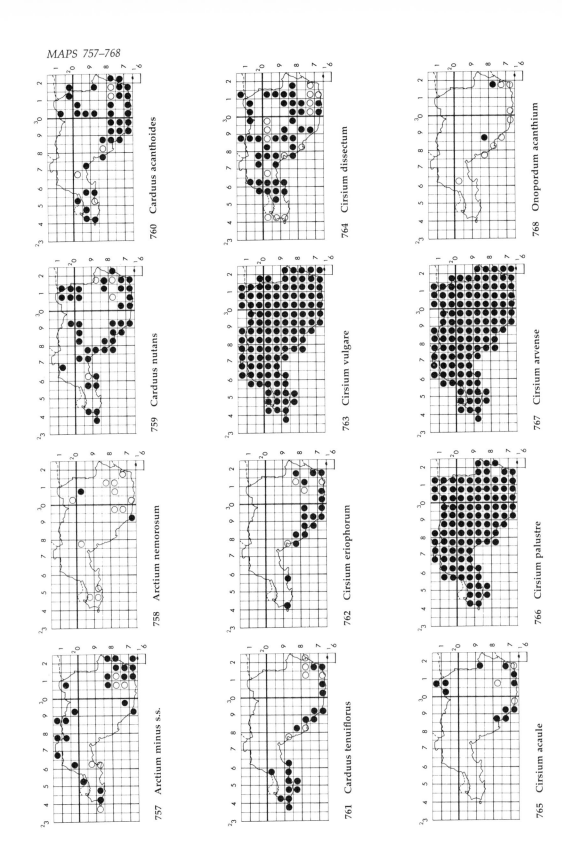

757 Arctium minus s.s.

758 Arctium nemorosum

759 Carduus nutans

760 Carduus acanthoides

761 Carduus tenuiflorus

762 Cirsium eriophorum

763 Cirsium vulgare

764 Cirsium dissectum

765 Cirsium acaule

766 Cirsium palustre

767 Cirsium arvense

768 Onopordum acanthium

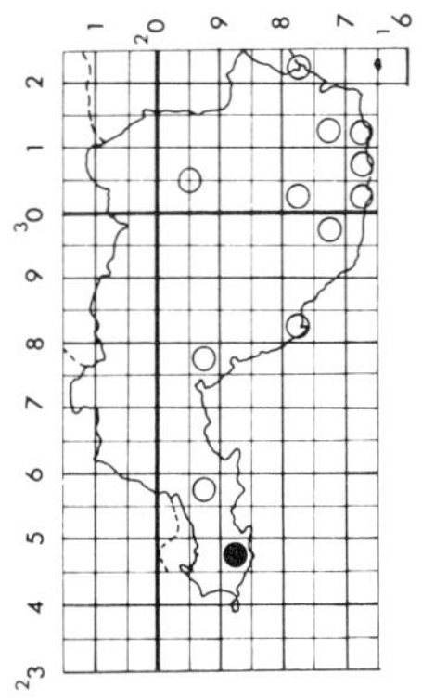

772 Centaurea nigra

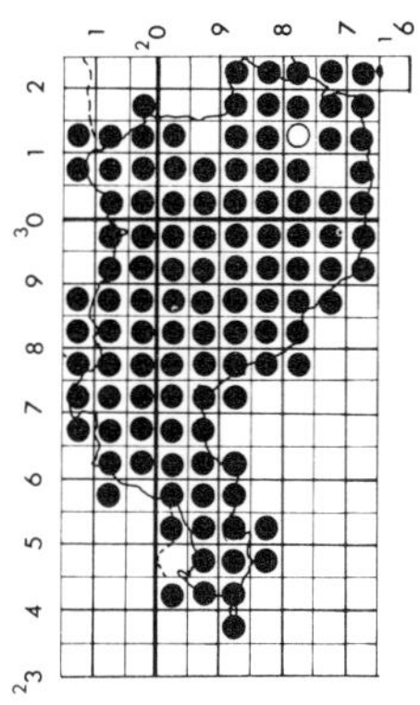

776 Leontodon autumnalis

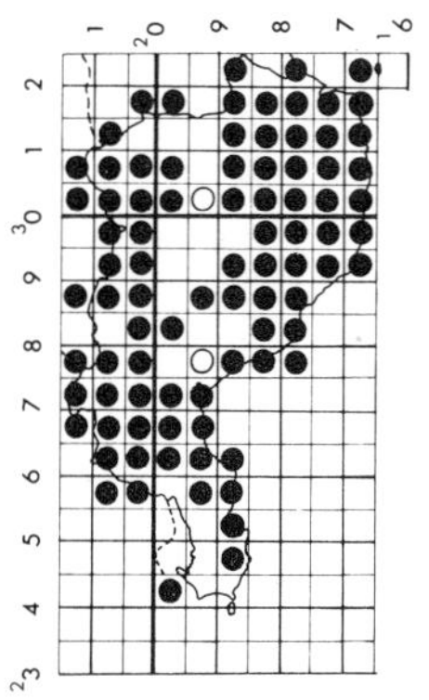

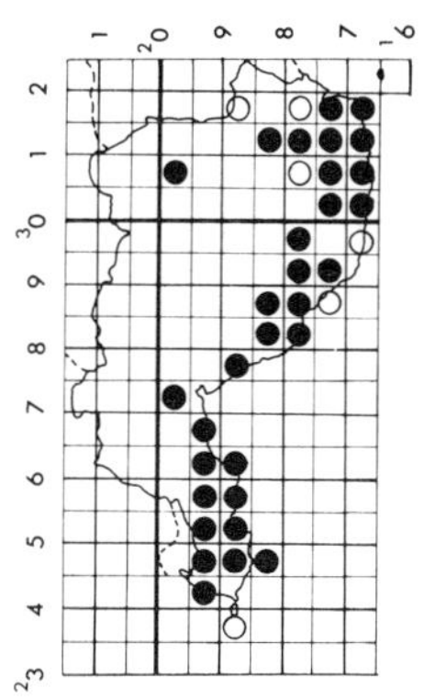

780 Picris hieracioides

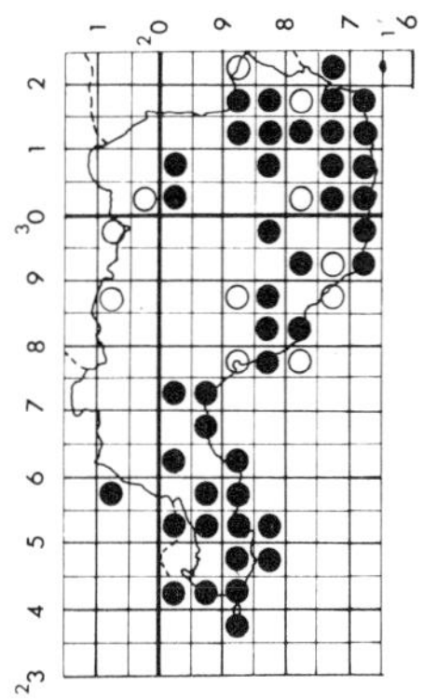

771 Centaurea scabiosa

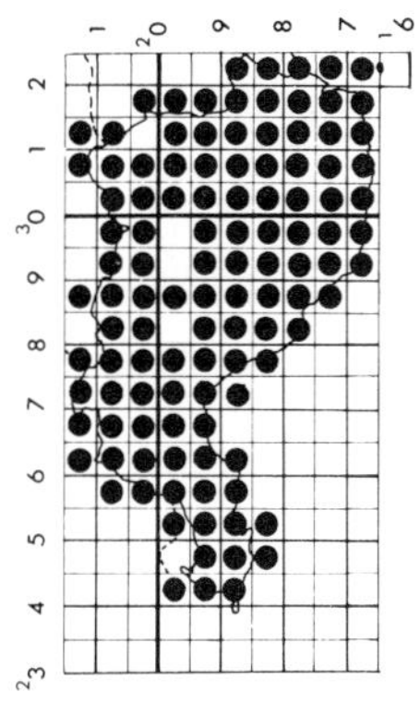

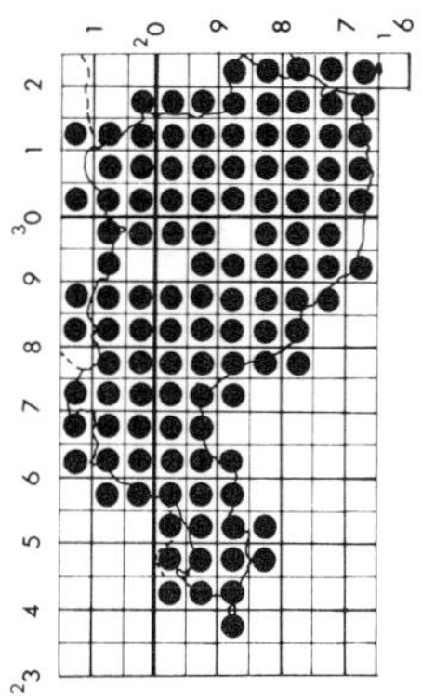

775 Hypochoeris radicata

779 Picris echioides

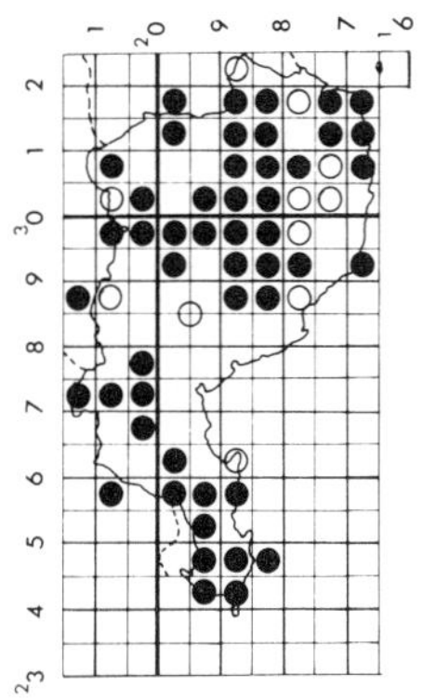

770 Serratula tinctoria

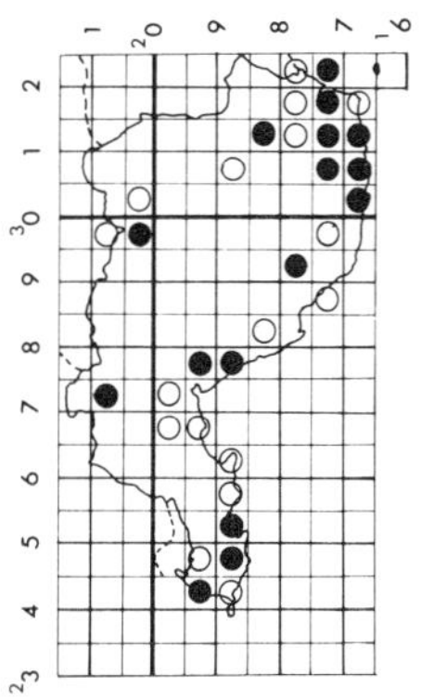

774 Cichorium intybus

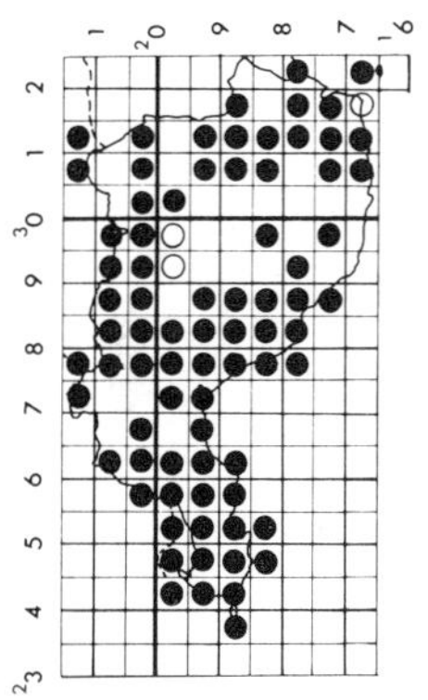

778 Leontodon taraxacoides

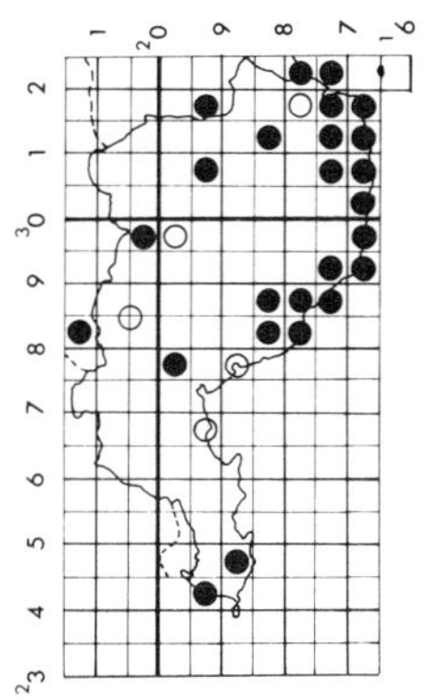

769 Silybum marianum

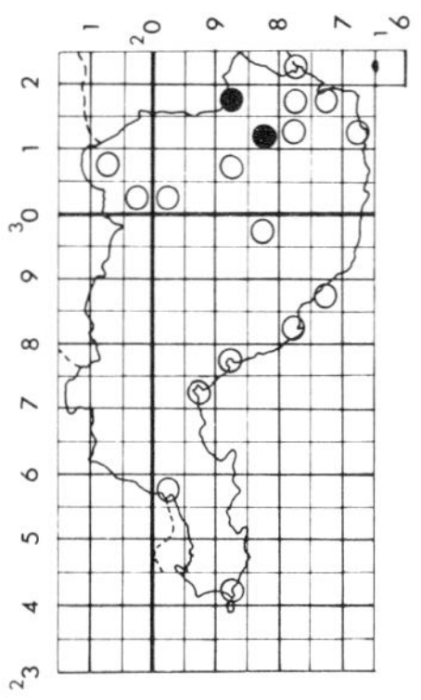

773 Centaurea cyanus

777 Leontodon hispidus

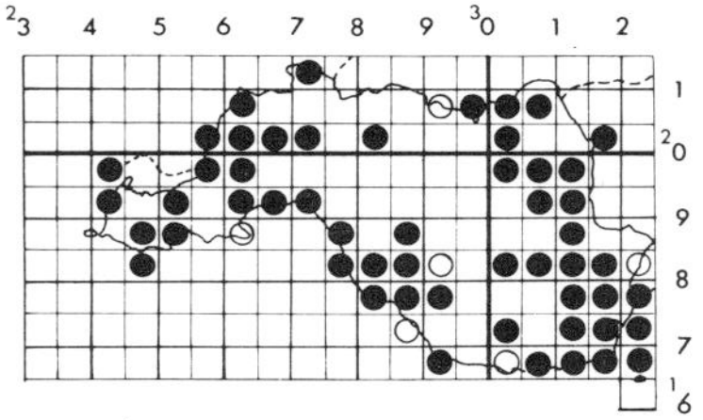

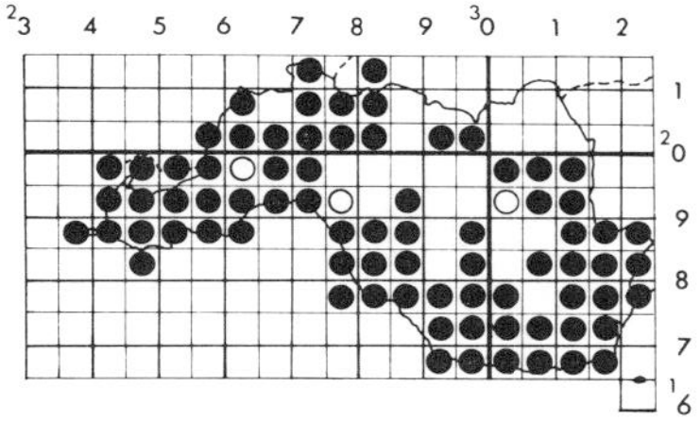

784 Sonchus arvensis

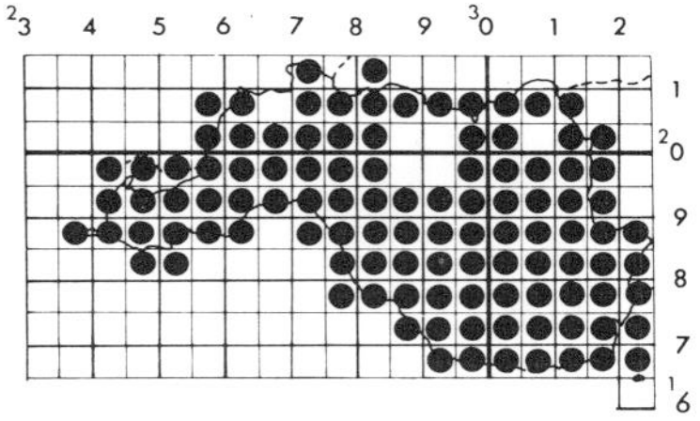

783 Sonchus oleraceus

782 Sonchus asper

781 Tragopogon pratensis
subsp. minor

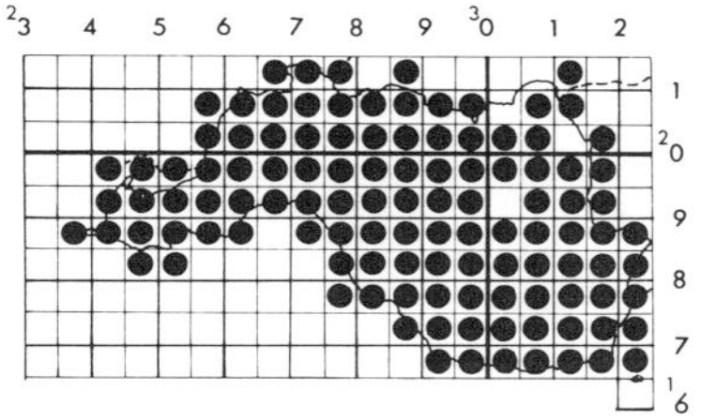

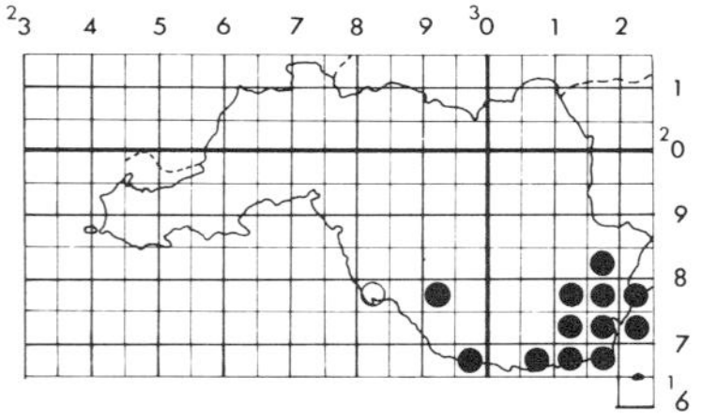

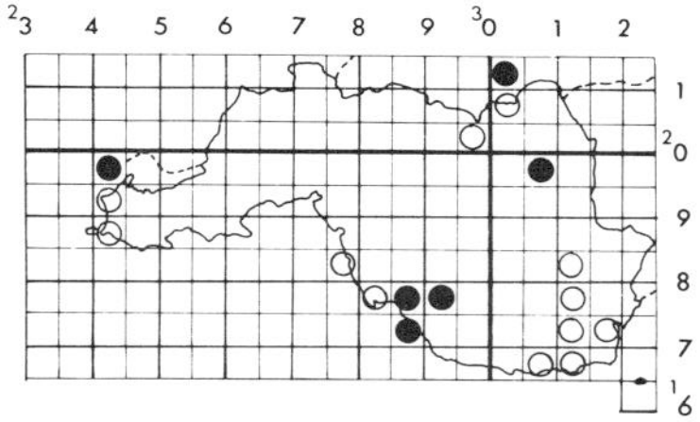

788 Taraxacum oxoniense

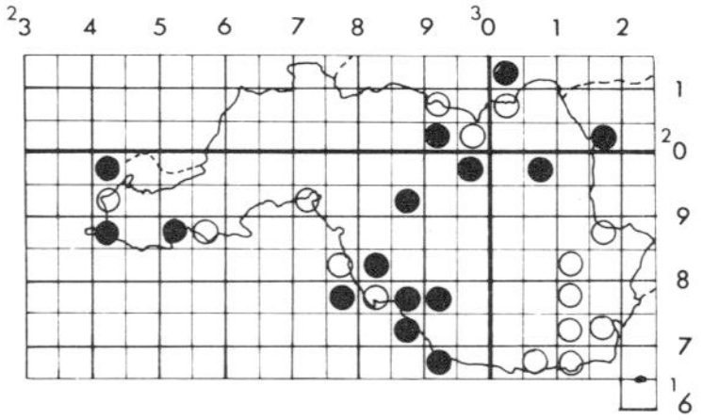

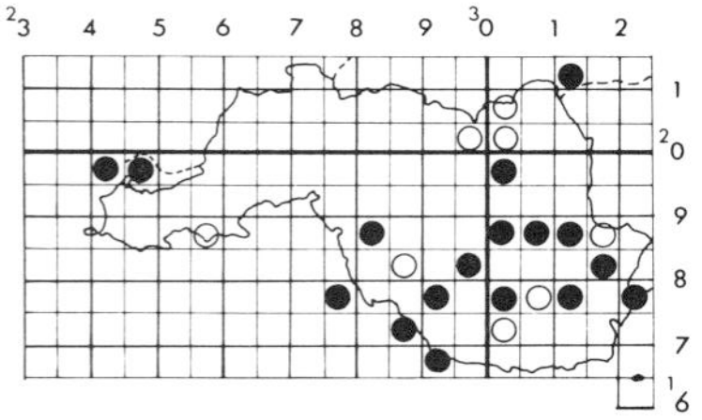

787 Taraxacum
Sect. Erythrosperma

786 Mycelis muralis

785 Lactuca serriola

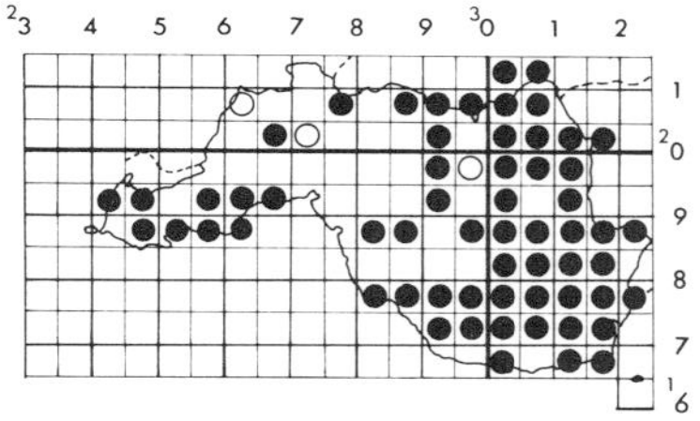

792 Taraxacum hamatum

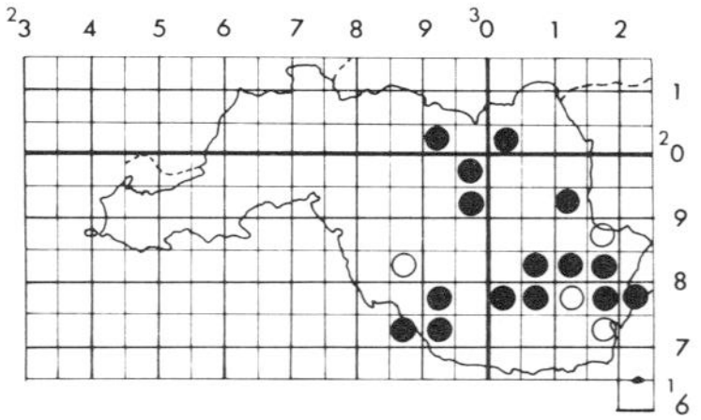

791 Taraxacum Sect. Hamata

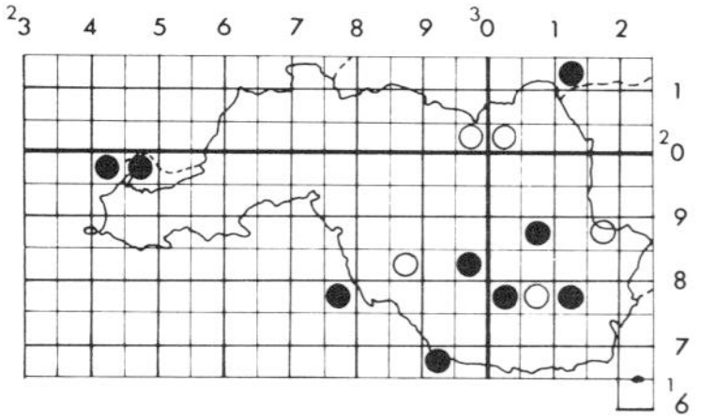

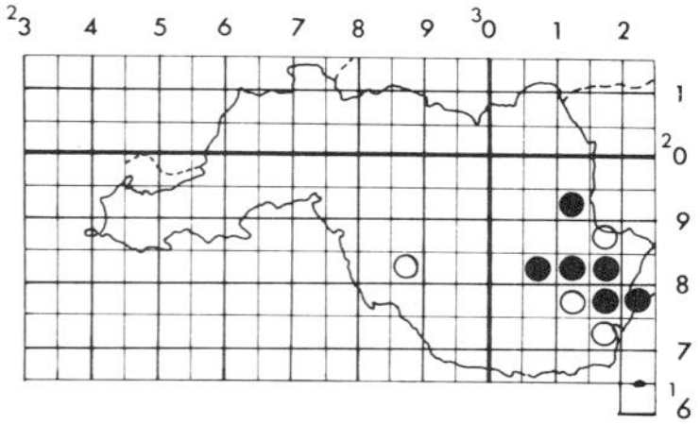

790 Taraxacum nordstedtii

789 Taraxacum Sect. Celtica

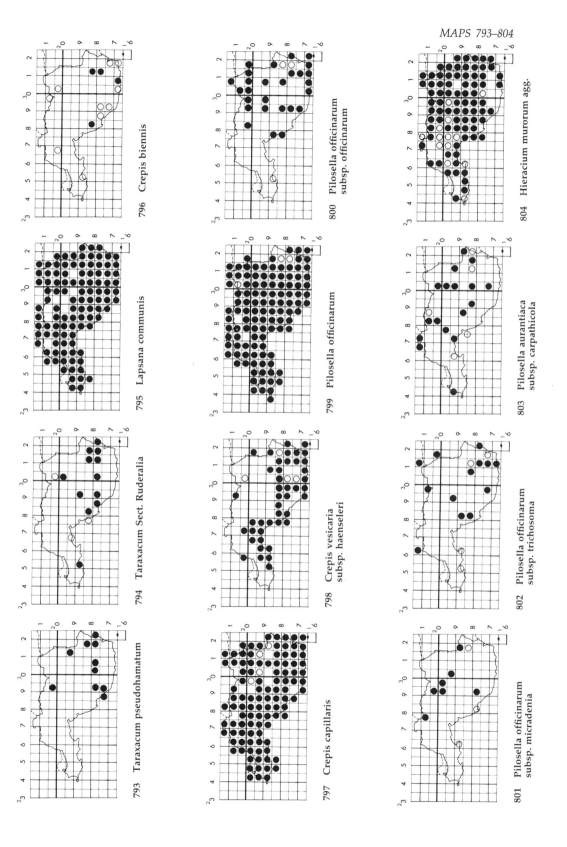

793 Taraxacum pseudohamatum

794 Taraxacum Sect. Ruderalia

795 Lapsana communis

796 Crepis biennis

797 Crepis capillaris

798 Crepis vesicaria subsp. haenseleri

799 Pilosella officinarum

800 Pilosella officinarum subsp. officinarum

801 Pilosella officinarum subsp. micradenia

802 Pilosella officinarum subsp. trichosoma

803 Pilosella aurantiaca subsp. carpathicola

804 Hieracium murorum agg.

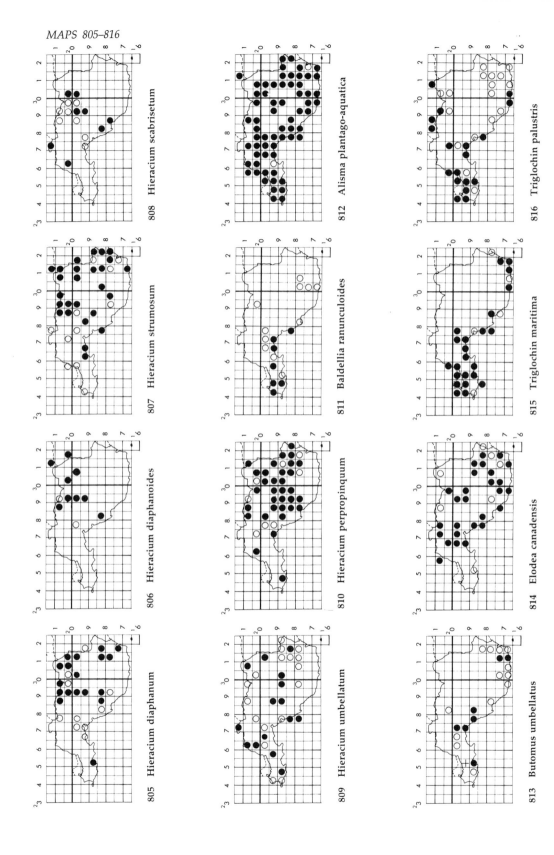

805 Hieracium diaphanum

806 Hieracium diaphanoides

807 Hieracium strumosum

808 Hieracium scabrisetum

809 Hieracium umbellatum

810 Hieracium perpropinquum

811 Baldellia ranunculoides

812 Alisma plantago-aquatica

813 Butomus umbellatus

814 Elodea canadensis

815 Triglochin maritima

816 Triglochin palustris

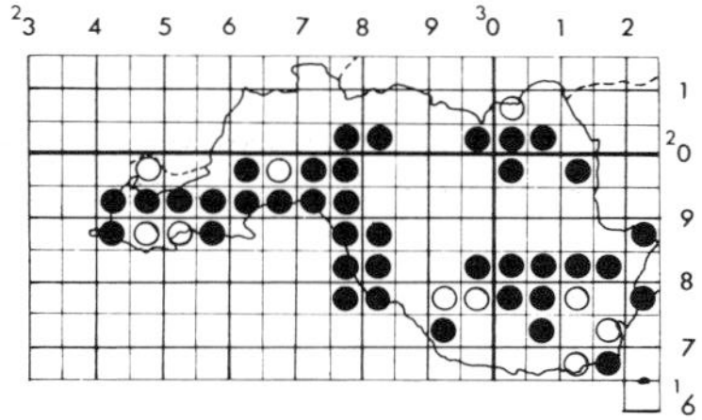

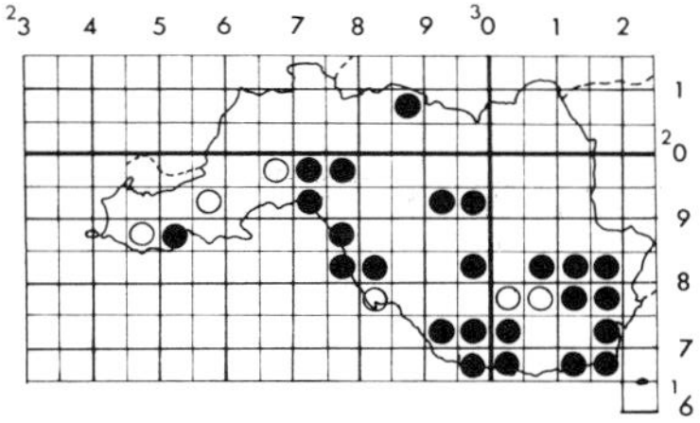

820 Potamogeton crispus

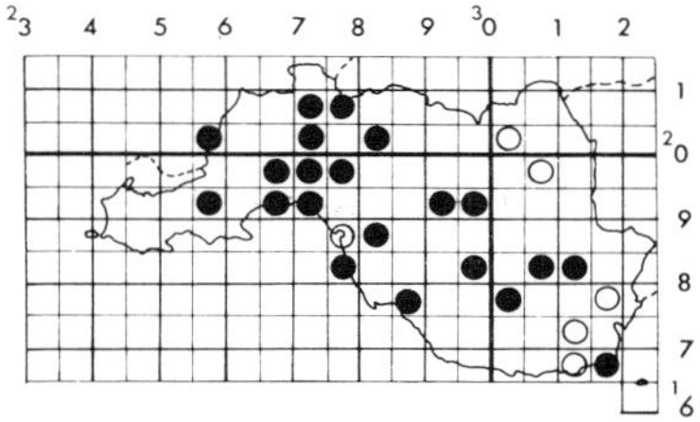

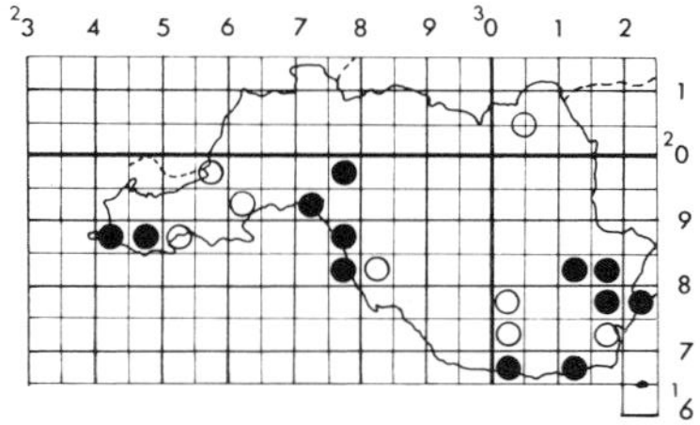

819 Potamogeton berchtoldii

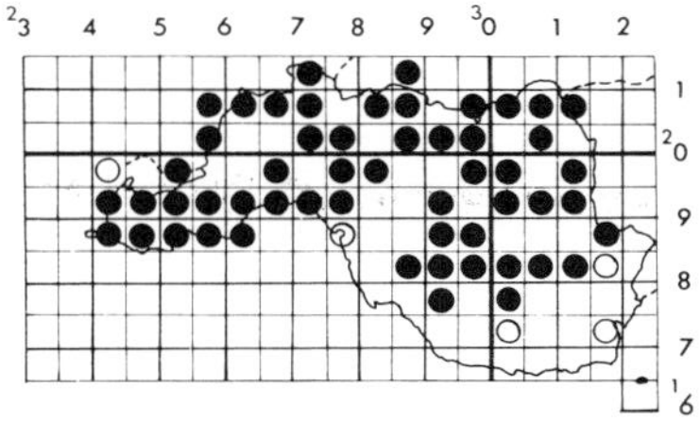

818 Potamogeton polygonifolius

817 Potamogeton natans

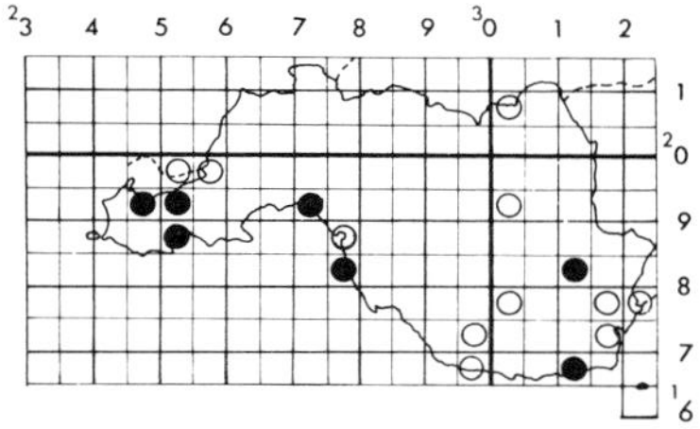

824 Zannichellia palustris

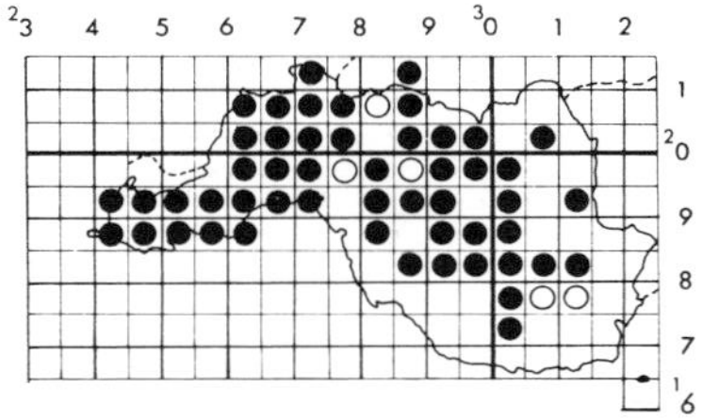

823 Ruppia maritima

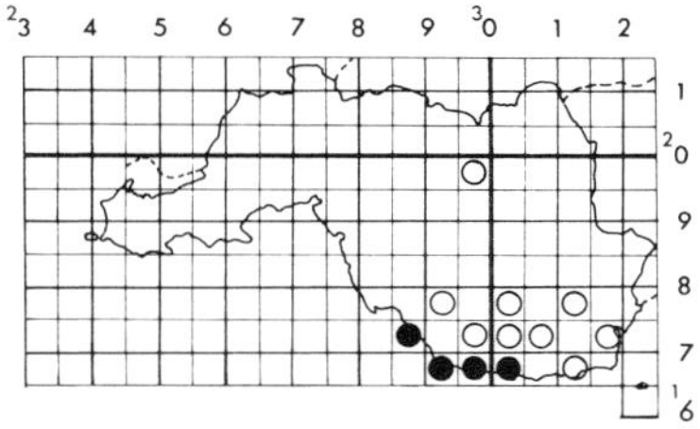

822 Groenlandia densa

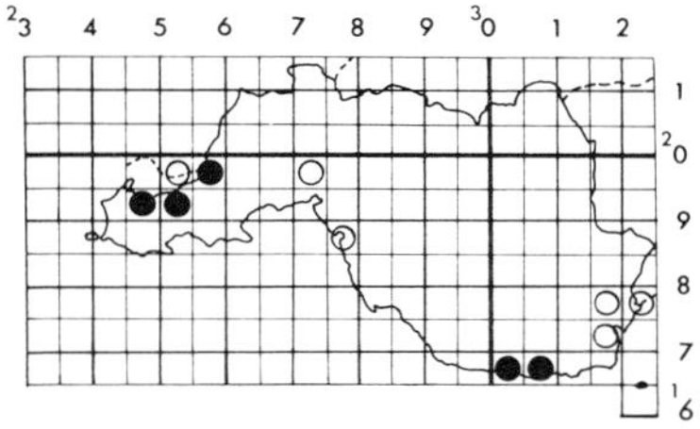

821 Potamogeton pectinatus

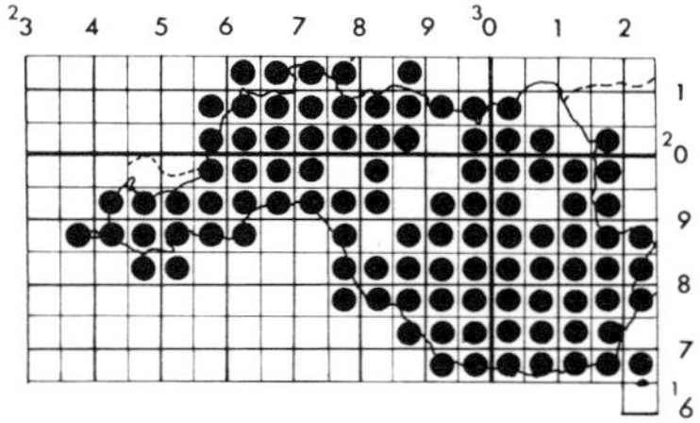

828 Hyacinthoides non-scripta

827 Scilla verna

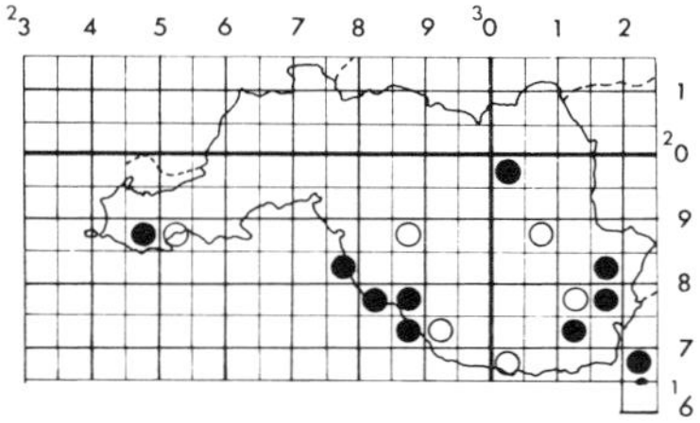

826 Ornithogalum umbellatum

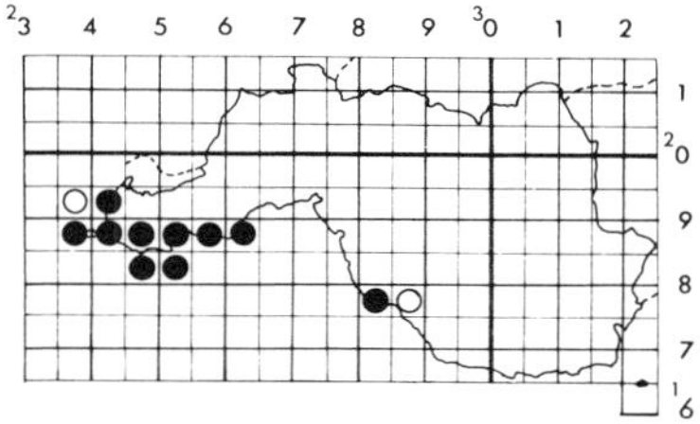

825 Narthecium ossifragum

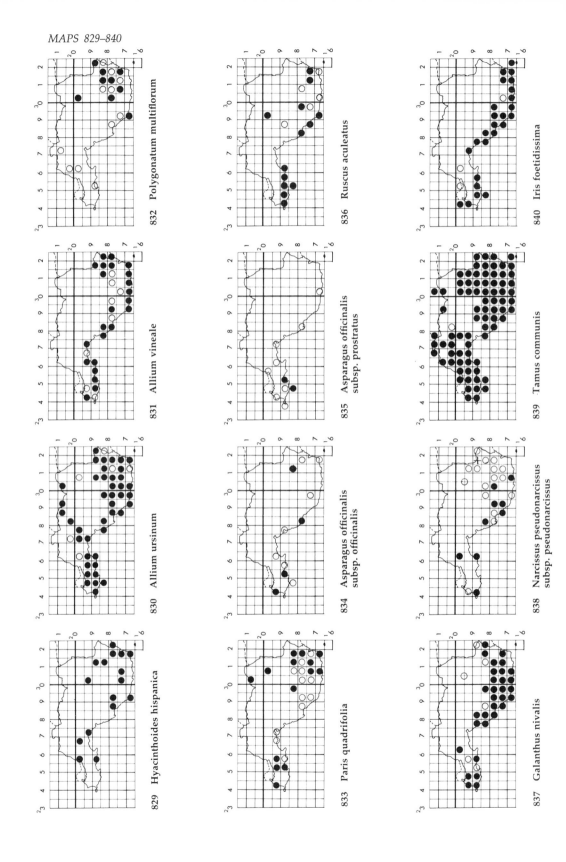

832 Polygonatum multiflorum

831 Allium vineale

830 Allium ursinum

829 Hyacinthoides hispanica

836 Ruscus aculeatus

835 Asparagus officinalis subsp. prostratus

834 Asparagus officinalis subsp. officinalis

833 Paris quadrifolia

840 Iris foetidissima

839 Tamus communis

838 Narcissus pseudonarcissus subsp. pseudonarcissus

837 Galanthus nivalis

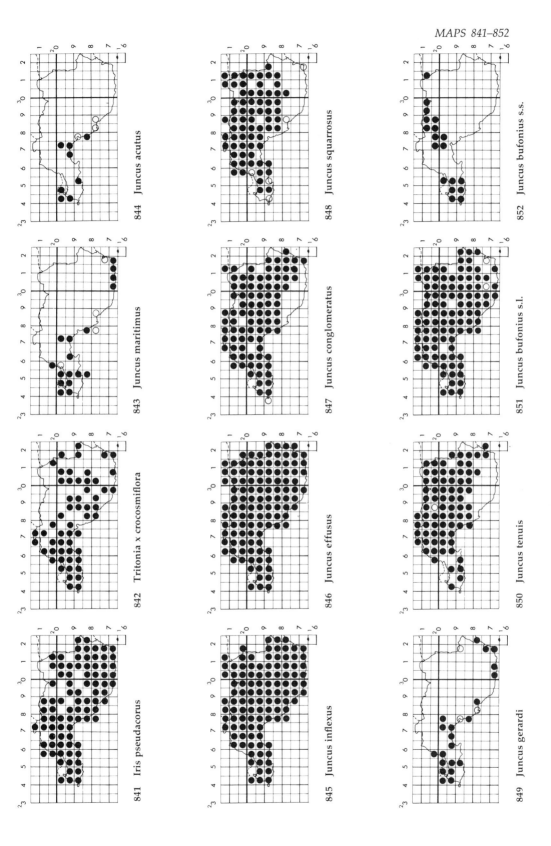

841 Iris pseudacorus

842 Tritonia x crocosmiflora

843 Juncus maritimus

844 Juncus acutus

845 Juncus inflexus

846 Juncus effusus

847 Juncus conglomeratus

848 Juncus squarrosus

849 Juncus gerardi

850 Juncus tenuis

851 Juncus bufonius s.l.

852 Juncus bufonius s.s.

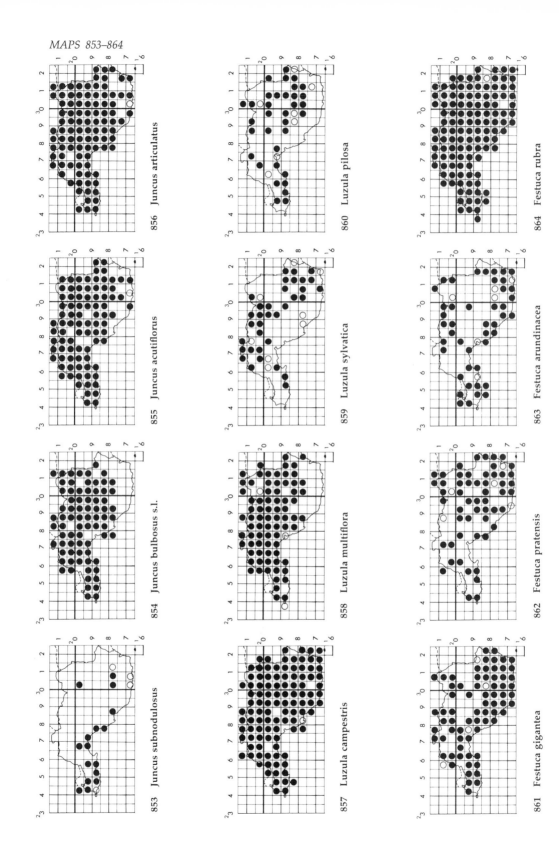

853 Juncus subnodulosus
854 Juncus bulbosus s.l.
855 Juncus acutiflorus
856 Juncus articulatus
857 Luzula campestris
858 Luzula multiflora
859 Luzula sylvatica
860 Luzula pilosa
861 Festuca gigantea
862 Festuca pratensis
863 Festuca arundinacea
864 Festuca rubra

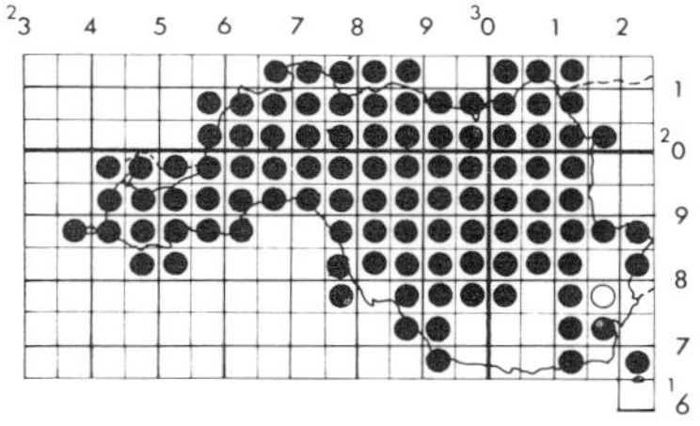

865 Festuca ovina

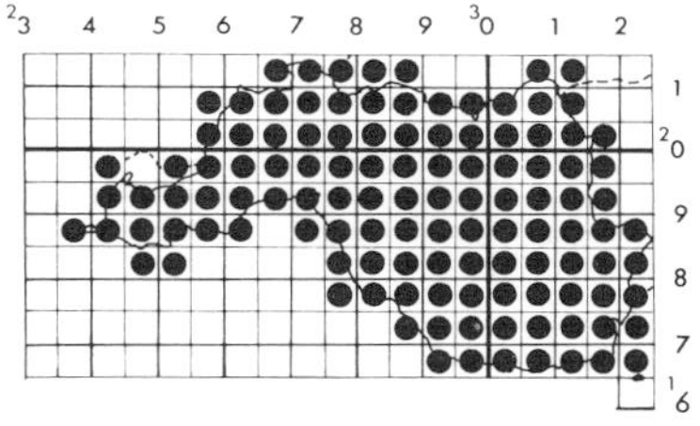

866 Lolium perenne

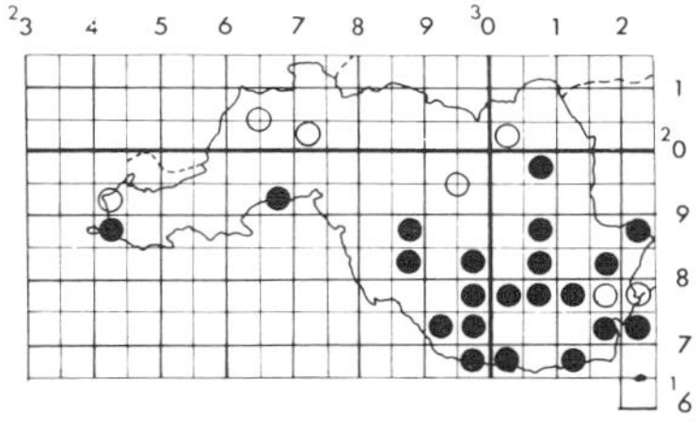

867 Lolium multiflorum

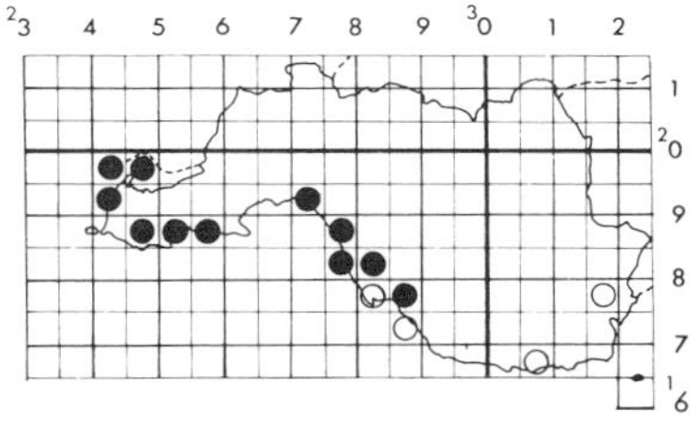

868 Vulpia fasciculata

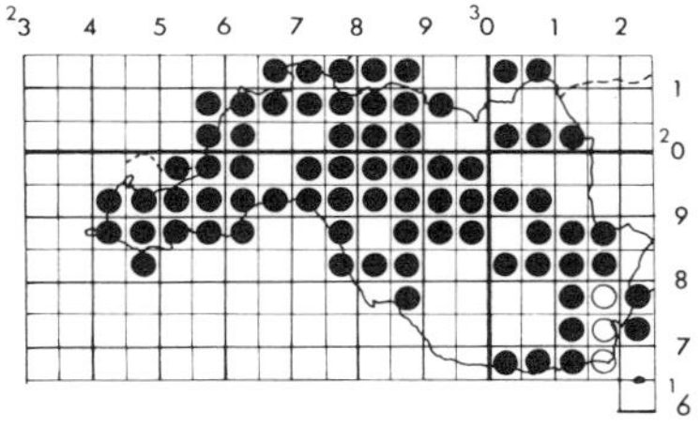

869 Vulpia bromoides

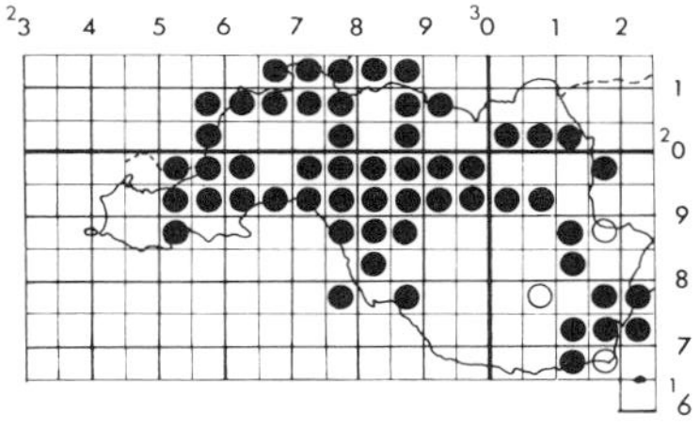

870 Vulpia myuros

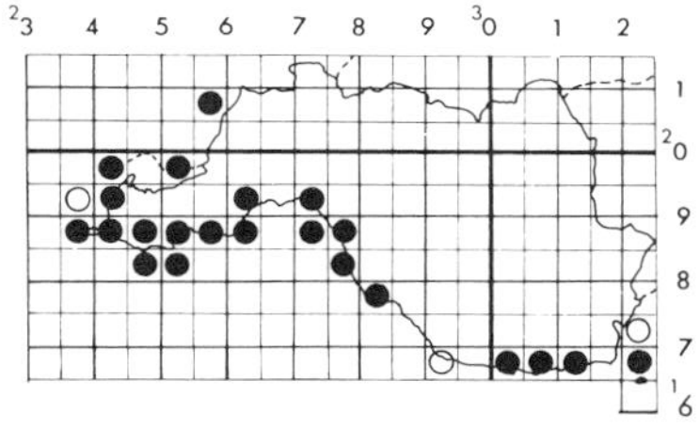

871 Desmazeria marina

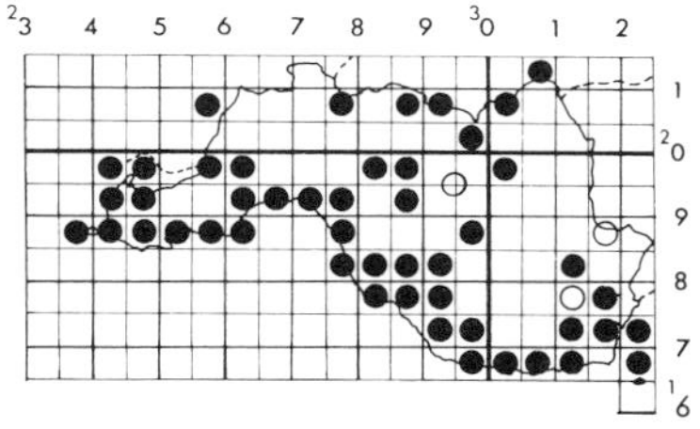

972 Desmazeria rigida

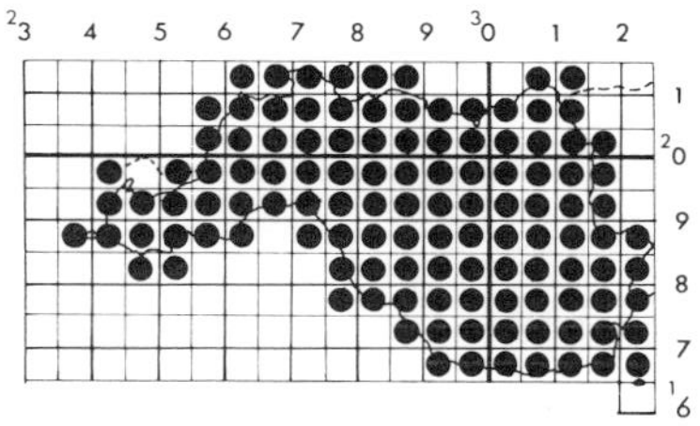

873 Poa annua

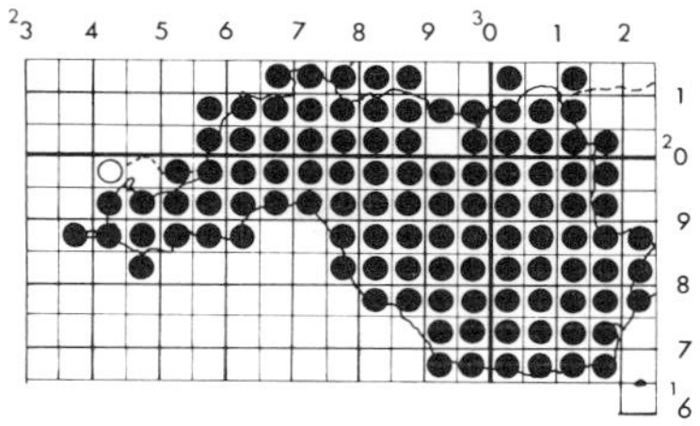

874 Poa trivialis

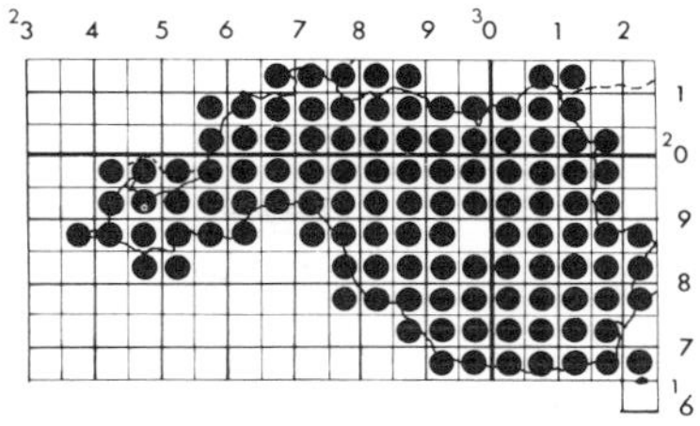

875 Poa pratensis s.l.

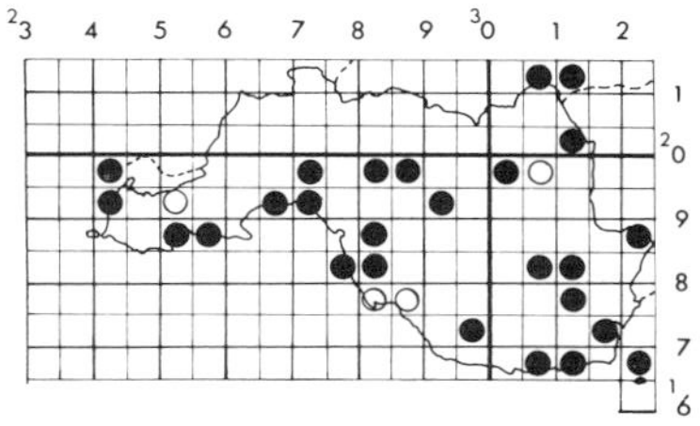

876 Poa subcaerulea

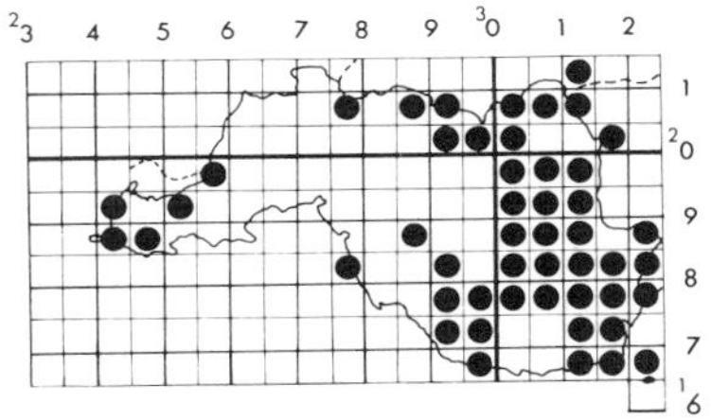

880 Poa nemoralis

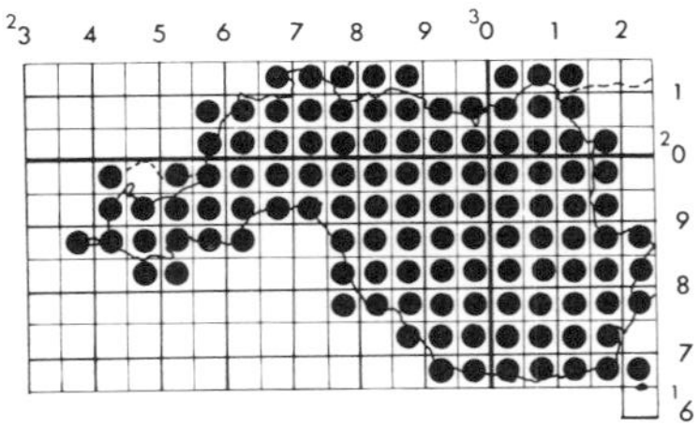

884 Cynosurus cristatus

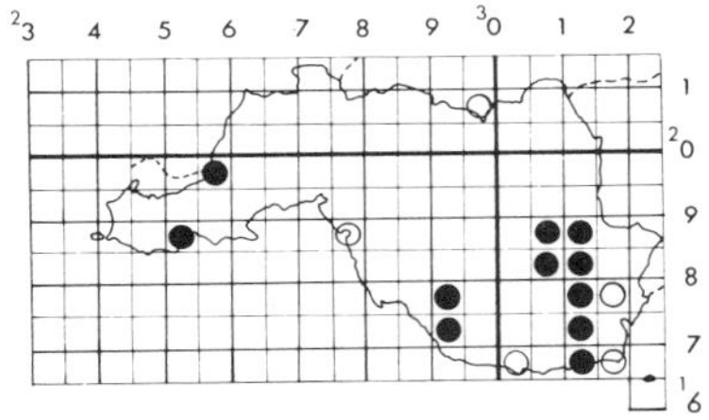

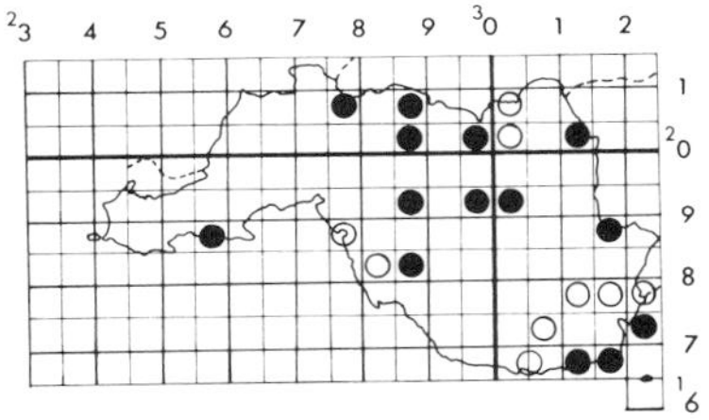

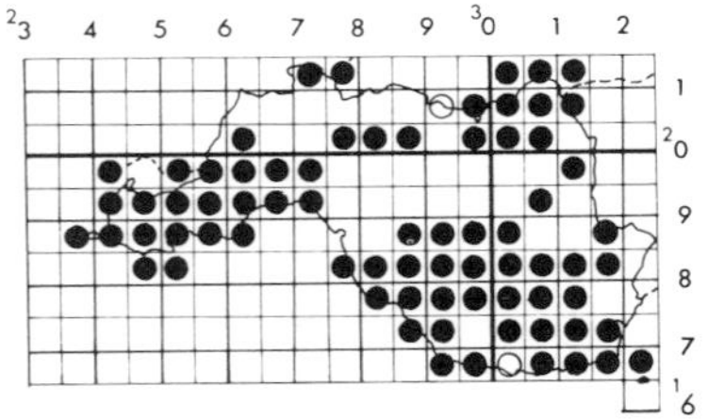

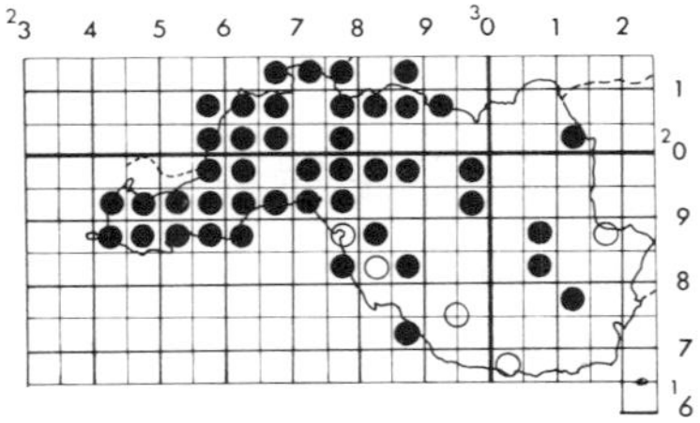

888 Glyceria declinata

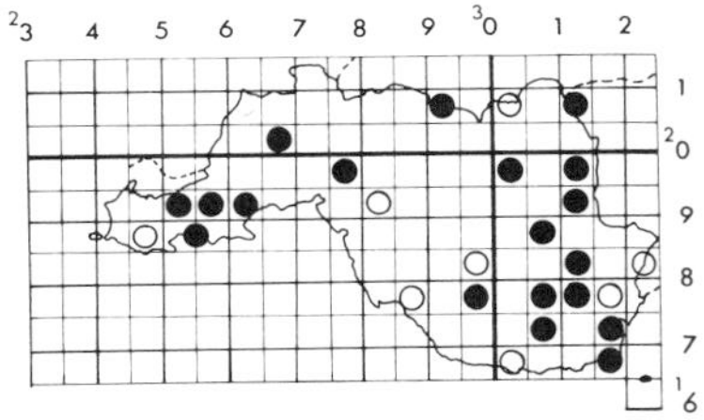

879 Poa compressa

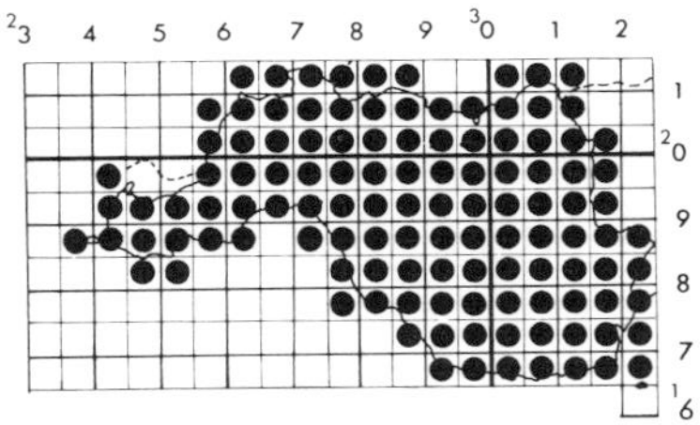

883 Dactylis glomerata

887 Glyceria maxima

878 Poa angustifolia

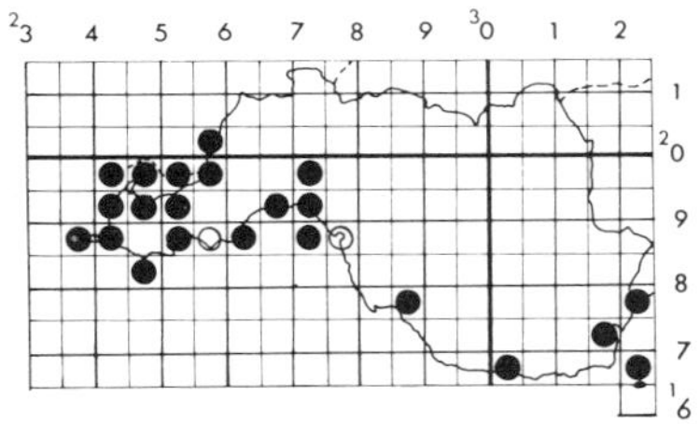

882 Puccinellia maritima

886 Melica uniflora

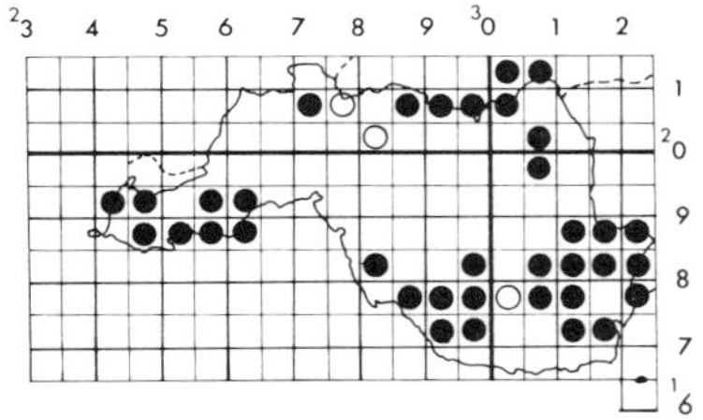

877 Poa pratensis s.s.

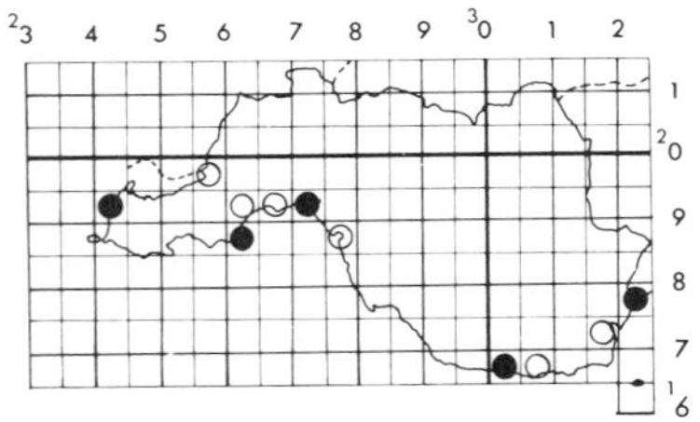

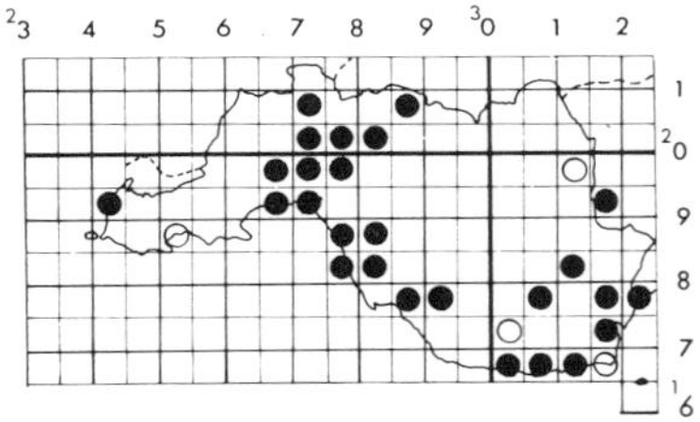

881 Puccinellia distans

885 Briza media

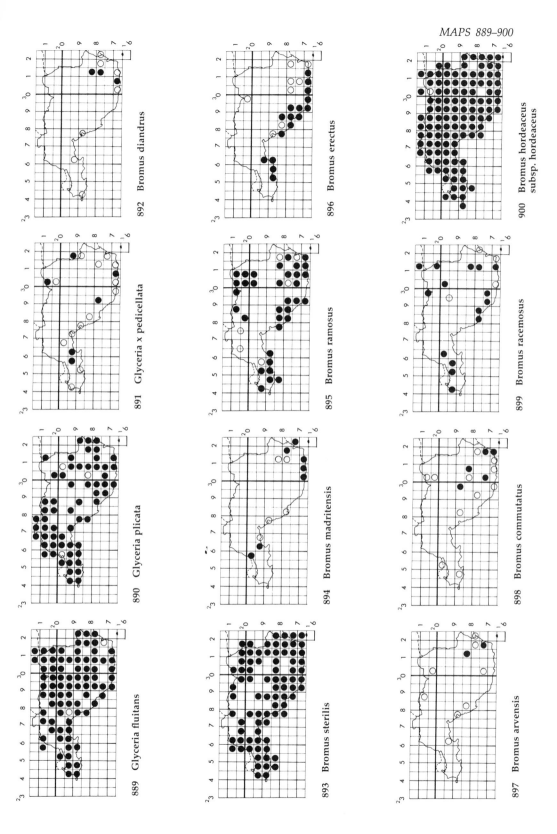

889 Glyceria fluitans

890 Glyceria plicata

891 Glyceria x pedicellata

892 Bromus diandrus

893 Bromus sterilis

894 Bromus madritensis

895 Bromus ramosus

896 Bromus erectus

897 Bromus arvensis

898 Bromus commutatus

899 Bromus racemosus

900 Bromus hordeaceus subsp. hordeaceus

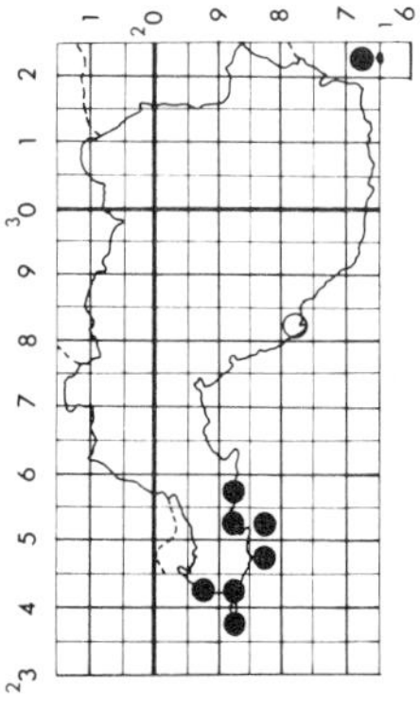

901 Bromus hordeaceus
subsp. ferronii

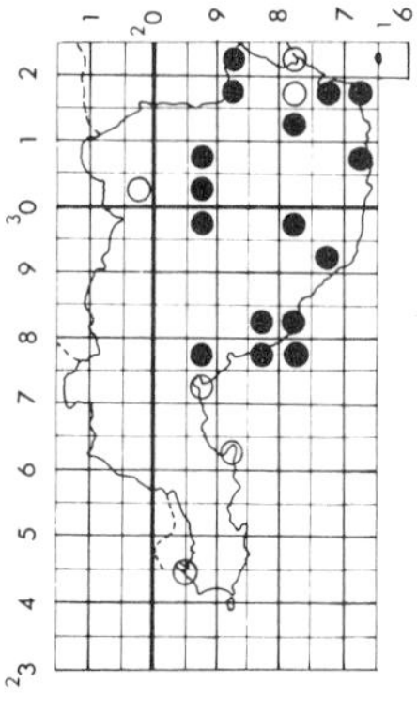

902 Bromus x pseudothominii

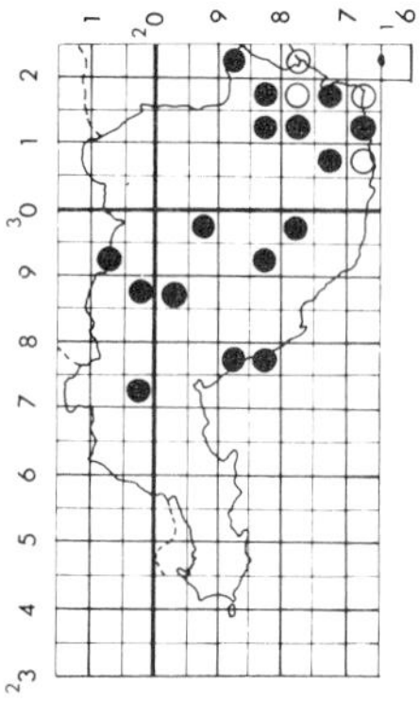

903 Bromus lepidus

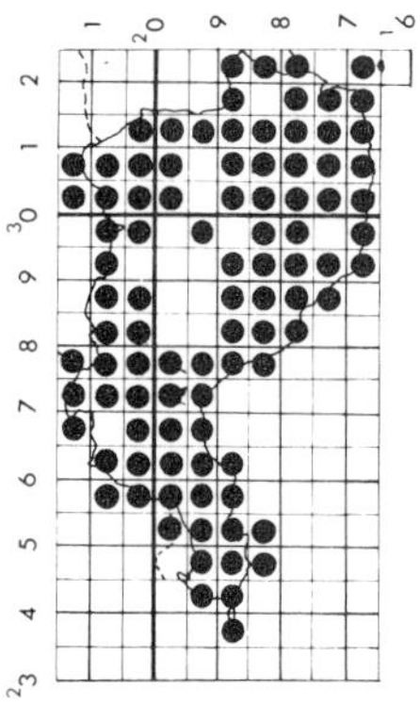

904 Brachypodium sylvaticum

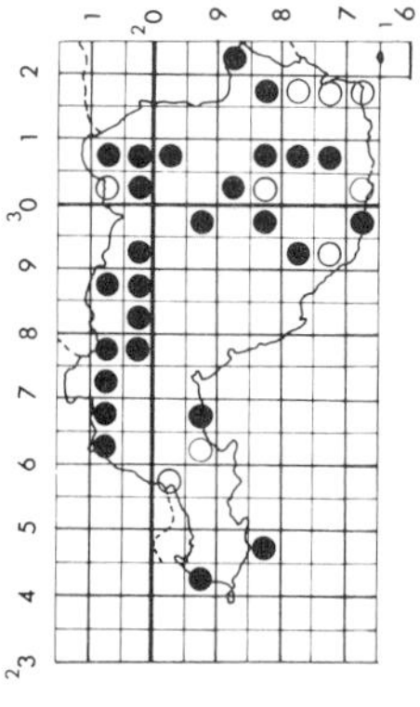

905 Elymus caninus

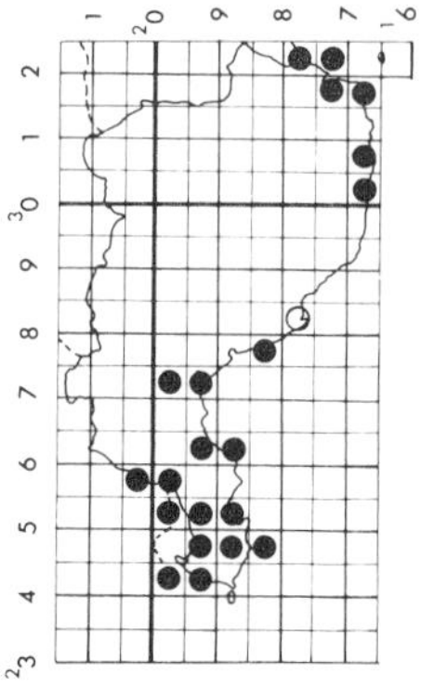

906 Elymus pycnanthus

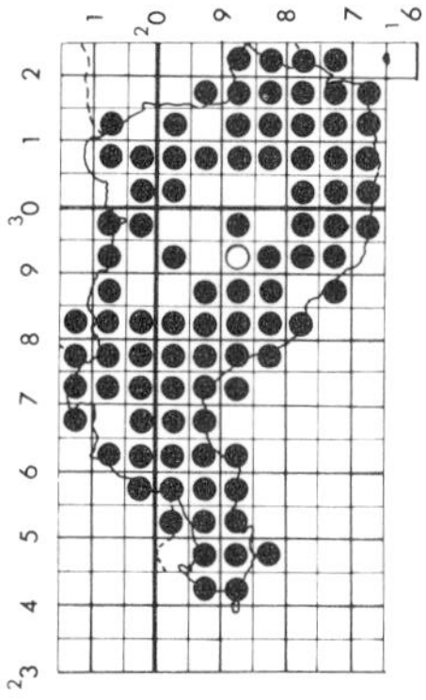

907 Elymus repens

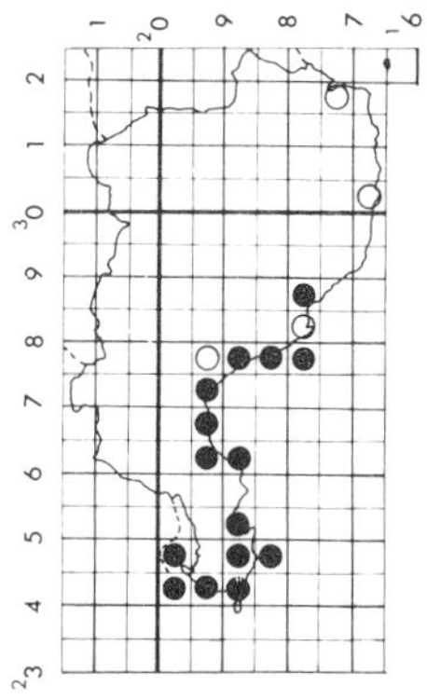

908 Elymus farctus subsp.
boreali-atlanticus

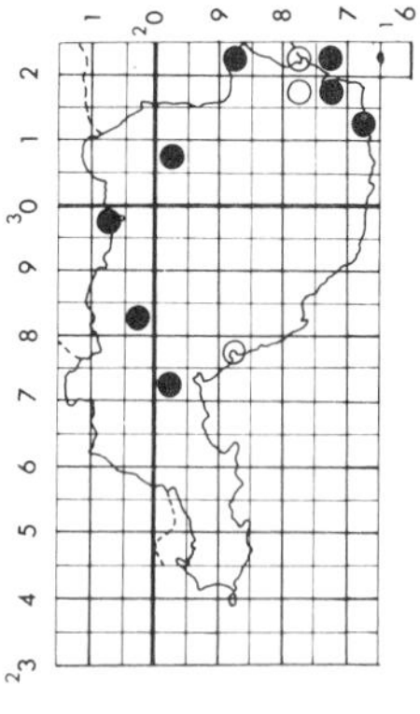

909 Triticum aestivum

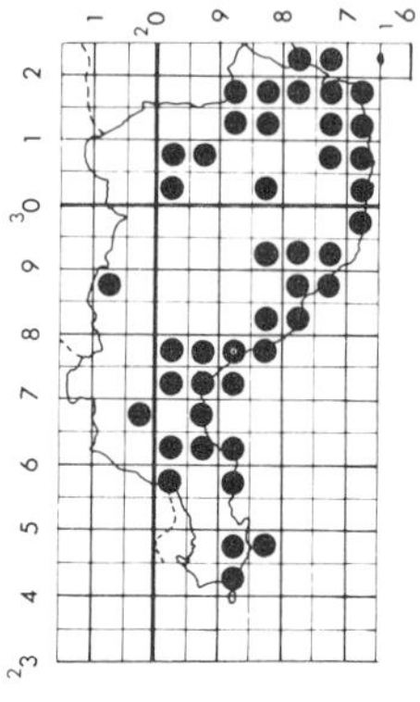

910 Hordeum murinum

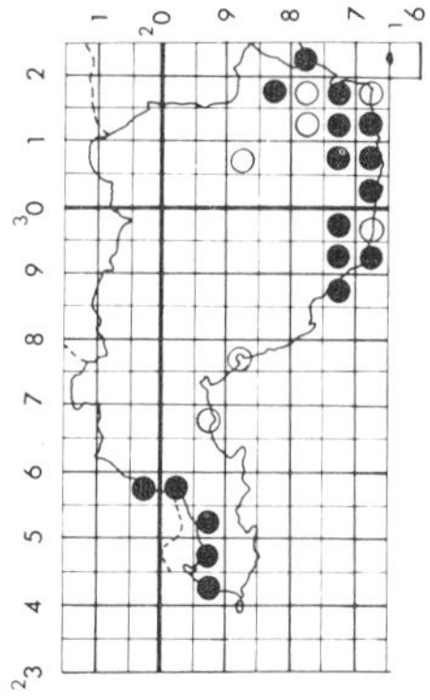

911 Hordeum secalinum

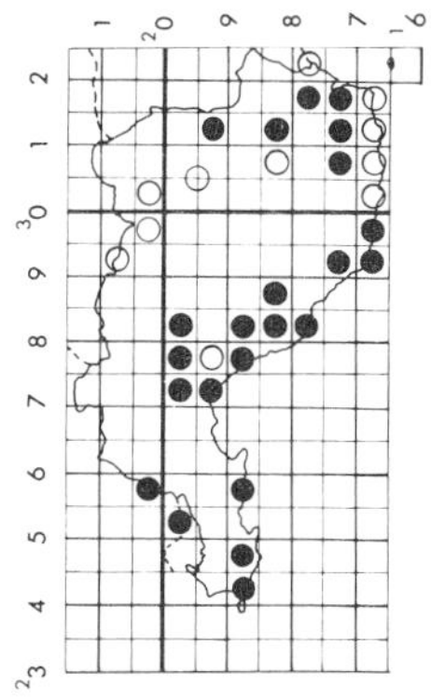

912 Avena fatua

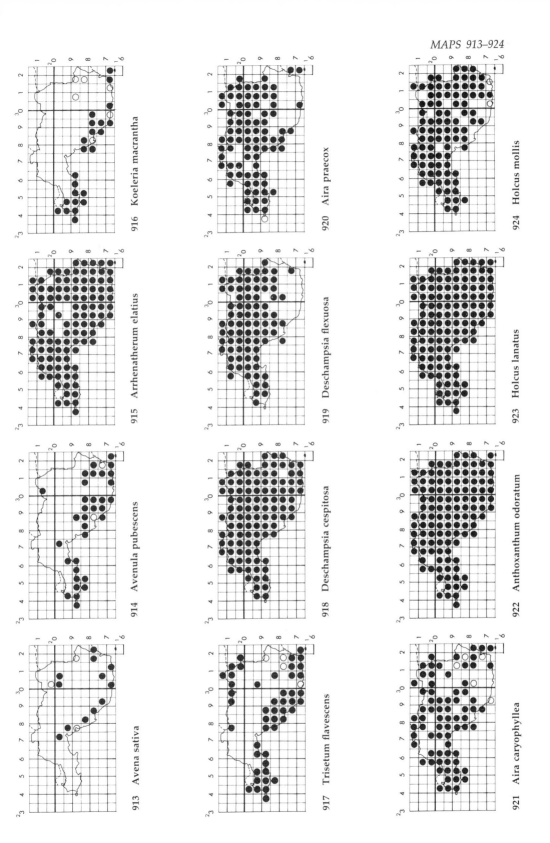

913 Avena sativa

914 Avenula pubescens

915 Arrhenatherum elatius

916 Koeleria macrantha

917 Trisetum flavescens

918 Deschampsia cespitosa

919 Deschampsia flexuosa

920 Aira praecox

921 Aira caryophyllea

922 Anthoxanthum odoratum

923 Holcus lanatus

924 Holcus mollis

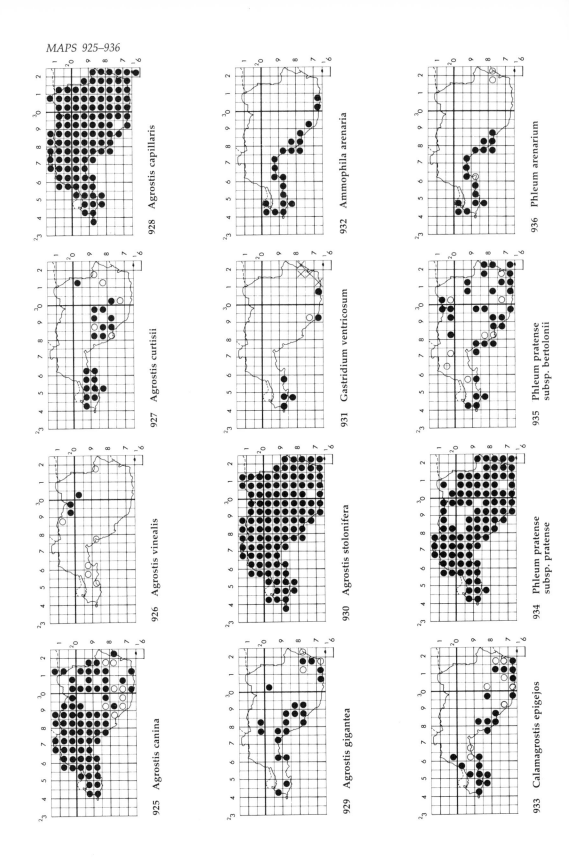

928 Agrostis capillaris

932 Ammophila arenaria

936 Phleum arenarium

927 Agrostis curtisii

931 Gastridium ventricosum

935 Phleum pratense
subsp. bertolonii

926 Agrostis vinealis

930 Agrostis stolonifera

934 Phleum pratense
subsp. pratense

925 Agrostis canina

929 Agrostis gigantea

933 Calamagrostis epigejos

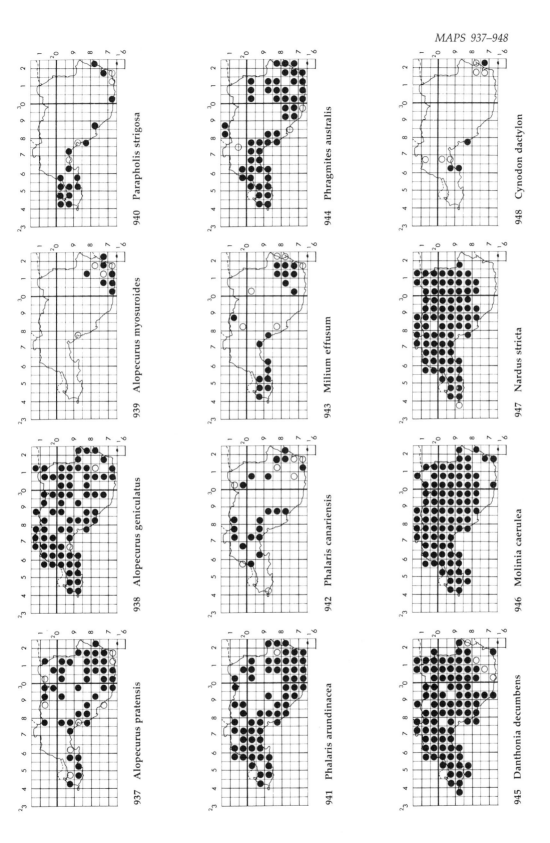

937 Alopecurus pratensis

938 Alopecurus geniculatus

939 Alopecurus myosuroides

940 Parapholis strigosa

941 Phalaris arundinacea

942 Phalaris canariensis

943 Milium effusum

944 Phragmites australis

945 Danthonia decumbens

946 Molinia caerulea

947 Nardus stricta

948 Cynodon dactylon

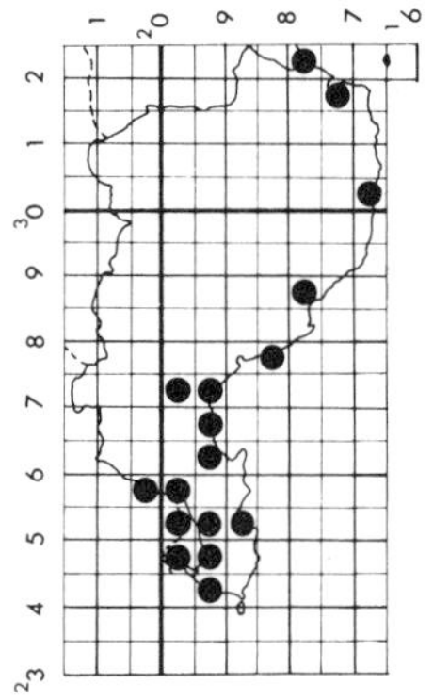

952 Arum maculatum

951 Setaria viridis

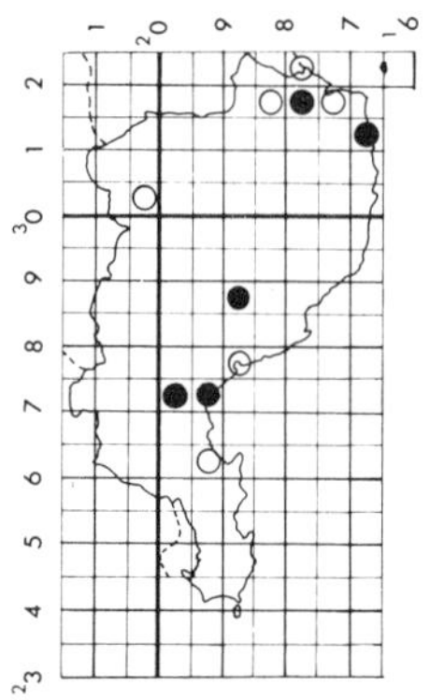

950 Echinochloa crus-galli

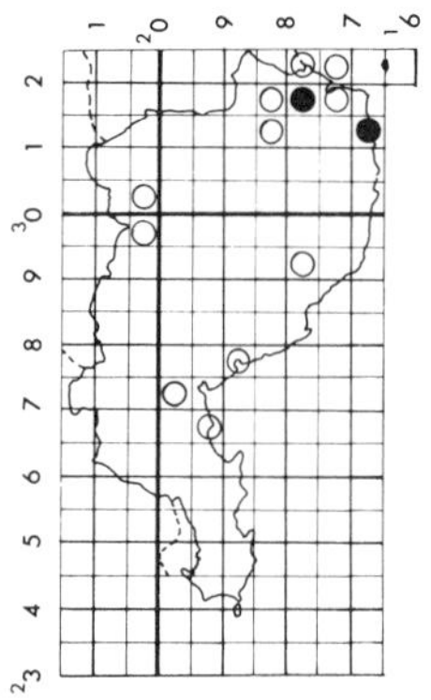

949 Spartina anglica

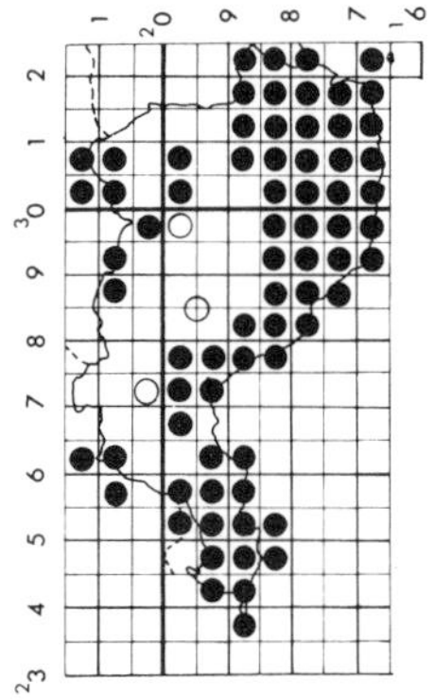

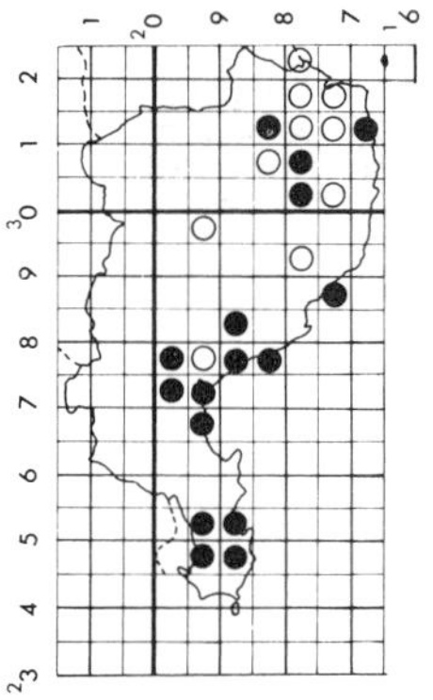

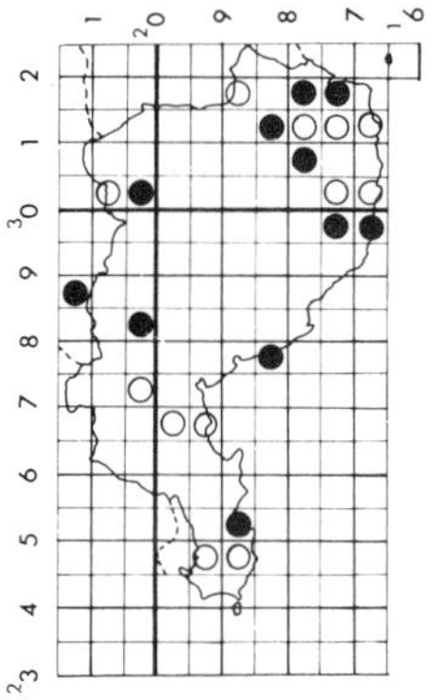

956 Sparganium erectum subsp. erectum

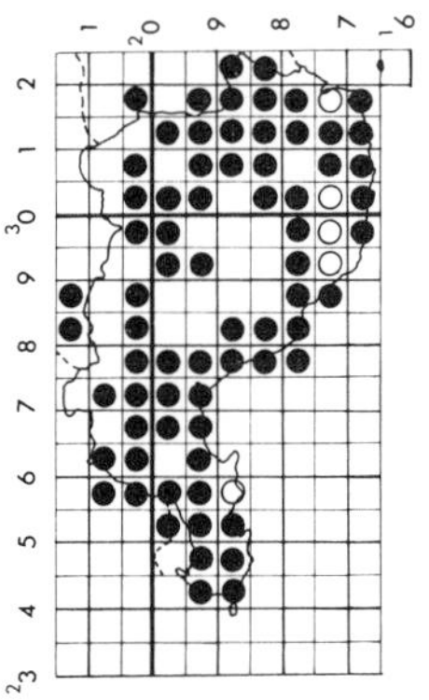

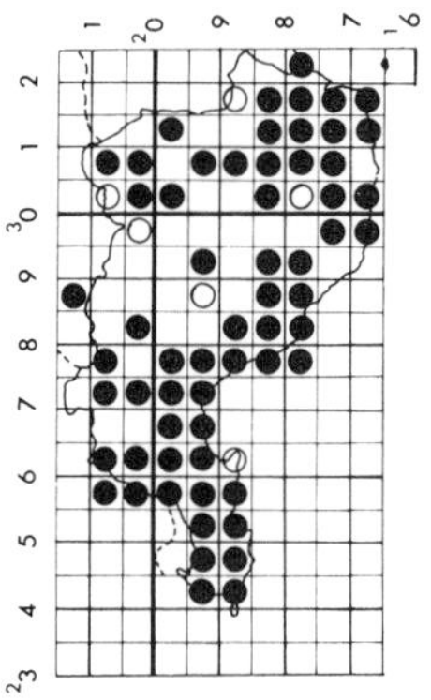

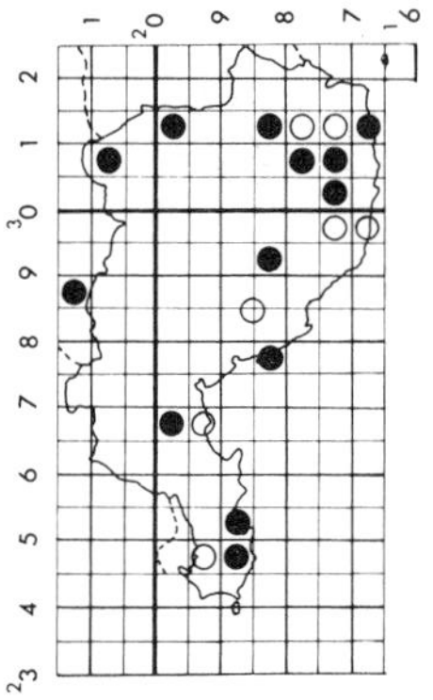

955 Sparganium erectum

954 Lemna minor

953 Lemna trisulca

960 Typha latifolia

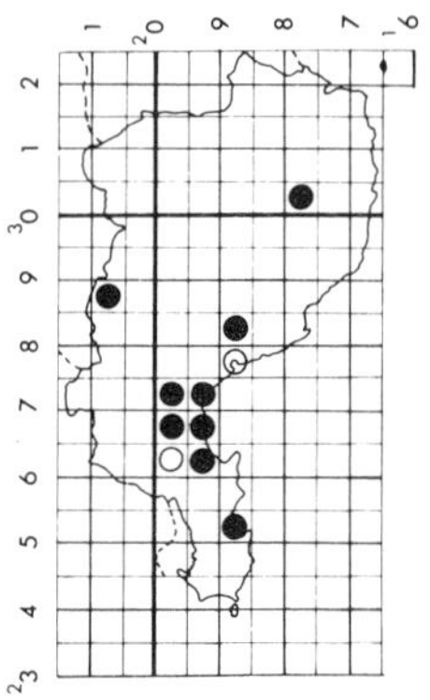

959 Typha angustifolia

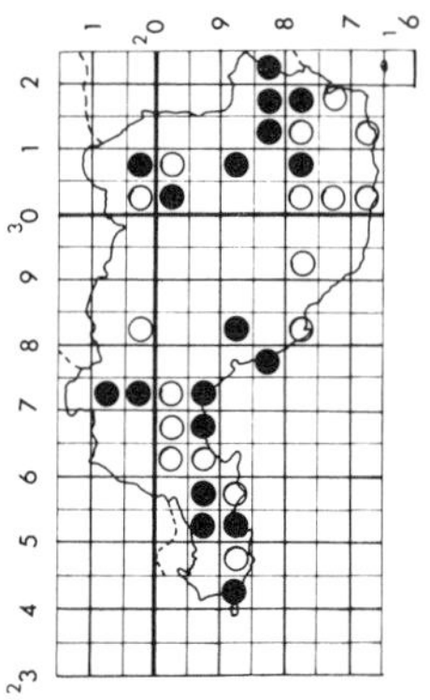

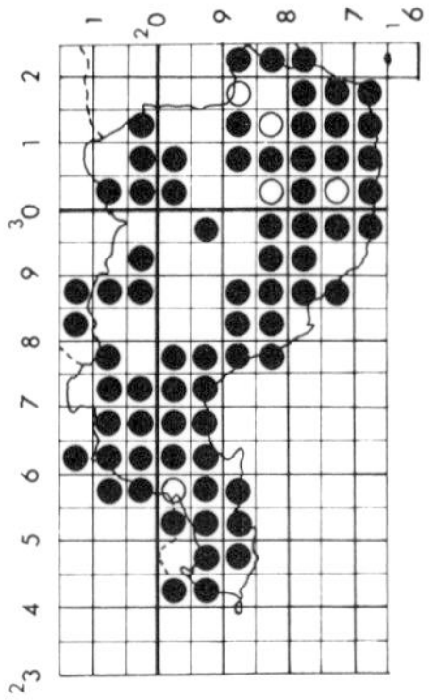

958 Sparganium emersum

957 Sparganium erectum subsp. neglectum

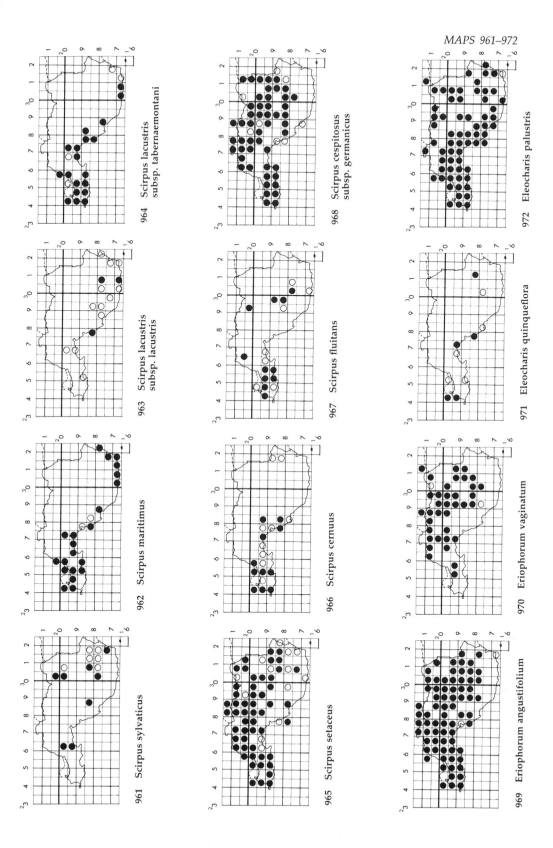

961 Scirpus sylvaticus

962 Scirpus maritimus

963 Scirpus lacustris subsp. lacustris

964 Scirpus lacustris subsp. tabernaemontani

965 Scirpus setaceus

966 Scirpus cernuus

967 Scirpus fluitans

968 Scirpus cespitosus subsp. germanicus

969 Eriophorum angustifolium

970 Eriophorum vaginatum

971 Eleocharis quinqueflora

972 Eleocharis palustris

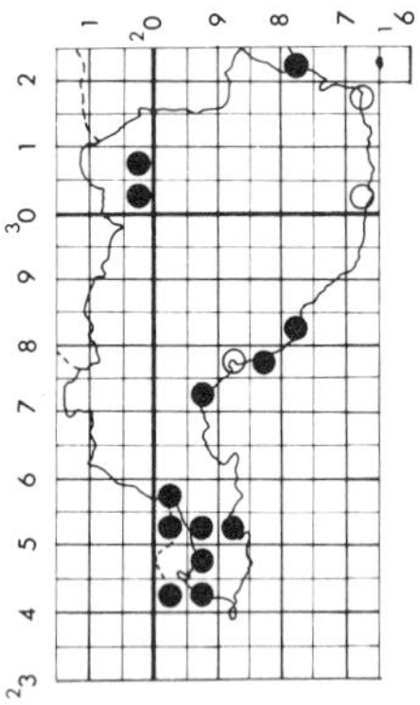

976 Carex otrubae

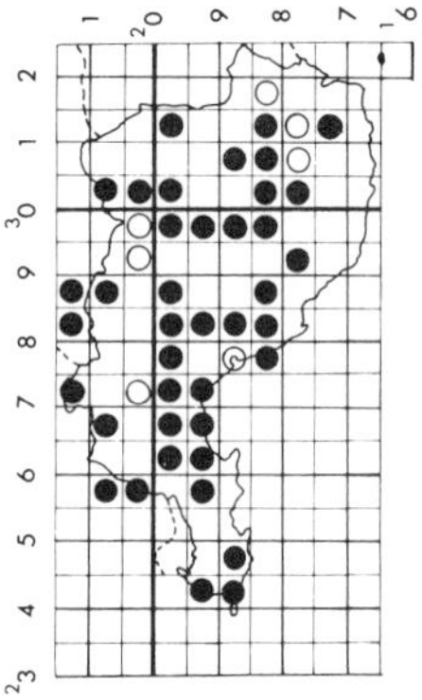

975 Carex paniculata

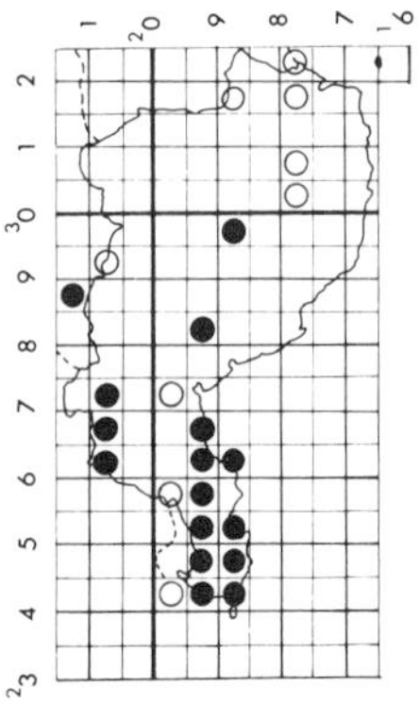

974 Eleocharis multicaulis

973 Eleocharis uniglumis

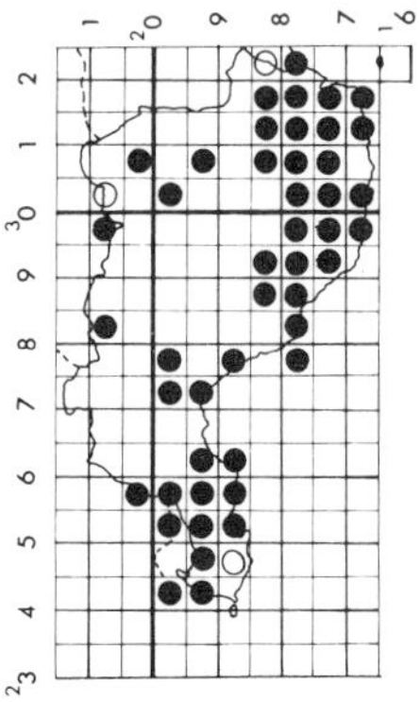

980 Carex remota

979 Carex arenaria

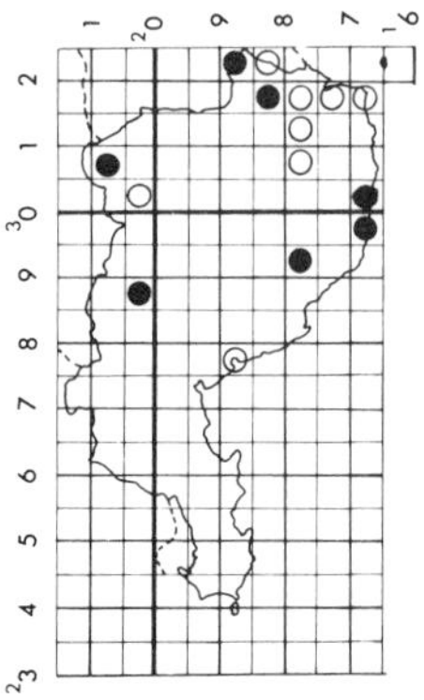

978 Carex spicata

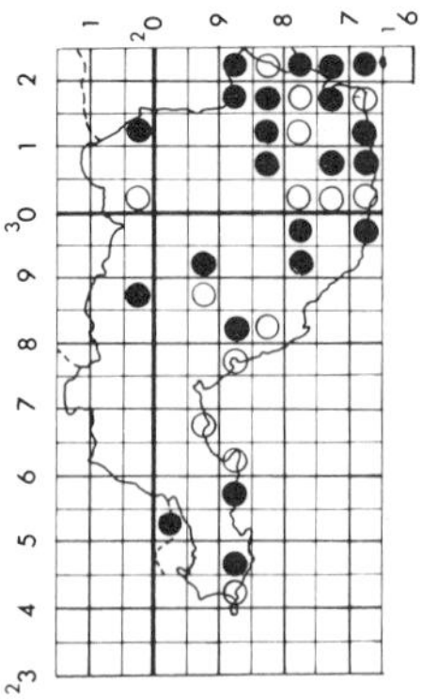

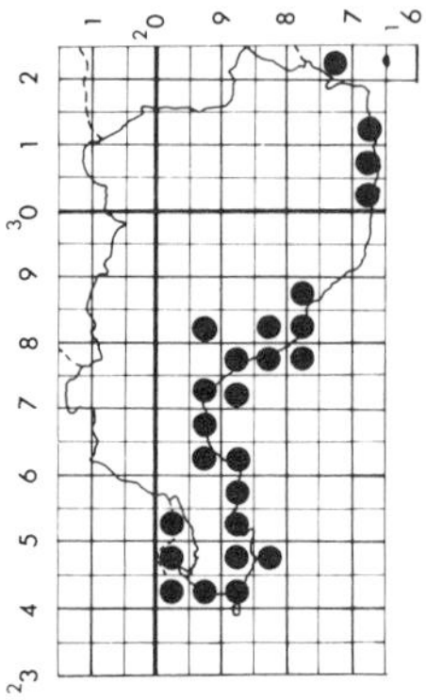

977 Carex muricata agg.

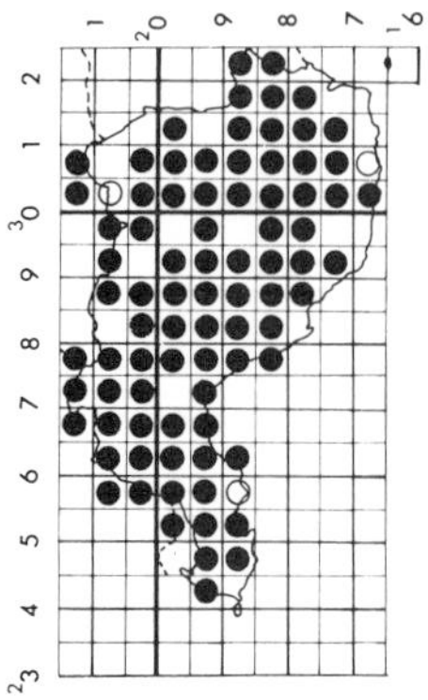

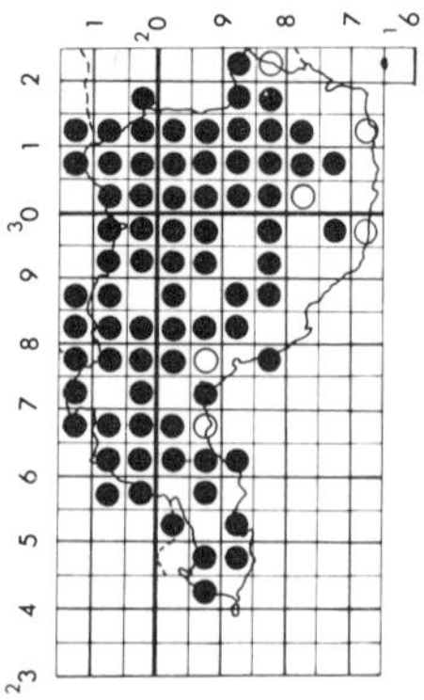

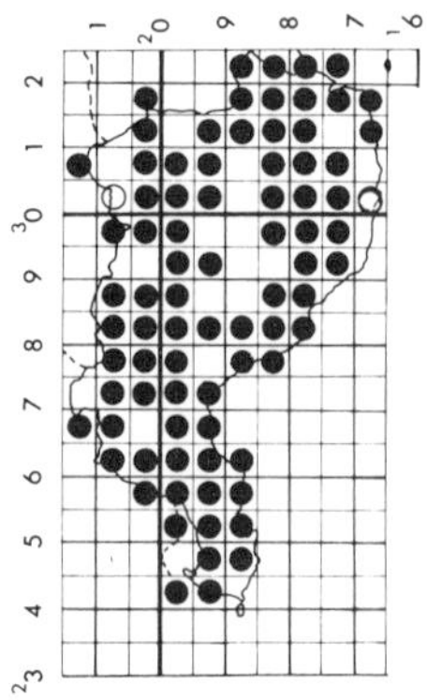

984 Carex hirta

983 Carex curta

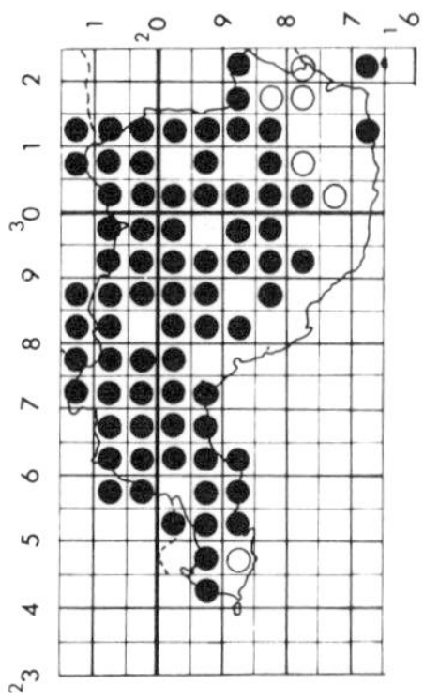

982 Carex echinata

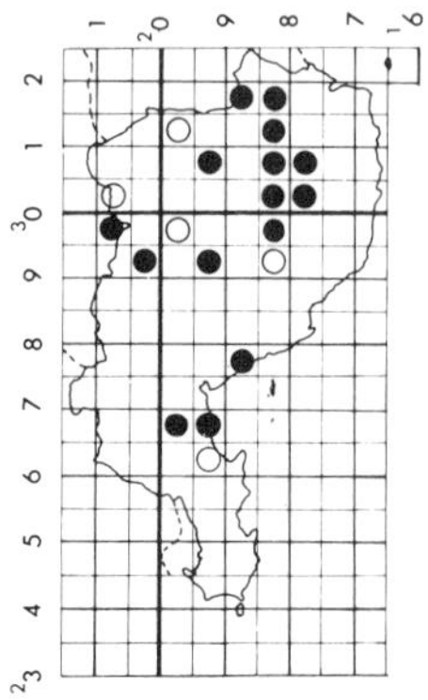

981 Carex ovalis

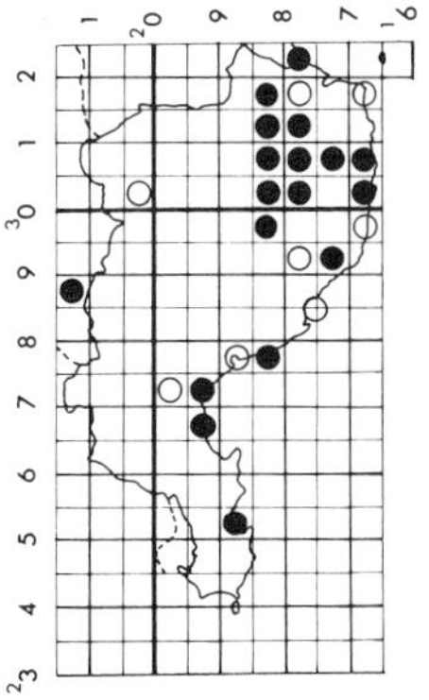

985 Carex acutiformis

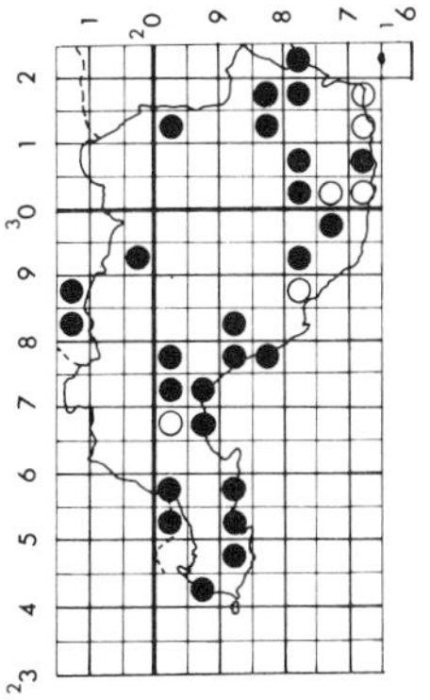

986 Carex riparia

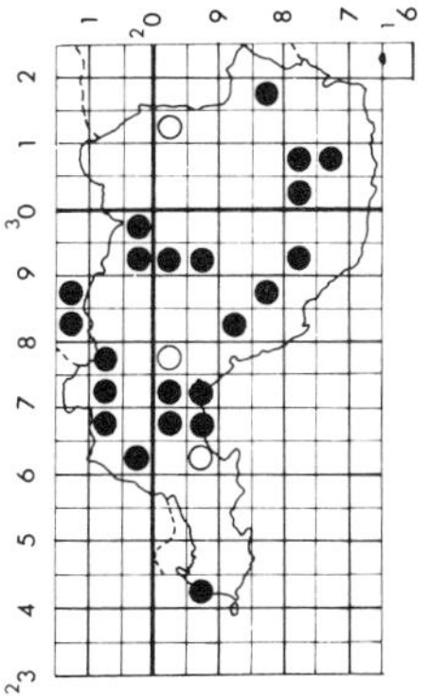

987 Carex rostrata

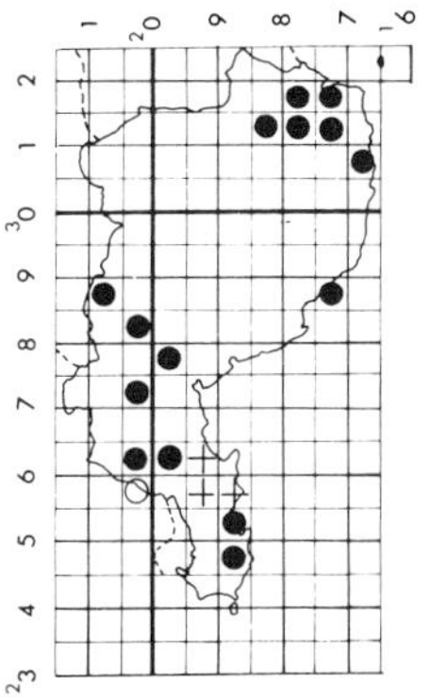

988 Carex pendula

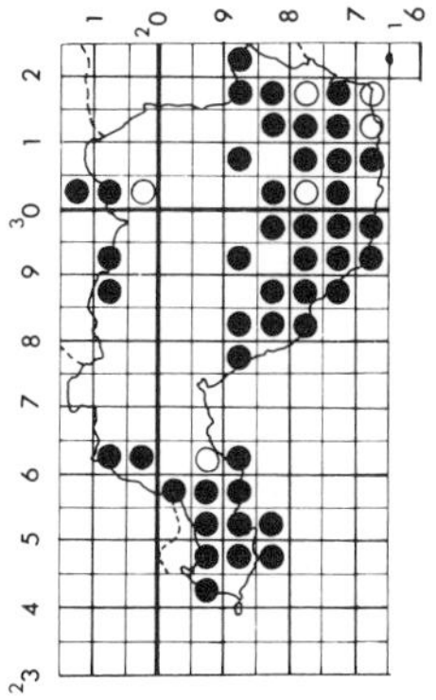

989 Carex sylvatica

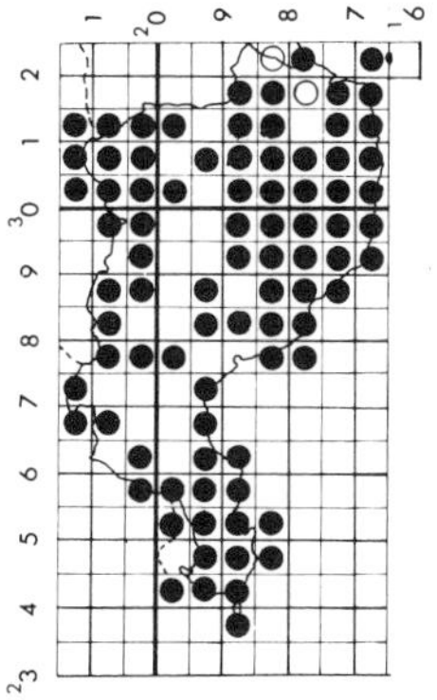

990 Carex flacca

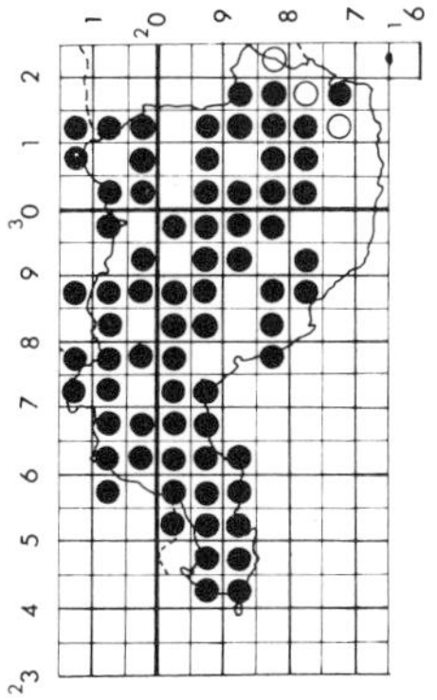

991 Carex panicea

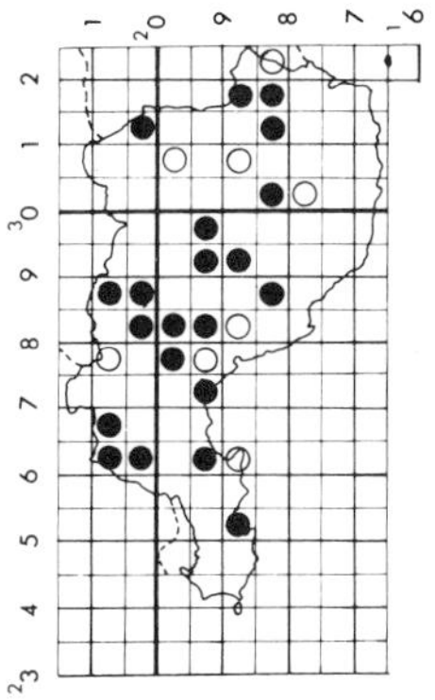

992 Carex laevigata

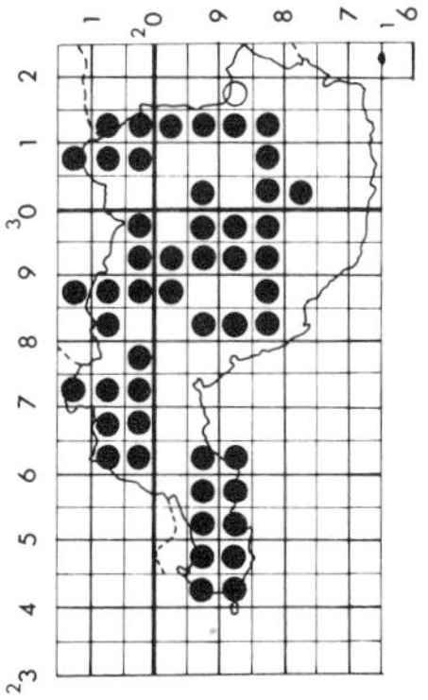

993 Carex binervis

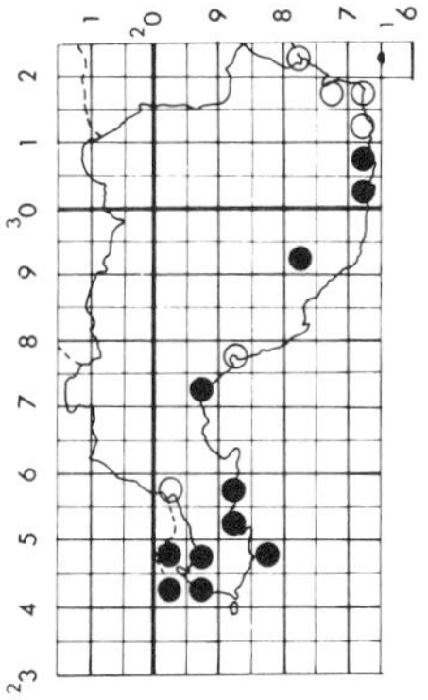

994 Carex distans

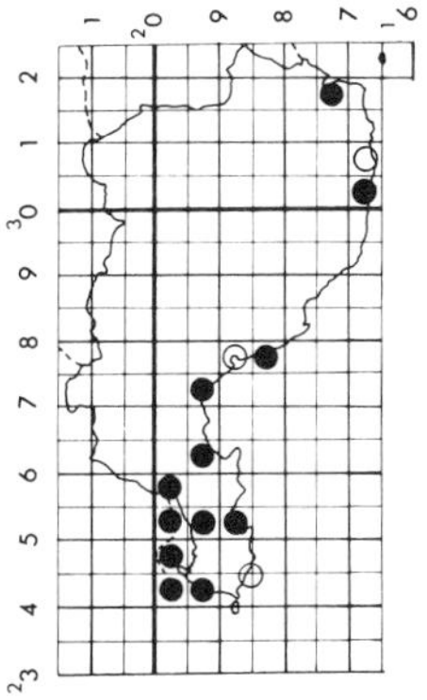

995 Carex extensa

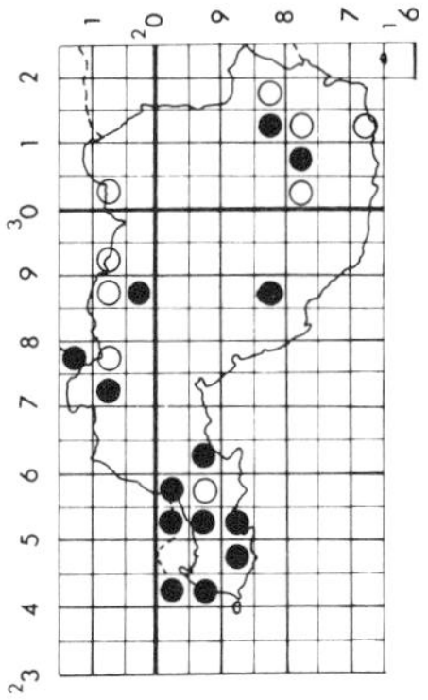

996 Carex hostiana

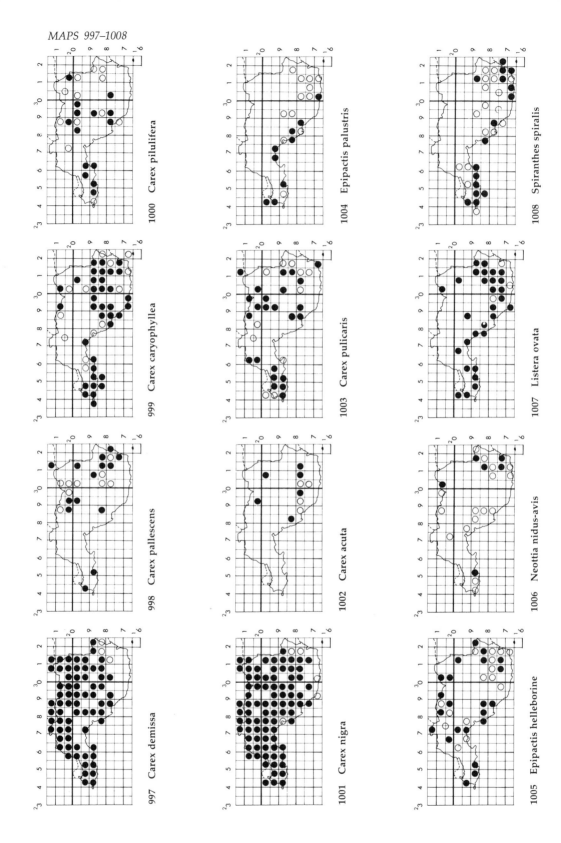

997 Carex demissa

998 Carex pallescens

999 Carex caryophyllea

1000 Carex pilulifera

1001 Carex nigra

1002 Carex acuta

1003 Carex pulicaris

1004 Epipactis palustris

1005 Epipactis helleborine

1006 Neottia nidus-avis

1007 Listera ovata

1008 Spiranthes spiralis

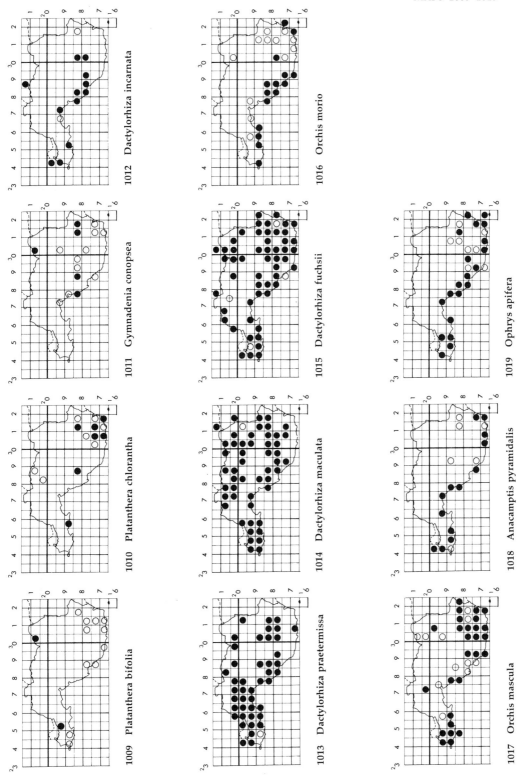

1012 Dactylorhiza incarnata

1016 Orchis morio

1011 Gymnadenia conopsea

1015 Dactylorhiza fuchsii

1019 Ophrys apifera

1010 Platanthera chlorantha

1014 Dactylorhiza maculata

1018 Anacamptis pyramidalis

1009 Platanthera bifolia

1013 Dactylorhiza praetermissa

1017 Orchis mascula

7

Hornworts, Liverworts and Mosses of Glamorgan

A. R. Perry

It is thirty years since Smith started collecting data for his *A Bryophyte Flora of Glamorgan* (Smith, 1964). Since then much taxonomic revision has taken place, for example in *Metzgeria, Sphagnum, Aloina, Gymnostomum* and *Bryum*, several new species have been described, for example *Plagiochila britannica, Dicranella staphylina* and *Fissidens celticus*, and many new records have been made in the county. So a modern account is needed and the present flora is an attempt to fill that need. At the end of 1990 bryophytes recorded (with numbers in Smith's Flora in parentheses) were 107 species and 3 varieties (92 + 5) of liverworts, 12 species and 3 varieties (11 + 2) of Sphagna and 340 species and 31 varieties and subspecies (295 + 33) of mosses.

This account incorporates all records in Smith's *Flora*. However, the identification of many of the specimens on which his records were based has been revised and these revisions have been incorporated here. But, because of the need for compactness these re-identifications are not highlighted: re-identified taxa are simply assigned new positions. It follows, therefore, that any records in Smith's *Flora* that are not in evidence here have either been transferred or, because there was no supporting specimen where one was needed (for example in *Racomitrium heterostichum*), have been rejected as dubious. A manuscript listing all Glamorgan bryophyte records in full, including all re-identifications, is available for consultation from the author in the Department of Botany of the National Museum of Wales, Cardiff.

During the early stages in the preparation of this flora the three main botanical divisions of the vice-county that were distinguished by Trueman (Trueman, 1936), namely the Gower Peninsula, the Uplands and the Vale of Glamorgan, were used as the botanical divisions of the area. Descriptions of these districts are given by Smith (1964). During the course of the present work, however, it became increasingly evident that a fourth district, the Upland ridge, was a necessary geographical segregate because it appeared to be bryologically distinct from the districts that it separates, namely the Vale and the Uplands. So the 'Upland ridge' figures quite prominently in this account. It is a relatively narrow strip of land that rises up from the northern edge of the Vale of Glamorgan, separating the Vale from the Uplands and which runs, from west to east, from Neath through Baglan, Margam, Tondu, Blackmill, Llanharan, Llantrisant, Creigiau and Tongwynlais to Draethen.

Vice-county 41 (Glamorgan) covers fully or in part 41 10 km grid squares. All or part of the following 10 km grid squares lie in the Gower Peninsula: SS 38, 39, 48, 49, 58, 59, 68 and 69. The small fragment of land in SS 39 has not been visited. All or part of the following lie in the Uplands: SS 59, 69, 78, 79, 88, 89, 98, 99; SN 50, 60, 61, 70, 71, 80, 81, 90; ST 08, 09, 18, 19, 28; SO 00, 01, 10 and 11. The small fragments of land in SN 50 and SO 11 have not been visited. All or part of the following lie in the Vale of Glamorgan: SS 77, 78, 79, 87, 88, 96, 97, 98; ST 06, 07, 08, 16, 17, 18, 26, 27 and 28. The fragment of land in SS 77 has not been visited. The Upland ridge passes from west to east through grid squares SS 79, 78, 88, 98; ST 08, 18 and 28.

The taxa are arranged according to Corley & Hill (1981) and, with few exceptions, their nomenclature is also used. In the account of each taxon an attempt has been made to indicate the plant's frequency in vice-county 41 by the use of the terms 'very rare', 'rare', 'uncommon', 'occasional', 'frequent', 'common' and 'very common'. These are rather vague and subjective terms and are used simply to give the author's impression of the frequency of the plant. The substrata and habitats in which the plant has been recorded *in Glamorgan* are listed and there is usually a statement about the plant's distribution in the vice-county as a whole.

If a taxon has been recorded in only a few grid squares, these are often listed in full (with the habitat details that have been recorded on the herbarium specimen), together with the source of the record. All such records are given a grid square number and these are listed in numerical order starting with 00 and ending with 99, their prefix letters being omitted. None of the records made prior to Smith's *Flora* (Smith, 1964) supply grid references, so for the purpose of this list these have been given. Smith (Smith, *l.c.*) has been followed in placing those records made by H.H. Knight by the River Perddyn in grid square **80**; those he made by the River Taff Fechan, Morlais Hill I have placed in **01**. Similarly, those made by A.E. Wade on Clyne Common without precise details have been placed in **58** though they could equally well have been made in **59** or in **69**.

Records made by collectors other than the author which have been refound in that locality by the author are shown by a refinding date followed by !. Specimens checked by the author are indicated by ! following the collector's name or, where the specimen is housed in a herbarium, the herbarium abbreviation (for example Davies! or NMW!).

Finally, unless the plant is so rare as to have its records listed in full, there is a list of Glamorgan 10km grid squares in which it has been recorded. Those records in roman type were made in 1960 onwards: those in italics were made before 1960. Records in ( ) are thought to be extinct. Species in [ ] are doubtfully recorded.

Abbreviations used, including those of collectors' names, are as follows:

Appleyard	Mrs J. Appleyard
Armitage	Miss E. Armitage
Banwell	A.D. Banwell
BBS	British Bryological Society
BBSUK	specimen in Herbarium of the British Bryological Society
Binstead	Rev C.H. Binstead
Blockeel	T.L. Blockeel
BM	specimen in Herbarium of the Natural History Museum, London
c., ca.	circa
Corley	M.F.V. Corley
Crundwell	A.C. Crundwell
Duncan	Miss U.K. Duncan
Gardiner	J.C. Gardiner
Garlick	G.W. Garlick
Harrison	S.G. Harrison
Kay	Q.O.N. Kay
Knight	H.H. Knight
Hb.	specimen in Herbarium of
NMW	specimen in Herbarium of the National Museum of Wales, Cardiff
Orange	A. Orange
OXF	specimen in Herbarium of the Botany School, Oxford University
Paton	Mrs J.A. Paton
Perry	A.R. Perry
Peterken	J.H.G. Peterken
Pettifer	A.J. Pettifer

Richards	P.W. Richards
s.d.	undated
Smith	A.J.E. Smith
Sowter	F.A. Sowter
UCSA	specimen in Herbarium of the University College of Swansea
Wade	A.E. Wade
Wallace	E.C. Wallace
Warburg	E.F. Warburg
Whitehouse	H.L.K. Whitehouse

HORNWORTS

Anthoceros agrestis Paton
Very rare. 49: muddy track, Burry Pill, 1963, *Apple-yard* (BBSUK).

Phaeoceros laevis (L.) Prosk. ssp. **laevis**
Uncommon, on damp soil, in arable land, on roadside banks and on tracks and ditch banks, in woodland, with a few records from Gower and the eastern part of the Upland ridge and in one locality in the western Uplands, absent from the Vale. *18, 28*, 48, 58, 70.

LIVERWORTS

Reboulia hemisphaerica (L.) Raddi
Frequent but usually in small quantity, on soil in rock crevices and at the base of walls, especially limestone, by rivers, in lane banks, on sea cliffs and in sheltered valleys, in all districts but absent from the mountains and the western Uplands. 00, 01, *07, 17, 18*, 48, 49, 58, 87, 97, *98*.

Conocephalum conicum (L.) Lindb.
Common and often abundant, on wet rocks and sheltered banks and bases of walls by culverts, streams, rivers and ponds, rarely on shaded soil over limestone, in woodland, wooded valleys and at roadsides, in all districts but absent from the mountains. 00, 01, 06, 07, 17, 18, 28, 48, 49, 58, 59, 60, 69, 70, 80, 87, 97, 98, 99.

Lunularia cruciata (L.) Dum. ex Lindb.
Frequent, on damp, shaded soil in walls, on stream banks, in dune slacks and in gardens and green-houses, and on soil over limestone, in all districts but absent from the central Uplands. 00, 06, 07, 08, *16*, 17, 18, *28*, 48, 49, 58, 59, 69, 70, 87, 88, 97.

Preissia quadrata (Scop.) Nees
Occasional, in calcareous conditions, amongst and on shaded rocks, in crevices in walls, and in dune slacks, in all districts but commonest in Gower with only three records from the northern and the eastern Uplands and two from the Vale, absent from the Upland ridge. 00, 01, *48*, 49, *58*, 68, 78, *80*, 87.

Marchantia polymorpha L.
Uncommon, on cinders and clinker and garden soil, between paving stones, on the bank of a drainage channel and in a bog, in a very few scattered localities throughout the county but absent from the Upland ridge. 07, 17, *19*, 49, 69, 89, *(99)*.

Marchantia alpestris (Nees) Burgeff
Very rare, but possible overlooked as *Marchantia polymorpha*. 17: between paving stones, Cathays, Cardiff, 1988, *Perry* (NMW).

Riccia cavernosa Hoffm.
Rare, on damp, sandy soil in dune slacks. 78: Kenfig Burrows, 1933, *E. Vachell* (NMW), Kenfig Pool, 1934, *Knight* (NMW); Margam Burrows, 1980, *M.E. Gillham* (NMW!).

Riccia fluitans L.
Rare, on damp soil in marshy ground and sand dunes. 07: muddy margin, St y-Nyll Pond, 1921, *Wade* (NMW); Pysgodlyn Mawr, 1958, *Wade* (NMW), 1975!, 1979!, absent 1985. 58: in ditches and *Phragmites* fen, Oxwich, 1968–88, *Kay* (Kay, 1989). 78: damp soil by path, Kenfig Burrows, 1986, *Orange* (Hb. Orange). 88: on mud beneath *Acorus, Iris, Glyceria maxima*, etc., side of lake SE of Abbey ruins, Margam Park, 1984, *Perry* (NMW).

Riccia glauca L.
Uncommon, on moist soil on banks, tracks and in fallow fields, fairly frequent in Gower but otherwise known only from two localities in the eastern Uplands. 00, 19, 48, 49, 69.

Riccia beyrichiana Hampe ex Lehm.
Very rare. 00: rock crevices under viaduct, Taff Fechan, 1963, *Smith* (NMW).

Riccia sorocarpa Bisch.
Occasional, on hard-packed soil on sea cliffs, on cultivated soil in fields, on reclaimed pit waste and on soil-covered limestone slabs in an old quarry, with a few scattered records, frequent on Gower, rare elsewhere. 00, *18*, 48, 58, 70, *87*.

Metzgeria fruticulosa (Dicks.) Evans
Uncommon, on living tree trunks and dead branches in woodland, recorded only from the Vale. 06, 07, 87, 97.

Metzgeria temperata Kuwah.
Occasional, on shaded tree trunks, with a few records from the Vale, Gower, the central Upland ridge and the western Uplands. 06, 07, 59, 60, 97, 98.

Metzgeria furcata (L.) Dum.
Common, on the bark of several species of trees, on stones, rocks and walls, more rarely on soil and on sand hillocks, in woodland, on limestone cliffs, on sea cliffs and in sand dunes, in all districts but commonest in the Vale and Gower, absent from the central Uplands. 00, 01, 06, 07, 08, 16, 17, 18, *28*, 48, 49, 58, 59, 60, 69, 78, 87, 88, 96, 97, 98.

Metzgeria conjugata Lindb.
Very rare, not seen recently. 18: wet rocky bank of lane, Craig Lysfaen, 1933, *Wade* (NMW!). A dubious record for 90: Craig-y-Llyn, *Trow* (Trow, 1899), needs confirmation.

Aneura pinguis (L.) Dum.
Common, on wet rocks and banks, in bogs and boggy ground, in dune slacks, by rivers, on mountain sides and in woodland, recorded from all districts but

commonest in the Uplands. 00, 18, 28, 49, 58, 60, 69, 70, 78, 79, 80, 81, 88, 89, 90, 97, 98, 99.

Riccardia multifida (L.) S.Gray
Occasional, in wet boggy ground and flushes and on wet soil and rocks, on dripping banks, on mountain sides, by a waterfall, on sea cliffs and in a dune slack, with a few scattered records. 07, *17*, 49, 58, 60, 80, 87, 90, 99.

Riccardia chamaedryfolia (With.) Grolle
Frequent, on wet banks, on wet cliffs, on lakeside mud, in dune slacks, on a sandy road in woodland and amongst other bryophytes on dripping rocks, with scattered records from all districts. 00, *07, 17*, 49, 58, 60, 78, 79, 80, 88, 90, 98, 99.

Pellia epiphylla (L.) Corda
Very common, on base-poor soils, on sheltered banks, on sides of streams and ditches, more rarely on wet rocks, in woodland, at roadsides, on hills and mountain sides, throughout all districts. 00, 01, 06, 07, 08, 09, 10, 18, 19, *48*, 49, 58, 59, 60, 69, 70, 78, 79, 80, 81, 87, 88, 89, 90, 98, 99.

Pellia neesiana (Gott.) Limpr.
Occasional, in marshy fields, at the margin of a lake and in boggy ground, often amongst *Juncus* tussocks, with a few scattered records in all districts. 00, 07, 08, 49, 60, 79, 88, 98.

Pellia endiviifolia (Dicks.) Dum.
Common, on moist, base-rich soil at the sides of streams, on sheltered banks, on wet limestone, in woodland, old quarries and on sand dunes, in all districts but rare in the mountains. 00, 01, 06, 08, 09, 17, 18, *48*, 49, 58, 59, 69, 78, 79, 80, 87, 96, 97, 98, 99.

Moerckia hibernica (Hook.) Gott.
Rare, in dune slacks. **49**: Whiteford Burrows, 1932, *D.A. Jones & Knight* (NMW), 1963, *Harrison* (NMW), 1974, *Kay* (Kay, 1989); Llangennith Burrows, 1943, *Richards* (NMW). **78**: Kenfig Burrows, 1961, *Smith* (NMW), 1986, *Orange*. **87**: Merthyr Mawr Warren, 1987, *Perry* (NMW).

Blasia pusilla L.
Very rare. **89**: ditch bank and clinkers beside R. Pelena above Ton Mawr, 1963, *Smith* (NMW).

Fossombronia husnotii Corb.
Very rare, on soil on sea cliffs in Gower. **48**: near Overton, 1961, *Smith* (NMW); cliff top south-west of Rhossili, 1963, *Crundwell* (Smith, 1964). **58**: between Caswell Bay and Pwlldu Bay, 1961, *Smith* (NMW, BBSUK).

Fossombronia pusilla (L.) Nees var. **pusilla**
Uncommon, on damp soil on ditch banks, in sheltered lanes and in turf, except for one record in Gower confined to the eastern Vale and the eastern end of the Upland ridge. 07, *08, 17*, 18, 28, 48.

Fossombronia wondraczekii (Corda) Dum. ex Lindb.
Very rare. **18**: bare damp clay, *Molinia* heath, Rudry Common, Rudry, 1978, *Garlick* (BBSUK).

Petalophyllum ralfsii (Wils.) Nees. & Gott.
Rare, in dune slacks. **49**: Whiteford Burrows, *E.J. Salisbury* (Smith, 1964), 1924, *H.A. Hyde* (NMW), 1932, *D.A. Jones & Knight* (NMW), 1961, *Smith* (NMW, UCSA), 1963, *Harrison* (NMW), 1968–88 *Kay* (Kay, 1989). **78**: Kenfig Burrows, 1954, *K. Benson-Evans* (NMW), 1960, *Richards* (NMW), 1963, *Harrison* (NMW). **87**: Merthyr Mawr dunes, *Sowter* (Wade, 1959).

Barbilophozia floerkei (Web. & Mohr) Loeske
Very rare, in turf. **49**: Rhosili Down, 1967, *Perry* (Kay, 1989). **90**: above Llyn Fach, 1961, *Smith* (NMW), 1976!; Craig y Llyn, 1973, *Perry*.

Barbilophozia attenuata (Mart.) Loeske
Uncommon, on rotting wood and tree bases in woodland, and on dry soily acidic rocks, in one locality in Gower and four scattered localities in the Uplands. 10, 58, 60, 70, *99*.

Barbilophozia barbata (Schmid. ex Schreb.) Loeske
Very rare. **99**: soil beside stream, Craig Ogwr, 1961, *Smith* (BBSUK).

Lophozia ventricosa (Dicks.) Dum. var. **ventricosa**
Probably frequent but under-recorded, in turf on acidic banks, on soily rock ledges and in boggy ground, in an old quarry and on moorland. **10**: roadside bank, Blaenllwynau, 1963, *Smith* (NMW). **18**: soil on ledges, Caerphilly Common, 1974, *Perry*; soil, acidic bank in old quarry on SW-facing slope, Mynydd Meio, Taff Vale, 1986, *Perry* (NMW). **80**: bog by road 1 mile SE of Dyffryn Cellwen, 1987, *K.L. Davies*!. **89**: stream bank, Mynydd Fforch-Dwm (Smith, 1964).

Lophozia ventricosa (Dicks.) Dum. var. **silvicola** (Buch) E.W. Jones ex Schust.
Frequent but probably under-recorded, in similar situations to var. *ventricosa*, recorded from Gower where it is fairly frequent in heathy grassland, and the Uplands and Upland ridge, absent from the Vale. 00, 10, 48, 58, 59, 60, 70, 71, 88, 90, 99.

Lophozia excisa (Dicks.) Dum. var. **excisa**
Very rare. **58**: soil on cliff above Pwll-du Bay, 1961, *Smith* (NMW).

Lophozia incisa (Schrad.) Dum.
Uncommon, in turf on banks and in moorland, rarely on wood, in the central and western Uplands and Upland ridge, absent elsewhere. 60, *80*, 88, 89, 90.

Lophozia bicrenata (Schrad. ex Hoffm.) Dum.
Occasional, on acidic banks, old anthills, in an old quarry and on moorland, with a few scattered records in the Uplands, Upland ridge and southern

Gower, absent from the Vale. 00, 09, *17*, 18, 19, 48, *69, 90.*

Leiocolea turbinata (Raddi) Buch
Occasional, on moist calcareous soil and rocks, on banks and in turf, in old quarries, on cliffs and in dune slacks, recorded from all districts but absent from the western and central Uplands. 00, 06, *16, 17,* 18, 58, 68, 78, *96.*

Leiocolea badensis (Gott.) Jørg.
Occasional, on moist calcareous soil, on banks, below limestone cliffs, in woodland and in dune slacks, recorded from the Vale and the eastern Uplands and Upland ridge, absent elsewhere. *00*, 17, 18, 28, 78, 97.

Leiocolea alpestris (Schleich. ex Web.) Isov.
Very rare, amongst moist, shaded calcareous rocks. **00**: Morlais Hill, 1932, *Knight* (NMW). **18**: shaded limestone rubble in woodland, Coed y bedw 1976, *E.W. Jones & Perry.*

Leiocolea bantriensis (Hook.) Jørg.
Very rare, on river and stream banks in the north. **01**: streamside detritus, Taff Fechan, 1963, *Smith* (NMW). **80**: soil on rocks, W bank of R. Neath NE of Glyn Neath, 1963, *Paton* (NMW).

Gymnocolea inflata (Huds.) Dum. var. **inflata**
Very common, in acidic situations, on peaty soil and boggy ground, sometimes amongst *Sphagnum*, on moorland, in crevices in acidic rocks, on dry stone walls and industrial waste, rarely on tree bases, common in the Uplands and in Gower but rare in the Vale and absent from the coastal strip of the southern Vale. 00, 01, 07, 08, 09, 10, 18, 19, 48, 49, 58, 59, 60, 69, 70, 71, 78, 79, 89, 81, 88, 89, 90, 98, 99.

Tritomaria quinquedentata (Breidl.) Loeske
Very rare, amongst rocks on mountain slope. **90**: above Llyn Fach, 1976, *Perry.*

Jungermannia atrovirens Dum.
Widespread but uncommon, on wet, shaded, basic rocks, often on limestone, sometimes near rivers, recorded from all districts. 00, 01, 18, 58, 68, 80, 87, 98.

Jungermannia pumila With.
Uncommon, on wet rocks by rivers and streams in the central and western Uplands. **60**: Cwm Clydach, 1988, *Orange.* **80**: by first waterfall, Afon Pyrddin (Smith, 1964). **90**: below Llyn Fawr reservoir, Craig-y-Llyn (Smith, 1964). **99**: by waterfall, Craig Ogwr, 1961, *Smith* (NMW).

Jungermannia gracillima Sm.
Frequent, on clay banks and damp soil, often by ditches, in old quarries, on moorland and in woodland, in all districts. 07, 08, 18, 49, 59, 60, *69*, 70, 79, 80, 90, 98, 99.

Jungermannia paroica (Schiffn.) Grolle
Very rare. **80**: R bank of R. Perddyn, 1932, *Knight* (NMW).

Jungermannia obovata Nees
Very rare. **60**: with *J. pumila* on damp rocks near river in woodland, Cwm Clydach, 1988, *Orange* (BBSUK).

Nardia compressa (Hook.) S. Gray
Occasional, on rocks in and by the upper reaches of streams, most frequent in the central Uplands and western Upland ridge, otherwise known only in one locality in the western Uplands. 60, 88, 89, 90, 98, 99.

Nardia scalaris S.Gray
Common, especially in the mountains, rare in the southern Vale, on raw soil, clayey stream banks, peaty soil, boggy ground, wet rocks and screes, often amongst other bryophytes, in old quarries, woodland and on moorland and mountain sides. 00, 01, *07*, 08, 09, 10, 17, 18, 19, *28*, 48, 49, 58, 59, 60, 70, 71, 78, 79, 80, 81, 87, 88, 89, 90, 98, 99.

Nardia geoscyphus (De Not.) Lindb.
Very rare. **18**: on acidic ledges in disused Pennant Sandstone quarry, Caerphilly Common, 1974, *Perry* (NMW).

Marsupella emarginata (Ehrh.) Dum. var. **emarginata**
Uncommon, on damp rocks and rocks in and about streams and rivers in the Uplands, absent elsewhere. 09, 60, *70, 80*, 90, 98, 99.

Marsupella emarginata (Ehrh.) Dum. var. **aquatica** (Lindenb.) Dum.
Very rare, on wet rocks by streams in the mountains. **90**: Craig-y-Llyn, 1932, *Knight* (NMW). **98**: Cwm Dimbath, 1951, *Wade* (NMW).

Southbya tophacea (Spruce) Spruce
Very rare and possibly decreasing, on wet sheltered tufaceous Liassic cliffs on the coast. **06**: between Aberthaw and Fontigary, 1949, *Wade* (NMW), 1963 *Harrison* (NMW); Porthkerry, 1950, *Wade* (NMW), 1974!; East Aberthaw, 1960, *Richards* (NMW).

Plagiochila porelloides (Torrey ex Nees) Lindenb.
Frequent, on banks, rocks and walls, and tree bases, especially in calcareous regions, with scattered records throughout the county. 00, 01, *16*, 17, 18, 27, 28, 48, 49, 58, 60, 69, 80, 87, 90, 98.

Plagiochila asplenioides (L.) Dum.
Frequent in humid conditions, on shaded walls and rocks, on grassy banks and on the floor of woodlands, especially in calcareous regions. 00, 01, 07, 17, 18, 28, 48, 49, 58, 59, 69, 80, 87, 90, 97, 98.

Plagiochila britannica Paton
Very rare. **58**: Oxwich, 1932, *Knight* (NMW); boulder in limestone woodland, Oxwich Bay, 1983, *Blockeel* (BBSUK).

Plagiochila spinulosa (Dicks.) Dum.
Very rare. **98**: Dimbath Valley, 1932, *E.M. Thomas* (NMW); shaded siliceous rock face, Daren y Dimbath, 1974, *Perry* (NMW).

Lophocolea bidentata (L.) Dum.
Widespread and very common, amongst grass on banks and in boggy ground, on tree and wall bases and occasionally in dune slacks, throughout the county. 00, 01, 06, 07, 08, 09, 10, 17, 18, 19, 28, 48, 49, 58, 59, 60, 68, 69, 70, 78, 79, 80, 87, 88, 89, 90, 97, 98, 99.

Lophocolea cuspidata (Nees) Limpr.
Widespread and very common, on tree bases, rotting logs, damp shaded rocks, on decaying vegetation and soil in woodland, throughout the county. 00, 01, 06, 07, 08, 09, 10, 17, 18, 19, 28, 48, 49, 58, 59, 60, 68, 69, 70, 71, 80, 87, 90, *96*, 97, 98, 99.

Lophocolea heterophylla (Schrad.) Dum.
Widespread and common in damp or humid places, on tree bases, logs and stumps in woodland, throughout the county. 00, 06, 07, 08, 09, 17, 18, 19, 27, 28, 48, 58, 59, 60, 68, 69, 70, 80, 89, 96, 97, 98, 99.

Lophocolea fragrans (Moris & De Not) Gott., Lindenb.& Nees
Very rare. **49**: stone in woodland with *Lejeunea* sp., Burry Pill, 1963, *Appleyard* (BBSUK).

Chiloscyphus polyanthos (L.) Dum.
Frequent, on stones in streams, in flushes and marshy ground, and on wet banks in woods, absent from the mountains and the north-eastern Uplands. 08, *17*, 18, 19, *27*, *48*, 49, 58, 59, 60, 69, *70*, 80, 98.

Chiloscyphus pallescens (Ehrh. ex Hoffm.) Dum.
Frequent in marshy ground, in bogs and wet banks in woods, and in dune slacks, with a few scattered records throughout the Vale and occasional records from Gower, the Upland ridge and the eastern Uplands. 00, 06, 09, *17*, 18, 27, *28*, 58, 69, 78, 80, 97.

Saccogyna viticulosa (L.) Dum.
Uncommon, on humid rocky banks by streams, especially in woodland, rarely on soil on sea cliffs, most frequent in the Uplands, with single records from Gower and from the Upland ridge. 18, 48, 60, 80, 98, 99.

Diplophyllum albicans (L.) Dum.
Very common, one of the commonest liverworts in the county, on acidic banks, dry exposed rocks and tree bases, and on cliffs, in woodland, lanes and old quarries, moorland and bogs, in all districts but absent from the coastal strip in the Vale. 00, 01, 07, 08, 09, 10, 17, 18, 19, 48, 49, 58, 59, 60, 69, 70, 71, 78, 79, 80, 87, 88, 89, 90, 98, 99.

Diplophyllum obtusifolium (Hook.) Dum.
Very rare. **28**: sandy bank, woodland, Coed Cefn Pwll Ddu, Rhydygwern, Machen, 1979, *Garlick* (BBSUK).

Scapania cuspiduligera (Nees) K.Muell.
Very rare. **68**: short turf on limestone, Mumbles Head near Swansea, 1969, *J.G. Duckett* (BBSUK).

Scapania scandica (H. Arn.& Buch) Macv.
Very rare, on soil on grassy slope. **00**: E-facing Pennant sandstone outcrop, Mynydd Cilfach-yr-encil, Troedyrhiw, 1986, *Perry* (NMW).

Scapania nemorosa (L.) Dum.
Frequent, on shaded rocks and stony tracks, earthy banks, mossy boulders and trees, in woodland, in quarries and occasionally on river banks and moorland margins, with most records from the Uplands and the Upland ridge, and very few from Gower and the Vale. 00, 01, 08, 09, 18, *58*, 59, 60, 70, 71, 78, 79, 80, 81, 89, 90, 98.

Scapania irrigua (Nees) Nees
Occasional, amongst grass in marshy or flushed ground, commonest in the Uplands and the eastern part of the Upland ridge with one record from Gower, absent from the Vale. 00, 01, 08, 10, 49, 60, 70, 80, 88, 89, 90, 98.

Scapania compacta (A.Roth) Dum.
Occasional, on acidic soil, amongst rocks and on walls and mossy boulders, with a few scattered localities in the Uplands and one in Gower, absent from the Vale, and the Upland ridge. 00, 01, 49, 70, *98*, 99.

Scapania undulata (L.) Dum.
Common, on stones, rocks and tree roots in and near water and in boggy ground, rarely on dry, stony ground, in Gower, the Uplands and Upland ridge, absent from the Vale. 00, 01, 08, 09, 10, 18, *48*, 49, 58, 59, 60, 70, 78, *79*, 80, 81, *88*, 89, 90, 98, 99.

Scapania aspera H.Bern.
Uncommon, on soil or in turf or moss over limestone, with one record from the Vale, one from the north-eastern Uplands and three from Gower. 00, 48, 49, 97.

Scapania gracilis Lindb.
Uncommon, in turf in old quarry, on siliceous boulders and on wet rocks by mountain stream, rarely on wood and soil, in scattered localities in Gower and the Uplands, absent from the Vale and from the Upland Ridge. 00, 09, 48, 49, 58, 60, 98, *99*.

Odontoschisma sphagni (Dicks.) Dum.
Rare, amongst *Sphagnum*. **58**: Cefn Bryn near Penmaen (Kay, 1989). **59**: in bog N of Cefn Bryn, 1963, *Smith* (NMW).

Odontoschisma denudatum (Mart.) Dum.
Very rare. **60**: on decaying log by stream in woodland, Cwm Clydach, 1988, *Orange* (BBSUK).

Cephaloziella rubella (Nees) Warnst.
Uncommon, on heathland, in turf on sea cliffs, on an ant-hill and on decaying stump and tree, with a very few scattered records throughout the county. *08, 48,* 59, 60, 87.

Cephaloziella hampeana (Nees) Schiffn.
Rare but possibly overlooked. **07**: in *Campylopus flexuosus* by pond, Mynydd y Glew, 1962, *Smith* (NMW). **48**: Rhossili, 1963, *Paton* (Smith, 1964). **59**: heath near Broad Pool, N of Cefn Bryn, 1963, *Paton* (Smith, 1964). **78**: sands, Kenfig, 1908, *Knight* (NMW).

Cephaloziella divaricata (Sm.) Schiffn.
Occasional, on soil or peaty ground and tree trunks, on ant-hills, on overhanging banks on hillside, and on moorland and cliff tops, in Gower, the Uplands and the Upland ridge, absent from the Vale. 00, 18, 49, 58, 60, *90,* 99.

Cephaloziella stellulifera (Tayl.) Schiffn.
Very rare. **48**: amongst turf on limestone cliffs, Rhossili, 1963, *Paton* (BBSUK).

Cephaloziella calyculata (Dur. & Mont.) K.Muell.
Very rare and local. **48**: on soil on limestone rock ledges, Rhossili Down, 1963, *J.W. Fitzgerald* (NMW), turf, limestone cliffs SW of Rhossili, 1963, *Crundwell* (BBSUK).

Cephalozia bicuspidata (L.) Dum. agg.
Very common, on soil on banks, stony tracks and tree bases, in woods and on moorland, amongst rocks, on decaying wood and in swamps and bogs, throughout all districts. 00, 06, 07, 08, 09, 10, 17, 18, 19, *28,* 48, 49, 58, 59, 60, 69, 70, 71, 78, 79, 80, 87, 88, 89, 90, 98, 99.

Cephalozia bicuspidata (L.) Dum. ssp. **bicuspidata**
Common and widespread in habitats as described for the aggregate species. 07, 18, 49, 59, 70, 88, 89, 90.

Cephalozia bicuspidata (L.) Dum. ssp. **lammersiana** (Hueb.) Schust.
Occasional, in similar habitats to var. *bicuspidata* with all records but one from the western half of the county. 07, 49, 58, 56, 70, 88, 89.

Cephalozia connivens (Dicks.) Lindb.
Very rare. **59**: on wet peat and mosses on wet heath, near Cillibion, 1963, *Paton, Appleyard & Wallace* (NMW).

Cephalozia catenulata (Hueb.) Lindb.
Very rare. **98**: rotting log in gully, Daren y Dimbath, 1979, *Perry* (NMW).

Nowellia curvifolia (Dicks.) Mitt.
Uncommon but often abundant where it occurs and possibly spreading, on decaying logs, often in sheltered woodland and valleys, restricted to a very few localities in the Upland ridge and the Uplands. 09, 18, 60, 88, 90, 98.

Cladopodiella francisci (Hook.) Buch ex Jørg.
Very rare. **58**: edge of drainage channel, Clyne Common, 1944, *Wade* (NMW). **60**: peaty ground by stream, near Upper Lliw Reservoir, 1963, *Smith* (NMW).

Kurzia pauciflora (Dicks.) Grolle
Uncommon, on peaty soil and amongst *Sphagnum* on heathland, with two records from Gower, one from the northern Uplands and one from the Upland ridge. 08, 58, 59, *81.*

Lepidozia reptans (L.) Dum.
Frequent to common, in Gower, the eastern Upland ridge and the Uplands, on peaty soil and acidic banks, on tree stumps and trunks, absent from the Vale. 00, 01, 08, 09, 10, 18, 19, *28,* 58, 59, 60, 69, 71, 80, 88, 89, 90, 98, 99.

Bazzania trilobata (L.) S.Gray
Very rare. **98**: moss-covered grit boulder, Daren y Dimbath, 1974, *Perry* (NMW), 1981 (NMW).

Bazzania tricrenata (Wahlenb.) Lindb.
Very rare. **90**: amongst rocks above Llyn Fach, 1976, *Perry.*

Calypogeia muelleriana (Schiffn.) K.Muell.
Occasional, on rotten logs in woodland, wet soil banks, and peaty ground in bogs and on moorland, in scattered localities in Gower, the Uplands and the Upland ridge, with one record from the western Vale. 09, 10, 18, 49, *58,* 60, 79, 90, 98.

Calypogeia fissa (L.) Raddi
Common throughout the county except in the coastal strip of the southern Vale, on shaded soil on banks and cliff ledges, and in bogs. 00, 01, 07, 08, 09, 10, 17, 18, 19, 28, 48, 58, 59, 60, 69, 70, 71, 78, 79, 80, 87, 88, 89, 90, 98, 99.

Calypogeia arguta Mont.& Nees
Common, on acidic soil on banks and ditch sides, rarely on bare rock, in Gower, the Uplands and the Upland ridge, very rare in the Vale. 01, 07, 08, 09, 17, 18, 19, 27, 48, 49, *58,* 59, 60, 69, 80, 90, 98, 99.

Blepharostoma trichophyllum (L.) Dum.
Very rare and not seen recently. **80**: river bank, right bank of R.Perddyn, 1932, *Knight* (NMW), by first waterfall, Afon Pyrddin, *Smith* (Smith, 1964).

Trichocolea tomentella (Ehrh.) Dum.
Very rare. **18**: calcareous springs in deciduous woodland, Coed-y-bedw, Taffs Well, 1985, *Blockeel* (BBSUK). **48**: stream bank and damp ground in S part of Penrice Woods, 1989, *Kay* (NMW).

Ptilidium ciliare (L.) Hampe
Uncommon, on wet acidic heath and amongst siliceous boulders on moorland, with one record from the eastern Uplands, three from the eastern Upland ridge and six from Gower. *09, 18,* 48, 49, 58.

Ptilidium pulcherrimum (G.Web.) Vainio
Rare. **07**: bole of sallow in carr, SW side of Pysgodlyn Mawr, 1983, *Perry* (NMW). **60**: on ash and oak, Cwm Clydach, 1988, *Orange*. **80**: fallen oak trunk by waterfall, Afon Perddyn, 1961, *Smith* (NMW, UCSA).

Radula complanata (L.) Dum.
Rare. **58**: tree beside stream, Bishopston Valley, 1963, *Smith* (BBSUK). **97**: on trees by Afon Alun, N of Castle-upon-Alun, 1974, *Perry*. **98**: on trunks of trees, carr woodland near Tir Eithin Farm West, 1988, *R.D. Pryce* (NMW!).

Porella platyphylla (L.) Pfeiff.
Frequent, on limestone rocks, cliffs and walls, less frequently on tree boles in calcareous areas, in all districts but commonest in Gower and the Vale. 00, 06, 07, 08, 16, 17, 18, *28*, 48, 49, 58, 59, *80*, 87, 97.

Frullania tamarisci (L.) Dum.
Occasional, on trees, walls, cliffs and acidic boulders, and in turf, fairly frequent in Gower but much rarer elsewhere, with four scattered records from the Uplands, one from the Upland ridge and one from the southern Vale. 00, *18*, 48, 49, 58, 59, 60, *87*, 90, 99.

Frullania dilatata (L.) Dum.
Common in Gower and the Vale with scattered records from the Upland ridge and the eastern and southern Uplands, on a variety of tree species, rarely on acidic boulders. 00, 01, 06, 07, *16*, *17*, 18, 48, 49, 58, 59, 60, 78, 87, *88*, 96, 97, 98.

Marchesinia mackaii (Hook.) S.Gray
Rare, on limestone. **16**: Sully Island, 1949, *Wade* (NMW). **48**: Paviland, 1934, *Wallace* (Smith, 1964). **59**: between Llethrid and Parkmill, 1961, *Smith* (NMW).

Lejeunea cavifolia (Ehrh.) Lindb.
Common, on a variety of tree species in woods, on soil on roadside banks and in quarries, and on stonework, calcareous rocks and stones in stream, in all districts, common in Gower and the Vale, rare in the Upland ridge and Uplands. 00, 01, 06, 07, 17, 18, 48, 49, 58, 59, 69, 80, 87, 97, 98.

Lejeunea lamacerina (Steph.) Schiffn.
Occasional, on rocks (often by water), soil banks, and on an oak trunk, in Gower, the Uplands and Upland ridge, absent from the Vale. 10, 48, 49, 58, 59, 79, 80, 98.

Lejeunea patens Lindb.
Very rare. **90**: on moist shaded cleft of rock at 1500 ft. alt., on *Campylopus atrovirens*, Craig y Llyn, 1963, *R.D. Fitzgerald* (Smith, 1964).

Lejeunea ulicina (Tayl.) Gott., Lindenb. & Nees
Frequent, on the bark of a variety of tree species in Gower, the Vale and the Upland ridge, absent from the Uplands except for one locality in the west. 07, 08, 17, 18, 48, 49, 58, 59, 60, 87, 97, 98.

Colura calyptrifolia (Hook.) Dum.
Very rare. **07**: trunk of *Salix* in wet carr, Hensol Forest, Welsh St Donats, S of Llantrisant, 1985, *Blockeel* (BBSUK), 1985, *Perry* (NMW).

Cololejeunea calcarea (Lib.) Schiffn.
Rare, on limestone. **01**: rocks by R. Taff Fechan, Morlais Hill, 1932, *Knight* (NMW). **80**: dry perpendicular rock by R. Neath near Pont-nedd-Fechan, 1963, *Duncan* (NMW).

MOSSES

Sphagnum papillosum Lindb.
Frequent, in bogs and in wet heaths, throughout the county where these occur. 07, 08, 09, 18, 19, 48, 58, 59, 60, 69, 70, 81, 89, 90, 98.

Sphagnum palustre L.
Frequent, in boggy or flushed ground in wet woodland and on moors, in scattered localities but very rare in the Vale. 07, 09, 18, 49, 58, 59, 60, 69, 71, 79, *90*, 98, 99.

Sphagnum squarrosum Crome
Occasional, in bogs and marshy hollows, rarely in slightly brackish marshes, in a few scattered localities throughout the county, sometimes in very small quantity. 00, 08, 09, *18*, 19, 49, 59, 60, 69, 70, 79.

Sphagnum teres (Schimp.) Ångstr.
Very rare. **09**: boggy field on Mynydd Eglwysilan, just S of Cilfynydd, 1975, *B. Scotter* (NMW).

Sphagnum fimbriatum Wils.
Common, in hollows in moorland and wet heath, at the edges of bogs and in flushes in woodland and hillsides, rare in the Vale. 00, 07, 08, 09, 18, 19, 49, 59, 60, 69, 70, 71, 79, 80, *88*, 89, 90, 98, 99.

Sphagnum capillifolium (Ehrh.) Hedw.
Frequent, in boggy places on heaths, in scattered localities throughout the county but rare in the Vale. *07*, *08*, 09, *18*, 48, 49, 60, 71, 78, 81, 88, 89, 90, *98*.

Sphagnum subnitens Russ.& Warnst.
Common, in boggy ground on wet heaths, wet woodland, open hillsides and moorland, throughout the county but rare in the Vale. 07, 08, 09, 10, 18, 48, 49, 58, 59, 60, 70, 78, 80, 81, 89, 90, *98*, 99.

Sphagnum compactum DC.
Very rare, on heaths. **49**: northern end of Rhossili Down (Smith, 1964). **98**: Hirwaun Common, 1973, *Wade*!.

Sphagnum auriculatum Schimp. var. **auriculatum**
Very common, in bogs, boggy ground and wet hollows in wet woodland, hillsides, heaths and moorland, occasionally on dripping rocks in the mountains, throughout the county but absent from

most of the Vale. 00, 07, 08, 09, 10, 18, 19, 48, 49, *58*, 59, 60, 69, 70, 78, 79, 80, 81, 88, 89, 90, 98, 99.

Sphagnum auriculatum var. **inundatum** (Russ.) M.O.Hill
Frequent in boggy ground in wet woodland, hillsides, heaths and moorland, throughout the county but rare in the Vale. 07, 08, 09, 18, *19*, 49, 59, 60, 80, *88*, 89, 90, *98*, 99.

Sphagnum cuspidatum Hoffm.
Frequent, in bogs and moorland, often submerged in acidic pools, scattered through Gower and the Uplands but absent from the Vale except for one record in the west. 08, 09, *18*, 19, 49, 59, 60, 69, 70, 79, 80, 81, 88, 89, *90*, *98*.

Sphagnum tenellum (Brid.) Brid.
Uncommon, in bogs and wet heaths, in a very few scattered localities in Gower and the Uplands, absent from the Upland Ridge and the Vale. 19, 48, 49, 81, *90*.

Sphagnum recurvum P.Beauv. var. **amblyphyllum** (Russ.) Warnst.
Occasional but under-recorded, in bogs, swampy hollows, heaths and wet woodland and on flushed hillsides. Recorded so far only from the Upland ridge and the Uplands in the eastern half of the county but almost certainly present in the western Uplands and in Gower. 08, 09, 10, 18, 19, *90*, 99.

Sphagnum recurvum P.Beauv. var. **tenue** Klinggr.
Very rare. 19: flush on N-facing slope at ca. 1050 ft. alt., Mynydd Eglwysilan, 1973, *Perry* (NMW).

Sphagnum recurvum P.Beauv. var. **mucronatum** (Russ.) Warnst.
Occasional but under-recorded, in bogs, by acidic pools, in boggy fields and on flushed banks on hillsides, in a few scattered localities in the central and eastern Uplands and from single localities in the Vale, Upland ridge and Gower. Almost certainly the plant is commoner in the west than the records indicate. 00, 07, 08, 09, 19, 49, 80, 90.

[**Andreaea alpina** Hedw.
90: about Llyn Vach near Aberpergam, 1805, *G. Sowerby* (Turner & Dillwyn, 1805), about Llyn Vach, near Aberpergwm (Gutch, 1842). Dubious record in need of confirmation.]

Andreaea rupestris Hedw. var. **rupestris**
Very rare, on siliceous rocks in the mountains. 90: Old Red Sandstone rock at top of Craig-y-Llyn, above Llyn Fach, 1961, *Smith* (BBSUK).

Andreaea rothii Web.& Mohr var. **rothii**
Very rare or doubtful, on acidic mountain rocks. 00: Cnwc, Troedyrhiw, 1959, *Wade* (NMW, det. as 'probably this – no capsules', B.M. Murray 1985).

Andreaea rothii Web. & Mohr var. **falcata** (Schimp.) Lindb.
Frequent, on hard siliceous rocks in the central and eastern Uplands, with one record from the eastern Vale. 00, 09, 18, 90, 98, 99.

Tetraphis pellucida Hedw.
Common, on rotting logs and stumps in shaded places, less commonly on tree bases, throughout the county but less frequent in the Vale. 00, 06, 07, 08, 09, 10, 17, 18, 19, 48, 58, 59, 60, 69, 70, 71, 78, 80, 81, 88, 89, 90, 98.

Tetrodontium brownianum (Dicks.) Schwaegr.
Very rare, in deeply shaded rock crevices. 80: on Millstone Grit by waterfall, Afon Perddyn, Pont-nedd-fechan, 1963, *Duncan* (BBSUK).

Polytrichum alpinum Hedw. var. **alpinum**
Very rare, on acidic rock ledges. 19: between Gellygaer and Bargoed, 1951, *T.L. Thomas* (Smith, 1964). 90: Craig-y-Llyn, 1906, *Knight* (NMW), 1932, *Knight* (NMW); above Llyn Fach, 1962, *Smith* (NMW).

Polytrichum longisetum Sw. ex Brid.
Uncommon, on peaty soil on a hillside in the western Vale, and on an ant-hill and on boggy ground, in four localities in the central Uplands, absent elsewhere. 79, 80, 81, 89, 90.

Polytrichum formosum Hedw.
Very common, on well-drained banks, on stumps and mossy boulders in woodland, often persisting after felling, on old ant-hills and on moorland and bogs, rarely amongst sandstone rocks, in all districts, common except in the Vale. 00, 01, 07, 08, 09, 10, 17, 18, 19, 48, 49, 58, 59, 60, 69, 70, 71, 78, 79, 80, 81, 88, 90, 98, 99.

Polytrichum commune Hedw. var. **commune**
Very common, in the wetter parts of bogs and heaths, occasionally in wet woodland, in all districts, common except in the Vale. 00, 01, 07, 08, 09, 10, 18, 19, 38, 48, 49, 58, 59, 60, 68, 69, 70, 71, 78, 79, 80, 81, 87, 88, 89, 90, 98, 99.

Polytrichum piliferum Hedw.
Very common, on dry gravelly acidic soil on banks, heaths and moorland, on ant-hills, in old quarries, and amongst acidic rocks in the mountains, in all districts, common except in the Vale. 00, 01, 07, 08, 09, 10, 18, 19, 48, 49, 58, 59, 60, 68, 69, 70, 71, 78, 79, 80, 81, 88, 89, 90, 98, 99.

Polytrichum juniperinum Hedw.
Common, on dry gravelly acidic soil on banks, walls and rocks, and ant-hills, in all districts, common except in the Vale. 00, 01, *07*, 08, 09, 10, 18, 19, 27, 48, 49, 58, 59, 60, 68, 69, 70, 71, 79, 80, 87, 88, 90, 98, 99.

Polytrichum alpestre Hoppe
Very rare and not seen recently, in *Sphagnum* bog. **81**: Gors Llwyn near Colbren Junction, 1908, *Knight* (NMW).

Pogonatum nanum (Hedw.) P.Beauv.
Uncommon, on acidic/peaty soil of stream banks and heathland, with two records from the Uplands and three from Gower. 09, 49, *69*, 90.

Pogonatum aloides (Hedw.) P.Beauv. var. **aloides**
Very common, on acidic soil on vertical earthy banks and cuttings, in all districts, common except in the Vale. 00, 07, 08, 09, 10, 18, 19, 28, 48, 49, 58, 59, 60, 69, 70, 71, 78, 80, 87, 88, 89, 90, 98, 99.

Pogonatum urnigerum (Hedw.) P.Beauv.
Common, on dry exposed gravelly soil on banks and rocks and in boggy ground, in all districts but uncommon in the Vale and very rare in Gower. 00, 01, 07, 08, 10, 18, 19, 49, 60, *69*, 70, 71, 78, 79, 80, 81, 88, 89, 90, 98, 99.

Oligotrichum hercynicum (Hedw.) Lam.& DC.
Uncommon, on clayey gravelly soil in the mountains. 60, 88, 89, 90, 98, 99.

Atrichum crispum (James) Sull.& Lesq.
Uncommon, on sandy detritus at the side of streams and rivers, especially in the mountains, with one record from the lowlands (Crymlyn Bog), the others from the Uplands and Upland ridge. 08, 09, 69, 70, 89, 90, 99.

Atrichum undulatum (Hedw.) P.Beauv. var. **undulatum**
Very common and widely distributed, on banks in woodland, in clearings, beside streams, amongst rocks, at the bases of trees, and on ant-hills and heaths, commonest in Gower and the Vale (except the southern coastal strip where it is absent), somewhat rarer in the Uplands. 00, 01, 07, 08, 09, 10, 17, 18, 19, 27, 28, 48, 49, 58, 59, 60, 69, 70, 71, 79, 80, 87, 88, 89, 90, 97, 98, 99.

Diphyscium foliosum (Hedw.) Mohr
Occasional, on dry acidic shaded soil, rock ledges and rocky banks, in a few scattered localities in the central and western Uplands. 60, 70, 79, 80, 89, 90, 99.

Archidium alternifolium (Hedw.) Mitt.
Occasional, on compacted soil on paths, in gateways, in arable fields, in turf on sea cliffs, on moorland and on mole-hills, in scattered localities throughout the county. 08, 17, 18, 48, 49, 58, 68, 69, 70, 87.

Pleuridium acuminatum Lindb.
Occasional, in turf in arable fields, on damp soil on paths, on river and lane banks, and on ant- and mole-hills, in Gower, the Uplands and the Upland ridge, absent from the Vale. 00, 18, 19, 48, 49, 58, 68, 78, 88.

Pleuridium subulatum (Hedw.) Lindb.
Rare, in turf in fields, on cliff tops and mole-hills. **00**: ant-hill, slope below Morlais Castle, 1976, *Perry* (NMW). **18**: Llanedeyrn Road, Pen-y-lan, 1949, *Wade* (Smith, 1964); Nant Glandulas, Lisvane reservoir, 1949, *Wade* (Smith, 1964); Mynydd Rudry, 1952, *Wade* (Smith, 1964). **28**: Cefn Mably Park, 1944, *Wade* (NMW). **48**: cliff tops near Paviland (Smith, 1964). **87**: mole-hills, Ogmore Down *Smith 1962* (NMW).

Pseudephemerum nitidum (Hedw.) Reim.
Occasional, on bare base-deficient soil on ditch sides, in wet fields and in turf, in a few scattered localities throughout the county. 07, *18*, 19, *58*, 70, *87*, 98.

Ditrichum cylindricum (Hedw.) Grout
Uncommon, on base-deficient soil on footpaths, in arable fields, on a stream bank, at the side of a lake and in grassland, in the Vale and the Uplands, absent from Gower. 07, 08, 18, 70, 87, 98.

Ditrichum flexicaule (Schimp.) Hampe
Frequent, in grassland, amongst rocks and on sand-dunes, in calcareous districts throughout the county, very rare elsewhere. 00, 01, *18*, 48, 49, 58, 78, 87, 88, 90, 97.

Ditrichum heteromallum (Hedw.) Britt.
Occasional, on acidic sandy, clayey or gravelly soil on banks in the Upland ridge and the Uplands, absent elsewhere. 00, 08, 70, 78, *79*, 89, 90, 98.

Distichium inclinatum (Hedw.) Br.Eur.
Rare and restricted to the Morlais Hill area and Kenfig Burrows, on limestone rocks and in calcareous grassland and dune. **00**: Morlais Castle Hill, 1923, *Wade* (NMW), 1932, *Knight* (NMW), 1963, *Smith* (NMW), 1968, *Harrison* (NMW), 1973, 1987, *Perry*; by Tâf Fechan (Smith, 1964). **01**: railway bridge over road c. ½ mile S of Pontsticill, 1973, *Perry* (NMW). **78**: sand dunes near Kenfig Pool, 1961, *Smith* (NMW), 1986, *Orange* (Hb. Orange).

Seligeria pusilla (Hedw.) Br.Eur.
Very rare, on shaded limestone. **00**: rock crevices under railway arches by Tâf Fechan (Smith, 1964). **58**: shaded rock, Bishopston Valley, s.d., *Binstead* (NMW).

Blindia acuta (Hedw.) Br.Eur.
Uncommon to rare, and in small quantity, on wet mountain rocks in the central Uplands. 70, 80, 90, 99.

Ceratodon purpureus (Hedw.) Brid. var. **purpureus**
Very common, occurring in every part of the county, on acidic soil, peat, industrial waste, amongst acidic rocks, on rocky banks, walls and tree trunks on roadsides, waste places, fields and grassland, moorland, cliff tops, old quarries and sand dunes. 00, 01, 06, 07, 08, 09, 10, 16, 17, 18, 19, 27, 28, 38, 48, 49, 58, 59, 60, 68, 69, 70, 71, 78, 79, 80, 81, 87, 88, 89, 90, 96, 97, 98, 99.

Rhabdoweisia fugax (Hedw.) Br.Eur.
Very rare, in rock crevices in the Uplands. **00:** E-facing Pennant sandstone outcrop, Mynydd Cilfach-yr-encil, Troedyrhiw, 1986, *Perry* (NMW). **80:** by R. Perddyn, near Pont-nedd-Fechan, 1932, *Wade* (NMW). **90:** Craig-y-Llyn, 1890, *A. Ley* (Armitage, 1921).

Rhabdoweisia crispata (With.) Lindb.
Very rare, in rock crevices in the Uplands. **80:** rocks by waterfall, Afon Perddyn, 1961, *Smith* (NMW). **90:** Craig-y-Llyn, 1932, *Knight* (NMW); sandstone rocks, cliffs above SE corner of Llyn Fach, 1976, *Perry* (NMW).

Rhabdoweisia crenulata (Mitt.) Jameson
Very rare, in mountain rocks. **90:** under large rock on cliffs above Llyn Fach, 1961, *Smith* (BBSUK, NMW, UCSA).

Cynodontium bruntonii (Sm.) Br.Eur.
Very rare, on exposed mountain rocks. **00:** E-facing Pennant sandstone outcrop, Mynydd Cilfach-yr-encil, Troedyrhiw, 1986, *Perry* (NMW). **90:** Craig-y-Llyn, 1932, *Knight* (NMW); sandstone rocks, cliffs above S side of Llyn Fach, 1976, *Perry* (NMW).

Dichodontium pellucidum (Hedw.) Schimp.
Occasional, on moist rocks, rarely ground, on banks, by streams and rivers and in a limestone quarry, in all districts but rare in the Vale and Gower. 00, 01, 18, 58, 59, 80, 90, 97, 98, 99.

Dichodontium flavescens (With.) Lindb.
Rare, on wet rocks by rivers. **01:** by R. Taff Fechan, Morlais Hill, 1932, *Knight* (NMW). **80:** by first waterfall, Afon Pyrddin (Smith, 1964). **90:** by R. Neath, Pont-nedd-Fechan, 1963, *Harrison* (NMW).

Dicranella palustris (Dicks.) Crundw. ex Warb.
Occasional, in flushes and on dripping rocks in the mountains, only in the Uplands. 09, 60, 70, 89, 90, 98, 99.

Dicranella schreberiana (Hedw.) Dix.
Uncommon, on soil in arable land and gardens and in open places in woods and roadsides, in a few localities throughout the county. 09, 17, 48, 58, 69, 70, 96.

Dicranella subulata (Hedw.) Schimp.
Very rare. **90:** bare clay on vertical face of bank, SE side of Llyn Fawr, 1973, *Perry* (NMW, BBSUK).

Dicranella rufescens (With.) Schimp.
Occasional, on clayey banks of waterways and ponds in the Uplands and Upland ridge, absent from the Vale and from Gower. 00, 08, 10, 70, 78, 79, 89, 90, 98.

Dicranella varia (Hedw.) Schimp.
Common, on calcareous soil on banks and amongst limestone and calcareous rocks, throughout the county, commonest in the Vale. 00, 01, *06*, 08, 09, 16, 17, 18, 28, 49, 58, 60, 87, 89, 96, 97, 98, 99.

Dicranella staphylina Whitehouse
Rare but probably under-recorded, on acidic soil. **07:** near lake margin, Hensol Forest, Welsh St Donats, 1985!, *Paton & Blockeel* (BBSUK). **17:** bare muddy bank of R. Ely, St Fagans, 1988, *Perry*. **98:** compacted sandy bank of stream, Nant Iechyd, Cwm Dimbath, 1974, *Perry*; on top of wall, Allt y Rhiw, Blackmill, 1987, *Orange* (Hb. Orange).

Dicranella heteromalla (Hedw.) Schimp.
Very common, on acidic soil on banks, more rarely on wood, in moorland and heathland, in old quarries, on roadsides and in woodland, in every part of the county. 00, 01, 06, 07, 08, 09, 10, 17, 18, 19, 28, 38, 48, 49, 58, 59, 60, 68, 69, 71, 78, 79, 80, 81, 87, 88, 89, 90, 97, 98, 99.

Dicranoweisia cirrata (Hedw.) Lindb. ex Milde
Very common and widespread, on trees throughout the county but especially in the lowlands, occasionally on exposed rocks in the mountains and on the coast. 00, 01, 06, 07, 08, 09, 10, 17, 18, 19, 28, 38, 48, 49, 58, 59, 60, 68, 69, 70, 71, 80, 87, 88, 90, 97, 98, 99.

Dicranum bonjeanii De Not.
Uncommon, in a few localities on moorland, sometimes amongst *Sphagnum*, and in calcareous grassland, on Gower and in the eastern Upland ridge, absent elsewhere. 18, 28, 48, 49, *58*, 98.

Dicranum scoparium Hedw.
Common, in neutral to strongly acidic conditions, in heathland, moorland and bogs, on banks, in woods at tree bases and on tree trunks, rarely on rocks, in all districts. 00, 01, 06, 07, 08, 09, 10, 18, 19, 28, 38, 48, 49, 58, 59, 60, 68, 69, 70, 71, 80, 87, 88, 90, 97, 98, 99.

Dicranum majus Sm.
Occasional, on banks in woodland and by a river, on rock ledges in old quarries and amongst rocks in the mountains, two records from Gower, the others from the Uplands and the Upland ridge. 00, 08, 18, 48, 59, 80, 90, 98, 99.

Dicranum fuscescens Sm. var. **fuscescens**
Very rare. **07:** old willow trunk in shallow water, heathland lake near Welsh St Donats, 1968, *A.G. Side* (BBSUK).

Dicranum scottianum Turn.
Very rare. **48:** Old Red Sandstone conglomerate, W-facing hillside, Rhossili Down, 1967, *Perry & Corley* (BBSUK).

Dicranum montanum Hedw.
Very rare. **18:** silver birch, shady damp woodland, Copi Gwythi, Rudry, 1980, *Garlick* (BBSUK).

Dicranum tauricum Sapehin
Rare but possibly spreading, on rotting wood in marshy ground and in woodland. 07: near Pysgodlyn Mawr, 1977, *Perry*. 18: in woodland, Castell Coch, Tongwynlais, 1976, *Perry* (NMW); on SE-facing wooded hillside on W side of Gwaelod y Garth, 1982, *Perry* (NMW); wet deciduous woodland, Coed y bedw, 1983, *Perry* (NMW). 19: in marsh, valley N of A472, *c.* ¾ mile E of Nelson, 1976, *Perry* (NMW, BBSUK).

Dicranodontium denudatum (Brid.) Broth. var. **denudatum**
Rare. 60: on wood and tree bases, Cwm Clydach 1988, *Orange*. 98: fallen bole of sallow in marsh by Nant Iechyd, Cwm Dimbath, 1974, *Perry* (BBSUK); on log, Allt y Rhiw, Blackmill, 1987, *Orange* (Hb. Orange).

Campylopus subulatus Schimp.
Very rare. 98: gravelly track by Nant Iechyd, Cwm Dimbath, 1974, *Perry* (BBSUK).

Campylopus fragilis (Brid.) Br.Eur.
Rare. 48: wall, Rhossili Down, 1932, *Knight* (NMW). 80: rocks near waterfall, Afon Perddyn, 1961, *Smith* (NMW).

Campylopus pyriformis (Schultz) Brid. var. **pyriformis**
Frequent, on peaty soil and tree bases in woodland and on moorland, in all districts but rare in the Vale. 00, 07, 09, 18, 19, 27, 48, 49, 58, 59, 60, 79, 88, 98.

Campylopus paradoxus Wils.
Common, on peaty soil, acidic rocks, tree bases and rotting stumps, sometimes in boggy ground and in turf, in moorland, open hillsides and woodland, commonest in Gower, the Uplands and the Upland ridge, rare in the Vale. 00, 01, 07, 08, 09, 18, 19, 48, 49, 58, 59, 60, 68, 69, 70, 71, 80, 81, 89, 90, 98, 99.

Campylopus atrovirens De Not. var. **atrovirens**
Uncommon, on wet siliceous rocks in Gower and the western and central Uplands. 48, 49, 58, 60, 90, 99.

Campylopus introflexus (Hedw.) Brid.
Frequent, amongst acidic rocks, on peaty ground, on heathlands and dry acidic banks, on the shelving edge of a reservoir, in woodland and on sand dunes. First noted in Glamorgan in 1963 this species is now scattered throughout the county in all districts, presumably having spread rapidly in the last thirty years. 00, 07, 09, 17, 18, 48, 49, 60, 70, 71, 79, 80, 87, 98.

Campylopus brevipilus Br. Eur.
Rare and in small quantity, on damp peat and flushed rocks on heath. 07: Mynydd-y-Glew, 1922, *Richards* (NMW). 48: Rhossili Down, 1934, *Wallace* (NMW), 1974, *Perry* (NMW). 49: Rhossili Down (N), 1974, *Perry* (NMW).

[**Leucobryum glaucum** (Hedw.) Ångstr.
Recorded for Glamorgan but all specimens that I have examined (NMW, BBSUK) are *L. juniperoideum*.]

Leucobryum juniperoideum (Brid.) C.Muell.
Uncommon and often in small quantity, on tree trunks, more rarely on acidic soil in woodland, in a very few localities in the Upland ridge and in one in the Vale and one in the western Uplands. 07, 18, 28, 60, 98.

Fissidens viridulus (Sw.) Wahlenb.
Occasional, on base-rich soil on banks, amongst rocks and in woods, rarely on mortar, in scattered localities in all districts but absent from the central Uplands. 00, 06, 17, 18, 58, 60, 87, 98.

Fissidens pusillus (Wils.) Milde var. **pusillus**
Occasional, on soft, usually slightly basic, boulders and stones and on tufa, often by streams, in shaded woodland, in a few scattered localities but absent from the central Uplands. 01, 07, 17, 18, 49, 58, 60, 69, 80, 97.

Fissidens pusillus (Wils.) Milde var. **tenuifolius** (Boul.) Podp.
Very rare. 49: on piece of limestone at edge of wood near Cheriton, 1962, *Smith* (*Trans.br.Bryol.Soc.* **6**, 196 (1970), as *F. minutulus* Sull. var. *minutulus*).

Fissidens limbatus Sull.
Uncommon, on soil: 17: clay bank, roadside near Old Cogan, 1950, *Wade* (NMW). 18: roadside bank NE of Lisvane, 1949, *Wade* (NMW). 59: wood near Llethrid, 1961, *Smith* (NMW). 87: soil crevices in walls, Ogmore Down, 1962, *Smith* (NMW).

Fissidens incurvus Starke ex Roehl.
Occasional but probably overlooked, on calcareous soil on sea cliffs, more rarely on calcareous banks inland, in Gower and the Vale, absent from the Uplands and Upland ridge. 06, 16, 17, 18, 48, 49, 58, 59, 87, 96, 97.

Fissidens bryoides Hedw.
Very common, on soil in woodlands and in gardens, probably favouring slightly acidic conditions, rarely on rocks in streams and on wet banks, in all districts but rare in the mountains. 00, 01, 06, 07, 08, 10, 16, 17, 18, 19, 27, 28, 48, 49, 58, 59, 60, 68, 69, 70, 71, 80, 88, 90, 96, 97, 98.

Fissidens curnovii Mitt.
Very rare. 08: on silty tree roots and rocks by stream by Ely River near Ynysmaerdy, 1987, *Orange* (Hb. Orange). 98: dripping rock outcrops by Nant Iechyd, Cwm Dimbath, 1974, *Perry* (NMW).

Fissidens crassipes Wils.ex Br.Eur.
Rare, on calcareous rocks and tree roots in streams. 17: tree roots by stream, Roath Park, Cardiff, 1990, *Orange*. 18: Castell Coch, 1932, *Richards* (NMW). 87: in

Afon Alun, Pont Alun near Ewenny, 1962, *Smith* (NMW). **97**: Llanblethian, 1928, *Richards* (NMW); at side of Afon Alun, *c*. ¾ mile NW of Castle-upon-Alun, 1976, 1983, *Perry* (NMW).

Fissidens rufulus Br.Eur.
Very rare, on rocks in streams and rivers. **58**: Bishopston Valley, 1961, *Smith* (UCSA), 1962, *Smith* (BBSUK), 1963, *Peterken* (BBSUK). **80**: Afon Pyrddin, 1963, *Duncan* (Smith, 1964).

Fissidens exilis Hedw.
Uncommon, on shaded calcareous clay banks, frequently in woodland, on Gower and on the coast of the southern Vale, absent elsewhere. 06, 48, 49, 58, 59, 96.

Fissidens celticus Paton
Uncommon, often on sticky, smooth clay on banks. **08**: on shaded soil bank by stream by Ely River near Ynysmaerdy, 1987, *Orange* (BBSUK, Hb. Orange). **58**: acid soil bank by stream, near Ilston, near Cartersford, 1990, *Orange*. **59**: soil bank near stream, among liverworts, Gelli-hîr Wood, 1989, *Kay* (NMW!). **70**: acid soil bank in fragment of wood by stream, near Tan-y-rhiw, Cilfrew, near Neath, 1990, *Orange*. **98**: spoil heap of opencast mine, NW of Llanharry, 1990, *Orange*.

Fissidens osmundoides Hedw.
Uncommon, in very few localities on wet vertical rock faces and ledges, in the central to eastern Uplands. 00, 80, 90, 99.

Fissidens taxifolius Hedw. ssp. **taxifolius**
Very common, on shaded soil, usually slightly basic, on woodland floors, banks, in lanes, fields and gardens, in all districts but rare in the mountains. 00, 01, 06, 07, 08, 16, 17, 18, 19, 28, 48, 49, 58, 59, 68, 70, 80, 87, 88, 90, 96, 97, 98, 99.

Fissidens cristatus Wils.ex Mitt.
Frequent, on calcareous rocks, mortar in walls and on sand dunes, throughout the county but absent from much of the Uplands. 00, 01, 07, 17, 18, 48, 49, 58, 59, 60, 68, 80, 97, 98.

Fissidens adianthoides Hedw.
Frequent, on damp rocky banks and flushed ground in woods and roadsides, on walls, mortar and on cliffs, frequently in quarries, throughout the county but absent from much of the Uplands. 00, 07, 17, 18, 48, 49, 58, 59, 60, 68, 80, 87, 98, 99.

Encalypta vulgaris Hedw.
Frequent, in crevices of calcareous rocks, on soily limestone slabs in old quarries, on mortar and on sand dunes, in scattered localities but absent from much of the Vale and the Uplands. 00, 01, 07, *18*, 48, 49, 58, 68, 78, 80, 87.

Encalypta streptocarpa Hedw.
Common, on the mortar of walls, on soily limestone slabs in old quarries and in crevices of calcareous rocks, in all districts but rare in the mountains. 00, 01, 07, 08, 17, 18, 48, 49, 58, 59, 60, 68, 69, 80, 87, 88, 90, *96*, *97*, 98.

Tortula ruralis (Hedw.) Gaertn. ssp. **ruralis**
Frequent, on sand dunes, on soil on sea cliffs, rarely on roofs and walls, absent from the Uplands. 06, *07*, 16, *17*, 18, 38, 48, 49, 58, 59, 78, 79, 87, 88, 97.

Tortula ruralis (Hedw.) Gaertn. ssp. **ruraliformis** (Besch.) Dix.
Common on sand dunes, sandy soil and concrete paths near the sea, rare or absent elsewhere, with one record, from a tarmac path, in the Uplands. 07, 09, 16, 48, 49, 58, 69, 78, 79, 87, 88, 96.

Tortula intermedia (Brid.) De Not.
Frequent, on calcareous rocks and walls in all districts but commonest in the Vale. 00, 06, 08, *17*, *18*, 48, 49, 58, *87*, 90, *96*, 97.

Tortula laevipila (Brid.) Schwaegr. var. **laevipila**
Occasional, on the trunks of trees in the Vale and Gower, with more records from the former, absent from the Uplands and Upland ridge. 06, *07*, *16*, *17*, 48, 58, 87, 96.

Tortula laevipila (Brid.) Schwaegr. var. **laevipiliformis** (De Not) Limpr.
Uncommon, on trees in the Vale and Gower, absent from the Uplands and Upland ridge. 06, 49, 58, 87, 97.

Tortula muralis Hedw. var. **muralis**
Very common, on walls and stonework, rarely on rocks and on soil in a quarry, in all districts but most frequent in the lowlands, rare in the mountains. 00, 01, 06, 07, 08, 09, 10, 16, 17, 18, 19, 27, 28, 48, 49, 58, 59, 60, 68, 69, 70, 71, 79, 87, 88, 89, 90, 96, 97, 98.

Tortula subulata Hedw. var. **subulata**
Uncommon, on sheltered limestone rocks and walls and calcareous soil, on sea cliffs, in a disused quarry and at a roadside, confined to southern Gower, the central Vale and the north-eastern Uplands. 00, 01, 07, 48, *58*.

Tortula subulata Hedw. var. **graeffii** Warnst.
Rare. **00**: limestone in cutting of disused railway E of Morlais Hill, 1973, *Perry* (NMW, BBSUK).

Tortula latifolia Bruch ex Hartm.
Rare. **78**: fallen log in alder wood by Afon Kenfig, Kenfig Burrows, 1963, *Smith* (BBSUK).

Aloina aloides (Schultz) Kindb. var. **aloides**
Frequent, on limestone and calcareous soil, amongst rocks, on shaded banks, walls and dunes, at roadsides, in old quarries and on sea cliffs, in Gower, the Vale, the eastern Uplands and the eastern half of the

Upland ridge. 00, 06, *16*, 17, *18*, 48, 49, *58*, 68, *96*, 97, 98.

Aloina aloides (Schultz) Kindb. var. **ambigua** (Br.Eur.) Craig
Very rare. **69**: Swansea, 1882, *T. Barker* (MANCH, teste BBS Recorder's Book).

Desmatodon convolutus (Brid.) Grout
Very rare, on coastal soil in Gower. **38**: soil patches, Inner Head, Worms Head, 1962, *Smith* (NMW). **58**: stony bare ground on coast near sea near Caswell Bay, 1929, *Binstead* (BBSUK).

Pottia starkeana (Hedw.) C.Muell. ssp. **starkeana** var. **starkeana**
Very rare. **58**: Caswell Bay, 1929, *Binstead* (BBSUK).

Pottia starkeana (Hedw.) C.Muell. ssp. **starkeana** var. **brachyodus** C.Muell.:
Very rare. **48**: soil on limestone cliff, Rhossili, 1965, *Paton* (Hill, M.O. (1980). *Bull.br.Bryol.Soc.* **35**, 9).

Pottia starkeana (Hedw.) C.Muell. ssp. **conica** (Schleich.ex Schwaegr.) Chamberlain: Very rare. **48**: soil near quarry on Cefn Bryn, near Arthur's Stone, 1962, *Smith* (NMW).

Pottia crinita Wils. ex. Br. Eur.
Very rare, on soil on coastal cliffs in Gower. **38**: Inner Worm, Worms Head, 1962, *Smith* (NMW, BBSUK). **58**: Pwll du Head, 1963, *Crundwell* (Smith, 1964); Pennard cliffs, 1967, *Perry* (Kay, 1989).

Pottia lanceolata (Hedw.) C.Muell.
Uncommon, on compacted soil on cliffs on the south coast of Gower. 38, 48, 58, 68.

Pottia intermedia (Turn.) Fuernr.
Occasional, on soil, frequently near the sea, in quarries and on cliffs, and in arable or waste land, most frequent in Gower and the Vale with a single record from the eastern Upland ridge. 06, 08, *16*, *17*, 48, 49, 58, 87.

Pottia truncata (Hedw.) Fuernr.
Occasional, on base-deficient soil in arable and waste land and gardens, in Gower, the Vale, the eastern Upland ridge and the south-eastern Uplands. 06, 08, 17, 18, 19, 48, 49, 58, 87, 98.

Pottia heimii (Hedw.) Fuernr.
Uncommon, on or near the coast on muddy saline soil or tracks, in a very few scattered localities from western Gower to eastern Vale. 06, 49, *58*, 79, *87*.

Pottia bryoides (Dicks.) Mitt.
Very rare, on calcareous soil on Gower. **38**: among rocks, Inner Head, Worms Head, 1962, *Smith* (BBSUK). **49**: Burry Pill, 1963, *Appleyard and K. Lye* (Smith, 1964).

Pottia recta (With.) Mitt.
Uncommon, on bare calcareous soil in fields and on sea cliffs, in a few localities in Gower and in one in the Vale. *17*, 48, 49, 58, 68.

Phascum cuspidatum Hedw. var. **cuspidatum**
Frequent, on soil in arable fields, on a soily bank, on gravelly paths, in gardens and waste places and in turf on sea cliffs, in all districts but very rare in the Uplands and Upland ridge. 00, 06, 16, 17, 18, *28*, 38, 48, 49, 58, 68, 69, 70, 87.

Phascum cuspidatum Hedw. var. **piliferum** (Hedw.) Hook.& Tayl.
Rare, on soil near the sea on Gower. **48**: Rhossili, 1963, *Paton* (Smith, 1964); Port Eynon Point, 1963, *Paton* (Smith, 1964). **49**: sea wall, Cheriton (Smith, 1964). **58**: soil on cliffs above Pwll du Bay, 1961, *Smith* (BBSUK).

Acaulon muticum (Brid.) C.Muell.
Rare and local, on base soil on limestone cliffs in Gower. **38**: top of eroded mole hill, Inner Head, Worms Head, 1962, *Smith* (NMW, BBSUK). **48**: field near Paviland (Smith, 1964); turf on cliff top near Paviland, 1962, *Smith* (NMW); mole-hill on cliffs south-west of Rhossili, 1963, *Crundwell* (Smith, 1964). **58**: ant-hills at base of Carboniferous limestone cliffs by the sea, Pwll Du Head, 1967, *Perry* (Hb. Perry).

Scopelophila cataractae (Mitt.) Broth.
Very rare, and possibly exterminated by cosmetic landscaping. **69**: old zinc waste, Lower Swansea Valley, 1967, *Perry*, new to Europe (NMW); at edge of site of Dillwyn spelter works, at bottom of bank of zinc waste which forms the N-facing side of the embankment of London-Swansea railway line, Lower Swansea Valley, 1983, *Perry* (NMW) (*J.Bryol.* **13**, 323–328, 1985), not re-found 1985.

Barbula convoluta Hedw. var. **convoluta**
Very common throughout the county, on bare soil, paths and waste ground, amongst rocks, in crevices in walls and on sand dunes. 00, 01, 06, 08, 09, 16, 17, 18, 27, 28, 38, 48, 49, 58, 59, 60, 68, 69, 71, 78, 79, 80, 87, 88, 89, 90, 97, 98.

Barbula convoluta Hedw. var. **commutata** (Jur.) Husn.
Frequent, on walls and soil on paths and in waste places, in scattered localities throughout the county. 00, 07, 08, 09, 10, 18, 19, 48, 49, 58, 87, 89, 97, 98.

Barbula unguiculata Hedw.
Very common throughout the county, on soil on banks and paths, on stony ground, in fields and in crevices in rocks and on mortar in walls. 00, 01, 06, 07, 08, 09, 10, 16, 17, 18, 19, 28, 48, 49, 58, 59, 60, 68, 69, 70, 71, 78, 79, 80, 87, 88, 90, 96, 97, 98, 99.

Barbula hornschuchiana Schultz
Frequent, on calcareous soil and sand, and amongst fine calcareous rubble, on concrete, in turf on

limestone or in dune slacks, on walls and in quarries, recorded from all districts, most frequent in the Vale, rare in the Upland ridge, in the Uplands restricted to the limestone region of the north-east. 00, 01, 06, 08, 16, 17, 19, 48, 49, *87*, 88, 97.

Barbula revoluta Brid.
Uncommon, on limestone walls and rocks, and on mortar and concrete, with a few scattered records from all districts, but absent from the central and western Uplands. 00, *17*, 18, 58, *96*, 97.

Barbula acuta (Brid.) Brid.
Rare, on calcareous soil on the coast. **48**: cliffs, Kitchen Corner, 1963, *Appleyard* (Smith, 1964). **58**: on soil among rocks on cliffs between Caswell Bay and Brandy Cove, 1960, *Smith* (BBSUK). **78**: dune slack, Kenfig Burrows, 1963, *Smith & Paton* (NMW).

Barbula fallax Hedw.
Common, on bare base-rich soil in fields, amongst rocks, on banks, in turf on sea cliffs, in quarries, on rides in woods and damp tracks, and in dune slacks, throughout the county but uncommon in the Uplands. 00, 01, 06, 07, 08, 16, 17, 18, 27, 28, 48, 49, 58, *60*, 68, 71, 78, 79, 87, 88, 89, 96, 97, 98.

Barbula reflexa (Brid.) Brid.
Very rare, on gravelly soil over limestone. **00**: Morlais Hill, 1968, *Pettifer, Gardiner & Peterken* (BBSUK), 1973!; Taf Fechan, 1976, *Perry*.

Barbula spadicea (Mitt.) Braithw.
Rare, on soil among rocks. **17**: roadside bank near Cyncoed, Penylan, Cardiff, 1925, *Wade* (NMW). **27?**: marly rocks between Penylan and Coed-y-Gores, Cardiff, 1925, *Wade* (NMW). **58**: earthy stone, Bishopston Valley, 1963, *Crundwell* (Smith, 1964).

Barbula rigidula (Hedw.) Mitt.
Common, on moist calcareous rocks and walls, brickwork and concrete, often in an ooze, throughout the county. 00, 01, 07, 08, 16, *17*, 18, 19, 48, 49, *58*, 60, 69, 70, 80, *87*, 90, 96, 97, 98, 99.

Barbula trifaria (Hedw.) Mitt.
Frequent, on humid calcareous walls and rocks, rarely on soil, in every district but uncommon in the Uplands and Upland ridge. 00, 07, 08, 18, 48, 49, 58, 68, 70, 87, 97, 99.

Barbula tophacea (Brid.) Mitt.
Frequent, and often forming tufa, on wet limestone and damp calciferous rocks, on damp calcareous soil, on walls and sea cliffs, and in sand dunes, frequent in the Vale, infrequent in Gower, rare in the Uplands and Upland ridge. 00, *06*, 17, 48, 49, 87, 88, 90, *96*, 97, 98.

Barbula cylindrica (Tayl.) Schimp.
Very common and widely distributed throughout the county, but probably less frequent in the mountains, on humid walls and stonework, mortar, gravelly

paths, shaded clayey soil, earthy rock crevices and on silt on river banks. 00, 01, 06, 07, 08, 09, 10, 17, 18, 19, 28, 48, 49, 58, 60, 68, 69, 71, 78, 80, 87, 89, 90, 97, 98, 99.

Barbula recurvirostra (Hedw.) Dix.
Very common, in sheltered crevices in brickwork and rocks, rarely on soil, preferring calcareous conditions, throughout the county. 00, 01, 06, 08, 16, 17, 18, 48, 49, 58, 60, 68, 69, 70, 71, 78, 87, 88, 89, 90, 97, 98, 99.

Barbula ferruginascens Stirt.
Rare, only in Gower and the Uplands. **49**: among limestone rocks, Tor-gro, Landimore, 1963, *Warburg* (Smith, 1964). **80**: wet shale rocks by waterfall, R. Perddyn, 1963, *Richards* (NMW). **90**: on exposed rock by stream below Llyn Fawr Reservoir, 1961, *Smith* (BBSUK).

Gymnostomum luisieri (Sérgio) Sérgio ex Crundw.
Rare, on shaded limestone rocks. **18**: Gelli Quarry, N of Rhiwbina, 1933, *Wade* (NMW), 1987, *Whitehouse*; roadside near the Black Cock Inn, near Thornhill, 1950, *Wade* (NMW). **97**: old limestone quarry, W of the R. Alun, opposite Ewenny Down, 1976, *E.W. Jones & Perry* (NMW).

Gymnostomum calcareum Nees & Hornsch.
Rare, on compacted calcareous soil and limestone. **18**: calcareous soil on exposed bank, woodland, Coed y bedw, Gwaelod y Garth, 1976, *Perry* (NMW). **58**: on rock, Bishopston Valley, s.d., *Binstead* (NMW).

Gymnostomum aeruginosum Sm.
Uncommon, restricted to calcareous habitats, on thin soil over rocks, on mortar in walls, on flat soily limestone slabs in an old quarry and on wet rocks by waterfalls and rivers, with very few records from the north and east Uplands, and the eastern Upland ridge. 00, 01, 18, 80, 90.

Gymnostomum recurvirostrum Hedw.
Rare, on damp base-rich rocks. **08**: Pontypridd, 1878, *E.M. Holmes* (Smith, 1964). **68**: road cutting, Mumbles Head (Smith, 1964). **90**: wet rocks, Craig-y-Llyn, near Hirwaun, 1950, *Crundwell* (BBSUK).

Gyroweisia tenuis (Hedw.) Schimp.
Rare, on shaded calcareous rock. **08**: base of stone ornament in garden, Talygarn, S of Pontyclun, 1985, *Whitehouse* (BBSUK). **18**: with *Eucladium verticillatum*, crevice in limestone cliff face under a beech at the eastern end of Gelli Quarry, 1987, *Whitehouse*.

Anoectangium aestivum (Hedw.) Mitt.
Rare, in crevices in acidic rocks in the Uplands. **80**: right bank of R. Perddyn, 1932, *Knight* (NMW), rocks on river bank, Afon Pryddin falls, Pont-nedd-Fechan, 1963, *Harrison* (NMW). **90**: Craig-y-Llyn (Smith, 1964).

Eucladium verticillatum (Brid.) Br.Eur.
Occasional, on dripping or moist calcareous rock faces in old quarries and on sea cliffs, often forming tufa, with a few scattered records from the southern Vale, Gower, the Upland ridge and the north-east Uplands. 00, 06, *16*, *17*, 18, 58, 59, 87, 90, 96.

Weissia controversa Hedw. var. **controversa**
Frequent, on neutral or slightly acidic bare soil on banks, in all districts but most common in the Vale and Gower, rare in the Uplands and Upland ridge. 00, *06*, 07, 08, *17*, 18, 28, 48, 49, 58, 68, 87, 97, 99.

Weissia controversa Hedw. × **Weissia longifolia** Mitt. var. **angustifolia** (Baumg.) Crundw. & Nyholm
Very rare. **49**: earth banks, Cwm Ivy, Cheriton, 1963, *R.D. Fitzgerald* (Hb. Perry).

Weissia perssonii Kindb.
Very rare. **48**: Carboniferous limestone cliffs south of Port Eynon, 1963, *Warburg* (OXF, *fide* Crundwell, A.C. (1971). *Trans.br.Bryol.Soc.* **6**, 222).

Weissia rutilans (Hedw.) Lindb.
Very rare. **18**: on sandy woodland road, Copi Gwythi, Rudry, Caerphilly, 1980, *Garlick* (BBSUK).

Weissia microstoma (Hedw.) C.Muell. var. **microstoma**
Uncommon, on calcareous soil on sea cliffs in Gower, also recorded from the north-eastern Uplands on a roadside bank. 00, 38, 48, 58, 68.

Weissia longifolia Mitt. var. **longifolia**
Very rare. **18**: Dolomite boulder in W bank of R. Taff, Gwaelod y Garth, Taff Vale, 1979, *Perry* (BBSUK).

Weissia longifolia Mitt. var. **angustifolia** (Baumg.) Crundw. & Nyholm
Rare, in turf over limestone near the sea, with three records from Gower and two from the western Vale. 48, 49, 58, 87.

Weissia longifolia Mitt. var. **angustifolia** (Baumg.) Crundw. & Nyh. × **Weissia controversa** Hedw.
Very rare. **49**: earth banks, Cwm Ivy, Cheriton, 1963, *R.D. Fitzgerald* (Hb. Perry). Probably also this hybrid at **48**: Kitchen Corner, Rhossili, 1963, *Paton* (Smith, 1964).

Weissia levieri (Limpr.) Kindb.
Rare and very local, only on Gower, in calcareous turf on sea cliffs and in rock crevices. **38**: Inner Head, Worms Head (Smith, 1964). **48**: bank near Overton, 1961, *Smith* (NMW); soil on cliffs, and rock crevices in quarry, Port Eynon Point, 1962, *Smith* (NMW); bare patches in limestone grassland, cliffs near Culver Hole, Port Eynon, 1963, *Richards* (NMW).

Oxystegus sinuosus (Mitt.) Hilp.
Occasional, on shaded humid calcareous rocks and stones, often in woodland, rarely on trees, through-

out the Vale and in Gower, absent from the Uplands and Upland ridge. 06, *16*, 17, 48, 49, 58, 87, 96, 97.

Oxystegus tenuirostris (Hook.& Tayl.) A.J.E.Smith var. **tenuirostris**
Occasional, on damp or shaded rocks often by water, in scattered localities from Gower to the eastern Uplands, absent from the Vale and the Upland ridge. 01, 48, 59, 60, 80, 90, 98, 99.

Trichostomum crispulum Bruch
Frequent, in crevices and on ledges in outcropping limestone, on limestone walls and boulders, and on sea cliffs, rarely on other basic rocks or in calcareous turf and sand dunes, in all districts but most frequent in Gower and the Vale, absent from the western and central Uplands. 00, 01, 06, 07, 18, 38, 48, 49, 58, 68, 78, 87, 96, 97.

Trichostomum brachydontium Bruch
Frequent, on rocks and walls, usually calcareous, occasionally on calcareous soil, in all districts. 00, 06, 07, *16*, 18, 48, 49, 58, 59, 68, 87, 96, 97, 98, 99.

Tortella tortuosa (Hedw.) Limpr.
Frequent, on limestone rocks and walls, rarely on soil, in all districts, but probably commonest in Gower, very rare in the central Uplands, absent from the eastern Vale and the western Uplands. 00, 01, 18, 28, 48, 49, 58, 59, 68, 87, 90, 97.

Tortella flavovirens (Bruch) Broth. var. **flavovirens**
Occasional, on sand-dunes and soil, rocks and cliffs, chiefly near the sea, in Gower and the southern Vale, absent elsewhere. *16*, 38, 48, 49, 58, 68, 87, 96.

Tortella flavovirens (Bruch) Broth. var. **glareicola** (Christens.) Crundw.& Nyholm
Rare, in sand-dunes. **49**: Llangennith Burrows, 1963, *Crundwell* (Smith, 1964). **68**: Langland Bay, 1963, *Warburg* (Smith, 1964). **78**: Kenfig Burrows, 1963, *Crundwell et al.* (BBSUK), 1989, *Perry* (NMW).

Tortella inclinata (Hedw.f.) Limpr.
Very rare and local, near the sea in west Gower. **49**: Spaniard Rocks near Burry Holmes, 1934, *Wallace* (Smith, 1964); dunes by track at SW end of Whiteford Burrows, 1963, *Smith* (NMW).

Tortella nitida (Lindb.) Broth.
Occasional, on exposed limestone and dolerite, frequent in Gower in suitable habitats with two records from the Vale and one from the Upland ridge. 18, 48, 49, 58, 59, 87, 97.

Pleurochaete squarrosa (Brid.) Lindb.
Rare, on sandy or calcareous soil. **49**: by track at southern end of Whiteford Burrows, 1963, *Paton, Smith & Wallace* (Smith, 1964); Llangennith Burrows, 1967, *Perry* (Kay, 1989). **58**: Pennard, 1934, *Wallace* (Smith, 1964); edge of dune slack, Oxwich Burrows (Smith, 1964). **97**: calcareous grassland on hillside NW of Castle-upon-Alun, 1975, *Perry* (NMW).

Trichostomopsis umbrosa (C.Muell.) Robins.
Very rare but probably under-recorded. 08: near base of N-facing brick wall in garden, Talygarn, S of Pontyclun, 1985, *Whitehouse & C.D. Preston* (BBSUK).

Leptodontium flexifolium (With.) Hampe ex Lindb.
Very rare. **18**: soily crevice in sandstone rock outcrops, Mynydd Rudry, Rudry Common, 1974, *Perry* (NMW, BBSUK).

Cinclidotus fontinaloides (Hedw.) P.Beauv.
Rare, on calcareous rocks periodically submerged in streams and rivers and in a canal, in a very few widely scattered localities. 01, 17, 58, 97.

Cinclidotus mucronatus (Brid.) Mach.
Rare, on calcareous rocks and tree bases and roots periodically submerged in or near rivers and on damp humid limestone, in a very few localities, in southern Gower and the western Vale. 48, *58*, 87, 97.

Schistidium apocarpum (Hedw.) Br.Eur. var. **apocarpum**
Common, on calcareous rocks, walls and concrete, very rarely on trees, very common in Gower, common in the Vale, the Uplands and the Upland ridge but rare in the mountains where it is mostly confined to man-made substrata. 00, 01, 06, 07, 08, 09, 10, 16, 18, 19, 48, 49, 58, 59, 60, 68, 70, 71, 79, 80, 81, 87, 88, 89, 90, 97, 98, 99.

Grimmia donniana Sm. var. **donniana**
Rare, on acidic rocks in the Uplands, absent elsewhere. 70: iron ore tips near Carreg yr Afon Park, Godre'r-graig, 1987, *K.L. Davies* (NMW!). 98: wall, Daren y Dimbath, Cwm Dimbath, 1951, *Wade* (NMW). 99: Pennant Grit boulder, Craig Fach near Cwm Parc, 1953, *Wade* (NMW); Pennant Grit boulders, scree below waterfall, Tarren Rhiw-maen, 1974, *Perry* (NMW).

Grimmia pulvinata (Hedw.) Sm. var. **pulvinata**
Common, on calcareous rocks, brickwork and concrete, throughout the county but rarer in the mountains. 00, 01, 07, 08, 18, 19, 48, 49, 58, 59, 68, 69, 70, 71, 87, 88, 89, 90, 97, 98, 99.

Grimmia orbicularis Bruch ex Wils.
Rare, confined to Carboniferous limestone sea cliffs in Gower. 48: Port Eynon, 1934, *Wallace* (NMW); near Overton, 1961, *Smith* (NMW); Kitchen Corner, Rhossili, 1962, *Smith* (NMW). 58: Pwll du, near Caswell Bay, 1929, *Binstead* (NMW), by path near Pwll du Bay, 1961, *Smith* (NMW); near Oxwich, 1932, *Knight* (NMW).

Grimmia trichophylla Grev. var. **trichophylla**
Very rare. 18: on sandstone boulders forming the shelving edge of Llanishen Reservoir, 1988, *Perry* (NMW). 49: on moist rocks in flush, Rhossili Down, 1967, *H. & M. Corley* (BBSUK), Old Red Sandstone boulders, N end Rhossili Down, 1968, *Kay* (UCSA!), 1974!.

Racomitrium aciculare (Hedw.) Brid.
Common, on acidic boulders, walls and tree roots in and near streams and a reservoir, mostly in the open, rarely in woodland, very common in the Uplands, especially in the mountains, in one locality in Gower, one in the eastern Vale. 00, 01, 08, 09, 10, 18, 19, 48, 49, 60, 61, 70, 80, 81, 89, 90, 98, 99.

Racomitrium aquaticum (Schrad.) Brid.
Uncommon, on damp acidic rocks, occasionally in streams and by water, confined to the Uplands and to one locality in the western Upland ridge. 09, 60, 78, 89, 90, 98, 99.

Racomitrium fasciculare (Hedw.) Brid.
Frequent, on bare siliceous rocks, boulders and walls in the Uplands, mainly in the mountains, but with one locality in the eastern Vale. 00, 08, 09, 10, 18, 19, 60, 80, 88, 89, 90, 98, 99.

Racomitrium heterostichum (Hedw.) Brid.
On bare siliceous rocks and boulders, probably occasional to frequent but confused until recently with other species in the *heterostichum* complex and at present recorded with certainty only from the Upland ridge and the eastern Uplands. 09, 18, 19, 88, 98.

Racomitrium affine (Web.& Mohr) Lindb.
On siliceous rocks, probably frequent but confused until recently with other species in the *heterostichum* complex and almost certainly under-recorded. 00: boulder by Taf Fechan E of Cefn-coed-y-cymmer, 1976, *Perry* (NMW). 18: on sandstone boulders, Llanishen Reservoir, Cardiff, 1988, *Perry* (NMW).

Racomitrium lanuginosum (Hedw.) Brid.
Frequent, mostly on acidic rocks and walls and boulder scree, more rarely in boggy ground on heaths and acid dune-heath, mainly in the Uplands, scarce in Gower and the Upland ridge, in one locality in the eastern Vale. 00, 08, 18, 48, 49, 58, 70, 90, 99.

Racomitrium canescens (Hedw.) Brid. var. **canescens**
Uncommon, in grassland and on soily limestone slabs, in old quarries and sand dunes, in Gower, the western Vale and the north-eastern Uplands. 00, *58*, 78, 87.

Racomitrium canescens (Hedw.) Brid. var. **ericoides** (Hedw.) Hampe
Occasional but often abundant where it occurs, amongst chippings, on soily limestone slabs, in old quarries, on an iron-ore tip, by tracks and amongst rocks in the mountains, recorded only from the Uplands and the eastern Upland ridge. 00, 01, *18*, 70, 71, 89, 99.

Ptychomitrium polyphyllum (Sw.) Br.Eur.
Occasional, on exposed siliceous rocks and walls, mainly in the Uplands, with one locality in the

eastern Vale, absent from Gower. 00, 01, 08, 18, 61, 70, 89, 90, 98, 99.

Funaria hygrometrica Hedw.
Common, on bare soil in a wide variety of habitats, on burnt and waste ground, cultivated soil and garden paths, pond margins and moorland, throughout the county. 00, 01, 06, 07, 08, 09, 10, 17, 18, 19, 27, 28, 38, 48, 49, 58, 59, 60, 68, 69, 70, 78, 79, 80, 87, 89, 90, 98, 99.

Funaria muhlenbergii Turn.
Rare, on calcareous soil. 18: among the ruins of Caerphilly Castle, 1842, *J.B. Woods* (Gutch, 1842); thin soil over limestone rocks, Cefn Onn, 1952, *Wade* (NMW). 48: Port Eynon Point (Smith, 1964); Kitchen Corner (Smith, 1964). 58: thin soil on limestone cliff by Nicholaston Wood, 1967, *M. Corley* (Crundwell & Nyholm, 1974).

Funaria pulchella Philib.
Very rare. 48: soil on cliffs near Overton, 1961, *Smith* (NMW). 58: S-facing limestone outcrop, Nicholaston Woods, 1967, *Perry* (Hb. Perry).

Funaria fascicularis (Hedw.) Lindb.
Very rare. 00: soil on bank by path along Taf Fechan c. 1 mile NE of Cefn-coed-y-Cymmer, 1976, *Perry* (NMW). 18: field behind Lisvane chapel, 1953, *Wade* (NMW).

Funaria obtusa (Hedw.) Lindb.
Very rare. 09: on soil of bank of ditch, St Gwynno Forest, Llanwonno, 1990, *Perry* (NMW). 59: on wet soil near High Pool, Clyne Common, 1944, *Wade* (NMW, BBSUK). 80: rocks beside Afon Perddyn, 1961, *Smith* (NMW).

Physcomitrium pyriforme (Hedw.) Brid.
Occasional, in arable land and on ditch and stream banks, in a few places in Gower and in the eastern Vale and eastern Uplands, absent from the whole of the centre of the county. 07, 17, 19, *28*, 48, 49, *68*, 69.

Physcomitrella patens (Hedw.) Br.Eur.
Very rare, on mud by water. 16: on dried mud of pond, Lavernock, 1954, *Wade* (NMW), 1975, *Wade* (NMW). 87: on side of river, Ewenny, 1944, *Miss M.G. Winkles* (NMW).

Ephemerum serratum (Hedw.) Hampe var. **serratum**
Very rare but possibly overlooked. 28: on barish clay soil at side of ride, mixed forestry plantation, Coed Cadwgan, ca. 8 km NE of Cardiff, 1988, *Perry* (NMW).

Ephemerum serratum (Hedw.) Hampe var. **minutissimum** (Lindb.) Grout
Uncommon, on bare soil in arable land and mole-hills in the Vale, and on sea-cliffs in Gower. 17, *18*, 48, 58, 87.

Tetraplodon mnioides (Hedw.) Br.Eur.
Very rare. 49: boggy ground, N end of Rhossili Down, 1963, *Smith* (UCSA). 90: bones, Craig-y-Llyn, 1957, *K. Benson-Evans & Wade* (NMW); sheep carcass by Llyn Fawr (Smith, 1964); dung in scree below cliffs, above SE corner of Llyn Fach, 1976, *Perry* (NMW).

Splachnum ampullaceum Hedw.
Very rare. 49: boggy ground, northern end of Rhossili Down, 1963, *Smith* (NMW). 69: bog about 1½ miles east of Swansea [Crymlyn Bog], 1928, *R.L. Smith* (NMW).

Orthodontium lineare Schwaegr.
Uncommon, on rotting wood and tree bases in seven localities in Gower, the extreme western Vale and the eastern Uplands, absent elsewhere. 00, 09, 58, 60, 69, 79.

Leptobryum pyriforme (Hedw.) Wils.
Uncommon or rare, on soil on waste ground and flower pots, once on a wooden stake, with a very few scattered records from Gower, the eastern Vale and the western Uplands, absent elsewhere. 06, *17*, *58*, 69, 70.

Pohlia elongata Hedw. ssp. **elongata** var. **elongata**
Rare, on damp mossy ledges in the central Uplands. 80: right bank of R. Perddyn, 1932, *Knight* (NMW), on ORS by waterfall, Afon Pyrddin, Pont-nedd-Fechan, 1961, *Smith* (NMW). 90: sandstone rocks, cliffs above SE corner of Llyn Fach, 1976, *Perry* (NMW). 99: crevice in Pennant Grit outcrop by waterfall, Tarren Rhiw-maen, near Craig Ogwr, 1974, *Perry* (NMW).

Pohlia nutans (Hedw.) Lindb.
Very common, on peaty and gravelly soil, among acidic rocks and on walls, less often on rotting logs and tree bases, on banks on heaths and moorlands, on ant-hills, in old quarries, in bogs and woodland, throughout the county but uncommon in the southern Vale. 00, 01, 06, 07, 08, 09, 10, 16, 17, 18, 19, 48, 49, 58, 59, 60, 69, 70, 71, 78, 79, 80, 87, 88, 89, 90, 98, 99.

Pohlia proligera (Kindb.ex Breidl.) Lindb.ex Arnell
Frequent, on gravelly soil on tracks and banks and in rock crevices, on waste ground, in plantations and woodland, in old quarries and on moorland, usually in small quantity, in all districts but most frequent in the Uplands. 00, 07, 18, 19, 59, 70, 79, 80, 89, 90, 98.

Pohlia camptotrachela (Ren.& Card.) Broth.
Very rare. 08: Llantrisant Common, 1977, *anon.* (BBSUK).

Pohlia muyldermansii Wilcz.& Demar. var. **pseudomuyldermansii** T. Arts, G. Nordhorn-Richter & A.J.E. Smith
Very rare. 90: rock crevice on cliffs above Llyn Fach, Craig-y-Llyn, 1962, *Smith* (NMW).

Pohlia lutescens (Limpr.) Lindb.f.
Very rare. **07**: clay bank in sunken lane, S of Pendoylan, 1976, *Perry* (NMW, BBSUK).

Pohlia lescuriana (Sull.) Andrews
Very rare. **98**: compacted sandy bank of stream, Nant Iechyd, Cwm Dimbath, 1974, *Perry* (BBSUK).

Pohlia carnea (Schimp.) Lindb.
Frequent, on clayey soil on vertical and sloping banks of streams, ditches and roadsides, and on wet ground, in all districts but rare in Gower and in the mountains. 06, 07, 08, 10, 17, 18, 19, 49, 60, 88, 96, 97, 98.

Pohlia wahlenbergii (Web.& Mohr) Andrews var. **wahlenbergii**
Common and widespread throughout the county but almost always in small quantity, on gravelly soil on waste ground, tracks and roadsides, in woodland, in old quarries, on moorland and in the mountains, on damp soil on ditch banks and in fields. 00, 01, 07, 08, 09, 10, 17, 18, 19, 28, 48, 49, 58, 60, 69, 70, 78, 79, 80, 88, 89, 90, 98, 99.

Epipterygium tozeri (Grev.) Lindb.
Very rare. **07**: clay bank in sunken lane, S of Pendoylan, 1976, *Perry* (NMW). **48/49**: damp soil (ORS), Rhossili Downs, 1932, *Knight* (NMW).

Plagiobryum zieri (Hedw.) Lindb.
Very rare. **00**: Morlais Castle Hill, 1932, *Knight* (NMW!).

Anomobryum filiforme (Dicks.) Solms var. **filiforme**
Very rare, in crevices in wet mountain rocks. **90**: Craig y Llyn near Hirwaun, 1950, *Crundwell* (BBSUK). **99**: Craig Ogwr, 1961, *Smith* (NMW).

Bryum warneum (Roehl.) Bland.ex Brid.
Very rare, on sand dunes. **49**: Whiteford Burrows, 1932, *D.A. Jones & Knight* (NMW).

Bryum pallens Sw. var. **pallens**
Common but usually in small quantity, on damp peaty or gravelly soil in waste places and on tracks, in old quarries, on moorland, marshy ground and in the mountains, on stream sides, dune slacks and sea cliffs and in fields, in every district but rare in the east. 00, 01, 10, 18, 38, 48, 49, 58, 59, 68, 69, 70, 78, 79, 87, 89, 90, 97.

Bryum algovicum Sendtn. ex C.Muell. var. **rutheanum** (Warnst.) Crundw.
Occasional, common on sand dunes and on sandy soil in Gower and the Vale, otherwise recorded only from a wall in Gower, gravelly soil in an old quarry in the north-eastern Uplands, and in a lawn in the eastern Vale. 00, 17, *48*, 49, 58, 78, 79, *87*, *96*.

Bryum inclinatum (Brid.) Bland.
Rare, on soil and on log. **00**: Morlais Hill, 1932, *Knight* (NMW), bank at roadside, E side of Morlais Hill, 1973, *Perry* (NMW). **17**: on log in alder and sallow carr, Culverhouse Cross, Cardiff, 1990, *Orange* (NMW). **58**: Oxwich Burrows, 1934, *Wallace*; Great Tor (Kay, 1989). **78**: old loading platform of disused mine, Cwm Dyffryn near Port Talbot, 1963, *Smith* (NMW).

Bryum donianum Grev.
Very rare. **49**: loamy roadside hedgebank, Landimore, 1963, *Wallace* (BBSUK).

Bryum capillare Hedw. var. **capillare**
Very common throughout the county, on rocks, walls, roofs, trees and fences and rarely on soil. 00, 01, 06, 07, 08, 09, 10, 16, 17, 18, 19, 27, 28, 38, 48, 49, 58, 59, 60, 68, 69, 70, 71, 78, 79, 80, 81, 87, 88, 89, 90, 96, 97, 98, 99.

Bryum flaccidum Brid.
Rare, but probably more frequent than the records indicate. **07**: bole of *Salix alba* ssp. *vitellina* by lake, Talygarn Estate, 1976, *Perry* (NMW, BBSUK). **18**: elder in woodland, W bank of R.Taff, Gwaelod y Garth, 1979, *Perry* (NMW); on elder at edge of forestry plantation, Coed Coesau-whips, ESE of Caerphilly, 1988, *Perry* (NMW). **97**: on ash bole by Afon Alun, N of Castle-upon-Alun *Perry 1983* (NMW).

Bryum torquescens Bruch ex De Not.
Very rare. **48**: cliffs, south of Port Eynon, 1963, *Warburg* (OXF, teste Syed, 1973).

Bryum canariense Brid.
Rare, in limestone rock crevices on the coast of Gower. **48**: cliffs south-west of Rhossili, 1963, *Crundwell* (Smith, 1964); quarry, Port Eynon Head, 1963, *Richards* (NMW). **49**: Llangennith Burrows, 1963, *Warburg* (Smith, 1964). **58**: Pwll du Head, 1963, *E.W. Jones, Warburg & Crundwell* (BBSUK).

Bryum pallescens Schleich. ex Schwaegr.
Very rare, in waste ground. **27**: derelict industrial site, East Moors, Cardiff, 1990, *Orange* (BBSUK).

Bryum pseudotriquetrum (Hedw.) Schwaegr.
Frequent to common, in boggy or marshy ground, in dune slacks and old quarries, on moorland, amongst damp rocks and on dripping rocks by waterfalls, in all districts but very rare in the eastern Vale. 00, 01, 08, 17, 48, 49, 58, 60, 70, 78, 79, 80, *87*, 88, 89, 90, 97, 98, 99.

Bryum pseudotriquetrum (Hedw.) Schwaegr. var. **bimum** (Brid.) Lilj.
Very rare but possibly under-recorded. **89**: rotting log under *Acer pseudoplatanus*, Bwlch-y-Cywion, near Ton Mawr, 1963, *Smith* (NMW, BBSUK).

Bryum caespiticium Hedw. var. **caespiticium**
Uncommon, in a dune slack and on exposed soil and walls, recorded only from the Vale and the eastern Uplands and Upland ridge. *06, 08, 16, 17, 18,* 78.

Bryum alpinum Huds. ex With.
Occasional, on flushed rocks often by water, on heath and moorland and in the mountains, with a few records from the Uplands and Gower and one from the Vale. *07,* 48, 49, *70, 90, 98.*

Bryum riparium Hagen
Very rare, and not seen recently. **98**: muddy flushed vertical bank by track by Nant Iechyd, Cwm Dimbath, 1974, *Perry* (NMW, BBSUK).

Bryum bicolor Dicks.
Common, on bare soil and amongst gravel in waste places, amongst rocks, on banks, paths and old walls, in old quarries and on cliffs, on sides of roads and paths and in arable land, rarely on wood, throughout the county, often in small quantity. 00, 01, 06, 08, *16, 17,* 18, 19, 28, 38, 48, 49, 58, 60, 68, 69, 70, 79, 87, 96, 97, 98, 99.

Bryum dunense Smith & Whitehouse
Very rare. **58**: soil on cliffs, Pwll du Head, 1963, *Crundwell & Warburg* (Smith & Whitehouse, 1978).

Bryum argenteum Hedw.
Very common but often in small quantity, on bare soil, amongst rocks and on gravelly ground in waste places and on tracks, cultivated ground, in old quarries, on roadsides and paths, between paving stones, occasionally on tarmac, on roofs and in crevices in brickwork, throughout the county. 00, 01, *06,* 08, 10, *17,* 18, 19, 27, 28, 38, 48, 49, 58, 59, 60, 68, 69, 70, 71, 78, 79, 80, 87, 88, 89, 90, 96, 97, 98, 99.

Bryum radiculosum Brid.
Occasional, on old mortar of walls, and on compacted calcareous soil between limestone rocks, in a few scattered localities in all districts except the Upland ridge. 28, 58, *60,* 80, *87,* 96.

Bryum ruderale Crundw.& Nyholm
Uncommon, on calcareous soil in a few localities near the coast in Gower, in a quarry in the north-eastern Uplands and on a wall in the south-eastern Vale. 00, 17, 49, 58.

Bryum klinggraeffii Schimp.
Very rare, but probably under-recorded. **48**: soil in stubblefield, Llandewi, Reynoldston, 1985, *Blockeel* (BBSUK).

Bryum microerythrocarpum C.Muell.& Kindb.
Uncommon, on acidic soil on a sea cliff in Gower, on a heath in the Upland ridge and on ant-hills and a riverbank in the Vale. 08, 18, 58, 87.

Bryum bornholmense Winkelm. & Ruthe

Very rare, but probably under-recorded. **49**: dunes, Llangennith Burrows, 1963, *Crundwell* (Smith, 1964).

Bryum rubens Mitt.
Frequent, on soil on sea cliffs and heathland, in arable fields, gardens and waste ground, in bare patches in turf, on mole-hills, and on banks and paths in woodland, in all districts but absent from much of the Uplands. 00, 17, 18, 28, 48, 49, 58, 69, 87, 96, 98.

Rhodobryum roseum (Hedw.) Limpr.
Rare, confined to Gower where it occurs on calcareous soil and in sandy turf. **49**: limestone ledges, Tor-gro, near Landimore, 1963, *Harrison* (NMW); Cwm Ivy Tor, 1971, 1989, *Kay* (Kay, 1989). **58**: Oxwich Burrows, 1934, *Wallace* (BBSUK), fixed dunes, Oxwich Burrows, 1963, *Smith* (NMW); Great Tor (Kay, 1989); Pennard Burrows (Kay, 1989).

Mnium hornum Hedw.
Very common, on acidic soil, rotting wood and stumps, tree bases, in rock crevices and on walls, occasionally on peat, in woodland, moorland and heath, in old quarries and among mountain rocks, throughout the county. 00, 01, 06, 07, 08, 09, 10, 17, 18, 19, 28, 48, 49, 58, 60, 68, 69, 70, 71, 80, 87, 88, 89, 90, 97, 98, 99.

Mnium marginatum (With.) P.Beauv. var. **marginatum**
Very rare. **00**: rock crevices beside Taff Fechan, near Merthyr Tydfil, 1963, *Smith* (NMW).

Mnium stellare Hedw.
Occasional, on soil-covered shaded limestone and basic rocks and walls and in a dune slack, in Gower and the eastern and southern Vale and eastern Upland ridge, absent from the Uplands except for one record in the north. 00, 01, 07, 17, 18, *28,* 48, 49, 58, 80, 87.

Rhizomnium punctatum (Hedw.) Kop.
Common, on wet ground by streams and rotting wood in woodland, on wet rocks in the mountains, and on marshy or flushed ground, throughout the county. 00, 01, 06, 07, 08, 09, 17, 18, 19, 27, 48, 49, 58, 59, 60, 68, 69, 80, 87, 90, 98, 99.

Plagiomnium cuspidatum (Hedw.) Kop.
Uncommon, on soil-covered limestone rocks, in limestone rock crevices and on a tree trunk, in a few localities on Gower and in one locality in the north-east Uplands. 00, *49,* 58, 59.

Plagiomnium affine (Funck) Kop.
Uncommon, on the ground in woodland, on turf on sea-cliffs and on mine spoil, in a very few scattered localities in all districts. 08, 18, 48, 60.

Plagiomnium elatum (Br.Eur.) Kop.
Rare, in calcareous turf and marsh. **07**: rich marsh at side of Nant Dyfrgi, *c.* 1 mile NE of Ystradowen, 1981,

Perry (NMW). **38**: shaded limestone turf, N side of Inner Worms Head, 1967, *H. & M. Corley* (BBSUK).

Plagiomnium ellipticum (Brid.) Kop.
Rare. **58**: Pwll du Head, 1963, *Paton* (Smith, 1964). **88**: Margam Park, 1941, *Banwell* (*Rep.Br.bryol.Soc.* **1940–43** (1944)).

Plagiomnium undulatum (Hedw.) Kop.
Very common, on soil, sometimes more or less flushed, in woodland and shady places and amongst grass on shaded banks, throughout the county but rare in the mountains. 00, 01, 06, 07, 08, 09, 10, 17, 18, 19, 28, 48, 49, 58, 59, 60, 68, 69, 70, 80, 87, 88, 89, 90, 96, 97, 98, 99.

Plagiomnium rostratum (Schrad.) Kop.
Frequent, on woodland floors, in sand dunes, on banks and among rocks, especially in calcareous areas, in all districts, but in the Uplands present only in the north-east. 00, 01, 06, 07, 08, 17, *18*, 28, 48, 49, 58, 68, 87, 88, 97.

Aulacomnium palustre (Hedw.) Schwaegr. var. **palustre**
Common, in bogs and boggy ground, often amongst *Sphagnum*, on heaths and wet acidic moorland, in all districts but absent from the southern Vale. 07, 08, 09, 17, 18, 19, 48, 49, 58, 59, 60, 69, 71, 79, 80, 81, 90, 98.

Aulacomnium androgynum (Hedw.) Schwaegr.
Uncommon, on soft-barked living tree boles and branches and rotten wood in damp woodland, in a few localities in the eastern Vale, Upland ridge and Uplands, with one record from near Swansea, but absent from Gower, the western Vale and the central and western Uplands. 00, 17, 18, 19, 28, 69.

Amblyodon dealbatus (Hedw.) Br.Eur.
Very rare, in dune slack. **49**: Whiteford Burrows, 1964, *Richards* (BBSUK), 1981, *Blockeel* (pers.comm. 1988)

Plagiopus oederi (Brid.) Limpr.
Very rare. **00**: Morlais Hill, 1932, *Knight* (NMW).

Bartramia pomiformis Hedw.
Occasional, in crevices and ledges usually in acidic rocks and walls but once recorded from limestone rock crevices, only in the Uplands and the eastern Upland ridge. 00, 19, *28*, *60*, *69*, 80, 90, 98, 99.

Bartramia ithyphylla Brid.
Very rare, in crevices of mountain rocks. **90**: Llyn-vach, near Pont-nedd-vechan, 1803, *G. Sowerby* (Sowerby, 1807, f.1710), Llyn-vach, 1803, Herb. Sowerby (BM), teste Orange 1987; Craig-y-Llyn, 1906, *Knight* (Smith, 1964).

Philonotis caespitosa Wils.
Uncommon, amongst wet rocks in rill, on a river bank and in marshy ground, with three records from

Gower and three from the Uplands. *18*, *49*, *59?*, *80*, *98*.

Philonotis fontana (Hedw.) Brid.
Common, on wet ground and rocks by streams and waterfalls, especially in the mountains, in bogs and on damp gravelly paths in woodlands and on moorland, in all districts but very rare in the Vale. 00, 01, *07*, 08, 09, 10, 18, 19, 48, 49, 58, 59, 60, 69, 70, 80, 81, 89, 90, 98, 99.

Philonotis calcarea (Br.Eur.) Schimp.
Very rare, in calcareous flushes. **01**: by road W of and below Twynau Gwynion, Merthyr Common, 1973, *Perry* (NMW, BBSUK). **59**: beside pool to north of Cefn Bryn [Broad Pool?], 1961, *Smith* (NMW).

Breutelia chrysocoma (Hedw.) Lindb.
Rare, on moist rock ledges in the north-central Uplands. **80**: right bank of R. Perddyn, 1932, Knight (NMW), vertical rocks by waterfall, Afon Pyrddin, 1963, *Smith* (NMW). **90**: Craig y Llyn, 1973, *Perry*.

Amphidium mougeotii (Br.Eur.) Schimp.
Uncommon, in shaded vertical rock crevices in the mountains, often in a continuous drip, in a few localities in the central Uplands. 80, 90, 98, 99.

Zygodon viridissimus (Dicks.) R.Br. var. **viridissimus**
Frequent, on a variety of tree species, in woodlands, gardens and at roadsides, in every district but absent from much of the Uplands. 00, 06, *07*, 08, 17, 18, 58, 68, 69, 80, 87, *88*, 96, 97, 98.

Zygodon viridissimus (Dicks.) R.Br. var. **stirtonii** (Schimp.ex Stirt.) Hagen
Rare, in calcareous rocks. **01**: limestone stonework of railway bridge over road *c.* ½ mile S of Pont Sticill, 1973, *Perry* (NMW). **16**: among stones of fixed shingle, Sully Is., 1961, *Wade* (NMW). **87**: soil in wall, Ogmore Down, 1962, *Smith* (NMW).

Zygodon conoideus (Dicks.) Hook. & Tayl.
Rare, but probably overlooked for Z. *viridissimus* as it rarely fruits. In October and November it becomes conspicuous, not only because of the more humid conditions which allow it to unfold, but then its gemmae start to germinate amongst the colonies of plants, forming wefts of protonemata. **07**: on elder in field, Tre-Dodridge, NE of Welsh St Donats, 1988, *Perry* (NMW). **18**: branches of elder, mixed woodland SW of Black Cock Inn, Thornhill, 1988, *Perry* (NMW). **58**: elder near stream, Bishopston Valley, 1963, *Paton* (NMW, BBSUK).

Orthotrichum striatum Hedw.
Rare and in small quantity, on trees. **07**: Garn-lwyd Mill, near Llancarfan, 1947, *Wade* (NMW). **48**: grounds of Penrice Castle, 1932, *Knight* (NMW). **58**: Oxwich, 1907, *Knight* (Smith, 1964); Bishopston Valley, 1963, *Duncan* (Smith, 1964).

Orthotrichum lyellii Hook.& Tayl.
Uncommon, on trunks and branches of well-lit trees, only in Gower and the western Vale. *48, 49, 58, 87, 97.*

Orthotrichum affine Brid.
Frequent, on a variety of tree species, but commonest on elder, in woodland, thickets and hedgerows and by streams and rivers, in all districts but absent from the western and northern Uplands. 01, 07, 09, 17, 18, 28, 48, 49, 58, 59, 87, 96, 97, 98.

Orthotrichum anomalum Hedw.
Frequent, on walls, bridges and other stonework, rocks and stones, chiefly limestone, in woodland, old quarries and on sea cliffs, in all districts but absent from most of the Uplands. 00, 01, 06, 07, 08, *16, 17,* 18, 48, 49, 58, 97, 98.

Orthotrichum cupulatum Brid. var. **cupulatum**
Rare, on rocks. **00:** by Taff Fechan, near Merthyr Tydfil, 1963, *Smith* (NMW). **01:** by R. Taff Fechan, Morlais Hill, 1932, *D.A. Jones & Knight* (NMW); by Tâf Fechan (Smith, 1964). **48:** rocks, Kitchen Corner (Smith, 1964).

Orthotrichum stramineum Hornsch.ex Brid.
Uncommon, on trees in a very few localities in the Vale, Gower and eastern Upland ridge. 28, 58, 87, 97.

Orthotrichum tenellum Bruch ex Brid.
Very rare. **58:** on elder, Oxwich Bay, 1985, *Blockeel* (BBSUK).

Orthotrichum diaphanum Brid.
Frequent, usually on soft-barked trees such as elder where it often grows associated with other species such as *Zygodon viridissimus*, less often on walls and concrete and on old wooden posts, in woodland, thickets and hedgerows, in old quarries, in all districts but absent from the central Uplands. 00, 01, 06, 07, 08, 09, 17, 18, 48, 49, 58, 60, 87, 96, 97.

Ulota crispa (Hedw.) Brid. var. **crispa**
Frequent, on shaded branches and trunks of living trees in humid conditions, in woodland and once on heathland, in all districts, but absent from the northern and western Uplands. 00, 01, 07, 18, 48, 49, 58, 59, 87, 97, 98.

Ulota crispa (Hedw.) Brid. var. **norvegica** (Groenvall) Smith & Hill
Occasional, on shaded branches of living trees in humid conditions, in woodland and on heathland, in all districts but in the Uplands restricted to three very distant localities and absent from the western Vale. 00, 06, 07, 48, 49, 58, 59, 60, 69, 97, 98.

Ulota phyllantha Brid.
Uncommon, on sheltered trees in woodland and garden, in a few localities in Gower and the Vale, absent elsewhere. 07, *48,* 49, 58, 87, *96,* 97.

Hedwigia ciliata (Hedw.) P.Beauv.
Very rare, and in small quantity. **49:** four small tufts on one large Devonian sandstone boulder, *c.* 200 ft.alt., associated with *Grimmia trichophylla*, Rhossili Down N. end, 1969, *Kay* (BBSUK, UCSA!).

Fontinalis antipyretica Hedw. var. **antipyretica**
Frequent, on rocks, stones and tree roots in streams, springs, rivers and pools, with scattered records in all districts, but with only one record from Gower. 00, 01, 07, 10, 58, 78, 80, 87, 90, 97.

Fontinalis squamosa Hedw. var. **squamosa**
Rare. **60:** on rocks, submerged at margin of river, shaded, Cwm Clydach, 1988, *Orange* (BBSUK). **89:** wet rocks by stream, Nant Cwm-du, Maesteg, 1989, *Orange* (NMW).

Climacium dendroides (Hedw.) Web.& Mohr
Occasional, amongst grass in dune slacks and fixed dunes and in calcareous to mildly acidic grassland, in scattered localities in all districts but rare in the Uplands and absent from the Upland ridge. 00, *07,* 17, 49, 58, 78, 80, 87, 88, 90.

Cryphaea heteromalla (Hedw.) Mohr
Occasional, usually in very small quantity, on broadleaved and coniferous tree trunks and exposed roots, and on stonework of bridges, often near water, in a few scattered localities in all districts but absent from the eastern Vale and from the western and central Uplands. 06, 07, 08, 18, 49, 58, 97.

Leucodon sciuroides (Hedw.) Schwaegr. var. **sciuroides**
Rare, on trees. **18:** Cefn-onn, 1932, *Wade* (Smith, 1964). **58:** Pennard, 1907, *Knight* (Smith, 1964); Bishopston Valley, 1963, *Duncan* (Smith, 1964). **87:** boles of poplar at edge of dunes near Candleston Castle, Merthyr Mawr Warren, 1973, *Perry* (NMW).

Pterogonium gracile (Hedw.) Sm.
Rare, on exposed rocks. **18:** Cefn-on, 1932, *Wade* (NMW). **49:** sandstone rocks at northern end of Rhossili Down, 1963, *Smith* (NMW); rare on exposed Conglomerate boulders, Rhossili Down (N) 1974, 1986, *Perry*. **58:** rocks in Ilston Valley (Smith, 1964).

Leptodon smithii (Hedw.) Web.& Mohr
Extinct. **07:** wall near Llancarfan, 1933, *Wade* (NMW), not refound 1975. **58:** plentiful on the trees between the house and gardens of Penrice Castle (Gutch, 1842).

Neckera crispa Hedw.
Occasional, on limestone rock faces, in one locality on dolerite, in a few localities in all districts but absent from the central and western Uplands and much of Gower, most frequent in the Vale. 00, 01, *07,* 17, *18,* 59, 87.

Neckera pumila Hedw.
Uncommon, on trunks of smooth-barked trees, in a very few localities in the Vale and Gower, absent from the Uplands and the Upland ridge. 06, *48*, 58, 87, 97.

Neckera complanata (Hedw.) Hueb.
Frequent, on shaded rocks, walls and trees in calcareous regions, in all districts, but in the Uplands and Upland ridge restricted to the east. 00, 01, 06, 07, 08, 16, 17, 18, 48, 49, 58, 59, 87, 88, 96, 97.

Homalia trichomanoides (Hedw.) Br.Eur.
Frequent, on tree bases and shaded calcareous rocks, occasionally extending onto the ground, usually in woodland, in all districts but in the Uplands very rare in the centre and absent from the west. 00, 01, 06, 07, 17, 18, 49, 58, 80, 87, 96, 97, 98.

Thamnobryum alopecurum (Hedw.) Nieuwl.
Common, on shaded calcareous rocks and occasionally wet, sometimes wet, in woodland and on bases of calcareous walls, in all districts, but most common in the Vale. 00, 01, 06, 07, 08, 16, *17*, 27, 48, 49, 58, 59, 80, 87, 96, 97.

Hookeria lucens (Hedw.) Sm.
Occasional, on banks of streams in woodland and in humid rock crevices, in a few localities in Gower, the Uplands and the Upland Ridge, absent from the Vale. 18, *28*, 48, 49, 58, 59, 60, 69, 80, 90.

Habrodon perpusillus (De Not.) Lindb.
Very rare and in small quantity, on trunks of sycamore, ash and elm, but probably decreasing owing to encroaching ivy. 87: with *Zygodon viridissimus* on ash trunks round Candleston Castle near Merthyr Mawr, 1962, *Smith* (NMW, BBSUK), 1973, *Perry* (NMW).

Leskea polycarpa Hedw.
Occasional, on roots and trunks of trees by streams and rivers and in marshy ground, recorded once on slag in a pond, always where liable to submergence, in all districts but absent from the mountains. 00, *07*, 09, *28*, 49, 58, 78, 87, 97.

Heterocladium heteropterum (Bruch ex Schwaegr.) Br.Eur. var. **heteropterum**
Occasional, on shaded and humid rocks and rocky banks, both acidic and calcareous, occasionally by streams and waterfalls and on tree bases, in scattered localities in Gower and the Uplands and the Upland ridge, absent from the Vale. 08, 09, 18, 58, 59, 60, 80, 90, 98, 99.

Heterocladium heteropterum (Bruch ex Schwaegr.) Br.Eur. var. **flaccidum** Br.Eur.
Very rare. 18: wall, Cwm Nofydd, near Rhiwbina, 1951, *Wade* (NMW!).

Anomodon viticulosus (Hedw.) Hook.& Tayl.
Frequent in calcareous districts, on sheltered rocks, walls and tree base, commonest in Gower and the Vale, otherwise found only in the north-eastern Uplands and the eastern Upland ridge. 00, *07*, 16, 17, 18, 48, 49, 58, 59, 87, 97.

Thuidium abietinum (Hedw.) Br.Eur. ssp. **abietinum**
Rare, in calcareous turf in sand-dunes. **49**: Llangennith Burrows, 1966, *Kay*, 1967, *Perry* (Kay, 1989). **78**: amongst *Salix repens*, Kenfig dunes, 1963, *Smith* (NMW). **87**: Porthcawl, 1908, *Knight* (NMW); Candleston, 1933, *Wade* (NMW), 1973, 1979, *Perry* (NMW). **88**: Kenfig Burrows (Smith, 1964).

Thuidium tamariscinum (Hedw.) Br.Eur.
Common, on rocks, hedge banks and in turf, in woodland, on hillsides, sea cliffs and roadsides, recorded from all districts but rare in the mountains. 00, 01, 07, 08, 09, 18, 19, 28, 48, 49, 58, 59, 69, 87, 97, 98.

Thuidium philibertii Limpr.
Rare, in calcareous grassland over limestone and in fixed dunes. **00**: amongst turf in quarry, Morlais Hill, Merthyr Tydfil, 1963, *Smith* (NMW), 1968, *Harrison* (NMW), 1987, *Perry* (NMW). **49**: limestone cliff, Tor-gro, Landimore, 1963, *Crundwell* (Smith, 1964). **78**: amongst *Salix repens*, Kenfig Burrows, 1963, *Smith & Paton* (NMW, BBSUK).

Cratoneuron filicinum (Hedw.) Spruce var. **filicinum**
Very common and widespread, in a wide range of wet habitats, especially (but not always) in calcareous conditions, in turf, on wet rocks, on concrete, banks and ditch-sides and boggy ground, in wet fields, woodland, dune slacks and on hillsides, on peaty soil in moorland and in wet gravelly ground in gateways, on tracks and at road-sides, evenly spread throughout all districts. 00, 01, 06, 07, 08, 09, 16, 17, 18, 28, 48, 49, 58, 59, 60, 68, 71, 78, 79, 80, 87, 88, 89, 90, 97, 98, 99.

Cratoneuron commutatum (Hedw.) Roth var. **commutatum**
Frequent, in wet calcareous (and usually slightly basic) habitats, on wet, often dripping, rocks, in flushes, springs and ditches, in woodland, old quarries and on hillsides, in all districts but uncommon in the Uplands. 00, 01, *06*, *07*, *10*, 17, 18, 48, 58, 80, 87, 90.

Cratoneuron commutatum (Hedw.) Roth var. **falcatum** (Brid.) Moenk.
Frequent, in wet, flushed or boggy ground, on wet moorland and hillside, in old quarries and in dune slacks, recorded from all districts but absent from the central Uplands. This variety appears to have a wider range of habitat tolerance than the variety *commutatum*, but even so it avoids very acidic habitats. 00, 01, *17*, *18*, 58, 60, *70*, 79, 87, 88.

Campylium stellatum (Hedw.) J.Lange & C.Jens. var. **stellatum**
Occasional, in damp, open, calcareous habitats, in turf and flushes and on dripping rocks, in dune slacks, on wet moorland and heath, in all districts but absent from the Upland ridge and the western Uplands. 01, *07*, *48*, 49, 58, *68*, 78, 88, 90.

Campylium stellatum (Hedw.) J.Lange & C.Jens. var. **protensum** (Brid.) Bryhn
Uncommon, in dry calcareous habitats, occurring on soil and in turf, in old quarries, dune slacks and on sea cliffs, in all districts but in the Uplands confined to the Morlais Hill area. 00, 01, 06, 17, 18, *58*, 78.

Campylium chrysophyllum (Brid.) J.Lange
Frequent, in dry calcareous habitats, on soil, rocks and rocky ground and in turf, in old quarries, sand dunes, on sea cliffs, once on basic rocks in the mountains. It is commonest in Gower and the Vale, with few records from the Uplands and Upland ridge districts and only one from the mountains. 00, *16*, 17, 18, 48, 49, 58, 59, 68, 79, 87, 90, 97.

Campylium polygamum (Br.Eur.) J.Lange & C.Jens.
Uncommon to rare, in base-rich, marshy ground, on sand-dunes, hillsides and once on rocks by a waterfall, with a few scattered records from all districts in the western half of the county. 49, 68, 78, 80, 88.

Campylium calcareum Crundw. & Nyholm
Very rare, amongst limestone rubble in shaded woodland. 18: Coed-y-bedw, near Pentyrch, 1985, *Appleyard et al.* (BBSUK).

Amblystegium serpens (Hedw.) Br.Eur. var. **serpens**
Very common, on soil, rocks, damp brickwork, mortar and concrete, on tree bases, roots and rotting wood and as an epiphyte on the branches and trunks of soft-barked trees such as elder, on roadside banks, in woodland and thickets, marshy ground, by streams and rivers, on sea-cliffs and in sand-dunes, throughout all districts. 00, 06, 07, 08, 16, 17, 18, 19, 27, 28, 48, 49, 58, 59, 60, 68, 69, 70, 78, 79, 87, 88, 89, 90, 96, 97, 98.

Amblystegium serpens (Hedw.) Br.Eur. var. **salinum** Carringt.
Very rare, on fixed dunes. 78/88: Kenfig Sands, 1908, *Knight* (NMW), 1950, *Wade* (NMW). 88: Kenfig Burrows (Smith, 1964).

Amblystegium fluviatile (Hedw.) Br.Eur.
Rare, in or near running water. 08: on old wooden post on bank of stream with *Rhynchostegium riparioides*, by Ely River near Ynysmaerdy, 1987, *Orange* (Hb. Orange). 18: boulders in stream, Cwm Nofydd near Rhiwbina, 1951, *Wade* (NMW). 80: rocks in R. Perddyn, 1932, *Knight* (NMW).

Amblystegium tenax (Hedw.) C.Jens.
Occasional, on shaded boulders in streams and river, with single records from Gower and the Vale, two from the eastern Upland ridge and two from the central Uplands. 18, *59*, 89, 90, 97.

Amblystegium varium (Hedw.) Lindb.
Very rare. **97**: blackthorn bole by stream, side of Afon Alun N of Castle-upon-Alun, 1974, *Perry* (NMW, BBSUK).

Amblystegium riparium (Hedw.) Br.Eur.
Occasional, on rotting or submerged wood, tree bases, rocks, concrete and soil, in marshy ground, in or by streams and rivers, in a water tank and in a reservoir, recorded from Gower, the Vale and the eastern end of the Upland ridge. *07*, 17, *18*, 58, 79, 87, 96.

Platydictya confervoides (Brid.) Crum
Very rare. **97**: lumps of limestone on steep wooded slope, Ewenni Park near Ewenny, 1965, *Smith* (BBSUK).

Drepanocladus aduncus (Hedw.) Warnst.
Occasional, in marshy ground, in dune slacks, at pond margins and on the floor of a disused limestone quarry, in a few scattered localities in Gower, the Vale and the eastern Uplands. 00, *18*, *48*, 59, 78, 97.

Drepanocladus sendtneri (Schimp.ex H.Muell.) Warnst.
Rare, confined to dune slacks where it is often abundant. 49: edge of pool, Whiteford Burrows, 1961, *Smith* (NMW). 78: dried up pool, Kenfig Burrows, 1948, *Wade* (NMW), 1961, *Smith* (NMW, UCSA), 1986, *Orange*.

Drepanocladus lycopodioides (Brid.) Warnst.
Very local and restricted to one system of dune slacks where it is often found in abundance. 78: Kenfig Burrows, 1961, *Smith* (NMW, BBSUK, UCSA), 1986, *P. Jones* (NMW).

Drepanocladus fluitans (Hedw.) Warnst. var. **fluitans**
Rare, in marshy pools and wet boggy ground. 49: waterlogged ground near Arthur's Seat, Cefn Bryn, 1962, *Smith* (UCSA). 79: N side of Crymlyn Bog, 1977, *Perry* (NMW). 89: Mynydd Fforch-Dwm (Smith, 1964).

Drepanocladus fluitans (Hedw.) Warnst. var. **falcatus** (Sanio ex C.Jens.) Roth
Occasional, in wet heaths and boggy ground, with scattered records in the Uplands, Upland ridge and Gower, absent from the Vale. 08, 19, 58, *59*, 70, 80, 88, 89.

Drepanocladus exannulatus (Br.Eur.) Warnst. var. **exannulatus**
Occasional, on wet heaths, on the edges of pools and in marshy ground, mainly in the Uplands and Upland ridge, but with one record from Gower, absent from the Vale. 08, 09, *18*, 48, 60, *69*, 70, 90, *99*.

Drepanocladus exannulatus (Br.Eur.) Warnst. var. **rotae** (De Not.) Loeske
Rare, on boggy ground, recorded only from Gower and the Vale. 07: Mynydd-y-Glew, Welsh St Donats, 1923, *R.Melville* (NMW!). 49: boggy ground, northern end of Rhossili Down, 1963, *Smith* (NMW). 59: by Broad Pool, N of Cefn Bryn (Smith, 1964).

Drepanocladus revolvens (Sw.) Warnst.
Occasional, on marshy ground on flushed heathy slopes and in bogs, in all districts except the Upland ridge, in a few scattered places, mainly in the west, but with one record from the eastern Vale. *07*, 49, 58, 69, 70, 79, 90.

Drepanocladus vernicosus (Mitt.) Warnst. 45
Very rare, in marshy ground. **49**: near pool, Llangennith Burrows, 1963, *Warburg* (BBSUK).

Drepanocladus uncinatus (Hedw.) Warnst.
Uncommon, on rocks and trees by rivers and on wet heath, recorded from the western and eastern Uplands and Gower but absent elsewhere. 00, 01, 49, 60.

Hygrohypnum ochraceum (Turn.ex Wils.) Loeske
Uncommon, on rocks in and by streams, in four scattered localities in the Uplands and eastern Upland ridge, absent elsewhere. 01, 08, 60, *80*.

Hygrohypnum luridum (Hedw.) Jenn. var. **luridum**
Occasional, on shaded tree roots and rocks in and by streams, often partly submerged, preferring base-rich habitats, in woodland or on open hillsides, most frequent in the Uplands, rare in Gower and the Upland ridge, absent from the Vale. 00, 01, 18, *58*, 60, 90, 99.

Hygrohypnum luridum (Hedw.) Jenn. var. **subsphaericarpon** (Schleich.ex Brid.) C.Jens.
01: approaching this variety but material without sporophytes, on sandstone in stream, sloping woodland on roadside W of Twynau Gwynion, NE of Merthyr Tydfil, 1987, *Perry* (NMW).

Scorpidium scorpioides (Hedw.) Limpr.
Rare, on wet heath in the lowlands. 07: Mynydd-y-Glew, 1920, *Wade* (NMW!), 1922, *Richards* (Smith, 1964). 49 Cefn Bryn, 1974, *Perry*.

Calliergon stramineum (Brid.) Kindb.
Frequent, in wet ground, often amongst *Sphagnum*, in bogs and heaths, moorland, hillsides and mountains, confined to the Upland and Upland ridge districts. 00, 08, 09, 18, 19, 60, 71, 79, 80, 90, 98, *99*.

Calliergon cordifolium (Hedw.) Kindb.
Occasional, in shallow water or on waterlogged ground, in fen and bog or at pool margins, often amongst reeds, in all districts but absent from most of the Vale and the Uplands. 07, *18*, 19, 49, 69, 79, 98.

Calliergon giganteum (Schimp.) Kindb.
Very rare. 49: with *Drepanocladus aduncus*, Whiteford Burrows, 1932, *Knight* (NMW!).

Calliergon cuspidatum (Hedw.) Kindb.
Very common and often abundant, in turf in lawns, grassland and marshes, on tracks, banks and rocks, in bogs, ditches and flushes, beside pools, streams and rivers, in old quarries, on moorland and on sea cliffs throughout the county. 00, 01, 06, 07, 08, 09, 10, 16, 17, 18, 19, 28, 48, 49, 58, 60, 68, 69, 70, 71, 78, 79, 80, 81, 88, 89, 90, 97, 98, 99.

Isothecium myurum Brid.
Uncommon, on tree bases in woodland, rarely on rocks, with a very few scattered records from the eastern Uplands and Upland ridge, the southern Vale and Gower. 00, 01, *06*, 18, 49.

Isothecium myosuroides Brid. var. **myosuroides**
Common, on tree trunks, tree bases, rotting logs, less frequently on rocks and walls, in all districts but mainly in the lowlands and Upland ridge, very rare in the Uplands. 01, 06, 07, 08, 09, 10, 17, 18, 48, 49, 58, 59, 60, 69, 90, 96, 97, 98.

Isothecium striatulum (Spruce) Kindb.
Uncommon, on limestone and dolerite rock outcrops, cliffs and walls, often exposed, in Gower, the Vale and the Upland ridge. 18, 48, 58, 97.

Scorpiurium circinatum (Brid.) Fleisch.& Loeske
Occasional, on limestone outcrops and walls and in calcareous turf, on sea-cliffs and in wooded valleys, only in Gower and the Vale, common in parts of the Gower coast, rare and sporadic elsewhere. 06, 38, 48, 49, 58, 59, 97.

Homalothecium sericeum (Hedw.) Br.Eur.
Very common, on a wide range of substrata, on rock outcrops and walls, boulders, roofs and tree trunks, rarely on sand dunes and in turf, in all districts but most common in limestone regions. 00, 01, 06, 07, 08, 09, 10, 16, 17, 18, 19, 28, 48, 49, 58, 59, 60, 68, 69, 70, 71, 78, 80, 87, 88, 90, 96, 97, 98, 99.

Homalothecium lutescens (Hedw.) Robins.
Occasional, in calcareous turf, especially in sand dunes and on sandy sea cliffs where it is particularly common, also rarely in turf over inland limestone, commonest in Gower and the Vale, rare in the Uplands, absent from the Upland ridge. 00, 06, 48, 49, 58, 78, 87, 90.

Brachythecium albicans (Hedw.) Br.Eur.
Common, in grass on sandy soils and on bare patches of sand or gravel, in sand dunes (where it is often

abundant), on sides of tracks and roads and on waste ground, in all districts but most frequent in the Vale and Gower. 00, 06, 08, 16, 18, 19, 28, 48, 49, 58, 68, 70, 71, 78, 79, 87, 88, 90, 99.

Brachythecium glareosum (Spruce) Br.Eur.
Occasional, on rocks, concrete, tracks and gravelly ground, in old quarries on open hillsides, and in sand dunes, in all districts but rare in Gower and in the mountains. 00, 01, 08, 16, *17*, 18, 19, 28, 58, *78*, 87, 90.

Brachythecium mildeanum (Schimp.) Milde
Occasional, in sand dunes, rarely on sheltered rocks and soil on sea cliffs and heathland, in Gower and the Vale, absent from the Uplands and Upland ridge. 49, 58, 59, 69, 78, 79, 87.

Brachythecium rutabulum (Hedw.) Br.Eur.
Very common, almost ubiquitous in the lowlands, often in disturbed sites, on banks, soil, rocks, walls, trees and decaying wood, in woodland, fields, lawns, marshes and old quarries, on roadsides, in heaths, moorland and sea cliffs, throughout the county but uncommon in the mountains. 00, 01, 06, 07, 08, 09, 10, 16, 17, 18, 19, 27, 28, 48, 49, 58, 59, 60, 68, 69, 70, 78, 79, 80, 87, 88, 89, 90, 96, 97, 98.

Brachythecium rivulare Br.Eur.
Common, on damp or waterlogged soil in marshes, moorland and wet woodland, on rocks and banks by streams and rivers, on the margins of ponds and reservoirs, and in dune slacks, throughout the county. 00, 01, 06, 07, 08, 09, 17, 18, 19, 28, 49, 58, 60, 69, 79, 80, 89, 90, 97, 98, 99.

Brachythecium velutinum (Hedw.) Br.Eur.
Frequent, on rocks, walls, banks, stumps and tree bases, in woodland, wooded valleys and on roadsides, in all districts, scattered throughout the county but absent from the mountains, commonest in the Vale, rarer in the west. 00, 01, 06, 09, 17, 18, 28, 58, 59, *60*, 78, 80, 87, 90, 96.

Brachythecium populeum (Hedw.) Br.Eur.
Common, on rocks and stones, walls, tree branches and trunks, in woodland, on roadsides and in other places, in all districts. 00, 01, 06, 08, 10, 17, 18, 27, 48, 49, 58, 59, 60, 69, 71, 80, 90, 97, 98.

Brachythecium plumosum (Hedw.) Br.Eur.
Common, on rocks, boulders and trees in and beside rivers, streams and a reservoir, in woodland, on roadsides and on mountains, rarely on rocks away from water, in woodland and in old quarries, recorded from all districts but uncommon in the Vale. 00, 01, 06, 08, 10, 18, *19*, 49, 58, 59, 60, 69, 70, 80, 89, 90, 96, 98, 99.

Pseudoscleropodium purum (Hedw.) Fleisch.
Very common, on banks in woodland and roadsides, and in turf in grassland, heath, moorland and sand dunes, in all districts but commonest in the lowlands.

00, 01, 06, 07, 08, 09, 10, 17, 18, 19, 28, 48, 49, 58, 59, 60, 68, 78, 80, 87, 88, 89, 90, 96, 97, 98, 99.

Scleropodium tourettii (Brid.) L.Koch
Rare, in limestone turf on sea cliffs in south-western Gower, and in dunes and an inland locality in the Vale. **17**: clay bank, near Penylan, Cardiff, 1949, *Wade* (Smith, 1964). **48**: path on cliff top near coastguard station, Rhossilli, 1963, *Paton* (NMW, BBSUK); mainland near Worms Head, 1974, *Perry*. **58**: Pennard Cliffs, 1967, *Kay* (Kay, 1989). **78**: small breccia outcrop in dunes S of Kenfig Pool, 1980, *Perry* (NMW).

Cirriphyllum piliferum (Hedw.) Grout
Frequent, often intricately mixed with other mosses, in slightly calcareous conditions, on woodland floors and banks and on roadside banks, scattered throughout the county in all districts, but nowhere common. 00, 06, 07, 08, 17, 18, 19, *48*, *58*, 59, 97, 98, 99.

Cirriphyllum crassinervium (Tayl.) Loeske & Fleisch.
Frequent, on rocks, walls and tree bases, mainly in calcareous habitats, occasionally by streams and rivers where liable to submergence, most frequent in the Vale, absent from the western and central Uplands. 00, 07, 08, *17*, 18, *48*, 49, 58, 70, 87, 96, 97.

Rhynchostegium riparioides (Hedw.) C.Jens.
Common, usually submerged, on stones and rocks and occasionally wood, in streams and rivers and on dripping rocks, tolerating both acidic and calcareous conditions, in all districts, from the coast to the mountains. 00, 01, 06, 07, 08, 09, 10, *17*, 18, 19, 49, 58, 59, 60, 70, 80, 87, 90, 96, 97, 98, 99.

Rhynchostegium murale (Hedw.) Br.Eur.
Frequent to common, on sheltered calcareous rocks and stones, in walls and bridges, in woodland and wooded valleys, and on calcareous rocks in tracks, from the coast to the mountains, recorded from all districts but absent from the western Uplands, thus showing a very similar distribution to that of *R. riparioides* though not as common as that species. 00, 01, 06, 07, 17, 18, 19, 28, 48, 49, 58, 59, 69, 80, 87, 89, 90, 96, 97.

Rhynchostegium confertum (Dicks.) Br.Eur.
Common to very common, on tree trunks, stumps, walls, concrete and rocks, often in woodland but also in the open, in all districts but commonest in the lowlands, absent from the mountains. 00, 06, 07, 08, 16, 17, 18, 19, 27, 28, 48, 49, 58, 59, 60, 68, 69, 70, 79, 80, 87, 96, 97, 99.

Rhynchostegium megapolitanum (Web.& Mohr) Br.Eur.
Occasional, on sand dunes and in turf on sandy soil on sea cliffs, in Gower and the Vale, along the coast from Merthyr Mawr to Whiteford. 48, 49, 58, 69, 78, 79, 87, 88.

Eurhynchium striatum (Hedw.) Schimp.
Common, on neutral to basic substrata, on rocks and soil in woodland, more rarely on roadside banks and in turf in sea cliffs, in all districts but commonest in the lowlands and Upland ridge. 00, 01, 06, 07, 08, 17, 18, 28, 38, 48, 49, 58, 59, 60, 70, 71, 80, 87, 88, 89, 90, 97, 98.

Eurhynchium pumilum (Wils.) Schimp.
Occasional, on bare soil and on stones and rocks in damp and often heavily shaded places, on woodland banks, occasionally on sea cliffs, in all districts but in the Uplands and Upland ridge restricted to the east. 01, 06, 07, 16, 18, 49, 58, 96, 97.

Eurhynchium praelongum (Hedw.) Br.Eur. var. **praelongum**
Very common, on clay soil, rocks and stones, on rotting wood and tree bases, in wooded areas where often heavily shaded, on banks in quarries, by streams and rivers, in turf in marshes and pastures, on sea cliffs and moorland, widespread and common in all districts. 00, 01, 06, 07, 08, 09, 10, 16, 17, 18, 19, 27, 28, 38, 48, 49, 58, 59, 60, 68, 69, 70, 71, 78, 79, 80, 87, 88, 89, 90, 96, 97, 98.

Eurhynchium praelongum (Hedw.) Br.Eur. var. **stokesii** (Turn.) Dix.
Uncommon, in turf, on the shaded base of a wall and on stones and mossy ground in woodland, restricted to the lowlands of the Vale and eastern Gower. 07, 69, 87, 97.

Eurhynchium swartzii (Turn.) Curn. var. **swartzii**
Very common, on soil in woods and in arable land, and in turf on banks, usually (always?) in calcareous or slightly calcareous conditions, in all districts, extending from the coast to all but the highest hills. 00, 01, 06, 07, 08, 09, 10, 16, 17, 18, 19, 28, 48, 49, 58, 59, 60, 68, 69, 70, 80, 87, 88, 89, 96, 97, 98.

Eurhynchium swartzii (Turn.) Curn. var. **rigidum** (Boul.) Thér.
Very rare. 16 on soil over Liassic limestone, Cosmeston Lakes, S of Penarth, 1987, *Perry* (NMW, BBSUK).

Eurhynchium schleicheri (Hedw.f.) Milde
Uncommon, on shaded calcareous banks and stones, restricted to Gower and to the eastern Upland ridge and Vale. 18, *28, 48*, 58, *69*.

Eurhynchium speciosum (Brid.) Jur.
Rare or overlooked, on marshy ground. 17: sterile, around tree bases, alder and sallow carr, Culverhouse Cross, Cardiff, 1990, *Orange*. 48: boggy ground in wood beside stream near Penrice Castle, 1961, *Smith* (BBSUK). 78: marshy ground in sallow scrub, Kenfig Burrows, 1986, *Orange* (Hb. Orange). 87: Pont Alun, 1956, *Sowter* (Wade, 1960).

Rhynchostegiella tenella (Dicks.) Limpr. var. **tenella**
Common, on limestone rocks, walls and mortar, in all districts but commonest in the Vale and Gower, rare

in the Uplands. 00, 06, 07, 08, *16*, 17, 18, 28, 48, 49, 58, 59, 60, 80, 87, 96, 97, 98.

Rhynchostegiella teesdalei (Br.Eur.) Limpr.
Very rare, on shaded limestone by rivers. **00**: bank of Taf Fechan N of Merthyr Tydfil, 1963, *Paton* (NMW, BBSUK). **97**: at edge of Afon Alun N of Castle-upon-Alun, 1974, *Perry* (NMW).

Entodon concinnus (De Not.) Paris
Very rare, amongst grass on sandy soil. **58**: grassy bank, Oxwich Bay, 1932, *Knight* (NMW). **87**: mossy slope, sand hill at back of dunes, Merthyr Mawr Warren, 1973, *Perry* (NMW).

Plagiothecium latebricola Br.Eur.
Very rare. **88**: decaying vegetation by the brook leading to the lake, Margam Park, 1955, *J.G. Hughes* (NMW, BBSUK).

Plagiothecium denticulatum (Hedw.) Br.Eur. var. **denticulatum**
Occasional, on moist shaded banks, in lanes and roadsides, and on tree roots and stumps, rarely in boggy ground, in all districts but absent from the central Uplands and western Vale. *00*, 07, 09, 17, 18, 19, 28, 58, 60.

Plagiothecium ruthei Limpr.
Very rare. **59**: peaty ground, heath near Llethrid, 1961, *Smith* (BBSUK).

Plagiothecium succulentum (Wils.) Lindb.
Common, on shaded banks and tree bases and on a mortared wall, in woodland, by streams and rivers, and among rocks on moorland and mountains, in all districts. 00, 01, 06, 07, 08, 09, 18, 28, 48, 49, 58, 59, 60, 71, 80, 87, 90, 97, 98, 99.

Plagiothecium nemorale (Mitt.) Jaeg.
Occasional, on shaded banks in woodland, by streams and rivers and by roads, and amongst rocks in the mountains, in all districts but absent from the coastal strip in the Vale and from much of the Uplands. 09, 10, 17, 18, 19, 48, 49, 59, 60, 98, 99.

Plagiothecium undulatum (Hedw.) Br.Eur.
Frequent, on moist acidic banks, rocks and rarely on wood, in woodland and on grassy banks by streams, in a few scattered localities in all districts. *00*, 07, 08, *18*, 48, 49, 60, 80, 90, 98, 99.

Isopterygium elegans (Brid.) Lindb.
Very common, on acidic boulders, rocks and rocky banks and ledges, on tree bases and on peaty soil, in woodland, heath and moorland and on cliffs, in all districts but absent from the coastal strip of the Vale. 00, 01, 07, 08, 09, 10, 17, 18, 19, 48, 49, 58, 59, 60, 69, 70, 71, 78, 79, 80, 88, 89, 90, 97, 98, 99.

Taxiphyllum wissgrillii (Garov.) Wijk & Marg.
Uncommon, on shaded limestone rocks and stones in sheltered woodland and wooded valleys, known

only from a few localities in Gower and the Vale and from one in the Upland ridge, absent from the Uplands. 07, 17, 18, 58, 59, 97.

Hypnum cupressiforme Hedw. var. **cupressiforme**
Very common, on trunks and branches of trees and on rocks and soil in woodland, heath, moorland and bog, and on rock exposures and sea cliffs, widespread in a wide range of conditions, from acidic to calcareous, throughout the county. 00, 01, 06, 07, 08, 09, 10, 16, 17, 18, 19, 28, 38, 48, 49, 58, 59, 60, 68, 69, 70, 71, 78, 79, 80, 87, 88, 89, 90, 96, 97, 98, 99.

Hypnum cupressiforme Hedw. var. **resupinatum** (Tayl.) Schimp.
Frequent, on tree trunks and branches and on rotting wood, in woodland and hedgerows, on rocks on heaths and sea cliffs and in buildings and other man-made structures, common in Gower and the Vale, otherwise known only from the north-east Uplands and the eastern Upland ridge. 00, 06, 07, 17, 18, 48, 49, 58, 59, 87, 96, 97.

Hypnum cupressiforme Hedw. var. **lacunosum** Brid.
Frequent, in turf on heaths, cliffs and sand dunes, and on exposed rocks, usually in calcareous regions, common in Gower and the Vale, otherwise known only from the north-east Uplands and the eastern Upland ridge. 00, 01, 08, *16, 18*, 48, 49, 58, 59, 68, 78, 87, 96, 97.

Hypnum mammillatum (Brid.) Loeske
Frequent, on trunks and branches of trees in woodland and wooded valleys and by rivers and streams, more rarely on rocks and walls, in wooded valleys and hillsides, in several scattered localities throughout the county, often abundant where it occurs, rare in the Vale. 00, 01, 07, 08, 09, 48, 49, 58, 59, 60, 69, 71, 89, 90, 98, 99.

Hypnum jutlandicum Holmen & Warncke
Common, always in acidic conditions, on moist ground, tree stumps and bases of trees, and on boulders in deciduous woodland, in conifer plantations, in moorland, heaths and old quarries, at road sides and on cliffs, spread throughout the county. 00, 01, 07, 08, 09, 10, 18, 19, 48, 49, 58, 59, 60, 68, 69, 70, 71, 78, 80, 81, 88, 89, 90, 98, 99.

Hypnum lindbergii Mitt.
Occasional, amongst gravel or grass, at track sides, in old quarries and in gateways, in woodland and on moist roadside banks, confined to the Uplands and the Upland ridge. 01, 08, 09, 18, 19, 28, 60, 98, 99.

Ctenidium molluscum (Hedw.) Mitt. var. **molluscum**
Common, in calcareous districts on soil, in turf and on rocks, walls and limestone outcrops, in woodland,

on rocky coasts, heaths, sand dunes and base-rich ledges in the mountains, widespread in all districts. 00, 01, 06, 07, 16, 17, 18, 48, 49, 58, 59, 60, 68, 79, 80, 87, 90, 96, 97, 98, 99.

Ctenidium molluscum (Hedw.) Mitt. var. **fastigiatum** (Bosw.ex Hobk.) Braithw.
Very rare. **48/49**: Rhossili cliffs, 1963, *Appleyard* (BBSUK).

Hyocomium armoricum (Brid.) Wijk & Marg.
Frequent, on acidic rocks, boulders, and on rocky banks by streams, in wooded valleys and on open hillsides, mostly in the Uplands and Upland ridge but with one record from Gower, absent from the Vale. 09, 10, 49, 60, 70, 80, 88, 89, 90, 98, 99.

Rhytidiadelphus triquetrus (Hedw.) Warnst.
Occasional, in calcareous turf in sand dunes and on banks and hillsides, and amongst boulders in woodland and wooded valleys, in all districts but absent from the central and western Uplands. 00, 07, 18, *28*, 49, 58, 87, 88.

Rhytidiadelphus squarrosus (Hedw.) Warnst.
Very common, on the ground, amongst grass in lawns, pastures, marshes and bogs, by tracks, on roadsides, on river and stream banks, in woodland, on moorland, heath and sea cliffs and in sand dunes, widespread in all districts. 00, 01, 06, 07, 08, 09, 10, 17, 18, 19, 28, 48, 49, 58, 59, 60, 68, 69, 70, 71, 78, 79, 80, 87, 88, 89, 90, 98, 99.

Rhytidiadelphus loreus (Hedw.) Warnst.
Uncommon, on the ground, amongst boulders in woodland and in turf, in the Uplands and the Upland ridge and in one locality in Gower, absent from the Vale. 18, 48, 60, 80, 90, 98.

Pleurozium schreberi (Brid.) Mitt.
Frequent to common, in acidic turf, in heaths, moorlands and mountain sides, on stream banks, in old quarries and in sand dunes, in all districts but in only one locality in the Vale. 00, 07, 08, 09, 10, 18, 19, 48, 49, 58, 59, 80, 88, 89, 90, 98, 99.

Hylocomium brevirostre (Brid.) Br.Eur.
Very rare, on limestone scree in wood. **00**: Morlais Hill, 1986, *Orange*, W side of Morlais Hill, N of Merthyr Tydfil, 1987, *Perry & Orange* (BBSUK).

Hylocomium splendens (Hedw.) Br.Eur.
Uncommon, in turf on heathy hillsides, in old limestone quarry, in woodland and on sea cliff, tolerating both acidic and calcareous situations, recorded from a few widely scattered localities in all districts but absent from the mountains. 00, *18*, 48, 49, *58, 87*.

REFERENCES

Armitage, E. (1921). Glamorganshire bryophyta. *J. Bot.* **59**, 49–50.

Crundwell, A.C. & Nyholm, E. (1974). Funaria muhlenbergii and related European species. *Lindbergia* **2**, 222–229.

Corley, M.F.V. & Hill, M.O. (1981). *Distribution of Bryophytes in the British Isles. A Census Catalogue of their Occurrence in Vice-Counties.* Cardiff: British Bryological Society.

Gutch, J.W.G. (1842). A list of plants met with in the neighbourhood of Swansea, Glamorganshire. *Phytologist* **1**, 180–187.

Kay, Q.O.N. (1989). Mosses and liverworts of the Gower Peninsula: an annotated list. 16 pp. Unpublished manuscript.

Knight, H.H. (1936). Mosses and hepatics of Glamorgan. In Tattersall, W.M. (ed.), *Glamorgan County History* **1**, 179–182. Cardiff: William Lewis Limited.

Smith, A.J.E. (1964). A bryophyte flora of Glamorgan. *Trans.br.Bryol.Soc.* **4**, 539–596.

Smith, A.J.E. & Whitehouse, H.L.K. (1978). An account of the British species of the *Bryum bicolor* complex including *B. dunense* sp.nov. *J.Bryol.* **10**, 29–47.

Sowerby, J. (1807). *English Botany; or, coloured figures of British Plants . . .*, **24**, f.1710. London: for the author.

Syed, H. (1973). A taxonomic study of *Bryum capillare* Hedw. and related species. *J.Bryol.* **7**, 265–326.

Trow, A.H. (1899). Liverworts found in the neighbourhood of Cardiff. *Rep. Trans. Cardiff Nat.Soc.* **30**, 57–59.

Trueman, A.E. (1936). A general survey of Glamorgan. In Tattersall, W.M. (ed.), Glamorgan County History **1**, 1–17. Cardiff: William Lewis Limited.

Turner, D. & Dillwyn, L.W. (1805). *The Botanist's Guide through England and Wales.* **1**, 305. London: Phillips & Fardon.

Wade, A.E. (1959). Glamorgan botanical notes, 1957. *Rep.Trans.Cardiff Nat.Soc.* **86**, 22–23.

Wade, A.E. (1960). Glamorgan botanical notes, 1958. *Rep.Trans.Cardiff Nat.Soc.* **87**, 27–28.

8

Lichens of Glamorgan

Alan Orange

The following account includes lichenized fungi (lichens), lichenicolous fungi, and fungi which have traditionally been regarded as lichenized, but which are now thought not to form stable associations with algae. The list comprises 518 species and 6 infraspecific taxa of lichens and 38 non-lichenized species. The county boundaries used are those of the botanical vice-county of Glamorgan (V.C. 41). No systematic survey of the county's lichen flora has yet been attempted, and the present account must be regarded as little more than a compilation of the available records.

The earliest record of a lichen from Glamorgan appears to be that of a species of *Lepraria* (as *Conferva pulverea*) in Dillwyn (1809): 'on the ruins of the Chapter House at Margam, and the walls of Oystermouth Castle, Glamorgan, Mr. Young'. Gutch (1842 a & b) listed several lichen records from the Swansea area, including *Dermatocarpon miniatum*, *Toninia coeruleonigricans* and *Squamarina cartilaginea*.

In this century, at least 22 people have contributed at least one lichen record. Many records were collected by A.E. Wade, who was employed at the National Museum of Wales in Cardiff from 1920 to 1961, and who collected records until 1975. A list of lichens recorded in the county was produced by Wade & Watson (1936): this comprised 324 species and 146 infraspecific taxa, with brief habitat details for each species, and localities for some of the rarer species. A visit to Gower by the British Lichen Society in September 1990 added more than 30 new species to the county list.

The main source of records is the herbarium of the National Museum of Wales, supplemented by published records and unpublished field notes and record cards. Nomenclature generally follows Cannon *et al.* (1985). Non-lichenized species are listed in parentheses.

For each species all 10 km squares from which it has been recorded are listed; records dating only from before 1960 are listed in parentheses. The vice-county of Glamorgan includes 41 ten-kilometre squares, namely: 21/38, 39, 48, 49, 58, 59, 68, 69, 77, 78, 79, 87, 88, 89, 96, 97, 98, 99, 22/50, 60, 61, 70, 71, 80, 81, 90, 31/06, 07, 08, 09, 16, 17, 18, 19, 26, 27, 28, 32/00, 01, 10, 11.

Abbreviations and symbols:

BLS: record collected by British Lichen Society in 1990.
MS: record derived from information held by the British Lichen Society mapping scheme.
!:   specimen identified, or an identification confirmed, by A. Orange.
TLC: thin-layer chromatography.

LICHENS

(**Abrothallus parmeliarum** (Sommerf.) Arnold)
21/48: Penrice Castle, on *Parmelia saxatilis*, 1986
Orange (NMW). 21/49: Rhossili Down, on *Parmelia discordans*, 1986 *Orange* (NMW). Probably frequent.

Acarospora fuscata (Nyl.) Arnold
On acidic rocks, frequent. 21/49, 87, 88, 96, 97, 98,
31/06, 07, 09, 16, 18, (19), 32/00, (01).

A. heppii (Naeg.) Naeg.
Rare. (31/06): The Leys, West Aberthaw, on Liassic
Limestone boulder, 1953 Wade (NMW)! 21/97:
Llansannor church, on pebble, 1989 *Orange*.

A. smaragdula (Wahlenb.) Massal.
On acidic, nutrient-enriched rocks; occasional. 21/48,
49, 87, 96, 97, 31/18.

Acrocordia conoidea (Fr.) Körber
Limestone rocks and walls, frequent. 21/48, 49, 58,
59, 87, 97, 31/(06), 07.

A. gemmata (Ach.) Massal.
On bark, occasional; records from Gower and coastal
parts of the Vale. 21/48, 87, 97, 31/(06), 07.

A. salweyi (Leighton ex Nyl.) A.L. Sm.
31/06: West Aberthaw, on soft mortar of limestone
wall, 1988 *Orange* (NMW).

(**Adelococcus cladoniae** (Anzi) Keissler)
21/49: Whiteford Burrows, on *Cladonia pocillum*, 1988
Q.O.N. Kay (UCSA)! 21/49: Bovehill, on *Cladonia
pocillum*, 1990 BLS.

Agonimia tristicula (Nyl.) Zahlbr.
On calcareous soil and over mosses, on limestone
rocks and walls. 21/49, 58, 59, 87, 97, 22/70, 31/07,
17.

Anaptychia ciliaris (L.) Körber
(21/87): Ewenny Downs, on hawthorn, Ivimey-Cook
(1959).

A. runcinata (With.) Laundon
Acidic rocks near the coast, and one record on wood;
rare. 21/48, 49.

Anisomeridium biforme (Borrer) R.C. Harris
On bark, few records. 21/49, 58, 97, 31/(07, 16).

A. nyssaegenum (Ell. & Ev.) R.C. Harris (*A. juistense*
(Erichsen) R.C. Harris)
Frequent on *Sambucus* bark, one record from bone.
21/49, 58, 97, 22/60, 31/06, 07, 17, 18, 26, 28.

Arthonia didyma Körber
31/07: Beaupre Castle, on bark, 1990 *B.J. Coppins*.

A. impolita (Hoffm.) Borrer
31/06: Penmark churchyard, sheltered bark of *Acer
pseudoplatanus*, 1988 *Orange* (NMW). 31/07: Beaupre
Castle, 1990 *B.J. Coppins*.

A. lapidicola (Taylor) Branth & Rostrup
On limestone. 21/48, 49, 31/07, (17).

A. leucodontis (Poelt & Döbb.) Coppins
21/48: Mewslade Bay, on *Sambucus*, 1990 BLS.

A. radiata (Pers.) Ach.
On bark, frequent. 21/49, 58, 59, 87, 88, 96, 97,
31/(06), 07, 16, (17).

A. spadicea Leighton
On shaded bark, frequent. 21/58, 97, 98, 22/60, 70,
31/07, 17, 18, 28.

A. vinosa Leighton
Rare. Castle-upon-Alun, *M.G. Winkles* (Wade 1950).
21/48: Penrice Castle, on oaks, 1986 *Orange*.

Arthopyrenia antecellans (Nyl.) Arnold
Ewenny, H.H. Knight (Watson 1930). Record requires
confirmation.

(**A. cinereopruinosa** (Schaerer) Massal.)
21/48: Penrice Castle, on old *Platanus*, 1986 *Orange*
(NMW).

A. elegans R. Sant.
21/58: Southgate, limestone rocks on sea shore, 1990
A. Fletcher.

A. halodytes (Nyl.) Arnold *sensu lato*
Limestone rocks, barnacles and limpet shells between
tide levels; frequent. 21/48, 58, 87, 96, 31/16, 26.

(**A. lapponina** Anzi)
On bark, frequent. 21/48, 49, 58, 59, 87, 31/07, 17.

A. monensis (Wheldon) Zahlbr.
31/06: Porthkerry Park, limestone stones by small
stream, 1987 *Orange* (NMW).

A. orustensis Erichsen
21/58: Southgate, limestone rocks on shore, 1990 *A.
Fletcher*.

(**A. punctiformis** Massal.)
Occasional on smooth bark. 21/58, 31/16.

A. ranunculospora Coppins & P. James
Rare. 21/48: Penrice Castle, on oak, 1986 *Orange*
(NMW).

A. salicis Massal.
21/59: near Ilston, on *Corylus*, 1990 *Orange* (NMW).

A. sublitoralis (Leighton) Arnold
On limestone rocks on sea shore. 21/48, 58.

Arthrorhaphis citrinella (Ach.) Poelt
21/99: Tarren Rhiw-maen, on mosses on sandstone cliff, alt. 400m, 1987 *Orange*.

(A. grisea Th. Fr.)
22/90: below Craig y Llyn, 1932 *Wade* (NMW)! Host usually *Baeomyces rufus*, but scarcely determinable here.

Aspicilia caesiocinerea (Nyl. ex Malbr.) Arnold
Acidic rocks. (31/19): Mynydd Eglwysilan, 1943 *Wade* (NMW). 32/00: Troedyrhiw, 1960 *Wade* (NMW)!

A. calcarea (L.) Mudd
Limestone rocks and walls, also concrete; common. 21/48, 49, 58, 59, (77), 87, (88), 96, 97, 98, 06, 07, 08, 16, 17, 26, 27, 22/90, 32/01.

A. cinerea (L.) Körber
21/48: Mewslade Bay, 1990 BLS. 21/49: Llanrhidian church, 1990 BLS.

A. contorta (Hoffm.) Krempelh.
Limestone rocks and concrete, frequent. 21/48, 49, 87, 96, 97, 98, 31/06, 07, 16, 17, (18), 26, 27, 32/00.

A. leprosescens (Sandst.) Havaas
21/49: Rhossili Down, 1990 *A. Fletcher*.

(Athelia arachnoidea (Berk.) Jülich)
On *Lecanora conizaeoides*, forming conspicuous lesions. Several records from the Cardiff area, probably frequent.

Bacidia arceutina (Ach.) Arnold
On bark. 21/48, 31/06, (07), 16.

B. arnoldiana Körber
31/18: Coed y Bedw, on shaded limestone, 1986 *Orange* (NMW).

B. bagliettoana (Massal. & de Not.) Jatta (*B. muscorum* (Ach.) Mudd)
Over mosses on calcareous soil and on dunes, probably frequent. 21/48, 49, 58, 78, 87, 98, 31/(17), (18), 32/(00).

B. caligans (Nyl.) A.L. Sm.
31/17: Bute Park, Cardiff, on base of *Aesculus* trunk, 1986 *Orange*, det. B.J. Coppins 1987 (NMW).

B. chloroticula (Nyl.) A.L. Sm.
On recently erected fenceposts and on urban wasteland, frequent. 21/48, 69, 31/17, 18, 28.

B. cuprea (Massal.) Lettau
21/48: Penrice Castle, on limestone rocks, 1990 BLS.

B. egenula (Nyl.) Arnold
31/17: Slanney Woods, on shaded mortar of brick wall, 1989 *Orange*.

B. friesiana (Hepp) Körber
Listed by Watson (1953). Record requires confirmation.

B. incompta (Borrer ex Hooker) Anzi
Rare. (21/87): Bridgend, on *Ulmus*, 1945 *M.G. Winkles* (NMW)! Bridgend, on *Ulmus*, 1957 *Wade* (NMW)! 31/07: Dyffryn Gardens, on bark, 1967 *Wade* (NMW)!

B. inundata (Fr.) Körber
Damp acidic rocks, often by streams, few records. 32/00.

B. laurocerasi (Delise ex Duby) Zahlbr.
Rare. (31/16): Swanbridge Halt, on *Ulmus*, 1957 *Wade* (NMW)! (31/17): near Llandough, Cardiff, on *Hedera* stem, 1959 *Wade* (NMW)! 21/48: Mewslade Bay, on *Sambucus*, 1990 BLS.

B. naegelii (Hepp) Zahlbr.
On bark, rare. 21/48, 97, (31/16).

B. phacodes Körber
Listed by Wade & Watson (1936).

B. rubella (Hoffm.) Massal.
On bark, rare. 21/48, 58, 97.

B. sabuletorum (Schreber) Lettau
On limestone and mortar, usually growing over mosses; frequent. 21/48, 49, 58, 59, 77, (87), 97, 98, 22/70, 32/07, (16, 17), 18, 26, 32/(00).

B. saxenii Erichsen
21/69: Swansea, Hafod, on urban wasteland (Gilbert 1990). 21/69: Swansea, Prince of Wales Dock, on urban wasteland (Gilbert 1990). 31/28: Draethen, calcareous soil in quarry, 1990 *Orange*.

B. vezdae Coppins & P. James
31/07: Pysgodlyn Mawr, on *Salix cinerea*, 1988 *Orange* (NMW). 31/07: near Bonvilston, on shaded *Quercus* and *Fagus*, 1988 *Orange* (NMW). 31/17: Leckwith, on shaded *Fraxinus* trunk, 1990, *Orange* (NMW).

Baeomyces roseus Pers.
On acidic soil, frequent. 21/89, 98, 22/60, (90), 31/(18), 22/(00).

B. rufus (Huds.) Rebent.
On acidic soil and stones, frequent. 21/49, 98, 99, 22/60, 70, 31/07, (08), (18), 28, 32/(00).

Belonia nidarosiensis (Kindt.) P.M. Jørg. & Vězda
Shaded limestone. 31/26: Flat Holm, 1986 *Orange* (NMW). 32/00: Morlais Hill, 1987 *Orange*. Both records of sterile plants.

Biatorina atropurpurea (Schaerer) Massal.
Rare. (31/07): between Peterston-super-Ely and Pendoylan, on bark, 1933 *Wade* (NMW)! 21/48: Penrice Castle, 1990 BLS.

Bryoria fuscescens (Gyelnik) Brodo & D. Hawksw.
Rare. (31/18): Coed Coesau-whips, on *Quercus*, 1933
Wade (NMW)! (21/99): record from MS. 22/60: Cwm
Clydach, on *Betula*, 1988 *Orange*.

Buellia aethalea (Ach.) Th. Fr.
On acidic rocks, frequent. 21/48, 49, 69, 96, 97, 31/07,
(08), 09, 32/00.

B. disciformis (Fr.) Mudd
'Benarth' (Leighton 1879).

B. griseovirens (Turner & Borrer ex Sm.) Almb.
On fence posts. 21/49, 97, 31/07, 17.

B. ocellata (Flotow) Körber
On acidic rocks. 21/48, 49, 96, 97, 31/07.

B. pulverea Coppins & P. James
(31/18): Cefn Onn (31/18) on *Crataegus*, 1932 *Wade*
(NMW)! (31/18): Craig Llanishen, on *Fraxinus*, 1932
Wade (NMW)! (31/07): between Bonvilston and
Llancarfan (31/07) on *Pinus*, 1933 *Wade* (NMW)!

B. punctata (Hoffm.) Massal.
On nutrient-rich bark and fence-posts, frequent.
21/48, 49, 97, 22/60, 31/06, 07, 16, 17, 18, 26, 27.

B. stellulata (Taylor) Mudd
31/19: Mynydd Eglwysilan, 1973 *Wade*. 21/49:
Llanrhidian church, 1990 BLS.

Calicium glaucellum Ach.
On fence-posts and decorticate trunks, few records.
22/60, 31/(07, 18).

C. viride Pers.
On bark, few records. 21/48, 31/(06), 07.

Caloplaca alociza (Massal.) Migula
Frequent on Carboniferous Limestone near the coast.
21/48, 58, 77, (87), 31/(26).

C. aurantia (Pers.) Steiner
On limestone rocks and walls, common. 21/48, 49, 58,
77, 87, 88, 96, 97, 98, 31/06, 07, 16, 17, (18), 26, 28.

C. cerina (Ehrh. ex Hedw.) Th. Fr.
On nutrient-rich bark (rarely on wood), including
dead stems of *Helianthemum canum*, several records
from near the coast. 21/48, 87, 31/06, (16).

C. cirrochroa (Ach.) Th. Fr.
On limestone, several records from near the coast.
21/48, 49, 58, 31/16.

C. citrina (Hoff.) Th. Fr.
On limestone, mortar, concrete etc.; very common.
21/48, 49, 58, 59, 69, (78), 87, 88, 96, 97, 98, 99, 31/06,
07, 08, 16, 17, 18, 26, 27, 28, 32/00.

C. crenularia (With.) Laundon
Acidic rocks, frequent. 21/59, 87, 96, 97, 31/07, 08,
18.

C. decipiens (Arnold) Blomb. & Forss.
(31/17): Penarth, on pales, *Wade* (Watson 1933).
31/17: Penarth, on cemented wall, 1967 *Wade*
(NMW)!

C. flavescens (Huds.) Laundon
On limestone rocks and walls, common. 21/48, 49, 58,
59, 77, 87, 88, 96, 97, 98, 31/06, 07, 08, 16, 17, 18, 26,
32/00, 01.

C. flavovirescens (Wulfen) Dalla Torre & Sarnth.
On limestone and concrete, occasional. 21/48, 49, 58,
69, 96, 97, 99, 31/06, 07, 16, 17, (18), 26, 32/00.

C. granulosa (Müll. Arg.) Jatta
On limestone outcrops on the coast, fairly frequent in
Gower. 21/48, 49, 58, 31/16.

C. holocarpa (Hoffm.) Wade
On limestone and concrete, rarely on wood; common.
21/48, 58, 69, 77, 79, 87, 96, 97, 98, 99, 31/06, 07, 08, 09,
16, 17, 18, 26, 28, 32/00, 10.

C. isidiigera Vĕzda
21/97: Llandow churchyard, on flat sandstone tomb,
1986 *Orange*. 31/07: St Mary Church churchyard,
sandstone flagstones on path, 1988 *Orange*. 21/49:
Cwm Ivy, sandstone paving stones in garden, 1990
BLS!

C. lactea (Massal.) Zahlbr.
On limestone, occasional. 21/(48), 49, 58, 87, 31/06,
16, 26.

C. marina (Wedd.) Zahlbr.
On limestone rocks on the sea shore, common. 21/48,
49, 58, 77, 87, 96, 31/16, 17, 26.

C. microthallina (Wedd.) Zahlbr.
On limestone rocks on the sea shore, few records.
21/48, 77, 31/16.

C. ochracea (Schaerer) Flagey
On limestone. 21/48, 49, 58, 87.

C. saxicola (Hoffm.) Nordin
On limestone, mortar and concrete, often on dry
vertical surfaces; frequent. 21/49, 58, 87, 96, (97),
31/06, 07, (17), 26.

C. thallincola (Wedd.) Du Rietz
On limestone rocks on the sea shore, common. 21/38,
48, 58, 77, 87, 31/16, 26.

C. teicholyta (Ach.) Steiner
21/49: Landimore, on limestone outcrop, 1990
Orange. 21/96: Llantwit Major, in churchyard, 1990
B.J. Coppins. 31/07: St Hilary church, 1990 *B.J.
Coppins*.

C. ulcerosa Coppins & P. James
31/07: Beaupre Castle, on bark, 1990 *B.J. Coppins.*

C. variabilis (Pers.) Müll. Arg.
On limestone rocks. 21/48, 58.

C. velana (Massal.) Du Rietz
On limestone rocks and walls, frequent in the Vale and Gower, many records from churches. 21/48, 49, 59, 77, 87, 97, 98, 31/06, 07, 16, (26).

Candelaria concolor (Dickson) B. Stein
21/49: Whiteford Burrows, on *Sambucus*, 1990 *D. McCutcheon*!

Candelariella aurella (Hoffm.) Zahlbr.
On limestone and concrete, frequent. 21/69, 77, 96, (97), 98, 31/06, 08, 16, 17, 26, 27.

C. coralliza (Nyl.) Magnusson
21/49: Rhossili Down, 1990 BLS.

C. medians (Nyl.) A.L. Sm.
On limestone rocks and walls, also on concrete; occasional. 21/48, 49, 58, 87, 96, 97, 31/06, 07, 16, 26.

C. reflexa (Nyl.) Lettau
Nutrient-rich bark, probably frequent. 21/48, 49, 58, 59, 97, 31/06, 07, 17.

C. vitellina (Ach.) Müll. Arg.
On nutrient-enriched acidic rocks, trees, wood, slag etc; common. 21/49, 69, 79, 87, 88, 96, 97, 98, 99, 22/70, 31/06, 07, 08, 09, 16, 17, 18, 26, 27, 28, 32/00, 10.

Catapyrenium pilosellum O. Breuss
On soil amongst limestone rocks and on walls. 21/48, 49, 58, 97. (Old records of *Catapyrenium* from 21/87, 31/06, (07, 16, 17), 26 need to be redetermined).

Catillaria atomarioides (Müll. Arg.) Kilias
32/00: Dowlais, Merthyr Tydfil, on blocks of acidic slag, 1986 *Orange* (NMW). 21/96: Llantwit Major, in churchyard, 1990 *B.J. Coppins.*

C. chalybeia (Borrer) Massal.
On acidic rocks, pebbles and brick, less frequently on calcareous rock; frequent. 21/48, 87, 96, 97, 98, 31/07, 16, (17), (18), 26, (28).

(C. episema (Nyl.) H. Olivier)
On *Aspicilia calcarea*. (21/48): Mewslade Bay, 1943 *Wade* (NMW). 31/06: West Aberthaw, 1988 *Orange* (NMW). 21/48: Mewslade Bay, 1990 BLS.

C. lenticularis (Ach.) Th. Fr.
On limestone and other calcareous substrata, common. 21/(48), 49, 58, (87), 96, 97, 31/06, 07, (18), 27.

C. pulverea (Borrer) Lettau
21/59: near Ilston, on *Salix cinerea*, 1990 *Orange* (NMW).

C. sphaeroides (Dickson) Schuler
On bark, rare. 21/48, 22/(80), 32/(00).

Catinaria grossa (Pers. ex Nyl.) Vainio
(21/87): Candleston Castle, on *Ulmus*, 1942 *F.A. Sowter* (Wade 1959).

Cetraria chlorophylla (Willd.) Vainio
On acidic bark and boulders, scattered records. 21/48, (99), 22/(90), 31/07, (18), 32/00.

Chaenotheca ferruginea (Turner ex Ach.) Mig.
Acidic bark, usually on *Quercus*, few records. 21/98, 31/07, 17, 18.

C. hispidula (Ach.) Zahlbr.
21/58: Parkmill, 1990 *D. McCutcheon*!

Chrysothrix candelaris (L.) Laundon
On bark in dry recesses, uncommon. 21/48, 58, 69, 31/07.

C. chrysophthalma (P. James) P. James & Laundon
21/58: Penrice Castle, on wood, 1986 *Orange* (NMW). 31/07: Dyffryn Gardens, dry *Quercus* trunk, 1988 *Orange* (NMW). 31/07: Beaupre Castle, 1990 *B.J. Coppins.*

Cladonia arbuscula (Wallr.) Rabenh.
Rare. (21/09): Taren y Gigfran, Aberfan, 1958 *Wade* (NMW)!

C. bacillaris auct.
On the ground, upland, few records. Specimens contain barbatic acid. 22/60, 31/(18, 19).

C. caespiticia (Pers.) Flörke
Acidic soil banks, tree bases and logs, frequent. 21/48, 98, 22/60, 70, 32/08, 09, 18.

C. cervicornis (Ach.) Flotow subsp. **cervicornis**
On rocks and stony banks, scattered records. 21/48, 49, 58, 31/(18).
  subsp. **verticillata** (Hoffm.) Ahti
31/18: Mynydd Meio, acidic soil bank, 1986 *Orange*. 21/98: Llanharry, on mine spoil heap, 1990 *Orange*. 21/49: Rhossili Down, 1990 BLS.

C. ciliata Stirton var. **ciliata**
Rare. (31/09): Taren y Gigfran, Aberfan, in rocky hollow, 1958 *Wade* (NMW)! 21/87: Merthyr Mawr, 1960 *M. Percival*, conf. Wade.
  var. **tenuis** (Flörke) Ahti
Rare. 21/49, 58, 98.

C. coccifera (L.) Willd.
On acidic soil on banks and amongst rocks, mainly upland, frequent. 21/49, 69, 98, 99, 22/60, 90, 31/09, 18, 19, 32/00, 01. Glamorgan material in NMW contains zeorin and usnic acid, with or without porphyrilic acid.

C. coniocraea auct.
On tree bases, acidic rocks and soil, common. 21/48, 49, 58, (87), 88, 96, 97, 98, 99, 22/60, 70, 31/06, 07, 08, 09, 17, 18, 32/00, 10.

C. crispata (Ach.) Flotow
On acidic soil, rarely on wood, upland, occasional. 21/(99), 22/60, 31/(18), 32/(00).

C. cryptochlorophaea Asah.
On acidic soil, peat, tree trunks and decaying plant debris, mainly upland; frequent. 21/48, 22/60, 90, 31/(07), 09, 18.

C. digitata (L.) Hoffm.
On tree trunks and rotting wood, few records. 21/58, 99, 22/60, 31/17, 28.

C. fimbriata (L.) Fr.
On acidic soil, wood, tree bases and rocks; frequent. 21/48, 49, 58, 87, 88, 96, 97, 98, 99, 22/70, 90, 32/06, 07, 08, 09, 16, 17, 18, (28), 32/00, 01, 10.

C. firma (Nyl.) Nyl.
21/49: Rhossili Down, on wet rock, 1990 *O.W. Purvis* (BM).

C. floerkeana (Fr.) Flörke
On peaty soil, tree stumps; frequent. 21/49, 58, (87), 98, 99, 22/60, 90, 31/07, 18, (19), (28), 32/(00), 01.

C. foliacea (Huds.) Willd.
On stony calcareous soil and on dunes, frequent. 21/48, 49, 58, 78, 87, 98.

C. fragilissima Østh. & P. James
21/49: Rhossili Down, 1990 *O.W. Purvis* (BM).

C. furcata (Huds.) Schrader subsp. **furcata**
On soil and over rocks, frequent. 21/49, 58, 69, 88, 89, 97, 98, 99, 22/90, 31/(07), 08, (17), 18, 19, 26, 28, 32/00.
  subsp. **subrangiformis** (Scriba ex Sandst.) Pisút
On calcareous soil and on dunes, frequent. 21/48, 49, 58, (78), (87), 31/(17).

C. glauca Flörke
22/60: Cwm Clydach, on wood and tree trunks, 1988 *Orange* (NMW). 21/48: Rhossili Down, 1990 BLS.

C. gracilis (L.) Willd.
On acidic soil and over rocks, occasional. 21/49, 58, 99, 22/90, 32/00.

C. humilis (With.) Laundon
On soil, also on mossy tree trunks; probably frequent. All specimens seen contained atranorin. 21/69, 87, 98, (07), (17), (18), 27, 32/00.

C. luteoalba A. Wilson & Wheldon
21/49: Rhossili Down, 1990 BLS.

C. macilenta Hoffm.
On acidic soil and on tree bases, frequent. 21/49, 58, 98, 22/70, 31/07, (17), 18, (19), 28.

C. merochlorophaea Asah.
On acidic soil and over decaying vegetation, upland; few records but probably frequent. All specimens seen contained merochlorophaeic acid by TLC. 21/99, 22/60, 31/(08), (18), 32/(00).

C. ochrochlora Flörke
On stumps and tree bases, frequent. 21/87, 07, 98, 99, 31/(08), 09, (18).

C. parasitica (Hoffm.) Hoffm.
On stumps and logs, occasional. 21/88, 98, 22/60, (80).

C. pocillum O.-J. Rich.
On calcareous soil and in dunes, frequent. 21/48, 49, 58, 78, 88, 98, 31/(17, 18), 26, 32/00.

C. polydactyla (Flörke) Sprengel
On acidic soil and on stumps, frequent. 21/48, 99, 22/60, 90, 31/(07), 32/(00).

C. portentosa (Dufour) Coem.
On acidic soil and amongst rocks, common and widespread. 21/49, 58, 69, 87, 89, 96, 98, 99, 22/60, 70, 90, 31/(07), 08, 09, 18, 19, 32/00, 01, 10.

C. pyxidata (L.) Hoffm.
On acidic soil, rotting wood, tree bases; frequent. 21/48, 58, 69, 87, 97, 98, 22/60, 31/07, 08, (16, 17, 18), 26, 32/01.

C. ramulosa (With.) Laundon
On soil, dunes and on tree bases, rather few records. 21/48, 49, 58, 88, 98, 31/(18), (28).

C. rangiformis Hoffm.
On more or less calcareous soil and on dunes, locally common. 21/48, 49, 58, 69, 77, (78), (87), 88, 89, 98, 31/06, (07), 16, (17), (18), 26, 28, 32/00, 10.

C. scabriuscula (Delise) Leighton
Rare. (31/18); below Craig Llanishen, 1932 *Wade* (NMW)! Dimbath Valley, 1933 *E.M. Thomas* (NMW)! 21/48: Cefn Bryn, on acidic bank, 1986 *Orange* (NMW)!

C. squamosa (Scop.) Hoffm. var. **squamosa**
On acidic soil, tree bases, and stumps; frequent. 21/48, 49, 98, 99, 22/60, (90), 31/(09, 18, 19), 32/00.
  var. **subsquamosa** (Nyl. ex Leighton) Vainio
(31/18): Craig Llysfaen (Wade & Watson 1936). 31/08: between Efail Isaf and Creigiau, 1974 *Wade*.

C. subcervicornis (Vainio) Kernst.
On acidic soil amongst rocks, upland, uncommon. 21/49, 99, 22/90, 32/(00).

C. subulata (L.) Wigg.
On acidic soil. 21/69, 98, 99, 31/06, 08, 18, 32/00.

C. uncialis (L.) Wigg.
On peaty soil, upland, frequent. 21/49, 58, 99, 22/90, 31/09, 18, 32/00.

Cliostomum graniforme (Hagen) Coppins
21/49: Llanmadoc, 1933 *D.A. Jones* (Wade 1950). Record requires confirmation.

C. griffithii (Sm.) Coppins
On bark, frequent. 21/48, 58, 87, 96, 97, 31/06, 07, 16, 17, (28).

Coelocaulon aculeatum (Schreber) Link.
On stony acidic soil and on soil over rocks, frequent. 21/49, 58, 22/(90), 31/18, 32/00.

C. muricatum (Ach.) Laundon
Similar habitats to the last species, occasional. 21/49, 99, 22/(90), 32/00.

Collema auriforme (With.) Coppins & Laundon
On limestone rocks, mortar, and calcareous soil, common. 21/48, 49, 58, 96, 97, 98, 22/70, 31/06, 17, 18, 32/00.

C. crispum (Huds.) Wigg.
On limestone, concrete and mortar, common. 21/48, 58, 49, 77, 87, 88, 96, 97, 31/06, 07, 08, 16, 17, 18, 26, 32/00, 01.

C. cristatum (L.) Wigg.
Calcareous rocks and soil, frequent. 21/48, 77, 87, 31/(18, 16).

C. fragrans (Sm.) Ach.
Listed by Watson (1953). Record requires confirmation.

C. fuscovirens (With.) Laundon
Limestone rocks. 21/48, 49, 97, 31/06, 07, (18).

C. limosum (Ach.) Ach.
Rare. (31/16): Cold Knap, Barry, on soil, 1958 *Wade* (NMW)! 21/69: Llandore, Lower Swansea Valley, soil on demolition site, 1989 *O.L. Gilbert* (NMW)!

C. multipartitum Sm.
21/48: Mewslade Bay, on limestone, 1990 BLS.

C. polycarpon Hoffm.
On limestone, 21/48, 49.

C. subflaccidum Degel.
Rare. (21/48): The Cross, Reynoldston, on *Ulmus*, 1932 *D.A. Jones* (NMW)!

C. tenax (Schwartz) Ach.
On calcareous soil, mortar, and on dunes; common. 21/48, 49, 58, 87, 88, 96, 97, 98, 31/06, 07, 08, 16, 17, (18).

C. undulatum Laurer ex Flotow var. *granulosum* Degel.
21/49: Llanrhidian, 1990 BLS.

(Cornutispora lichenicola D. Hawksw. & Sutton)
(31/18): Mynydd Rudry, on a sterile crustose lichen, 1932 *Wade* (NMW)!

Cystocoleus ebeneus (Dillwyn) Thwaites
On shaded acidic rocks in underhangs; frequent. 21/59, 98, 22/(90), 32/00.

(Dactylospora parasitica (Flörke) Zopf)
31/07: Dyffryn Gardens, on *Pertusaria hymenea*, 1988 *Orange* (NMW). 21/97: Llanmihangel, 1990 *B.J. Coppins*.

Dermatocarpon luridum (With.) Laundon
Rare. 22/70: Tre-Forgan, periodically submerged rocks, 1975 *Q.O.N. Kay* (UCSA)! 22/60: Cwm Clydach, rocks by river, 1988 *Orange*.

D. miniatum (L.) Mann
On dry shaded limestone rocks; occasional. 21/48, 49, 58, (87), 97.

(Didymella sphinctrinoides (Zwackh) Berl. & Vogl.)
31/27: Cardiff East Moors, on *Leptogium schraderi* on wasteland, 1990 *Orange* (NMW).

Dimerella diluta (Pers.) Trevisan
Frequent on bark and wood in shady woodland. 21/48, (87), 89, 97, 22/60, 70, 31/07, 17, 18, 28, 32/01.

D. lutea (Dickson) Trevisan
Rare. 21/98: Cwm Dimbath, on the base of *Quercus* trunk, 1986 *Orange*. 21/48: Penrice Castle, on *Quercus* trunk, 1986 *Orange*. 22/60: Cwm Clydach, slightly sheltered bark on two large ash trees, 1988 *Orange*.

(Dinemasporium strigosum (Pers. ex Fr.) Sacc.)
21/58: Pennard Burrows, on dying parts of *Peltigera canina*, 1990 *Orange*, det. *V. Alstrup* (NMW).

Diploicia canescens (Dickson) Massal.
On nutrient-rich bark and on dry limestone, common. 21/48, 49, 58, 87, 88, 96, 97, 98, 31/06, 07, 16, 17, (28).

Diploschistes muscorum (Scop.) R. Sant.
Calcareous dunes and soil, usually overgrowing *Cladonia pocillum*, also on moss; locally frequent. 21/48, 49, 58, 78, (87).

D. scruposus (Schreber) Norman
On acidic rocks and walls, common. 21/48, 49, 98, 99, 22/60, 70, (90), 31/09, (18, 19), 32/00, (10).

Diplotomma alboatrum (Hoffm.) Flotow (including *D. epipolium* (Ach.) Arnold)
Frequent on limestone and mortar, rare on bark. 21/48, 49, 59, 87, 96, 97, 31/06, 07, (16), 17, 26.

D. chlorophaeum (Hepp ex Leighton) Szat.
On acidic rocks, rare. (21/48): Penrice, *Wade* (Watson 1933). Rhossili, 1932 *D.A. Jones* (NMW)! 21/77: near Sker Point, on wall by sea, 1988 *Orange* (NMW).

Dirina massiliensis Durieu & Mont. forma **sorediata** (Müll. Arg) Tehler
Shaded limestone and plaster on church walls, locally frequent, rarely on natural outcrops of Carboniferous Limestone. 21/48, 59, 87, 96, 97, 31/06, 07.

Endocarpon pusillum Hedw.
Rare. (31/17): St Fagans Castle, stones by fish pond, 1947 *Wade* (NMW)! 21/97: Afon Alun valley, soil on limestone rocks in railway cutting, 1986 *Orange* (NMW).

(Endococcus alpestris D. Hawksw.)
22/60: Cwm Clydach, on *Usnea subfloridana*, 1988 *Orange* (NMW). 21/58: Penrice, on *Usnea florida*, 1990 *Orange*.

(E. rugulosus Nyl.)
21/48: Cefn Bryn, on *Huilia macrocarpa*, 1986 *Orange* (NMW).

Enterographa crassa (DC.) Fée
Uncommon on bark near the coast. 21/48, 58, 31/06, 07.

Ephebe lanata (L.) Vainio
21/99: Tarren Rhiw-maen, wet rocks by stream, alt. 400 m, 1987 *Orange* (NMW).

Evernia prunastri (L.) Ach.
On bark, stems of ericaceous shrubs, sometimes on the ground in dunes; common. 21/48, 49, 58, 59, 78, 87, 88, 89, 96, 97, 98, 22/60, 31/07, 09, 16, 17, 18, 28, 32/01.

Foraminella ambigua (Wulfen) Friche Meyer
On acidic bark, fence posts and logs, common. 21/58, 59, 88, 89, 98, 99, 22/60, 70, 80, 31/07, 08, 09, 18, 28, 32/00, 01, 10.

Fuscidea cyathoides (Ach.) V. Wirth & Vězda
Acidic rocks. 21/49, 31/(09), (18), 32/00.

F. kochiana (Hepp.) V. Wirth & Vězda
Listed by Watson (1953). 21/49: Rhossili Down, 1990 *A. Fletcher*.

F. lightfootii (Sm.) Coppins & P. James
Humid woodland, often on *Salix cinerea*. 21/48, 22/60, 97, 31/07, 17.

F. tenebrica (Nyl.) V. Wirth & Vězda
21/97: record from MS.

F. viridis Tønsb.
31/09: Cwm Clydach, Ynysybwl, on shaded *Corylus* trunk, 1988 *Orange* (NMW). 21/48: Penrice Castle, on shaded *Salix cinerea* and *Fagus sylvatica*, 1990 *Orange* (NMW).

Graphis elegans (Borrer ex Sm.) Ach.
On smooth bark, common. 21/48, 49, 58, 59, 89, 97, 98, 22/60, 70, (80), 31/07, 08, 17, 18, 28.

G. scripta (L.) Ach.
On smooth bark, common. 21/48, 49, 58, 59, 96, 97, 98, 22/60, 70, 31/07, 28, 32/00.

Gyalecta flotowii Körber
(21/87): Newbridge Fields, Bridgend, on *Ulmus*, *F.A. Sowter* (Wade 1959). Record requires confirmation; the record in Watson (1933) is apparently based on Penrice (21/48) 1932 *D.A. Jones* (NMW) which now contains no spores.

G. jenensis (Batsch.) Zahlbr.
On shaded limestone, frequent. 21/48, 49, 58, 87, 31/18, 32/00.

G. truncigena (Ach.) Hepp
On relatively basic bark on mature trees in areas of low pollution. 21/(48), 58.

Gyalideopsis anastomosans P. James & Vězda
On bark and wood in sheltered humid woodland, widespread. Often fertile. 21/59, 89, 22/60, 70, 31/07, 09, 17.

(Guignardia olivieri (Vouaux) Sacc.)
21/58: Southgate, on *Xanthoria parietina*, 1990 *Orange* (NMW).

Haematomma caesium Coppins & P. James
22/81: Near Coelbren, on sallow, 1987 *R.G. Woods* (NMW)!

H. ventosum (L.) Massal.
On acidic rocks, upland, rare. (32/00): Cnwc, Troedyrhiw, 1959 *Wade* (NMW)!, usnic acid absent by TLC. 32/00; Mynydd Cilfach-yr-encil, 1959 *Wade* (NMW)! and same locality, 1971 *A.R. Perry* (NMW)! 21/99: Craig y Geifr, 1987 *Orange*. The last three collections contained usnic acid by TLC.

(Homostegia piggotii (Berk. & Br.) P. Karsten)
21/49: Rhossili Down, on *Parmelia saxatilis*, 1986 *Orange* (NMW).

Huilia cinereoatra (Ach.) Hertel
Rare; records need confirmation. (31/18): Mynydd Rudry, *Wade* (Wade & Watson 1936). Dimbath Valley, 1942 *M.G. Winkles* (Wade 1950). 31/09: record from MS.

H. crustulata Ach.) Hertel
On acidic rocks, often on pebbles, common. 21/49, 69, 88, 99, 22/(80, 90), 31/(07, 17, 18, 19), 28, 32/00.

H. macrocarpa (DC.) Hertel
On acidic rocks, common. 21/48, 49, 59, 87, 88, 89, 98, 99, 22/60, 70, (80), 31/(07), 08, 09, 17, 18, 32/00, 01, 10.

H. platycarpoides (Bagl.) Hertel
On acidic rocks, few records. Specimens with norstictic acid and variable amounts of stictic acid by TLC. 21/87, 31/(18).

H. soredizodes (Lamy) Hertel
On acidic stones, few records. The two specimens in NMW contain stictic acid by TLC. 21/99, 31/28, 32/(00).

H. tuberculosa (Sm.) P. James
On acidic rocks and stones, common and widespread. 21/48, 49, 59, 69, 87, 88, 96, 97, 98, 99, 22/60, 70, 90, 31/07, 08, 09, (16, 17), 18, 19, 26, 27, 32/00, 10.

Hymenelia lacustris M. Choisy
On acidic rocks at stream margins, mainly upland, occasional. 21/98, 99, 22/60, 70, (80), 31/09, 32/(00).

Hyperphyscia adglutinata (Flörke) Mayrh. & Poelt
Rare. (31/07): near Llancarfan, on *Pinus sylvestris*, 1933 *Wade* (NMW)! 31/16: record from MS. 31/06: Penmark churchyard, on large trunk of *Acer pseudoplatanus*, 1988 *Orange* (NMW).

Hypocenomyce caradocensis (Leighton ex Nyl.) P. James & G. Schneider
On bark, rare. 22/60, 31/07, (18, 28).

H. scalaris (Ach. ex Liljeblad) M. Choisy
On acidic bark. 22/70, 31/07, (09), 17, (18).

Hypogymnia physodes (L.) Nyl.
On acidic bark, wood and rocks; common. 21/48, 49, 58, 69, 87, 88, 89, 96, 97, 98, 99, 22/60, 70, 90, 31/07, 08, 09, 16, 17, 18, 28, 32/00, 10.

H. tubulosa (Schaerer) Havaas
On acidic bark, wood and rocks; common. 21/48, 96, 97, 98, 99, 22/60, 70, 90, 31/07, 08, 09, 17, 18, 28.

(Illosporium corallinum Rob.)
Recorded on *Parmelia saxatilis* and *Lecanora conizaeoides*. 21/48, 89, 98, 22/70, 31/17, 18.

Ionaspis epulotica (Ach.) Blomb. & Forss.
21/48: The Knave, on limestone, 1990 BLS.

Lasallia pustulata (L.) Mérat
On acidic rocks in the uplands, rare. 31/09, 32/(00).

Lecanactis amylacea (Ehrh. ex Pers.) Arnold
Rare. 21/48: Penrice Castle, sheltered bark on a large *Quercus*, 1986 *Orange* (NMW). Specimen contains confluentic acid by TLC.

L. premnea (Ach.) Arnold
On oaks, rare. (31/07): between Peterston-super-Ely and Pendoylan, 1933 *Wade* (NMW)! (31/07): Peterston-super-Ely, 1944 *Wade* (NMW)! 21/48: Penrice Castle, on one *Quercus* trunk, 1986 *Orange* (NMW).

L. subabietina Coppins & P. James
Dry sheltered bark, rare. 21/48: Penrice Castle, on *Quercus* (NMW) and on *Platanus*, 1986 *Orange*. 21/58: Nicholaston Woods, on one *Quercus*, 1986 *Orange*.

Lecania cyrtella (Ach.) Th. Fr.
On nutrient-rich bark, few records. 21/48, 58, 31/(07, 17).

L. erysibe (Ach.) Mudd *sensu lato*
On limestone, concrete and bark; common. 21/48, (87), 22/60, 31/(06, 07), 17, (18), 26. forma *sorediata* Laundon 21/96, 97, 31/26.

L. fuscella (Schaerer) Körber
On nutrient-rich bark, uncommon. 31/(06, 07, 17, 28).

L. rabenhorstii (Hepp) Arnold
21/48: Mewslade Bay, on limestone, 1990 BLS. 21/48: Port Eynon, 1990 BLS.

L. suavis (Müll. Arg.) Migula
21/48: Port Eynon, 1990 BLS.

L. turicensis (Hepp) Müll. Arg.
21/48: The Knave, 1990 BLS.

Lecanora actophila Wedd.
Rocks on the sea shore, apparently uncommon. (21/77): near Sker Point, on hard acidic pebble, Wade (NMW, under *Rhizocarpon obscuratum*)! 21/58: Oxwich Bay, on limestone, 1986 *Orange*. 21/48: Port Eynon, on limestone, 1986 *Orange*.

L. albescens (Hoffm.) Branth. & Rostrup
On limestone, frequent. 21/48, 49, 58, (87), 96, 97, 31/07, (16), 18, 26, 32/00.

L. argentata (Ach.) Malme (L. *subfuscata* Magnusson)
On bark, few records. 21/49, 96, 32/17.

L. atra (Huds.) Ach.
On acidic rocks, occasionally on bark, common. 21/48, 49, 58, 59, 77, 87, 88, 96, 97, (98), 31/06, 07, 08, 09, 16, 17, 18, 26.

L. badia (Pers.) Ach.
On acidic rocks, upland, frequent. 21/99, 22/(90), 31/08, (09), 18, (19), 32/(00, 10).

L. campestris (Schaerer) Hue
On calcareous or acidic rocks,and on concrete; frequent. 21/48, 49, 87, 88, 96, 97, 98, 22/70, 90, 31/06, 07, 08, 16, 17, (18), (26), 28.

L. carpinea (L.) Vainio
On bark, several records from lowland sites. 21/(87), 31/06, (07), 16, (28).

L. chlarotera Nyl. (including *L. rugosella* Zahlbr.)
On relatively basic bark, common. 21/48, 49, 58, 59, 87, 96, 97, 31/06, 07, 16, (17, 18), 26, (28).

L. confusa Almb.
On bark, often on twigs, including *Calluna* and *Rosa pimpinellifolia* stems, rarely on wood; occasional. 21/49, 58, 78, 87, 31/17, 26.

L. conizaeoides Nyl. ex Crombie
On acidic bark, wood and rocks; very common. 21/58, 59, 77, 79, 87, 88, 89, 96, 97, 98, 99, 22/60, 70, 31/06, 07, 08, 09, (16), 17, 18, 28, 32/00, 01, 10.

L. crenulata Hook.
On limestone rocks, often on dry vertical surfaces; common. 21/48, 49, 58, 77, 87, 96, 97, 31/06, 07, 16, 26.

L. dispersa (Pers.) Sommerf.
On limestone, concrete and wood; common. 21/48, 49, 58, 69, 77, 87, 88, 96, 97, 98, 99, 22/(70), 90, 31/06, 07, 08, 09, 16, 17, 18, 26, 27, 28, 32/00, 10.

L. epanora (Ach.) Ach.
Rare. (32/00): below Cnwc, Troedyrhiw, on acidic rocks on wall, 1959 *Wade* (NMW)!

L. expallens Ach.
On bark, common. 21/48, 49, 58, 87, 88, 96, 97, 98, 99, 31/06, 07, 16, 17, 18. Sterile crusts apparently belonging to this species are frequent on church walls.

L. fugiens Nyl.
On acidic rocks on the coast, rare. 21/48, 49, 58, 77.

L. gangaleoides Nyl.
On acidic rocks, mainly upland, frequent. 21/48, (58), 96, 31/07, (09, 18, 28).

L. helicopis (Wahlenb.) Ach.
On limestone rocks on the sea shore, frequent. 21/48, 58, 77, 87, 31/16, 26.

L. intricata (Ach.) Ach.
On acidic rocks, common. 21/48, 49, 97, 99, 22/60, 70, 90, 31/(18, 19), 32/00.

L. intumescens (Rebent.) Rabenh.
Listed by Watson (1953). Record requires confirmation.

L. jamesii Laundon
On bark, occasional. Usually sterile; found fertile in 21/48. 21/48, 49, 58, 87, 97, 22/60, 31/07, 17.

L. muralis (Schreber) Rabenh.
On limestone, concrete, asbestos-cement, sandstone, and other substrata; common. 21/69, 79, 87, 96, 97, 98, 99, 31/(06), 07, 16, 17, 18, 26, 27, 28, 32/00, 10.

L. pallida (Schreber) Rabenh.
(21/97): Castle-upon-Alun, *M.G. Winkles* (Herb. Sowter) (Wade 1950). Record requires confirmation.

L. polytropa (Hoffm.) Rabenh.
On acidic rocks, common. 21/48, 49, (69), 87, 88, 96, 97, 98, 99, 22/(90), 06, 07, 09, 16, 17, 18, 32/00, 10.

L. pulicaris (Pers.) Ach.
Probably overlooked. (31/07): between Bonvilston and Mynydd y Glew, on *Quercus*, 1933 *Wade* (NMW)! (21/87): Ewenny Down, on *Crataegus* (Ivimey-Cook 1959).

L. rupicola (L.) Zahlbr.
On acidic rocks, frequent. 21/49, 96, 97, 31/07, (08), 09, 17, (18).

L. saligna (Schrader) Zalbr.
21/69: Swansea, Prince of Wales Dock, wood on urban wasteland (Gilbert 1990).

L. sambuci (Pers.) Nyl.
Rare. Oxwich, on bark, 1932 *D.A. Jones* (NMW)!

L. soralifera (Suza) Räsänen
On acidic rocks, especially upland, frequent. 21/48, 87, 88, 96, 97, 98, 99, 22/70, 90, 31/08, 09, 17, 18, 19, 28, 32/00, 10.

L. stenotropa Nyl.
On rock and man-made materials, overlooked but probably frequent in urban areas. 21/69, 31/17, 27.

L. symmicta (Ach.) Ach.
On acidic bark and wood, probably frequent. 21/58, 59, 88, 96, 97, 31/07, 16, 17.

L. varia (Hoff.) Ach.
Listed by Wade & Watson (1936). Near Pentyrch, on ash and oak, 1952 *Wade*. Records require confirmation.

Lecidea erratica Körber
On acidic rocks, often on low rocks and stones, one record on wood; frequent. 21/69, 22/90, 31/18, (19), 27, 28.

L. furvella Nyl. ex Mudd
Watson (1953).

L. fuscoatra (L.) Ach.
On acidic rocks, especially upland; frequent. 21/48, 87, 88, 96, 99, 22/90, 31/07, (09), 17, 18, (19), 32/00, (10).

L. hypnorum Lib. *sensu lato*
22/80: Aber-pergwm Wood, on moss on *Quercus* trunk, 1986 *Orange*, det. B.J. Coppins 1987 (NMW). 22/60: Cwm Clydach, moss on *Quercus* trunk, 1988 *Orange* (NMW).

L. immersa (Hoffm.) Ach.
On limestone rocks, occasional. 21/48, (87, 97).

L. lactea Flörke ex Schaerer
21/49: Rhossili Down, 1990 BLS.

L. leucophaea (Flörke ex Rabenh.) Nyl.
On acidic rocks in the uplands, occasional. 21/99, 22/(90), 31/(09, 19), 32/00.

L. lithophila (Ach.) Ach.
On acidic rocks; few records but probably frequent in the uplands. 21/89, 99, 31/(09), 18, 32/(00).

L. metzleri (Körber) Th. Fr.
21/87: Merthyr Mawr Warren, on limestone stone, 1979 *A.R. Perry* (NMW)! 21/58: Southgate, limestone cliffs, 1990 BLS.

L. monticola (Ach.) Schaerer
On limestone, concrete and mortar, frequent. 21/49, 87, 88, 22/70, 31/06, 07, 16, (17), 18, 26, 28, 32/01.

L. orosthea (Ach.) Ach.
On acidic rocks, often on dry vertical faces, widespread. 21/49, 87, 89, 96, 97, 99, 31/(09), 32/00.

L. phaeops Nyl.
22/60: Cwm Clydach, shaded rocks near river, 1988 *Orange*.

L. speirea (Ach.) Ach.
On acidic rocks, few records, mainly upland. 21/(87), 88, 99, 31/08, (18), 32/(00).

L. sulphurea (Hoffm.) Wahlenb.
On acidic rocks, few records. 21/96, 97, 31/(09), 17.

L. valentior Nyl.
22/60: Cwm Clydach, rocks by river, 1988 *Orange* (NMW).

Lecidella elaeochroma (Ach.) M. Choisy forma **elaeochroma**
On bark, common. 21/48, 49, 58, 59, (78), 87, 88, 96, 97, 31/06, 07, 16, 17, (18, 28), 32/00.
    forma **soralifera** (Erichsen) D. Hawksw.
On bark and wood, rare. 21/(48), 31/18.

L. scabra (Taylor) Hertel
On acidic rocks and slag, occasional. 21/48, 49, (58), 87, 96, 97, 98, 22/70, 31/07, 17, 32/00.

L. stigmatea (Ach.) Hertel & Leuckert
On calcareous rocks, frequent. 21/79, 97, 31/(06), 07, (16, 18), 28, 32/00

L. subincongrua (Nyl.) Hertel & Leuckert
21/49: Llanrhidian church, 1990 BLS.

Lemmopsis arnoldianum (Hepp) Zahlbr.
21/58: Pwlldu Bay, limestone scree on coast, 1990 *Orange*, det. O.L. Gilbert (NMW).

Lempholemma botryosum (Massal.) Zahlbr.
21/49: Bovehill, 1990 BLS.

L. chalazanum (Ach.) B. de Lesd.
31/27: Cardiff East Moors, mossy ground over old road on wasteland, 1990 *Orange* (NMW).

Lepraria incana (L.) Ach.
On dry sheltered bark, wood, acidic rocks and decaying vegetation; very common. Pollution-tolerant, frequent in urban areas. Contains divaricatic acid and zeorin; material from Rhossili Down had patches also containing parietin. Confirmed records from 21/48, 49, 96, 99, 22/70, 31/07, 18.

L. lesdainii (Hue) R.C. Harris
Shaded dry limestone rocks and walls, locally frequent. 21/48, 49, 58, 59, 87, 97, 22/60, 31/06, 07, 17.

L. lobificans Nyl.
On sheltered bark, moss, decaying vegetation and rocks, common. Pollution-tolerant. Contains stictic acid, zeorin and atranorin. 21/48, 49, 58, 97, 99, 22/70, 31/17, 18, (19).

L. neglecta Vainio *sensu lato*
On acidic rocks in the uplands, often on moss-cushions; occasional. 21/99, 22/90, 31/(09, 18), 32/00.

Lepraria sp.
A greenish species containing thamnolic acid. 22/60: Cwm Clydach, Craig-cefn-parc, dry base of *Quercus*, 1988 *Orange* (NMW). 31/09: Cwm Clydach, Ynysybwl, dry base of *Alnus*, 1988 *Orange* (NMW).

Lepraria sp.
A yellowish species with usnic acid as the major secondary substance. 21/49: Rhossili Down, acidic rocks below overhang, with *L. incana*, 1990 *Orange* (NMW).

Lepraria sp.
Contains roccellic and other fatty acids and atranorin. 22/60: Cwm Clydach, Craig-cefn-parc, dry bases of *Quercus* trunks, 1988 *Orange* (NMW). 21/48: Penrice Castle, on *Fraxinus* and *Platanus*, 1990 *Orange* (NMW).

Leproloma diffusum Laundon var. **diffusum**
32/00: Morlais Hill, on dry limestone, 1986 *Orange*. 21/58: Pwlldu Bay, moss on side of limestone

boulder on coast, 1990 *Orange* (NMW). 31/17: Llandaff Cathedral, limestone wall, 1990 *Orange*.

L. vouauxii (Hue) Laundon
On mosses, bark, sandstone and limestone, probably frequent. 21/48, 87, 96, 31/07, 09, 17, (18).

Leproplaca chrysodeta (Vainio ex Räsäen) Laundon
On shaded dry limestone walls and rocks. 21/48, 49, 59, 97, 31/17.

L. xantholyta (Nyl.) Hue
On shaded dry limestone, rare. 21/48, 49, 87.

Leptogium biatorinum (Nyl.) Leighton
On soil. 21/48, 31/16.

L. britannicum P. Jørg. & P. James
Amongst moss or in turf on rocky banks, coastal. 21/48, 49, 58.

L. diffractum Krempelh. ex Körber
21/48: Mewslade Bay, 1990 BLS.

L. gelatinosum (With.) Laundon
On calcareous soil and limestone. 21/48, 49, 97, 31/09, (18), 32/00.

L. lichenoides (L.) Zahlbr.
On limestone and calcareous dunes, often amongst mosses. 21/48, 58, (78), 31/(06, 18), 32/(00).

L. massiliense Nyl.
Rare. 31/06: Porthkerry Bay, on limestone, 1963 *Wade* (NMW).

L. minutissimum (Flörke) Fr.
Rare. 32/00: Taf Fechan, Merthyr Tydfil, on *Sambucus*, 1986 *Orange* (NMW).

L. plicatile (Ach.) Leighton
On limestone rocks and walls and on concrete, occasional. 21/49, 58, 87, 98, 31/(06), 07, (17).

L. schraderi (Ach.) Nyl.
On soil and amongst mosses, over limestone and mortar. 21/(49), 58, 97, 31/(17, 18), 27.

L. teretiusculum (Wallr.) Arnold
On limestone rocks and mortar, and on basic bark; occasional. 21/48, 49, 58, 87, 31/06, 07, (17).

L. turgidum (Ach.) Crombie
Listed by Wade & Watson (1936). 21/69: Swansea, Prince of Wales Dock, soil on urban wasteland (Gilbert 1990). 21/97: Llanmihangel church, 1990 *B.J. Coppins*.

(Lichenoconium erodens M.S. Christ. & D. Hawksw.)
31/17: Bute Park, Cardiff, on *Lecanora conizaeoides*, 1986 *Orange* (NMW). Probably frequent.

(L. lecanorae (Jaap) D. Hawksw.)
31/18: Mynydd Rudry, on *Lecanora conizaeoides*, 1986 *Orange* (NMW). Probably frequent.

Lichina confinis (Müller) Agardh
Limestone rocks on the sea shore, frequent. 21/(48), 58, 87, 31/16, 26.

L. pygmaea (Lightf.) Agardh
Limestone rocks on the sea shore, apparently rare. 21/(38), 48, 58.

Macentina stigonemoides A. Orange
Frequent and widespread on *Sambucus* bark, one record from *Acer pseudoplatanus*. 21/48, 58, 88, 31/07, 17, 18.

(Merismatium lopadii (Anzi) Zopf)
31/06: West Aberthaw, on thallus of *Protoblastenia calva* on shingle, 1986 *Orange* (NMW).

Micarea alabastrites (Nyl.) Coppins
On bark and decaying wood, particularly in sheltered woodland; occasional. 21/(48), 89, 98, 22/60, 70, 80, 31/09, 32/(00).

M. bauschiana (Körber) V. Wirth & Vĕzda
Shaded acidic rocks, occasional. 22/48, 60, 70, 80, 31/(18).

M. botryoides (Nyl.) Coppins
On shaded acidic rocks and on bryophytes, also recorded from *Pinus* bark. 22/60, 70, 80, 31/18.

M. cinerea (Schaerer) Hedl.
Rare. 22/80: Afon Pyrddin, on moss on *Fraxinus* trunk, 1986 *Orange* (NMW).

M. denigrata (Fr.) Hedl.
21/69: Swansea, Prince of Wales Dock, urban wasteland (Gilbert 1990). 31/27: Cardiff East Moors, leather of old shoe on wasteland, 1990 *Orange*.

M. lignaria (Ach.) Hedl. var. **lignaria**
On wood, acidic soil, over bryophytes and on decaying vegetation; frequent. 21/89, 98, 99, 22/80, 90, 31/09, 32/00.

M. melaena (Nyl.) Hedl.
Acidic wood and bark in hilly areas, occasional. 21/89, 98, 22/60, 80, (90), 31/09.

M. misella (Nyl.) Hedl.
22/60: Cwm Clydach, decorticate *Betula* trunk in humid woodland, 1988 *Orange* (NMW).

M. nitschkeana (Lahm ex Rabenh.) Harm.
21/97: Afon Alun Valley, on *Corylus* twigs, 1986 *Orange* (NMW). 31/07: near Bonvilston, on shaded wooden gate, 1988 *Orange* (NMW). 21/89: Cwm Du, on fence post, 1989 *Orange*.

M. peliocarpa (Anzi) Coppins & R. Sant.
22/60: Cwm Clydach, on log in woodland, 1988 *Orange* (NMW).

M. prasina Fr.
On acidic bark and wood, few records but probably frequent. 21/48, 58, 59, 69, 99, 22/60, 31/07, 08, 09, 17.

M. sylvicola (Flotow) Vezda & V. Wirth
On acidic rocks and stones. 22/60, (90), 31/28, 32/00.

Microglaena muscorum (Fr.) Th. Fr.
Rare or overlooked. 21/78: Kenfig Burrows, on moss on dunes, 1986 *Orange*. 21/58: Southgate, 1990 BLS. 21/48: Penrice Castle, on moss on *Fraxinus* trunk, 1990 BLS!

(Milospium graphideorum (Nyl.) D. Hawksw.)
Rare. 21/48: Penrice Castle, on *Lecanactis amylacea*, 1986 *Orange* (NMW).

(Minutophoma chrysophthalmae D. Hawksw.)
21/58: Penrice Castle, on thallus of *Chrysothrix chrysophthalma*, 1986 *Orange* (NMW, with the host).

Mniacea jungermanniae (Nees ex Fr.) Boud.
21/98: Brynnau Gwynion, Pencoed, over dying hepatic on bank of ditch, 1974 *A.R. Perry* (NMW)!

Moelleropsis nebulosa (Hoffm.) Gyelnik
Rare on calcareous soil over limestone. 21/48, 58, 32/00.

Mosigia illita (Nyl.) R. Sant.
(22/80): River Perddyn, acidic rocks in river, 1932 *Wade* (NMW)!

(Muellerella hospitans Stizenb.)
Rare. (21/48): Penrice, on *Bacidia rubella*, 1932 *D.A. Jones* (NMW, with the host)!

(M. lichenicola (Sommerf.) D. Hawksw.)
(32/00); Morlais Hill, on *Catillaria chalybeia* on limestone, 1932 *Wade* (NMW)! (31/28): Cefn Mably Park, on *Lecanora campestris*, 1932 *Wade* (NMW)! Near Port Talbot, on *Caloplaca flavescens* and *Lecanora campestris* (Pyatt & Harvey 1973).

Mycoblastus sanguinarius (L.) Norman
21/99: Craig y Geifr, acidic rocks on wall, 1987 *Orange*.

M. sterilis Coppins & P. James
On acidic bark, particularly in humid sites, frequent. 21/48, 59, 89, 99, 22/60, 70, 80, 97, 31/07, 09, 17, 28, 32/00.

Mycosphaerella ascophylli Cotton
On the seaweed *Pelvetia canaliculata*, forming a 'mycophycobiosis'; probably common. 21/58: Oxwich, 1986 *Orange* (NMW). 21/16: Swanbridge, 1987 *Orange* (NMW).

Myxomphalia maura (Fr.) Hora
21/69: Swansea, Prince of Wales Dock, soil on urban wasteland (Gilbert 1990).

(Nectriella robergei (Mont. & Desm.) Weese)
21/78: Kenfig Burrows, on *Peltigera* sp., 1932 *Wade* (NMW)! Same locality and host, 1986 *Orange*. Present as the anamorph *Illosporium carneum* Fr.

(Neolamya peltigerae (Mont.) Theiss & H. Sydow)
Glamorganshire, on old *Peltigera* thalli (Hawksworth 1983).

(Niesslia cladoniicola D. Hawksw. & W. Gams)
21/87: Merthyr Mawr, on *Cladonia rangiformis* in dunes, 1973 *M.C. Clark* (Hawksworth 1975). 21/78: Kenfig Burrows, on *Cladonia foliacea* in dunes, 1986 *Orange* (NMW).

Normandina pulchella (Borrer) Nyl.
On bark, often over mosses; occasional records in Gower and the Vale of Glamorgan. 21/48, 49, 58, 59, 87, 97, 31/(06), 07, (16).

Ochrolechia androgyna (Hoffm.) Arnold
On bark and acidic rocks, frequent. 21/48, 49, 88, 22/60, 31/(09), 32/(00), 01.

O. parella (L.) Massal.
On acidic rocks, rarely on bark, frequent. 21/48, 49, 87, 88, 96, 97, 98, 31/06, 07, 08, 16, (17, 28).

O. subviridis (Hoeg) Erichsen
21/97: Llanmihangel, 1990 *B.J. Coppins*.

O. tartarea (L.) Massal.
Rare. (31/09): Craig Evan Leyshon Common, on acidic boulders, 1957 *Wade* (NMW)! 21/49: Rhossili Down, on acidic boulders, 1974 *Q.O.N. Kay* (UCSA)!

O. turneri (Sm.) Hasselrot
On bark. 21/48, 31/06, 07, 17.

Omphalina ericetorum (Fr. ex Fr.) M. Lange
22/90: Llyn Fach, acid soil bank, 1988 *Orange*. The sterile thallus of some *Omphalina* spp., known as '*Botrydina vulgaris*', is frequent on acid soil and on rotten wood in humid habitats, and in most cases is likely to belong to O. *ericetorum*.

O. hudsoniana (Jenn.) Bigelow (*Coriscium viride* (Ach.) Vainio)
22/90: Craig y Llyn, 1973 *Wade*. 22/60: Cwm Clydach, on stump, 1988 *Orange*.

O. pararustica Clémençon
22/80: near Afon Pyrddin, on soil of ant hill, 1986 *Orange* (NMW).

Opegrapha atra Pers.
On bark. 21/(48, 87), 31/06, (07, 16, 17).

O. calcarea Turner ex Sm.
On limestone. 21/48, 96, 97, 31/07.

O. gyrocarpa Flotow
On sheltered acidic rocks, frequent. 21/49, 96, 97, 99, 31/07, 09, 18.

O. herbarum Mont.
On bark. 21/(48), 97, 31/(07, 28).

O. lichenoides Pers.
Wade & Watson (1936). 21/58: Oxwich, on *Fraxinus*, 1965 *P.W. James* (BM).

O. mougeotii Massal.
Rare or overlooked. 21/96: St Donats Church, limestone on east-facing wall, 1983 *T. Morakinyo* (NMW)! 21/97: Llanmihangel church, 1990 *B.J. Coppins*. 31/07: Beaupre Castle, 1990 *B.J. Coppins*.

O. niveoatra (Borrer) Laundon
On bark, rare. (31/16); near Lavernock Station, on *Fraxinus* sapling, 1932 *Wade* (NMW)! Oxwich, 1932 *D.A. Jones* (NMW)!

(O. parasitica (Massal.) H. Olivier)
On various crustose lichens on limestone. 21/48, 49, 58, 97, 31/07, 26.

O. pseudovaria Coppins *ad int.*
21/97: Llanmihangel, 1990 *B.J. Coppins*.

O. saxatilis DC.
On limestone, rarely on siliceous rock; frequent. 21/48, 49, 58, 96, (97), 31/07, (16, 17), 26.

O. saxicola Ach.
32/00: Morlais Hill, on shaded limestone, 1987 *Orange* (NMW). 21/48: Port Eynon, 1990 BLS.

O. sorediifera Coppins & P. James
21/48: Penrice Castle, on trunk of *Fagus sylvatica*, 1990 BLS!

O. varia Pers.
On bark; few records, all from Gower. 21/48, 58, 59.

O. vermicellifera (Kunze) Laundon
21/97: Llanmihangel, 1990 *B.J. Coppins*.

O. vulgata auct.
On bark, common. 21/48, 58, 87, 31/06, 07, (16), 17.

Opegrapha sp.
Parasitic on *Aspicilia calcarea*, frequent on Carboniferous Limestone in Gower. 21/48, 58, 49.

Pannaria conoplea (Ach.) Bory
Rare, probably extinct. (21/48): Penrice, mossy bark, 1932 *D.A. Jones* (NMW)!

P. pezizoides (Weber) Trevisan
Rare. Rhossili, on rocks, 1932 *D.A. Jones* (NMW)! 21/58: Penmaen, limestone cliffs, 1990 BLS!

P. rubiginosa (Ach.) Bory
Listed by Watson (1953). Record needs confirmation.

Parmelia britannica D. Hawksw. & P. James
Cefn Bryn, 1945 *J.G. Rutter* (NMW). 21/48: Rhossili Down, 1990 BLS.

P. caperata (L.) Ach.
On bark and acidic rocks, frequent, rarely on limestone slopes by sea. 21/48, 49, 58, 59, 87, 88, 96, 97, 98, 22/70, 31/06, 07, 08, (17, 18, 28).

P. conspersa (Ehrh. ex Ach.) Ach.
On acidic rocks, mainly upland; frequent. 21/58, 98, 99, 22/(80, 90), 31/(08, 09), 18, (19), 32/(00).

P. discordans Nyl.
Acidic rocks, rare or overlooked. (31/18): Mynydd Rudry, 1932 *Wade* (NMW)! (31/18): Caerphilly Common, 1934 *Wade* (NMW)! 21/49: Rhossili Down, 1986 *Orange* (NMW).

P. exasperata de Not.
On bark, often on twigs; frequent. 21/48, 49, 58, 59, 87, 96, 97.

P. exasperatula Nyl.
On nutrient-enriched acidic bark. 21/97, 31/07, 17.

P. glabratula (Lamy) Nyl. subsp. **glabratula**
On bark, rarely on acidic rocks, common. 21/48, 49, 58, 59, 87, 88, 89, 96, 98, 99, 22/60, 70, (80), 31/07, 08, 09, 17, (18), 28, 32/00, 01.
  subsp. **fuliginosa** (Fr. ex Duby) Laundon
On acidic rocks, common. 21/48, 49, 58, 59, 79, 87, 88, 89, 96, 97, 98, 99, 22/60, 70, 90, 31/06, 07, 08, 09, 16, 17, 18, 32/00, 01, 10.

P. incurva (Pers.) Fr.
(31/09): near Garth Hall, west of Llanfabon, on wall, 1959 *Wade* (NMW)!

P. laciniatula (Flagey ex H. Olivier) Zahlbr.
21/97: Llanmihangel, 1990 *B.J. Coppins*. 31/07: Beaupre Castle, 1990 *B.J. Coppins*.

P. mougeotii Schaerer ex D. Dietr.
(21/88): near Cwrt Colman, Bridgend, acidic rocks, *Wade* (NMW).

P. omphalodes (L.) Ach.
On acidic rocks, occasional. 21/49, 58, 31/(18), 32/(00).

P. pastillifera (Harm.) R. Schubert & Klem.
Rare. 21/49: Cwm Ivy churchyard, small thallus on gently sloping sandstone tomb, 1990 *Orange*. 21/97: Wick Churchyard, on sandstone memorial stone, 1976 *Wade* (NMW)! 21/48: Penrice Castle, branch of *Juglans regia*, 1990 BLS.

P. perlata (Huds.) Ach.
On bark, rarely on acidic rocks, frequent. 21/48, 49, 58, 59, 96, 97, 22/70.

P. revoluta Flörke
On bark, rarely on acidic rocks; possibly frequent. 21/48, 49, 58, (87), 97, 22/60, 31/07, (16, 18).

P. saxatilis (L.) Ach.
On bark, wood and acidic rocks; common. 21/48, 49, 58, 59, 79, 87, 88, 89, 97, 98, 99, 22/60, 70, 90, 31/07, 08, 09, 17, 18, 19, (28), 32/00, 01, 10.

P. subaurifera Nyl.
On bark, frequent. 21/48, 58, 59, (87), 96, 97, 98, 22/70, 31/06, 07, 16, 17, (18, 28).

P. subrudecta Nyl.
On bark, frequent. 21/48, 49, 58, 59, 87, 96, 97, 31/06, 07, 17, (18).

P. sulcata Taylor
On bark, wood and acidic rocks; common. 21/48, 49, 58, 59, 87, 88, 96, 97, 98, 99, 22/60, 70, 31/06, 07, 08, 09, 16, 17, 18, 26, (27), 28, 32/00.

P. verruculifera Nyl.
Rare. (31/09): Mynydd Goitre-coed, Abercynon, on sandstone boulder, 1959 Wade (NMW)!

Parmeliella triptophylla (Ach.) Mül. Arg.
21/48: Penrice Castle, shaded trunk of old *Platanus* and on *Quercus*, 1990 BLS!

Peltigera canina (L.) Willd.
Uncommon, on dunes and on calcareous soil near the coast. 21/49, 58, (78).

P. didactyla (With.) Laundon
Occasional on basic soil on spoil heaps and waste ground. 21/48, 58, 79, 98, 31/27.

P. horizontalis (Huds.) Baumg.
Rare. (21/58): Crawley, 1940 *J.A. Webb* (NMW)! 21/48: Penrice Castle, on old *Acer platanoides*, 1986 *Orange*. 21/58: Pwlldu Bay, coastal rocks, 1990 *Orange*.

P. lactucifolia (With.) Laundon
On soil in turf, by tracks, etc.; frequent. 21/48, 49, 58, 59, 97, 98, 99, 22/60, 70, 90, 31/(17), 26.

P. membranacea (Ach.) Nyl.
On mossy boulders and in turf. 21/48, 49, 58, (87).

P. neckeri Hepp ex Mül.Arg.
21/49: Whiteford Burrows, on acid dunes, 1990 *Orange*(NMW). 21/58: Pennard Burrows, 1990 BLS.

P. praetextata (Flörke ex Sommerf.) Zopf
On walls, boulders, tree bases and logs; frequent. 21/48, 58, 59, 97, 98, 31/07, 09, 18, (28), 32/00.

P. rufescens (Weis) Humb.
On stony ground and in dunes, frequent. 21/49, 58, 69, (78), 87, 97, 22/90, 31/06, (16, 17), 32/00.

Pertusaria albescens (Huds.) M. Choisy & Werner
var. **albescens**
On bark, rarely on rock; occasional on limestone slopes facing sea in South Gower. 21/48, 58, 96, 97, 31/(18).
   var. **corallina** (Zahlbr.) Laundon
On bark, frequent. 21/48, 49, 87, 96, 97, 31/07, (17, 18).

P. amara (Ach.) Nyl.
On bark, less often on acidic rock; frequent. 21/48, 49, 58, 59, 87, 88, 96, 97, 98, 31/06, 07, 16, 17, 18, (28).

P. corallina (L.) Arnold
On acidic rocks, mainly upland; frequent. 21/49, 58, 98, 99, 22/90, 31/ 09, (18).

P. dealbescens Erichsen
On acidic rocks, mainly upland; frequent. 21/49, 88, 98, 22/(90), 31/(08, 09, 18), 32/(00).

P. hymenea (Ach.) Schaerer
On bark, few records. 21/48, 58, 97, 31/07.

P. lactea (L.) Arnold
Between Abercynon and Llanfabon, on acidic rock on wall, 1957 *Wade* (NMW)! 32/00: Cwm Ddu, Troedyr-hiw, on acidic rocks, 1960 *Wade* (NMW)!

P. leioplaca DC.
On smooth bark, frequent. 21/48, 49, 58, 59, (87), 88, 96, 97, 31/06, 07, (16, 17).

P. multipuncta (Turner) Nyl.
On bark, rather few records. 21/48, 59, 87, 31/(07).

P. pertusa (Weigel) Tuck.
On bark, frequent. 21/48, 49, (87), 96, 97, 31/06, 07, 17, (28).

P. pseudocorallina (Liljeblad) Arnold
On siliceous rocks and tombstones, a few records near the coast. 21/49, 31/06.

P. pupillaris (Nyl.) Th. Fr.
22/60: Cwm Clydach, on *Quercus* and *Acer pseudo-platanus*, 1988, *Orange* (NMW).

Petractis clausa (Hoffm.) Krempelh.
21/58: Southgate, limestone rocks on coast, 1990 *Orange*. 21/48: The Knave, 1990 BLS.

Phaeographis dendritica (Ach.) Müll. Arg.
On bark, rare. (21/48): Penrice, 1932 *D.A. Jones* (NMW). 21/48: near Penrice Castle, on *Platanus*, 1965 *Wade* (NMW). Both det. B.J. Coppins 1978–79.

Phaeophyscia orbicularis (Necker) Moberg.
On nutrient-enriched substrata including limestone, concrete and bark; common. 21/48, 49, 58, 69, 77, 87, 88, 96, 97, 98, 99, 22/(70), 90, 31/06, 07, 08, 09, 16, 17, 18, 26, 27, 32/00.

Phlyctis argena (Sprengel) Flotow
On bark, frequent. 21/48, 49, 58, 87, 88, 97, 22/60, 31/06, 07, 08, 16, 17, (18), 32/00.

Physcia adscendens (Fr.) H. Olivier
On nutrient-enriched substrata including limestone, concrete, sandstone and bark; common. 21/48, 49, 58, 87, 88, 96, 97, 98, 22/70, 31/06, 07, 08, (16), 17, 18, 26, 28, 32/00.

P. aipolia (Ehrh. ex Humb.) Fürnrohr
On nutrient-rich bark, frequent. 21/48, 49, 58, 59, 87, 88, 96, 97, 31/06, 07, (17, 28).

P. caesia (Hoffm.) Fürnrohr
On nutrient-enriched substrata, including limestone and concrete, rarely wood and sandstone; frequent. 21/48, 49, 58, 79, 87, 88, 97, 98, 22/70, 90, 31/07, (09), 17, 18, 26, 32/00, 01.

P. semipinnata (Gmelin) Moberg
On bark, most records from coastal areas; occasional. 21/49, 58, 87, 31/(06, 17).

P. stellaris (L.) Nyl.
On bark, rare. 21/48, (49), 31/07, (16).

P. tenella (Scop.) DC.
On nutrient-enriched substrata, including limestone, concrete, sandstone and bark; common. 21/48, 49, 58, 87, 88, 96, 97, 31/06, 07, 16, 17, 18, 32/00, 01, 10.

P. tribacia (Ach.) Nyl.
21/49: Llanrhidian, on church wall, 1990 BLS!

Physconia distorta (With.) Laundon
On bark, few records. 21/49, 87, 31/(17).

P. grisea (Lam.) Poelt
On bark; occasional. 21/(49, 87), 96, 31/06, 07, 16, 17.

Placidiopsis custnanii (Massal.) Körber
Occasional and in small quantity on soil on Carboniferous Limestone cliffs along south coast of Gower. 21/48, 58.

Placopsis gelida (L.) Lindsay
On acidic rocks, upland, rare. (22/90): Craig y Llyn, 1932 *Wade* (NMW)!

(32/10): lane from Bargoed to Cefn Brithdir, 1958 *Wade* (NMW)! (32/00); Cnwc, Troedyrhiw, 1959 *Wade* (NMW)!

Placynthiella icmalea (Ach.) Coppins & P. James
On acidic substrata including peaty soil, wood, bark, moss, and plant debris; frequent. 21/69, 89, 98, 99, 22/60, (90), 31/09, 18, 26, 32/00.

P. uliginosa (Schrader) Coppins & P. James
On peaty soil. 22/69, (90), 31/(18).

Placynthium garovaglii (Massal.) Malme
21/48: sheltered limestone, 1990 BLS.

P. nigrum (Huds.) Gray
On limestone and concrete, common. 21/48, 49, 58, 59, 87, 96, 97, (98), 31/ 06, 07, 08, 16, (17, 18), 26, 32/01.

Platismatia glauca (L.) Culb. & C. Culb.
On acidic bark, rocks, and posts; common. 21/48, 49, 58, 59, 97, 98, 99, 22/60, 70, 90, 31/07, 08, 09, 17, 18, 28.

Polyblastia albida Arnold
(32/00): Morlais Hill, on limestone, 1932 *Wade* (NMW)!

P. dermatodes Massal.
(31/17): near Dinas Powis, on calcareous stones, 1934 *Wade* (NMW)!

P. gelatinosa (Ach.) Th. Fr.
On calcareous soil on cliffs and in quarries. 21/48, 58, 31/06, 16, 17.

(Polysporina dubia (H. Magn.) Vězda)
21/49: Llanrhidian church, on ?*Lecidea fuscoatra*, 1990 BLS.

P. simplex (Davis) Vězda
On rock. 21/49, 96, 97, 31/07.

Porina aenea (Wallr.) Zahlbr.
On smooth, relatively basic bark, frequent. 21/49, 59, 87, 97, 31/07, 17, 18, 28.

P. borreri (Trevisan) D. Hawksw. & P. James
On bark, few records. 21/58, 97.

P. chlorotica (Ach.) Müll. Arg. forma **chlorotica**
Shaded acidic rocks, under-recorded. 21/(69), 22/60, 31/(18).

P. lectissima (Fr.) Zahlbr.
Damp shaded acidic rocks, upland. 22/60, 80, 32/00.

P. leptalea (Durieu & Mont.) A.L. Sm.
Shaded bark, probably frequent. 21/(48), 31/07, 17, 28.

P. linearis (Leighton) Zahlbr.
On limestone rocks, rare. (21/97): Old Castle Down, 1955 *Wade* (NMW)! (21/87): Ewenny Downs, (Ivimey-Cook 1959). 21/49: Cwm Ivy Tor, 1990 BLS.

Protoblastenia calva (Dickson) Zahlbr.
On limestone rocks, few records. 21/48, 49, 58, (87, 97), 31/16.

P. incrustans (DC.) Steiner
On Carboniferous Limestone, occasional in Gower. 21/48, 58.

P. rupestris (Scop.) Steiner
On limestone and concrete, common. 21/48, 49, 58, 77, 87, 96, 97, 98, 99, 22/60, (70), 90, 31/06, 07, 08, 16, 17, (18), 26, 28, 32/00.

Psilolechia leprosa Coppins & Purvis
31/17: Heath Cemetery, below overhang on tombstone, receiving run-off from copper plaque, 1990 *Orange*.

P. lucida (Ach.) M. Choisy
On acidic rocks, rarely on mortar, in dry crevices, common. 21/48, 49, 59, (87, 96), 97, 98, 99, 22/60, (80, 90), 31/06, 07, 08, 09, (16), 17, 18, 32/00, 10.

Psora lurida (Ach.) DC.
Uncommon on soil amongst limestone rocks. 21/48, 58, 87.

Pyrenula chlorospila (Nyl.) Arnold
On smooth bark, few records, mostly from coastal sites. 21/48, 58, 87, 96, 97.

P. macrospora (Degel.) Coppins & P. James
On smooth bark, occasional in Gower and the south of the Vale. 21/48, 49, 58, 87, 96, 31/(06, 07, 17).

Pyrrhospora quernea (Dickson) Körber
On bark, few records. 21/48, 58, 97, 31/06, 07, 16, 17.

Racodium rupestre Pers.
On acidic rocks below overhangs, upland. 21/99, 22/60, 70, 80, (90).

Ramalina calicaris (L.) Fr.
Rare. 21/97: Castle-upon-Alun, 1974 *Wade & A.R. Perry*. Same locality, on *Fraxinus* twigs, 1979 *A.R. Perry* (NMW)!

R. cuspidata (Ach.) Nyl.
On limestone rocks on the coast, rare. (31/16): Sully Island, 1926 *Wade* (NMW)! (21/48): Overton Mere, 1933 *R.C. McLean* (NMW)! Both collections containing stictic and norstictic acid by TLC.

R. farinacea (L.) Ach.
On bark, rarely on sandstone headstones, frequent in Gower and the Vale. Chemotypes with protocetraric acid and with hypoprotocetraric acid occur in a proportion of approximately 2: 1. 21/48, 49, 58, 59, 87, 88, 96, 97, 98, 31/06, 07, 16, 17.

R. fastigiata (Pers.) Ach.
On bark, occasional in Gower and the south part of the Vale. 21/48, 49, 58, 87, 96, 97, 31/06, (07, 16).

R. fraxinea (L.) Ach.
(21/87): Candleston Castle, on *Salix*, 1942 *F.A. Sowter* (Wade 1959). 21/49: record by *Q.O.N. Kay*. 21/96: St Donats, on *Acer pseudoplatanus* and *Quercus*, 1983 *T. Morakinyo* (NMW)!

R. polymorpha (Liljeblad) Ach.
(21/48): Penrice (Wade & Watson 1936). Record requires confirmation.

R. siliquosa (Huds.) A.L. Sm
On acidic rocks on or near the coast, uncommon, occasionally on tombstones in churchyards near coast. Chemotypes with protocetraric acid, with hypoprotocetraric acid and with salazinic acid occur. 21/48, 49, (58), 96.

R. subfarinacea (Nyl ex Crombie) Nyl.
On acidic rocks, rare. Chemotypes with norstictic acid and with protocetraric acid occur. 21/49, 58, (87), 31/18.

Rhizocarpon concentricum (Davies) Beltr.
On limestone and siliceous rocks, occasional. 21/(48), 87, 22/(70), 31/(06, 08, 16, 17, 18, 28).

R. constrictum Malme
21/49: Rhossili Down, 1990 BLS.

R. geographicum (L.) DC.
On acidic rocks, mainly upland, frequent. 21/87, 96, 97, 98, 99, 22/90, 31/(09, 18), 32/(00), 01.

R. hochstetteri (Körber) Vainio
On acidic rocks, upland, few records. 21/(98, 99), 22/60, (90), 31/(09).

R. lavatum (Fr.) Hazslin
22/70: above Tre-forgan, acidic rocks by stream, 1975 *Q.O.N. Kay* (UCSA)!

R. lecanorinum Anders
Rare. (31/09): Mynydd Goitre-coed, Abercynon, on acidic rock, 1957 *Wade* (NMW)!

R. obscuratum (Ach.) Massal.
On acidic rocks and pebbles, common. 21/49, 59, 69, 77, 87, 88, 96, 97, (98), 99, 22/70, (90), 31/07, (09, 17, 18, 19), 27, 28.

R. oederi (Web.) Körber
On acidic rocks, upland, frequent. 21/99, 22/(90), 31/08, 09, 32/00.

R. riparium Räsänen
On acidic rocks, upland, few records. 21/58, 22/(90), 31/(18).

R. umbilicatum (Ramond) Flagey
(32/00): Morlais Hill, on rock, 1932 *Wade* (NMW).

Rinodina atrocinerea (Dickson) Körber
Rare. Cefn Bryn, Gower, on acidic rocks, 1932 *D.A. Jones* (NMW)! (31/16): Swanbridge (Wade & Watson 1936). 21/49: Rhossili Down, on acidic boulders, 1974 *Q.O.N. Kay* (UCSA)! 21/49: Rhossili Down, 1990 BLS.

R. bischoffii (Hepp) Massal.
Rare. Port Eynon (Watson 1933). (31/06): Porthkerry Park, on Liassic limestone on shingle, 1932 *Wade* (NMW)!

R. exigua (Ach.) Gray
(31/07): between Peterston-super-Ely and Pendoylan, on *Ulmus*, 1933 *Wade* (NMW)!

R. gennarii Bagl.
On limestone, concrete, occasionally acidic rocks; common. 21/48, 77, 87, 96, 31/06, 07, (16), 17, 26.

R. immersa (Körber) Arnold
21/48: Mewslade Bay, on limestone, 1990 *Orange* (NMW).

R. roboris (Dufour ex Nyl.) Arnold
(31/08): record from MS.

R. sophodes (Ach.) Massal.
(21/96): St Donats (Smith 1918). (21/48): record from MS. 21/48: Penrice Castle, on twigs, 1990 BLS.

R. teichophila (Nyl.) Arnold
Rare. (31/17): St Fagans Castle, stones at margin of fish pond, 1947 *Wade*, det. J.W. Sheard 1964 (NMW). Record for 21/96 (Sheard 1967). 21/96: Llantwit Major, churchyard, 1990 *B.J. Coppins*.

Sarcogyne privigna (Ach.) Massal.
Rare. (31/18): Craig Llanishen (Wade & Watson 1936). 32/00: Cwm Ddu, Troedyrhiw, on acid rock on wall, 1960 *Wade* (NMW)!

S. regularis Körber
On limestone rocks and pebbles, and on concrete; common. 21/48, 49, 69, 87, 97, 98, 99, 22/(70), 31/06, (07), 16, 17, 18, 26, 27, 32/00.

Sarcosagium campestre (Fr.) Poetsch. & Schied.
21/69: Swansea, Landore, soil on urban wasteland (Gilbert 1990). 31/28: Draethen, calcareous soil in quarry, 1990 *Orange*.

Schaereria cinereorufa (Schaerer) Th. Fr.
On acidic rocks, upland; occasional. 21/99, 22/90, 31/09, 32/00.

Schismatomma decolorans (Turner & Borrer ex Sm.) Clauz. & Vězda
On bark, one record from church wall, rare. 21/48, 97, 31/07.

(**Sclerococcum sphaerale** (Ach. ex Ficinus & Schub.) Fr.)
On *Pertusaria corallina*. 21/90: Mynydd Blaenrhondda, 1986 *Orange* (NMW). 21/49: Rhossili Down, 1986 *Orange*. Probably frequent.

Scoliciosporum chlorococcum (Graewe ex Stenhammar) Vězda
Recorded from twigs of various trees, and on wood of gate. 21/48, 78, 31/07, 18.

S. umbrinum (Ach.) Arnold
On acidic rocks and slag, frequent, also on bark and wood. 21/48, 59, 69, 87, 89, 97, 98, 22/60, 70, 31/(07, 08), 09, 17, 18, 27, 32/00.

Solenopsora candicans (Dickson) Steiner
On limestone, frequent. 21/948), 49, 58, 87, 96, 31/06, (16), 17, (18).

Solorina spongiosa (Ach.) Anzi
Rare. (32/00): Morlais Hill, calcareous soil, 1932 *Wade* (NMW)!

Sphaerophorus globosus (Huds.) Vainio
On acidic boulders, a few scattered records. 21/49, (87, 99), 22/(90), 32/(18).

(**Sphinctrina turbinata** (Pers.) de Not.)
Rare. 21/58: Oxwich Burrows, on *Pertusaria* sp. on *Acer pseudoplatanus*, 1965 *P. James & Wade* (NMW)!

Squamarina cartilaginea (With.) P. James
Crevices of limestone rocks. 21/48, 49, 58, (87), 97, 31/(17, 18, 26), 32/00. Confirmed records of var. *cartilaginea* from 21/48, 49, 58, 31/(26), 32/00; confirmed records of var. **pseudocrassa** (Mattick) D. Hawksw. from 31/(17, 18).

Staurothele caesia (Arnold) Arnold
21/48: Mewslade Bay, 1990 BLS.

S. fissa (Taylor) Zwackh.
Rare. (22/80): Afon Pyrddin, acidic rocks in stream, 1932 *Wade* (NMW)!

S. guestphalica (Lahm ex Körb.) Arnold
31/06: West Aberthaw, Liassic limestone pebble on storm beach, 1988 *Orange* (NMW). 21/49: Bovehill, on Carboniferous Limestone outcrop, 1990 *Orange* (NMW).

S. rufa (Massal.) Zsch.
31/06: Porthkerry, on Liassic limestone in old quarry, 1964 *Wade* (NMW). The only British record.

S. rupifraga (Massal.) Arnold
On limestone. 21/48, 49, (87), 31/(18), 21/(00).

Steinia geophana (Nyl.) B. Stein
21/69: Swansea, Hafod, urban wasteland (Gilbert 1990). 31/27: Cardiff East Moors, dying moss on wasteland, 1990 *Orange* (NMW). 31/28: Draethen, decaying leaf on calcareous soil in quarry, 1990 *Orange.*

(Stenocybe pullulata (Ach.) B. Stein)
On *Alnus* twigs, widespread. 21/48, 22/60, 31/09, 18.

Stereocaulon dactylophyllum Flörke
On acidic rocks, often on walls, upland; occasional. 31/(09), 18, 32/(00, 10).

S. pileatum Ach.
On acidic rocks and slag, upland, few records. 21/99, 31/(18), 32/(00).

S. vesuvianum Pers. var. **vesuvianum**
On acidic rocks, upland, occasional. 21/99, 22/70, 31/(18, 19), 32/00.
    var. **symphycheilioides** Lamb
21/69: Swansea, Landore, urban wasteland (Gilbert 1990). 21/69: Swansea, Hafod, urban wasteland (Gilbert 1990).

Sticta limbata (Sm.) Ach.
21/48: Penrice castle, on old *Platanus,* 1990 BLS!

S. sylvatica (Huds.) Ach.
Rare. 21/48: Penrice Castle, on old *Platanus,* 1986 *Orange.*

(Stigmidium marinum (Deakin) Swinscow)
(31/16): Sully Island, on marine *Verrucaria* sp., 1932 *D.A. Jones* (NMW).

Strangospora moriformis (Ach.) Stein
Swansea (Leighton 1879).

Thelidium decipiens (Nyl.) Krempelh.
On limestone rocks and shingle. 21/49, 31/(06), 32/00.

T. incavatum Mudd
On limestone. 21/48, 87, 31/(17).

T. microcarpum (Davies ex Leighton) A.L. Sm.
On usually shaded calcareous stones, on soil, and on damp siliceous stones by streams, frequent. 21/48, 69, 31/07, 18, 29.

T. minutulum Körber
On shaded calcareous pebbles, occasional. 31/06, 07, 09, 18.

T. papulare (Fr.) Arnold
21/49: Bovehill, on Carboniferous Limestone, 1990 BLS!

T. pluvium A. Orange
Cwm Dimbath, stones in stream, 1958 Wade (NMW)! 31/09: Cwm Clydach, Ynysybwl, moist shaded rock outcrop, 1988 *Orange.*

(Thelocarpon impressellum Nyl.)
21/98: Llanharry, calcareous soil on spoil heap, 1990 *Orange* (NMW).

(T. lichenicola (Fuckel) Poelt & Hafellner)
On shaded stones with a thin covering of alga. 31/18: Garth Wood, Taffs Well, on dolomite pebbles, 1986 *Orange* (NMW). 21/98: Allt y Rhiw, Blackmill, on sandstone stone, 1987 *Orange* (NMW).

Thelotrema lepadinum (Ach.) Ach.
Rare. (21/58): record from MS. 21/93: St Donats, on beech, 1983 *T. Morakinyo* (NMW)! 22/80: Afon Pyrddin, on *Fraxinus* in wooded valley, 1986 *Orange.*

(Tomasiella gelatinosa (Chev.) Zahlbr.)
21/59: near Ilston, on *Corylus,* 1990 *Orange* (NMW).

Toninia aromatica (Turner ex Sm.) Massal.
On limestone rocks, mortar and concrete, frequent. 21/48, 49, 58, 87, 96, 97, 31/06, 07, 16, 17, 18, 26, (28).

T. coeruleonigricans (Lightf.) Th. Fr.
On limestone and on calcareous dunes, occasional. 21/48, 49, 58, 69, (78), 31/06, (16, 26), 32/00.

T. cervina Lönnr.
Local and in small quantities on Carboniferous Limestone in Gower. 21/49, 58.

T. lobulata (Sommerf.) Lynge
On calcareous rocks and soil, rare. 21/48, 31/(18, 26), 32/(00).

Trapelia coarctata (Sm.) M. Choisy
On acidic rocks and pebbles, common. 21/48, 69, 98, 99, 22/60, 70, (80), 31/09, 18, 28, 32/00, 10.

T. corticola Coppins & P. James
Occasional on acidic bark in humid woodland. 21/98, 22/60, 80, 31/09, 18, 32/01.

T. involuta (Taylor) Hertel
On acidic rocks, frequent. 21/48, 49, 69, 87, 89, 99, 22/60, 70, (90), 31/09, (18), 32/00.

T. obtegens (Th. Fr.) Hertel
Acidic stones and low rocks, few records. 21/69, 89, 98, 99, 22/60.

T. placodioides Coppins & P. James
On acidic rocks. 21/49, 99, 22/60, 32/00.

Trapeliopsis flexuosa (Fr.) Coppins & P. James
(31/18): near Castell Coch, on decorticated stump
(Wade 1959). 21/69: Swansea, Haford, urban wasteland (Gilbert 1990). 21/69: Swansea, Prince of Wales
Dock, urban wasteland (Gilbert 1990). Overlooked.

T. glaucolepidea (Nyl.) G. Schneider
32/00: Mynydd Cilfach-yr-encil, on peaty bank, 1986
Orange (NMW). 21/99: Tarren Rhiw-maen, on lip of
peaty bank, 1987 *Orange*.

T. granulosa (Hoffm.) Lumbsch.
On acidic soil and wood, frequent. 21/69, 98, 99,
22/60, 90, 31/17, (18), 32/00. 10.

T. pseudogranulosa Coppins & P. James
On acidic soil, wood and over bryophytes; frequent.
21/48, 49, 98, 99, 22/60, 70, 90, 31/07, 09, 18, 32/00.

Umbilicaria polyphylla (L.) Baumg.
On acidic rocks, upland. 21/99, 22/60, 90, 31/(09),
32/(00), 01.

Usnea articulata (L.) Hoffm.
Rare. 21/49: Whiteford Burrows, *Q.O.N. Kay.* 21/58:
Pennard Burrows, on the ground in dunes, 1979
Q.O.N. Kay. 21/78: Kenfig Burrows, on mossy hillock
in dunes, 1985 *A.R. Perry* (NMW)!

U. flammea Stirton
Rare. 21/49: Rhossili Down, on acidic boulders, 1974
Q.O.N. Kay, det. P. James (UCSA).

U. florida (L.) Wigg.
On bark, rare. 21/48, 58, 59, 31/07, 32/00.

U. fragilescens Havaas ex Lynge
Rare. 21/98: Cwm Dimbath, on *Larix*, 1974 *Q.O.N.
Kay*, det. P. James (UCSA). 21/49: Rhossili Down, on
the ground in heathland, 1974 *Q.O.N. Kay*, det. P.
James (UCSA).

U. subfloridana Stirton
On bark, frequent. All specimens examined contained thamnolic acid. 21/48, 58, 59, 87, 89, 97, 98, 99,
22/60, 70, 80, (90), 31/07, (09), 17, 18, (28), 32/00, 01.

Verrucaria aethiobola Wahlenb.
On acidic rocks in streams, rare. 32/00.

V. amphibia Clemente
On limestone rocks on sea shore. 21/48, 58.

V. aquatilis Mudd
31/06: Porthkerry Park, submerged limestone in
shaded stream, 1987 *Orange* (NMW).

V. aspiciliicola R. Sant.
On Carboniferous Limestone, parasitic on *Aspicilia
calcarea* when young. 21/48: Mewslade Bay, 1990
Orange (NMW). 21/58: Penmaen, 1990 BLS. 21/49:
Bovehill, 1990 BLS.

V. baldensis Massal.
On limestone rocks and walls, common. 21/48, 49, 58,
87, 88, 96, 97, 31/06, 07, (08), 16, 17, 18, 26, (28), 32/00,
(01).

V. bryoctona (Th. Fr.) A. Orange
21/69: Swansea, Prince of Wales Dock, soil on waste
ground, 1989 *O.L. Gilbert* (NMW)! 31/27: Cardiff East
Moors, mossy soil on waste ground, 1990 *Orange*
(NMW).

V. bulgarica Szat.
31/07: Dyffryn Gardens, calcareous pebbles on
shaded garden bed, 1988 *Orange*, det. P.M. McCarthy
(NMW). 31/07: St Mary Church, damp calcareous
stones by church wall, 1988 *Orange* (NMW). 31/17:
Leckwith, shaded calcareous stone, 1990, *Orange*
(NMW).

V. caerulea DC.
On limestone, rare. 31/(06, 07, 17, 18).

V. calciseda DC.
Limestone rocks and walls, rare. 21/97, 31/(07),
32/(00).

V. ditmarsica Erichsen
21/48: Mewslade Bay, limestone rocks on sea shore,
1990 *A. Fletcher.* 21/58: Southgate, limestone rocks on
sea shore, 1990 *A. Fletcher.*

V. dolosa Hepp
Recorded from siliceous and calcareous stones on
woodland path, on a river bank, and on wasteland;
probably common. 22/70, 31/06, 17, 27.

V. dufourii DC.
On Carboniferous Limestone, rare. 21/48, 58, 87, 97.

V. funckii (Sprengel) Zahlbr.
On rocks by streams. 21/48, 59, 31/06.

V. fusconigrescens Nyl.
21/49: Rhossili Down, 1990 BLS.

V. glaucina Ach.
On limestone rocks and walls, frequent. 21/48, 49, 58,
59, 87, 96, 97, 31/06, 07, 16, 17, 26.

V. halizoa Leighton
On rocks between tide-levels, few records. 21/48, 58,
31/16.

V. hochstetteri Fr.
On limestone and mortar, frequent. 21/48, 49, 58, 77,
22/60, 31/(07, 16, 18), 26.

V. hydrela Ach.
On acidic rocks in streams, few records. 21/(69),
22/80, 31/(18).

V. latericola Erichs.
21/58: Penmaen Burrows, on *Caloplaca flavescens*, 1990 *H. Fox* (NMW).

V. macrostoma Dufour ex DC.
21/96: Llantwit Major, in churchyard, 1990 *B.J. Coppins*. 31/07: Beaupre Castle, 1990 *B.J. Coppins*. 31/07: St Hilary church, 1990 *B.J. Coppins*. These are records of the sorediate form, which is probably widespread.

V. margacea (Wahlenb.) Wahlenb.
Siliceous rocks by streams. 22/80: Aber-pergwm Wood, 1986 *Orange* (NMW). 21/98: Cwm Dimbath, 1986 *Orange* (NMW). 22/60: Cwm Clydach, 1988 *Orange*.

V. maura Wahlenb.
On limestone between tide-levels, frequent. 21/48, 49, 58, 77, 87, 31/16, 17, 26.

V. mucosa Wahlenb.
On limestone or siliceous rocks between tide levels, low on the shore, frequent. 21/49, 58, 87, 31/(16), 26.

V. muralis Ach.
On limestone rocks, frequent. 21/49, 58, 69, 87, 97, 22/70, 31/06, (07), 08, 16, (17. 18, 26), 28.

V. murina Leighton
On shaded limestone pebbles and fragments of brick. 21/49, 58, 22/70, 31/06, 07.

V. nigrescens Pers.
On limestone rocks and walls, also concrete, common. 21/48, 49, 58, 87, 88, (96), 97, 98, 31/06, 07, 08, 16, (17), 18, 26, (27), 28, 32/00, 01.

V. praetermissa (Trevisan) Anzi
On acidic rocks in streams, rare. (22/80): Afon Pyrddin, 1932 *Wade* (NMW)! (32/00): Morlais Hill, 1932 *Wade* (NMW)! (32/00): Nant Cwm-du, Troedyrhiw, 1959 *Wade* (NMW)!

V. rheitrophila Zsch.
(31/18): Nant Cwm Nofydd, boulder in stream, 1933 *Wade* (NMW)!

V. simplex McCarthy
31/17: Grangetown, urban wasteland (Gilbert 1990).

V. striatula Wahlenb.
Limestone rocks on sea shore. 21/48, 58.

V. viridula (Schrader) Ach.
On limestone and concrete, frequent. 21/(48), 49, (87), 96, 31/(07, 17), 18. Records need confirmation.

Vezdaea aestivalis (Ohl.) Tsch.-Woess & Poelt
Over mosses on ± calcareous soil on dunes, spoil heaps and wasteland, and on a wall. 21/49, 98, 22/60, 31/27.

V. leprosa (P. James) Vezda
21/69: Swansea, Prince of Wales Dock, soil on urban wasteland (Gilbert 1990). 22/81: near Coelbren, below galvanised fencing, 1987 *R.G. Woods* (NMW).

V. retigera Poelt & Döbb.
On soil on urban wasteland and soil amongst limestone outcrops. 21/49, 69, 31/17.

V. rheocarpa Poelt & Döbb.
31/28: Draethen, amongst moss on calcareous soil in quarry, 1990 *Orange* (NMW).

(Vouauxiella lichenicola (Lindsay) Petrak & Sydow)
31/16: Swanbridge, on *Lecanora chlarotera*, 1971 *A.R. Perry* (NMW).

(Wedellomyces epicallopismum (Weddell) D. Hawksw.)
On thallus of *Caloplaca flavescens*, frequent. 21/48, 49, 97, 31/06.

Xanthoria calcicola Oxner
On limestone and concrete, frequent. 21/48, 58, 96, (97), 31/06, 07, 17, 18, 26, 32/00.

X. coralligera (With.) Laundon
21/48: Mewslade Bay, 1990 BLS.

X. elegans (Link.) Th. Fr.
Rare. 31/16: Cold Cnap Park, on concrete, 1967 *Wade* (NMW).

X. parietina (L.) Th. Fr.
On nutrient-rich limestone, concrete and bark; common. 21/48, 49, 58, 59, 69, 79, 87, 88, 96, 97, 98, 99, 22/(70), 31/06, 07, 08, 09, 16, 17, 18, 19, 26, 27, 28, 32/00, 10.

X. polycarpa (Hoffm.) Rieber
On nutrient-rich bark, usually on twigs, frequent. 21/48, 49, 58, 87, 96, 97, 31/06, 07, 16, 17, 26, 32/00.

DOUBTFUL RECORDS

The following published records are doubtfully correct, or refer to species whose present identity is uncertain. Supporting specimens have not been traced.

Bacidia ? assulata (Körber) Vězda (Wade 1959, (as *B. effusa* (Sm.)Arn.).
Cladonia incrassata Flörke (Watson 1933).
C. pleurota (Flörke) Schaerer (Watson 1953).
Lecanora agardhiana Ach. Watson (1933).
L. vernalis (L.) Ach. Watson (1953).
Lepraria spp.: (As *Crocynia superba* Hue (Watson 1933); *C. finkii* B. de Lesd., *C.fragilis* B. de Lesd., *C. lanuginosa* (Ach.) Hue var. *inactiva* B. de Lesd. (Wade & Watson 1936); *C. flava* (Ach.) Hue (Watson 1953)).

(*Leptorhaphis epidermidis* (Ach.) Th. Fr.) (Wade & Watson 1936).

Micarea subnigrata (Nyl.) Coppins & Kilias (Watson 1953).

Ochrolechia inversa (Nyl.) Laundon (Watson 1933).

Peltigera venosa (L.) Hoffm. (Gutch 1842a).

Phaeographis inusta (Ach.) Müll. Arg. (Watson 1953).

Physcia lithotea (Ach.) Nyl. Watson (1933).

Rhizocarpon confervoides DC. Wade & Watson (1936).

R. plicatile (Leighton) A.L. Sm. Watson (1953).

Toninia turneri (Leight.) Oliv. (Leighton 1879).

Usnea lapponica Vainio (Watson 1953, as *U. fulvoreagens*).

Verrucaria papillosa Ach. (Watson 1953).

EXCLUDED SPECIES

The list includes published records found to be incorrect, usually by the examination of specimens referred to in the place of publication, and records of species now known not to occur in the British Isles. Some of the former category have since been correctly recorded from the county.

Aspicilia laevata (Ach.) Arnold

A. verrucigera Hue

Bacidia adjuncta Th. Fr. (Watson 1933).

Calicium abietinum Pers. (Wade & Watson 1936).

Caloplaca luteoalba (Turner) Th. Fr. (Watson 1933).

Catapyrenium cinereum (Pers.) Körber (Wade & Watson 1936, as *Dermatocarpon hepaticum*).

Catapyrenium lachneum (Ach.) R. Sant. (Wade & Watson 1936).

Catillaria erysiboides (Nyl.) Th. Fr. (Wade & Watson 1936).

Collema flaccidum (Ach.) Ach. (Wade & Watson 1936).

C. nigrescens (Huds.) DC. (Wade & Watson 1936).

Huilia albocaerulescens (Wulfen) Hertel

Hypogymnia vittata (Ach.) Parr.Wade & Watson (1936).

Lecanora allophana (Ach.) Nyl. (Wade & Watson 1936).

L. coilocarpa (Ach.) Nyl. (Watson 1953).

Lecidea confluens (Web.) Ach. (Wade & Watson 1936).

L. scotinodes Nyl. (Watson 1953).

Lecidella anomaloides (Massal.) Hertel & Kilias (Wade & Watson 1936).

L. pulveracea (Schaerer) Sydow (Watson 1933).

Leptogium andegavense Hy (Wade & Watson 1936).

(*Microthelia micula auct.*) (Watson 1933).

(*Muellerella pygmaea* (Körber) D. Hawksw.) (Wade 1950).

Parmelia taylorensis Mitchell (Wade & Watson 1936, as *P. revoluta* var. *rugosa*).

P. tiliacea (Hoffm.) Ach. (Wade & Watson 1936).

Pertusaria amarescens Nyl. (Watson 1953).

P. ceuthocarpoides Zahlbr. (Wade & Watson 1936).

Phaeophyscia ciliata (Hoffm.) Moberg. (Watson 1953).

P. sciastra (Ach.) Moberg (Watson 1953).

P. phaea (Tuck.) Thomson (Wade & Watson 1936).

Polyblastia deminuta Arnold (Watson 1933).

P. terrestris Th. Fr. (Watson 1933).

Polychidium muscicola (Swartz) Gray (Wade & Watson 1936).

Pyrenula nitida (Weigel.) Ach. (Wade & Watson 1936).

P. nitidella (Flörke ex Schaerer) Mül. Arg. (Wade & Watson 1936).

Rhizocarpon grande (Flörke ex Flotow) Arnold (Wade 1959).

Rinodina albidorimulosa (Harm.) Zahlbr. (Watson 1953).

Stereocaulon condensatum Hoffm. (Wade & Watson 1936).

Thelidium cataractarum (Hepp) Lönnr. (Watson 1953).

Trapeliopsis viridescens (Schrader) Coppins & P. James (Wade & Watson 1936).

Umbilicaria deusta (L.) Baumg. (Wade 1962).

Usnea ceratina Ach. (Wade & Watson 1936).

U. hirta (L.) Wigg. (Wade & Watson 1936).

Verrucaria maculiformis Kremp. (Watson 1933).

V. prominula Nyl. (Wade & Watson 1936).

REFERENCES

Cannon, P.F., Hawksworth, D.L. & Sherwood-Pike, M.A. (1985). *The British Ascomycotina. An annotated checklist.* Slough: Commonwealth Mycological Institute.

Dillwyn, L.W. (1809). *British Confervae.* London: W. Philips.

Gilbert, O.L. (1990). The lichen flora of urban wasteland. *Lichenologist* **22**, 87–101.

Gutch, J.W.G. (1842a). A list of plants met with in the neighbourhood of Swansea, Glamorganshire. *Phytologist* **1**, 180–187.

Gutch, J.W.G. (1842b). Additions to the list of plants met with in the neighbourhood of Swansea. *Phytologist* **1**, 377–380.

Hawksworth, D.L. (1975). Notes on British lichenicolous fungi, I. *Kew Bulletin* **30**, 183–203.

Hawksworth, D.L. (1983). A key to the lichen-forming, parasitic, parasymbiotic and saprophytic fungi occurring on lichens in the British Isles. *Lichenologist* **15**, 1–44.

Hawksworth, D.L., James, P.W. & Coppins, B.J. (1980). Checklist of British lichen-forming, lichenicolous and allied fungi. *Lichenologist* **12**, 1–115.

Hertel, H. & Knoph, J. G. (1984). *Porpidia albocaerulescens*, eine weit verbreitete, doch in Europa seltene und vielfach verkannte Krustenflechte. *Mitteilungen der Botanische Staatsammlung München* **20**, 467–488.

Ivimey-Cook, R.B. (1959). The lichens of the Ewenny Downs, Glamorganshire. *Lichenologist* **1**, 97–103.

Leighton, W.A. (1879). *The lichen-flora of Great Britain, Ireland, and the Channel Islands.* 3rd edition. Shrewsbury: privately printed.

Pyatt, F.B. & Harvey, R. (1973). *Tichothecium erraticum* Mass. on *Lecanora campestris* (Schaer.) Hue and *Caloplaca heppiana* (Müll. Arg.) Zahlbr. *Israel Journal of Botany* **22**, 139–143.

Rose, F. (1976). Lichenological indicators of age and continuity in woodlands. In: Brown, D.H., Hawksworth, D.L. & Bailey, R.H. (eds.) *Lichenology: Progress and Problems.* London: Academic Press.

Sheard, J.W. (1967). A revision of the lichen genus *Rinodina* (Ach.) Gray in the British Isles. *Lichenologist* **3**, 328–367.

Smith, A. L. (1918). *A Monograph of the British Lichens* Vol. I. 2nd edition. London: British Museum (Natural History).

Swinscow, T.D.V. (1965). Pyrenocarpous lichens: 9. Notes on various species. *Lichenologist* **3**, 72–83.

Wade, A.E. (1950). Botanical Notes, 1947–48. *Transactions of the Cardiff Naturalists' Society* **79**, 52–54.

Wade, A.E. (1959). Glamorgan botanical notes. 1957. *Transactions of the Cardiff Naturalists' Society* **86**, 22–23.

Wade, A.E. (1960a). Glamorgan botanical notes. 1958. *Transactions of the Cardiff Naturalists' Society* **87**, 27.

Wade, A.E. (1960b). The British Anaptychiae and Physciae. *Lichenologist* **1**, 127–144.

Wade, A.E. (1962). Glamorgan botanical notes. 1960. *Transactions of the Cardiff Naturalists' Society* **89**, 31.

Wade, A.E. & Watson, W. (1936) Lichens of Glamorgan. In: W.M. Tattersall, ed. *Glamorgan County History.* Vol. 1. Cardiff: W. Lewis.

Watson, W. (1930). Lichenological notes. V. *Journal of Botany* **68**, 265–270.

Watson, W. (1933). Lichenological notes. VII. *Journal of Botany* **71**, 314–318, 327–338.

Watson, W. (1953). *Census catalogue of British lichens.* London: Cambridge University Press.

Watson, W. (1942). Lichenological notes. X. *Journal of Botany* **80**, 137–149.

Appendix I
Key to Abbreviations of Recorders' Initials

This list includes those recorders whose names, in abbreviated form, are referred to in the catalogue of flowering plants and ferns. It does not include those names given there in full. The list is arranged alphabetically in order of the first initial.

ABP	Mrs A. B. Pinkard	JBe	Mrs J. Berney
AHT	Prof. A. H. Trow	JEL	J. E. Lousley
AJES	Dr A. J. E. Smith	JIL	Mrs J. I. Littley
AL	Rev. A. Ley	JM	J. Motley
ALe	Dr A. Lees	JPC	J. P. Curtis
AMP	Mrs A. M. Pell	JS	J. Storrie
AN	A. Newton	JTa	Mrs J. Tann
ARP	A. R. Perry	JWD	J. W. Davies
BAM	B. A. Miles	JWGG	J. W. G. Gutch
BAW	B. A. Williams	KA	Mrs K. Adams
BS	Dr B. Seddon	LWD	L. W. Dillwyn
BSBI	Botanical Society of the	MG	Dr M. E. Gillham
	British Isles	MSP	Dr M. S. Percival
CCB	C. C. Babington	NCC	Nature Conservancy Council
CH	Dr C. R. Hipkin	PJo	P. S. Jones
DMcC	D. McClintock	PS	Miss P. Simons
EF	E. Forster	PWR	Prof. P. W. Richards
EFL	Rev. E. F. Linton	RAH	R. A. Henderson
EMT	Miss E. Mary Thomas	RE & FC	Misses R. E. & F. Cundall
EV	Miss E. Vachell	RLS	R. L. Smith
GCD	Dr G. C. Druce	RM	Dr R. Melville
GH	Dr G. Hutchinson	SGH	S. G. Harrison
GMB	Miss G. M. Barrett	TBF	T. B. Flower
GSW	G. S. Walters	TGE	T. G. Evans
GTG	Dr G. T. Goodman	THT	T. H. Thomas
H & CH	H. & C. R. Hipkin	VCB	Mrs V. C. Baldwin
HAH	Dr H. A. Hyde	WAS	Dr W. A. Shoolbred
HJR	Rev. H. J. Riddelsdell	WT	Dr W. Turton
HOB	Miss H. O. Booker	WWB	W. W. Boucher
JAW	J. A. Webb		

Appendix II

List of Contributors to the Flora of Glamorgan, 1969–1989

This list includes the names of the vast majority of botanists who helped, in one way or another, to record plants for this Flora. The authors accept full responsibility for the list and apologise for any omissions.

An exclamation mark (!) following a name indicates a member of the Flora Committee, and an asterisk (*) a contribution of over 100 records.

Ackers, B. J.
Adams, D. & Mrs S.
Adams, H. F. & Mrs K. F.*
Allen, Mrs M. J.
Asprey, Prof. G. F.!
Atkins, Mrs K. A.*
Baker, Mrs K.*
Baker, R. M.
Baldwin, Mrs V. C.*
Barling, D. M.*
Barrett, Miss G. M.*
Barrow, M. D.
Baxter, Miss C. F.*
Baxter, D. A. & Mrs C. M.*
Benson-Evans, Dr K.!
Berney, Mrs J.*
Bevan, J.
Blackler, Mrs D. K.*
Boorman, Dr L. A.
Bowen, D. B.
Bowen, Dr H. J. M.
Burn, A.
Cahn, Dr M.
Campbell, Mrs G.
Carey, I. M.
Castell, Mrs R.
Cawood, Mrs E.*
Clarke, Mrs G. M.
Cohen, Mrs M.
Collings, Mrs K. I.!
Collings, R.!
Coniber, R. A.
Coombe, Dr D. E.
Cotton, R.

Cropper, R. S.
Curtis, Mrs C. A.*
Curtis, J. P.*!
Cussack, I. B.
David, R. W.
David, T. G.
Davies, C.
Davies, D.
Davies, H. T.
Davies, J.
Davies, Mrs J.
Davies, J. W. & Mrs R.*
Davies, M.
Davies, Mrs M.
Davies, P. M. & R.*
Davies, Mrs R. M.
Davies, T.
Davis, R.
Dawson, Mrs C.
Dawson, Dr H.
Doe, M.*
Driscoll, R. J.
Dunn, Miss J.
Dunn, P. H.
Elias, D. O.
Ellis, R. G.*! (Recorder)
Ellis, Mrs V. G.
Elward, E.
Etherington, Dr J.
Evans, Miss D. E.
Evans, L.
Evans, T. G.*
Fearn, Dr G. M.
Fenton, Miss E. J.

Fielding, J. L.*
Fitter, R. A.
FitzGerald, Lady R.
Fraser-Jenkins, C. R. & C. D.
Fry, Mrs A.
Gillham, Dr M. E.*!
Goode, Dr D. A.
Goodman, Dr G. T.*
Gordon, Miss V.
Grant, G.
Green, D.
Grenfell, A. L.
Grubb, Dr P.
Hammond, P. E.*
Hankinson, Miss M.
Hankinson, W.
Hare, A. R.*
Harris, B. G.
Harrison, S. G.*! (Chairman)
Harvey, Dr R.!
Heathcote, Miss
Henderson, Mr & Mrs R. A.*
Henshaw, Miss E. B.
Hipkin, Dr C. R. & Mrs H.*
Hoddle, Mrs K.
Holland, Miss D. G.
Howarth, D. M.*
Howells, M.
Hughes, M. R.
Hutchinson, Dr G.*!
Hyde, Dr H. A.!
Jakeman, Mr & Mrs K.
James, Miss M.
Jarvis, P.
Jepson, P.
John, Mrs J.
John, Mrs R. F.*
John, Miss S.
Jones, Mrs B. E.*
Jones, Mrs D.
Jones, E.
Jones, H
Jones, I.
Jones, J. R.
Jones, P. S.
Kay, Mrs B.
Kay, Dr Q. O. N.*! (Recorder)
Killick, H. J.
Kilpatrick, Mrs J.
Lack, Dr A. J.
Lanigan, Mrs J.*

Lees, Dr A. J.
Lehane, J.
Lewis, Miss I. D.
Lewis, Mrs J.*
Lewis, J. W.
Littley, Mrs J. I.*
Llewellyn, G.*
Lord, J.
Lovell, Mr & Mrs J. W. A.*
Mapleson, Dr W. W.*
Marsden, Mrs M. P.
Matt, Miss
McClintock, D.
McKenzie, H.
Meade, Dr R.
Miller, J. B.
Mock, P.
Moon, S.
Morgan , I. K.
Morgan, Miss V.
Morris, G.
Morton, M. P.
Moseley, Mrs J.
Nelson, W.
Newton, A.
Nicholls, W. H.*
Noltie, H. J.
Orange, A.
Oxenham, Mrs S.
Page, Dr M.
Pankhurst, R. J.
Panison, Dr B.
Pawson, Dr B.
Peddle, L.
Pell, Mrs A. M.*!
Percival, Dr M. S.*!
Perring, Dr & Mrs F. H.*
Perry, A. R.*!
Petch, Miss E.*
Phillips, Miss N.
Pinkard, Mrs A. B.*! (Secretary)
Popham, J.
Powley, A. R.
Price, Mrs J. M.
Price, Mrs P.*
Proctor, Dr M. C. F.
Pryce, R. D.
Randall, G.
Rees, I.
Richards, Dr A. J.
Roberts, Miss E. G.

Roderick, Mrs N.
Rostanski, K.
Rotheray, G. O.
Rowlands, A.
Russell, Mrs M. E.*
Russell, Mrs V. J.
Salmon, Col. H. M.
Sanday, D. B.
Scotter, Mrs B.*
Scotter, Dr C.
Shanahan, Mrs M. R.*
Sheldon, T. H.
Smith, Prof. A. G.!
Staddon, Ms J.
Stark, Mrs O.
Stevens, Dr D. P.
Stevens, Dr J.
Stewart, Mrs O.
Swan, Dr M.*
Sykes, Mrs H.
Syrett, Prof. P. J.!
Tann, Mrs J.*
Taylor, P.
Taylor, Mrs S.
Tew, I.*
Thomas, Dr B. A.
Thomas, D. J.

Thomas, Miss H.
Thomas, J. M.*
Thomas, P. L.
Tidswell, R. J.
Tipper, A. D.
Turner-Ettlinger, D. M.
Tyrer, Dr F. H.*
Vaughan, Mrs I. M.
Vincent, Mrs M.
Voyce, M.
Wade, A. E.*! (Treasurer)
Wade, Dr M.
Wainwright, Dr S. J.
Waldren, Dr S.
Walker, C.
Wall, W.
Walsh, D. B.
Walters, G. S.
Ward, Dr L. K.
Webb, A. M.
Weston, Mrs W.!
Wiley, E.
Wilson, Mrs L.
Woodell, Dr S. R. J.
Worrall, Dr D.
Zehetmayr, J. W.

Appendix III

Gazetteer of Place Names and Geographical Features

This Gazetteer lists most of the place names and geographical features mentioned in this book, but, since there is no universally accepted spelling of many place and locality names, and Welsh and English forms may exist side by side, many alternative spellings are also included. Spellings which are so similar as to be unambiguous, including some archaic spellings found in the chapter 'A short history of botany and botanists in Glamorgan' (e.g. Swanzey / Swansea), are omitted. Names for the same locality which can exist as separate or hyphenated words (e.g. Blaenycwm / Blaen-y-cwm or Cefn-onn / Cefn Onn) are only listed once. The entries are generally listed in strict alphabetical order.

With each entry a four-figure grid reference is given (i.e. to a 1km square), as these references do not overlap in Glamorgan, 100km prefix letters are given only for those few localities outside the county, which are also followed, in parentheses, by an abbreviated vice-county name. A '?' following any reference indicates uncertainty.

It is important to note that the four-figure reference given is usually to the square in which the printed name occurs on the Ordnance Survey map consulted, which need not necessarily be in the same square as the actual locality. Where two or more places have the same name, each place is qualified by placing after it (in parentheses) the name of the nearest town or village. Rivers are listed under the name of the river (afon in Welsh) and where two references are given, the first is to the source and the second to the mouth of the river (or valley).

Aberafan	7590	Allt y Rhiw	9285
Aberavon	7590	Alps Quarry	1374
Aberavon Marshes	7590	Alun, Afon	9274–8977
Abercynon	0795	Alun, Valley	9076
Aberdare	0002	Aman, River	9900
Aberdare Country Park	9702	Ammanford (Carms.)	SN6212
Aberdovey (Merioneth)	SN6196	Arthur's Stone	4990
Aberdulais	7699		
Aberfan	0600	Bacon Hole	5586
Abergavenny (Mons.)	SO3013	Baglan	7492
Abernant	0103	Baglan Bay	7292
Abernant Park	0103	Baglan Burrows	7391
Aberpergam	8606	Baglan Moors	7491
Aberpergwm	8606	Baglan Sands	7291
Aber-pergwm Wood	8505	Bannau Sir Gaer (Carms.)	SN8121
Aberthaw	0366	Bargoed	1499
Afan, Afon	8295	Bargoed Rhymney	0907–1500
Afan, River	8295	Bargoed Taf	0906–0996
Afan Valley	8796–8195	Barland	5789
Afon – see under name of river		Barland Common	5789

Barry	1068	Burry Estuary	4497
Barry Docks	1167	Burry Green	4691
Barry Island	1166	Burry Holmes	3992
Beaupre	0073	Burry Holms	3992
Bendrick Rock	1366	Burry Inlet	4397
Berges Island	4596	Burry Pill	4389–4597
Berry Wood	4788	Burton Bridge	0367
Birchgrove (Cardiff)	1779	Bute Dock	1975
Birchgrove (Swansea)	7098	Bute Docks	1975
Bishop's Wood	5987	Bute Park	1777
Bishopston	5788	Bute Town	1009
Bishopston Churchyard	5789	Bwlch y Clawdd	9494
Bishopston Valley	5789–5787	Bwlch-y-Cywion	8298
Black Cock Inn	1484		
Blackmill	9286	Cadoxton (Neath)	7598
Blackpill	6190	Cadoxton (Barry)	1269
Blackpill Burrows	6290	Cadoxton Mill	1368
Blackweir	1778	Caerau	1375
Blackwood (Mons.)	1797	Caerffili Mountain	1585
Blaenafan	9096	Caerphilly	1586
Blaenllechau	0097	Caerphilly Castle	1587
Blaenllwynau	1002	Caerphilly Common	1585
Blaenrhondda	9299	Candleston	8777
Blaen-y-cwm (Monknash)	9070	Candleston Castle	8777
Blaen-y-Cwm (Treherbert)	9298	Canton	1676
Blue Anchor	5495	Canton Common	1676?
Bonvilston	0673	Cardiff	1877
Bonvilston Church	0674	Cardiff Castle	1876
Boverton	9868	Cardiff Docks	1974
Braich-y-cymer	9090	Cardiff Marshes (East Moors)	2076
Brandy Cove	5887	Cardiff railway station	1875
Brecon (Brecs.)	SO0428	Cardigan (Cards.)	SN1846
Brecon Beacons (Brecs.)	SO0121	Carmarthenshire Vans (Carms.)	SN8121
Bridgend	9180	Carn Caca	8200
Bristol (W. Gloucs.)	ST5872	Carnedd Lwydion	1092
Brithdir	1402	Carn y Bugail	1003
Brithdir Wood	7198	Carreg yr Afon Park	7607
Briton Ferry	7494	Castell Coch	1383
Briton Ferry Docks	7393	Castell Coch Woods	1383
Broad Pool	5191	Castle-upon-Alun	9174
Broughton Burrows	4192	Caswell	5987
Broughton Farm	4192	Caswell Bay	5987
Bryn	8192	Caswell Corner	5987
Brynamman (Carms.)	SN7013	Caswell valley	5988–5987
Brynau	6089	Cathays Park	1777
Bryn-cous (Cadoxton)	7598?	Cefn Brithdir	1203
Brynmill	6392	Cefn Bryn	5089
Brynnau Gwynion	9782	Cefn Carnau	1684
Brynsadler	0280	Cefn-coed-y-cymmer	0208
Bull Cliff	0966	Cefn Cribwr	8582
Bulwarks, The	0866	Cefn Goleu	5795
Burry	4590	Cefn Mably	2284

Cefn Mably Park	2284	Cold Knap	1066
Cefn Morfudd	7896	Cold Knap Park	1066
Cefn On	1885	Colwinston	9475
Cefn Onn	1784	Cooper's Fields	1776
Cefn Pennar	0402	Copi Gwythi	1987
Cefn y Brithdir	1203	Cornelly	8280
Chepstow (Mons.)	ST5393	Corntown	9177
Cheriton	4593	Corrwg, Afon	8903–8696
Cheriton Green	4593	Cosmeston	1869
Church Village	0885	Cosmeston Lakes	1769
Cilfynydd	0992	Court Colman	8882
Cillibion	5191	Cowbridge	9974
Cimla	7696	Coychurch Higher	9585
Clifton (W. Gloucs.)	ST5673	Craig Blaen Rhondda	9100
Clun, Afon	0783–0382	Craig-cefn-parc	6703
Clydach	6801	Craig Cerrig-gleisiad (Brecs.)	SN9621
Clydach Vale	9793	Craig Evan Leyshon Common	0894
Clyne (Swansea)	6190	Craig fach	9395
Clyne (Vale of Neath)	8000	Craig fawr	9296
Clyne Castle	6190	Craig Llysfaen	1884
Clyne Common	5990	Craig lysfaen	1884
Clyne Valley	5893–6190	Craig Llanishen	1684
Clyne Wood	6091	Craig Ogwr	9394
Cnwc	0601	Craig Ruperra	2286
Cockett	6394	Craig y Geifr	9494
Coedarhydyglyn	1075	Craig-y-bwlch	9403
Coed Cadwgan	2184	Craig-y-llyn	9003
Coed Cefn Pwll Ddu	2287	Craig-y-llyn Fach	9003
Coed Coesau-whips	1985	Craig-y-Pant	8903
Coed Gelli (Coed Gelli-draws)	0588	Craig y Parcau	8978
Coed Leyshon	0476	Craig-yr-Allt	1284
Coed Morgannwg	8087–8508	Crawley	5288
Coedrhiglan	1075	Crawley Cliff	5288
Coedrhiglan Woods	1075	Creigiau	0881
Coedrhydyglyn	1075	Crockherbtown (Cardiff)	1876
Coedrhydyglyn Woods	1075	Crofty	5294
Coedriglan	1075	Cross, The	4789
Coed-y-Bedw	1182	Crug-yr-afan	9295
Coed-y-goras	2080	Crymlyn	7093
Coed-y-gores	2080	Crymlyn Bay	7092
Coedymwster	9481	Crymlyn Bog	6995
Coelbren (Brecs.)	SN8511	Crymlyn Burrows	7093
Coety Green	4291	Crymlyn Dingle	7096
Cogan	1772	Crymlyn Fen	6995
Cogan Farm	1670	Crynant	7904
Cogan Hall	1670	Culver Hole	4684
Cogan Pill	1773	Culverhouse Cross	1174
Cogan Plantation	1769	Cwmafan	7892
Cog Moors	1569	Cwm Alun	9175–9076
Coity	9281	Cwm Aman	6913
Coity Green	9281	Cwmbwrla	6494
Colbren Junction (Brecs.)	SN8511	Cwmbwrla Bog	6494

Llwchwr, Afon (R. Loughor)	6210–5697	Melin-y-cwrt Glen	8300–8201
Llwchwr Valley	6210–5697	Melin-y-cwrt Waterfalls	8201
Llwydcoed	9904	Merthyr	0607
Llwyn-y-grant	1979	Merthyr Common	0709
Llwyn-y-pia	9993	Merthyr Dyfan	1169
Llwynypia Mountain	9893	Merthyr Mawr	8877
Llyn Fach	9003	Merthyr Mawr Burrows	8676
Llyn Fawr	9103	Merthyr Mawr Church	8877
Llynfi, River	8894–8983	Merthyr Mawr Dunes	8676
Llynfi Valley	8694–8983	Merthyr Mawr Warren	8676
Llyn Teivy (Cards.)	SN7867	Merthyr Tydfil	0607
Llyn vach (Llyn Fach)	9003	Mewslade	4287
Llysworney	9674	Mewslade Bay	4186
Llyswyrney	9674	Michaelston y Fedw (Mons.)	ST2484
Locks Common	8077	Mill Lane	6190
Longhole	4585	Mill Wood (Penrice)	4888
Longhole Cliff	4485	Mill Wood (Porthkerry)	0967
Loughor	5797	Miskin	0480
Loughor, River	6210–5697	Mitchin Hole	5586
Lower Clydach, River	6709–6801	Monknash	9270
Lower Gnoll Pond	7697	Morfa Dyffryn (Merioneth)	SH5525
Lower Sketty	6191	Morfa Ponds	7985
Lower Swansea Valley	6695	Morfa Pools	7985
		Morfa Road	6694
Machen	2189	Morganstown	1281
Maendy	1778	Morlais Castle	0409
Maendy Pool	1778	Morlais Castle Hill	0509
Maerdy	9798	Morlais Glen	0409
Maesteg	8591	Morlais Hill	0509
Maindy	1778	Morriston	6698
Marcross	9269	Mountain Ash	0599
Mardy	9798	Mouth of Ogmore	8675
Margam	7887	Mumbles	6287
Margam Abbey	8086	Mumbles Head	6387
Margan Burrows	7784	Mumbles Hill	6287
Margam Lakes	8086	Murton	5888
Margam Moors	7885	Mynydd (= Mountain)	
Margam Park	8086	Mynydd Blaenrhondda	9100
Margam Pond	8086	Mynydd Bwllfa	9503
Margam Steelworks	7886	Mynydd Cilfach-yr-encil	0703
Margam Woods	8086	Mynydd Dinas	7691
Mawdlam	8081	Mynydd Drumau	7299
Mawdlam Church	8081	Mynydd Eglwysilian	1292
Mayals	6090	Mynydd Gellionen	7004
Mayals Green	6091	Mynydd Goetre-coed	0895
Mead Moor	4788	Mynydd Llangeinor	9192
Melin-court	8201	Mynydd Maendy	9689
Melin Court Brook	8300	Mynydd Maendy	9886
Melin-cwrt Waterfall	8201	Mynydd Maerdy	9697
Melingriffith Works	1480	Mynydd Maes-teg	9690
Melin-y-cwrt	8201	Mynydd Maesteg	9690
Melin-y-cwrt Falls	8201	Mynydd Maio	1188

Mynydd Marchywel	7603	Ogmore	8876
Mynydd Margam	8188	Ogmore by Sea	8674
Mynydd Mayo	1188	Ogmore Castle	8876
Mynydd Meio	1188	Ogmore Down	8876
Mynydd Merthyr	0402	Ogmore, River	9386–8675
Mynydd Resolfen	8503	Ogmore Vale	9390
Mynydd Rudry	1886	Ogmore Valley	9184
Mynydd Troed y rhiw	0192	Ogwr	9386–8675
Mynydd Uchaf	7110	Ogwr District	8785
Mynydd-y-drum	8110	Ogwr Fach	9592–9386
Mynydd y Gaer	9585	Ogwr Fawr	9294–9386
Mynydd-y-glew	0376	Ogwr, River (Ogmore)	9386–8675
Mynydd y Glyn	0389	Olchfa	6193
Mynydd y Gwair	6609	Old Castle Down	9076
		Onllwyn	8410
		Overton	4685
Nant Clydach	0693	Overton Cliff	4584
Nant Cwm-du	8789	Overton Mere	4684
Nant Dyfrgi	0278	Oxwich	4986
Nant Iechyd	9488	Oxwich Bay	5186
Nant-y-moel	9392	Oxwich Burrows	5087
Nant yr Aber	1388	Oxwich Green	4986
Nash Point	9168	Oxwich Marsh	5087
National Museum of Wales site	1877	Oxwich National Nature Reserve	5087
Neath	7496	Oxwich Point	5185
Neath Canal	7395	Oxwich Wood	5086
Neath, River	9007–7292	Oystermouth	6188
Neath Technical College	7497	Oystermouth Castle	6188
Neath Valley	9007–7298	Oystermouth Road	6592
Nedd, Afon (R. Neath)	9007–7292		
Nedd-fechan	8908–9007	Padell-y-bwlch	9403
Nedd valley	9007–7292	Pant St Bride's	8976
Nelson	1195	Pant-y-Sais	7194
Nelson Bog	1295	Parc le Breos	5289
Newborough Warren (Anglesey)	SH4263	Park Farm	9386
Newbridge Fields	8979	Park le Breos	5289
Newport (Mons.)	ST3386	Parkmill	5489
Newport Road	1977	Parkmill Valley	5389
Newton (Mumbles)	6088	Parkmill Wood	5389
Newton (Porthcawl)	8277	Park Place	1877
Newton Burrows	8477	Paviland	4385
Newtown (Monts.)	SO1091	Paviland Farm	4486
Nicholaston	5288	Paviland Slade	4385
Nicholaston Burrows	5287	Pelina, River	8399–7994
Nicholaston Pill	5187	Pembrey (Carms.)	SN4201
Nicholaston Woods	5188	Pembrey Burrows (Carms.)	SN4000
N. M. W. site	1877	Pembrey Forest (Carms.)	SN3703
North Cornelly	8281	Penarth	1871
North-hill Wood	5488	Penarth Docks	1872
Norton (Mumbles)	6088	Penarth Ferry	1872
Norton (Ogmore by Sea)	8775	Penarth Ferry Road	1773
Nottage	8278	Penarth Head	1971

Reynoldston	4889	St Fagans	1277
Reynoldston Common	4890	St Fagans Castle	1177
Rhaslas Pond	0907	St George's	1075
Rheola	8304	St Helen's	6492
Rheola Forest	8004	St Hilary	0173
Rhigos	9205	St Lythan's	1172
Rhigos-Hirwaun gap	9105–9505	St Mary Church	0071
Rhiwbina	1581	St Mary's Well Bay	1767
Rhondda	9596	St Nicholas	0974
Rhondda Fach	9302–0291	St-y-Nyll	0978
Rhondda Fawr	9201–0790	St-y-Nyll Ponds	0978
Rhondda Valley	9302–0291	Salthouse Point	5295
Rhoose	0666	Sarn	9083
Rhoose Point	0765	Scurlage	4687
Rhoose Quarry	0665	Scwd Einon Gam	8909
Rhosili	4188	Sellack (Herefs.)	SO5627
Rhosili Down	4290	Seven Sisters	8208
Rhosili Dunes	4188	Seven Slades	5887
Rhossili	4188	Sgwd Einon Gam	8909
Rhossili Cliffs	4087	Sgwd Gwladys	8909
Rhossili Down	4290	Singleton	6392
Rhyd-y-fro	7105	Singleton Bog	6291
Rhydygwern	2188	Singleton Burrows	6391
Rhyd-y-gwern Wood	2188	Singleton Marsh	6291
Rhyd-y-pandy	6601	Sker	7879
Rhymney (Mons.)	SO1107	Sker Point	7879
Rhymney Estuary	2277	Sketty	6293
Rhymney, River	0911–2277	Sketty Bog	6191
Rhymney Valley	0911–2277	Sketty Cwm	6191
Rhymni	2177	Skewen	7296
Rhymni Valley	0911–2277	Slade	4885
River – see under name of river		Slanney woods	1178
Roath	2077	South Cornelly	8180
Roath Brook	1978	South Cornelly Quarry	8280
Roath Dock	2074	Southerndown	8874
Roath Park	1879	Southgate	5588
Rockwood Hospital	1478	Spaniard Rocks	4092
Rudry	1986	Splott	2076
Rudry Common	1886	Stanleytown	0194
Rumney (Mons.)	ST2279	Steep Holm (N. Somerset)	ST2260
Rumney hill (Mons.)	ST2179	Stoke Newington (Middlesex)	TQ3286
Ruperra	2286	Stonemill	5589
Ryer's Down	4592	Stoneylands Farm	1670
Rymney, River	0911–2277	Stormy Down	8480
		Stouthall	4789
St Athan	0167	Stouthall Woods	4789
St Bride's-Major	8974	Strata Florida Abbey (Cards.)	SN7465
St Brides-super-Ely	0977	Sully	1568
St Donat's	9368	Sully House Copse	1667
St Donat's Bay	9367	Sully Island	1667
St Donat's Castle	9368	Sully Moors	1468–1569
St Donat's Church	9368	Summerhouse Point	9966

Swanbridge	1767	Thistledown	6287
Swanbridge Halt	1668	Thornhill	1584
Swansea	6594	Three Cliffs Bay	5387
Swansea Bay	6688	Three Crosses	5794
Swansea Canal	7203	Thurba	4286
Swansea Docks	6792	Tinkinswood	0973
Swansea Pier	6692	Tir Eithin Farm West	9781
Swindon (N. Wilts.)	SU1484	Tir Hester	6194
		Tir John	6893
Taf, Afon	0307–1972	Tir John Power Station	6893
Taf Fawr, Afon	0307–0307	Tir John tip	6893
Taf Fechan	0611–0307	Tir-y-beth	1596
Taf-fechan Gorge	0409–0307	Tondu	8984
Taf-fechan Valley	0611–0307	Tongwynlais	1382
Taff-Ely District	0385	Ton Mawr	8096
Taff Fechan, River	0611–0307	Tonplanwydd	8709
Taff Gorge	0409–0307	Tonteg	0986
Taff, River	0307–1972	Tonypandy	9992
Taff's Well	1283	Tonyrefail	0187
Taff Vale	1086	Tor-gro	4593
Taff Valley	0307–1972	Tottenham (Middlesex)	TQ3491
Tai-bach	7789	Townhill	6393
Talbot Green	0482	Townhill Bog	6394
Talygarn	0380	Towyn Burrows (Carms.)	SN3605
Talygarn Estate	0379	Trecynon	9803
Tankeylake Pond	4391	Tre-Dodridge	0577
Tareni Gleision	7506	Tre-forgan	7905
Taren Maerdy	9897	Treharris	0997
Taren y Gigfran	0698	Treherbert	9498
Tarren Rhiw-maen	9294	Treorchy	9697
Tarren y Gigfran	0698	Treorci	9697
Tawe, Afon	7808–6692	Tresilian	9467
Tawe, River	7808–6692	Tresilian Bay	9467
Tawe Valley	7808–6692	Troedyrhiw	0702
Tears Point	4186	Trwyn-y-Witch	8872
Temperance Town	1876	Twmpath Valley (Cwm Dimbath)	9589
Tennant Canal	7093	Twrch, Afon	7511–7808
Thaw, River (Afon Ddaw)	9880–0265	Twynau Gwynion	0610
Thaw Valley	9778–0367	Twyn Brynbychan	0698
The Bulwarks	0866	Twyn Howel	1091
The Cross	4789	Ty-coch	6293
The Dams	0665	Tythegston	8578
The Ferry (Swansea)	6694	Tywi, Afon (Cards.-Carms.)	SN8063–3507
The Garth	1083	Tywyn Burrows (Carms.)	SN3605
The Groose	4494		
The Heath	1779	University College of Swansea	6291
The Knap	5787	Upper Clydach River	6909–7203
The Knave	4386	Upper Lliw Reservoir	6606
The Leys	0266	Usgod Eynon Gam	8909
The Little Garth	1282		
The Rest	8078	Vale of Neath	9007–7298
The Wenallt	1583	Vaynor (Brecs.)	SO0410

Victoria Park	1576	Whiteford Point	4496
		Whitford (see Whiteford)	
Walthamstow (S. Essex)	TQ3788	Whitmore Bay	1166
Warren Mill Pond	0575	Wick	9272
Watford Fawr	1485	Wick Churchyard	9272
Wattstown	0293	Widegate	5688
Waunarlwydd	6095	Witch's Point	8872
Waycock, River	0872–0668	Witford Point	7392
Welsh St Donats	0276	Worm's Head	3887
Wenallt	1583	Wyndcliff (Mons.)	ST5297
Wentloog levels (Mons.)	ST2479		
Wenvoe	1272		
Weobley	4792	Y Drummau	7299
Weobley Castle	4792	Y Graig	0483
Werfa	9194	Ynysboeth	0796
Wergan-rhos	5790	Ynyscynon	0102
Wern-ffrwd	5193	Ynysforgan	6899
West Aberthaw	0266	Ynyshir	0292
West Cliff	5587	Ynys Maerdy	0383
West Cross	6189	Ynystawe	6700
Whitchurch	1579	Ynysybwl	0594
Whitchurch Canal	1480	Ystalyfera	7608
Whiteford	4394	Ystrad	9795
Whiteford Burrows	4495	Ystrad Mynach	1493
Whiteford Marsh	4494	Ystradowen	0177
Whiteford National Nature		Ystradowen Bog	0278
Reserve	4394	Ystradowen Moor(s)	0278
Whiteford Plantation	4394	Ystrad Rhondda	9795

Appendix IV
Bibliography

This bibliography includes:

1. References which include a vascular plant taxon mentioned at a specified site in Glamorgan (v.c. 41), but excluding records reported in BSBI 'Plant Records', as published in Watsonia and its predecessors, unless considered relevant by the authors, or referred to in the text.   Also excluded are references in newspapers, popular magazines, radio or television programmes and video recordings.
2. References cited in the chapter on 'A short history of botany and botanists in Glamorgan'.
3. References cited in the vascular plant catalogue.

References to works in other chapters are listed at the end of the relevant chapter.

Abbreviations and symbols.

* * Not seen by authors.
* ** Included in Simpson (1960) but not seen by Simpson or the authors of this Flora.

Anon. – Anonymous or author unknown.
Ed. – Edited.
ed(s). – Editor(s).
Edn (edn) – Edition (edition).
et al. – and other authors.
illus. – illustration.
MS – Manuscript.
n.d. – no date.
n.s. – new series.
obit. – obituary.
p.(pp.) – Page(s).
q.v. – which see.
sp.(p). – species, sing. (plural).
suppl. – supplement.
trans. – transcribed.

The authors are indebted to Dr G. Hutchinson for compiling the initial expanded draft of this bibliography (Hutchinson, G., 1990).

Ab-Shukor, N.A., Kay, Q.O.N., Stevens, D.P. & Skibinski, D.O.F. (1988). Salt tolerance in natural populations of *Trifolium repens* L. *New Phytologist*, **109**, 483–490.

Ackers, B.J. (1978). *The distribution of* Ranunculus *L. subgenus* Batrachium *(DC.) A. Gray in West Glamorgan*. B.Sc. project report, Department of Botany, University College of Swansea.

Ahmad, I. & Wainwright, S.J. (1976). Ecotype differences in leaf surface properties of *Agrostis stolonifera* from salt marsh, spray zone and inland habitats. *New Phytologist*, **76**, 361–366.

Al-Gharbi, A.S. (1984). *A study of nitrate reductase activity in British angiosperms*. Ph.D. thesis. University College of Swansea.

Angel, H. (1977). *The Countryside of South Wales*. Jarrold Colour Publications, Norwich.

Anon. (1859a). *A Week's Walk in Gower*. R. Mason, Tenby.

Anon. (1859b) *Sparganium minimum* [from Crymbyn [sic] Bog, near Swansea, Rev. J.H. Thompson]. *Phytologist [n.s.]*, **3**, 306.

Anon. (1869). Barry Field Meeting. *Transactions of the Cardiff Naturalists' Society 1867–8*, **1**, 74–76.

Anon. (1869–70). Exhibit. T.B. Flower. *Proceedings of the Linnean Society*, p. xxxiii.*.

Anon. (1872). An old Oak Tree with Elm growing on the top, Tydee Farm, Llanharry, nr. Cowbridge. *Report and Transactions of the Cardiff Naturalists' Society 1870–71*, **3**, [opp. p. 80 Illustration].

Anon. (1877). The President's Botany Prize No. 1. [Collection of dried specimens of plants sent in by C.T. Vachell, M.D. Lond., with locations]. *Report and Transactions of the Cardiff Naturalists' Society*, **8**, 146.

Anon. (1879). Manchester Literary and Philosophical Society. Ref. to Mr T. Rogers reading a paper on, and exhibiting many specimens of, ballast plants collected at Cardiff in September 1878. *Nature*, **19**, 571.

Anon. (1887). A collection of wild flowers, chiefly local, collected, arranged, mounted and named by the donor [Mrs Fisher]. *Annual Report of the Swansea Scientific Society 1886–7*, p. 11.

Anon. (1893–1894). Field meetings in *Swansea Scientific Society Annual Report*, 1892–3, pp. 17–24 (Paviland to Port Eynon); pp. 30–32 (Crymlyn Burrows and Bog); 1893–4, pp. 16–21 (Penmaen Burrows).

Anon. (1903–1911). Field meetings and plant records in *Royal Institution of South Wales Reports*, 1902–3, pp. 12–13 (Bicheno and Motley collections); 1904–5, pp. 117–119 (Pont-Neath-Vaughan); 1905–6, p. 86 (Ogmore and Ewenny); 1906–7, p. 80 and pp. 123–128 (Orchids; Port Talbot ballast species); 1907–8, pp. 170–175; 1908–9, pp.127–134; 1909–10, pp. 147–155 (Port Talbot ballast species); 1910–11, p.158 (Kenfig and Porthcawl); pp. 160–162 (Swansea).

Anon. (1912). [Wild Plants: *Claytonia perfoliata* nr. Cardiff]. *The Country-side Monthly*, **4**, 282.

Anon. (1913a). *A Descriptive Catalogue of Trees and Shrubs in the Aberdare Park 1913*. Copy in Library, National Museum of Wales, Cardiff.

Anon. (1913b). Notices of work on British Vegetation: Sand Dunes in Glamorgan, Wales. [Abstract of Orr, M.Y. (1912), 'Kenfig Burrows'. *Scottish Botanical Review*, **1**, 209–216]. *Journal of Ecology* , **1**, 63.

Anon. (1966a). Rainfall as a factor in determining plant distribution in Wales. *Botanical Society of the British Isles Welsh Region Bulletin No. 9*, pp. 1–6.

Anon. (1966b). [Note of prize awarded for photograph taken by Dr J.P. Savidge, of *Orobanche minor* Sm. growing on *Eryngium maritimum* L. at Jersey Marine, Glamorgan]. In: Report of Autumn Meeting, 1966, in: *Botanical Society of the British Isles Welsh Region Bulletin No. 9*, p. 10.

Anon. (1968). Study Courses [Ref. to trips to Overton Mere, Whitford, Rhossili Down and Whitemoor]. *Glamorgan County Naturalists' Trust, Bulletin No. 7*, pp. 23–24.

Anon. (1973). Notes from the National Museum of Wales, Department of Botany: Flora of Glamorgan. *Nature in Wales*, **13**, 193.

Anon. (1981). Let's visit a reserve: Broad Pool and Bog. *GNAT No. 3*, [p. 3]. (Newsletter of the Glamorgan Naturalists' Trust).

Anon. (1982). *The Vegetation of Merthyr Mawr*. Nature Conservancy Council.*.

Anon. (1985a). Cowslips return. *GNAT No. 12*, [p. 5]. (Newsletter of the Glamorgan Trust for Nature Conservation).

Anon. (1985b). *Glamorgan Canal Local Nature Reserve. Guided Walks Summer 1985*.

Anon. (1988) Gower gets a new reserve [Kilvrough Manor Woods]. *Glamorgan Wildlife No. 5*, [p. 1]. (Newsletter of the Glamorgan Wildlife Trust).

Anon. (1989). Trust gets first up valley wood [Blaenant-y-Gwyddyl Nature Reserve, Glyn-neath]. *Glamorgan Wildlife No. 7*, [p. 1]. (Newsletter of the Glamorgan Wildlife Trust).

Anon. (1990). A bright future for the Taf Fechan LNR. *Glamorgan Wildlife No. 12*, p. 3. (Newsletter of the Glamorgan Wildlife Trust).

Ashcroft, H. (1983). *Phyllophane studies of* Fagus sylvatica *and* Tsuga heterophylla *in relation to the air spora of a mixed woodland*. Ph.D. thesis. University College of Cardiff.

Atherton, V. (1975). *The distribution and ecology of the diploid and tetraploid cytotypes of* Chrysanthemum leucanthemum L. *in the Swansea area*. B.Sc. project report, Department of Botany, University College of Swansea.

[Babington, A.M.] (1897). *Memorials, journals, and botanical correspondence of C.C. Babington,* pp. 83, 145, 190, 281. Macmillan and Bowes, Cambridge.

Babington, C.C. (1841). *Brassicella monensis* (errore *Sinapis cheiranthus*). *Annals and Magazine of Natural History,* **6**, 314*.

Baker, C.R.B., Blackman, R.L. & Claridge, M.F. (1972). Studies in *Haltica carduorum* Guerin (Coleoptera: Chrysomelidae) an alien beetle released in Britain as a contribution to the biological control of creeping thistle, *Cirsium arvense* (L.) Scop. *Journal of Applied Ecology* , **9**, 819–830.

Baker, J.M. (1971). *The effects of oil pollution and cleaning on the ecology of salt marshes* . Ph.D. thesis. University College of Swansea.

Baker, J.M. (1973a). Effects of refinery effluents on the plants of the Crumlyn Bog. In: *Annual Report, Oil Pollution Research Unit, Orielton, Pembs.,* pp. 29–35.

Baker, J.M. (1973b). Recovery of salt-marsh vegetation from successive oil spillages. *Environmental Pollution,* **4**, 223–230.

Baker, R.M. (1987). The Growth and Survival of *Aconitum anglicum* (Monkshood) at Gwern Rhyd Nature Reserve, South Glamorgan. In: Notes, *Nature in Wales (n.s.) 1986,* **5**, 53–57.

Baker, R.M. (1988). Mechanical control of Japanese Knotweed in an S.S.S.I. *Aspects of Applied Biology,* **16**, 189–192.

Baker, R.M. & Wislocka, G.M.S. de K. (1990). Expansions of *Hippophae rhamnoides* in a South Wales Dune System. In: *The Biology and Control of Invasive Plants.* Industrial Ecology Group of the British Ecological Society at University of Wales College of Cardiff, Conference Report, pp. 34–46.

Balchin, W.G.V. [ed.] (1971). *Swansea and its Region.* University College, Swansea.

Ball, J. (1849). Contributions to the flora of South Wales. *Botanical Gazette* , **1**, 107–109.

Ball, P.W. & Tutin, T.G. (1959). Notes on annual species of *Salicornia* in Britain. *Watsonia,* **4**, 193–205.

Bannon, N. et al. (1972). *Aspects of the vegetation and soils of Oxwich Bay Nature Reserve – the saltmarsh.* Nature Conservancy (Oxwich National Nature Reserve). Unpublished.

Barker, T.W. (1905). *Handbook to the Natural History of Carmarthenshire.* W. Spurrell, Carmarthen.

Barling, D.M. (1955). Some population studies in *Ranunculus bulbosus* L. *Journal of Ecology,* **43**, 207–218.

Barling, D.M. (1957). *Poa pratensis* subsp. *subcaerulea* in north Glamorgan and south Brecon. *Nature in Wales,* **3** , 429–433.

Barling, D.M. (1959). Biological studies in Welsh *Poa subcaerulea. Nature in Wales,* **5**, 774–780.

Barling, D.M. (1962). Studies in the biology of *Poa subcaerulea* Sm. *Watsonia,* **5**, 163–173.

Barrett, G.M. (1984). Cwm George in Summer. *GNAT No. 8,* p. 9. (Newsletter of the Glamorgan Naturalists' Trust).

Barrow, D.A. (1983). *Ecological studies on bumblebees in South Wales with special reference to resource partitioning and bee size variation.* Ph.D. thesis. University College of Cardiff.

Beanland, W.A. (1935). *The History of The Royal Institution of South Wales, Swansea 1835–1935.* Royal Institution, Swansea.

Beerling, D.J. (1990). The use of non-persistent herbicides to control riparian stands of Japanese Knotweed (*Reynoutria japonica* Houtt.). In: *The Biology and Control of Invasive Plants.* Industrial Ecology Group of the British Ecological Society at University of Wales College of Cardiff, Conference Report, pp. 121–129.

Bennett, A. (1903). James Motley and his herbarium at Swansea. *The Naturalist* [n.s.], p. 4.

Bennett, A. (1905). Supplement to Topographical Botany. Edn 2. *Journal of Botany,* **43**, (supplement).

Bennett, A. (1907). *Liparis loeselii,* Rich., An Extension of Range. *Transactions of the Norfolk and Norwich Naturalists' Society, 1906–1907,* **8**, 340–343.

Bennett, A. (1918). Watson Exchange Club Report, 1916–1917: *Liparis loeselii* [A.] Rich. *Journal of Botany,* **56** , 111–112.

Bennett, A., Salmon, C.E. & Matthews, J.R. (1929–1930). Second Supplement to Watson's Topographical Botany. *Journal of Botany,* **67** & **68** (supplements).

Bennett, A.W. (1886). [Report of Indoor Meeting] April 17th, 1884. *Proceedings of the Linnean Society of London* , p. 10.

Bennett, L. et al. (1989). Glamorgan. In: *A Guide to the Nature Reserves of Wales and the West Midlands,* pp. 86–97. Macmillan Press Ltd., London.

Benoit, P.M. (1962). *Erodium glutinosum* in Wales. *Nature in Wales,* **8**, 2–6.

Berney, J. (1987). Gower's rock-garden. *The Countryman*, **92**, 104–109.

[Bevan, J.] [ed.] (1985). [Notice of] Field Meetings 1985: '. . . Glamorgan, vc 41'. *Hieracium Group Notes No. 7*, p. 1. [Botanical Society of the British Isles, Hieracium Study Group].

[Bevan, J.] (1985). Reports of Meetings: Glamorgan, vc 41, Saturday 8th June 1985. *Hieracium Group Notes No. 8*, p. 6. [Botanical Society of the British Isles, Hieracium Study Group].

Bigwood, M.H. (1953). Exhibition Meeting 1952: *Ononis reclinata*. *Botanical Society of the British Isles Year Book*, p. 52.

Blackburn, K.B. (1949). Exhibits: *Limosella* [Morfa Pools]. In: *Botanical Society of the British Isles Conference Report* [Ed. by A.J. Wilmott], p. 81.

Boorman, L.A. (1986). *Chief Scientist Directorate Report No. 632: A survey of sand dunes in relation to grazing (interim report)*. Nature Conservancy Council, Peterborough.

Boswell Syme, J.T. & in part Brown, N.E. eds. (1902). *[Sowerby's] English Botany* , 3rd edn, 13 vols. London.

Bowen, E.J. (1930). A Survey of the flora of the North Gower coast in relation to marine ecological conditions. *Proceedings of the Swansea Scientific and Field Naturalists' Society*, **1** , 109–111.

Bowra, J.C. (1989). *Oenothera* L. in South Wales. *Botanical Society of the British Isles News No. 53*, p. 20.

Bridges, E.M. (1973). *An investigation into the origin and development of selected soils of the Gower Peninsula* . Ph.D. thesis. University College of Swansea.

Bridges, E.M. (1976). The Evolution of the Burry Inlet. *Gower*, **27**, 84–89.

Bridges, E.M. ed. (1979). *Problems of Common Land: The Example of West Glamorgan*. Proceedings of a Symposium on Commons. University College of Swansea 1978. Department of Geography, University College of Swansea.

Bridges, M. (1987). *Classic Landform Guides No. 7: Classic Landforms of the Gower Coast*. The Geographical Association in conjunction with the British Geomorphological Research Group. The Geographical Association, Sheffield.

Brinn, P.J. [1976]. *An Ecological Assessment of Suburban Expansion within the City of Cardiff*. Nature Conservancy Council.

British Association (1960). *Kenfig Burrows*. [Plant list produced for British Association Meeting 1960].

British Ecological Society – Industrial Ecology Group (1990). *The Biology and Control of Invasive Plants*. University of Wales College of Cardiff, Conference Report.

Britten, J. (1887). Notices of Books: The Flora of Cardiff. A descriptive list of the Indigenous Plants found in the District of the Cardiff Naturalists' Society by John Storrie. *Journal of Botany*, **25**, 349–351.

Britten, J. ed. (1891). Short Notes: Plants of the Flat Holme. *Journal of Botany*, **29**, 345.

Britten, J. ed. (1905). [Editor's notes on *Cotoneaster microphyllus* Lindley]. *Journal of Botany*, **43**, 244, 274.

Britten, J. (1917). *Liparis liliifolia* and *L. loeselii*. *Journal of Botany*, **55**, 246–250.

Britten, J. & Boulger, G.S. (1893). *A Bibliographical Index of British and Irish Botanists*. West, Newman & Co., London.

Britten, J. & Boulger, G.S. (1931). *A Bibliographical Index of deceased British and Irish Botanists*. 2nd edn revised and completed by A.B. Rendle. Taylor and Francis, London.

Bryan, A. (1989). Cwmtalwg Pocket Park. *Glamorgan Wildlife No. 9*, pp. 4, 8. (Newsletter of the Glamorgan Wildlife Trust).

Bryan, A.M. (1988). Towards a pocket park. [Cwmtalwg Woodlands, Barry]. *Glamorgan Wildlife No. 6*, [p. 6]. (Newsletter of the Glamorgan Wildlife Trust).

Bunce, R.G.H. (1982). *A Field Key for Classifying British Woodland Vegetation Part 1*. Institute of Terrestrial Ecology, Cambridge.

Burd, F. (1989). *Research and Survey in nature conservation. No. 17 – The Saltmarsh Survey of Great Britain (An inventory of British saltmarshes)*. Nature Conservancy Council, Peterborough.

Butcher, R.W. (1947). Biological Flora of the British Isles: *Atropa belladonna* L. *Journal of Ecology*, **34**, 345–353.

Cahn, M. (1984). Cwmllwyd Nature Reserve. *Fair Country (Conservation Volunteers Wales)* (Spring Issue), pp. 3–4.

Camden, W. (1695). *Britannia*. Ed. by E. Gibson, London.

Camden, W. (1789). *Britannia*. Ed. by R. Gough, London.

Camden, W. (1806). *Britannia*. Ed. by R. Gough, London.

Campion, P.S.A. (1979). *Factors affecting the establishment of vegetation on colliery spoil heaps in South Wales*. Ph.D. thesis. University of Wales Institute of Science and Technology.

Cardiff City Council (1983a). *Notification of sites of Special Scientific Interest (SSSI) under Section 28 of the Wildlife and Countryside Act 1981: Flatholm. [Allium ampeloprasum]*.

Cardiff City Council (1983b). *Notification of sites of Special Scientific Interest (SSSI) under Section 28 of the Wildlife and Countryside Act 1981: Garth Wood. [Neottia]*.

Cardiff City Council (1983c). *Wildlife and Countryside Act 1981: Section 28: Notification of sites of Special Scientific Interest: Castle Coch Woodland. [Neottia, Monotropa, Platanthera]*.

Cardiff City Council [c.1987]. *Glamorgan Council Local Nature Reserve*.

Cardiff City Council, Chief Executive (1985). *Wildlife and Countryside Act 1981– Section 28 Fforest Ganol: Notification of sites of Special Scientific Interest*. MS. Cardiff City Council, Leisure and Amenities Department.

Cardiff City Council, Chief Executive (1989). *Severn Estuary Site of Special Scientific Interest. Notification under Section 28 of the Wildlife and Countryside Act 1981 as amended*. [Pengam Moors – List of taxa]. MS. Cardiff City Council, Leisure and Amenities Department.

Cardiff City Council, Leisure and Amenities Department (n.d.1). *Glamorganshire Canal Local Nature Reserve: Mammals Found in the Reserve*.

Cardiff City Council, Leisure and Amenities Department (n.d.2). *Glamorganshire Canal Local Nature Reserve: Birds*.

Cardiff City Council, Leisure and Amenities Department [c.1980–]. *Howardian Nature Reserve: List of Species*. MS.

Cardiff City Council, Leisure Services Department (n.d.). *Wildlife in Cardiff*.

Cardiff City Council, Parks Department (n.d.1). *Bute Park*. [Booklet in the Nature Trail Series].

Cardiff City Council, Parks Department (n.d.2). *Caerphilly Common*. [Booklet in the Nature Trail Series].

Cardiff City Council, Parks Department (n.d.3). *Cefn-on*. [Booklet in the Nature Trail Series].

Cardiff City Council, Parks Department (n.d.4). *Coed Coesau-whips*. [Booklet in the Nature Trail Series].

Cardiff City Council, Parks Department (n.d.5). *Glamorgan Canal*. [Booklet in the Nature Trail Series].

Cardiff City Council, Parks Department (n.d.6). *Howardian Nature Reserve*. [Booklet in the Nature Trail Series].

Cardiff City Council, Parks Department (n.d.7). *Identifying Trees Series: Trees Company: Tree Trail I, II, III*.

Cardiff City Council, Parks Department (n.d.8). *Wenallt*. [Booklet in the Nature Trail Series].

Cardiff City Council, Parks Department [c.1970a]. *Take a walk in the park: Canton to Llandaff Walk*.

Cardiff City Council, Parks Department [c.1970b]. *Take a walk in the park: Rhymney walk*.

Cardiff City Council, Parks Department [c.1970c]. *Take a walk in the park: Taff Valley Walk*.

Cardiff Wildlife Group (1984). *Cardiff Wildlife Survey*.

Carter, P.W. (1955). Some Account of the History of Botanical Exploration in Glamorganshire. *Reports and Transactions of the Cardiff Naturalists' Society 1952–3*, **82**, 5–31.

Chambers, F.M. (1980). *Aspects of vegetational history and blanket peat initiation in upland South Wales*. Ph.D. Thesis. University College of Cardiff.

Chambers, F.M., Dresser, P.Q. & Smith, A.G. (1979). Radiocarbon dating evidence on the impact of atmospheric pollution on upland peats. *Nature*, **282**, 829–831.

Chapman, V.J. (1941). Studies in Salt-marsh ecology Section VIII. *Journal of Ecology*, **29**, 69–82.

Charman, K., Fojt, W. & Penny, S. (1986). C. Flora: iv Wales. In: *Research and Survey in nature conservation No. 3: Saltmarsh survey of Great Britain: Bibliography*. Nature Conservancy Council, Peterborough.

Chase, D.S. (1978). *Botanical and economic aspects of revegetating parts of the Lower Swansea Valley*. Ph.D. thesis. University College of Swansea.

Chatfield, J.E. (1981). *Nature Guide to Wales; Usborne Regional Guides*. Usborne Publishing Ltd., London.

Clapham, A.R., Tutin, T.G. & Moore, D.M. (1987). *Flora of the British Isles*, 3rd edn. Cambridge University Press, Cambridge.

Clapham, A.R., Tutin, T.G. & Warburg, E.F. (1962). *Flora of the British Isles*, 2nd edn. Cambridge University Press, Cambridge.

Clapham, A.R., Tutin, T.G. & Warburg, E.F. (1981). *Excursion Flora of the British Isles*, 3rd edn. Cambridge University Press, Cambridge.

Claridge, M.F., Blackman, R.L., Baker, C.R.B. (1970). *Haltica carduorum* Guerin introduced into Britain as a potential control agent for creeping thistle *Cirsium arvense* (L.) Scop. *The Entomologist*, **103**, 210–212.

Clark, J.H. (1853). *Cardiff and its neighbourhood*. [List of plants, p. 83].

Cliffe, C.F. (1847). *The Book of South Wales, The Bristol Channel, Monmouthshire and The Wye*. Hamilton, Adams & Co., London.

Condry, W.M. (1981). *The Natural History of Wales* (The New Naturalist No. 66). Collins, London.

Conolly, A.P. (1977). The distribution and history in the British Isles of some alien species of *Polygonum* and *Reynoutria*. *Watsonia*, **11**, 291–311.

Cooper, A. (1972). *An ecological study on some carboniferous limestone soils in South Wales*. Ph.D thesis. University College of Cardiff.

Cooper, A. (1976). The vegetation of carboniferous limestone soils in South Wales. *Journal of Ecology*, **64**, 147–155.

Cooper, A. & Etherington, J.R. (1974). The vegetation of carboniferous limestone soils in South Wales. 1. Dolomitization, Soil Magnesium Status and Plant Growth. *Journal of Ecology*, **62**, 179–190.

Coppock, J.T. (1964). *An Agricultural Atlas of England and Wales*. Faber and Faber Ltd., London.

Cornish, V. (1946). *The Churchyard Yew and immortality*. Muller, London.

County Borough of Cardiff [c.1923]. Wild Garden. In: *Guide to Roath Park and Catalogue*, pp. 57–58. Western Mail, Cardiff.

Coxhead, A. (1989). Cowslips, Cabbages and a Celtic Hill Fort – Cwm Colhuw. *Glamorgan Wildlife No. 9*, [p. 1]. (Newsletter of the Glamorgan Wildlife Trust).

Cowie, R.J. & Hinsley, S.A. (1987). Breeding success of Blue and Great Tits in suburban gardens [Trees, North Cardiff]. *Ardea*, **75**, 81–90.

Crampton, C.B. (1961). The influence of ice and wind on some Glamorgan soils. In: *Welsh Soils Discussion Group Report No. 2: Glaciation in Wales as related to soil profile development* (Ed. by J.A. Taylor), pp. 43–48. Geography Department, University College of Wales, Aberystwyth.

Crampton, C.B. (1963). The Development and Morphology of Iron Pan Podzols in Mid and South Wales. *Journal of Soil Science*, **14**, 282–302.

Crampton, C.B. (1966). An Interpretation of the Pollen and Soils in Cross-Ridge Dykes of Glamorgan. *Bulletin of the Board of Celtic Studies*, **21**, 376–390.

Crampton, C.B. & Webley, D.P. (1963). The Correlation of Prehistoric Settlement and Soils: Gower and the South Wales Coalfield. *Bulletin of the Board of Celtic Studies*, **20**, 326–337.

Crampton, C.B. & Webley, D.P. (1964). Preliminary Studies of the Historic Succession of Plants and Soils on Selected Archaeological Sites in South Wales. *Bulletin of the Board of Celtic Studies*, **20**, 440–449.

Crouch, C. (1891). *Lathyrus palustris* L. in Glamorganshire. *Journal of Botany*, **29**, 251.

Culver, S.J. (1976). The development of the Swansea Bay area during the past 20,000 years. *Gower*, **27**, 58–63.

Cundall, R.E. & F. (1902). Glamorganshire Plants. *Journal of Botany*, **29**, 251.

Curtis, J.P. (1981–1989). [Numerous articles and field meeting reports in *Cardiff Naturalists' Society Newsletter* and *Botanical Section Newsletter*].

Curtis, J.P. (1989). Report of Field Meetings, 1988: Porth, Glamorgan, 3rd July. *Botanical Society of the British Isles News No. 52*, pp. 43–44.

Dallman, A.A. (1941). *Crepis taraxacifolia* Thuill. in Wales. *North West Naturalist*, **16**, 203–206.

Dandy, J.E. ed. (1958). *The Sloane Herbarium*. British Museum (Natural History), London.

Dandy, J.E. (1969). *Watsonian Vice-counties of Great Britain*. Ray Society, London. [Includes maps of the vice-counties].

Dapre, P. (1970). Pond pollution in the Lower Swansea Valley. *Bios (Swansea)*, **4**, 71–82.

Davidson, D. & Withers, E. (1937). Reclamation of upland peat in Glamorgan. *Welsh Journal of Agriculture*, **13**, 256–259.

Davies, D. ed. (1979). Y daith i flaenau Cwm Nedd 20th April 1979. *Y Naturiaethwr*, **2**, 8–11.

Davies, D. ed. (1979). Traith Bro Gwyr Mehefin I af, 1979. Rhosili i Ogof Pafiland. *Y Naturiaethwr*, **2**, 12–15.

Davies, D. ed. (1980). Taith ym Mro Gwyr 11th April. Arweinydd: Ken Maddocks. *Y Naturiaethwr*, **4**, 19.

Davies, D. ed. (1983). Y Daith i Benrhyn-gwyr 6 Awst 1982. Arweinydd: Ken Maddocks. *Y Naturiaethwr*, **9**, 16–18.

Davies, D., Roberts, E. G. & Jones, A.O. (n.d.) ['Welsh Names of Plants' – in preparation].

Davies, M.S. & Singh, A.K. (1983). Population differentiation in *Festuca rubra* L. and *Agrostis stolonifera* L. in response to soil waterlogging. *New Phytologist*, **94**, 573–583.

Davies, P.W. (1963). Kenfig Dunes and Pool. A proposed local Nature Reserve in Glamorgan. *Glamorgan County Naturalists' Trust, Bulletin No. 2*, [pp. 3–5].

Davies, R. [c.1912]. Flora & Fauna of Merthyr Tydfil. In: *The Democrat's Handbook to Merthyr*, pp. 24–35.

Davies, W. (1815). *General View of the Agriculture and Domestic Economy of South Wales Vols. I & II*. Sherwood, Neely & Jones, London.

Davis, B.C. (1974). *A study of Salicornia species in the North Gower saltmarshes*. B.Sc. Thesis, University College of Wales, Swansea.

Davis, T.A.W., (1969). Report of the Welsh Region Autumn Meeting, 1967 at Department of Botany, University College, Swansea. *Botanical Society of the British Isles Welsh Region Bulletin No. 11*, pp. 5–6.

Dawson, H.J. (1986). *Petrorhagia nanteuilii* (Burnat) P.W. Ball & Heywood in Mid Glamorgan. *Watsonia*, **16**, 174–175.

Derrick, L.N., Jermy, A.C., & Paul, A.M. (1987). Checklist of European Pteridophytes. *Sommerfeltia*, **6** , i–xx, 1–94.

Dillwyn, L.W. (1802). Catalogue of the more rare plants found in the environs of Dover. *Transactions of the Linnean Society*, **6** , 177.

Dillwyn, L.W. (1828). *Rarer plants of Swansea.***.

Dillwyn, L.W. (1840). *Contributions towards a History of Swansea*. [Botany, Chapter 14]. Murray and Rees, Swansea.

Dillwyn, L.W. (1848). *Materials for a Fauna and Flora of Swansea and the Neighbourhood*. Swansea. (Unpublished).

Doe, M. (1976). Coed-yr-ysgol Nature Reserve (Howardian Nature Reserve). *Glamorgan Naturalists' Trust, Bulletin No. 15*, p. 39.

Donovan, E. (1805). *Descriptive excursions through South Wales*, **1**, 302, 378–379; **2**, 5, 13, 116, 129.

Doody, P. ed. (1984). *Focus on nature conservation No. 5:* Spartina anglica *in Great Britain*. Interpretative Branch, Nature Conservancy Council, Shrewsbury.

Doody, P. ed. (1985). *Focus on nature conservation No. 13: Sand dunes and their management – A report of a meeting held at the University of Swansea*. Nature Conservancy Council, Peterborough.

Douglas-Jones, N. (1972). Another New Reserve: Deborah's Hole, Gower. *Glamorgan County Naturalists' Trust, Bulletin No. 11*, p. 26.

Douglas-Jones, N. (1979). Ilston Reserve, Gower. *Glamorgan Naturalists' Trust, Bulletin and Annual Report No. 18*, pp. 18–19.

Drabble, E. (1927). Notes on the distribution of pansies in England and Wales. *Report of the Botanical Society and Exchange Club of the British Isles*, **8**, 191–204.

Druce, G.C. (1897). Short Notes: *Vicia villosa* Roth. var. *glabrescens* Koch. *Journal of Botany*, **35**, 362.

Druce, G.C. (1924). Plant Notes etc., for 1923 : 2210 R[umex] acetosella L. var. nova McLeanii, *Report of the Botanical Society and Exchange Club of the British Isles*, **7**, 60.

Druce, G.C. (1932). *The Comital Flora of the British Isles*. T. Buncle & Co., Arbroath.

Druery, C.T. ed. (1912). New Ferns: *Ceterach officinarum ramocristatum* Kirby. *British Fern Gazette*, **2**, 50.

Duckett, D.P. (1980). Let's Visit a Reserve. Coed-y-Bedw. *GNAT No. 1*, [pp. 3–4]. (Newsletter of the Glamorgan Naturalists' Trust).

Dunn, S.T. (1905). *Alien Flora of Britain*. West, Newman & Co., London.

Dyce, J.W. (1988). *Polypodium australe* 'Cambricum'. *Pteridologist*, **1**, 217–220.

Edees, E.S. & Newton, J. (1988). *Brambles of the British Isles*. The Ray Society, London.

Edington, J.M. (1968). New reserves in E. Glamorgan. *Glamorgan County Naturalists' Trust, Bulletin No. 7*, pp. 6–7.

Edington, J.M. & Douglas-Jones, N. (1970). New Reserves: Ogmore Down. *Glamorgan County Naturalists' Trust, Bulletin No. 9*, p. 25.

Edlin, H.L. ed. (1961). *Glamorgan Forests*. Forestry Commission Guide. Her Majesty's Stationery Office. London.

Edwards, E.E. (1938). Plant Pests in Glamorgan in 1936. *Report and Transactions of the Cardiff Naturalists' Society 1936*, **69**, 82–89.

Edwards, E.E. (1939). Plant Pests in Glamorgan in 1937. *Report and Transactions of the Cardiff Naturalists' Society 1937*, **70**, 46–52.

Elias, D.O. (1981). A provisional account of the flowering plants and ferns of Oxwich, 78pp. [Report to Nature Conservancy Council].

Ellis, R.G. (1972–1974). Plant Hunting in Wales I-III. *Amgueddfa*, **10; 13; 16**.

Ellis, R.G. (1973). The Flora of Glamorgan. *Botanical Society of the British Isles Welsh Region Bulletin No. 18*, pp. 8–9.

Ellis, R.G. [1974]. *Plant Hunting in Wales*. National Museum of Wales, Cardiff.

Ellis, [R.]G. (1976a). *Botany in Wales*. National Museum of Wales, Cardiff.

Ellis [R.]G. (1976b). A.E. Wade, M.Sc., F.L.S. *Nature in Wales*, **15**, 22–23.

[Ellis, R.G.]. (1980a). Report of field meeting Cardiff area, (leader M.E. Gillham) In: Annual General Meeting 1979. *Botanical Society of the British Isles Welsh Bulletin No. 32*, p. 4.

Ellis, [R.]G. (1980b). Botanical Recording in Wales – the last 500 years. *Botanical Society of the British Isles Welsh Bulletin No. 32*, pp. 6–7.

Ellis, R.G. (1983). *Flowering Plants of Wales*. National Museum of Wales, Cardiff.

Ellis, [R.]G., Ellis, P.G., Humphrey, D.R. & Wade, A.E. (1970). *Helianthemum apenninum* (L.) Mill. in Wales. In: Field Notes: Plants, in: *Nature in Wales*, **12**, 40–41.

Ellis, R.G. & Perry, A.R. (1990). Arthur Edward Wade (1895–1989) [obit.]. *Watsonia*, **18**, 113–114.

Ellis, R.G. & Waldren, S. (1984). *Frankenia laevis*, The Sea-Heath, in Glamorgan. (Exhibition Meeting, 1981), *Watsonia*, **14**, 229.

Emery, F.V. (1954). The value of Nicholaston Wood. *Gower*, **7**, 6–8.

Emery, F.V. (1965). Edward Lhuyd and some of his Glamorgan correspondents: A view of Gower in the 1690's. *Transactions of the Honourable Society of Cymmrodorion 1965*, 59–114.

Emery, F.V. (1975). The early cultivation of Clover in Gower. *Gower*, **26**, 45–49.

Etherington, J.R. (1967). Studies of nutrient cycling and productivity in oligotrophic ecosystems 1. Potassium and wind blown sea spray in a South Wales dune grassland. [Kenfig Burrows]. *Journal of Ecology*, **55**, 743–752.

Etherington, J.R. (1977). The effect of limestone quarrying dust on a limestone heath in South Wales. *Nature in Wales*, **15**, 218–223.

Etherington, J.R. (1981). Limestone heaths in South-West Britain: their soils and the maintenance of their calcicole-calcifuge mixtures. *Journal of Ecology*, **69**, 277–294.

Etherington, J.R. (1983). Control of germination and seedling morphology by ethene: differential responses, related to habitat of *Epilobium hirsutum* L. and *Chamerion angustifolium* (L.) J. Holub. *Annals of Botany*, **52**, 653–658.

Etherington, J.R. (1984). Comparative studies in plant growth and distribution in relation to waterlogging. X. Differential formation of adventitious roots and their experimental excision in *Epilobium hirsutum* and *Chamerion angustifolium*. *Journal of Ecology*, **72**, 389–404.

Etherington, J.R. (1988). Limestone heaths in Britain. *Plants Today*, **1**, 177–182.

Evans, C.E. (1986). *Effect of soil water potential and pH on germination and establishment of some British plants*. Ph.D thesis. University College of Cardiff.

Evans, C.J.O. (1938). *Glamorgan: Its History and Topography*. William Lewis (Printers) Ltd, Cardiff. 2nd edn revised, 1943.

Evans, J. (1804). *Letters written during a tour through South Wales*. C. and R. Baldwin, London.

Evans, J. (1831). Notes made by Rev. J. Evans. Llandough*.

Evans, J.O. [1985]. A Site Worth Preserving. [Cyfarthfa Canal, Cwm Woods and Cwm Pit Area, Merthyr Tydfil]. *Merthyr Tydfil and District Naturalists' Society Newsletter [No. 10]*, pp. 12–15 and map.

Evans, L. (1973). An Ecological Study of Broad Pool in 1960. *Gower*, **24**, 50–53.

Evans, T.C. (1887). *History of Llangynwyd Parish* : The Old Castle, pp. 34–36. Llanelly.

Evans, T.G. (1984). *Draba aizoides* (Yellow Whitlowgrass). *Botanical Society of the British Isles Welsh Bulletin No. 40*, p. 3.

Evans, W.F. (1905). Short Notes: '*Cotoneaster microphylla*'. *Journal of Botany*, **43**, 244.

Feachem, R.W. (1973). Ancient agriculture in the highlands of Britain. *Proceedings of the Prehistoric Society*, **39**, 332–353.

Fearn, G.M. (1972). The distribution of intraspecific chromosome races of *Hippocrepis comosa* L. and their phytogeographical significance. *New Phytologist*, **71**, 1221–1225.

Fearn, G.M. (1977). A morphological and cytological investigation of *Cochlearia* populations on the Gower Peninsula, Glamorgan. *New Phytologist*, **79**, 455–458.

Ferns, P.N. (1987). The Taff Estuary and its Birds. *Transactions of the Cardiff Naturalists' Society 1979–1986*, **100**, 13–25.

Firth, J.N.M. & Caine, C.V. (1978). *Environmental Bibliography of Wales 1900–1976*. Her Majesty's Stationery Office, Cardiff.

Fisher, J. (1987). *Wild Flowers in Danger*. H.F. & G. Witherby Ltd., London (pbk edn 1989).

Flatholm Project (n.d.). *Flatholm*. County of South Glamorgan & The Manpower Services Commission.

Flower, T.B. (1839). Catalogue of Swansea plants. *Loudon's Magazine of Natural History [n.s.]*, **3**, 561–564.

Flower, T.B. (1852). On the Glamorganshire locality for *Cnicus tuberosus* [*Cirsium tuberosum*]. *Phytologist*, **4**, 519.

Flower, T.B. (1877). Plants of Glamorganshire. *Journal of Botany*, **15**, 180.

Flower, T.B. (1886). *Helleborus foetidus* in Glamorganshire. *Journal of Botany*, **24**, 83.

Flower, T.B. (1890). Short Notes: *Lepidium draba* L., in South Wales. *Journal of Botany*, **28**, 218.

Flower, T.B. & Lees, E. (1842). Additions to the list of plants met with in the neighbourhood of Swansea. *Phytologist*, **1**, 377–380.

Ford, B.J. (1962). *The Thaw Estuary, an ecological Survey*. [On behalf of Glamorgan County Naturalists' Trust acting for the Nature Conservancy].

Ford, M.A. & Kay, Q.O.N. (1985). The genetics of incompatibility in *Sinapis arvensis*. *Heredity*, **54**, 99–102.

Forestry Commission (n.d.1). *The Forestry Commission offer a walk in the Tair Onen Forest*.

Forestry Commission (n.d.2). *Mill Wood Forest Walks*. South Wales Conservancy, Resolven, Neath.

Forestry Commission (1949). *Britain's Forests : Rheola*. His Majesty's Stationery Office, London.

Forestry Commission (1983). *Census of Woodlands and Trees 1979–1982*. Forestry Commission Conservancy South Wales. Forestry Commission, Edinburgh.

Fox, C. (1943). The Climate, Flora and Fauna of Britain: (B) The Flora of Britain. In: *The Personality of Britain*, pp. 53–62. 4th edn. [1st edn (1932)]. National Museum of Wales, Cardiff.

Fraser, J. (1886). *Helleborus foetidus* in Glamorganshire. *Journal of Botany*, **24**, 23.

Fry, D. (1888). Glamorganshire plants. *Journal of Botany*, **26**, 57.

Gadgil, P.D. (1965). The soil biology of the Lower Swansea Valley. *Lower Swansea Valley Project, Study Report No. 10*. University College of Swansea.

Gadgil, R.L. (1969). Tolerance of heavy metals and the reclamation of industrial waste [Lower Swansea Valley]. *Journal of Applied Ecology*, **6**, 247–259.

Gamwell, S.C. (1880). Some of the rarer flowering plants and ferns of the district. In: *British Association Handbook for Swansea*, pp. 165–166.

[Garner, R.] (1867). *Holiday excursions of a naturalist* [Gower, pp. 119–142]. Robert Hardwicke, London.

Garrard, I. & Streeter, D. (1983). *The Wild Flowers of the British Isles*. Macmillan, London.

Garrett-Jones, R. (1961). Bracken grazing by sheep. *Agriculture*, **68**, 510. (Journal of the Ministry of Agriculture).

Gemmell, R.P. (1976). The maintenance of grassland on smelter wastes in the Lower Swansea Valley. I. Blast furnace slag. *Journal of Applied Ecology*, **13**, 285–294.

Gemmell, R.P. (1977). *Colonisation of industrial wasteland. Studies in Biology No. 80*. Edward Arnold (Publishers) Ltd., London.

Gemmell, R.P. & Goodman, G.T. (1980). The maintenance of grassland on smelter wastes in the Lower Swansea Valley, Wales, U.K. 3. Fine smelter waste. *Journal of Applied Ecology*, **17**, 461–468.

George, T.N. (1930). The Submerged Forest of Gower. *Proceedings of the Swansea Scientific Society*, **1**, 100–108.

Gillham, M.E. (1964). The vegetation of local coastal gull colonies. *Transactions of the Cardiff Naturalists' Society*, **91**, 23–33.

Gillham, M.[E.] (1965). Pysgodlyn Mawr, Welsh St. Donats. *Glamorgan County Naturalists' Trust, Bulletin No. 4*, pp. 10–12.

Gillham, M.[E.] (1968). Cwm George. In: New Reserves in E. Glamorgan. *Glamorgan County Naturalists' Trust, Bulletin No. 7*, p. 8.

Gillham, M.E. (1969a). A New Nature Reserve: Glamorganshire Canal between Tongwynlais and Whitchurch and the adjacent Long Wood. *Transactions of the Cardiff Naturalists' Society*, **94**, 4–40.

Gillham, M.E. (1969b). Glamorganshire Canal: Tongwynlais to Whitchurch. *Glamorgan County Naturalists' Trust, Bulletin No. 8*, pp. 18–20.

Gillham, M.E. (1969c). Taf Fechan Gorge. Proposed Nature Reserve N.E. of Merthyr Tydfil. *Glamorgan County Naturalists' Trust, Bulletin No. 8*, pp. 27–29.

Gillham, M.E. [c.1971]. Aliens in the Taff Vale. [Notes by Dr. M.E. Gillham]. Library, National Museum of Wales, Cardiff.

Gillham, M.E. (1972). The Llyn Fach Reserve, Craig y Llyn, between Rhigos and Blaen Rhondda. *Glamorgan County Naturalists' Trust, Bulletin No. 11*, pp. 11–14.

Gillham, M.E. (1973a). Eglwys Nunydd Reservoir. *Glamorgan County Naturalists' Trust, Bulletin No. 12*, pp. 19–22.

Gillham, M.E. (1973b). Biological Notes from the Scientific Secretary – East Glamorgan. *Glamorgan County Naturalists' Trust, Bulletin No. 12*, pp. 35–38.

Gillham, M.E. (1974). *Flatholm Island*. Extracted from notes loaned by Dr M.E. Gillham. MS in Cardiff Central Reference Library.

Gillham, M.E. (1975). Taf Fechan Gorge Local Nature Reserve. *Glamorgan Naturalists' Trust, Bulletin & Annual Report No. 14*, pp. 29–31.

Gillham, M.E. (1976). Botanical Society of the British Isles Excursion on the Vale of Glamorgan coast, 22nd May, 1976. *Botanical Society of the British Isles Welsh Region Bulletin No. 25*, pp. 6–8.

Gillham, M.E. (1977a). *The Natural History of Gower*. D. Brown & Sons Ltd., Cowbridge and Bridgend.

Gillham, M.E. (1977b). Birthworthy Birthwort in Mid Glamorgan. *Glamorgan Naturalists' Trust, Bulletin and Annual Report No. 16*, p. 10.

Gillham, M.E. (1977c). Environs of Kenfig Pool at the start of 1977. *Glamorgan Naturalists' Trust, Bulletin and Annual Report No. 16*, pp. 20–24.

Gillham, M.E. (1978a). From fresh water to salt: sand dune pools at Merthyr Mawr. *Glamorgan Naturalists' Trust, Bulletin and Annual Report No. 17*, pp. 15–16.

Gillham, M.E. (1978b). From salt water to fresh: Aberthaw Lagoon. *Glamorgan Naturalists' Trust, Bulletin and Annual Report No. 17*, pp. 30–31.

Gillham, M.E. (1979a). Plants of Glamorgan. *Botanical Society of the British Isles Welsh Bulletin No. 30*, pp. 5–8.

Gillham, M.E. (1979b). A new nature reserve on the River Ely. *Glamorgan Naturalists' Trust, Bulletin and Annual Report No. 18*, pp. 19–22.

Gillham, M.E. (1980a). New nature Reserve at East Aberthaw, South Glamorgan. *Glamorgan Naturalists' Trust, Bulletin and Annual Report No. 19*, pp. 18–21.

Gillham, M.E. (1980b). Bob-tailed Drug Addicts of Glamorgan's Heritage Coast [*Helleborus foetidus*]. *Glamorgan Naturalists' Trust, Bulletin and Annual Report No. 19*, pp. 28–29.

Gillham, M.E. (1982a). *Swansea Bay's Green Mantle*. D. Brown & Sons Ltd., Cowbridge.

Gillham, M.E. (1982b). A Walk on Merthyr Mawr Dunes. *GNAT, No. 4*, p. 7. (Newsletter of the Glamorgan Naturalists' Trust).

Gillham, M.E. (1983–1984). Let's Visit a Reserve: Aberdare Canal north of Cwmbach. *GNAT No. 8*, pp. 5–6. (Newsletter of the Glamorgan Naturalists' Trust).

Gillham, M.E. (1984a). Botanical skirmish on Flatholm. *GNAT No. 9*, pp. 6–7. (Newsletter of Glamorgan Naturalists' Trust).

Gillham, M.E. [1984b]. Taf Fechan in the drought of August 1983. *Merthyr Tydfil and District Naturalists' Society Newsletter [No. 4]*, pp. 10–11.

Gillham, M.[E.] (1985), Equinoctual springs. *GNAT No. 12*, [p. 8]. (Newsletter of the Glamorgan Trust for Nature Conservation).

Gillham, M.E. (1987). *The Glamorgan Heritage Coast Wildlife Series. Volume 1: Sand Dunes*. Heritage Coast Joint Management and Advisory Committee.

Gillham, M.E. (1989a). *The Glamorgan Heritage Coast Wildlife Series. Volume 2: Rivers*. Glamorgan Heritage Coast Project, Southerndown, Mid Glamorgan.

Gillham, M.E. (1989b). Melincourt Falls. *Glamorgan Wildlife No. 8*, [p. 1, 5]. (Newsletter of the Glamorgan Wildlife Trust).

Gillham, M.[E.], Perkins, J. & Thomas, C. (1979). *A Guide to the Historic Taff Valley – from Quakers' Yard to Aberfan*. Merthyr Tydfil & District Naturalists' Society. D. Brown & Sons Ltd., Cowbridge and Bridgend.

Gilmour, J.S.L. (1933a). Notes from Glamorgan. *Journal of Botany*, **71**, 16–17.

Gilmour, J.S.L. (1933b). A new variety of *Prunella vulgaris*. *Journal of Botany*, **71**, 320–321.

Gilmour, J.S.L., Thomas, E.M. & Vachell, E. (1936). Short Notes: *Rumex cuneifolius* Campd. *Journal of Botany*, **74**, 56.

Glading, P.R. (1984). *Ecological Studies upon Carboniferous Limestone Vegetation in South Wales*. Ph.D. thesis. University College of Cardiff.

Glamorgan County Naturalists' Trust (n.d.). *Nature and Conservation in Glamorgan*.

Glamorgan County Naturalists' Trust (1973). *Four Nature Walks in Glamorgan*.

Glamorgan Heritage Coast (n.d.1). *1. Nature Walk; 2. Country Walk; 3. Cliff Walk; 4. Beach Walk*.

Glamorgan Heritage Coast (n.d.2). *The Story of Dunraven*.

Glamorgan Heritage Coast (n.d.3). *The Story of Summerhouse Point*.

Glamorgan Heritage Coast (n.d.4). *The Wildlife of Merthyr Mawr*.

Glamorgan Heritage Coast (n.d.5). *Wild Flowers of the Coast*.

Glamorgan Heritage Coast [c.1975]. *Glamorgan Heritage Coast: Plan Statement*.

Glamorgan Heritage Coast [c.1981]. *Cwm Marcross Nature Trail*.

Glamorgan Naturalists' Trust (n.d.1). *Naturalists' Trust Reserves in Gower and Wild Orchids of Gower*. Reprint from *Gower*, December 1968.

Glamorgan Naturalists' Trust (n.d.2). *Nature Walk in Glamorgan: Four walks in Gower*.

Glamorgan Naturalists' Trust (n.d.3). *Nature Walks in Glamorgan: Three Walks in the Coalfield*.

Glamorgan Naturalists' Trust (n.d.4). *[Nature] Reserve Cards*.

[GLAMORGAN PLANT RECORDS]. *Nature in Wales* 1956, **2**, 230–231, 353–356; 1957, **3**, 397–400; 1964, **9**, 69; 1965, **9**, 216, 223; 1970, **12**, 35; 1971, **12**, 171; 1972, **13**, 42; 1974, **14**, 49; 1976, **15**, 28; 1977, **15**, 143; 1978, **16**, 58; 1979, **16**, 210; *BSBI Welsh Regional Bulletin No. 33* 1980, p.3; *Nature in Wales (n.s.)* 1982, **1** Supplement, p.3; *BSBI Welsh Regional Bulletin No. 35* 1981, pp. 2–4; *BSBI Welsh Bulletin No. 39 Supplement* 1984, pp. 2–6; *No. 41 Supplement* 1985, pp. 6–8; *BSBI Welsh Bulletin No. 42* 1985, pp. 18–23; *No. 44* 1986, pp. 17–19; *No. 49* 1990, pp. 19–21.

Glamorgan Trust for Nature Conservation (n.d.). *Visitor Pack*.

Glamorgan Wildlife Trust (n.d.). *Coed y Bedw Nature Reserve, Pentyrch*.

Glamorgan – Gwent Archaeological Trust [1979]. Flat Holm – A Preliminary Survey. In: *The Glamorgan – Gwent Archaeological Trust Ltd. Annual Report 1978–1979*, pp. 94–97.

[Glyncornel Environmental Centre] (n.d.1). *Cefn Glas Short Walk*.

[Glyncornel Environmental Centre] (n.d.2). *Glyncornel Lake Walk*.

[Glyncornel Environmental Centre] (n.d.3). *Graig Llyn Ridge Walk*.

[Glyncornel Environmental Centre] (n.d.4). *Lluest Wen/Cefn Bryn Gelli*.

[Glyncornel Environmental Centre] (n.d.5). *Woodlands Walk*.

Godwin, H. (1940a). A boreal transgression of the sea in Swansea Bay. (Data for the study of Post-glacial history VI). *New Phytologist*, **39**, 308–321.

Godwin, H. (1940b). Pollen Analysis and Forestry History of England and Wales. *New Phytologist*, **39**, 370–400.

Godwin, H. (1943). Coastal Peat Beds of the British Isles and North Sea. *Journal of Ecology*, **31**, 199–247.

Godwin, H. (1975). *History of the British Flora*, 2nd edn. Cambridge University Press. [1st edn, 1956; Pbk. 2nd edn 1984].

Golding, H.F.A.H. (1919). *The nature and history of the vegetation of part of the cliffs on the northern shore of the Severn Estuary. The halophytic vegetation of the pitching of the left bank of the Ely tidal harbour, Penarth Dock*. M.Sc. thesis, University College of Cardiff.

Goodman, G.T. [1955a]. *The Vegetation of Pwll-du Head and Bishopston Valley, Gower, Glamorgan.* MS. University College of Swansea. Unpublished. Copy of MS in Library, National Museum of Wales, Cardiff.

Goodman, G.T. [1955b]. *The Vegetation of the country between Oxwich and Nicholaston in the Gower Peninsula, Glamorgan.* MS. University College of Swansea. Unpublished. Copy of MS in Library, National Museum of Wales, Cardiff.

Goodman, G.T. (1957a). Whiteford Burrows [Field Meeting]. *Proceedings of the Botanical Society of the British Isles,* **2,** 414–415.

Goodman, G.T. (1957b). Kenfig Burrows, [Field Meeting]. *Proceedings of the Botanical Society of the British Isles,* **2,** 416.

Goodman, G.T. (1961). Plants and Man at Oxwich. *Gower,* **13,** 32–40.

Goodman, G.T. (1962). Broadpool. *Gower,* **15,** 71.

Goodman, G.T. (1963). *Plant life in Gower.* Gower Society, Swansea.

Goodman, G.T. (1965a). Flowers and trees. In: *A Guide to Gower,* pp. 80–84. The Gower Society.

Goodman, G.T. (1965b). Whitford Burrows. *The Glamorgan County Naturalists' Trust, Bulletin No. 4,* pp. 7–9.

Goodman, G.T. (1969). The Lower Swansea Valley – a text for the conservationist. *The Glamorgan County Naturalists' Trust, Bulletin No. 8,* pp. 10–16.

Goodman, G.T. & Bray, S.A. (1975). *Ecological Aspects of the Reclamation of Derelict and Disturbed Land – an annotated bibliography.* National Environment Research Council, London.

Goodman, G.T., Edwards, R.W. & Lambert, J.M. eds. (1965). *Ecology and the Industrial Society,* 5th Symposium of the British Ecological Society. Blackwell Scientific Publications, Oxford.

Goodman, G.T. & Gemmell, R.P. (1978). The maintenance of grassland on smelter wastes in the Lower Swansea Valley: II. Copper smelter waste. *Journal of Applied Ecology,* **15,** 875–884.

Goodman, G.T., Pitcairn, C.E.R. & Gemmell, R.P. (1973). Ecological factors affecting growth on sites contaminated with heavy metals. In: *Ecology and Reclamation of Devastated Land,* Vol. 2 (Ed. by R.J. Hutnik and G. Davis), pp. 149–173. Gordon and Breach, New York and London.

Goodman, G.T. & Roberts, T.M. (1971). Plants and soils as indicators of metals in the air. *Nature,* **231,** 287–292.

Gower Field Education Project (n.d.). *Caswell Valley and Bishops Wood Teachers Guide.* West Glamorgan County Council Education Department & Nature Conservancy Council.

Gray, A.J. & Benham, P.E.M. [eds.] (1990). *Spartina anglica* - a research review. Institute of Terrestrial Ecology research publication no. 2. Her Majesty's Stationery Office, London.

Gray, T. (1909). *The Buried City of Kenfig,* pp. 13–15, 32, 184. T. Fisher Unwin, London and Leipsic.

Gregory, R.P.G. & Bradshaw, A.D. (1965). Heavy metal tolerance in populations of *Agrostis tenuis* Sibth. and other grasses. *New Phytologist,* **64,** 131–143.

Gregson, D. (1979). Vegetational succession and climax communities. *Glamorgan Naturalists' Trust, Bulletin and Annual Report No. 18,* pp. 23–25.

Gregson, D. (1980). Lavernock Point Nature Reserve, Penarth. In: Vegetational succession and climax communities. *Nature in Wales,* **17,** 23–28.

Grenfell, H.E. (1978). Winter Trees. *Gower,* **29,** 27–30.

Grenfell, H.E. & Hayward, J. (1973). Some Gower Wildflowers. *Gower,* **24,** 37–43.

Grimes, W.F. & Hyde, H.A. (1937). A prehistoric hearth at Radyr, Glamorgan and its bearings on the nativity of beech (*Fagus sylvatica* L.) in Britain. *Transactions of the Cardiff Naturalists' Society 1935,* **168,** 46–54.

Groenhof, A.C. (1984). *Physiological and biochemical responses to water stress in British vascular plants.* Ph.D. thesis. University College of Cardiff.

Grove, R.H. ed. (1985). *The SSSI Handbook, Volume 1. The Provisional Lists for Great Britain.* Cambridge Friends of the Earth, Cambridge.

Groves, E.W. (1958). *Hippophae rhamnoides* L. in the British Isles. *Proceedings of the Botanical Society of the British Isles,* **3,** 1–21.

Groves, H. & J. (1907). *Ononis reclinata* L. in Glamorgan. *Journal of Botany,* **45,** 280–281.

Gunn, I.D.M. (1986). *Biology and Control of Japanese Knotweed.* Unpublished thesis, University College of Cardiff.*

Gunther, R.T. (1945). Life and Letters of Edward Lhwyd. *Early Science in Oxford,* **14.** Oxford. Reprinted 1968, London.

Gutch, J.W.G. (1839). On the medical topography, statistics, climatology and natural history of Swansea, Glamorganshire. *Transactions of The Provincial Medical and Surgical Association, 7*, (for 1838) 249–282.

Gutch, J.W.G. (1844a). A list of plants met with in the neighbourhood of Swansea, Glamorganshire, *The Phytologist*, **1**, 104–108 (1841), 118–121, 141–145, 180–187 (1842). Additions and corrections, **1**, 380 (1842).

Gutch, J.W.G. (1844b). [*Polygonum raii* not *P. maritimum*, Swansea]. *Phytologist*, **1**, 205 (1842).

Hallett, H.M. (1912). Index to Flora of Glamorgan (Ed. by A.H. Trow). *Report and Transactions of the Cardiff Naturalists' Society*, Appendix, **44**, 195–209.

Hallett, H.M. (1947). General Index to The Transactions of the Cardiff Naturalists' Society, **1–70**, 1867–1868 to 1937. [Including field meetings of the Society arranged alphabetically, pp. 120–125]. *Transactions of the Cardiff Naturalists' Society 1939–1945*, **72–78**. [126pp].

Halliday, G. & Beadle, M. (1983). *Consolidated Index to Flora Europaea*. Cambridge University Press, Cambridge, New York & Melbourne.

Hambury, J. (1963). Practical Conservation work at Broad Pool. *Glamorgan County Naturalists' Trust, Bulletin No. 2*, [pp. 12–13].

Hambury, H.J. [1971]. Botanical Highlights in Glamorgan 1961–1971. *Glamorgan County Naturalists' Trust, Bulletin No. 10*, pp. 11–14.

Harris, C. (1974a). Oxwich Burrows. *Gower*, **25**, 48–56.

Harris, C. (1974b). Wind speed and sand movement in a coastal dune environment [Oxwich]. *Area*, **6**, 243–249.

Harris, H. (1905). *The Flora of the Rhondda*. Compiled for the Rhondda Naturalists' Society. Burleigh Ltd., Bristol.

Harrison, S.G. (1968a). Field Notes: Plants: A New Zealand Willow-herb in Wales. *Nature in Wales*, **11**, 74–78.

Harrison, S.G. (1968b). Our Nameless Cultivars [*Polypodium australe* 'Cambricum']. *British Fern Gazette*, **10**, 36–37.

Harrison, S.G. (1970). Flora of Glamorgan. *Botanical Society of the British Isles Welsh Region Bulletin No. 12*, pp. 3–5.

Harrison, S.G. (1974). Harold Augustus Hyde (1892–1973) [obit.]. *Watsonia*, **10**, 113–114.

Harrison, S.G. (1977). The National Museum of Wales, Department of Botany and Welsh National Herbarium. *Botanical Society of the British Isles Welsh Region Bulletin No. 27*, pp. 2–10.

Harrison, S.G. (1980). Ferns in Glamorgan. *Glamorgan Naturalists' Trust, Bulletin and Annual Report No. 19*, pp. 34–35.

Harrison, S.G. (1983). The Collections of the Welsh National Herbarium – National Museum of Wales. *Nature in Wales (n.s.) 1982*, **1**(2), 39–55.

Harrison, S.G. (1985). *Index of Collectors in the Welsh National Herbarium*. National Museum of Wales. Cardiff.

Hatton, R.H.S. (1963). Oxwich National Nature Reserve. *Glamorgan County Naturalists' Trust, Bulletin No. 2*, [pp. 5–6].

Hatton, R.H.S. (1968). Some Problem Species. *Glamorgan County Naturalists' Trust, Bulletin No. 7*, pp. 10–13.

Hatton, R.H.S. (1969a). Broad Pool. Glamorgan County Naturalists' Trust, Bulletin No. 8, p. 30.

Hatton, R.H.S. (1969b). Visit to Port Eynon/Overton Cliffs. *Glamorgan County Naturalists' Trust, Bulletin No. 8*, p. 34.

Hatton, R.H.S. (1972a). *Saltmarshes of Gower*. [Glamorgan County Naturalists' Trust Booklet].

Hatton, R.H.S. (1972b). Three Glamorgan County Naturalists' Trust Nature Reserves. *Nature in Wales*, **13**, 132–134.

Haworth, C.C. (1988). *An Annotated list of British and Irish Dandelions*. C.C. Haworth.

Haworth, C.C. & Rundle, A.J. (1986). *An Annotated List of British and Irish Dandelions*. C.C. Haworth.

Hayward, J. (1972). Ferns of Gower. *Gower*, 24–27.

Hayward, J. (1979). Present-day flora of West Glamorgan commonland. In: *Problems of Commonland* (Ed. by E.M. Bridges), 3:2/1–3:2/6. University College of Swansea.

Healey, I.N. (1967). An ecological study of temperatures in a Welsh Moorland Soil, 1962–1963. *Journal of Animal Ecology*, **36**, 425–434.

Hemsley, W.B. (1908). The history of three casual dodders. *Journal of Botany*, **46**, 241–247.

[Henderson, H.] [1986]. *The Flowers of Llandaff*, Llandaff Society Occasional Paper: I. Llandaff Society, Cardiff.

Henderson, H. ed. (1988). *Cardiff Naturalists' Society, Botanical Section Newsletter No. 7*.

Henderson, H. ed. (1988–1989). *Cardiff Naturalists' Society Newsletter Nos. 1–4*.

Henderson, R. (1979). A little-known wood. [Forest Ganol]. *Glamorgan Naturalists' Trust, Bulletin and Annual Report No. 18*, pp. 28–29.

Henderson, R.A. (1972). *The Wenallt, an ecological survey*. Conservation Group, Cardiff Parks and Allotments Department, Cardiff City Council.

Henderson, R.A. (1974). Glamorganshire Canal Nature Reserve: Additional Information, 1971–1972. *Transactions of the Cardiff Naturalists' Society*, **96**, 29–34.

Hepburn, I. (1952). *Flowers of the Coast* (The New Naturalist No. 24). Collins, London.

Heron, T. (1968). The Wild Orchids of Gower. *Gower*, **19**, 25–29.

Hiern, W.P. (1901). *Limosella aquatica* L. var. *tenuifolia* Hook. fil. *Journal of Botany*, **39**, 336–339.

Hill, J. (1775). *Notes in Ray's Synopsis*, **3**, Library, Department of Botany, Oxford.

Hilton, K.J. (1963). The Lower Swansea Valley Project. *Geography*, **48**, 296–299.

Hilton, K.J. ed. (1967). *The Lower Swansea Valley Project*. Longmans, London.

Hilton, K.J., Goodman, G.T. & Ward, W.J. (1971). The Lower Swansea Valley Project. In: *Swansea and its Region* (Ed. by W.G.V. Balchin), pp. 365–380. University College, Swansea.

Hinde, T. (1985). *Forests of Britain*. V. Gollancz. [Sphere Books Ltd. (1987)].

Hipkin, C.R. (1975). The river and waterfall flora of the Neath Valley. *Bios (Swansea)* **6**, 47–56.

Hoare, F. & Davis, G. [c.1989]. *Coed-y-bwl Nature Reserve, Castle-upon-alun*. Glamorgan Wildlife Trust.

Hodges, J. (1981). *Land North of Howardian Nature Reserve and Llanedeyrn Interchange*. MS. Cardiff City Council Leisure and Amenities Department.

Hodges, J.E. (1981). *The Glamorgan Canal Nature Reserve – Draft Management Plan*. Leisure and Amenities Department, Cardiff City Council.

Holmes, N.T.H. (1983). *Focus on nature conservation No. 4: Typing British Rivers according to their flora*. Nature Conservancy Council, Attingham Park, Shrewsbury.

Holmes, N.[T.H.] & Newbold, C. (1984). *Focus on nature conservation No. 9*: River plant communities – reflectors of water and substrate chemistry. Nature Conservancy Council, Shrewsbury.

Hood, C. & Williams, M. (n.d.). *Rosehill Quarry*. [Swansea]. Rosehill Quarry Project.

Hooker, W.J. ed. (1831–1834). *Supplement to the English Botany of the late Sir J.E. Smith & Dr Sowerby*. 2 vols, London.

Hooper, L.J. & Jones, R.G. (1966). *Lower Swansea Valley project. Report on grass seeding trials conducted by National Agricultural Advisory Service, 1962–1966*. N.A.A.S. Ministry of Agriculture Fisheries and Food, Cardiff.

Hope-Simpson, J.F. & Jefferies, R.L. (1966). Observations relating to vigour and debility in marram grass (*Ammophila arenaria* (L.) Link). *Journal of Ecology*, **54**, 271–274.

Hopkins, N. (1971). An Investigation of heavy metal tolerance of *Rhizobium trifolii* isolated from clover taken from various sources. *Bios (Swansea)*, **5**, 83–88.

Hopkins, T.J. (1961). Welsh St. Donat's: A Border Vale Parish. In: *The Garden of Wales* (Ed. by S. Williams). *The Vales Series vol. III*. D. Brown & Sons Ltd., Cowbridge.

Howarth, W.O. (1919). *Festuca rubra* near Cardiff. *New Phytologist*, **18**, 263–286.

Howarth, W.O. (1921). Notes on the habitats and ecological characters of three sub-varieties of *Festuca rubra*. *Journal of Ecology*, **8**, 216–231.

Hubbard, C.E. (1954). *Grasses: A Guide to their Structure, Identification, Uses and Distribution in the British Isles*. Penguin, Harmondsworth, Baltimore and Ringwood. 2nd edn (1968).

Hubbard, C.E. & Hubbard, J.C.E. (1984). *Grasses: A guide to their Structure, Identification, Uses and Distribution in the British Isles*. 3rd edn. Penguin Books, Harmondsworth, New York, Ringwood, Markham and Auckland.

Hubbard, J.C.E. & Stebbings, R.E. (1967). Distribution, Dates of Origin and Acreage of *Spartina townsendii* (s.l.) Marshes in Great Britain. *Proceedings of the Botanical Society of the British Isles*, **7**, 1–7.

Hudson, W. (1762). *Flora Anglica*. London.

Hughes, M. (1983). National Nature Reserves in Wales: a systematic survey. 3. Oxwich National Nature Reserve, West Glamorgan. *Nature in Wales (n.s.)* 1982, **1**, 27–33.

Hughes, M.R. (1983). The Dune Slacks of Whiteford National Nature Reserve, *Gower*, **34**, 19–29.

Hussien, F. (1949). Exhibits: Chromosome races of *Cardamine pratensis* in the British Isles. In: *British Flowering Plants and modern systematic methods* (Ed. by A.J. Wilmott), p. 77 and Plate XV. Botanical Society of the British Isles, London.

Hutchinson, G. ed. (1981–1987). *Cardiff Naturalists' Society, Botanical Section Newsletter Nos. 1–6.*

Hutchinson, G. (1985). Glamorgan Canal Nature Reserve, Whitchurch, Cardiff. *Llanelli Naturalists' Newsletter – March 1985.* [p. 15].

Hutchinson, G. (1988). Report of Field Meetings, 1987: Fontygary, Glamorgan, 30th May. *Botanical Society of the British Isles News No. 48*, p. 49.

Hutchinson, G. (1990). Expanded draft of bibliography for *Flora of Glamorgan*. MS in Library, National Museum of Wales, Cardiff.

Hyde, H.A. (1934). Some notable trees in Glamorgan. *Quarterly Journal of Forestry*, **28**, 101–108.

Hyde, H.A. (1936). Trees and Shrubs. In: *Glamorgan County History*, **1**, 217–231. Cardiff.

Hyde, H.A. (1937). Plant distribution in the Cardiff district. In: *The Book of Cardiff* (Ed. by C.J. Evans), pp. 28–38. National Association of Head Teachers.

Hyde, H.A. (1938a). Report on Plant Remains. In: Grimes, W.F. : A Barrow on Breach Farm, Llanbleddian, Glamorgan. Appendix II. In: *Proceedings of the Prehistoric Society (n.s.)*, **4**, 120–121.

Hyde, H.A. (1938b). Trees and Plantations at Margam Abbey, Glamorgan. *Quarterly Journal of Forestry*, **32**, 224–225.

Hyde, H.A. (1938c). On a peat bed at the East Moors, Cardiff. *Transactions of the Cardiff Naturalists' Society 1936*, **69**, 39–48.

Hyde, H.A. (1938d). Report on Plant Remains. In: Grimes, W.F. : A Barrow on Breach Farm, Llanbleddian, Glamorgan. Appendix II. In: *Transactions of the Cardiff Naturalists' Society 1936*, **69**, 68.

Hyde, H.A. (1939). Part 2: The Palaeobotanical Evidence for the date of the Hoard. In: A Second Cauldron and an Iron Sword from the Llyn Fawr Hoard, Rhigos, Glamorganshire. *The Antiquaries Journal*, **19**, 391–401.

Hyde, H.A. (1940). On a peat-bog at Craig-y-Llyn, Glamorgan. *New Phytologist*, **39**, 226–233.

Hyde, H.A. (- 1950). MS notes in his annotated copy of Trow's '*Flora of Glamorgan*' with Welsh orthography corrected by I.C. Peate. Library, National Museum of Wales, Cardiff.

Hyde, H.A. (1950). Eleanor Vachell (1879–1948) [obit.]. *Watsonia*, **1**, 325–327.

Hyde, H.A. (1950–1952). Mynydd Llysworney. In: *46th Annual Report of the National Museum of Wales.*

Hyde, H.A. (1952a). Studies in Atmospheric Pollen IVa. Pollen deposition at two Cardiff Stations in 1943 compared. *Reports and Transactions of the Cardiff Naturalists' Society*, **80**, 3–6.

Hyde, H.A. (1952b). Studies in Atmospheric Pollen V. A daily census of patterns at Cardiff for the six years 1943–1948. *New Phytologist*, **51**, 281–293.

Hyde, H.A. (1953). Appendix. In: Savory, H.N. The excavation of a Neolithic dwelling and a Bronge Age cairn, *Reports and Transactions of the Cardiff Naturalists' Society*, **81**, 91–92.

Hyde, H.A. (1961). Plant life in and around the Glamorgan Forests. In: *Glamorgan Forests* (Ed. by H.L. Edlin), pp. 49–66. Forestry Commission Guide, HMSO, London.

Hyde, H.A. (1963). Lewis Weston Dillwyn as a botanist. In: *South Wales and Monmouth Record Society, Publication No. 5: Diary of Lewis Weston Dillwyn* (Ed. by H.J. Randall & W. Rees), pp. 6–8. R.H. Johns Ltd., Newport.

Hyde, H.A. (1977). *Welsh Timber Trees*, 4th edn. revised by S.G. Harrison. 1st edn (1931), 2nd edn (1935), 3rd edn (1961). National Museum of Wales, Cardiff.

Hyde, H.A. & Wade, A.E. (1934). *Welsh Flowering Plants*. 2nd edn (1957). National Museum of Wales, Cardiff.

Hyde, H.A. & Wade, A.E. (1940). *Welsh Ferns*. 2nd edn (1948), 3rd edn (1954), 4th edn (1962). National Museum of Wales, Cardiff.

Hyde, H.A., Wade, A.E. & Harrison, S.G. (1969). *Welsh Ferns including Clubmosses, Quillworts and Horsetails*, 5th edn (1978). National Museum of Wales, Cardiff.

Hyde, H.A. & D.A. Williams (1944). Studies in Atmospheric Pollen I: A daily census of pollens at Cardiff, 1942. *New Phytologist*, **43**, 49–61.

Hylander, N. (1943). Die Grassameneinkömmlinge schwedischer Parke. *Symbolae botanicae upsalienses*, **7**, 125–274.

Hywel-Davies, J., Thom, V. & Bennett, L. (1989). *The Macmillan Guide to Britain's Nature Reserves.* 1st edn (1986). Macmillan, London.

Ingrouille, M.J. (1986). The *Limonium binervosum* aggregate (Plumbaginaceae) in the British Isles. *Botanical Journal of the Linnean Society*, **92**, 177–217.

Ingrouille, M.J. & Pearson, J. (1987). The pattern of morphological variation in the *Salicornia europaea* L. aggregate (Chenopodiaceae). *Watsonia*, **16**, 269–291.

Irvine, A. ed. (1855–1856). An Epitome of Botanical Tours in Wales from the earliest period. *Phytologist* (n.s.), **1**, 264–269. [contd from p. 219].

Isaac, W.E. (1930). *The rate of humidification in relation to soil types and vegetative covering in Glamorgan.* Ph.D. thesis. University College of Cardiff.

Ivimey-Cook, R.B. (1955). *The ecology of a limestone heath at Ewenny, Glamorgan with special reference to the occurrence of* Agrostis setacea. Ph.D. thesis, Wales.

Ivimey-Cook (1959). Biological Flora of the British Isles: *Agrostis setacea* Curt. *Journal of Ecology*, **47**, 697–706.

Jackett, R. (1892–1987). Field meetings and plant records in *Swansea Scientific Society Annual Report and Transactions*, 1891–2, pp. 40–43 (Crymlin Bog); pp. 45–59 (Flora of Swansea); 1892–3, pp. 20–24 (S. Gower); 1893–4, pp. 26–28 (Pontardawe); 1894–5, pp. 51–55 (Clyne and Killay); 1895–6, pp. 73–78 (Penllegaer); pp. 94–97 (Parc-le-Breos).

Jackson, C. (1983–1984). *Aberdare Park. Tree Identification and Location Survey 1983–1984.* Cynon Valley Borough Council.

Jackson, N. & Eversham, B. eds. (1989). *Pan/Ordnance Survey Nature Atlas of Great Britain, Ireland and the Channel Isles.* Duncan Petersen Publishing Ltd., London.

Jermy, A.C., Arnold, H.R., Farrell, L. & Perring, F.H. eds. (1978). *Atlas of Ferns of the British Isles.* Botanical Society of the British Isles and British Pteridological Society, London.

Jermy, A.C., Chater, A.O., & David, R.W. (1982). *Sedges of the British Isles. Botanical Society of the British Isles Handbook No. 1.* 2nd edn. Botanical Society of the British Isles, London.

Jermy, A.C. & Tutin, T.G. (1968). *British Sedges.* Botanical Society of the British Isles, London.

Jewell, P.S. (1934a). Flowers. In: Notes and Records. *Proceedings of the Swansea Scientific and Field Naturalists' Society*, **1**, 220.

Jewell, P.S. (1934b). A Note on some common Gower Orchids. *Proceedings of the Swansea Scientific and Field Naturalists' Society*, **1**, 226–228.

John, A.D. (1978). *An ecological survey of the Glamorgan Canal Nature Reserve, Forest Farm, Whitchurch.* Parks and Baths Department, Cardiff City Council.

Jones, A. (1984). Notes on Coed-y-felin. *Forest Farm News No. 2.* [pp. 19–20].

Jones, E.J. [trans. 1937]. Welsh names of plants collected in Glamorgan and recorded in his copy of Skene's 'A Flower Book for the Pocket'. MS in Library, National Museum of Wales, Cardiff.

Jones, G.E. (1972). An investigation into the possible causes of poor growth of Sitka Spruce, Margam Forest, Glamorgan. In: *Research Papers in Forestry Meteorology No. 13*, 1970, pp. 147–156.

Jones, G.M. & Scourfield, E. (1986). Sully Island. In: *Sully, A village and parish in the Vale of Glamorgan*, p. 138. Caerphilly and London.

Jones, H.C. (1976). *Place Names in Glamorgan.* The Starling Press Ltd., Risca, Newport, Gwent. 2nd edn (1978).

Jones, H.E. (1968). *Some chemical effects of waterlogging in soils of low pH on the growth of* Erica cinerea *and* Erica tetralix. Ph.D thesis. University College of Cardiff.

Jones, H.E. & Etherington, J.R. (1970). Comparative studies of plant growth and distribution in relation to waterlogging. 1. The survival of *Erica cinerea* L. and *E. tetralix* L. and its apparent relationship to iron and manganese uptake in waterlogged soil. *Journal of Ecology*, **58**, 487–496.

Jones, H.M. (1971). *Crymlyn Bog: a pollen diagram and its interpretation.* M.Sc. thesis. University College of Swansea.

Jones, P.S. (1983a). Adiantum capillus-veneris L. Survey. MS in Library, National Museum of Wales, Cardiff.

Jones, P.S. (1983b). *Notes on the Glamorgan (v.c. 41) distribution of* Cirsium tuberosum (L.) All. MS in Library, National Museum of Wales, Cardiff.

Jones, P.S. & Etherington, J. (1985–1986). Preliminary Account of the Vegetation Ecology of Dune Slacks at Kenfig Pool and Dunes Local Nature Reserve, Mid Glamorgan. *GNAT No. 13*, [p.5]. (Newsletter of the Glamorgan Trust for Nature Conservation).

Jones, R. (n.d.). *Thirty Walks in Gower*. Privately published.

Jones, R. (1967). *The relationship of dune-slack plants to soil moisture and chemical conditions*. [Kenfig]. Ph.D. thesis. University College of Cardiff.

Jones, R. (1972a). Comparative studies of plant growth and distribution in relation to waterlogging V. The uptake of iron and manganese by dune and dune slack plants. [Kenfig Burrows]. *Journal of Ecology*, **60**, 131–139.

Jones, R. (1972b). Comparative studies of plant growth and distribution in relation to waterlogging VI. The effect of manganese on the growth of dune and dune slack plants. *Journal of Ecology*, **60**, 141–145.

Jones, R. (1973). Comparative studies of plant growth and distribution in relation to waterlogging VII. The influence of water-table fluctuations on iron and manganese availability in dune slack soils. *Journal of Ecology*, **61**, 107–116.

Jones, R. & Etherington, J.R. (1971). Comparative studies of plant growth and distribution in relation to waterlogging IV. The growth of dune and dune slack plants. *Journal of Ecology*, **59**, 793–801.

Jones, T.G. & ap Gwynn, A. (1950). *Geiriadur Cymraeg-Saesneg a Saesneg-Cymraeg*. Hughes & Son, Cardiff.

Jory, R.M. (1967). Flat Holm in the Bristol Channel. In: *Glamorgan Historian* [Ed. by S. Williams]. Vol. 4. pp. 162–171. D. Brown and Sons Ltd., Cowbridge.

Judd, D.R. (1983). *Intraspecific variation in response to shade*. Ph.D. thesis. University College of Cardiff.

Kay, Q.O.N. (1971a). *A preliminary survey of the vegetation of Crymlyn Bog. Scientific report to the Nature Conservancy 1971*.

Kay, Q.O.N. (1971b). Botany. In: *Swansea and its Region* (Ed. by W.G.V. Balchin), pp. 85–100. University College, Swansea.

Kay, Q.O.N. (1971c). Biological Flora of the British Isles: *Anthemis cotula* L. *Journal of Ecology*, **59**, 623–636.

Kay, Q.O.N. (1971d). Biological Flora of the British Isles: *Anthemis arvensis* L. *Journal of Ecology*, **59**, 637–648.

Kay, Q.O.N. (1972). The Dune Gentian in the Gower Peninsula. *Nature in Wales*, **13**, 81–85.

Kay, Q.O.N. (1974a). Botany. In: *Neath and District. A Symposium* (Ed. by E. Jenkins), pp. 305–318. Elis Jenkins, Neath.

Kay, Q.O.N. (1974b). Diploid *Isoetes echinospora* in Britain. *Fern Gazette*, **11**, 56–57.

Kay, Q.O.N. (1975). J.A. Webb's account of 'The presumably extinct plants of West Glamorgan' reconsidered after forty-five years. *Transactions of the Cardiff Naturalists' Society*, **96**, 23–28.

Kay, Q.O.N. (1976). Preferential pollination of yellow-flowered morphs of *Raphanus raphanistrum* by *Pieris* and *Eristalis* spp. *Nature*, **261**, 230–232.

Kay, Q.O.N. (1978). The role of preferential and assortative pollination in the maintenance of flower colour polymorphisms. In: *Linnean Society Symposium Series No. 6 – The Pollination of Flowers by Insects*. (Ed. by A.J. Richards, pp. 175–190.) Academic Press Inc., London & New York.

Kay, Q.O.N. (1979). Post-glacial history of commonland in West Glamorgan. In: *Problems of Commonland* (Ed. by E.M. Bridges), 3:1/1–3:1/19. University College of Swansea.

Kay, Q.O.N. (1982). Botanical Society of the British Isles Field Meeting: Crymlyn Burrows and Crymlyn Fen, Glamorgan, 19th July, 1981. *Botanical Society of the British Isles, Welsh Bulletin No. 36*, pp. 6–8.

Kay, Q.O.N. (1985a). Hermaphrodites and subhermaphrodites in a reputedly dioecious plant, *Cirsium arvense* (L.) Scop. *New Phytologist*, **100**, 457–472.

Kay, Q.O.N. (1985b). Nectar from willow catkins as a food source for Blue Tits. *Bird Study*, **32**, 40–44.

Kay, Q.O.N. & Ab-Shukor, N.A. (1988). *Trifolium occidentale* D.E. Coombe, new to Wales. *Watsonia*, **17**, 168–170.

Kay, Q.O.N. & Harrison, J. (1970). Biological Flora of the British Isles: *Draba aizoides* L. *Journal of Ecology*, **58**, 877–888.

Kay, Q.O.N., Lack, A.J., Bamber, F.C. & Davies, C.R. (1984). Differences between sexes in floral morphology, nectar production and insect visits in a dioecious species, *Silene dioica*. *New Phytologist*, **98**, 515–529.

Kay, Q.O.N. & Page, J. (1985). Dioecism and pollination in *Ruscus aculeatus*. *Watsonia*, **15**, 261–264.

Kay, Q.O.N., Roberts, R.H. & Vaughan, I.M. (1974). The spread of *Pyrola rotundifolia* L. subsp. *maritima* (Kenyon) E.F. Warb. in Wales. *Watsonia*, **10**, 61–67.

Kay, Q.O.N. & Rojanavipart, P. (1977). Salt marsh ecology and trace-metal studies. In: *Problems of a small estuary* (Ed. by A. Nelson-Smith and E.M. Bridges), 2.1–2.16. University College of Swansea, Institute of Marine Studies.

Kay, Q.O.N. & Stevens, D.P. (1986). The frequency, distribution and reproductive biology of dioecious species in the native flora of Britain and Ireland. *Botanical Journal of the Linnean Society*, **92**, 39–64.

Kay, Q.O.N. & Woodell, S.R.J. (1976). The vegetation of anthills in West Glamorgan saltmarshes. *Nature in Wales*, **15**, 81–87.

Kent, D.H. (1968). *Senecio squalidus* L. in the British Isles – 7. Wales. *Nature in Wales*, **8**, 175–178.

Kent, D.H. & Allen, D.E. (1984). *British and Irish Herbaria - An index to the location of herbaria of British and Irish Vascular plants*. Botanical Society of the British Isles, London.

King, T.J. (1972). *The plant ecology of anthills in grassland*. D. Phil. thesis, Oxford.

King, T.J. (1981). Anthill vegetation in acidic grasslands in the Gower peninsula, South Wales. *New Phytologist*, **88**, 559–571.

Kirby, A. (1913). An amateur's notes and observations to amateurs. *British Fern Gazette*, **2**, 86–89.

Knight, F.A. (1902). *The Sea-Board of Mendip*. J.M. Dent & Co., London.

Lacey, W.S. ed. (1970). *Welsh Wildlife in Trust*. North Wales Naturalists' Trust, Bangor.

Lack, A., Llewellyn, P. & Thomas, D. (1985). Heavy Plant in Broad Pool. *GNAT No. 11*, [p.6]. (Newsletter of the Glamorgan Trust for Nature Conservation).

Lack, A.J. & Kay, Q.O.N. (1986). Phosphoglucose isomerase (EC 5.3.1.9) isoenzymes in diploid and tetraploid *Polygala* species: Evidence for gene duplication and diversification. *Heredity*, **56**, 111–118.

Lack, A.J. & Kay, Q.O.N. (1987). Genetic structure, gene-flow and reproductive ecology in sand-dune populations of *Polygala vulgaris*. *Journal of Ecology*, **75**, 259–276.

Lack, A.J. & Kay, Q.O.N. (1988). Allele frequencies, genetic relationships and heterozygosity in *Polygala vulgaris* populations from contrasting habitats in southern Britain. *Biological Journal of the Linnean Society*, **34**, 119–148.

Lack, A.J. & Llewellyn, P.J. (1986). Broad Pool. *Gower*, **37**, 9–18.

Lang, D.C. (1987). *The Complete Book of British Berries*. Threshold Books Ltd., London.

Langford, J. (1984). Coppicing. [Effect of Coppicing at Glamorgan Canal and Coed Garn Llwyd Nature Reserves]. *Forest Farm News No. 1*, [pp. 9–10, 15–16].

[Langhelt, L., *et al. c.*1978]. *Flora and Fauna of the Rhymney Valley*. Manpower Services Commission, Rhymney Valley District Council.

Lankester, E. ed. (1846). *Memorials of John Ray*. Ray Society, London.

Lansdown, A.R. (1972). *Spartina* in Swansea Bay. *Glamorgan County Naturalists' Trust, Bulletin No. 11*, pp. 24–25.

Lavender, S.J. (1981). *New Land for Old, The Environmental Renaissance of the Lower Swansea Valley*. Adam Hilger Ltd., Bristol.

Leadlay, E.A. & Heywood, V.H. (1990). The biology and systematics of the genus *Coincya* Porta & Rigo ex Rouy (Cruciferae). *Botanical Journal of the Linnean Society*, **102**, 313–398.

Lees, E. (1842). *The Botanical Looker-out*. 2nd edn (1851). London.

Lees, E. (1843). Note on *Osmunda regalis*, near Swansea, Glamorganshire. *Phytologist*, **1**, 748–749.

Legard, P.H. & Harris, E.H.M. (1964). Foresting in the S. Wales coalfield. *Forestry*, **37**, 13–20.

Leonard, P. (1982). Flatholm: an island nature reserve. *GNAT No. 4*, p. 11. (Newsletter of the Glamorgan Naturalists' Trust).

Lewis, E.R. ed. (1962). *Glamorgan County Naturalists' Trust Directory of Scientific Areas used by Schools for Educational Purposes*.

Lewis-Jones, L.J. (1980). Development and Status of the Royal Institution of South Wales Herbarium at Swansea Museum. *Biological Curators' Group Newsletter*, **2**, 368–369.

Lewis-Jones, L.J. & Kay, Q.O.N. (1977). The cytotaxonomy and distribution of water starworts (*Callitriche* spp.) in West Glamorgan. *Nature in Wales*, **15**, 180–183.

Ley, A. (1900). Some Welsh Hawkweeds. *Journal of Botany*, **38**, 3–7.

Lightfoot, J. (1773). *Journal of a botanical excursion in Wales*. Printed in *Journal of Botany* (1905), **43**, 297–307.

Lightfoot, J. (1788). *Notes in Ray's Synopsis*, **3**. Library, Department of Botany, Oxford.

Linnard, W. (1978). Nodiadau Amrywiol 1. Miscellaneous Notes: ffawydd fel elfen mewn enwau lleoedd. *Bulletin of the Board of Celtic Studies*, **28**, 83–86.

Linnard, W. (1979). Historical Distribution of Beech in Wales. *Nature in Wales*, **16**, 154–159.

Linnard, W. (1982). *Welsh Woods & Forests – History and utilisation*. National Museum of Wales, Cardiff.

Linton, E.F. (1886a). Hieracium Notes. *Journal of Botany*, **24**, 84–85.

Linton, E.F. (1886b). New Glamorgan Plants. *Journal of Botany*, **24**, 112.

Linton, E.F. (1890). Glamorgan Plants. *Journal of Botany*, **28**, 157.

Linton, E.F. (1892). Glamorgan notes and records. *Journal of Botany*, **30**, 296–297.

Linton, W.R. (1886). New records [Glamorganshire]. *Journal of Botany*, **24**, 376–377.

Lloyd, J.B. [pre 1907]. MS lists of Porthcawl and Mumbles Plants, by the late Rev. J.B. Lloyd of Liverpool.* [Note in Riddelsdell (1907), p. 4].

Lloyd, J.E. & Jenkins, R.T. eds. (1959). *The Dictionary of Welsh Biography down to 1940*. The Honourable Society of Cymmrodorion, London.

Lloyd, O.C. [1978]. Fern-hunting in South Wales. *The Red Dragon (Y Ddraig Goch) 1978–1979*, **5**, 64–65. (The Annual Journal of the Cambrian Caving Council).

Lloyd, O.C. [1979]. Fern-hunting in South Wales. *Merthyr Tydfil and District Naturalists' Society Newsletter [No. 3]*, pp. 11–13.

Llwyd, E. (1707). *Archaeologia Britannica*. Oxford.

Lockley, R.M. (1941). *The Way to an Island*. [autobiography]. J.M. Dent & Sons Ltd., London.

Lockley, R.M. (1970). *The Naturalist in Wales*. David & Charles, Newton Abbot.

Lockley, R.M. (1979). *Myself when young*. [autobiography]. Andre Deutsch, London.

Longman, L. (n.d.). *Trees*. Glyncornel Environmental Centre.

Lousley, J.E. [1948]. Some autumn wild flowers of chalk and limestone. In: *Birds, beasts and flowers*. [Ed. by A.W. Coysh], p. 11.*.

Lousley, J.E. (1950a). South Wales. In: *Wild Flowers of Chalk and Limestone*. (The New Naturalist No. 16), pp. 142–147. 2nd edn (1969). Collins, London.

Lousley, J.E. (1950b). [Tribute to] Eleanor Vachell. *Watsonia*, **1**, 327.

Lousley, J.E. & Kent, D.H. (1981). *Docks and Knotweeds of the British Isles. Botanical Society of the British Isles Handbook No. 3*. Botanical Society of the British Isles, London.

Lovell, D. (1950). I went to Flatholm. *Glamorgan County Magazine No. 6*, pp. 38–42.

Lucas, C.C. (1854). *Viola lactea* and *Viola flavicornis* [*V. canina*] (?). *Phytologist*, **5**, 216.

Lucas, J. & Turton, W. (1802). The rarer vegetables. In: *The Swansea Guide* by J. Oldisworth, pp. 165–172.

Lyon, A.G. (1956). Field Meeting: [Cwm. George, Dinas Powis]. *Proceedings of the Botanical Society of the British Isles*, **2**, 71.

MacGillivray, W. (1830). *A Systematic arrangement of British Plants by W. Withering M.D. Corrected and condensed by W. MacGillivray*. London.

Majeed, A. (1984). *Nitrogen metabolism and salt tolerance in higher plants*. Ph.D. thesis. University College of Swansea.

Malkin, B.H. (1804). *Scenery, Antiquities and Biography of South Wales from materials of two excursions in the year 1803*. London.

Manpower Services Commission (1988). *Glyncornel Walks 1. Graig Pont Rhondda*.

Marchant, N.G. (1970). *Experimental taxonomy of* Veronica *section* Beccabungae *Griseb.* Ph.D. thesis. University of Cambridge.

Marshall, E.S. (1907). A Glamorganshire Sedge. *Journal of Botany*, **45**, 163.

Marshall, E.S. & Shoolbred, W.A. (1902). Glamorganshire plants. *Journal of Botany*, **40**, 248–250.

Matthews, J.R. (1937). Geographical relationships of the British Flora. *Journal of Ecology*, **25**, 1–90.

Maxwell, J.F. (1989). Gower – June 30th, July 1st and 2nd. [Field Meeting, Wild Flower Society]. *Wild Flower Magazine No. 416*, pp. 24–25.

Mayhead, G.J., Broad, K. & Marsh, P. (1974). *Tree Growth on the South Wales Coalfield. Research and Development Paper No. 108*. Forestry Commission, London.

McAllister, H. (1980). The problem of *Hedera hibernica* resolved. *Ivy Exchange Newsletter*, **3**, 26–27.

McCain, S. (1981). *Physiological adaptations to soil factors in natural populations of certain grass species.* Ph.D. thesis. University College of Cardiff.

McColl, R.H.S. (1969). *The inorganic element regimes of mire habitats in relation to the ecology of selected mire species.* Ph.D. thesis. University College of Swansea.

[McLaggan, M. & Gillham, M. eds.] (1972). Coed y bwl Reserve, Castle-upon-alun, nr. Bridgend. *Glamorgan County Naturalists' Trust, Bulletin No. 11,* pp. 7–10.

McLean, R.C. (1935). An ungrazed grassland on limestone in Wales [Worm's Head]. *Journal of Ecology,* 23, 436–442.

McLean, R.C. (1936a). Botany: Introduction. In: *Glamorgan County History,* 1, 121–122.

McLean, R.C. (1936b). Notes on the ecology of Glamorgan vegetation. *Glamorgan County History,* 1, 242–251.

[McLean, R.C.] (1937 –). Survey maps of selected areas in Glamorgan by students. Botany Department, University College of Cardiff.

McLean, R.C. & Salisbury, E.J. (1925). British Ecological Society, Summer Excursion, 1924 [Gower Peninsula]. *Journal of Ecology,* 13, 167–168.

Meade, R. (1982). *Botanical and hydrological Survey of Oxwich Marsh, 1981.* Nature Conservancy Council (South Wales Region). Unpublished.

Mears, M.A. (1987). *Vegetation and soil conditions on unreclaimed colliery spoil heaps in South Wales.* M.Sc. thesis. University College of Cardiff.

Meikle, R.D. (1984). *Willows and Poplars of Great Britain and Ireland. Botanical Society of the British Isles Handbook No. 4.* Botanical Society of the British Isles, London.

Merthyr Teachers Centre Group (1981). *Merthyr Tydfil, A Valley Community.* D. Brown & Sons Ltd., Cowbridge.

Mettam, C.J. (1982). Tidal Power from the Severn Estuary: an energy review. *Nature in Wales* (n.s.), 1, 50–61.

Mid Glamorgan County Council (n.d.1). *A Walk in Quaker's Yard.*

[Mid Glamorgan County Council] (n.d.2). *Brecon Mountain Railway.*

Mid Glamorgan County Council (n.d.3). *Darren Valley Walk.*

Mid Glamorgan County Council (n.d.4). *Discover Mid Glamorgan. (Tourist Guide).*

Mid Glamorgan County Council (n.d.5). *Lluest Wen/Cefn Bryn Gelli (Long Walk).*

[Mid Glamorgan County Council] (n.d.6). *Trees Trail, Cyfarthfa Castle.*

Mid Glamorgan County Council (n.d.7). *Walks in Mid Glamorgan. No. 1. Blaenrhondda Waterfalls Walk.*

Mid Glamorgan County Council (n.d.8). *Walks in Mid Glamorgan No. 2: Rudry Common Walk.*

Mid Glamorgan County Council (1977). *Mid Glamorgan County Structure Plan: Landscape and Conservation. Report of Survey.*

Mid Glamorgan County Council & Rhymney Valley District Council (n.d.). *Parc Cwm Darran.*

Mid Glamorgan County Council, Education Department (1988). *Mid Glamorgan Meadows – A Guide for Teachers.*

Mid Glamorgan Education Authority (n.d.). *Darren Park, Ferndale.*

Mid Glamorgan Education Authority, Rhondda Educational Centre. [1977a]. *Environmental Studies Research Project (Job Creation Scheme 1976–1977) Rhondda District: Sites of Interest in Rhondda Area.*

Mid Glamorgan Education Authority, Rhondda Education Centre [1977b]. Points of interest at Mynydd Maendy In: *Environmental Studies Research Project (Job Creation Scheme 1976–1977) Rhondda District: Penrhys; Mynydd Maendy.*

Miles, B. (1960). Aberthaw, Glamorgan [Report of Field Meeting]. *Proceedings of the Botanical Society of the British Isles,* 3, 457.

Miller, W.D. (1928). The Botanical Excursion in Somerset and Glamorgan. *Report of the Botanical Society and Exchange Club of the British Isles,* 8, 458–459.

Milward, J. & Storrie, J. (1877). Southerndown field meeting. *Report and Transactions of the Cardiff Naturalists' Society 1876,* 8, 38–40.

Mitchell, A. (1897). The Pine-wood wasp (*Sirex juvencus*) and its occurrence at Dunraven. *Report and Transactions of the Cardiff Naturalists' Society 1896–7,* 29, 69–73.

Mitchell, A.F. & Hallett, V.E. (1987). *Champion Trees in the British Isles. Forestry Commission Research and Development Paper 138.* 2nd edn. Forestry Commission, Edinburgh.

Mitchell, R., Probert, P.K., McKirdy, A.P. & Doody, J.R. (1981). *Severn Tidal Power: Nature Conservation.* Nature Conservancy Council.

Mockeridge, F.A. (1929a). Reversion in *Primula vulgaris*. *Proceedings of the Swansea Scientific and Field Naturalists' Society*, **1**, 77–79.

Mockeridge, F.A. (1929b). An abnormality in *Ulmus campestris*. *Proceedings of the Swansea Scientific and Field Naturalists' Society*, **1**, 75–77.

Mockeridge, F.A. (1938). Botany in Glamorgan [Review of 'Glamorgan County History (Vol. 1)']. *Proceedings of the Swansea Scientific and Field Naturalists' Society*, **2**, 20–22.

Moger, G.M. (n.d.). *Report No. 4: Tree Planting in the Lower Swansea Valley*. Conservator's Office [Lower Swansea Valley Project].

Mogford, D.J. (1974). Flower colour polymorphism in *Cirsium palustre*. *Heredity*, **33**, 241–256.

Mogford, D.J. (1978). Pollination and flower colour polymorphism, with special reference to *Cirsium palustre*. In: *Linnaean Society Symposium Series No. 6 – The Pollination of Flowers by Insects* (Ed. by A.J. Richards), pp. 191–199. Academic Press Inc., London & New York.

Moggridge, M. (1844a). A statement of the times of flowering of the following plants in the neighbourhood of Swansea on an average of the last six years. *Report, Royal Institution of South Wales 1843–4*, **9**, 28–31.

Moggridge, M. (1844b). List. In: *Proceedings of Royal Institution of South Wales*.* [See Carter (1955), p. 13].

Moggridge, M. (1852). List of plants in flower at Caswell Gardens.* [See Carter (1955), p. 13].

Moon, S. [J.] (1989). Kenfig National Nature Reserve. *Plant Press No. 7*, [p. 3]. (Plant Conservation News from the Conservation Association of Botanical Societies [CABS]).

[Moon, S.J., Nelson, W. & Wykes, J.R. (1980)]. *Kenfig Pool and Dunes (Local Nature Reserve, Mid Glamorgan)*. Mid Glamorgan County Council. D. Brown and Sons Ltd., Bridgend.

[Moon, S.J., Nelson, W., Wykes, J.R.] (1982). *A Checklist of The Flora of Kenfig*. Mid Glamorgan County Council.

Moore, D.M. (1982). *Flora Europaea Check-list and Chromosome Index*. Cambridge University Press. Cambridge, New York & Melbourne.

Moore, T. (1848). *A Handbook of British Ferns*. London.

More, A.G. (1884). Plants gathered in the counties of Pembroke and Glamorgan. *Journal of Botany*, **22**, 43–46 [Glamorgan, p. 46].

Morgan, A. (1974). About Plants. In: *Legends of Porthcawl and the Glamorgan Coast*, pp. 111–113. 2nd edn (1978), D. Brown and Sons, Cowbridge & Bridgend.

Morgan, C.D. (1886). *Wanderings in Gower*. The Cambrian Office, Swansea.

Morgan, T. (1901). *Glamorganshire Place-names with names of mountains, fields, historic places etc.* J.E. Southall, Newport, Gwent.

Motley, J. (1847). Remarks on *Centaurea nigrescens*, *Malva verticillata*, and *Euphorbia peplis*. *Phytologist*, **2**, 972–974.

Mott, F.B. (1919). Short Notes: *Impatiens glandulifera* Royle. *Journal of Botany*, **57**, 69.

National Museum of Wales (n.d.). *Glamorganshire Canal Nature Reserve*. [Plan of Botany Galleries showing where the plants were displayed in the National Museum of Wales, Cardiff].

Nature Conservancy (1958). New Nature Reserves: The Gower Coast Nature Reserve. *Nature in Wales*, **4**, 629.

Nature Conservancy (1964). New National Nature Reserves: Whiteford National Nature Reserve, Glamorgan. *Nature in Wales*, **9**, 65.

Nature Conservancy (1967). *Whiteford National Nature Reserve*. Nature Conservancy, Plas Gogerddan, Aberystwyth. (Revised 1971).

Nature Conservancy (1968). *Oxwich National Nature Reserve*. Nature Conservancy, Plas Gogerddan, Aberystwyth. (2nd printing 1969).

Nature Conservancy (1969). *Oxwich Sand Trail*. Nature Conservancy, Plas Gogerddan, Aberystwyth. (2nd printing 1971).

Nature Conservancy (1970). *A limestone nature trail : Gower Coast National Nature Reserve*. Nature Conservancy, Plas Gogerddan, Aberystwyth.

Nature Conservancy Council (1975). *Gower Coast National Nature Reserve – Limestone Nature Trail*. Nature Conservancy Council, Bangor, Gwynedd.

Nature Conservancy Council (1977a). *Oxwich National Nature Reserve*. Nature Conservancy Council, Bangor, Gwynedd.

Nature Conservancy Council (1977b). *Oxwich National Nature Reserve: The Sand Dune Nature Trail*. Nature Conservancy Council, Bangor, Gwynedd.

Nature Conservancy Council (1979). *Oxwich National Nature Reserve: A Woodland Walk*. Nature Conservancy Council, Bangor, Gwynedd.

Nature Conservancy Council (1980). *Nature Conservation and Field Studies in Gower*.

Nature Conservancy Council (1981). *National Nature Reserves in Wales. Nature Conservancy Council Information and Library Services Information sheet No. 5*. Nature Conservancy Council, Banbury, Oxon.

Nature Conservancy Council (1984). *Glamorgan Canal and Long Wood, Site of Special Scientific Interest: Notifiable operations. Proposed work*. MS. Cardiff City Council, Leisure and Amenities Department. [*Sagittaria, Nuphar, Sparganium*].

Nature Conservancy Council (1987). Award for Cardiff Community Nature Area [Radyr Woods]. *Urban Wildlife News*, **4**, 2–3.

Nature Conservancy Council & City of Swansea (1981). *Crymlyn Proposals*. Swansea City Planning Department.

Nature Conservancy Council with The National Trust and Glamorgan Wildlife Trust (1988). *South West Gower Coast Nature Reserves*. Nature Conservancy Council in Wales.

Nature Conservancy Council – South Wales Region (1976). *South Wales Region – A Review*. Nature Conservancy Council, South Wales Region, Cardiff.

Nature Conservancy Council – South Wales Region (1977). *A Nature Conservation Study of the Neath Estuary Complex (West Glamorgan)*. Nature Conservancy Council, South Wales Region, Cardiff.

Nature Conservancy Council – South Wales Region (1979). *Whiteford National Nature Reserve*. Nature Conservancy Council.

Nature Conservancy Council – South Wales Region (1981). *Oxwich National Nature Reserve Naturalists' Guide*. Nature Conservancy Council.

Nature Conservancy Council – South Wales Region (1983a). *Oxwich National Nature Reserve Information Sheet Number Two: Butterflies*. Nature Conservancy Council Interpretative Branch.

Nature Conservancy Council – South Wales Region (1983b). *Oxwich National Nature Reserve Information Sheet Number Three: Flowering Plants and Ferns*. Nature Conservancy Council Interpretative Branch.

Nature Conservancy Council – South Wales Region (1984, 1986). *Worms Head, Gower Coast National Nature Reserve*. Nature Conservancy Council Interpretative Services Branch.

Nature Conservancy Council – South Wales Region (1986a). *Crymlyn Bog National Nature Reserve*. Nature Conservancy Council Interpretative Services Branch.

Nature Conservancy Council – South Wales Region (1986b). *Oxwich National Nature Reserve Information Sheet. Number Five: The Flowering Plants of Oxwich Bay*. Nature Conservancy Council Interpretative Services Branch.

Nature Conservancy Council – South Wales Region (1988). *Oxwich National Nature Reserve Explorers' Guide*. Nature Conservancy in Wales.

Neale, M.H. & Morgan, D.H. (1937). Kenfig Pool. In: Report of the Council. *Report and Transactions of the Cardiff Naturalists' Society 1935*, **68**, p. 80.

Neama, J.D. (1982). *Physiological and ecological studies on altitudinal distribution in the genus* Salix. Ph.D. thesis. University College of Cardiff.

Nelson, W. (1981). Let's Visit a Reserve: Eglwys Nunydd Reservoir. *GNAT No. 2*, [p. 3]. (Newsletter of the Glamorgan Naturalists' Trust).

Nelson, W. (1984). Kenfig Pool and Dunes. *Forest Farm News No. 2*. [pp. 20–22].

Nelson, W. (1985). Field meeting: Kenfig Pool and Dunes, 6th July 1985. *Llanelli Naturalists' Newsletter, September 1985*, [p. 13].

Nelson-Smith, A. (1984). Swansea Bay: Green mantle and toxic sludge – an essay review. *Nature in Wales* (n.s.) 1983, **2**, 80–89.

Nelson-Smith, A. & Bridges, E.M. eds. (1977). *Problems of a small estuary: proceedings of the symposium on the Burry Inlet. (South Wales) Swansea, 1976*. University College of Swansea, Institute of Marine Studies, Swansea.

Neve, P.J. (1976). *Grazing on the North Gower saltmarshes*. B.Sc. project report, Department of Botany, University College of Swansea.

Newport Museum and Forestry Commission (1970). *Draethen Forest Nature Trail*.

Newton, A. (1974). Five Brambles from South Wales. *Nature in Wales*, **14**, 24–34.

Newton, A. (1976). Brambles in Glamorgan and neighbouring areas. *Transactions of the Cardiff Naturalists' Society*, **97**, 44–49.

Newton, A. & Porter, M. (1990). Five brambles from Wales. *Watsonia*, **18**, 189–198.

Nicholson, G. (1808). *The Cambrian Traveller's Guide*. George Nicholson, Stourbridge. [2nd edn 1813, 3rd edn 1840].

Nock, A.I., Wade, A.E. & Webb, J.A. (1952). The Clyne Common Survey of 1944. *Proceedings of the Swansea Scientific and Field Naturalists' Society*, **2**, 329–342.

North, F.J. (1930). The river scenery at the head of the Vale of Neath. *Report and Transactions of the Cardiff Naturalists' Society 1928*, **61**, 12–54.

North, F.J. (1933). From Giraldus Cambrensis to the Geological Map. *Report and Transactions of the Cardiff Naturalists' Society 1931*, **64**, 20–97.

North, F.J. (1935). *The Coal Forests – and David Davies, Gilfach Goch*. South Wales Institute of Engineers, Cardiff.

Ogwr Borough Council Planning Department (n.d.). *Bryngarw County Park*. D. Brown & Sons Ltd., Bridgend.

[Oldisworth, J.]. (1802). *The Swansea Guide*. Z.B. Morris, Swansea. (A New Edn, 1823).

Ollivant, C.E. (1874). Six-monthly flora. *Transactions of the Cardiff Naturalists' Society 1873*, **5**, 89–97.

Orr, M.Y. (1912). Kenfig Burrows: an ecological study. *Scottish Botanical Review*, **1**, 209–216.

Orr, M.Y. (1914). *Pilularia globulifera* L. in Glamorgan. *Transactions and Proceedings of the Botanical Society of Edinburgh*, **26**, 281–285.

Osbourne, T.M. (1988). *Gower Management Plan*. Swansea City Council.

Palmer, J.P. (1990). Japanese Knotweed (*Reynoutria japonica*) in Wales. In: *The Biology and Control of Invasive plants*. Industrial Ecology Group of the British Ecological Society at University of Wales College of Cardiff, Conference Report, pp. 96–109.

Parkinson, E. & Plant, A.J. (1978). *Cosmeston Country Park Development and Management Plan*. South Glamorgan County Council (amended 1979).

Parry, M. (1969). *Casgliad o Enwau Blodau, Llysiau a Choed*. [Welsh Plant names]. Gwasg Prifysgol Cymru, Caerdydd.

Paul, A.M. (1987). The status of *Ophioglossum azoricum* (Ophioglossaceae: Pteridophyta) in the British Isles. *Fern Gazette*, **13**, 173–187.

Peate, I.C. (n.d.). See Hyde, H.A. (- 1950).

Penford, N., Francis, I.S., Hughes, E.J. & Aitchison, J.W. (1990). *Biological survey of Common Land, No. 13: West Glamorgan*. Nature Conservancy Council, Northminster House, Peterborough.

Percival, M. (1981). Raising wild seed for road verges. *GNAT No. 3* [p. 6] (Newsletter of the Glamorgan Naturalists' Trust).

Percival, M.S. (1960). Botany. In: *The Cardiff Region* (Ed. by J.F. Rees *et al.*) [British Association Handbook for the Cardiff Meeting], pp. 45–57. University of Wales Press, Cardiff.

Perkins, J., Evans, J. & Gillham, M. (1982). *The Historic Taf Valleys Vol. 2*. In: *The Brecon Beacons National Park*. Merthyr Tydfil and District Naturalists' Society in conjunction with D. Brown & Sons Limited, Cowbridge.

Perkins, J., Thomas, C. & Evans, J. (1986). *The Historic Taf Valleys Vol. 3 : From the Taf confluence at Cefn-Coed-y-Cymmer to Aberfan*. Merthyr Tydfil and District Naturalists' Society in conjunction with D. Brown & Sons Limited, Cowbridge.

Perring, F.H. ed. (1968). *Critical Supplement to the Atlas of the British Flora*. Botanical Society of the British Isles, London.

Perring, F.H. & Farrell, L. (1977). *British Red Data Books, 1: Vascular Plants*. The Society for the Promotion of Nature Conservation.

Perring, F.H. & Farrell, L. (1983). *British Red Data Books, 1: Vascular Plants*. 2nd edn. Royal Society for Nature Conservation.

Perring, F.H., Perring, M.D., Kay, Q.[O.N.] & Weston, W. (n.d.). *Wild Plants in Glamorgan*. Glamorgan Naturalists' Trust.

Perring, F.H. & Walters, S.M. eds. (1962). *Atlas of the British Flora*.2nd edn (1976), 3rd edn (1982). Botanical Society of the British Isles, London.

Perry, A.R. ed. (1986). *Glamorgan Trust For Nature Conservation 1961–1986*. Glamorgan Trust For Nature Conservation, Tondu, Bridgend.

Peterken, G.F. & Allison, H. (1989). *Focus on conservation No. 22: Woods, trees and hedges (a review of changes in the British countryside)*. Nature Conservancy Council, Peterborough.

Petiver, J. (1712). An account of divers rare plants. *Philosophical Transactions of the Royal Society*, **27**, 375.

Pettigrew, A. (1875). On British plants. *The Gardener.**

Pettigrew, A.A. (1927). Welsh Vineyards. *Report and Transactions of the Cardiff Naturalists' Society 1926*, **59**, 25–34.

Pettigrew, W.W. (1895). The Botanical Garden, Roath Park. *Report and Transactions of the Cardiff Naturalists' Society 1894–1895*, **27**, 56–62.

Phillips, O. (1956). Part 2: Natural History. In: *Gower* (Regional Book Series), pp. 145–164. Robert Hale Ltd., London.

Pigott, C.D. & Walters, S.M. (1954). On the interpretation of the discontinuous distributions shown by certain British species of open habitats. [Gower]. *Journal of Ecology*, **42**, 95–116.

Pitcairn, C.E.R. (1969). *An ecological study of the factors influencing revegetation of industrial waste heaps contaminated with heavy metals.* Ph.D. thesis. University College of Swansea.

Plummer, B.A.G. (1960). *An investigation into human influence on marsh development in the Burry Estuary, South Wales.* M.Sc. thesis. University College of Swansea.

Potts, E.A. (1968). *The geomorphology of the sand dunes of South Wales with special reference to Gower.* Ph.D. thesis. 3 vols. University College of Swansea.

Poulton, S.M.C. (1978). *The ecology of two areas of saltmarsh and mudflats on the east and west coasts of Britain.* University College of Cardiff. Unpublished.

Preston, C.D. (1988). The spread of *Epilobium ciliatum* Raf. in the British Isles. *Watsonia*, **17**, 279–288.

Preston, T.A. (1884). Plants flowering in January and February, 1884. *Journal of Botany*, **22**, 257–261.

Price, W.R. (1944). Rev. H.J. Riddelsdell [obit. and photo]. *The Botanical Society and Exchange Club Report for 1941–1942*, **12**, 460–462.

Prime, C.T., Buckle, O. & Lovis, J.D. (1960). The distribution and ecology of *Arum neglectum* in southern England and Wales. *Proceedings of the Botanical Society of the British Isles*, **4**, 26–33.

Pritchard, N.M. (1959). *Gentianella* in Britain. *Watsonia*, **4**, 69–192.

Pryce, R.D. (1985). Joint Meeting of Llanelli Naturalists' and Cardiff Naturalists' Botany Section to Whitchurch, Cardiff, 4th August 1985. *Llanelli Naturalists' Newsletter, September 1985*, [pp. 14–15].

Pugsley, D.J. (1941). The Study of the colonisation and subsequent flora of coal dumps at Cwmbach, Aberdare. *Proceedings of the Swansea Society and Field Naturalists' Society*, **2**, 159–177.

Pugsley, H.W. (1948). A Prodromus of the British Hieracia. *Journal of the Linnean Society of London (Botany)*, **54**.

Pyatt, F.B. (1973a). Plant sulphur content as an air pollution gauge in the vicinity of a steel works [Port Talbot]. *Environmental Pollution*, **5**, 103–115.

Pyatt, F.B. (1973b). Some aspects of plant contamination by airborne particulate pollutants [Port Talbot]. *International Journal of Environmental Studies*, **5**, 215–220.

Pyatt, F.B. (1975). An appraisal of tree sulphur content as a longterm air pollution gauge [Port Talbot]. *International Journal of Environmental Studies*, **7**, 103–106.

Qureshi, H.U.A. (1983). *Ecophysiological studies of seed germination strategies in sand-dune plants.* Ph.D. thesis. University College of Swansea.

Randall, H.J. (1961). *The Vale of Glamorgan, Studies in Landscape History.* R.H. Johns Ltd., Newport.

Randall, H.J. & Rees, W. eds. (1963). *South Wales and Monmouth Record Society, Publication No. 5: Diary of Lewis Weston Dillwyn.* R.H. Johns Ltd., Newport.

Ranwell, D. (1955). *Notes on the vegetation of Worm's Head, Inner region, 1949.* Report by King's College, London, Biological Society.

Ratcliffe, D.A. ed. (1977). *A Nature Conservation Review* Vol. I and II. Cambridge University Press, Cambridge, London, New York, Melbourne.

Ratcliffe, J.B. & Hattey, R.P. (1982). *Welsh Lowland Peatland Survey.* Nature Conservancy Council.

Raum, J. & Lees, A. (1981). Crymlyn Bog – don't let Swansea City Council waste it!. *GNAT No. 2*, [pp. 1 and 5] (Newsletter of the Glamorgan Naturalists' Trust).

Raven, P.H. (1963). *Circaea* in the British Isles. *Watsonia*, **5**, 262–272.

Ray, J. (1670). *Catalogus plantarum Angliae.* London.

Ray, J. (1688). *Historia Plantarum Vol II.* London.

Ray, J. (1690). *Synopsis methodica stirpium Britannicarum.* 1st edn London, 2nd edn (1696), 3rd edn (Ed. by J. Dillenius) (1724).

Ray Society (1973). *John Ray Synopsis methodica stirpium Britannicarum edition tertia 1724. Carl Linnaeus Flora Anglica, 1754 and 1759.* Facsimiles with introduction by W.T. Stearn. No. 148. The Ray Society, London.

Read, J.C. (1977). *The distribution of diploid and tetraploid cytotypes and variation in leaf markings in* Ranunculus ficaria. B.Sc. project report. Department of Botany, University College of Swansea.

Reed, A.E. (1881). The Manufacture of Paper as carried on at Ely. *Transactions of the Cardiff Naturalists' Society 1880,* **12**, 41–49.

Rees, D.I. (1977). *A nature conservation study of the Neath Estuary complex.* Nature Conservancy Council (South Wales Region). Unpublished.

Rees, F.E. (1935). Some effects of the blizzard of May 17th, 1935, on the growth of ash and oak. *Proceedings of the Swansea Scientific and Field Naturalists' Society,* **1**, 325–330.

Rees, I. (1979). Nature conservation and the commons of West Glamorgan. In: *Problems of Commonland* (Ed. by E.M. Bridges), 4:3/1–4:3/23. University College of Swansea.

Rees, J. (1928). Sainfoin or French grass in S. Wales. *Welsh Journal of Agriculture,* **4**, 170–183.

Rees, J.F. *et al.* eds. (1960). *The Cardiff Region, A Survey.* [British Association Meeting]. University of Wales Press, Cardiff.

Rees, T.K. (1935). Competition and mortality amongst seedlings. *Proceedings of the Swansea Scientific and Field Naturalists' Society,* **1**, 300–305.

Reynolds, L.D. (1955). The flowering plants and Pteridophyta of the Caerphilly basin. *North Western Naturalist,* **26**, 35–57.

Reynolds, N. (1948). *The morphological development of certain epiphytic algae from ponds in Glamorgan.* Ph.D. thesis. University College of Cardiff.

Reynolds, W.J. (1975). *Ecological investigations of some Leaf-hoppers* (Hemiptera: Homoptera, Cicadellidae) *of woodland canopy.* M.Sc. thesis. University College of Cardiff.

Rhys (1925). *Renant Valley in Wales, Neath* *.

Rhys, E. (1911). *The South Wales Coast from Chepstow to Aberystwyth.* T. Fischer Unwin, London and Leipsic.

Rich, T.C.G. & Rich, M.D.B. (1988). *Plant Crib.* Botanical Society of the British Isles, London.

Richards, A.J. (1972). The *Taraxacum* Flora of the British Isles. *Watsonia,* **9 suppl.**

Richens, R.H. (1987). The history of the elms of Wales. *Nature in Wales (n. s.) 1986,* **5**, 3–11.

Richens, R.H. & Jeffers, J.N.R. (1985). The Elms of Wales. *Forestry,* **58**, 9–25.

Ricks, G.R. & Williams, R.J.H. (1974). Effects of atmospheric pollution on deciduous woodland. II. Effects of particulate matter upon stomatal diffusion resistance in leaves of *Quercus petraea* (Mattuschka) Leibl. *Environmental Pollution,* **6**, 87–109.

Ricks, G.R. & Williams, R.J.H. (1975). Effects of atmospheric pollution on deciduous woodland III. Effects on photosynthetic pigments of leaves of *Quercus petraea* (Mattuschka) Leibl. *Environmental Pollution,* **8**, 96–106.

Riddelsdell, H.J. (n.d.). Papers in manuscript relating to the 'Flora of Glamorgan' [Two volumes, three note-books and one bundle of papers]. Library, National Museum of Wales, Cardiff.

Riddelsdell, H.J. (1902a). North of England plants in the Bicheno Herbarium at Swansea. *The Naturalist (n.s.),* pp. 337–342.

Riddelsdell, H.J. (1902b). North of England plants in the Motley Herbarium at Swansea. *The Naturalist (n.s.),* pp. 343–351.

Riddelsdell, H.J. (1903). Further notes on Yorkshire plants in the Bicheno herbarium at Swansea. *The Naturalist (n.s.),* pp. 167–168.

Riddelsdell, H.J. (1904). *Viola stagnina* Kit. and other plants in Glamorganshire. *Journal of Botany,* **42**, 312. [The record for *Viola stagnina* was an error for *V. lactea*].

Riddelsdell, H.J. (1905a). Notes on Mr Dunn's 'Alien Flora' with particular reference to Glamorganshire plants. *Journal of Botany,* **43**, 89–94.

Riddelsdell, H.J. (1905b). Glamorgan plants. *Journal of Botany,* **43**, 217.

Riddelsdell, H.J. (1905c). Short Notes: *Liparis loeselii* Rich. in Glamorganshire. *Journal of Botany,* **43**, 274.

Riddelsdell, H.J. (1905d). Lightfoot's journey to Wales in 1773. *Journal of Botany,* **43**, 290–307.

Riddelsdell, H.J. (1905e). Glamorgan Botany. In: Swansea Scientific Society Annual Report and Transactions 1904–1905. In: *Royal Institution of South Wales Reports,* p. 143.

Riddelsdell, H.J. (1906). *Rubi* of Glamorganshire. *Journal of Botany,* **44**, 90–99.

Riddelsdell, H.J. (1907). A Flora of Glamorganshire. *Journal of Botany* [Supplement] **45**, 1–88.

Riddelsdell, H.J. (1908). Monmouth and Glamorgan plants. *Journal of Botany*, **46**, 231.

Riddelsdell, H.J. (1909). Further Glamorganshire records. *Journal of Botany*, **47**, 397–412.

Riddelsdell, H.J. (1911). The Flora of the Worms Head and the nativity of certain disputed species. *Journal of Botany*, **49**, 89–92.

Riddelsdell, H.J. (1934). A Flora of Glamorganshire (Supplement to the *Journal of Botany*, 1907). *Botanical Society and Exchange Club Reports 1933*, **10**, 666–669.

Rilstone, F. (1933). Short notes: *Rumex rupestris* Le Gall. *Journal of Botany*, **71**, 107.

Riley, W. (1894). Second Field Walk. *Report and Transactions of the Cardiff Naturalists' Society 1892–1893*, **25**, 52–61.

Roberts, R.H. & Lacey, W.S. eds. (1979). Plants of Glamorgan: [Precis of a paper delivered to the Welsh Section of the Botanical Society of the British Isles AGM at Lampeter on 22nd July 1978, by Dr M.E. Gillham]. *Botanical Society of the British Isles Welsh Bulletin No. 30*, pp. 5–8.

Roberts, T.M. (1972). Accumulation of atmospheric metal pollution in the natural vegetation of the Swansea area. *Welsh Soils Discussion Group Report No. 13*, pp. 41–58.

Robinson, D.M. ed. (1985). The Castle's Woodland Setting. In: *Castell Coch*, pp. 19–20. Cadw: Welsh Historic Monuments, Cardiff.

Roblin, E. (1988). Chemical control of Japanese Knotweed (*Reynoutria japonica*) on river banks in South Wales. *Aspects of Applied Biology*, **16**, 1–7.*.

Ross, M. (1982). Let's visit a reserve: Overton, Longhole Cave and Deborah's Hole. *GNAT No. 4*, pp. 12–13. (Newsletter of the Glamorgan Naturalists' Trust).

Rostański, K. (1977). Some new taxa in the genus *Oenothera* L. subgenus *Oenothera*, III. *Fragmenta floristica et geobotanica*, **23**, 289–293.

Rostański, K. (1982). The species of *Oenothera* L. in Britain. *Watsonia*, **14**, 1–34.

Rostański, K. & Ellis, G. (1979). Evening Primroses (*Oenothera* L.) in Wales. *Nature in Wales*, **16**, 238–249.

Rousham, S. (1985). *Castell Coch*. Cadw: Welsh Historic Monuments, Cardiff.

Rowlands, S.P. (1934). *Adiantum capillus-veneris* in Glamorganshire. *British Fern Gazette*, **6**, 292.

Royal Commission on Land in Wales and Monmouthshire (1896). *Bibliographical, statistical and other miscellaneous memoranda, being Appendices to the Report of the Royal Commission on land in Wales and Monmouthshire*. Her Majesty's Stationery Office, London. Appendix B. – Bibliographical: Supplemental Notes on the Bibliography of the Natural History of Wales II. The Flora of Wales, pp. 140–144, and Rare Welsh Plants, pp. 145–146.

Russell, G.W.E. (1895). *Letters of Matthew Arnold*, **2**, 162.*

Ryle, G.B. (1932). The coastal sand dunes of South Wales. *Quarterly Journal of Forestry*, **26**, 140–157.

Ryle, G.B. (1950). Forestry in South Wales. *Report and Transactions of the Cardiff Naturalists' Society 1945–1948*, **79**, 23–30.

Saleh, H.A. (1984). *Chemical and physical techniques for sand-dune fixation and stabilisation*. M.Sc. thesis. University College of Swansea.

Salisbury, E. (1952). *Downs and Dunes - their plant life and its environment*. G. Bell & Sons, Ltd., London.

Salisbury, E. (1961). *Weeds & Aliens*. (The New Naturalist No. 43) Collins, London.

Salter, B.R. (1965). Afforestation in the lower Swansea Valley. *Lower Swansea Valley Project Study Report No. 11*. University College of Swansea.

Sampson, K. (1984). Spring Flowers in Coed Garn Llwyd. *Forest Farm News No. 2*, [pp. 16–17].

Sargent, C. (1982). *Chief Scientists Team Report No. 435: The Biological Survey of British Rail Property*. Nature Conservance Council, Banbury.

Sargent, C. (1984). *Britain's railway vegetation*: Institute of Terrestrial Ecology, Cambridge.

Saunders, L. (1990). *A study of Plant Succession on sand dunes at Delvid Burrows, Gower*. B.A. (Soc. St.) Exon. University of Exeter.

Scurfield, G. (1962). Biological flora of the British Isles : *Cardaria draba* (L.) Desv. *Journal of Ecology*, **50**, 489–499.

Seddon, B. (1964a). Aquatic plants in Welsh lakes. *Nature in Wales*, **9**, 3–8.

Seddon, B. (1964b). The 1963 Field Meetings: Kenfig Burrows & Aberthaw. *Botanical Society of the British Isles Welsh Region Bulletin No. 1*, p. 11.

Seddon, B. (1965). Schools and Field Work in Glamorgan. *Botanical Society of the British Isles Welsh Region Bulletin No. 4*, pp. 6–7.

Seddon, B. (1972). Aquatic macrophytes as limnological indicators. *Freshwater Biology*, **2**, 107–130.

Seddon, B. (1975). Water Plants in Wales: Their distribution & history. *Botanical Society of the British Isles Welsh Region Bulletin No. 22*, pp. 6–8.

Shakoer, M.A. (1984). *Biological techniques for sand-dune fixation and stabilisation*. M.Sc. thesis. University College of Swansea.

Sheen, A.W. (1928). Flora. In: The History of the Cardiff Naturalists' Society. *Report and Transactions of the Cardiff Naturalists' Society 1927*, **60**, 119.

Simpson, N.D. (1960). *A Bibliographical Index of the British Flora*. (Glamorgan; v.c. 41, pp. 308–310), privately printed, Bournemouth.

Singh, A.K. (1977). *Intraspecific variation within plant species in response to soil factors*. M.Sc. thesis. University College of Cardiff.

Smith, A. (1984). *Cardiff Wildlife Survey*. Nature Conservancy Council *et al*. Cardiff. Unpublished. Copy in Library, Nature Conservancy Council, Cardiff.

Smith, A.E. ed. (1982). *A Nature Reserves Handbook*. Royal Society for Nature Conservation, Lincoln.

[Smith, I.R. *c*.1981]. A botanical survey of Margam Moors, West Glamorgan, Wales Field Unit Project W80–4/2. Nature Conservancy Council, South Wales Region.

Smith, J.E. ed. (1832–1841). [*Sowerby's*] *English Botany*. 2nd edn. Vols. 1–8, London.

Smith, L.P. (1979). *A Survey of the Saltmarshes in the Severn Estuary*. Nature Conservancy Council.

Smith, R.L. & Wade, A.E. (1939). Notes on the adventive flora of the Cardiff District. *Report of The Botanical Society and Exchange Club (for 1938)*, **12**, 72–83.

Smith, R.S. (1976). *An ecological basis for land-use decisions on the South Wales coalfield*. Ph.D. thesis (2 vols.), University College of Cardiff.

Smith, R.S. and Edington, J.M. (1973). Broadleaved woodlands in South Wales and their conservation. *Nature in Wales*, **13**, 146–154.

Smollett, T. (1769). *The present state of all nations* Vol. 3. [*Polypodium cambro-Britannicum*].

Smyth, B. (1987). Gazetteer: Bristol and South Wales. In: *City Wildspace*. H. Shipman Ltd., London.

Somerville, C. (1989). Oxwich. In: *Britain Beside the Sea*, pp. 176–182. Grafton Books, London.

Sothern, E. (1986). *Glamorgan Inventory of Ancient Woodlands (Provisional)*. Nature Conservancy Council.

South Glamorgan County Council (1974). *Cardiff River Valleys recreation study*.

South Glamorgan County Council [1977a]. *County Treasurer's Survey: Marcross, Monknash, St. Donats*.

South Glamorgan County Council. [1977b]. *County Treasurer's Survey: Pendoylan, Ystradowen, Welsh St. Donats*.

South Glamorgan County Council (1979, amended 1980). *Flatholm – A Study*.

South Glamorgan County Council [1980]. *County Treasurer's Survey: Llancarfan*.

South Glamorgan County Council (1983). *Flatholm Project* [Brochure].

South Glamorgan County Council. [*c*.1983]. *County Treasurer's Report: Radyr/Tongwynlais*.

Sowerby, J. and in part Smith, J.E. (1790–1814). *English Botany*, Vols 1–19, London.

Spence, E.I. (1809). *Summer Excursion through parts of Oxfordshire. . . and South Wales*. Vol. 2. 2nd edn. Longman, Hurst Rees and Orme, London.

Spencer, N. (1771, 1773). *The complete English Traveller*.

Stace, C.A. ed. (1975). *Hybridization and the Flora of the British Isles*. Published in collaboration with The Botanical Society of the British Isles by Academic Press, London.

Stace, C.A. & Cotton, R. (1974). Hybrids between *Festuca rubra* L. sensu lato and *Vulpia membranacea* (L.) Dum. *Watsonia*, **10**, 119–138.

Stapledon, R.G. ed. (1936). *A Survey of the Agricultural and Waste Lands of Wales*. (Edited for the Cahn Hill Improvement Scheme) Faber and Faber Ltd, London.

Steers, J.A. (1946). *The Coastline of England and Wales*. University Press, Cambridge.

Stepney-Gulston, A. (1906). A Contribution towards an account of the Narcissi of South Wales. *Carmarthenshire Antiquarian Society and Field Club*, **1**, 113–115.

Stevens, D. (1988). Growing Strong. (Nature Conservancy Council Grassland Survey: progress report). *Glamorgan Wildlife No. 5*, [p. 2]. (Newsletter of the Glamorgan Wildlife Trust).

Stevens, J.P. & Kay, Q.O.N. (1989). The number, dominance relationships and frequencies of self-incompatibility alleles in a natural population of *Sinapis arvensis* L. in South Wales. *Heredity*, **62**, 199–206.

Stiles, B. (1984). Coed-y-Bedw Nature Reserve. *GNAT No. 10*, p. 5. (Newsletter of the Glamorgan Trust for Nature Conservation).

Still, A.L. (1935). Mints in Gower. *Botanical Society and Exchange Club Report for 1934*, **10**, 919–922.

Storrie, J. (1877). Notes on ballast plants of Cardiff and neighbourhood. *Transactions of the Cardiff Naturalists' Society 1876*, **8**, 141–145.

Storrie, J. (1886). *The Flora of Cardiff*. Cardiff Naturalists' Society, Cardiff.

Storrie, J. (1886–1901). Annotated copy of his *Flora of Cardiff* (1886). Library, National Museum of Wales, Cardiff.

Storrie, J. (1896a). Botany of the Island. In: *Notes on Excavations made during the Summers of 1894–1895 at Barry Island and Ely Race Course*, pp. 41–43. Cardiff.

Storrie, J. (1896b). Iolo Morganwg. In: *Notes on Excavations made during the Summer of 1984–1895 at Barry Island and Ely Race Course*, pp. 43–45. Cardiff.

Storrie, J. (1896c). List of plants found on Barry Island. In: *Notes on Excavations made during the Summers of 1894–1895 at Barry Island and Ely Race Course*, pp. 45–50. Cardiff.

Strahan, A. (1896). On submerged land-surfaces at Barry, Glamorganshire. With Notes on the Fauna and Flora by C. Reid. . . . *Quarterly Journal of the Geological Society of London*, **52**, 474–489.

Street, H.E. & Goodman, G.T. (1967). Revegetation techniques in the Lower Swansea Valley. In: *The Lower Swansea Valley Project*. (Ed. by K.J. Hilton), pp. 71–110. Longmans, London.

Summer, A. (1980). Main Field meetings, 1980: The Gower, May 16th–18th. *Wild Flower Magazine No. 389, Autumn 1980*, p. 13.

Summerhayes, V.S. (1951). *Wild Orchids of Britain*. (New Naturalist No. 19). 2nd edn (1968). Collins, London.

Swan, M.C. (1981). The orchids of Gower. *Bios (Swansea)*, **7**, 41–45.

Swan, M.C. (1986). *An investigation of pollen transfer by selected pollinating insects*. Ph.D. thesis. University College of Swansea.

Swansea City Council (n.d.1). *Bishop's Wood Nature Reserve.*

Swansea City Council (n.d.2). *Clyne Wood Trail.**

Swansea City Council (n.d.3). *Gower Heritage Coast Management Programme. Project No. 1: Port Eynon.*

Swansea City Council (n.d.4). *Gower Heritage Coast Management Programme Project No. 5: Hill End Burrows.*

Swansea City Council (n.d.5). *Nature Reclaimed. A Journey of discovery through the Lower Swansea Valley.*

Swansea City Council (n.d.6). *Nature Trail leaflet: Bishop's Wood Local Nature Reserve, Caswell Bay.*

Swansea City Council (1978a). *A conservation and management study: Fairwood Common.* Environmental Design Section, Environment Department, Swansea City Council.

Swansea City Council (1978b). *Lower Swansea Valley Facts Sheet – An introduction to the valley and the literature.*

[Swansea City Council] (1979). *Kilvey Hill*. Environmental Design Section. Environment Department, Swansea City Council.

Swansea City Council (1982a). *Clyne Valley, a new Country Park*. Swansea City Planning Department.

Swansea City Council (1982b). *Lower Swansea Valley – Legacy and Future.*

Swansea City Council Development Department (1986). *Clyne Valley Country Park: Clyne Wood Trail.*

Swansea City Council Development Department (1989). *A Strategy for Greening the City (Consultation Draft).*

Swansea City Council, Leisure Services Department (1988). *Clyne Gardens, Blackpill, Swansea.*

[Swansea Heritage Committee] (n.d.1). *Birds of the Lower Swansea Valley. (A Supplement to the nature trail).* [Swansea City Planning Department and the Conservator for the Lower Swansea Valley].

Swansea Heritage Committee (n.d.2). *Discovering the Lower Swansea Valley: Walk 1.*

Swansea Heritage Committee (n.d.3). *Wild Flowers in the Lower Swansea Valley. (A Supplement to the nature trail).* Swansea City Council Planning Department and the Conservator for the Lower Swansea Valley.

Sykes, M.H. & Webb, J.A. (1947). The Flora of the Bombed Areas and Slum-clearance Sites of Swansea. *Proceedings of the Swansea Scientific and Field Naturalists' Society,* 2, 291–306.

Symonds, W.S. (1872). Rare plants of Gower. In: *Records of the Rocks,* pp. 331–332. John Murray, London.

Talbot, R.J. (1983). *Biochemical, physiological and ecological studies of waterlogging tolerance in* Salix *species.* Ph.D. thesis. University College of Cardiff.

Tann, J. (1972). Operation Bog Bean. *Glamorgan County Naturalists' Trust, Bulletin No. 11,* pp. 15–18.

Tansley, A.G. (1939). *The British Islands and their vegetation.* Cambridge University Press.

Tattersall, W.M. ed. (1936). The Botany of Glamorgan. In: *Glamorgan County History, Vol. 1 Natural History,* pp. 121–254. W. Lewis (Printers) Ltd., Cardiff.

Taylor, C.J. (1974). *The physiology and ecology of the stolon in* Potentilla reptans. Ph.D. thesis. University College of Cardiff.

Teverson, R. (1980). *Saltmarsh ecology in the Severn Estuary.* Department of Energy Agreement, No. 1619. ST51. Contract No. UKEA. E/SA/CON/1619/51/054. Department of Zoology, University of Bristol.

Thomas, A.N. (1938). *The Land of Britain. The Report for the Land Utilisation Survey of Britain.* (Ed. by L.D. Stamp), Part 31: Glamorgan. Published for The Survey by Geographical Publications Ltd., London.

Thomas, B. (1984). Tall tale from Ron. [*Digitalis purpurea* x].*GNAT No. 10,* p. 7 (Newsletter of the Glamorgan Trust for Nature Conservation).

Thomas, C. (1950). The Kenfig *Epipactis. Watsonia,* 1, 283–288.

Thomas, D. (1963). *Agriculture in Wales during the Napoleonic Wars.* University of Wales Press.

Thomas, H. (1959). Flowers of the Limestone. *Gower,* 12, 40–41.

Thomas, H. ed. (1990). *Taff Trails.* The Taff Trail Project, Merthyr Tydfil.

Thomas, J.M. ed. (1965). Flowers and Trees. In: *A Guide to Gower.* Gower Society (5th edn 1982, reprinted 1985).

Thomas, J.K. (1983). *The pollen loads of honeybees foraging in rural and urban environments.* (Apis mellifera *L.).* M.Sc. thesis. University College of Cardiff.

Thomas, M. (1935). Members' Evening. 'The true and extraordinary story of the discovery of two rare Docks (*Rumex rupestris* and *Rumex cunefolius*) on Kenfig Dunes'. In: Report of the Forty-eighth session 1934–1935. Biological and Geological Section. *Cardiff Naturalists' Society Report and Transactions 1935,* 68, 67.

Thomas, P.H. (1967). Medical Men of Glamorgan: 4. The Vachells of Cardiff. *Glamorgan Historian,* 4, 136–161. Stewart William, Barry.

Thomas, P.H. [c.1972]. Medical Men of Glamorgan: Thomas Williams of Swansea. *Glamorgan Historian,* 9, 70–95. ['Wild chamomile' surviving copper smoke, p. 88]. Stewart Williams, Barry.

Thomas, T.H. (1881). Notes upon some fine specimens of Oak, Yew, Elm and Beech, chiefly in the Counties of Monmouth & Glamorgan. *Report and Transactions of the Cardiff Naturalists' Society 1880,* 12, 15–24.

Thomas, T.H. (1888). The Identity of some plants native to the Rocky Mountains with Local Species. *Transactions of the Cardiff Naturalists' Society 1888,* 20, 46–47.

Thomas, T.H. (1891). Botany. In: *Handbook for Cardiff and District* (Ed. by I. James), pp. 200–207. British Association.

Thomas, T.H. (1892). A field walk to Neath valley. *Report and Transactions of the Cardiff Naturalists' Society 1891–1892,* 24, 76–79.

Thomas, T.H. (1893). Botanical Notes during the Summer of 1892. *Report and Transactions of the Cardiff Naturalists' Society 1892–1893,* 25, 76–77.

Thomas, T.H. (1895). Notes upon . . . oak, elm and beech . . . Glamorgan and Monmouthshire. *Swansea Scientific Society,* p. 51.*

Thomas, T.H. (1897). Monstrous form of *Plantago major. Report and Transactions of the Cardiff Naturalists' Society 1896–1897,* 29, 58 and illus.

Thomas, T.H. (1903). Note upon *Meconopsis. Report and Transactions of the Cardiff Naturalists' Society 1901–1902,* 34, 63–64.

Thomas, T.H. (1906). Influence of farm and cottage gardens upon flora in part of the district of the Cardiff Naturalists' Society. *Report and Transactions of the Cardiff Naturalists' Society 1905*, **38**, 61–68.

Thompson, H.S. (1917). *Liparis loeselii* Rich. In: *Watson Botanical Exchange Club 33rd Annual Report (1916–1917)*, **3**, 33–34.

Thompson, H.W. (1933). Plant Pests in Glamorgan. *Report and Transactions of the Cardiff Naturalists' Society 1931*, **64**, 98–101.

Thompson, H.W. (1934). Plant Pests in Glamorgan II. *Report and Transactions of the Cardiff Naturalists' Society 1932*, **65**, 116–120.

Thompson, H.W. (1935). Plant Pests in Glamorgan III. *Report and Transactions of the Cardiff Naturalists' Society 1933*, **66**, 103–109.

Tidswell, R.J. (1983a). *List of Vascular Plants in Radyr Wood, Cardiff (ST 134 800) 30/11/83.* MS in Library, National Museum of Wales, Cardiff.

Tidswell, R.J. (1983b). *Vascular Plants in the Marshland Area of Bute West Dock, Cardiff (ST 193 754) August 1983.* MS in Library, National Museum of Wales, Cardiff.

Traherne (Mrs). (1883). An alphabetical list of some flowering plants found in the Vale of Glamorgan and not mentioned in Miss Ollivant's 'Six-monthly Flora' [*q.v.*]. *Transactions of the Cardiff Naturalists' Society 1882*, **14**, 21–23.

Trew, M.J. (1973). The effects and management of trampling on coastal sand-dunes [Port-Eynon]. *Journal of Environmental Planning and Pollution Control*, **1**, 38–49.

Trist, P.J.O (1986). The distribution, ecology, history and status of *Gastridium ventricosum* (Gowan) Schinz & Thell. in the British Isles. *Watsonia*, **16**, 43–54.

Trotman, D.M. (1963). *Data for Late-Glacial and Post-Glacial History in South Wales.* Ph.D. thesis. University College, Swansea.

Trow, A.H. (1893). Notes on the flora of the Society's district. *Transactions of the Cardiff Naturalists' Society 1891–1892*, **24**, 13–21.

Trow, A.H. ed. (1907–1912). *The Flora of Glamorgan.* Published in parts in *Transactions of the Cardiff Naturalists' Society*, Appendix, (*1906*, **39**, 1–44; *1907*, **40**, 45–78; *1908*, **41**, 79–132; *1909*, **42**, 133–146; *1910*, **43**, 147–194; *1911*, **44**, 195–209). Published as a single volume Trow, A.H. (1911).

Trow, A.H. (c.1908). Nature Study. *Transactions of the Guild of Graduates* [University of Wales] 1907–1908, pp. 12–27.

Trow, A.H. (1909). Forms of *Senecio vulgaris. Journal of Botany*, **47**, 304–306.

Trow, A.H. (1911). *The Flora of Glamorgan Vol. 1.* Cardiff Naturalists' Society. See also Hyde, H.A. (- 1950) and Vachell, E. (n.d.4).

Turner, D. & Dillwyn, L.W. (1805). *The Botanist's Guide through England and Wales.* (Glamorganshire, **1**, 298–306; Supplement to Glamorganshire, **2**, 753–755). Phillips & Fardon, London.

Tutin, T.G. *et al.* eds. (1964–1980). *Flora Europaea Vols. 1–5.* Cambridge University Press, London & Cambridge.

[Vachell, C.T.] (n.d.). Portfolio of notes and papers concerning the early history of Cardiff Museum. Vachell Collection, Library, National Museum of Wales, Cardiff.

Vachell, C.T. (1895a). Contribution towards an account of the 'Narcissi' of South Wales. *Report and Transactions of the Cardiff Naturalists' Society 1893–1894*, **26**, 81–94.

Vachell, C.T. (1895b). Note on a doubtful species of *Senecio. Report and Transactions of the Cardiff Naturalists' Society 1893–1894*, **26**, 110.

Vachell, C.T. (1897). Notes on the Flora of Wales. *Report and Transactions of the Cardiff Naturalists' Society 1896–1897*, **29**, 74–75.

Vachell, E. (n.d.1). Botanical notes in the autograph of E. Vachell. MS in Library, National Museum of Wales, Cardiff.

Vachell, E. (n.d.2). File of typed botanical records [Glamorgan]. Library, National Museum of Wales, Cardiff.

Vachell, E. (n.d.3). MS list of plants found by E. Vachell, arranged and numbered according to Bentham's Handbook of the British Flora to accompany Fitch & Smith's *Illustrations of the British Flora*, in which book the illustrations of the plants found by her *in situ* are coloured. Library, National Museum of Wales, Cardiff.

Vachell, E. (n.d.4). MS notes in her annotated copy of Trow's *Flora of Glamorgan* (1911). Library, National Museum of Wales, Cardiff.

Vachell, E. (n.d.5). MS notes relating to the localities of certain botanical rarities. Library, National Museum of Wales, Cardiff.

Vachell, E. (n.d.6). Portfolio of notes, prints, etc., concerning *Limosella* plants of Glamorgan. Library, National Museum of Wales, Cardiff.

Vachell, E. [1883–1948]. Diary, with botanical notes (2 volumes). Library, National Museum of Wales, Cardiff.

Vachell, E. [1891–1931]. Botanical diary. Library, National Museum of Wales, Cardiff.

Vachell, E. (1917–1919). Annotated copy of L.W. Dillwyn's *Materials for a fauna and flora of Swansea*. MS in Library, National Museum of Wales.

Vachell, E. [1919]. Flowers of the field: botanical notes. In MS, compiled during the summer of 1919. Library, National Museum of Wales, Cardiff.

Vachell, E. (1920). The Botany of the Cardiff District. In: *Handbook to Cardiff and the neighbourhood.* (Ed. by H.M. Hallet), pp. 218–227. British Association, Cardiff Meeting 1920.

Vachell, E. [1921–1948]. Wild flowers in Wales: Six volumes of mounted newspaper cuttings from the *Western Mail.* April 11th, 1921 – December 7th, 1948. Library, National Museum of Wales, Cardiff.

Vachell, E. (1922). The Leek – The National Emblem of Wales. *Report and Transactions of the Cardiff Naturalists' Society 1919,* **52,** 26–49.

Vachell, E. (1927). An unusual specimen of *Anagallis* [Talk to Linnean Society 17th February 1927]. *Proceedings of the Linnean Society 1926–1927,* pp. 26–28.

Vachell, E. (1928). Meeting of some of the Botanical Society of the British Isles in Glamorgan, 1927. *Report of the Botanical Society & Exchange Club of the British Isles,* 8, 455–458.

Vachell, E. (1931). Flowers and Climate. *Wild Flower Magazine No. 207,* pp. 7–8.

Vachell, E. [1931–1945]. Field Botanist's Diary. Library, National Museum of Wales, Cardiff.

Vachell, E. (1934). A List of Glamorgan Plants. *Report of the Botanical Society and Exchange Club 1933,* **10,** 686–743.

Vachell, E. (1936a). Flowering Plants and Ferns. In: *Glamorgan County History Vol. 1* (Ed. by W.M. Tattersall) pp. 123–178. W. Lewis (Printers) Ltd, Cardiff.

Vachell, E. (1936b). Glamorgan Botanists. In: *Glamorgan County History, Vol. 1* (Ed. by W.M. Tattersall), pp. 252–254. W. Lewis (Printers) Ltd., Cardiff.

Vachell, E. (1938). Botanical Notes, 1936. *Transactions of the Cardiff Naturalists' Society 1936,* **69,** 90–91.

Vachell, E. (1941a). Botanical Notes. *Transactions of the Cardiff Naturalists' Society,* **71,** 29–31.

Vachell, E. (1941b). *Limosella* plants of Glamorgan. *Transactions of the Cardiff Naturalists' Society,* **71,** 32–35.

Vachell, E. (1947a). Principal A.H. Trow [Obituary]. *Report and Transactions of the Cardiff Naturalists' Society 1939–1945,* **72–78,** 9–10.

Vachell, E. (1947b). Botanical Notes 1939–1946. *Report and Transactions of the Cardiff Naturalists' Society, 1939–1945,* **72–78,** 23–26.

Vachell, E. (1949). Exhibits: Some Interesting British Specimens, (a) Variation of *Senecio squalidus* in leaf form around Cardiff, (b) variations in *Carduus tenuiflorus* Curt. in Glamorgan, (c) Specimens of *Cirsium* from Monknash, Glamorgan. In: A.J. Wilmott ed. (1949), *q.v.*

Vachell, E. (1950). The disappearance of Morfa Pools (with list of flowering plants). *Transactions of the Cardiff Naturalists' Society,* **79,** 40–42.

Vachell, E. & Blackburn, K.B. (1939). The *Limosella* plants of Glamorgan, *Journal of Botany,* **77,** 65–71.

Vale of Glamorgan Borough Council Parks Department (n.d.). *Porthkerry Country Park.*

Vale of Glamorgan Borough Council Planning Services Department & Barry Preservation Society (1982). *Discovering Barry, 1 : Old Barry.*

Valentine, D.H. (1949). Vegetative and cytological variation in *Viola riviniana* Rchb. In: A.J. Wilmott ed. (1949), pp. 48–53, *q.v.*

Valentine, D.H. (1980). Ecotypic and polymorphic variation in *Centaurea scabiosa* L. *Watsonia,* **13,** 103–109.

Vaughan, I.M. (1973). Notes on the increase of *Pyrola rotundifolia* subsp. *maritima* (Kenyon) E.F. Warb. on Welsh Dune Systems. *Botanical Society of the British Isles Welsh Region Bulletin No. 18,* 7–8.

Vaughan, I.M. (1977). *Pyrola minor* L. in Pembrey Forest, Carmarthenshire. *Watsonia,* **11,** 388–389.

Vaughan, I.M., Donovan, J.W. & Davis, T.A.W. (1972). Notes on the Dune Gentian. *Nature in Wales*, **13**, 33–36.

Vaughan Thomas, W. (1976). *Portrait of Gower*. Robert Hale, London. 2nd edn (1983).

Vaughan Thomas, W. (1978). The Countryside in Autumn: Hambury Wood. *Glamorgan Naturalists' Trust Bulletin and Annual Report No. 17*, 29.

Wade, A.E. (1950–1966). Glamorgan Botanical Notes in *Transactions of the Cardiff Naturalists' Society*, 1947–8, **79**, 52–54; 1949–1950, **80**, 37–38; 1951–1952, **81**, 100–101; 1953–1954, **83**, 25–26; 1955, **84**, 18; 1956, **85**, 25–26; 1957, **86**, 22–23; 1958, **87**, 27; 1959, **88**, 21; 1960, **89**, 31; 1961, **90**, 25; 1961–3, **91**, 38; 1963–66, **93**, 47–48.

Wade, A.E. (1958). The History of *Symphytum asperum* Lepech. and *S.* x *uplandicum* Nyman in Britain. *Watsonia*, **4**, 117–118.

Wade, A.E. & Smith, R.L. (1926). The adventive flora of the Port of Cardiff. *The Botanical Society and Exchange Club, Report for 1925*, pp. 999–1027.

Wade, A.E. & Smith, R.L. (1927). Additions to the adventive flora of the Port of Cardiff. *The Botanical Society and Exchange Club, Report for 1926*, pp. 181–183.

Wade, A.E. & Webb, J.A. (1952). Clyne Common: A Summary of the Types of Habitats. *Proceedings of the Swansea Scientific and Field Naturalists' Society*, pp. 337–338.

Wade, G.W. & J.H. (1913). *South Wales*. [Flora section by H.J. Riddelsdell], pp. 18–20.

Wade, J.H. (1914). (Natural History. In: *Glamorganshire*, pp. 37–45.) [Cambridge County Geography Series]. Cambridge University Press, London and Edinburgh.

Wakefield, H.R. (1907). Alien plants collected at Port Talbot. *Report of the Swansea Field Naturalists' Society 1906–1907*, p. 128.

Wakefield, H.R. (1910). Flora of the Swansea district. *Report of the Swansea Field Naturalists' Society 1909–1910*, pp. 147–153.

Wakefield, H.R. (1915). Report of the botanical section. *Report of the Swansea Field Naturalists' Society 1914–1915*, pp. 26–27.

Wakefield, H.R. (1916). Botany. *Swansea Field Naturalists' Society 1915–1916*, p. 24.

Waldren, S. (1982). *Frankenia laevis* L. in Mid Glamorgan. *Watsonia*, **14**, 185–186.

Waldren, S. (1985). *Physiological and morphological studies of waterlogging tolerance in* Geum rivale *L., G. urbanum L., and their hybrids*. Ph.D. thesis. University College of Cardiff.

Waldron, C. (1887). Gall on the Creeping Thistle [in litt.]. *Report and Transactions of the Cardiff Naturalists Society 1886*, **18**, 89.

Wales Tourist Board (1984). *The Great Nature Trails of Wales*. Wales Tourist Board, Cardiff.

Walford, T. (1818). *The scientific tourist through England, Wales and Scotland*, 2 vols. London.

Waring, E. (1850). *Recollections and Anecdotes of Edward Williams, the bard of Glamorgan; or Iolo Morganwg, B.B.D.* Charles Gilpin, London.

Warren, R.G. (1973). Butterflies and Moths in Some Gower Reserves. *Glamorgan County Naturalists' Trust Bulletin No. 12*, pp. 28–31.

Waters, B. (1955). *The Bristol Channel* [Osmunda regalis]. J.M. Dent & Sons, Ltd., London.

Watkins, M. (1983). Let's Visit a Reserve: Coed Garnllwyd. *GNAT No. 6*, p. 4. (Newsletter of the Glamorgan Naturalists' Trust).

Watkins, M.G. ed. (1984). *Cardiff Wildlife Survey*. Cardiff Wildlife Group. Cardiff. (Unpublished).

Watson, C.J. (1892). Notes at the British Association, Cardiff 1891. *The Midland Naturalist*, **15**, 10–19.

Watson, H.C. (1835). *The New Botanist's Guide to the localities of the rarer plants of Britain, Vol. 1*. [Glamorganshire, pp. 216–218]. London.

Watson, H.C. (1837). *The New Botanist's Guide to the localities of the rarer plants of Britain, Vol. 2*. [Glamorgan (supplement), p. 630]. London.

Watson, H.C. (1847–1859). *Cybele Britannica; or British Plants and their Geographical Relations*. **1–4**. Longman Co. London. Suppl. (1860). Privately printed.

Watson, H.C. (1850). Explanatory Notes on certain British Plants for distribution by the Botanical Society of London, in 1850. *Phytologist*, **3**, 801–811. ['*Sinapis cheiranthus* (Koch)', from Gower, pp. 805–6].

Watson, H.C. (1868–1872). *Compendium of the Cybele Britannica. Parts 1–3 and suppl.* Thames Ditton, London.

Watson, H.C. (1883). *Topographical Botany*. Edn 2. Bernard Quaritch, London.

Webb, J.A. (1913–1926). Field meetings and plant records in *Royal Institution of South Wales*

Reports, 1912–3, pp. 104–114; 1913–4, pp. 116–129; 1914–5, pp. 95–102; 1918–9, pp. 24–26; 1920–21, pp. 35–37; 1921–22, pp. 37–38; 1922–23, p. 29; 1923–4, p. 41; 1924–25, pp. 37–38; 1925–26, pp. 34–35.

Webb, J.A. (1927–1956). Field meetings and plant records in *Proceedings of the Swansea Scientific and Field Naturalists' Society*, **1**, 6–8, 22–24, 44–46, 55–58, 62–64, 89–92, 93–95, 117–123, 123–126, 151–156, 156–159, 172–175, 176–179; **2**, 86–89, 125–128, 128–130, 182–187, 188–191, 251–256, 256–258, 258–260, 318, 319–320, 320–323, 323–328; **3**, 7–10.

Webb, J.A. (1929c). The presumably extinct plants of West Glamorgan, with notes on dubious and erroneous records. *Proceedings of the Swansea Scientific and Field Naturalists' Society*, **1**, 70–75.

Webb, J.A. (1932c). On *Polygonum cuspidatum* (Japanese knotweed) in West Glamorgan. *Proceedings of the Swansea Scientific and Field Naturalists' Society*, **1**, 165–166.

Webb, J.A. (1947d). Plant Records 1947. *Proceedings of the Swansea Scientific and Field Naturalists' Society*, **2**, 261–262.

Webb, J.A. (1948). Naturalists in Gower – Part 1 (1906–1920). *Gower*, **1**, 9–12.

Webb, J.A. (1949). Naturalists in Gower – Part 2 (1920–1949). *Gower*, **2**, 22–24.

Webb, J.A. (1950). Gower's Own Plant – *Draba aizoides montana*. *Gower*, **3**, 44–45.

Webb, J.A. (1954). The Trees and Shrubs of Gower – I. *Gower*, **7**, 12–16.

Webb, J.A. (1955a). A Year with Nature in an East Gower Parish. *Gower*, **8**, 12–14.

Webb, J.A. (1955b). Trees and Shrubs of Gower, Part 2, *Gower*, **8**, 37–40.

Webb, J.A. (1956b). The Flora of Gower. *Proceedings of the Swansea Scientific and Field Naturalists' Society*, **3 suppl.**, pp. 1–25. Including the Flora of the peninsula of Gower. Dicotyledons Part 1 – Ranunculaceae to Geraniaceae.

Webb, J.A. (1957). The Fern and Fern-allies of Gower. *Gower*, **10**, 75–78.

Webb, J.A. (1958). Ferns and Fern-allies of Gower Wallica. *Gower*, **11**, 47–48.

Webb, L.M. (1934). The Field Excursions, 1933. *Proceedings of the Swansea Scientific and Field Naturalists' Society*, **1**, 214–216.

Webb, L.M. & Allen, E.E. (1938). The Field Excursion for 1936 and 1937. *Proceedings of the Swansea Scientific and Field Naturalists' Society*, **2**, 7–10.

Webb, L.M. & Jewell, P.S. (1935). The Field Excursions, 1934, with ecological notes. *Proceedings of the Swansea Scientific and Field Naturalists' Society*, **1**, 262–265.

Welsh Development Agency (1987). *Working with nature* [Research into low cost reclamation].

Welsh Office (1975). *Report of a collaborative study on certain elements in air, soil, plants, animals and humans in the Swansea – Neath – Port Talbot area together with a report on a moss-bag study of atmospheric pollution across S. Wales*. Welsh Office, Cardiff, pp. 1–365.

Welsh Water Authority (1984). Report of herbicide trials on floodbanks at Treforest and Trehafod. *Scientific Services Report B/8/1984.**

Westcombe, T. (1844). Localities of a few plants lately observed: Glamorganshire (in addition to Mr Gutch's List. . .). *Phytologist*, **1**, 780–787 (1843).

West Glamorgan County Council (n.d.1). *Gower Field Education Project: Caswell Valley and Bishop's Wood Teacher's Guide*. West Glamorgan County Council & Nature Conservancy Council.

West Glamorgan County Council (n.d.2). *West Glamorgan County Series No. 3: A Pocket Guide to Lower Lliw Valley*.

West Glamorgan County Council (1980). *Cwmllwyd Wood Nature Reserve*. Revised 1984.

West Glamorgan County Council (1988a). *Forest Walks: Afon Argoed County Park*.

West Glamorgan County Council (1988b). *Gower Farm Trail*.

West Glamorgan County Council and Afan Borough Council (n.d.). *Margam Park*. Howard Jones Associates, Swansea.

West Glamorgan County Council and Forestry Commission (1979). *Afon Argoed Country Park*. A. McLay & Co. Ltd., Cardiff. [Reprinted with minor amendments 1984].

Weston, R.L. & Gadgil, P.D. (1965). The plant ecology and soil biology of the Lower Swansea Valley. *Lower Swansea Valley Project, Study Report No. 9*. University College of Swansea.

Weston, R.L., Gadgil, P.D., Salter, B.R. & Goodman, G.T. (1965). Problems of revegetation in the Lower Swansea Valley, an area of extensive industrial dereliction. In: *Ecology and the Industrial Society*, 5th Symp. Br. Ecol. Soc. (Ed. by G.T. Goodman, R.W. Edwards & J.M. Lambert), pp. 297–325. Blackwell Scientific Publications, Oxford.

Weston, W. (1968). An extremely rare wild flower rediscovered in Glamorgan. [*Matthiola sinuata*]. *Glamorgan County Naturalists Trust, Bulletin No. 7*, pp. 8–9.

Weston, W. (1969). Crymlyn Bog. *Glamorgan County Naturalists Trust, Bulletin No. 8*, pp. 24–26.

Westrup, A.W. (1957). Field meetings 1956: June 2nd 1956 Kenfig Burrows. *Proceedings of the Botanical Society of the British Isles*, **2**, 408.

Wheldon, J.A. (1915). Short Notes: *Polygala dunense* Dumort. *Journal of Botany*, **53**, 250.

White, E.J. and Smith, R.I. (1982). *Climatological Maps of Great Britain*. Institute of Terrestrial Ecology., Cambridge.

White, J. (1959). Afforestation of former opencast coal site in Coed Morgannwg. *Journal of Forestry Commission*, **28**, 69–74.

Whyte, R.O. & Sisam, J.W.B. (1949). Present interest and extent of the problem: Wales. In: *Commonwealth Agricultural Bureaux Joint Publication No. 14: The Establishment of Vegetation on Industrial Waste Land*, pp. 14–19. The Commonwealth Bureau of Pastures and Field Crops, Aberystwyth, Wales, and the Commonwealth Forestry Bureau, Oxford, England.

Wiley, M. (1989). A walk along the towpath in Summer. *Forest Farm Watch News*, [pp. 2–3].

Wilkinson, J. (1941). The *Salices* of Gower. *Proceedings of the Swansea Scientific and Field Naturalists' Society*, **2**, 133–141.

Wilkinson, J. (1946). Some factors affecting the distribution of the *Capreae* group of *Salix* in Gower. *Journal of Ecology*, **33**, 214–221.

Wilkinson, M.J. & Stace, C.A. (1988). The taxonomic relationships and typification of *Festuca brevipila* Tracey and *F. lemanii* Bastard (Poaceae). *Watsonia*, **17**, 289–299.

Willers, B.W. (1984). *An Investigation into the Dietary Habits of Fallow Deer* (Dama dama L.) *in relation to the available Forage on a Grassland Site in Margam Park (W. Glamorgan), and a study of some factors that may influence the Vegetational Composition of the Site*. Joint Honours thesis. University College of Cardiff.

Williamson, A.M. (1981). *Botanical studies in the Lower Ogmore system*. M.Sc. thesis. University College of Cardiff.

Williams, D. (1984). [Ref. to *Viola odorata* and *Helleborus foetidus*]. In: Short Notes, *Llanelli Naturalists' Newsletter*, September 1984, [p. 16].

Williams, D.A. (1984). *Studies in atmospheric pollen 1. Daily census of pollens at Cardiff, 1942*. M.Sc. thesis. University College of Cardiff.*

Williams, R.J.H., Lloyd, M.M. & Ricks, G.R., (1971). Effects of atmospheric pollution on deciduous woodland. I. Some effects on leaves of *Quercus petraea* (Mattuschka) Leibl. *Environmental Pollution*, **2**, 57–68.

Williams, T. (1854). *Report on The Copper Smoke*. [Lower Swansea Valley]. Herbert Jones, Swansea.

Williams, W. (1920). *Some woodlands of Glamorgan: an ecological study*. M.Sc. thesis, University College of Cardiff.

Wilmott, A.J. ed. (1949). *British flowering plants and modern systematic methods being the report of the Conference on the study of Critical British Groups. . . .* Botanical Society of the British Isles, London.

Wilson, A. (1949). The Altitude Range of British Plants 2nd edn. *North Western Naturalist*, Supplement. Reprinted 1956. T. Buncle & Co. Ltd., Arbroath.

Wilson, L. (1988). *Wild Flowers in their Seasons*, D. Brown & Sons, Cowbridge.

Wislocka, G.M.S.deK. (1989). *The synecology and growth of* Hippophae rhamnoides *at Merthyr Mawr S.S.S.I. South Wales*. Unpublished thesis, Polytechnic of Wales, Treforest, Mid Glamorgan.*

Wisniewski, P.J. (1982–1983). Let's visit a Nature Reserve: Cwm Risca Wood and Park Pond. *GNAT No. 5*, pp. 4–5. (Newsletter of the Glamorgan Naturalists' Trust).

Wisniewski, P.J. (1982–1983). West Woods, Atlantic College. *GNAT No. 5*, p. 11. (Newsletter of the Glamorgan Naturalists' Trust).

Wisniewski, P.J. (1983). Vegetational succession at The Nature Centre. *GNAT No. 7*, [pp. 3–4]. (Newsletter of the Glamorgan Naturalists' Trust).

Wisniewski, P.[J.] (1984). A Naturalist on the Road – In pursuit of the Common or Garden Weed. *GNAT No. 10*, pp. 7–8. (Newsletter of the Glamorgan Trust for Nature Conservation).

Withering, W. (1812). *A Systematic Arrangement of British plants*. (Corrected and enlarged by W. Withering Jr.). Birmingham. 5th edn, 4 vols.

Wood, M. (1978). Welsh Polypody is rediscovered in Glamorgan. *Glamorgan Naturalists' Trust, Bulletin and Annual Report No. 17*, pp. 32–33.

Woods, A. ed. (1985). *Broadleaved Woodlands in Wales: The Core Report*. Cynefin. c/o Countryside Commission Office for Wales, Newtown.

Woods, J. (1850). Botanical Notes, the result of a visit to Glamorgan and Monmouthshire, . . . 1850. *Phytologist*, **3**, 1053–1061.

Worrall, D. (1984). Flat Holm Project. *Forest Farm News No. 1*, [p. 21].

Worrall, D.H. & Surtees, P.R. (1984). *Flatholm*. South Glamorgan County Council.

Wotton, F.W. (1891). A short historical account of Flatholm and its natural history. [Incl. List of plants by J.G. Thomas and J. Storrie]. *Transactions of the Cardiff Naturalists' Society 1890*, **22**, 105–111.

Wotton, F.W. (1897). Appendix on Fauna and Flora. [Of shell bed deposits]. East Barry Docks. *Report and Transactions of the Cardiff Naturalists' Society 1895–1896*, **28**, 90–93.

Wyatt, N. (1989a). Flora: The Bogbean. *Glamorgan Wildlife No. 7*, [p. 6]. (Newsletter of the Glamorgan Wildlife Trust).

Wyatt, N. (1989b). Conservation: Where are our orchids?. *Glamorgan Wildlife No. 8*, [p. 3]. (Newsletter of the Glamorgan Wildlife Trust).

Yaqub, M. (1981). *The implication of climate in the control of the distribution of* Verbena officinalis *and* Eupatorium cannabinum. Ph.D. thesis. University College of Cardiff.

Young, E. (1856). *The Ferns of Wales*. T. Thomas, Neath.

Ysgol Addysg Prifysgol Cymru (1969). *Enwau planhigion*. [Welsh plant names]. Gwasg Prifysgol Cymru, Caerdydd.

Zehetmayr, J. (1984). Lavernock Point Nature Reserve. *Forest Farm News No. 10*, [p. 22].

Index

Accepted names are in **bold** type, synonyms are in *italics*. English and Welsh names are in ordinary type. During the preparation of the index a few minor errors were noted in the text and have been corrected in the index.

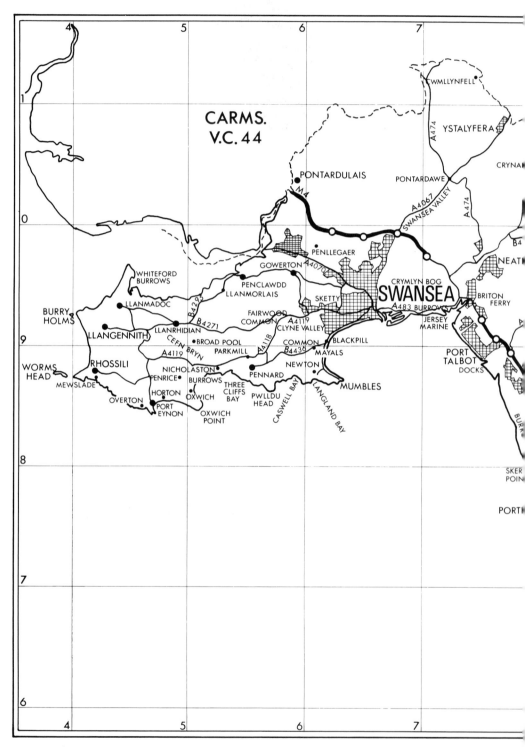

Map 1 Urban areas, main botanical site and road network.

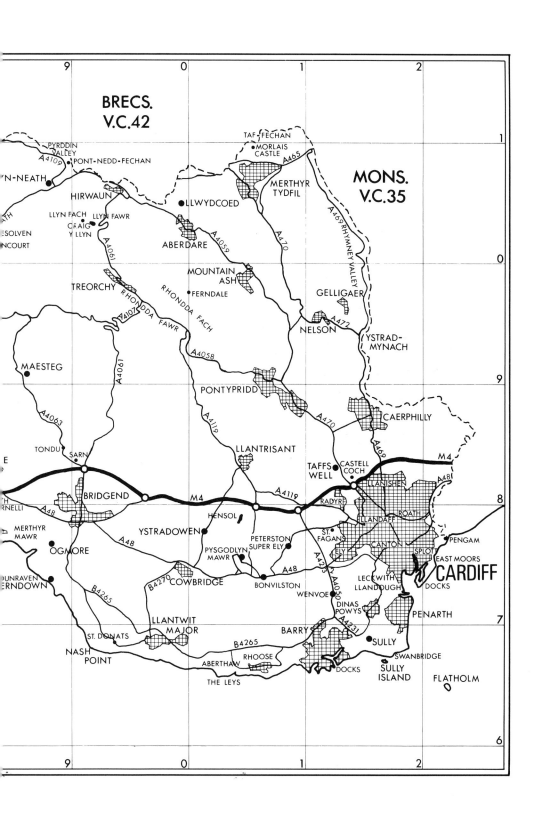

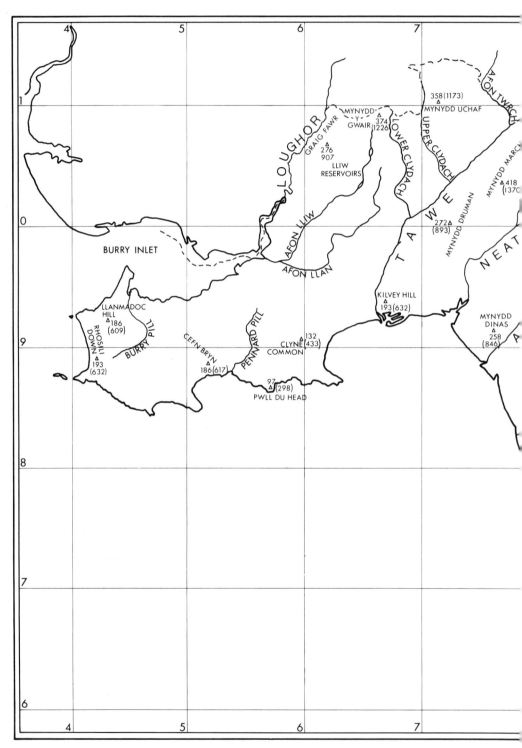

Map 2 Mountains and river systems (Δ major hill or mountain, altitude in metres (feet)).

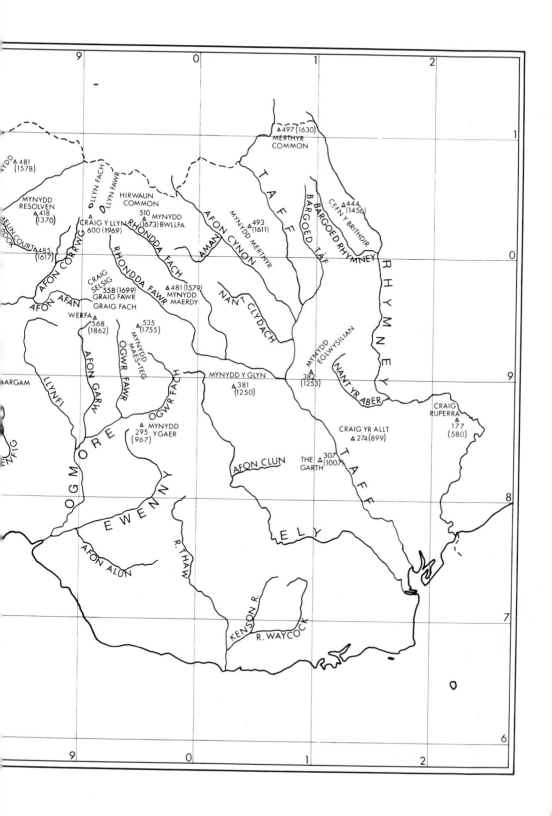